深入解析Android虚拟机

钟世礼◆编著

人民邮电出版社
北京

图书在版编目（CIP）数据

深入解析Android虚拟机 / 钟世礼　编著. -- 北京：人民邮电出版社, 2016.9
ISBN 978-7-115-42353-5

Ⅰ. ①深… Ⅱ. ①钟… Ⅲ. ①移动终端－应用程序－程序设计－虚拟处理机 Ⅳ. ①TP338

中国版本图书馆CIP数据核字(2016)第238207号

内 容 简 介

Android 系统从诞生到现在的短短几年时间里，凭借其易用性和开发的简洁性，赢得了广大开发者的支持。在整个 Android 系统中，Dalvik VM 一直是贯穿从底层内核到高层应用开发的核心。本书循序渐进地讲解了 Android 虚拟机系统的基本知识，并剖析了其整个内存系统的进程和运作流程，并对虚拟机系统优化和异常处理的知识进行了详细讲解。本书几乎涵盖了 Dalvik VM 系统的所有主要内容，并且讲解方法通俗易懂，特别有利于读者学习并消化。

本书适合 Android 初学者、Android 底层开发人员、源代码分析人员和虚拟机开发人员学习，也可以作为大专院校相关专业师生的学习用书和培训学校的教材。

◆ 编　　著　钟世礼
责任编辑　张　涛
责任印制　沈　蓉　焦志炜

◆ 人民邮电出版社出版发行　　北京市丰台区成寿寺路 11 号
邮编 100164　　电子邮件 315@ptpress.com.cn
网址 http://www.ptpress.com.cn
固安县铭成印刷有限公司印刷

◆ 开本：787×1092　1/16
印张：38　　2016 年 9 月第 1 版
字数：1004 千字　　2024 年 7 月河北第 6 次印刷

定价：99.00 元

读者服务热线：(010)81055410　印装质量热线：(010)81055316
反盗版热线：(010)81055315

前　言

Android 虚拟机技术——Dalvik VM 是通往 Android 高级开发的必备技术！为了让广大读者深入理解 Android 系统，不再停留在抽象的原理和概念之上，本书对 Android 虚拟机方面的知识进行了细致分析，这样做的目的是“提炼”出 Android 系统的本质，了解 Android 系统究竟是如何运作的，进程和线程之间是如何协调并进的，内存之间是如何分配并存的。并以此为基础，详细讲解了内存优化、垃圾收集和系统优化方面的基本原理和具体实现。

本书的内容

本书共 24 章，循序渐进地讲解了 Android 虚拟机系统的基本知识，从获取并编译 Android 源码开始，依次讲解了 Java 虚拟机基础、Android 虚拟机基础、分析 JNI、分析内存系统、Android 程序的生命周期管理、IPC 进程通信机制、init 进程、Dalvik VM 的进程系统、Dalvik VM 运作流程、DEX 文件、Dvlik VM 内存系统、Dalvik VM 垃圾收集机制、Dalvik VM 内存优化机制、Dalvik VM 的启动过程、注册 Dalvik VM 并创建线程、Dalvik VM 异常处理、JIT 编译、Dalvik VM 内存优化、Dalvik VM 性能优化等内容。

本书特色

在内容的编写上，本书具有以下特色。

（1）结构合理

从用户的实际需要出发，科学安排知识结构，详细讲解了 Android 虚拟机的各方面知识，内容循序渐进、由浅入深。

（2）遵循“基础讲解—源码分析—核心技术剖析”这一主线

为了使广大读者彻底弄清楚 Android 虚拟机中的各个知识点，剖析了与 Android 虚拟机相关的进程运行机制、内存系统、生命周期管理等核心知识，并讲解了读者关心的系统优化技术。

（3）易学易懂

本书内容条理清晰、语言简洁，可以帮助读者快速掌握每个知识点。使读者既可以按照本书编排的章节顺序进行学习，也可以根据自己的需求对某一章节进行有针对性地学习。

本书参考资料

由于 Android 虚拟机系统十分深奥，加上市面上的相关资料十分稀缺。作者在写作过程中对每一段文字都进行了深入研究和推敲，并参阅了国内外大师们的经典资料，对这些资料进行了深入地研读。在作者的写作过程中，从下面 4 部分资料中获得了帮助。

（1）Oracle 官方资料

http://docs.oracle.com/javase/7/docs/

http://docs.oracle.com/javase/6/docs/

http://www.oracle.com/technetwork/java/

上述资料是 Oracle 官方提供的 Java 虚拟机资料，这些资料也是国内外读者学习 Java 虚拟机的第一手资料。

（2）国外经典名著

《The Java Language Specification, Third Edition》

《The Java Virtual Machine Specification》

上述资料是国外大师们根据 Oracle 官方资料而著成的经典名著，也是国内外读者学习 Java 虚拟机的参考资料。在国内的一些开源论坛中，有很多热心网友进行了翻译。

（3）Google 官方资料

Google I/O 2010 - A JIT Compiler for Android's Dalvik VM

Dalvik VM Internals - Presentation from Google I/O 2008, by Dan Bornstein

Detailed Dalvik specifications documents

上述资料是 Google 公司《Google I/O 讲座系列》的内容，讲解了 Android 虚拟机优化和内存系统的知识，对广大初学者来说有很强的借鉴作用。当然，Google 提供的 Android 源码更是人们分析 Dalvik VM 的第一手资料。

（4）国内著作

《解析 Java 虚拟机开发：权衡优化、高效和安全的最优方案》清华大学出版社，张善香，2013-06-01。

这是国内技术高人的一本著作，可以说是讲解 Java 虚拟机方面较全的一本参考书。里面介绍的很多内容对写作本书有很大启发，想了解这方面内容的读者可以参考一下。

读者对象

- 初学 Android 编程的自学者
- Linux 开发人员
- 大中专院校的老师和学生
- 做毕业设计的学生
- Android 编程爱好者
- 相关培训机构的老师和学员
- 从事 Android 开发的程序员

本书在编写过程中，我的家人在我写作时给予了巨大支持，在此表示深深的感谢。另外，由于本人水平有限，书中如有纰漏和不尽如人意之处在所难免，诚请读者提出意见或建议，以便今后修订并使之更臻完善。另外为本书提供了售后支持网站：http://www.toppr.net/，读者如有疑问可以在此提出，一定会得到满意的答复。编辑联系邮箱：zhangtao@ptpress.com.cn。

作　者

目　录

第 1 章　获取并编译 Android 源码……1

1.1　获取 Android 源码……1

1.1.1　在 Linux 系统获取 Android 源码……1

1.1.2　在 Windows 平台获取 Android 源码……2

1.1.3　Windows 获取 Android L 源码……4

1.2　分析 Android 源码结构……6

1.3　编译 Android 源码……8

1.3.1　搭建编译环境……8

1.3.2　开始编译……9

1.3.3　在模拟器中运行……10

1.3.4　常见的错误分析……10

1.3.5　实践演练——演示两种编译 Android 程序的方法……11

1.4　编译 Android Kernel……14

1.4.1　获取 Goldfish 内核代码……14

1.4.2　获取 MSM 内核代码……17

1.4.3　获取 OMAP 内核代码……17

1.4.4　编译 Android 的 Linux 内核……17

第 2 章　Java 虚拟机基础……19

2.1　虚拟机的作用……19

2.2　Java 虚拟机概述……20

2.2.1　JVM 的数据类型……20

2.2.2　Java 虚拟机体系结构……21

2.2.3　JVM 的生命周期……25

2.3　JVM 的安全性……26

2.3.1　JVM 的安全模型……26

2.3.2　沙箱模型的 4 种组件……27

2.3.3　分析 Java 的策略机制……28

2.4　网络移动性……29

2.4.1　现实需要网络移动性……29

2.4.2　网络移动性……30

2.5　内存异常和垃圾处理……31

2.5.1　内存分配中的栈和堆……31

2.5.2　运行时的数据区域……33

2.5.3　对象访问……34

2.5.4　内存泄露……35

2.5.5　JVM 的垃圾收集策略……36

2.5.6　垃圾收集器……37

2.6　Java 内存模型……37

2.6.1　Java 内存模型概述……38

2.6.2　主内存与工作内存……38

2.6.3　内存间交互操作……39

第 3 章　Dalvik 和 ART 基础……40

3.1　Dalvik VM 和 JVM 的差异……40

3.2　Dalvik 虚拟机的主要特征……41

3.3　Dalvik VM 架构……42

3.3.1　Dalvik 虚拟机的代码结构……42

3.3.2　dx 工具……44

3.3.3　Dalvik VM 的进程管理……44

3.3.4　Android 的初始化流程……44

3.4　Dalvik VM 控制 VM 命令详解……45

3.4.1　基本命令……45

3.4.2　扩展的 JNI 检测……45

3.4.3　断言……46

3.4.4　字节码校验和优化……46

3.4.5　Dalvik VM 的运行模式……47

3.4.6　死锁预测……47

3.4.7　dump 堆栈追踪……48

3.4.8　dex 文件和校验……48

3.4.9　产生标志位……48

3.5　ART 机制基础……48

3.5.1　什么是 ART 模式……48

3.5.2　ART 优化机制基础……50

第 4 章　分析 JNI……52

4.1　JNI 的本质……52

4.2　分析 Java 层……54

4.2.1　加载 JNI 库……54

4.2.2　实现扫描工作……55

4.2.3　读取并保存信息……56

4.2.4　删除 SD 卡外的信息……58

4.2.5　直接转向 JNI……58

4.2.6　扫描函数 scanFile……59

4.2.7　JNI 中的异常处理……59

4.3　分析 JNI 层……60

4.3.1 将Native对象的指针保存到Java对象……60
4.3.2 创建Native层的MediaScanner对象……60
4.4 Native（本地）层……61
4.4.1 注册JNI函数……61
4.4.2 完成注册工作……63
4.4.3 动态注册……64
4.4.4 处理路径参数……65
4.4.5 扫描文件……66
4.4.6 添加TAG信息……66
4.4.7 总结函数JNI_OnLoad()与函数JNI_OnUnload()的用途……67
4.4.8 Java与JNI基本数据类型转换……67
4.4.9 JNIEnv接口……69
4.4.10 JNI中的环境变量……70

第5章 分析内存系统……71

5.1 分析Android的进程通信机制……71
5.1.1 Android的进程间通信（IPC）机制Binder……71
5.1.2 Service Manager是Binder机制的上下文管理者……72
5.1.3 Service Manager服务……86
5.2 匿名共享内存子系统详解……89
5.2.1 基础数据结构……89
5.2.2 初始化处理……90
5.2.3 打开匿名共享内存设备文件……91
5.2.4 内存映射……93
5.2.5 读写操作……94
5.2.6 锁定和解锁……95
5.2.7 回收内存块……100
5.3 C++访问接口层详解……101
5.3.1 接口MemoryBase……101
5.3.2 接口MemoryBase……108
5.4 Java访问接口层详解……111

第6章 Android程序的生命周期管理……115

6.1 Android程序的生命周期……115
6.1.1 进程和线程……115
6.1.2 进程的类型……116
6.2 Activity的生命周期……116
6.2.1 Activity的几种状态……117
6.2.2 分解剖析Activity……117
6.2.3 几个典型的场景……119
6.2.4 管理Activity的生命周期……119
6.2.5 Activity的实例化与启动……120
6.2.6 Activity的暂停与继续……120
6.2.7 Activity的关闭/销毁与重新运行……121
6.2.8 Activity的启动模式……121
6.3 进程与线程……122
6.3.1 进程……122
6.3.2 线程……123
6.3.3 线程安全的方法……123
6.3.4 Android的线程模型……123
6.4 测试生命周期……125
6.5 Service的生命周期……129
6.5.1 Service的基本概念和用途……129
6.5.2 Service的生命周期详解……129
6.5.3 Service与Activity通信……129
6.6 Android广播的生命周期……133
6.6.1 Android的广播机制……133
6.6.2 编写广播程序……133
6.7 ART进程管理……135

第7章 IPC进程通信机制……147

7.1 Binder机制概述……147
7.2 Service Manager是Binder机制的上下文管理者……148
7.2.1 入口函数……148
7.2.2 打开Binder设备文件……149
7.2.3 创建设备文件……149
7.2.4 管理内存映射地址空间……154
7.2.5 发生通知……156
7.2.6 循环等待……161
7.3 内存映射……162
7.3.1 实现内存分配功能……162
7.3.2 分配物理内存……164
7.3.3 释放物理页面……166
7.3.4 分配内核缓冲区……167
7.3.5 释放内核缓冲区……168
7.3.6 查询内核缓冲区……170

第8章 init进程详解……171

8.1 init基础……171
8.2 分析入口函数……172
8.3 配置文件详解……174
8.3.1 init.rc简介……174
8.3.2 分析init.rc的过程……176
8.4 解析service……179

8.4.1 Zygote 对应的 service action…179
8.4.2 init 组织 service…180
8.4.3 函数 parse_service 和 parse_line_service…181
8.5 字段 on…184
8.5.1 Zygote 对应的 on action…184
8.5.2 init 组织 on…185
8.5.3 解析 on 用到的函数…186
8.6 在 init 控制 service…186
8.6.1 启动 Zygote…186
8.6.2 启动 service…187
8.6.3 4 种启动 service 的方式…191
8.7 控制属性服务…194
8.7.1 引入属性…194
8.7.2 初始化属性服务…197
8.7.3 启动属性服务…197
8.7.4 处理设置属性的请求…200
第 9 章 Dalvik VM 的进程系统…202
9.1 Zygote（孕育）进程详解…202
9.1.1 Zygote 基础…202
9.1.2 分析 Zygote 的启动过程…203
9.2 System 进程详解…216
9.2.1 启动 System 进程前的准备工作…216
9.2.2 分析 SystemServer…217
9.2.3 分析 EntropyService…220
9.2.4 分析 DropBoxManagerService…222
9.2.5 分析 DiskStatsService…227
9.2.6 分析 DeviceStorageManager Service…231
9.2.7 分析 SamplingProfilerService…233
9.2.8 分析 ClipboardService…241
9.3 应用程序进程详解…247
9.3.1 创建应用程序…247
9.3.2 启动线程池…256
9.3.3 创建信息循环…257
第 10 章 Dalvik VM 运作流程详解…259
10.1 Dalvik VM 相关的可执行程序…259
10.1.1 dalvikvm、dvz 和 app_process 简介…259
10.1.2 对比 app_process 和 dalvikvm 的执行过程…260
10.2 初始化 Dalvik 虚拟机…262
10.2.1 开始虚拟机的准备工作…262
10.2.2 初始化跟踪显示系统…262
10.2.3 初始化垃圾回收器…263
10.2.4 初始化线程列表和主线程环境参数…263
10.2.5 分配内部操作方法的表格内存…264
10.2.6 初始化虚拟机的指令码相关的内容…264
10.2.7 分配指令寄存器状态的内存…264
10.2.8 分配指令寄存器状态的内存和最基本用的 Java 库…265
10.2.9 初始化使用的 Java 类库线程类…266
10.2.10 初始化虚拟机使用的异常 Java 类库…267
10.2.11 初始化其他对象…268
10.3 启动 Zygote…276
10.3.1 在 init.rc 中配置 Zygote 启动参数…276
10.3.2 启动 Socket 服务端口…276
10.3.3 加载 preload-classes…277
10.3.4 加载 preload-resources…277
10.3.5 使用 folk 启动新进程…278
10.4 启动 SystemServer 进程…278
10.4.1 启动各种系统服务线程…279
10.4.2 启动第一个 Activity…280
10.5 加载 class 类文件…281
10.5.1 DexFile 在内存中的映射…281
10.5.2 ClassObject——Class 在加载后的表现形式…283
10.5.3 加载 Class 并生成相应 ClassObject 的函数…283
10.5.4 加载基本类库文件…284
10.5.5 加载用户类文件…284
第 11 章 DEX 文件详解…285
11.1 DEX 文件介绍…285
11.2 DEX 文件的格式…285
11.2.1 map_list…286
11.2.2 string_id_item…288
11.2.3 type_id_item…291
11.2.4 proto_id_item…292
11.2.5 ield_id_item…293
11.2.6 method_id_item…293
11.2.7 class_def_item…294

11.3 DEX 文件结构……297
11.3.1 文件头（File Header）……297
11.3.2 魔数字段……298
11.3.3 检验码字段……298
11.3.4 SHA-1 签名字段……300
11.3.5 map_off 字段……300
11.3.6 string_ids_size 和 off 字段……301
11.4 DEXFile 接口详解……303
11.4.1 构造函数……303
11.4.2 公共方法……304
11.5 DEX 和动态加载类机制……306
11.5.1 类加载机制……306
11.5.2 具体加载……306
11.5.3 代码加密……308
11.6 动态加载 jar 和 DEX……309
第 12 章 Dvlik VM 内存系统详解……310
12.1 如何分配内存……310
12.2 内存管理机制详解……312
12.3 优化 Dalvik 虚拟机的堆内存分配……326
第 13 章 Dalvik VM 垃圾收集机制……328
13.1 引用计数算法……328
13.2 Mark Sweep 算法……328
13.3 和垃圾收集算法有关的函数……330
13.4 垃圾回收的时机……346
13.5 调试信息……347
13.6 Dalvik VM 和 JVM 垃圾收集机制的区别……348
第 14 章 Dalvik VM 内存优化机制详解……350
14.1 sp 和 wp 简介……350
14.1.1 sp 基础……350
14.1.2 wp 基础……351
14.2 智能指针详解……351
14.2.1 智能指针基础……352
14.2.2 轻量级指针……353
14.2.3 强指针……355
14.2.4 弱指针……365
第 15 章 分析 Dalvik VM 的启动过程……369
15.1 Dalvik VM 启动流程概览……369
15.2 Dalvik VM 启动过程详解……370
15.2.1 创建 Dalvik VM 实例……370
15.2.2 指定一系列控制选项……371
15.2.3 创建并初始化 Dalvik VM 实例……376
15.2.4 创建 JNIEnvExt 对象……378
15.2.5 设置当前进程和进程组 ID……382
15.2.6 注册 Android 核心类的 JNI 方法……382
15.2.7 创建 javaCreateThreadEtc 钩子……385
第 16 章 注册 Dalvik VM 并创建线程……387
16.1 注册 Dalvik VM 的 JNI 方法……387
16.1.1 设置加载程序……387
16.1.2 加载 so 文件并验证……387
16.1.3 获取描述类……392
16.1.4 注册 JNI 方法……392
16.1.5 实现 JNI 操作……394
16.2 创建 Dalvik VM 进程……395
16.2.1 分析底层启动过程……395
16.2.2 创建 Dalvik VM 进程……395
16.2.3 初始化运行的 Dalvik VM……398
16.3 创建 Dalvik VM 线程……399
16.3.1 检查状态值……399
16.3.2 创建线程……399
16.3.3 分析启动过程……402
16.3.4 清理线程……404
第 17 章 Dalvik VM 异常处理详解……407
17.1 Java 异常处理机制……407
17.1.1 方法调用栈……407
17.1.2 Java 提供的异常处理类……409
17.2 Java VM 异常处理机制详解……409
17.2.1 Java 语言及虚拟机的异常处理机制……410
17.2.2 COSIX 虚拟机异常处理的设计与实现……410
17.3 分析 Dalvik 虚拟机异常处理的源码……414
17.3.1 初始化虚拟机使用的异常 Java 类库……414
17.3.2 抛出一个线程异常……415
17.3.3 持续抛出进程……415
17.3.4 找出异常原因……416
17.3.5 找出异常原因……417
17.3.6 清除挂起的异常和等待初始化的异常……420
17.3.7 包装“现在等待”异常的不同例外……420
17.3.8 输出跟踪当前异常的错误信息……421

17.3.9 搜索和当前异常相匹配的方法……421
17.3.10 获取匹配的捕获块……423
17.3.11 进行堆栈跟踪……424
17.3.12 生成堆栈跟踪元素……425
17.3.13 将内容添加到堆栈跟踪日志中……426
17.3.14 将内容添加到堆栈跟踪日志中……427
17.4 常见异常的类型与原因……428
17.4.1 SQLException：操作数据库异常类……428
17.4.2 ClassCastException：数据类型转换异常……428
17.4.3 NumberFormatException：字符串转换为数字类型时抛出的异常……428
17.5 调用堆栈跟踪分析异常……429
17.5.1 解决段错误……429
17.5.2 跟踪 Android Callback 调用堆栈……431

第 18 章 JIT 编译……434

18.1 JIT 简介……434
18.1.1 JIT 概述……434
18.1.2 Java 虚拟机主要的优化技术……436
18.1.3 Dalvik 中 JIT 的实现……436
18.2 Dalvik VM 对 JIT 的支持……436
18.3 汇编代码和改动……438
18.3.1 汇编部分代码……438
18.3.2 对 C 文件的改动……438
18.4 Dalvik VM 中的 JIT 源码……439
18.4.1 入口文件……439
18.4.2 核心函数……447
18.4.3 编译文件……450
18.4.4 BasicBlock 处理……458
18.4.5 内存初始化……459
18.4.6 对 JIT 源码的总结……462

第 19 章 Dalvik VM 内存优化……463

19.1 Android 内存优化的作用……463
19.2 查看 Android 内存和 CPU 使用情况……464
19.2.1 利用 Android API 函数查看……464
19.2.2 直接对 Android 文件进行解析查询……464
19.2.3 通过 Runtime 类实现……465
19.2.4 使用 DDMS 工具获取……465
19.2.5 其他方法……469
19.3 Android 的内存泄露……472
19.3.1 什么是内存泄漏……472
19.3.2 为什么会发生内存泄露……473
19.3.3 shallow size、retained size……474
19.3.4 查看 Android 内存泄露的工具——MAT……475
19.3.5 查看 Android 内存泄露的方法……478
19.3.6 Android（Java）中常见的容易引起内存泄漏的不良代码……480
19.4 常见的引起内存泄露的坏习惯……480
19.4.1 查询数据库时忘记关闭游标……481
19.4.2 构造 Adapter 时不习惯使用缓存的 convertView……481
19.4.3 没有及时释放对象的引用……482
19.4.4 不在使用 Bitmap 对象时调用 recycle()释放内存……482
19.5 解决内存泄露实践……483
19.5.1 使用 MAT 根据 heap dump 分析 Java 代码内存泄漏的根源……483
19.5.2 演练 Android 中内存泄露代码优化及检测……489

第 20 章 Dalvik VM 性能优化……491

20.1 加载 APK/DEX 文件优化……491
20.1.1 APK 文件介绍……492
20.1.2 DEX 文件优化……493
20.1.3 使用类动态加载技术实现加密优化……493
20.2 SD 卡优化……496
20.3 虚拟机优化详解……497
20.3.1 平台优化——ARM 的流水线技术……497
20.3.2 Android 对 C 库优化……501
20.3.3 优化创建的进程……504
20.3.4 渲染优化……504

第 21 章 分析 ART 的启动过程……508

21.1 运行环境的转换……508
21.2 运行 app_process 进程……509
21.3 准备启动……512
21.4 创建运行实例……518

21.5 注册本地 JNI 函数……519
21.6 启动守护进程……520
21.7 解析参数……521
21.8 初始化类、方法和域……528

第 22 章 执行 ART 主程序……534

22.1 进入 main 主函数……534
22.2 查找目标类……535
22.2.1 函数 LookupClass()……535
22.2.2 函数 DefineClass()……537
22.2.3 函数 InsertClass()……540
22.2.4 函数 LinkClass()……541
22.3 类操作……543
22.4 实现托管操作……544

第 23 章 安装 APK 应用程序……549

23.1 PackageManagerService 概述……549
23.2 主函数 main……549
23.3 调用初始化函数……550
23.4 创建 PackageManagerService 服务……553
23.5 扫描并解析……554
23.6 保存解析信息……570

第 24 章 ART 环境安装 APK 应用程序……572

24.1 Android 安装 APK 概述……572
24.2 启动时安装……572
24.3 ART 安装……581
24.4 实现 dex2oat 转换……586
24.4.1 参数解析……586
24.4.2 创建 OAT 文件指针……588
24.4.3 dex2oat 准备工作……588
24.4.4 提取 classes.dex 文件……589
24.4.5 创建 OAT 文件……594
24.5 APK 文件的转换……595

第1章　获取并编译 Android 源码

在本章中，将详细讲解获取并编译 Android 源码的基本知识，介绍各个目录中主要文件的功能，为读者步入本书后面知识的学习打下基础。

1.1 获取 Android 源码

要想研究 Android 系统的源码，需要先获取其源码。目前市面上主流的操作系统有 Windows、Linux、Mac OS 的操作系统，由于 Mac OS 源自于 Linux 系统，因此本书将讲解分别在 Windows 系统和 Linux 系统中获取 Android 源码的知识。

1.1.1　在 Linux 系统获取 Android 源码

在 Linux 系统中，通常使用 Ubuntu 来下载和编译 Android 源码。由于 Android 的源码内容很多，Google 采用了 Git 的版本控制工具，并对不同的模块设置不同的 Git 服务器，可以用 repo 自动化脚本来下载 Android 源码，下面介绍获取 Android 源码的过程。

（1）下载 repo。

在用户目录下创建存放 repo 的 bin 文件夹，并把该路径设置到环境变量中去，具体命令如下所示：

```
$ mkdir ~/bin
$ PATH=~/bin:$PATH
```

下载用于执行 repo 的 repo 的脚本，具体命令如下所示：

```
$ curl https://dl-ssl.google.com/dl/googlesource/git-repo/repo > ~/bin/repo
```

设置可执行权限，命令如下所示：

```
$ chmod a+x ~/bin/repo
```

（2）初始化一个 repo 的客户端。

在用户目录下创建一个空目录，用于存放 Android 源码，命令如下所示：

```
$ mkdir AndroidCode
$ cd AndroidCode
```

进入到 AndroidCode 目录，并运行 repo 下载源码，下载主线分支的代码，主线分支包括最新修改的 bug，以及并未正式发布版本的最新源码，命令如下所示：

```
$ repo init -u https://android.googlesource.com/platform/manifest
```

下载其他分支，建议下载正式发布的版本，可以通过添加-b 参数来下载，例如下载 Android 4.3 正式版的命令如下所示：

```
$ repo init -u https://android.googlesource.com/platform/manifest -b
android-4.3_r1
```

在下载过程中会需要填写 Name 和 E-mail，填写完毕之后，选择 Y 进行确认。最后提示 repo 初始化完成，这时可以开始同步 Android 源码了。同步过程非常漫长，需要大家耐心等待。执行下面命令开始同步代码：

```
$ repo sync
```

经过上述步骤后，便开始下载并同步 Android 源码了，界面效果如图 1-1 所示。

```
Checking out files: 100% (2497/2497), done. out files:  31% (790/2497)
Checking out files: 100% (1654/1654), done. out files:  39% (649/1654)
Checking out files: 100% (3471/3471), done.
Checking out files: 100% (24607/24607), done.ut files:  21% (5269/24607)
Checking out files: 100% (2431/2431), done. out files:  32% (800/2431)
Checking out files: 100% (18696/18696), done.
Checking out files: 100% (4276/4276), done. out files:  48% (2058/4276)
Checking out files: 100% (2216/2216), done. out files:  49% (1093/2216)
Checking out files: 100% (857/857), done.ng out files:  13% (115/857)
Checking out files: 100% (1141/1141), done. out files:   2% (28/1141)
Checking out files: 100% (431/431), done.ng out files:  10% (46/431)
Checking out files: 100% (175/175), done.ng out files:  10% (19/175)
Checking out files: 100% (135/135), done.
Checking out files: 100% (378/378), done.
Checking out files: 100% (433/433), done.
Checking out files: 100% (2407/2407), done. out files:  30% (724/2407)
Checking out files: 100% (2489/2489), done.
Checking out files: 100% (2493/2493), done.
Checking out files: 100% (177/177), done.ng out files:  15% (27/177)
Checking out files: 100% (137/137), done.
Checking out files: 100% (40775/40775), done.ut files:   5% (2199/40775)
Checking out files: 100% (93/93), done.
Checking out files: 100% (450/450), done.
Checking out files: 100% (5265/5265), done. out files:  41% (2167/5265)
Syncing work tree: 100% (329/329), done.
```

图 1-1　下载同步界面

1.1.2　在 Windows 平台获取 Android 源码

在 Windows 平台获取源码与在 Linux 上原理相同，但是需要预先在 Windows 平台上搭建一个 Linux 环境，此处需要用到 Cygwin 工具。Cygwin 的作用是构建一套在 Windows 上的 Linux 模拟环境，下载 Cygwin 工具的地址如下所示：

```
http://cygwin.com/install.html
```

下载成功后会得到一个名为“setup.exe”的可执行文件，通过此文件可以更新和下载最新的工具版本，具体流程如下所示。

（1）启动 Cygwin，如图 1-2 所示。

（2）单击“下一步”按钮，选择第一个选项：从网络下载安装，如图 1-3 所示。

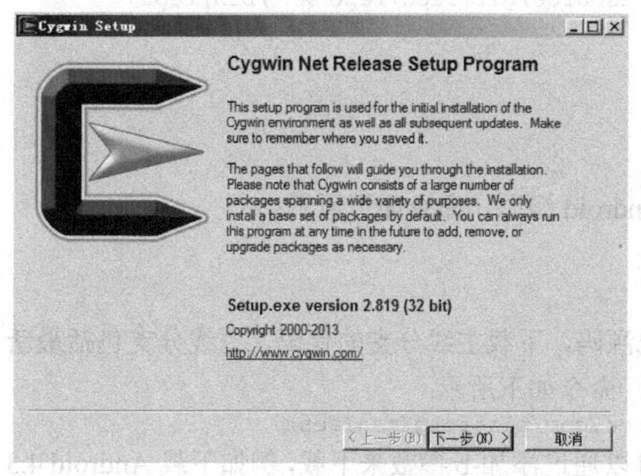

图 1-2　启动 Cygwin

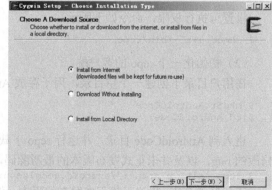

图 1-3　选择从网络下载安装

（3）单击“下一步”按钮，选择安装根目录，如图 1-4 所示。

（4）单击“下一步”按钮，选择临时文件目录，如图 1-5 所示。

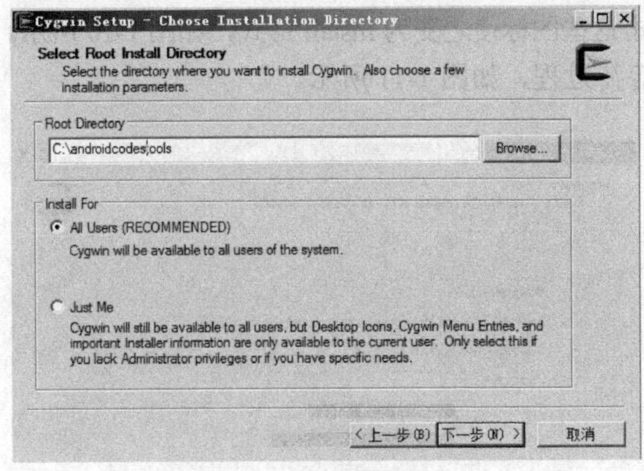

图 1-4 选择安装根目录

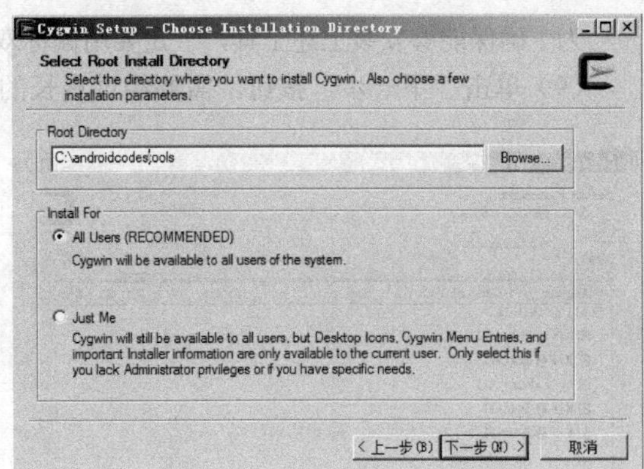

图 1-5 选择临时文件目录

（5）单击“下一步”按钮，设置网络代理。如果所在网络需要代理，则在这一步进行设置，如果不用代理，则选择直接下载，如图 1-6 所示。

（6）单击“下一步”按钮，选择下载站点。一般选择离得比较近的站点，速度会比较快，如图 1-7 所示。

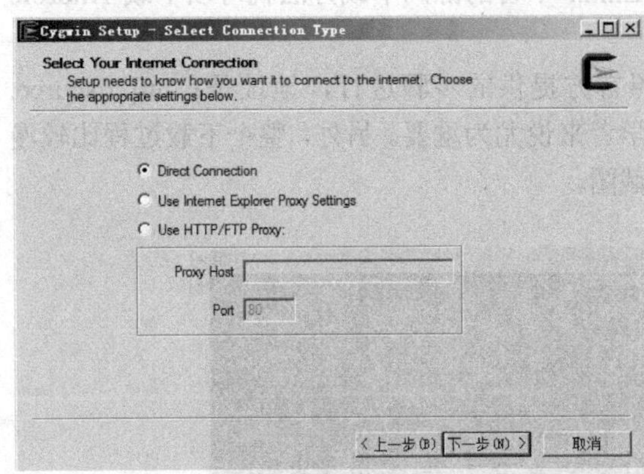

图 1-6 设置网络代理

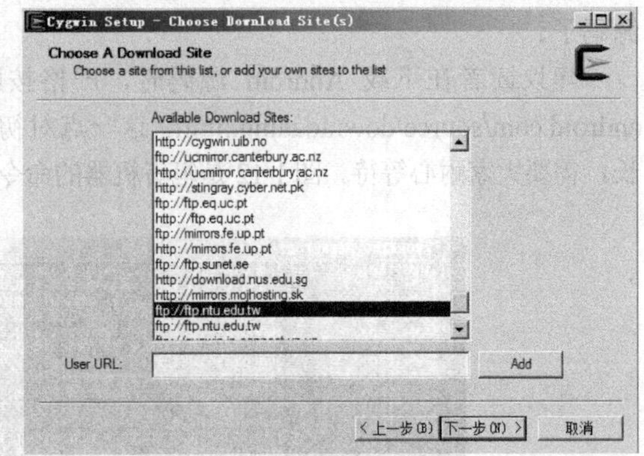

图 1-7 选择下载站点

（7）单击“下一步”按钮，开始更新工具列表，如图 1-8 所示。

（8）单击“下一步”按钮，选择需要下载的工具包。在此需要依次下载 curl、git、python 这些工具，如图 1-9 所示。

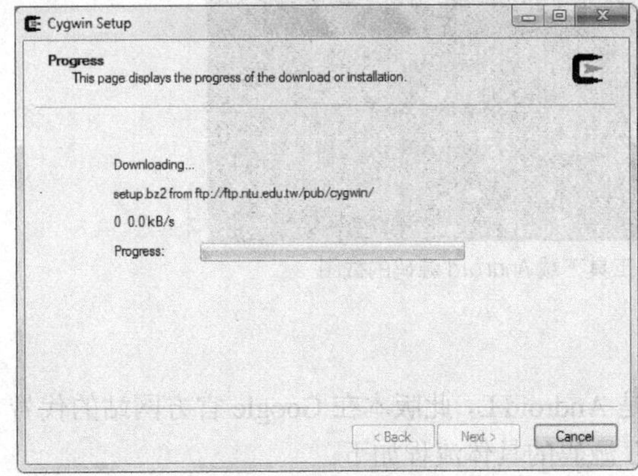

图 1-8 更新工具列表

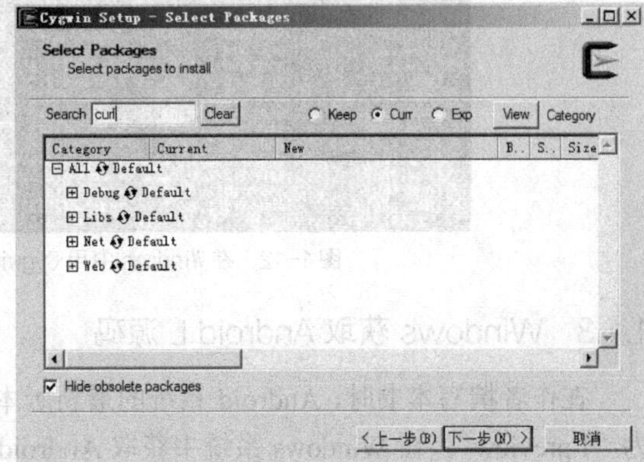

图 1-9 依次下载工具

为了确保能够安装上述工具，一定要用鼠标双击这些图标使之变为 Install 形式，如图 1-10 所示。

（9）单击“下一步”按钮，需要经过漫长的等待过程，如图 1-11 所示。

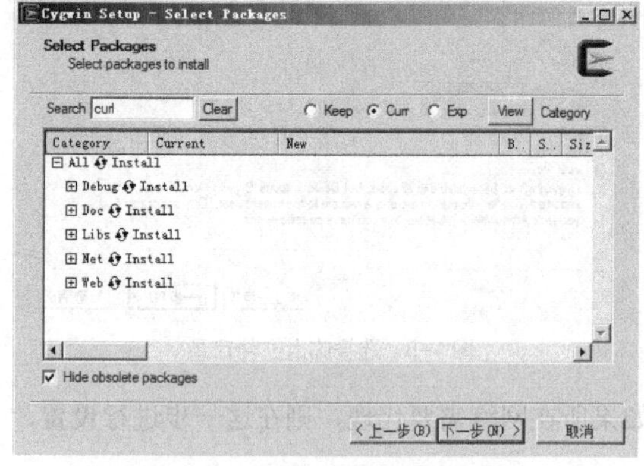

图 1-10　务必设置为 Install 形式

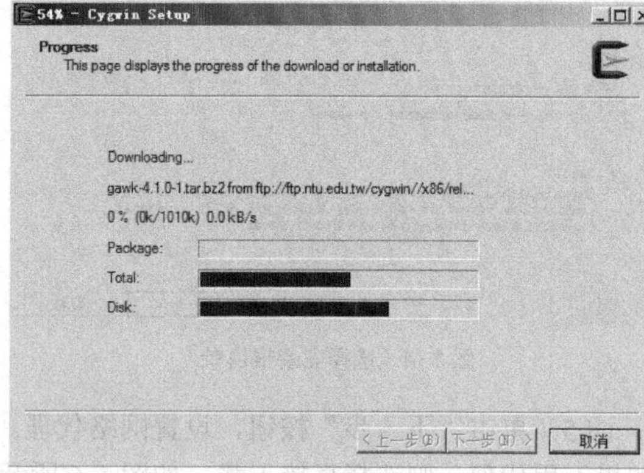

图 1-11　下载进度条

如果下载安装成功会出现提示信息，单击“完成”按钮即完成安装。打开安装好的 Cygwin 后，会模拟出一个 Linux 的工作环境，然后按照 Linux 平台的源码下载方法就可以下载 Android 源码了。

建议读者在下载 Android 源码时，严格按照官方提供的步骤进行，地址是：http://source.android.com/source/downloading.html，这一点对初学者来说尤为重要。另外，整个下载过程比较漫长，需要大家耐心等待。图 1-12 是笔者机器的命令截图。

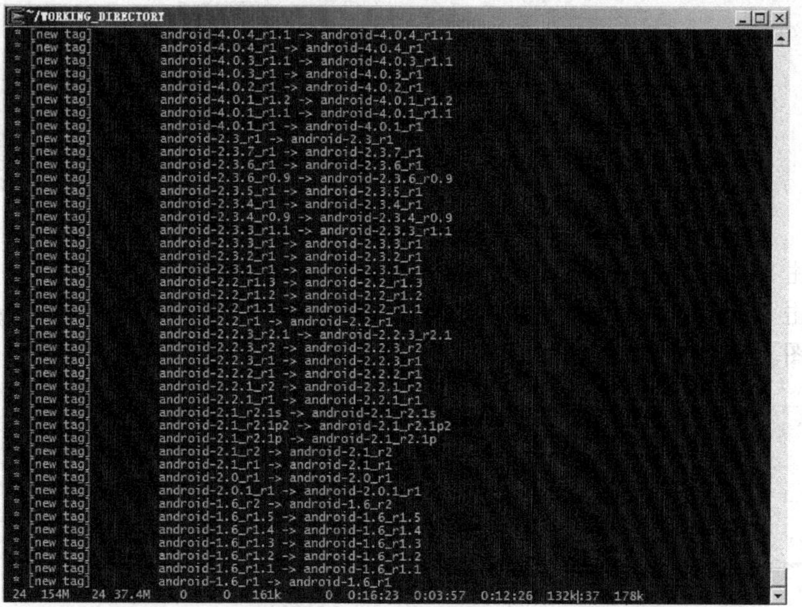

图 1-12　在 Windows 中用 Cygwin 工具下载 Android 源码的截图

1.1.3　Windows 获取 Android L 源码

在作者撰写本书时，Android 系统的最新版本是 Android L，此版本在 Google 官方网站的代号为“l-preview”。在 Windows 系统中获取 Android L 源码的具体流程如下。

（1）下载 Git 工具，其官方下载地址是 http://www.git-scm.com/downloads，如图 1-13 所示。

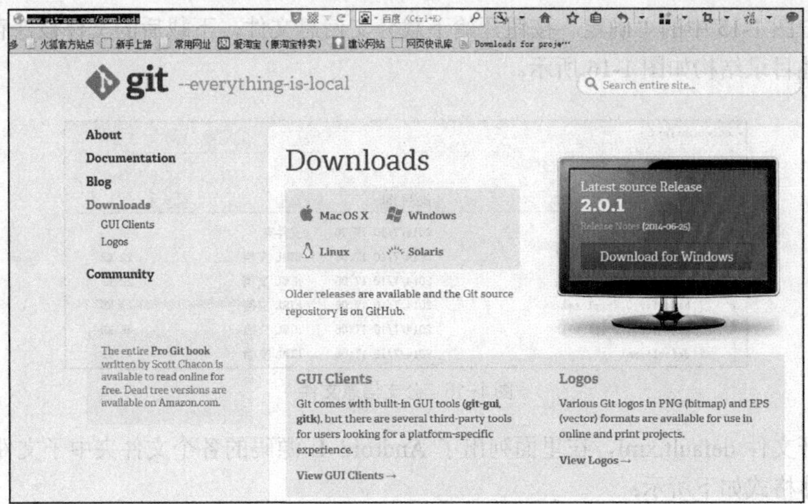

图 1-13 下载 Git 工具

下载后双击可执行文件进行安装，在安装过程按照默认选项安装即可。

（2）下载并安装 TortoiseGit 工具，下载地址是 https://code.google.com/p/tortoisegit/，如图 1-14 所示。

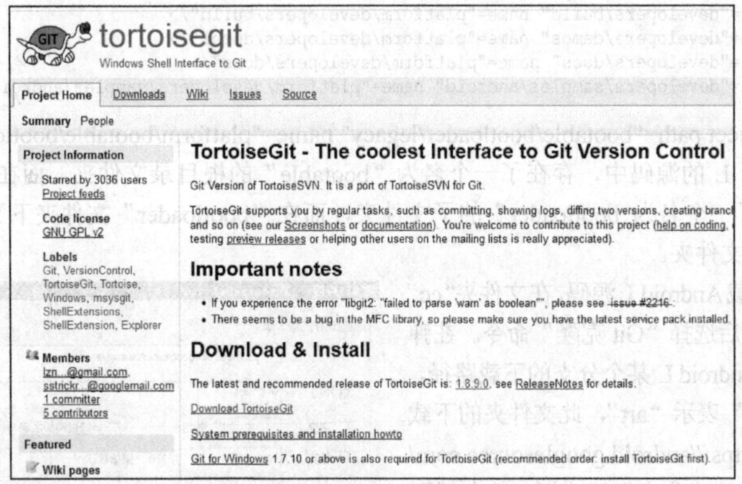

图 1-14 下载 TortoiseGit 工具

下载后双击下载后的可执行文件进行安装，安装过程按照默认选项安装即可。如果读者对英文不敢兴趣，可以下载 TortoiseGit 的中文版本，笔者安装的就是 TortoiseGit 中文版。

（3）新建一个保存源码的文件夹，例如“cc”，在文件夹上单击右键，然后选择“Git 克隆”命令。

（4）在弹出的“Git 克隆”对话框界面中，在“URL”后面的文本框中输入 Android L 项目下载路径：https://android.googlesource.com/platform/manifest.git，如图 1-15 所示。

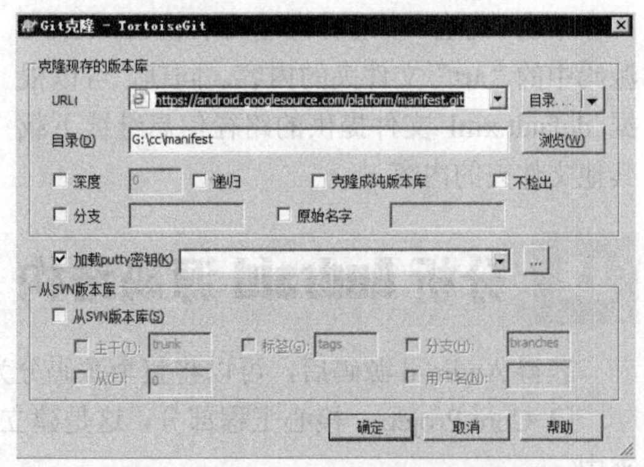

图 1-15 输入 Android L 项目下载路径

（5）单击图 1-15 中的“确定”按钮开始下载分支信息文件，下载后的文件被保存在“cc”文件夹中，具体目录结构如图 1-16 所示。

▸ cc ▸ manifest ▸

刻录　新建文件夹

名称	修改日期	类型	大小
.git	2014/7/10 17:06	文件夹	
base-for-3.0-gpl.xml	2014/7/10 17:06	HTML 文档	22 KB
base-for-3.1-gpl.xml	2014/7/10 17:06	HTML 文档	22 KB
base-for-3.2-gpl.xml	2014/7/10 17:06	HTML 文档	23 KB
base-for-4.2-gpl.xml	2014/7/10 17:06	HTML 文档	45 KB
default.xml	2014/7/10 17:06	HTML 文档	45 KB

图 1-16　分支信息文件

（6）打开文件 default.xml，在里面列出了 Android L 源码的各个文件夹中子文件夹目录的分支信息，具体格式如下所示：

```
<project path="abi/cpp" name="platform/abi/cpp" groups="pdk"/><project path="art" name="platform/art"/>
<project path="bionic" name="platform/bionic" groups="pdk"/>
<project path="bootable/bootloader/legacy" name="platform/bootable/bootloader/legacy"/>
<project path="bootable/diskinstaller" name="platform/bootable/diskinstaller"/>
<project path="bootable/recovery" name="platform/bootable/recovery" groups="pdk"/>
<project path="cts" name="platform/cts" groups="cts"/>
<project path="dalvik" name="platform/dalvik"/>
<project path="developers/build" name="platform/developers/build"/>
<project path="developers/demos" name="platform/developers/demos"/>
<project path="developers/docs" name="platform/developers/docs"/>
<project path="developers/samples/android" name="platform/developers/samples/android"/>
```

例如“<project path="bootable/bootloader/legacy" name="platform/bootable/bootloader/legacy"/>”表示在 Android L 的源码中，存在了一个名为“bootable”的根目录文件夹，而在“bootable”文件夹中又包含了一个名为 bootloader”的子文件夹，而在“bootloader”文件夹下又包含了一个名为“legacy”的文件夹。

（7）开始下载 Android L 源码，在文件夹“cc”上单击右键，然后选择“Git 克隆”命令。在弹出界面中输入 Android L 某个分支的下载路径，例如 path=“art”表示“art”，此文件夹的下载地址是：https://android.googlesource.com/a/art.git。然后勾选“分支”复选框，并在后面填写“l-prevew”分支，如图 1-17 所示。

单击“确定”按钮后将开始下载 Android L 源码中的“art”文件夹的内容。同理，可以根据 default.xml 文件提供的路径信息继续下载其他文件夹的内容。

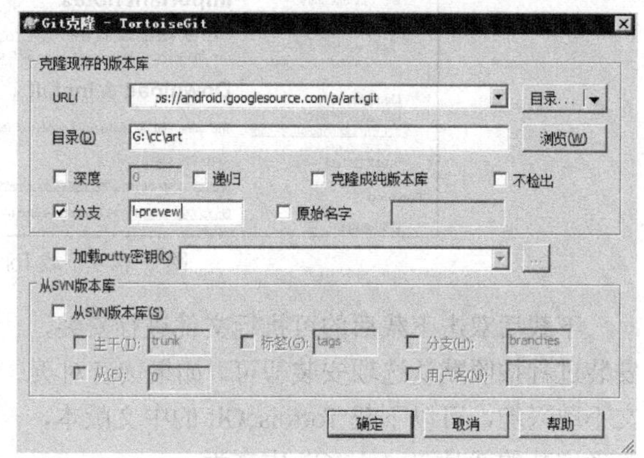

图 1-17　开始下载“art”文件夹的内容

1.2　分析 Android 源码结构

获得 Android 源码后，可以将整个源码分为如下 3 个部分。

- Core Project：核心工程部分，这是建立 Android 系统的基础，被保存在根目录的各个文件夹中。

❑ External Project：扩展工程部分，可以使其他开源项目具有扩展功能，被保存在“external”文件夹中。

❑ Package：包部分，提供了 Android 的应用程序、内容提供者、输入法和服务，被保存在“package”文件夹中。

无论是 Android 1.5 还是 Android 4.3 和 Android L，各个版本的源码目录基本类似。在里面包含了原始 Android 的目标机代码、主机编译工具和仿真环境。解压缩下载的 Android 4.3 源码包后，第一级别目录结构的具体说明如表 1-1 所示。

表 1-1　Android 源码的根目录

Android 源码根目录	描　述
abi	abi 相关代码，abi:application binary interface，应用程序二进制接口，在 Android L 中，此文件夹被修改为“api”
art	ART 运行环境文件夹，ART 机制是从 Android 4.4 开始推出的运行模式。和传统的 Dalvik 模式相比，ART 模式的运行速度更快，所需的内存更小。从 Android L 开始，ART 被设置为系统默认的运行环境
bionic	bionic C 库
bootable	启动引导相关代码
build	存放系统编译规则及 generic 等基础开发配置包
cts	Android 兼容性测试套件标准
dalvik	Dalvik Java 虚拟机
development	应用程序开发相关
device	设备相关代码
docs	介绍开源的相关文档
external	android 使用的一些开源的模组
frameworks	核心框架——Java 及 C++语言，是 Android 应用程序的框架
gdk	即时通信模块
hardware	主要是硬件适配层 HAL 代码
kernel	Linux 的内核文件
libcore	核心库相关
libnativehelper	是 Support functions for Android's class libraries 的缩写，表示动态库，是实现实现的 JNI 库的基础。
ndk	ndk 相关代码。Android NDK（Android Native Development Kit）是一系列的开发工具，允许程序开发人员在 Android 应用程序中嵌入 C/C++语言编写的非托管代码
out	编译完成后的代码输出在此目录
packages	应用程序包
pdk	Plug Development Kit 的缩写，是本地开发套件
prebuilts	x86 和 ARM 架构下预编译的一些资源
sdk	sdk 及模拟器
system	文件系统和应用及组件，是用 C 语言实现的
tools	工具文件夹
vendor	厂商定制代码
Makefile	全局的 Makefile

1.3 编译 Android 源码

编译 Android 源码的方法非常简单，只需使用 Android 源码根目录下的 Makefile，执行 make 命令即可轻松实现。因为 Android L 是一个 Preview 版本，官方并没有公布其完整的内核代码。所以本节中的编译内容将以正式版 Android 4.3 进行。当然在编译 Android 源码之前，首先要确定已经完成同步工作。进入 Android 源码目录使用 make 命令进行编译，使用此命令的格式如下所示：

```
$: cd ~/Android4.3（这里的“Android4.3”就是我们下载源码的保存目录）
$: make
```

编译 Android 源码可以得到“～/project/android/cupcake/out”目录，笔者的截图界面如图 1-18 所示。

```
意：external/doclava/src/com/google/doclava/Stubs.java 使用了未经检查或不安全
操作。
意：要了解详细信息，请使用 -Xlint:unchecked 重新编译。
otice file: external/doclava/NOTICE -- out/host/linux-x86/obj/NOTICE_FILES/src/
framework/doclava.jar.txt
nstall: out/host/linux-x86/framework/doclava.jar
arget Java: core (out/target/common/obj/JAVA_LIBRARIES/core_intermediates/class
s)
意：某些输入文件使用或覆盖了已过时的 API。
意：要了解详细信息，请使用 -Xlint:deprecation 重新编译。
意：某些输入文件使用了未经检查或不安全的操作。
意：要了解详细信息，请使用 -Xlint:unchecked 重新编译。
opying: out/target/common/obj/JAVA_LIBRARIES/core_intermediates/classes-jarjar.
ar
opying: out/target/common/obj/JAVA_LIBRARIES/core_intermediates/emma_out/lib/cl
sses-jarjar.jar
opying: out/target/common/obj/JAVA_LIBRARIES/core_intermediates/classes.jar
arget Java: conscrypt (out/target/common/obj/JAVA_LIBRARIES/conscrypt_intermedi
tes/classes)
意：某些输入文件使用或覆盖了已过时的 API。
意：要了解详细信息，请使用 -Xlint:deprecation 重新编译。
意：某些输入文件使用了未经检查或不安全的操作。
意：要了解详细信息，请使用 -Xlint:unchecked 重新编译。
ost Prebuilt: jarjar (out/host/common/obj/JAVA_LIBRARIES/jarjar_intermediates/javalib.jar)
nstall: out/host/linux-x86/framework/jarjar.jar
arJar: out/target/common/obj/JAVA_LIBRARIES/conscrypt_intermediates/classes-jarjar.jar
opying: out/target/common/obj/JAVA_LIBRARIES/conscrypt_intermediates/emma_out/lib/classes-jarjar.jar
opying: out/target/common/obj/JAVA_LIBRARIES/conscrypt_intermediates/classes.jar
arget Java: bouncycastle (out/target/common/obj/JAVA_LIBRARIES/bouncycastle_intermediates/classes)
意：某些输入文件使用或覆盖了已过时的 API。
意：要了解详细信息，请使用 -Xlint:deprecation 重新编译。
意：某些输入文件使用了未经检查或不安全的操作。
意：要了解详细信息，请使用 -Xlint:unchecked 重新编译。
arJar: out/target/common/obj/JAVA_LIBRARIES/bouncycastle_intermediates/classes-jarjar.jar
opying: out/target/common/obj/JAVA_LIBRARIES/bouncycastle_intermediates/emma_out/lib/classes-jarjar.jar
opying: out/target/common/obj/JAVA_LIBRARIES/bouncycastle_intermediates/classes.jar
arget Java: ext (out/target/common/obj/JAVA_LIBRARIES/ext_intermediates/classes)
```

图 1-18　编译过程的界面截图

整个编译过程也是非常漫长的，需要读者耐心等待。在本节的内容中，将详细讲解编译 Android 源码的基本过程。

1.3.1　搭建编译环境

在编译 Android 源码之前，需要先进行环境搭建工作。在接下来的内容中，以 Ubuntu 系统为例讲解搭建编译环境以及编译 Android 源码的方法。具体流程如下。

（1）安装 JDK，编译 Android 4.3 的源码需要 JDK1.6，下载 jdk-6u21-linux-i586.bin 后进行安装，对应命令如下所示：

```
$ cd /usr
$ mkdir java
$ cd java
$ sudo cp jdk-6u21-linux-i586.bin 所在目录 ./
$ sudo chmod 755 jdk-6u21-linux-i586.bin
$ sudo sh jdk-6u21-linux-i586.bin
```

（2）设置 JDK 环境变量，将如下环境变量添加到主文件夹目录下的.bashrc 文件中，然后用

source 命令使其生效，加入的环境变量代码如下所示：

```
export JAVA_HOME=/usr/java/jdk1.6.0_23
export JRE_HOME=$JAVA_HOME/jre
export CLASSPATH=.:$JAVA_HOME/lib:$JRE_HOME/lib:$CLASSPATH
export PATH=$PATH:$JAVA_HOME/bin:$JAVA_HOME/bin/tools.jar:$JRE_HOME/bin
export ANDROID_JAVA_HOME=$JAVA_HOME
```

（3）安装需要的包，读者可以根据编译过程中的提示进行选择，可能需要的包的安装命令如下所示：

```
$ sudo apt-get install git-core bison zlib1g-dev flex libx11-dev gperf sudo aptitude install git-core
gnupg flex bison gperf libsdl-dev libesd0-dev libwxgtk2.6-dev build-essential zip curl libncurses5-dev
zlib1g-dev
```

1.3.2 开始编译

当完成安装所依赖包的工作后，就可以开始编译 Android 源码了，具体步骤如下。

（1）首先进行编译初始化工作，在终端中执行以下命令：

```
source build/envsetup.sh
```

或：

```
.build/envsetup.sh
```

执行后将会输出以下内容。

```
source build/envsetup.sh
including device/asus/grouper/vendorsetup.sh
including device/asus/tilapia/vendorsetup.sh
including device/generic/armv7-a-neon/vendorsetup.sh
including device/generic/armv7-a/vendorsetup.sh
including device/generic/mips/vendorsetup.sh
including device/generic/x86/vendorsetup.sh
including device/samsung/maguro/vendorsetup.sh
including device/samsung/manta/vendorsetup.sh
including device/samsung/toroplus/vendorsetup.sh
including device/samsung/toro/vendorsetup.sh
including device/ti/panda/vendorsetup.sh
including sdk/bash_completion/adb.bash
```

（2）然后选择编译目标，具体命令如下：

```
lunch full-eng
```

执行后会输出如下所示的提示信息：

```
============================================
PLATFORM_VERSION_CODENAME=REL
PLATFORM_VERSION=4.3
TARGET_PRODUCT=full
TARGET_BUILD_VARIANT=eng
TARGET_BUILD_TYPE=release
TARGET_BUILD_APPS=
TARGET_ARCH=arm
TARGET_ARCH_VARIANT=armv7-a
HOST_ARCH=x86
HOST_OS=linux
HOST_OS_EXTRA=Linux-3.2.2-5-generic-x86_61-with-Ubuntu-10.01-lucid
HOST_BUILD_TYPE=release
BUILD_ID=JOP40C
OUT_DIR=out
============================================
```

（3）接下来开始编译代码，在终端中执行以下命令：

```
make -j4
```

其中“-j4”表示用 4 个线程进行编译。整个编译进度根据不同机器的配置而需要不同的时间。例如笔者电脑为 Intel i5-2300 四核 2.8 Hz，4 GB 内存，经过近 4 小时才编译完成。当出现下面的信息时表示编译完成：

```
target Java: ContactsTests (out/target/common/obj/APPS/ContactsTests_intermediates/classes)
target Dex: Contacts
Done!
Install: out/target/product/generic/system/app/Browser.odex
Install: out/target/product/generic/system/app/Browser.apk
Note: Some input files use or override a deprecated API.
Note: Recompile with -Xlint:deprecation for details.
Copying: out/target/common/obj/APPS/Contacts_intermediates/noproguard.classes.dex
target Package: Contacts (out/target/product/generic/obj/APPS/Contacts_intermediates/package.apk)
 'out/target/common/obj/APPS/Contacts_intermediates/classes.dex' as 'classes.dex'...
Processing target/product/generic/obj/APPS/Contacts_intermediates/package.apk
Done!
Install: out/target/product/generic/system/app/Contacts.odex
Install: out/target/product/generic/system/app/Contacts.apk
build/tools/generate-notice-files.py                          out/target/product/generic/obj/NOTICE.txt
out/target/product/generic/obj/NOTICE.html "Notices for files contained in the filesystem images in this
directory:" out/target/product/generic/obj/NOTICE_FILES/src
Combining NOTICE files into HTML
Combining NOTICE files into text
Installed file list: out/target/product/generic/installed-files.txt
Target system fs image: out/target/product/generic/obj/PACKAGING/systemimage_intermediates/system.img
Running:  mkyaffs2image -f out/target/product/generic/system out/target/product/generic/obj/PACKAGING/
systemimage_intermediates/system.img
Install system fs image: out/target/product/generic/system.img
DroidDoc took 5331 sec. to write docs to out/target/common/docs/doc-comment-check
```

1.3.3　在模拟器中运行

在模拟器中运行的步骤就比较简单了，只需在终端中执行下面的命令即可：

```
emulator
```

运行成功后的效果如图 1-19 所示。

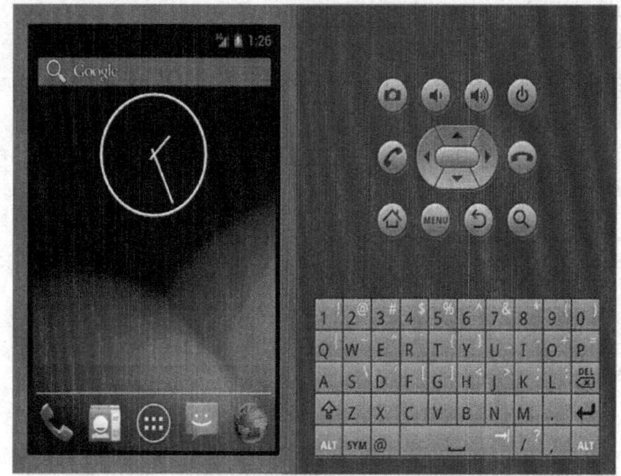

图 1-19　在模拟器中的编译执行效果

1.3.4　常见的错误分析

虽然编译方法非常简单，但是作为初学者来说非常容易出错，在下面列出了其中常见的编译错误类型。

（1）缺少必要的软件。

进入到 Android 目录下，使用 make 命令进行编译，可能会发现出现如下所示的错误提示。

```
host C: libneo_cgi <= external/clearsilver/cgi/cgi.c
external/clearsilver/cgi/cgi.c:22:18: error: zlib.h: No such file or directory
```

上述错误是因为缺少 zlib1g-dev，需要使用 apt-get 命令从软件仓库中安装 zlib1g-dev，具体命令如下所示：

```
sudo apt-get install zlib1g-dev
```

同理需要安装下面的软件，否则也会出现上述类似的错误：

```
sudo apt-get install flex
sudo apt-get install bison
sudo apt-get install gperf
sudo apt-get install libsdl-dev
sudo apt-get install libesd0-dev
sudo apt-get install libncurses5-dev
sudo apt-get install libx11-dev
```

（2）没有安装 Java 环境 JDK。

当安装所有上述软件后，运行 make 命令再次编译 Android 源码。如果在之前忘记安装 Java 环境 JDK，则此时会出现很多 Java 文件无法编译的错误，如果打开 Android 的源码，可以在如下目录中下发现有很多 Java 源文件。

```
android/dalvik/libcore/dom/src/test/java/org/w3c/domts
```

这充分说明在编译 Android 之前必须先安装 Java 环境 JDK，安装流程如下所示。

① 登录 Oracle 官方网站，下载 jdk-6u16-linux-i586.bin 文件并安装。

在 Ubuntu 8.04 中，"/etc/profile" 文件是全局的环境变量配置文件，它适用于所有的 shell。在登录 Linux 系统时应该先启动"/etc/profile"文件，然后再启动用户目录下的"～/.bash_profile"、"～/.bash_login" 或 "～/.profile" 文件中的其中一个，执行的顺序和上面的排序一样。如果"～/.bash_profile"文件存在，则还会执行"～/.bashrc"文件。在此只需要把 JDK 的目录放到"/etc/profile"目录下即可：

```
JAVA_HOME=/usr/local/src/jdk1.6.0_16
PATH=$PATH:$JAVA_HOME/bin:/usr/local/src/android-sdk-linux_x86-1.1_r1/tools:~/bin
```

② 重新启动计算机，输入 java –version 命令，输出下面的信息则表示配置成功：

```
ava version "1.6.0_16"
Java(TM) SE Runtime Environment (build 1.6.0_16-b01)
Java HotSpot(TM) Client VM (build 13.1-b01, mixed mode, sharing)
```

当成功编译 Android 源码后，在终端会输出如下提示：

```
Target system fs image: out/target/product/generic/obj/PACKAGING/systemimage_unopt_intermediates/system.img
Install system fs image: out/target/product/generic/system.img
Target ram disk: out/target/product/generic/ramdisk.img
Target userdata fs image: out/target/product/generic/userdata.img
Installed file list: out/target/product/generic/installed-files.txt
root@dfsun2009-desktop:/bin/android#
```

1.3.5 实践演练——演示两种编译 Android 程序的方法

Android 编译环境本身比较复杂，并且不像普通的编译环境那样只有顶层目录下才有 Makefile 文件，而其他的每个 Component 都使用统一标准的 Android.mk 文件。不过这并不是我们熟悉的 Makefile，而是经过 Android 自身编译系统的很多处理。所以说要真正理清楚其中的联系还比较复杂，不过这种方式的好处在于，编写一个新的 Android.mk 给 Android 增加一个新的 Component 会变得比较简单。为了使读者更加深入地理解在 Linux 环境下编译 Android 程序的方法，在接下来的内容中，将分别演示两种编译 Android 程序的方法。

1. 编译 Native C（本地 C 程序）的 helloworld 模块

编译 Java 程序可以直接采用 Eclipse 的集成环境来完成，实现方法非常简单，在这里就不再重复了。接下来将主要针对 C/C++进行说明，通过一个例子来讲解在 Android 中增加一个 C 程序的 Hello World 的方法。

（1）在“$(YOUR_ANDROID)/development”目录下创建一个名为“hello”的目录，并用“$(YOUR_ANDROID)”指向 Android 源代码所在的目录：

```
- # mkdir $(YOUR_ANDROID)/development/hello
```

（2）在目录“$(YOUR_ANDROID)/development/hello/”下编写一个名为“hello.c”的 C 语言文件，文件 hello.c 的实现代码如下所示：

```
#include <stdio.h>
int main()
{
   printf("Hello World!\n");//输出Hello World
return 0;
}
```

（3）在目录“$(YOUR_ANDROID)/development/hello/”下编写 Android.mk 文件。这是 Android Makefile 的标准命名，不能更改。文件 Android.mk 的格式和内容可以参考其他已有的 Android.mk 文件的写法，针对 helloworld 程序的 Android.mk 文件内容如下所示：

```
LOCAL_PATH:= $(call my-dir)
include $(CLEAR_VARS)
LOCAL_SRC_FILES:= \
   hello.c
LOCAL_MODULE := helloworld
include $(BUILD_EXECUTABLE)
```

上述各个内容的具体说明如下所示。

- ❑ LOCAL_SRC_FILES：用来指定源文件用。
- ❑ LOCAL_MODULE：指定要编译的模块的名字，在下一步骤编译时将会用到。
- ❑ include $(BUILD_EXECUTABLE)：表示要编译成一个可执行文件，如果想编译成动态库则可用 BUILD_SHARED_LIBRARY，这些具体用法可以在“$(YOUR_ANDROID)/build/core/config.mk”查到。

（4）回到 Android 源代码顶层目录进行编译。

```
# cd $(YOUR_ANDROID) && make helloworld
```

在此需要注意，make helloworld 中的目标名 helloworld 就是上面 Android.mk 文件中由 LOCAL_MODULE 指定的模块名。最终的编译结果如下所示：

```
target thumb C: helloworld <= development/hello/hello.c
target Executable: helloworld (out/target/product/generic/obj/EXECUTABLES/helloworld_intermediates/
LINKED/helloworld)
target Non-prelinked: helloworld (out/target/product/generic/symbols/system/bin/helloworld)
target  Strip:  helloworld  (out/target/product/generic/obj/EXECUTABLES/helloworld_intermediates/
helloworld)
Install: out/target/product/generic/system/bin/helloworld
```

（5）如果和上述编译结果相同，则编译后的可执行文件存放在如下目录：

```
out/target/product/generic/system/bin/helloworld
```

这样通过“adb push”将它传送到模拟器上，再通过“adb shell”登录到模拟器终端后就可以执行了。

2. 手工编译 C 模块

在前面讲解了通过标准的 Android.mk 文件来编译 C 模块的具体流程，其实可以直接运用 gcc 命令行来编译 C 程序，这样可以更好地了解 Android 编译环境的细节。具体流程如下。

（1）在 Android 编译环境中，提供了“showcommands”选项来显示编译命令行，可以通过打开这个选项来查看一些编译时的细节。

（2）在具体操作之前需要使用如下命令把前面中的 helloworld 模块清除：

```
# make clean-helloworld
```

上面的“make clean-$(LOCAL_MODULE)”命令是 Android 编译环境提供的 make clean 的方式。

（3）使用 showcommands 选项重新编译 helloworld，具体命令如下所示：

```
# make helloworld showcommands
build/core/product_config.mk:229: WARNING: adding test OTA key
target thumb C: helloworld <= development/hello/hello.c
prebuilt/linux-x86/toolchain/arm-eabi-4.3.1/bin/arm-eabi-gcc   -I  system/core/include   -I
hardware/libhardware/include   -I hardware/ril/include   -I dalvik/libnativehelper/include  -I
frameworks/base/include   -I external/skia/include   -I out/target/product/generic/obj/include
-I bionic/libc/arch-arm/include    -I bionic/libc/include    -I bionic/libstdc++/include   -I
bionic/libc/kernel/common    -I bionic/libc/kernel/arch-arm    -I bionic/libm/include    -I
bionic/libm/include/arch/arm    -I bionic/libthread_db/include     -I development/hello   -I
out/target/product/generic/obj/EXECUTABLES/helloworld_intermediates  -c  -fno-exceptions -Wno-multichar
-march=armv5te -mtune=xscale -msoft-float -fpic -mthumb-interwork -ffunction-sections -funwind-tables
-fstack-protector  -D__ARM_ARCH_5__ -D__ARM_ARCH_5T__  -D__ARM_ARCH_5E__  -D__ARM_ARCH_5TE__  -include
system/core/include/arch/linux-arm/AndroidConfig.h -DANDROID -fmessage-length=0 -W -Wall -Wno-unused
-DSK_RELEASE -DNDEBUG -O2 -g -Wstrict-aliasing=2 -finline-functions -fno-inline-functions-called-once
-fgcse-after-reload  -frerun-cse-after-loop  -frename-registers  -DNDEBUG  -UDEBUG  -mthumb  -Os
-fomit-frame-pointer -fno-strict-aliasing -finline-limit=64       -MD -o out/target/product/generic/
obj/EXECUTABLES/helloworld_intermediates/hello.o development/hello/hello.c

target Executable: helloworld (out/target/product/generic/obj/EXECUTABLES/helloworld_intermediates/
LINKED/helloworld)

prebuilt/linux-x86/toolchain/arm-eabi-4.3.1/bin/arm-eabi-g++  -nostdlib  -Bdynamic  -Wl,-T,build/core/armelf.x
-Wl,-dynamic-linker,/system/bin/linker -Wl,--gc-sections -Wl,-z,nocopyreloc -o out/target/product/generic/
obj/EXECUTABLES/helloworld_intermediates/LINKED/helloworld    -Lout/target/product/generic/obj/lib   -Wl,
-rpath-link=out/target/product/generic/obj/lib  -lc  -lstdc++  -lm   out/target/product/generic/obj/lib/
crtbegin_dynamic.o          out/target/product/generic/obj/EXECUTABLES/helloworld_intermediates/hello.o
-Wl,--no-undefined     prebuilt/linux-x86/toolchain/arm-eabi-4.3.1/bin/../lib/gcc/arm-eabi/4.3.1/interwork/libgcc.a
out/target/product/generic/obj/lib/crtend_android.o

target Non-prelinked: helloworld (out/target/product/generic/symbols/system/bin/helloworld)

out/host/linux-x86/bin/acp -fpt out/target/product/generic/obj/EXECUTABLES/helloworld_intermediates/
LINKED/helloworld out/target/product/generic/symbols/system/bin/helloworld

target Strip: helloworld (out/target/product/generic/obj/EXECUTABLES/helloworld_intermediates/helloworld)

out/host/linux-x86/bin/soslim --strip --shady --quiet out/target/product/generic/symbols/system/bin/
helloworld --outfile out/target/product/generic/obj/EXECUTABLES/helloworld_intermediates/helloworld

Install: out/target/product/generic/system/bin/helloworld

out/host/linux-x86/bin/acp -fpt out/target/product/generic/obj/EXECUTABLES/helloworld_intermediates/
helloworld out/target/product/generic/system/bin/helloworld
```

从上述命令行可以看到，Android 编译环境所用的交叉编译工具链如下所示：

```
prebuilt/linux-x86/toolchain/arm-eabi-4.3.1/bin/arm-eabi-gcc
```

其中参数“-I”和“-L”分别指定了所用的 C 库头文件和动态库文件路径分别是“bionic/libc/include ”和“out/target/product/generic/obj/lib”，其他还包括很多编译选项以及-D 所定义的预编译宏。

（4）此时就可以利用上面的编译命令来手工编译 helloworld 程序，首先手工删除上次编译得到的 helloworld 程序：

```
# rm out/target/product/generic/obj/EXECUTABLES/helloworld_intermediates/hello.o
# rm out/target/product/generic/system/bin/helloworld
```

然后再用 gcc 编译以生成目标文件：

```
# prebuilt/linux-x86/toolchain/arm-eabi-4.3.1/bin/arm-eabi-gcc -I bionic/libc/arch-arm/include -I
bionic/libc/include -I bionic/libc/kernel/common -I bionic/libc/kernel/arch-arm -c -fno-exceptions
-Wno-multichar -march=armv5te -mtune=xscale -msoft-float -fpic -mthumb-interwork -ffunction-sections
-funwind-tables -fstack-protector -D__ARM_ARCH_5__ -D__ARM_ARCH_5T__ -D__ARM_ARCH_5E__ -D__ARM_ARCH_5TE__
-include system/core/include/arch/linux-arm/AndroidConfig.h -DANDROID -fmessage-length=0 -W -Wall
-Wno-unused -DSK_RELEASE -DNDEBUG -O2 -g -Wstrict-aliasing=2 -finline-functions
-fno-inline-functions-called-once -fgcse-after-reload -frerun-cse-after-loop -frename-registers
-DNDEBUG -UDEBUG -mthumb -Os -fomit-frame-pointer -fno-strict-aliasing -finline-limit=64 -MD -o
out/target/product/generic/obj/EXECUTABLES/helloworld_intermediates/hello.o
development/hello/hello.c
```

如果此时与 Android.mk 编译参数进行比较，会发现上面主要减少了不必要的参数“-I”。

（5）接下来开始生成可执行文件：

```
# prebuilt/linux-x86/toolchain/arm-eabi-4.3.1/bin/arm-eabi-gcc -nostdlib -Bdynamic -Wl,-T,build/core/
armelf.x -Wl,-dynamic-linker,/system/bin/linker -Wl,--gc-sections -Wl,-z,nocopyreloc -o out/target/
product/generic/obj/EXECUTABLES/helloworld_intermediates/LINKED/helloworld -Lout/target/product/generic/obj/lib
-Wl,-rpath-link=out/target/product/generic/obj/lib -lc -lm out/target/product/generic/obj/EXECUTABLES/
helloworld_intermediates/hello.o out/target/product/generic/obj/lib/crtbegin_dynamic.o -Wl,--no-
undefined ./prebuilt/linux-x86/toolchain/arm-eabi-4.3.1/bin/../lib/gcc/arm-eabi/4.3.1/interwork/lib
gcc.a out/target/product/generic/obj/lib/crtend_android.o
```

在此需要特别注意的是参数“-Wl,-dynamic-linker,/system/bin/linker”，它指定了 Android 专用的动态链接器是“/system/bin/linker”，而不是平常使用的 ld.so。

（6）最后可以使用命令 file 和 readelf 来查看生成的可执行程序：

```
# file out/target/product/generic/obj/EXECUTABLES/helloworld_intermediates/LINKED/helloworld
out/target/product/generic/obj/EXECUTABLES/helloworld_intermediates/LINKED/helloworld: ELF 31-bit LSB
executable, ARM, version 1 (SYSV), dynamically linked (uses shared libs), not stripped
#  readelf -d out/target/product/generic/obj/EXECUTABLES/helloworld_intermediates/LINKED/helloworld
|grep NEEDED
 0x00000001 (NEEDED)                     Shared library: [libc.so]
 0x00000001 (NEEDED)                     Shared library: [libm.so]
```

这就是 ARM 格式的动态链接可执行文件，在运行时需要 libc.so 和 libm.so。当提示“not stripped”时表示它还没被 STRIP（剥离）。嵌入式系统中为节省空间通常将编译完成的可执行文件或动态库进行剥离，即去掉其中多余的符号表信息。在前面“make helloworld showcommands”命令的最后也可以看到，Android 编译环境中使用了“out/host/linux-x86/bin/soslim”工具进行 STRIP。

1.4 编译 Android Kernel

编译 Android Kernel 代码就是编译 Android 内核代码，在进行具体编译工作之前，需要先了解在 Android 开源系统中包含的以下 3 部分代码。

- 仿真器公共代码：对应的工程名是 kernel/common.get。
- MSM 平台的内核代码：对应的工程名是 kernel/msm.get。
- OMAP 平台的内核代码：对应的工程名是 kernel/omap.get。

在本节的内容中，将详细讲解编译上述 Android Kernel 的基本知识。

1.4.1 获取 Goldfish 内核代码

Goldfish 是一种虚拟的 ARM 处理器，通常在 Android 的仿真环境中使用。在 Linux 的内核中，Goldfish 作为 ARM 体系结构的一种“机器”。在 Android 的发展过程中，Goldfish 内核的版本也从 Linux

2.6.25 升级到了 Linux 3.4，此处理器的 Linux 内核和标准的 Linux 内核有以下 3 个方面的差别。

- Goldfish 机器的移植。
- Goldfish 一些虚拟设备的驱动程序。
- Android 中特有的驱动程序和组件。

Goldfish 处理器有两个版本，分别是 ARMv5 和 ARMv7，在一般情况下，只需使用 ARMv5 版本即可。在 Android 开源工程的代码仓库中，使用 git 工具得到 Goldfish 内核代码的命令如下所示：

```
$ git clone git://android.git.kernel.org/kernel/common.git
```

在其 Linux 源代码的根目录中，配置和编译 Goldfish 内核的过程如下所示：

```
$make ARCH=arm goldfish_defconfig .config
$make ARCH=arm CROSS_COMPILE={path}/arm-none-linux-gnueabi-
```

其中，CROSS_COMPILE 的 path 值用于指定交叉编译工具的路径。

编译结果如下所示：

```
LD vmlinux
SYSMAP system.map
SYSMAP .tmp_system.map
OBJCOPY arch/arm/boot/Image
Kernel: arch/arm/boot/Image is ready
AS arch/arm/boot/compressed/head.o
GZIP arch/arm/boot/compressed/piggy.gz
AS arch/arm/boot/compressed/piggy.o
CC arch/arm/boot/compressed/misc.o
LD arch/arm/boot/compressed/vmlinux
  OBJCONPY arch/arm/boot/zImage
  Kernel: arch/arm/boot/zImage is ready
```

- vmlinux：是 Linux 进行编译和连接之后生成的 Elf 格式的文件。
- Image：是未经过压缩的二进制文件。
- piggy：是一个解压缩程序。
- zImage：是解压缩程序和压缩内核的组合。

在 Android 源代码的根目录中，vmlinux 和 zImage 分别对应 Android 代码 prebuilt 中的预编译的 ARM 内核。使用 zImage 可以替换 prebuilt 中的“prebuilt/android-arm/”目录下的 goldfish_defconfig，此文件的主要片断如下所示：

```
CONFIG_ARM=y
#
# System Type
#
CONFIG_ARCH_GOLDFISH=y
#
# Goldfish options
#
CONFIG_MACH_GOLDFISH=y
# CONFIG_MACH_GOLDFISH_ARMV7 is not set
```

因为 GoldFish 是 ARM 处理器，所以 CONFIG_ARM 宏需要被使能，CONFIG_ARCH_GOLDFISH 和 CONFIG_MACH_GOLDFISH 宏是 GoldFish 处理器这类机器使用的配置宏。

在 gildfish_defconfig 中，与 Android 系统相关的宏如下所示：

```
#
# android
#
CONFIG_ANDROID=y
CONFIG_ANDROID_BUNDER_IPC=y #binder ipc 驱动程序
CONFIG_ANDROID_LOGGER=y #log 记录器驱动程序
# CONFIG_ANDROID_RAM_CONSOLE is not set
CONFIG_ANDROID_TIMED_OUTPUT=y #定时输出驱动程序框架
```

```
CONFIG_ANDROID_LOW_MEMORY_KILLER=y
CONFIG_ANDROID_PMEM=y #物理内存驱动程序
CONFIG_ASHMEM=y #匿名共享内存驱动程序
CONFIG_RTC_INTF_ALARM=y
CONFIG_HAS_WAKELOCK=y 电源管理相关的部分 wakelock 和 earlysuspend
CONFIG_HAS_EARLYSUSPEND=y
CONFIG_WAKELOCK=y
CONFIG_WAKELOCK_STAT=y
CONFIG_USER_WAKELOCK=y
CONFIG_EARLYSUSPEND=y
goldfish_defconfig 配置文件中，另外有一个宏是处理器虚拟设备的“驱动程序”，其内容如下所示：
CONFIG_MTD_GOLDFISH_NAND=y
CONFIG_KEYBOARD_GOLDFISH_EVENTS=y
CONFIG_GOLDFISH_TTY=y
CONFIG_BATTERY_GOLDFISH=y
CONFIG_FB_GOLDFISH=y
CONFIG_MMC_GOLDFISH=y
CONFIG_RTC_DRV_GOLDFISH=y
```

在 Goldfish 处理器的各个配置选项中，体系结构和 Goldfish 的虚拟驱动程序基于标准 Linux 内容的驱动程序框架，但是这些设备在不同硬件平台的移植方式不同；Android 专用的驱动程序是 Android 中特有的内容，非 Linux 标准，但是和硬件平台无关。

和原 Linux 内核相比，Android 内核增加了 Android 的相关驱动（Driver），对应的目录如下所示：

```
kernel/drivers/android
```

Android 的相关驱动主要分为以下几类驱动。

- ❑ Android IPC 系统：Binder (binder.c)。
- ❑ Android 日志系统：Logger (logger.c)。
- ❑ Android 电源管理：Power (power.c)。
- ❑ Android 闹钟管理：Alarm (alarm.c)。
- ❑ Android 内存控制台：Ram_console (ram_console.c)。
- ❑ Android 时钟控制的 gpio：Timed_gpio (timed_gpio.c)。

对于本书讲解的驱动程序开发来说，我们比较关心的是 GoldFish 平台下相关的驱动文件，具体说明如下所示。

（1）字符输出设备：

```
kernel/drivers/char/goldfish_tty.c
```

（2）图象显示设备(Frame Buffer)：

```
kernel/drivers/video/goldfishfb.c
```

（3）键盘输入设备文件：

```
kernel/drivers/input/keyboard/goldfish_events.c
```

（4）RTC 设备(Real Time Clock)文件：

```
kernel/drivers/rtc/rtc-goldfish.c
```

（5）USB Device 设备文件：

```
kernel/drivers/usb/gadget/android_adb.c
```

（6）SD 卡设备文件：

```
kernel/drivers/mmc/host/goldfish.c
```

（7）FLASH 设备文件：

```
kernel/drivers/mtd/devices/goldfish_nand.c
kernel/drivers/mtd/devices/goldfish_nand_reg.h
```

（8）LED 设备文件：

```
kernel/drivers/leds/ledtrig-sleep.c
```

（9）电源设备：

```
kernel/drivers/power/goldfish_battery.c
```

（10）音频设备：

```
kernel/arch/arm/mach-goldfish/audio.c
```

（11）电源管理：

```
kernel/arch/arm/mach-goldfish/pm.c
```

（12）时钟管理：

```
kernel/arch/arm/mach-goldfish/timer.c
```

1.4.2 获取 MSM 内核代码

在目前市面上，谷歌的手机产品 G1 是基于 MSM 内核的，MSM 是高通公司的应用处理器，在 Android 代码库中公开了对应的 MSM 的源代码。在 Android 开源工程的代码仓库中，使用 Git 工具得到 MSM 内核代码的命令如下所示：

```
$ git clone git://android.git.kernel.org/kernel/msm.git
```

1.4.3 获取 OMAP 内核代码

OMAP 是德州仪器公司的应用处理器，为 Android 使用的是 OMAP3 系列的处理器。在 Android 代码库中公开了对应的 OMAP 的源代码，使用 Git 工具得到 OMAP 内核代码的命令如下所示：

```
$ git clone git://android.git.kernel.org/kernel/omap.git
```

1.4.4 编译 Android 的 Linux 内核

了解了上述 3 类 Android 内核后，下面开始讲解编译 Android 内核的方法。在此以 Ubuntu 8.10 为例，完整编译 Android 内核的流程如下。

（1）构建交叉编译环境。

Android 的默认硬件处理器是 ARM，因此需要在自己的机器上构建交叉编译环境。交叉编译器 GNU Toolchain for ARM Processors 下载地址如下所示：

```
http://www.codesourcery.com/gnu_toolchains/arm/download.html
```

单击 GNU/Linux 对应的链接，再单击“Download Sourcery CodeBench Lite 5.1 2012.03-117”链接后直接下载，如图 1-20 所示。

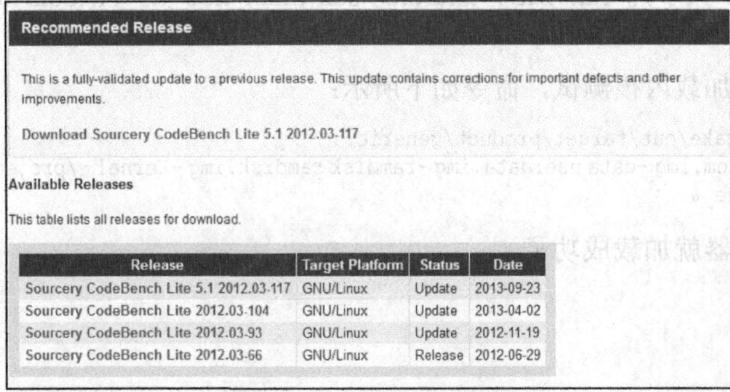

图 1-20　下载交叉编译器

把 arm-2008q3-71-arm-none-linux-gnueabi-i686-pc-linux- gnu.tar.bz2 解压到一目录下，例如“～/programes/”，并加入 PATH 环境变量：

```
vim ~/.bashrc
```

然后添加：

```
ARM_TOOLCHIAN=~/programes/arm-2008q3/bin/
export PATH=${PATH}:${ARM_TOOLCHIAN};
```

保存后并 source ～/.bashrc。

（2）获取内核源码，源码地址如下所示：

```
http://code.google.com/p/android/downloads/list
```

选择的内核版本要与选用的模拟器版本尽量一致。下载并解压后得到 kernel.git 文件夹。

```
tar -xvf ~/download/linux-3.2.5-android-4.3_r1.tar.gz
```

（3）获取内核编译配置信息文件。

编译内核时需要使用 configure，通常 configure 有很多选项，我们往往不知道需要那些选项。在运行 Android 模拟器时，有一个文件“/proc/config.gz”，这是当前内核的配置信息文件，把 config.gz 获取并解压到“kernel.git/”下，然后改名为.config。命令如下所示：

```
cd kernel.git/
emulator &
adb pull /proc/config.gz
gunzip config.gz
mv config .config
```

（4）修改 Makefile。

修改 195 行的代码：

```
CROSS_COMPILE    = arm-none-linux-gnueabi-
```

将 CROSS_COMPILE 值改为 arm-none-linux-gnueabi-，这是安装的交叉编译工具链的前缀，修改此处意在告诉 make 在编译的时候要使用该工具链。然后注释掉 562 和 563 行的如下代码：

```
#LDFLAGS_BUILD_ID = $(patsubst -Wl$(comma)%,%,/
#                 $(call ld-option, -Wl$(comma)--build-id,))
```

必须将上述代码中的 build id 值注释掉，因为目前版本的 Android 内核不支持该选项。

（5）编译。

使用 make 进行编译，并同时生成 zImage：

```
 LD      arch/arm/boot/compressed/vmlinux
 OBJCOPY arch/arm/boot/zImage
 Kernel: arch/arm/boot/zImage is ready
```

这样生成 zImage 大小为 1.23 MB，android- sdk-linux_x86-4.3_r1/tools/lib/images/kernel-qemu 是 1.24 MB。

（6）使用模拟器加载内核测试，命令如下所示：

```
cd android/out/cupcake/out/target/product/generic
emulator -image system.img -data userdata.img -ramdisk ramdisk.img -kernel ~/project/android/kernel.git/
arch/arm/boot/zImage &
```

到此为止，模拟器就加载成功了。

第 2 章　Java 虚拟机基础

Java 虚拟机和 Android 虚拟机十分相似，所以在本书中将以 Java 虚拟机开始，逐步引领广大读者步入 Android 虚拟机的世界。在本章的内容中，将简要讲解 Java 虚拟机技术的基本知识，为读者步入本书后面知识的学习打下基础。

2.1 虚拟机的作用

虚拟机（Virtual Machine）这一概念最初由波佩克与戈德堡定义，是指通过软件模拟的具有完整硬件系统功能的、运行在一个完全隔离环境中的完整计算机系统。由此可见，虚拟机是跟特定硬件无关的一个系统。在现实应用中，虚拟机最常见的情形便是双系统。例如计算机原装系统是 Windows，为了在这台机器上能够体验 Linux 系统，可以安装一个虚拟机环境，在这个虚拟机环境中运行 Linux 系统，这样就实现了“一机双系统”的功效。在现实应用中，通过虚拟机软件可以在一台物理计算机上模拟出一台或多台虚拟的计算机。这些虚拟机完全可以像真正的计算机那样进行工作，例如可以安装操作系统、安装应用程序、访问网络资源等等。对于使用用户而言，虚拟机只是运行在物理计算机上的一个应用程序。但是对于在虚拟机中运行的应用程序来说，虚拟机就是一台真正计算机。正因为如此，所以当在虚拟机中进行软件评测时，可能会发生系统崩溃的情形。但是这里崩溃的只是虚拟机上的操作系统，而不是物理计算机上的操作系统。可以使用虚拟机的“Undo”（恢复）功能，立即恢复虚拟机到安装软件之前的状态。

虚拟机根据它们的运用以及与直接机器的相关性分为两大类。系统虚拟机提供一个可以运行完整操作系统的完整系统平台；相反，程序虚拟机为运行单个计算机程序设计，这意谓它支持单个进程。虚拟机的一个本质特点是运行在虚拟机上的软件被局限在虚拟机提供的资源里——它不能超出虚拟世界。

在现实应用中，对于一般计算机用户来说，最常见的使用虚拟机的情形是安装双系统。例如在 Windows 平台上安装一个虚拟机，然后在这个虚拟机中安装 Linux 操作系统或 iOS 系统，这样就实现了双系统功能。

在当前流行的编程语言 Java 中，便是采用了虚拟机机制，Java 的虚拟机被称为 Java Virtual Machine，缩写为 JVM。用 Java 编写的程序可以通过对 Java 运行环境（JRE）软件发出命令获得服务，取得期望的结果。透过提供这种服务，JRE 起到了虚拟机的作用，程序不必为特定的操作系统或硬件编写。

Java 虚拟机和 Android 虚拟机十分相似，所以在本书中将以 Java 虚拟机开始，逐步引领广大读者步入 Android 虚拟机的世界。

2.2 Java 虚拟机概述

Java 虚拟机（JVM）是一个虚构出来的计算机，是通过在实际的计算机上仿真模拟各种计算机功能模拟来实现的。Java 虚拟机有自己完善的硬件架构，如处理器、堆栈、寄存器等，还具有相应的指令系统。JVM 虚拟机的运作结构如图 2-1 所示。

从该图中可以看到，JVM 是运行在操作系统之上的，与硬件没有直接的交互。JVM 的具体组成部分如图 2-2 所示。

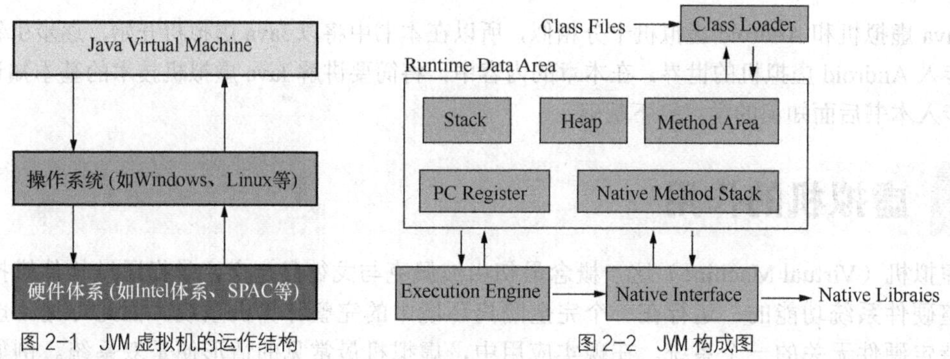

图 2-1 JVM 虚拟机的运作结构　　图 2-2 JVM 构成图

（1）使用 JVM 的原因。

Java 语言的一个非常重要的特点就是与平台的无关性。而使用 JVM 是实现这一特点的关键。一般的高级语言如果要在不同的平台上运行，至少需要编译成不同的目标代码。在引入 JVM 后，Java 语言在不同平台上运行时不需要重新编译。Java 语言使用模式 JVM 屏蔽了与具体平台相关的信息，使得 Java 语言编译程序只需生成在 JVM 上运行的目标代码（字节码），就可以在多种平台上不加修改地运行。当 JVM 执行字节码时，把字节码解释成具体平台上的机器指令执行。

（2）JVM 的作用。

JVM 是 Java 语言底层实现的基础，对 Java 语言感兴趣的读者来说，很有必要对 Java 虚拟机有一个大概的了解。因为这不但有助于理解 Java 语言的一些性质，而且也有助于使用 Java 语言。对于要在特定平台上实现 JVM 的软件人员、Java 语言的编译器作者以及要用硬件芯片实现 JVM 的人员来说，必须深刻理解 JVM 的规范。另外，如果你想扩展 Java 语言，或是把其他语言编译成 Java 语言的字节码，你也需要深入地了解 JVM。

在本节的内容中，将简要讲解和 JVM 相关的基本知识。

2.2.1 JVM 的数据类型

在 JVM 机制中，可以支持如下所示的基本数据类型。

- byte：1 字节有符号整数的补码。
- short：2 字节有符号整数的补码。
- int：4 字节有符号整数的补码。
- long：8 字节有符号整数的补码。
- float：4 字节 IEEE754 单精度浮点数。
- double：8 字节 IEEE754 双精度浮点数。
- char：2 字节无符号 Unicode 字符。

- object：对一个 Javaobject（对象）的 4 字节引用。
- returnAddress：4 字节，用于 jsr/ret/jsr-w/ret-w 指令。

几乎所有的 Java 类型检查工作都是在编译时完成的，上述列出的原始数据类型数据在 Java 执行时不需要用硬件标记。操作这些原始数据类型数据的字节码（指令）本身就已经指出了操作数的数据类型，例如 iadd、ladd、fadd 和 dadd 指令都是把两个数相加，其操作数类型分别是 int、long、float 和 double。虚拟机没有给 boolean（布尔）类型设置单独的指令。boolean 型的数据是由 integer 指令，包括 integer 返回来处理的。boolean 型的数组则是用 byte 数组来处理的。虚拟机使用 IEEE754 格式的浮点数，不支持 IEEE 格式的较旧的计算机，在运行 Java 数值计算程序时，可能会非常慢。

虚拟机的规范对于 object 内部的结构没有任何特殊的要求。在 Oracle 公司的实现中，对 object 的引用是一个句柄，其中包含一对指针：一个指针指向该 object 的方法表，另一个指向该 object 的数据。用 Java 虚拟机的字节码表示的程序应该遵守类型规定。Java 虚拟机的实现应拒绝执行违反了类型规定的字节码程序。Java 虚拟机由于字节码定义的限制似乎只能运行于 32 位地址空间的机器上。但是可以创建一个 Java 虚拟机，它自动地把字节码转换成 64 位的形式。从 Java 虚拟机支持的数据类型可以看出，Java 对数据类型的内部格式进行了严格规定，这样使得各种 Java 虚拟机的实现对数据的解释是相同的，从而保证了 Java 的与平台无关性和可移植性。

2.2.2 Java 虚拟机体系结构

JVM 由如下 5 个部分组成。

- 一组指令集。
- 一组寄存器。
- 一个栈。
- 一个无用单元收集堆（Garbage-collected-heap）。
- 一个方法区域。

这 5 部分是 Java 虚拟机的逻辑成分，不依赖任何实现技术或组织方式，但它们的功能必须在真实机器上以某种方式实现。在接下来的内容中，将简要介绍上述组成部分的基本知识，更加详细的知识读者可以参阅本书后面的内容。

1. Java 指令集

Java 虚拟机支持大约 248 个字节码，每个字节码执行一种基本的 CPU 运算，例如把一个整数加到寄存器，子程序转移等。Java 指令集相当于 Java 程序的汇编语言。

Java 指令集中的指令包含一个单字节的操作符，用于指定要执行的操作，还有 0 个或多个操作数，提供操作所需的参数或数据。许多指令没有操作数，仅由一个单字节的操作符构成。

虚拟机的内层循环的执行过程如下：

```
do{
取一个操作符字节；
根据操作符的值执行一个动作；
}while(程序未结束)
```

由于指令系统的简单性，使得虚拟机执行的过程十分简单，这样有利于提高执行的效率。指令中操作数的数量和大小是由操作符决定的。如果操作数比一个字节大，那么它存储的顺序是高位字节优先。假如一个 16 位的参数存放时占用两个字节，其值为：

```
第一个字节*256+第二个字节
```

字节码指令流一般只是字节对齐的，但是指令 tabltch 和 lookup 是例外，在这两条指令内部要求强制的 4 字节边界对齐。

2. 寄存器

Java 虚拟机的寄存器用于保存机器的运行状态，与微处理器中的某些专用寄存器类似，所有寄存器都是 32 位的。在 Java 虚拟机中有以下 4 种寄存器。

- pc：Java 程序计数器。
- optop：指向操作数栈顶端的指针。
- frame：指向当前执行方法的执行环境的指针。
- vars：指向当前执行方法的局部变量区第一个变量的指针。

Java 虚拟机是栈式的，它不定义或使用寄存器来传递或接收参数，其目的是为了保证指令集的简洁性和实现时的高效性，特别是对于寄存器数目不多的处理器。

3. 栈

Java 虚拟机中的栈有 3 个区域，分别是局部变量区、运行环境区、操作数区。

（1）局部变量区。

每个 Java 方法使用一个固定大小的局部变量集。它们按照与 vars 寄存器的字偏移量来寻址。局部变量都是 32 位的。长整数和双精度浮点数占据了两个局部变量的空间，却按照第一个局部变量的索引来寻址（例如，一个具有索引 n 的局部变量，如果是一个双精度浮点数，那么它实际占据了索引 n 和 n+1 所代表的存储空间）。虚拟机规范并不要求在局部变量中的 64 位的值是 64 位对齐的。虚拟机提供了把局部变量中的值装载到操作数栈的指令，也提供了把操作数栈中的值写入局部变量的指令。

（2）运行环境区。

在运行环境中包含的信息可以实现动态链接、正常的方法返回和异常、错误传播。

- 动态链接。

运行环境包括对指向当前类和当前方法的解释器符号表的指针，用于支持方法代码的动态链接。方法 clas 文件代码在引用要调用的方法和要访问的变量时使用符号。动态链接把符号形式的方法调用翻译成实际方法调用，装载必要的类以解释还没有定义的符号，并把变量访问翻译成与这些变量运行时的存储结构相应的偏移地址。动态链接方法和变量使得方法中使用的其他类的变化不会影响到本程序的代码。

- 正常的方法返回。

如果当前方法正常地结束了，在执行了一条具有正确类型的返回指令时，调用的方法会得到一个返回值。执行环境在正常返回的情况下用于恢复调用者的寄存器，并把调用者的程序计数器增加一个恰当的数值，以跳过已执行过的方法调用指令，然后在调用者的执行环境中继续执行下去。

- 异常和错误传播。

异常情况在 Java 中被称作 Error（错误）或 Exception（异常），是 Throwable 类的子类，在程序中的原因有如下两点。

- 动态链接错，如无法找到所需的 class 文件。
- 运行时出错，如对一个空指针的引用程序使用了 throw 语句。当发生异常时，Java 虚拟机采取如下措施解决。
    - 检查与当前方法相联系的 catch 子句表。每个 catch 子句包含其有效指令范围，能够处理的异常类型，以及处理异常的代码块地址。

- 与异常相匹配的 catch 子句应该符合下面的条件:造成异常的指令在其指令范围之内，发生的异常类型是其能处理的异常类型的子类型。如果找到了匹配的 catch 子句，那么系统转移到指定的异常处理块处执行。如果没有找到异常处理块，重复寻找匹配的 catch 子句的过程，直到当前方法的所有嵌套的 catch 子句都被检查过。
- 由于虚拟机从第一个匹配的 catch 子句处继续执行，所以 catch 子句表中的顺序是很重要的。因为 Java 代码是结构化的，因此总可以把某个方法的所有的异常处理器都按序排列到一个表中，对任意可能的程序计数器的值，都可以用线性的顺序找到合适的异常处理块，以处理在该程序计数器值下发生的异常情况。
- 如果找不到匹配的 catch 子句，那么当前方法得到一个"未截获异常"的结果并返回到当前方法的调用者，好像异常刚刚在其调用者中发生一样。如果在调用者中仍然没有找到相应的异常处理块，那么这种错误传播将被继续下去。如果错误被传播到最顶层，那么系统将调用一个缺省的异常处理块。

（3）操作数栈区。

机器指令只从操作数栈中取操作数，对它们进行操作，并把结果返回到栈中。选择栈结构的原因是：在只有少量寄存器或非通用寄存器的机器（如 Intel486）上，也能够高效地模拟虚拟机的行为。操作数栈是 32 位的。它用于给方法传递参数，并从方法接收结果，也用于支持操作的参数，并保存操作的结果。例如，iadd 指令将两个整数相加，相加的两个整数应该是操作数栈顶的两个字，这两个字是由先前的指令压进堆栈的，这两个整数将从堆栈弹出、相加，并把结果压回到操作数栈中。

每个原始数据类型都有专门的指令对它们进行必须的操作。每个操作数在栈中需要一个存储位置，除了 long 和 double 型，它们需要两个位置。操作数只能被适用于其类型的操作符所操作。例如压入两个 int 类型的数，如果把它们当作是一个 long 类型的数则是非法的。在 Sun 的虚拟机实现中，这个限制由字节码验证器强制实行。但是有少数操作（操作符 dupe 和 swap），用于对运行时数据区进行操作时是不考虑类型的。

4. 无用单元收集堆

Java 的堆是一个运行时数据区，类的实例（对象）从中分配空间。Java 语言具有无用单元收集能力，即它不给程序员显示释放对象的能力。Java 不规定具体使用的无用单元收集算法，可以根据系统的需求使用各种各样的算法。

5. 方法区

方法区与传统语言中的编译后代码或是 Unix 进程中的正文段类似。它保存方法代码（编译后的 java 代码）和符号表。在当前的 Java 实现中，方法代码不包括在无用单元收集堆中，但计划在将来的版本中实现。每个类文件包含了一个 Java 类或一个 Java 界面的编译后的代码。可以说类文件是 Java 语言的执行代码文件。为了保证类文件的平台无关性，Java 虚拟机规范中对类文件的格式也作了详细的说明。其具体细节请参考 Sun 公司的 Java 虚拟机规范。

在 Java 虚拟机规范中，一个虚拟机实例的行为是分别按照子系统、内存区、数据类型以及指令这几个术语来描述的。这些组成部分一起展示了抽象的虚拟机的内部抽象体系结构。但是规范中对它们的定义并非要强制规定 Java 虚拟机实现内部的体系结构，更多的是为了严格地定义这些实现的外部特征。规范本身通过定义这些抽象的组成部分以及它们之间的交互，来定义任何 Java 虚拟机实现都必须遵守的行为。

图 2-3 是 Java 虚拟机的结构框图，包括在规范中描述的主要子系统和内存区。前一章曾提到，

每个 Java 虚拟机都有一个类装载器子系统，它根据给定的全限定名类装入类型（类或接口），同样，每个 Java 虚拟机都有一个执行引擎，它负责执行那些包含在被装载类的方法中的指令。

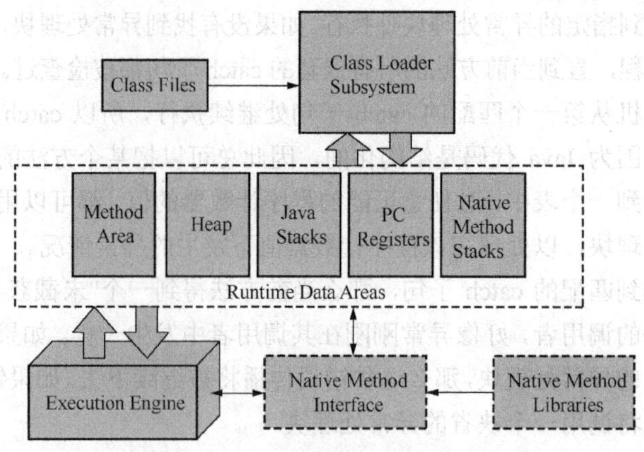

图 2-3　Java 虚拟机的内部体系结构

当 Java 虚拟机运行一个程序时，它需要使用内存来存储许多东西，例如下面所示的元素。

- 字节码。
- 从已装载的 class 文件中得到的其他信息。
- 程序创建的对象。
- 传递给方法的参数。
- 返回值。
- 局部变量。
- 运算的中间结果。

Java 虚拟机会把上述元素都组织到几个“运行时数据区”中，目的是便于管理。尽管这些“运行时数据区”都会以某种形式存在于每一个 Java 虚拟机实现中，但是规范对它们的描述却是相当抽象的。这些运行时数据区结构上的细节，大多数都由具体实现的设计者决定。

不同的虚拟机实现可能具有很不同的内存限制，有的实现可能大量的内存可用，有的可能只有很少的内存，有的实现可以利用虚拟内存，有的则不能。规范本身对“运行时数据区”只有抽象的描述，这就使得 Java 虚拟机可以很容易地在各种计算机和设备上实现。

某些运行时数据区是由程序汇总所有线程共享的，还有一些则只由一个线程拥有。每个 Java 虚拟机实例都有一个方法区以及一个堆，它们是由该虚拟机实例中所有线程共享的。当虚拟机装载一个 class 文件时，它会从这个 class 文件包含的二进制数据中解析类型信息，然后，它把这些类型信息放到方法区中。当程序运行时，虚拟机会把所有该程序在运行时创建的对象都放到堆中。图 2-4 对这些内存区域进行了描绘。

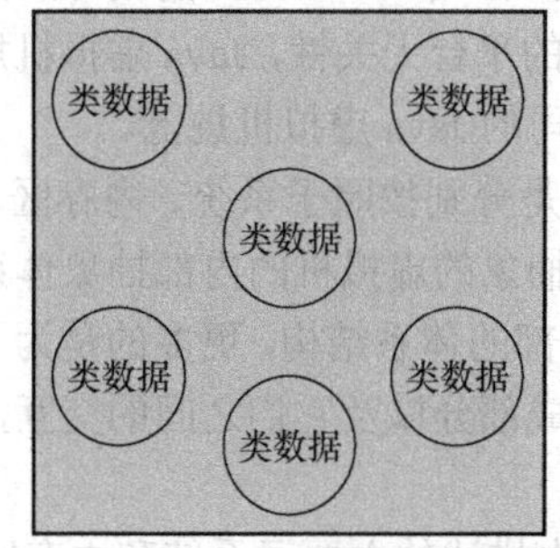

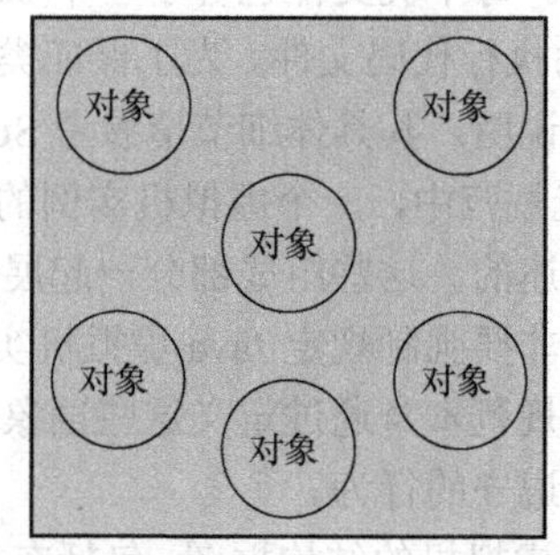

图 2-4　由所有线程共享的运行时数据区

当每一个新线程被创建时，它都将得到它自己的 PC 寄存器（程序计数器）以及一个 Java 栈：如果线程正在执行的是一个 Java 方法（非本地方法），那么 PC 寄存器的值将总是指示下一条将被执行的指令，而它的 Java 栈则总是存储该线程中 Java 方法调用的状态——包括它的局部变量、被调用时传进来的参数、它的返回值以及运算的中间结果等。而本地方法调用的状态，则是以某种依赖于具体实现的方式存储在本地方法栈中，也可能是在寄存器或者其他某些与特定实现相关的内存区中。

Java 栈是由许多栈帧（stackframe）或者说帧（frame）组成的，一个栈帧包含一个 Java 方法调用的状态。当线程调用一个 Java 方法时，虚拟机压入一个新的栈帧到该线程的 Java 栈中。当该方法返回时，这个栈帧被从 Java 栈中弹出并抛弃。

Java 虚拟机没有寄存器，其指令集使用 Java 栈来存储中间数据。这样设计的原因是为了保持 Java 虚拟机的指令集尽量紧凑，同时也便于 Java 虚拟机在那些只有很少通用寄存器的平台上实现，另外 Java 虚拟机的这种基于栈的体系结构，也有助于运行时某些虚拟机实现的动态编译器和即时编译器的代码优化。

图 2-5 描绘了 Java 虚拟机为每一个线程创建的内存区，这些内存区域是私有的，任何线程都不能访问另一个线程的 PC 寄存器或者 Java 栈。

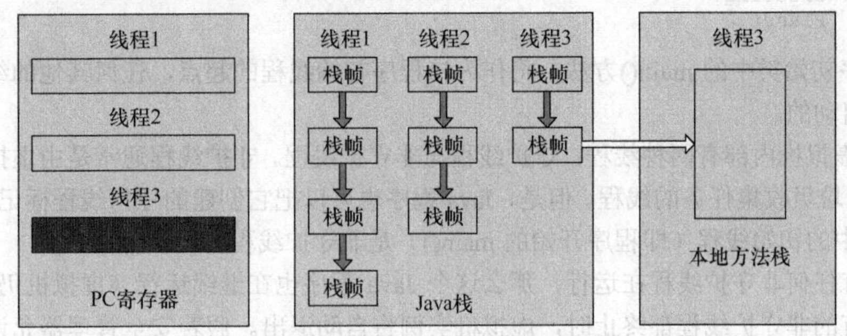

图 2-5　线程专有的运行时数据区

图 2-5 展示了一个虚拟机实例的快照，它有 3 个线程正在执行。线程 1 和线程 2 都正在执行 Java 方法，而线程 3 则正在执行一个本地方法。在图 5-3 中，和本书其他地方一样，Java 栈都是向下生长的，而栈顶都显示在图的底部，当前正在执行的方法的栈帧则以浅色表示，对于一个正在运行 Java 方法的线程而言，它的 PC 寄存器总是指向下一条将被执行的指令。在图 2-5 中，像这样的 PC 寄存器（比如线程 1 和线程 2 的）都是以浅色显示的。由于线程 3 当前正在执行一个本地方法，因此，它的 PC 寄存器（以深色显示的那个）的值是不确定的。

2.2.3　JVM 的生命周期

一个运行时的 Java 虚拟机实例的天职是：负责运行一个 Java 程序。在启动一个 Java 程序的同时会诞生一个虚拟机实例，当该程序退出时，虚拟机实例也随之消亡。如果在同一台计算机上同时运行 3 个 Java 程序，会得到 3 个 Java 虚拟机实例。每个 Java 程序都运行于它自己的 Java 虚拟机实例中。

Java 虚拟机实例通过调用某个初始类的 main()方法来运行一个 Java 程序。而这个 main()方法必须是公有的（public）、静态的（static）、返回值为 void，并且接受一个字符串数组作为参数。任何拥有这样一个 main()方法的类都可以作为 Java 程序运行的起点。假如存在这样一个 Java 程序，此程序能够打印出传给它的命令行参数：

```
package jvm.ext1;
public class Echo {
```

```
    public static void main(String[]args) {
        int length = args.length;
        for (int i = 0; i <length; i++) {
            System.out.print(args[i] +"");
        }
        System.out.println();
    }
}
```

上述代码必须告诉 Java 虚拟机要运行的 Java 程序中初始类的名字，整个程序将从它的 main()方法开始运行。现实中一个 Java 虚拟机实现的例子如 SunJava 2 SDK 的 Java 程序。比如，如果想要在 Windows 上使用 Java 运行 Echo 程序，需要键入如下命令。

```
java Echo Greeting, Planet
```

该命令的第一个单词“java”，告诉操作系统应该运行来自 Sun Java 2 SDK 的 Java 虚拟机。第二个词”Echo”则支持初始类的名字。Echo 这个初始类中必须有个公有的、静态的方法 main()，它获得一个字符串数组参数并且返回 void。上述命令行中剩下的单词序列“Greeting，Planet”，作为该程序的命令行参数以字符串数组的形式传递给 main()，因此，对于上面这个例子，传递给类 Echo 中 main()方法的字符串数组参数的内容就是：

```
args[0]为" Greeting,"
args[1]为 "Planet."
```

Java 程序初始类中的 main()方法，将作为该程序初始线程的起点，任何其他的线程都是由这个初始线程启动的。

在 Java 虚拟机内部有两种线程：守护线程与非守护线程。守护线程通常是由虚拟机自己使用的，比如执行垃圾收集任务的线程。但是，Java 程序也可以把它创建的任何线程标记为守护线程。而 Java 程序中的初始线程（即程序开始的 main()）是非守护线程。

只要还有任何非守护线程在运行，那么这个 Java 程序也在继续运行（虚拟机仍然存活）。当该程序中所有的非守护线程都终止时，虚拟机实例将自动退出。假若安全管理器允许，程序本身也能够通过调用 Runtime 类或者 System 类的 exit 方法来退出。

在上面的 Echo 程序中，方法 main()并没有调用其他的线程。所以当它打印完命令行参数后返回 main()方法。这就终止了该程序中唯一的非守护线程，最终导致虚拟机实例退出。

2.3 JVM 的安全性

除了平台无关性以外，Java 还必须解决的另一个技术难题就是安全。因为网络运行多台计算机共享数据和分布式处理，所以它提供了一条侵入计算机系统的潜在途径，使得其他人可能窃取信息、改变或破坏信息、盗取计算资源等。因此，将计算机联入网络产生了很多安全问题。为了解决由网络引起的安全问题，Java 体系结构采用了一个扩展的内置安全模型，这个模型随着 Java 平台的主要版本而不断发展。在本节的内容中，将简要讲解 JVM 安全性的基本知识，为读者步入本书后面知识的学习打下基础。

2.3.1　JVM 的安全模型

Java 安全模型侧主要用于保护终端用户免受从网络下载的、来自不可靠来源的、恶意程序的侵犯。为了达到这个目的，Java 提供了一个用户可配置的“沙箱”，在沙箱中可以放置不可靠的 Java 程序。沙箱对不可靠程序的活动进行了限制，程序可以在沙箱的安全边界内做任何事，但是不能进行任何跨越这些边界的举动。例如，原来在版本 1.0 中的沙箱对很多不可靠 Javaapplet 的活动做了限制，主要包括：

- 对本地硬盘的读写操作；
- 进行任何网络连接，但不能连接到提供者 applet 的源主机；
- 创建新的进程；
- 装载新的动态连接库。

由于下载的代码不可能进行这些特定的操作，这使得 Java 安全模型可以保护终端用户避免受到有漏洞的代码的威胁。在沙箱内有严格的限制，其安全模型甚至规定了对不可靠代码能做什么、不能做什么，所以用户可以比较安全地运行不可靠代码。但是对于 1.0 系统的程序员和用户来说，这个最初的沙箱限制太过严格，善意的代码常常无法进行有效的工作。所以在后来的 1.1 版本中，对最初的沙箱模型进行了改进，引入了基于代码签名和认证的信任模式。签名和认证使得接收端系统可以确认一系列 class 文件已经由某一实体进行了数字签名（有效，可被信赖），并且在经过签名处理以后，class 文件没有改动。这使得终端用户和系统管理员减少了对某些代码在沙箱中的限制，但这些代码必须已由可信任团体进行数字签名。

虽然 1.1 版本的安全 API 包含了对认证的支持，但是其实只是提供了完全信任和完全不信任策略。Java 1.2 提供的 API 可以帮助建立细粒度的安全策略，这种策略是建立在数字签名代码的认证基础上的。Java 安全模型的发展经历了 1.0 版本的基本沙箱，然后是 1.1 版本的代码签名和认证，最后是 1.2 版以后的细粒度访问控制。

2.3.2　沙箱模型的 4 种组件

在计算机系统中，个人电脑中运行一个软件的前提是必须信任它。普通用户只能通过小心地使用来自可信任来源的软件来达到安全性，并且定期扫描，检查病毒来确保安全性。一旦某个软件有权使用我们的系统，那么它将拥有对这台电脑的完全控制权。如果这个软件是恶意的，那么它就可以为所欲为。所以在传统的安全模式中，必须想办法防止恶意代码有权使用你的计算机。

沙箱安全模型使得工作变得容易，即使某个软件来自我们不能完全信任的地方，通过沙箱模型可以使我们接受来自任何来源的代码，而不是要求用户避免将来自不信任站点的代码下载到机器上。当运行来自不可靠来源的代码时，沙箱会限制它进行任何可能破坏系统的动作指令。并且在整个过程中，无需指出哪些代码可以信任，哪些代码不可以信任，也不必扫描查找病毒。沙箱本身限制了下载的任何病毒或其他恶意的，有漏洞的代码，使得它们不能对计算机进行破坏。

如果你还有疑问，在确信它能保护你之前，用户必需确认沙箱没有任何漏洞。为了保证沙箱没有漏洞，Java 安全模型对其体系结构的各方面都进行了考虑。如果在 Java 体系结构中有任何没有考虑到安全的区域，恶意的程序员很可能会利用这些区域来绕开沙箱。因此，为了对沙箱有一个了解，必须先看一下 Java 体系结构的几个不同部分，并且理解它们是怎样一起工作的。

下面列出了组成 Java 沙箱的基本组件。

- 类加载体系结构。
- class 文件检验器。
- 内置于 Java 虚拟机（及语言）的安全特性。
- 安全管理器及 Java API。

Java 的上述安全模型的前 3 个部分——类加载体系结构、class 文件检验器、Java 虚拟机（及语言）的安全特性一起达到一个共同的目的：保持 JVM 的实例和它正在运行的应用程序的内部完整性，使得它们不被下载的恶意代码或有漏洞的代码侵犯。相反，这个安全模型的第四个组成部分是安全管理器，它主要用于保护虚拟机的外部资源不被虚拟机内运行的恶意或有漏洞的代码侵犯。这个安全管理器是一个单独的对象，在运行的 Java 虚拟机中，它在对于外部资源的访问控制起中枢作用。

2.3.3 分析 Java 的策略机制

沙箱安全模型的最大优点之一是可以是用户自定义的，通过从 Java1.1 版本就已经引入的代码签名和认证技术，使正在运行的应用程序可以对代码区分不同的信任度。通过自定义沙箱，被信任的代码可以比不可靠的代码获得更多的访问系统资源的权限。这就防止了不可靠代码访问系统，但是却允许被信任的代码访问系统并进行工作。Java 安全体系结构的真正好处在于，它可以对代码授予不同层次的信任度来部分地访问系统。

Microsoft 提供了 ActiveX 控件认证技术，它和 Java 的认证技术相类似，但是 ActiveX 控件并不在沙箱中运行。这样使用了 ActiveX，一系列移动代码要么是被完全信任的，要么是完全不被信任的。如果一个 ActiveX 控件不被信任，则它将被拒绝执行。虽然这对于没有认证来说是一个很大的提高，但是如果一些恶意的或是有漏洞的代码得到了认证，这段危险的代码将拥有对系统的完全访问权。Java 的安全体系结构的优点之一就是，代码可以被授予只对它需要的资源进行访问的有限权限。即使一些恶意的或者有漏洞的代码得到了认证，它也很少有机会进行破坏。例如，一段恶意的或者有漏洞的代码可能只能删除一个固定目录下的为它设置的文件，而不是在本地硬盘上的所有文件。

从 1.2 版本开始的安全体系结构的主要目标是建立（以签名代码为基础的）细粒度的访问控制策略，这样不但过程更为简单而且更少出错。为了将不同的系统访问权限授予不同的代码单元，Java 的访问控制机制必须能确认应该给每个代码段授予什么样的权限。为了使这个过程变得容易，载入 1.2 版本或其他虚拟机的每一个代码段（每个 class 文件）将和一个代码来源关联。代码来源主要说明了代码从哪里来，如果它被某个人签名担保的话，是从谁那里来。在 1.2 版本以后的安全模型中，权限（系统访问权限）是授给代码来源的。因此如果代码段请求访问一个特定的系统资源，只有当这个访问权限是和那段代码的代码来源相关联时，Java 虚拟机才会把对那个资源的访问权限授予这段代码。

在 1.2 版本的安全体系结构中，对应于整个 Java 应用程序的一个访问控制策略是由抽象类 java.security.Policy 的一个子类的单个实例所表示的。在任何时候，每一个应用程序实际上都只有一个 Policy 对象。获得许可的代码可以用一个新的 Policy 对象替换当前的 Policy 对象，这是通过调用 Policy.setPolicy()并把一个新的 Policy 对象的引用传递给它来实现的。类装载器利用这个 Policy 对象来帮助它们决定，在把一段代码导入虚拟机时应该给它们什么样的权限。

安全策略是一个从描述运行代码的属性集合到这段代码所拥有的权限的映射。在 1.2 版本的安全体系结构中，描述运行代码的属性被总称为代码来源。一个代码来源是由一个 java.security.CodeSource 对象表示的，这个对象中包含了一个 java.net.URL，它表示代码库和代表了签名者的零个或多个证书对象的数组。证书对象是抽象类 java.security.Certificate 的子类的一个实例，一个 Certificate 对象抽象表示了从一个人到一个公钥的绑定，以及另一个为这个绑定作担保的人（以前提过的证书机构）。CodeSource 对象包含了一个 Certificate 对象的数组，因为同一段代码可以被多个团体签名（担保）。这个签名通常是从 Jar 文件中获得的。

从 1.2 版本开始，所有和具体安全管理器有关的工具和访问控制体系结构都只能对证书起作用，而不能对公钥起作用。如果附近没有证书机构，可以用私钥对公钥签名，生成一个自签名的证书。当使用 keytool 程序生成密钥时，总是会产生一个自签名的证书。例如在上一节的签名例子中，keytool 不仅产生了“公钥/私钥”对，而且还为别名 friend 和 stranger 产生了自签名的证书。

权限是用抽象类 java.security.Permission 的一个子类的实例表示的。一个 Permission 对象有 3 个属性，分别是类型、名字和可选的操作。权限的类型是由 Permisstion 类的名字指定的，例如

java.io.FilePermission、java.net.SocketPermission 和 java.awt.AWTPermission。权限的名字是封装在 Permission 对象内的。例如某个 FilePermission 的名字可能是“/my/finances.dat”，某个 SocketPermission 的名字可能是“applets.artima.com:2000”，某个 AWTPermission 的名字可能是“showWindowWithoutBannerWarning”。Permission 对象的第 3 个属性是它的动作。并不是所有的权限都有动作。例如，FilePermission 的动作是“read, write”，SocketPermission 的动作是“accept, connect”。如果一个 FilePermission 的名字为“/my/finances.dat”，并且有动作“read, write”，那么它就表示对文件“/my/finance.dat”可以进行读写操作。名字和动作都是由字符串来表示的。

Java API 有一个很大的权限层次结构，在里面表示了所有可能潜在危险的操作。可以根据自己的目的创建自己的 Permission 类来表示自定义的权限，例如可以创建一个 Permission 类来表示对属性数据库的特定记录的访问权限。定义自定义的 Permission 类也是一种扩展版本 1.2 的安全机制类满足自己需要的方法。如果创建了自己的 Permission 类，可以像使用 Java API 中的 Permission 类一样来使用它们。

在 Policy 对象中，每一个 CodeSource 是和一个或多个 Permission 对象相关联的。和一个 CodeSource 相关联的 Permission 对象被封装在 java.security.PermissionCollection 的一个子类实例中。类装载器可以调用 Policy.getPolicy()来获得一个当前有效的 Policy 对象的引用。然后它们可以调用 Policy 对象的 getPermission()方法，传入一个 CodeSource，从而得到和那个 CodeSource 对应的 Permission 对象的 PermissionCollection。然后类装载器可以使用这个从 Policy 对象中得到的 PermissionCollection 来帮助判断应该给导入的代码授予什么权限。

2.4 网络移动性

长久以来，如何开发网络软件是 Java 开发人员所面临的最大挑战之一。在网络领域需要实现平台无关性，因为同一网络中通常连接了多种不同的计算机和设备。除此之外，安全模式也是一个挑战，因为网络可以方便地传输病毒和其他形式的恶意代码。在本节将详细讲解 Java 如何把握网络所带来的巨大机遇，为步入本书后面知识的学习打下基础。

2.4.1　现实需要网络移动性

当个人计算机互联成网变得越来越普遍的时候，另一种软件模式日益重要起来，即“客户机/服务器”模式。“客户机/服务器”模式将任务分为两部分，分别运行在两种计算机上：客户端进程运行在终端用户的个人计算机上，而服务器端进程运行在同一网络的另一台计算机上。客户端和服务器端的进程通过网络来回发送数据进行传输。服务器端进程通常只是简单地接收网络中客户端发来的数据请求命令，从中央数据库中提取需要的数据，并将该数据发送给客户端。而客户端在接到数据后，进行处理，然后显示并允许用户操作数据。这样的模式允许个人计算机的终端用户读取并操作放在中央储藏库的数据，而不需强迫这些用户共享中央 CPU 来处理数据。终端用户地区是共享了运行服务器端进程的 CPU，但在一定程度上，数据处理是由客户端完成的，因此大大减轻了服务器端 CPU 的负载。

“客户机/服务器”模式最初被称作两层客户机/服务器模式，一层是客户端，另一层是服务器。更复杂一些的模型叫做 3 层（表示有 3 个进程）、4 层（4 个进程）或者 N 层结构，也就是说层次结构越来越多了。当更多的进程加入计算时，客户端和服务器的区别模糊了，于是人们开始使用“分布式处理”这个新名词来涵盖所有这些结构模式。

分布式处理模式综合了网络和处理器发展的优点，将进程分布在多个处理器上运行，并允许这些进程共享数据。尽管这种模式有许多大型计算机系统所无法比拟的优势，但它也有个不可忽视的缺点：分布式处理比大型计算机系统更难管理。在大型计算机系统中，软件应用程序存储在主机的磁盘上，虽然可以有多个用户使用该软件，但它只需在一个地方安装和维护。升级一个软件后，所有用户在下一次登录并启动该软件的时候可以得到这个新的版本。但是相反，在分布式系统中，不同组件的软件往往存储在不同的磁盘上，因此，系统管理员需要在分布式系统的不同组件上安装和维护软件。要升级一个软件时，管理员不得不分别升级每台计算机上的这个软件。所以，分布式处理的系统管理比大型计算机系统要困难得多。

Java 的体系结构使软件的网络移动性成为可能，同时也预示了一种新的计算模式的到来。这种新的模式建立在流行的分布式处理模式的基础上，并可以将软件通过网络自动传送到各台计算机上。这样就解决了分布式处理系统中系统管理的困难。例如在一个 C/S 系统中，客户端软件可以存储在网络中的一台中央计算机上，当终端用户需要用该软件的时候，这个中央计算机会通过网络将可执行的软件传送到终端用户的计算机上运行。

因此，软件的网络移动性标志着计算模式发展历程中的重要一步，尤其是它解决了分布式处理系统中系统管理的问题，简化了将软件分布在多台 CPU 上的工作，使数据可以和相关软件一起被传送。

2.4.2 网络移动性

平台无关性使得在网络上传送程序更加容易，因为不需要为每个不同的主机平台都准备一个单独的版本，因此也不需要判断每台计算机需要哪个特定的版本，一个版本就可以对付所有的计算机。Java 的安全特性促进了网络移动性的推广，因为最终用户就算从不信任的来源下载 class 文件，也可以充满自信。因此实际上，Java 体系结构通过对平台无关性和安全性的支持，更好地推广了其 class 文件的网络机动性。

除了平台无关性和安全性之外，Java 体系结构对网络移动性的支持主要集中在对在网络上传送程序的时间进行管理上。假若你在服务器上保存了一个程序，在需要的时候通过网络来下载它，这个过程一般都会比从本地执行该程序要慢。因此对于在网络上传送程序来说，网络移动性的一个主要难题就是时间。Java 体系结构通过把传统的单一二进制可执行文件切割成小的二进制碎片——Javaclass 文件——来解决这个问题。class 文件可以独立在网络上传播，因为 Java 程序是动态链接、动态扩展的，最终用户不需要等待所有的程序 class 文件都下载完毕，就可以开始运行程序了。第一个 class 文件到手，程序就开始执行。class 文件本身也被设计得很紧凑，所以它们可以在网络上飞快地传送。因此 Java 体系结构为网络移动性带来的直接主要好处就是把一个单一的大二进制文件分割成小的 class 文件，这些 class 文件可以按需装载。

Java 应用程序从某个类的 main()方法开始执行，其他的类在程序需要的时候才动态链接。如果某个类在一次操作中没有被用到，这个类就不会被装载。比如说，假若你在使用一个字处理程序，它有一个拼写检查器，但是在你使用的这次操作中没有使用拼写检查器，那么它就不会被装载。

除了动态链接之外，Java 体系结构也允许动态扩展。动态扩展是装载 class 文件的另一种方式，可以延迟到 Java 应用程序运行时才装载。使用用户自定义的类装载器，或者 Class 类的 forName()方法，Java 程序可以在运行时装载额外的程序，这些程序就会变成运行程序的一部分。因此，动态链接和动态扩展给了 Java 程序员一些设计上的灵活性，既可以决定何时装载程序的 class 文件——而这又决定了最终用户需要等待多少时间来从网络上装载 class 文件。

除了动态连接和动态扩展，Java 体系结构对网络移动性的直接支持还通过 class 文件格式体现。为了减少在网络上传送程序的时间，class 文件被设计得很紧凑。它们包含的字节码流设计得特别紧凑——之所以被称为“字节码”，是因为每条指令都只占据一个字节。除了两个例外情况，所有的操作码和它们的操作数都是按照字节对齐的，这使得字节码流更小。这两个例外是这样一些操作码，在操作码和它们的操作数之间会填上 1～3 个字节，一边操作数都按照字边界对齐。

class 文件的紧凑型隐含着另外一个含义，那就是 Java 编译器不会做太多的局部优化。因为二进制兼容性规则的存在，Java 编译器不能做一些全局优化，比如把一个方法调用转化为整个方法的内嵌（内嵌指把被调用方法的整个方法体都替换到发起调用的方法中去，这样在代码运行的时候，可以节省方法调用和返回的时间）。二进制兼容性要求，假若一个方法被现有 class 文件包括以后，那么改变这个方法的时候必须不破坏已有的调用方法。在同一个类中使用的方法可能使用内嵌，但是一般来说，Java 编译器不会做这种优化，部分原因是这样为 class 文件瘦身得不偿失。优化常常是在代码大小和执行速度间进行的折中。因此，Java 编译器通常会把优化工作留给 Java 虚拟机，后者在装载类之后，在解释执行，即时编译或者自适应编译的时候都可以优化代码。

除了动态链接、动态扩展和紧凑的 class 文件之外，还有一些并非体系结构必须的策略，可以帮助控制在网络上传送 class 文件的时间。因为 HTTP 需要单独为 Javaapplet 中用到的每一个 class 文件请求连接，所以下载 applet 的很大一部分时间并不是用来实际传输 class 文件的时间，而是每一个 class 文件请求的网络协议握手的时间。一个文件需要的总时间是按照需要下载的 class 文件的数目倍增的。为了解决这个问题，Java 1.1 包含了对 Jar 的支持，Jar 文件允许在一次网络传输过程中传输多个文件，这和一次传送一个个单独 class 文件相比，大幅度降低了需要的总体下载时间。更大的优点是，Jar 文件中的数据可以压缩，从而使下载时间更少。所以有时候通过一个大文件来传送软件。例如有些 class 文件是程序开始运行之前所必需的，这些文件可以很快地通过 Jar 文件一次性传递。

另外一个降低最终用户等待时间的策略就是不采取按需下载 class 文件的做法。有几种不同的技术，例如 MarimbaCastanet 使用的订阅模式，可以在需要 class 文件之前就已经把它们下载下来了，这样程序就可以更快地启动。

因此，除了平台无关性和安全性能够对网络移动性有利外，Java 体系结构的主要着眼点就是控制 class 文件在网络上传送的时间。动态链接和动态扩展允许 Java 程序按照小功能单元设计，在最终用户需要的时候才单独下载。Class 文件的紧凑性本身有助于减少 Java 程序在网络上传送的时间。Jar 文件允许在一次网络连接中传送多个文件，还允许数据压缩。

2.5 内存异常和垃圾处理

对于 C 和 C++的开发人员来说，在内存管理领域应该能够游刃有余。在计算机系统中，内存负责维护每一个对象生命的从开始到终结。Java 内存分配与管理是 Java 的核心技术之一，通常 Java 在内存分配时会涉及到以下区域。

- 寄存器：在程序中无法控制。
- 栈：存放基本类型的数据和对象的引用，但对象本身不存放在栈中，而是存放在堆中。
- 堆：存放用 new 产生的数据。
- 静态域：存放在对象中用 static 定义的静态成员。
- 常量池：存放常量。
- 非 RAM 存储：硬盘等永久存储空间。

2.5.1 内存分配中的栈和堆

1. 栈

在函数中定义的一些基本类型的变量数据，还有对象的引用变量都在函数的栈内存中分配。

当在一段代码块中定义一个变量时，Java 就在栈中为这个变量分配内存空间；当该变量退出该作用域后，Java 会自动释放掉为该变量所分配的内存空间，该内存空间可以立即被另作他用。

栈也称为栈内存，是 Java 程序的运行区，是在线程创建时创建，它的生命期跟随着线程的生命期，线程结束栈内存也就释放。对于栈来说不存在垃圾回收问题，只要线程一结束，该栈就被释放。问题出来了：栈中存的是那些数据呢？又什么是格式呢？

栈中的数据都是以栈帧（Stack Frame）的格式存在，栈帧是一个内存区块，是一个数据集，是一个有关方法（Method）和运行期数据的数据集。当一个方法 A 被调用时就产生了一个栈帧 F1，并被压入到栈中，A 方法又调用了 B 方法，于是产生栈帧 F2 也被压入栈；执行完毕后，先弹出 F2 栈帧，再弹出 F1 栈帧，遵循“先进后出”原则。

那栈帧中到底存在着什么数据呢？在栈帧中主要保存如下 3 类数据。

- ❑ 本地变量（Local Variables）：包括输入参数和输出参数以及方法内的变量。
- ❑ 栈操作（Operand Stack）：记录出栈、入栈的操作。
- ❑ 栈帧数据（Frame Data）：包括类文件、方法等。

光说比较枯燥，画个图来理解一下 Java 栈，如图 2-6 所示。

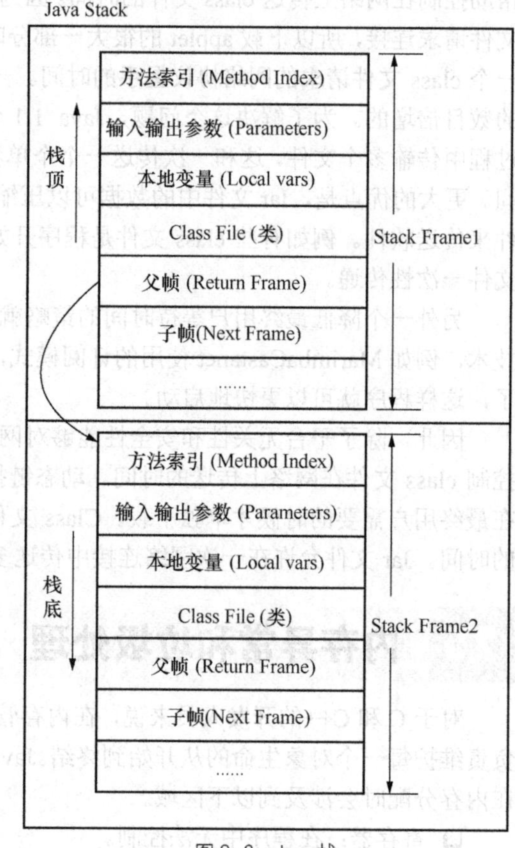

图 2-6　Java 栈

在图 2-6 中，一个栈中有两个栈帧，栈帧 2 是最先被调用的方法，先入栈，然后方法 2 又调用了方法 1。栈帧 1 处于栈顶的位置，栈帧 2 处于栈底，执行完毕后，依次弹出栈帧 1 和栈帧 2，线程结束，栈释放。

2. 堆

堆内存用来存放由关键字 new 创建的对象和数组。在堆中分配的内存，由 Java 虚拟机的自动垃圾回收器来管理。

在堆中产生了一个数组或对象后，还可以在栈中定义一个特殊的变量，让栈中这个变量的取值等于数组或对象在堆内存中的首地址，栈中的这个变量就成了数组或对象的引用变量。引用变量就相当于是为数组或对象起的一个名称，以后就可以在程序中使用栈中的引用变量来访问堆中的数组或对象。引用变量就相当于是为数组或者对象起的一个名称。

引用变量是普通的变量，定义时在栈中分配，引用变量在程序运行到其作用域之外后被释放。而数组和对象本身在堆中分配，即使程序运行到使用 new 产生数组或者对象的语句所在的代码块之外，数组和对象本身占据的内存不会被释放，数组和对象在没有引用变量指向它的时候，才变为垃圾，不能再被使用，但仍然占据内存空间不放，在随后的一个不确定的时间被垃圾回收器收走（释放掉）。这也是 Java 比较占内存的原因。

实际上，栈中的变量指向堆内存中的变量，这就是 Java 中的指针。

3. 常量池（constant pool）

常量池指的是在编译期被确定，并被保存在已编译的.class 文件中的一些数据。除了包含代码中所定义的各种基本类型（如 int、long 等等）和对象型（如 String 及数组）的常量值（final）还包含一些以文本形式出现的符号引用，比如：

- 类和接口的全限定名；
- 字段的名称和描述符；
- 方法和名称和描述符。

虚拟机必须为每个被装载的类型维护一个常量池。常量池就是该类型所用到常量的一个有序集合，包括直接常量（string, integer 和 floating point 常量）和对其他类型、字段和方法的符号引用。

对于 String 常量，它的值是在常量池中的。而 JVM 中的常量池在内存当中是以表的形式存在的，对于 String 类型，有一张固定长度的 CONSTANT_String_info 表用来存储文字字符串值，但是该表只存储文字字符串值，并不存储符号引用。在程序执行的时候，常量池会储存在 Method Area（方法区域）中，而不是堆中。

一个 JVM 实例只存在一个堆内存，堆内存的大小是可以调节的。类加载器读取了类文件后，需要把类、方法、常变量放到堆内存中，以方便执行器执行。堆内存分为 3 部分。

（1）永久存储区（Permanent Space）。

永久存储区是一个常驻内存区域，用于存放 JDK 自身所携带的 Class Interface 的元数据。也就是说，它存储的是运行环境必须的类信息，被装载进此区域的数据是不会被垃圾回收器回收掉的，关闭 JVM 才会释放此区域所占用的内存。

（2）新生区（Young Generation Space）。

新生区是类的诞生、成长、消亡的区域，一个类在这里产生、应用，最后被垃圾回收器收集，结束生命。新生区又分为伊甸区（Eden space）和幸存者区（Survivor pace）两部分。所有的类都是在伊甸区被 new（新建）出来的；幸存区有两个：0 区（Survivor 0 space）和 1 区（Survivor 1 space）。当伊甸园的空间用完时，程序又需要创建对象，JVM 的垃圾回收器将对伊甸园区进行垃圾回收，将伊甸园区中不再被其他对象所引用的对象进行销毁，然后将伊甸园中的剩余对象移动到幸存 0 区。若幸存 0 区也满了，再对该区进行垃圾回收，然后移动到 1 区。那如果 1 区也满了呢？再移动到养老区。

（3）养老区（Tenure Generation Space）。

养老区用于保存从新生区筛选出来的 Java 对象，一般池对象都在这个区域活跃。

上述 3 个区的示意图如图 2-7 所示。

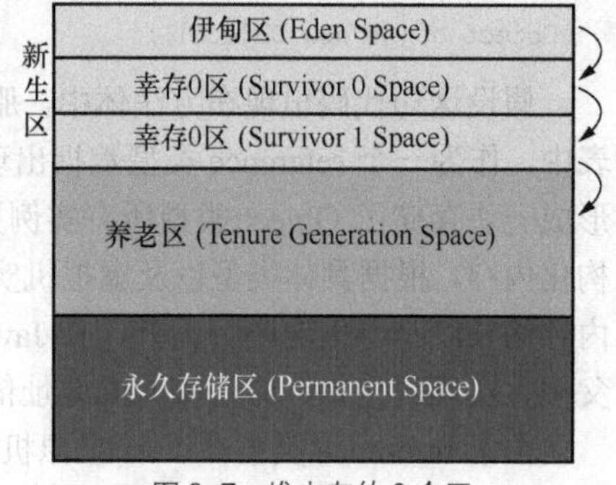

图 2-7 堆内存的 3 个区

2.5.2 运行时的数据区域

Java 通过自身的动态内存分配和垃圾回收机制，可以使 Java 程序员不用像 C++程序员那么头疼内存的分配与回收。对于这一点来说，相信熟悉 COM 机制的朋友对于引用计数管理内存的方式深有感触。通过 Java 虚拟机的自动内存管理机制，不仅降低了编码的难度，而且不容易出现内存泄露和内存溢出的问题。但是这过于理想的愿望正是由于把内存的控制权交给了 Java 虚拟机，一旦出现内存泄露和溢出，我们就必须翻过 Java 虚拟机自动内存管理这堵高墙去排查错误。

根据《Java 虚拟机规范》的规定，Java 虚拟机在执行 Java 程序时，即运行环境下会把其所管理的内存划分为几个不同的数据区域。有的区域伴随虚拟机进程的启动而创建，死亡而销毁；有些区域则是依赖用户线程的启动时创建，结束时销毁。所有线程共享方法区和堆，虚拟机栈、本地方法栈和程序计数器是线程隔离的数据区。Java 虚拟机运行时的数据区结构如图 2-8 所示。

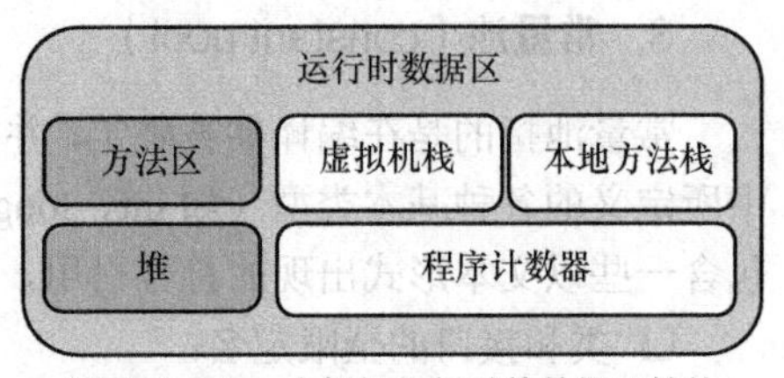

图 2-8 Java 虚拟机运行时的数据区结构

2.5.3 对象访问

JVM 的逻辑内存模型如图 2-9 所示。

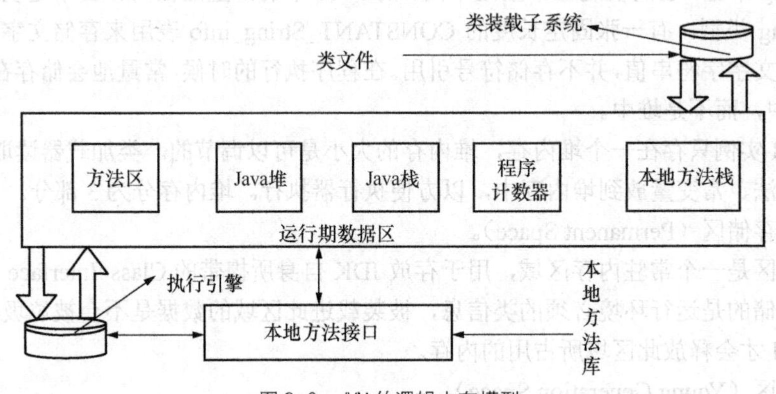

图 2-9 JVM 的逻辑内存模型

当建立一个对象时如何进行访问呢？在 Java 语言中，对象访问是如何进行的？对象访问在 Java 语言中无处不在，是最普通的程序行为，但即使是最简单的访问，也会涉及 Java 栈、Java 堆、方法区这 3 个最重要内存区域之间的关联关系，如下面的代码：

```
Object obj = new Object();
```

假设这句代码出现在方法体中，那“Object obj”这部分的语义将会反映到 Java 栈的本地变量表中，作为一个 reference 类型数据出现。而“new Object()”这部分的语义将会反映到 Java 堆中，形成一块存储了 Object 类型所有实例数据值（Instance Data，对象中各个实例字段的数据）的结构化内存，根据具体类型以及虚拟机实现的对象内存布局（Object Memory Layout）的不同，这块内存的长度是不固定的。另外，在 Java 堆中还必须包含能查找到此对象类型数据（如对象类型、父类、实现的接口、方法等）的地址信息，这些类型数据则存储在方法区中。

由于 reference 类型在 Java 虚拟机规范中只规定了一个指向对象的引用，并没有定义这个引用应该通过哪种方式去定位，以及访问到 Java 堆中的对象的具体位置，因此不同虚拟机实现的对象访问方式会有所不同，主流的访问方式有使用句柄和直接指针两种。

（1）如果使用句柄访问方式，Java 堆中将会划分出一块内存来作为句柄池，reference 中存储的就是对象的句柄地址，而句柄中包含了对象实例数据和类型数据各自的具体地址信息，如图 2-10 所示。

（2）如果使用直接指针访问方式，Java 堆对象的布局中就必须考虑如何放置访问类型数据的相关信息，reference 中直接存储的就是对象地址，如图 2-11 所示。

这两种对象的访问方式各有优势，使用句柄访问方式的最大好处就是 reference 中存储的是稳定的句柄地址，在对象被移动（垃圾收集时移动对象是非常普遍的行为）时只会改变句柄中的实例数据指针，而 reference 本身不需要被修改。

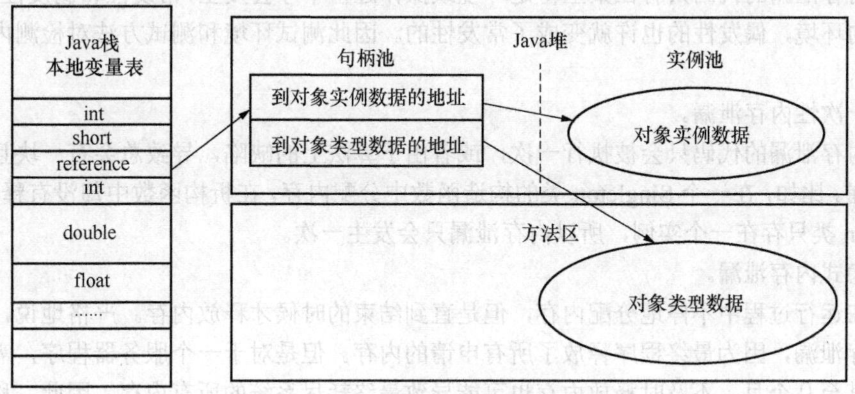

图 2–10　用句柄访问对象

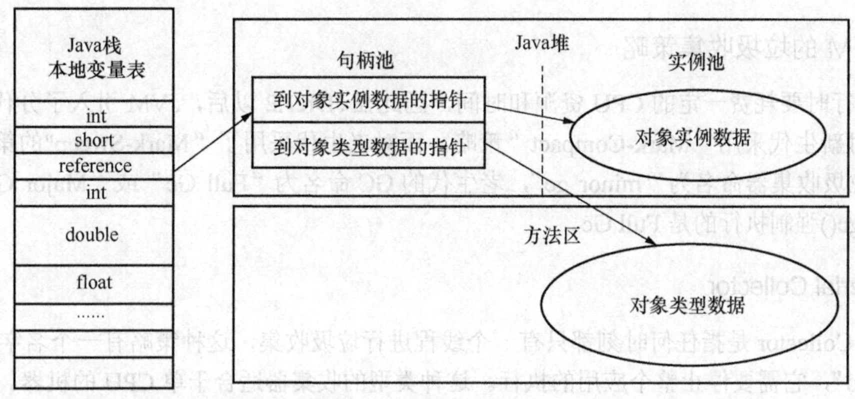

图 2–11　通过指针访问对象

使用直接指针访问方式的最大好处就是速度更快，它节省了一次指针定位的时间开销，由于对象的访问在 Java 中非常频繁，因此这类开销积少成多后也是一项非常可观的执行成本。从整个软件开发的范围来看，各种语言和框架使用句柄来访问的情况也十分常见。

2.5.4　内存泄露

在计算机科学中，内存泄漏（Memory Leak）是指由于疏忽或错误造成程序未能释放已经不再使用的内存的情况。内存泄漏并非指内存在物理上的消失，而是应用程序分配某段内存后，由于设计错误，失去了对该段内存的控制，因而造成了内存的浪费。内存泄漏与许多其他问题有着相似的症状，并且通常情况下只能由那些可以获得程序源代码的程序员才可以分析出来。然而，有不少人习惯于把任何不需要的内存使用的增加描述为内存泄漏，严格意义上来说这是不准确的。一般常说的内存泄漏是指堆内存的泄漏。堆内存是指程序从堆中分配的、大小任意的（内存块的大小可以在程序运行期决定）、使用完后必须显式释放的内存。应用程序一般使用 malloc、realloc、new 等函数从堆中分配到一块内存，使用完后，程序必须负责相应的调用 free 或 delete 释放该内存块，否则这块内存就不能被再次使用，一般就说这块内存泄漏了。

通常可以将内存泄露分为以下 4 类。

（1）常发性内存泄漏。

发生内存泄漏的代码会被多次执行到，每次被执行的时候都会导致一块内存泄漏。

（2）偶发性内存泄漏。

发生内存泄漏的代码只有在某些特定环境或操作过程下才会发生，常发性和偶发性是相对的。对于特定的环境，偶发性的也许就变成了常发性的。因此测试环境和测试方法对检测内存泄漏至关重要。

（3）一次性内存泄漏。

发生内存泄漏的代码只会被执行一次，或者由于算法上的缺陷，导致总会有一块且仅一块内存发生泄漏。比如，在一个 Singleton 类的构造函数中分配内存，在析构函数中却没有释放该内存。而 Singleton 类只存在一个实例，所以内存泄漏只会发生一次。

（4）隐式内存泄漏。

程序在运行过程中不停地分配内存，但是直到结束的时候才释放内存。严格地说，这里并没有发生内存泄漏，因为最终程序释放了所有申请的内存。但是对于一个服务器程序，需要运行几天、几周甚至几个月，不及时释放内存也可能导致最终耗尽系统的所有内存。因此，称这类内存泄漏为隐式内存泄漏。

2.5.5 JVM 的垃圾收集策略

GC 执行时要耗费一定的 CPU 资源和时间，因此在 JDK1.2 以后，JVM 引入了分代收集的策略，其中对新生代采用“Mark-Compact“策略，而对老生代采用了“Mark-Sweep"的策略。其中新生代的垃圾收集器命名为“minor gc”，老生代的 GC 命名为“Full Gc”或“Major GC”。其中用 System.gc()强制执行的是 Full Gc。

1. Serial Collector

Serial Collector 是指任何时刻都只有一个线程进行垃圾收集，这种策略有一个名字“stop the whole world”，它需要停止整个应用的执行。这种类型的收集器适合于单 CPU 的机器。

Serial Copying Collector

此种 GC 用-XX:UseSerialGC 选项配置，它只用于新生代对象的收集。1.5.0 以后-XX:MaxTenuringThreshold 来设置对象复制的次数。当 eden 空间不够时，GC 会将 eden 的活跃对象和一个名叫 From survivor 空间中尚不够资格放入 Old 代的对象复制到另外一个名字叫 To Survivor 的空间。而此参数就是用来说明到底 From survivor 中的哪些对象不够资格，假如这个参数设置为 31，那么也就是说只有对象复制 31 次以后才算是有资格的对象。

From Survivor 和 To Survivor 的角色是不断变化的，同一时间只有一块空间处于使用状态，这个空间就叫做 From Survivor 区，当复制一次后角色就发生了变化。

如果复制的过程中发现 To Survivor 空间已经满了，那么就直接复制到 Old Generation。

比较大的对象也会直接复制到 Old Generation，在开发中，应该尽量避免这种情况的发生：

Serial Mark-Compact Collector

串行的标记-整理收集器是 JDK5 update 6 之前默认的老生代的垃圾收集器，此收集使得内存碎片最少化，但是它需要暂停的时间比较长

2. Parallel Collector

Parallel Collector 主要是为了应对多 CPU，大数据量的环境。Parallel Collector 又可以分为以下两种。

（1）Parallel Copying Collector：此种 GC 用-XX:UseParNewGC 参数配置，它主要用于新生代

的收集，此 GC 可以配合 CMS 一起使用。

（2）在 1.4.1 版本以后用：

```
Parallel Mark-Compact Collector
```

此种 GC 用-XX:UseParallelOldGC 参数配置，此 GC 主要用于老生代对象的收集。1.6.0 后用：

```
Parallel scavenging Collector
```

此种 GC 用-XX:UseParallelGC 参数配置，它是对新生代对象的垃圾收集器，但是它不能和 CMS 配合使用，它适合于比较大新生代的情况，此收集器起始于 jdk 1.4.0。它比较适合于对吞吐量高于暂停时间的场合。

3. Concurrent Collector

Concurrent Collector 通过并行的方式进行垃圾收集，这样就减少了垃圾收集器收集一次的时间，这种 GC 在实时性要求高于吞吐量的时候比较有用。此种 GC 可以用参数-XX:UseConcMarkSweepGC 配置，此 GC 主要用于老生代和 Perm 代的收集。

2.5.6 垃圾收集器

如果说收集算法是内存回收的方法论，垃圾收集器就是内存回收的具体实现。Java 虚拟机规范中对垃圾收集器应该如何实现并没有任何规定，因此不同的厂商、不同版本的虚拟机所提供的垃圾收集器都可能会有很大的差别，并且一般都会提供参数供用户根据自己的应用特点和要求组合出各个年代所使用的收集器。这里讨论的收集器基于 Sun HotSpot 虚拟机 1.6 版 Update 22，这个虚拟机包含的所有收集器如图 2-12 所示。

图 2-12 展示了 7 种作用于不同分代的收集器（包括 JDK 1.6_Update14 后引入的 Early Access 版 G1 收集器），如果两个收集器之间存在连线，就说明它们可以搭配使用。

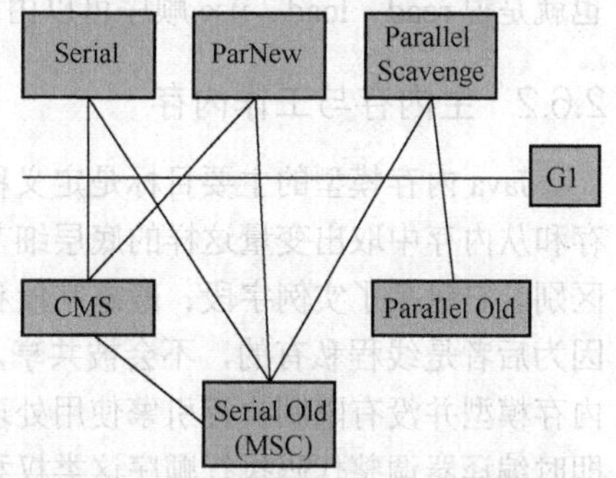

图 2-12 HotSpot JVM 1.6 的垃圾收集器

在介绍这些收集器各自的特性之前，先来明确一个观点：虽然是在对各个收集器进行比较，但并非为了挑选一个最好的收集器出来。因为直到现在为止还没有最好的收集器出现，更加没有万能的收集器，所以本书选择的只是对具体应用最合适的收集器。这点不需要多加解释就能证明：如果有一种放之四海皆准、任何场景下都适用的完美收集器存在，那 HotSpot 虚拟机就没必要实现那么多不同的收集器了。

2.6 Java 内存模型

不同的平台，内存模型是不一样的，但是 JVM 的内存模型规范是统一的。其实 Java 的多线程并发问题最终都会反映在 Java 内存模型上，所谓线程安全无非是要控制多个线程对某个资源的有序访问或修改。总结 Java 的内存模型，要解决两个主要的问题：可见性和有序性。

人们都知道计算机有高速缓存的存在，处理器并不是每次处理数据都是取内存的。JVM 定义了自己的内存模型，屏蔽了底层平台内存管理细节，对于 Java 开发人员，要清楚在 JVM 内存模型的基础上，如果解决多线程的可见性和有序性。

那么，何谓可见性？多个线程之间是不能互相传递数据通信的，它们之间的沟通只能通过共享变量来进行。Java 内存模型（JMM）规定了 JVM 有主内存，主内存是多个线程共享的。当新建一个对象的时候，也是被分配在主内存中，每个线程都有自己的工作内存，工作内存存储了主存的某些对象的副本，当然线程的工作内存大小是有限制的。当线程操作某个对象时，执行顺序如下。

（1）从主存复制变量到当前工作内存（read and load）。

（2）执行代码，改变共享变量值（use and assign）。

（3）用工作内存数据刷新主存相关内容（store and write）。

2.6.1　Java 内存模型概述

Java 平台自动集成了线程以及多处理器技术，这种集成程度比 Java 以前诞生的计算机语言要厉害很多。该语言针对多种异构平台的平台独立性而使用的多线程技术支持也是具有开拓性的一面，有时候在开发 Java 同步和线程安全要求很严格的程序时，往往容易混淆的一个概念就是内存模型。究竟什么是内存模型？内存模型描述了程序中各个变量（实例域、静态域和数组元素）之间的关系，以及在实际计算机系统中将变量存储到内存和从内存中取出变量这样的底层细节，对象最终是存储在内存里面的，这点没有错，但是编译器、运行库、处理器或者系统缓存可以有特权在变量指定内存位置存储或者取出变量的值。

JVM 规范定义了线程对主存的操作指令：read、load、use、assign、store、write。当一个共享变量在多个线程的工作内存中都有副本时，如果一个线程修改了这个共享变量，那么其他线程应该能够看到这个被修改后的值，这就是多线程的可见性问题。那么，什么是有序性呢？线程在引用变量时不能直接从主内存中引用，如果线程工作内存中没有该变量，则会从主内存中复制一个副本到工作内存中，这个过程为 read-load，完成后线程会引用该副本。当同一线程再度引用该字段时，有可能重新从主存中获取变量副本（read-load-use），也有可能直接引用原来的副本（use），也就是说 read、load、use 顺序可以由 JVM 实现系统决定。

2.6.2　主内存与工作内存

Java 内存模型的主要目标是定义程序中各个变量的访问规则，即在虚拟机中将变量存储到内存和从内存中取出变量这样的底层细节。此处的变量（Variable）与 Java 编程中所说的变量略有区别，它包括了实例字段、静态字段和构成数组对象的元素，但是不包括局部变量与方法参数，因为后者是线程私有的，不会被共享，自然就不存在竞争问题。为了获得较好的执行效能，Java 内存模型并没有限制执行引擎使用处理器的特定寄存器或缓存来与主内存进行交互，也没有限制即时编译器调整代码执行顺序这类权利。

Java 内存模型规定了所有的变量都存储在主内存（Main Memory）中（此处的主内存与介绍物理硬件时的主内存名字一样，两者也可以互相类比，但此处仅是虚拟机内存的一部分）。每条线程还有自己的工作内存（Working Memory，可与前面所讲的处理器高速缓存类比），线程的工作内存中保存了被该线程使用到的变量的主内存副本复制，线程对变量的所有操作（读取、赋值等）都必须在工作内存中进行，而不能直接读写主内存中的变量。不同的线程之间也无法直接访问对方工作内存中的变量，线程间变量值的传递均需要通过主内存来完成，线程、主内存、工作内存三者的交互关系如图 2-13 所示。

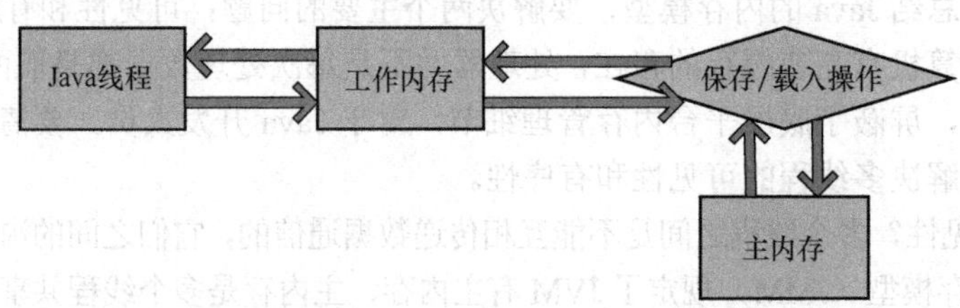

图 2-13　线程、主内存和工作内存之间的交互关系

2.6.3 内存间交互操作

关于主内存与工作内存之间具体的交互协议，即一个变量如何从主内存拷贝到工作内存、如何从工作内存同步回主内存之类的实现细节，在 Java 内存模型中定义了以下 8 种操作来完成内存间的交互操作。

- lock（锁定）：作用于主内存的变量，它把一个变量标识为一条线程独占的状态。
- unlock（解锁）：作用于主内存的变量，它把一个处于锁定状态的变量释放出来，释放后的变量才可以被其他线程锁定。
- read（读取）：作用于主内存的变量，它把一个变量的值从主内存传输到线程的工作内存中，以便随后的 load 动作使用。
- load（载入）：作用于工作内存的变量，它把 read 操作从主内存中得到的变量值放入工作内存的变量副本中。
- use（使用）：作用于工作内存的变量，它把工作内存中一个变量的值传递给执行引擎，每当虚拟机遇到一个需要使用变量值的字节码指令时将会执行这个操作。
- assign（赋值）：作用于工作内存的变量，它把一个从执行引擎接收到的值赋给工作内存的变量，每当虚拟机遇到一个给变量赋值的字节码指令时执行这个操作。
- store（存储）：作用于工作内存的变量，它把工作内存中一个变量的值传送到主内存中，以便随后的 write 操作使用。
- write（写入）：作用于主内存的变量，它把 store 操作从工作内存中得到的变量的值放入主内存的变量中。

如果要把一个变量从主内存复制到工作内存，那就要按顺序地执行 read 和 load 操作；如果要把变量从工作内存同步回主内存，就要按顺序地执行 store 和 write 操作。注意，Java 内存模型只要求上述两个操作必须按顺序执行，而没有保证必须是连续执行。也就是说 read 与 load 之间、store 与 write 之间是可插入其他指令的，如对主内存中的变量 a、b 进行访问时，一种可能出现的顺序是 read a、read b、load b、load a。除此之外，Java 内存模型还规定了在执行上述 8 种基本操作时必须满足如下规则。

- 不允许 read 和 load、store 和 write 操作之一单独出现，即不允许一个变量从主内存读取了但工作内存不接收，或者从工作内存发起回写操作但主内存不接收的情况出现。
- 不允许一个线程丢弃它最近的 assign 操作，即变量在工作内存中改变了之后必须把该变化同步回主内存。
- 不允许一个线程无原因地（没有发生过任何 assign 操作）把数据从线程的工作内存同步回主内存中。
- 一个新的变量只能在主内存中“诞生”，不允许在工作内存中直接使用一个未被初始化（load 或 assign）的变量，换句话说就是对一个变量实施 use 和 store 操作之前，必须先执行过了 assign 和 load 操作。
- 一个变量在同一个时刻只允许一条线程对其进行 lock 操作，但 lock 操作可以被同一条线程重复执行多次，多次执行 lock 后，只有执行相同次数的 unlock 操作时才会解锁变量。
- 如果对一个变量执行 lock 操作，将会清空工作内存中此变量的值。在执行引擎使用这个变量前，需要重新执行 load 或 assign 操作初始化变量的值。
- 如果一个变量事先没有被 lock 操作锁定，则不允许对它执行 unlock 操作，也不允许去 unlock 被其他线程锁定住的变量。
- 对一个变量执行 unlock 操作之前，必须先把此变量同步回主内存中（执行 store 和 write 操作）。

第3章　Dalvik和ART基础

Dalvik VM是Android虚拟机的称呼，从Android系统诞生之日起到Android L，一直是Google等厂商合作开发的Android移动设备平台的核心组成部分之一。从Android 4.4开始，谷歌推出了ART运行环境机制，这种机制的运行速度更快、效率更高，将完全取代Dalvik VM成为唯一的运行机制。Dalvik VM和ART都可以支持已转换为.dex（即Dalvik Executable）格式的Java应用程序的运行。.dex格式是专为Dalvik设计的一种压缩格式，适合内存和处理器速度有限的系统。Dalvik是由Dan Bornstein编写的，名字来源于他的祖先曾经居住过的名叫Dalvik的小渔村。大多数虚拟机包括JVM都是一种堆栈机器，而Dalvik虚拟机则是基于寄存器的。两种架构各有优劣，一般而言，基于栈的机器需要更多指令，而基于寄存器的机器指令更大。Dalvik VM和本书第2章中讲解的Java虚拟机十分相似。在本章的内容中，将详细讲解Dalvik VM和ART技术的基础性知识，为读者步入本书后面知识的学习打下基础。

3.1 Dalvik VM和JVM的差异

很多人认为Dalvik VM是一个Java虚拟机，因为Android的编程语言恰是Java语言。但是这种说法并不准确，因为Dalvik虚拟机并不是按照Java虚拟机的规范来实现的，两者并不兼容。其中有如下两个显著的不同点。

- Java虚拟机运行的是Java字节码，而Dalvik虚拟机运行的则是其专有的文件格式dex（Dalvik Executable）。
- 在Java SE程序中的Java类会被编译成一个或者多个字节码文件（.class）然后打包到JAR文件，而后Java虚拟机会从相应的CLASS文件和JAR文件中获取相应的字节码；Android应用虽然也是使用Java语言进行编程，但是在编译成CLASS文件后，还会通过一个工具（dx）将应用所有的CLASS文件转换成一个DEX文件，而后Dalvik虚拟机会从其中读取指令和数据。

Dalvik VM与大多数虚拟机和真正的Java VM不同，它们是栈机（Stack Machine），而Dalvik VM是基于寄存器的架构。此外，两种方法的相对优势取决于所选择的解释/编译策略。但总的来说，基于stack的机器必须使用指令来载入stack上的数据，或使用指令来操纵数据。所以与基于寄存器的机器相比，Dalvik VM需要更多的指令。然而，在寄存器的指令必须编码源和目的地寄存器，因此往往指令更大。

在Dalvik VM机制中，使用一个名为dx的工具来转换Java的.class文件到.dex格式。多个类文件可包含到单个的.dex文件中，重复的、可用于多个类的字符串和其他常量在转换到.dex格式时输出到保留空间。Java字节码还可转换成可选择的、Delvik VM使用的指令集。一个未经过压缩的.dex文件，在文件大小方面往往比从同样的.class文件压缩成的.jar文件更小。

当Dalvik可执行文件安装到移动设备时，它们是可以被修改的。为了进一步优化，在某些数据、简单数据结构和内联的函数库中的字节顺序可以互换，例如空类对象被短路。

为了满足低内存要求而不断优化，Dalvik VM 存在如下所示的独特特征。

（1）Dalvik VM 很小，使用的空间也小。

（2）Dalvik VM 没有 JIT 编译器。

（3）Dalvik VM 常量池已被修改为只使用 32 位的索引，以简化解释器。

（4）Dalvik VM 使用自己的字节码，而非 Java 字节码。

此外，Dalvik VM 被设计来满足可高效运行多种虚拟机实例。Dalvik VM 和标准 Java 虚拟机（JVM）之间的首要差别之一，就是 Dalvik VM 基于寄存器，而 JVM 基于栈。Dalvik VM 和 JVM 之间的另外一大区别就是运行环境——Dalvik VM 经过优化，允许在有限的内存中同时运行多个虚拟机的实例，并且每一个 Dalvik VM 应用作为一个独立的 Linux 进程执行。

具体来说，Dalvik 虚拟机和 Java 虚拟机的差异如下所示。

❑ Dalvik 虚拟机早期并没有使用 JIT（Just-In-Time）技术，从 Android 2.2 版本开始，Dalvik 虚拟机也支持 JIT。

❑ Dalvik 虚拟机有自己的 bytecode，并非使用 Java bytecode。

❑ Dalvik 虚拟机基于暂存器（register），而 JVM 基于堆栈（stack）。

❑ Dalvik VM 通过 Zygote 进行 Class Preloading，Zygote 会完成虚拟机的初始化，也是与 JVM 的不同之处。

概括来说，Dalvik 虚拟机和 Java 虚拟机的区别如表 3-1 所示。

表 3-1　Java 虚拟机和 Dalvik 虚拟机的区别

Java 虚拟机	Dalvik 虚拟机
Java 虚拟机基于栈，基于栈的机器必须使用指令来载入和操作栈上数据，所需指令更多	Dalvik 虚拟机是基于寄存器的
Java 虚拟机运行的是 Java 字节码。（java 类会被编译成一个或多个字节码.class 文件，打包到.jar 文件中，java 虚拟机从相应的.class 文件和.jar 文件中获取相应的字节码）	Dalvik 运行的是自定义的.dex 字节码格式（java 类被编译成.class 文件后，会通过一个 dx 工具将所有的.class 文件转换成一个.dex 文件，然后 dalvik 虚拟机会从其中读取指令和数据）
	常量池已被修改为只使用 32 位的索引，以简化解释器。Dalvik 的堆和栈的参数可以通过-Xms 和-Xmx 更改
	一个应用，一个虚拟机实例，一个进程（所有 Android 应用的线程都是对应一个 Linux 线程，都运行在自己的沙盒中，不同的应用在不同的进程中运行。每个 Android Dalvik 应用程序都被赋予了一个独立的 Lnux PID(app_*))

3.2 Dalvik 虚拟机的主要特征

在 Dalvik 虚拟机中，一个应用中会定义很多类，编译完成后即会有很多相应的 CLASS 文件，CLASS 文件间会有不少冗余的信息；而 DEX 文件格式会把所有的 CLASS 文件内容整合到一个文件中。这样，除了减少整体的文件尺寸和 I/O 操作，也提高了类的查找速度。在每个类文件中的常量池，都是在 DEX 文件中由一个常量池负责管理。

每一个 Android 应用都运行在一个 Dalvik 虚拟机实例里，而每一个虚拟机实例都是一个独立的进程空间。虚拟机的线程机制、内存分配和管理、Mutex 等都是依赖底层操作系统实现的。所有 Android 应用的线程都对应一个 Linux 线程，虚拟机因而可以更多地依赖操作系统的线程调度和管理机制。

不同的应用在不同的进程空间里运行，加之对不同来源的应用都使用不同的 Linux 用户来运行，可以最大程度地保护应用的安全和独立运行。

Zygote 是一个虚拟机进程，同时也是一个虚拟机实例的孵化器，每当系统要求执行一个 Android 应用程序，Zygote 就会 FORK（孕育）出一个子进程来执行该应用程序。这样做的好处如下所示：

- Zygote 进程是在系统启动时产生的，它会完成虚拟机的初始化、库的加载、预置类库的加载和初始化等操作。
- 当系统需要一个新的虚拟机实例时，Zygote 通过复制自身，最快速地提供一个系统。
- 对于一些只读性的系统库来说，所有虚拟机实例都和 Zygote 共享一块内存区域，大大节省了内存开销。

相对于基于堆栈的虚拟机实现，基于寄存器的虚拟机实现虽然在硬件通用性上要差一些，但是它在代码的执行效率上却更胜一筹。在基于寄存器的虚拟机里，可以更为有效地减少冗余指令的分发和减少内存的读写访问。

3.3 Dalvik VM 架构

在 Android 源码中，Dalvik 虚拟机的实现位于“dalvik/”目录下，其中“dalvik/vm”是虚拟机的实现部分，将会编译成 libdvm.so。而“dalvik/libdex”将会编译成 libdex.a 静态库，作为 dex 工具使用；“dalvik/dexdump”是.dex 文件的反编译工具，虚拟机的可执行程序位于“dalvik/dalvikvm”中，将会编译成 dalvikvm 可执行文件。

Dalvik 虚拟机的架构如图 3-1 所示。

Android 应用编译及运行流程如图 3-2 所示。

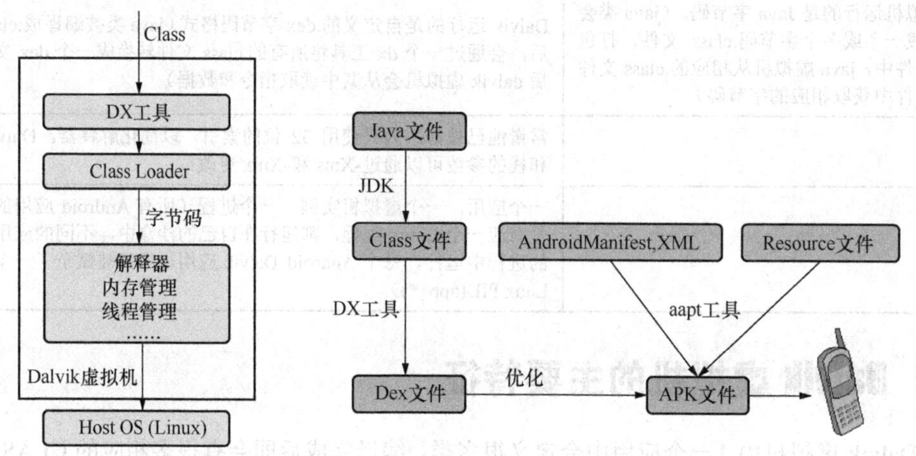

图 3-1 Dalvik 虚拟机的架构

图 3-2 Android 应用编译及运行流程

3.3.1 Dalvik 虚拟机的代码结构

Dalvik 是 Android 程序的 Java 虚拟机，代码保存在“dalvik/”目录下，目录的具体结构如下所示：

```
./
|-- Android.mk
|-- CleanSpec.mk
|-- MODULE_LICENSE_APACHE2
|-- NOTICE
```

```
|-- README.txt
|-- dalvikvm 虚拟机的实现库
|-- dexdump
|-- dexlist
|-- dexopt
|-- docs
|-- dvz
|-- dx
|-- hit
|-- libcore
|-- libcore-disabled
|-- libdex
|-- libnativehelper 使用 JNI 调用本地代码时用到这个库
|-- run-core-tests.sh
|-- tests
|-- tools
`-- vm
```

"dalvik/"目录的效果图如图 3-3 所示。

图 3-3 "dalvik/"目录的效果图

Dalvik 虚拟机各个目录的具体说明如下所示：

- ❑ android.mk：是虚拟机编译的 makefile 文件。
- ❑ dalvikvm：此目录是虚拟机命令行调用入口文件的目录，主要用来解释命令行参数，调用库函数接口等。
- ❑ dexdump：此目录是生成 dex 文件反编译查看工具，主要用来查看编译出来的代码文件是否正确，查看编译出来的文件结构如何。
- ❑ dexlist：此目录是生成查看 dex 文件里所有类的方法的工具。

- dexopt：此目录是生成 dex 优化工具。
- docs：此目录是保存 Dalvik 虚拟机相关帮助文档。
- dvz：此目录是生成从 Zygote 请求生成虚拟机实例的工具。
- dx：此目录是生成从 Java 字节码转换为 Dalvik 机器码的工具。
- hit：此目录是生成显示堆栈信息/对象信息的工具。
- libcore：此目录是 Dalvik 虚拟机的核心类库，提供给上层的应用程序调用。
- libcore-disabled：此目录是一些禁用的库。
- libdex：此目录是生成主机和设备处理 DEX 文件的库。
- libnativehelper：此目录是 Dalvik 虚拟核心库的支持库函数。
- MODULE_LICENSE_APACHE2：这个是 APCHE2 的版权声明文件。
- NOTICE：这个文件是说明虚拟机源码的版权注意事项。
- README.txt：这个文件是说明本目录相关内容和版权。
- run-core-tests.sh：这个文件是用来运行核心库测试。
- tests：此目录是保存测试相关测试用例。
- tools：此目录是保存一些编译/运行相关的工具。
- vm：此目录是保存虚拟机绝大部份代码，包括读取指令读取、指令执行等。

3.3.2 dx 工具

在 Android 虚拟机中，dx 工具是用来转换 Java Class 成为 DEX 格式，但不是全部。多个类型包含在一个 DEX 文件之中。多个类型中重复的字符串和其他常数包括会存放在 DEX 之中只有一次，以节省空间。Java 字节码（betecode）转换成 Dalvik 虚拟机所使用的替代指令集。一个未压缩 DEX 文件通常是稍稍小于一个已经压缩.Jar 文档。

当启动 Android 系统时，Dalvik VM 监视所有的程序（APK），并且创建依存关系树，为每个程序优化代码并存储在 Dalvik 缓存中。Dalvik VM 第一次加载后会生成 Cache 文件，以提供下次快速加载，所以第一次会变得很慢。

Dalvik VM 解释器采用预先算好的 Goto 地址，基于每个指令集 OpCode，都固定以 64 B 为 Memory Alignment。这样可以节省一个指令集 OpCode 后，要进行查表的时间。为了强化功能，Dalvik VM 还提供了 Fast Interpreter。

dx 是一套工具，可以将 Java 的.class 文件转换成.dex 格式。一个 dex 文档通常会有多个.class 文件。由于 dex 有时必须进行优化，会使文件大小增加 1～4 倍，以 ODEX 结尾。

3.3.3 Dalvik VM 的进程管理

Dalvik VM 进程管理是依赖于 Linux 的进程体系结构的，如要为应用程序创建一个进程，它会使用 Linux 的 fork 机制来复制一个进程（复制进程往往比创建进程效率更高）。

Zygote 是一个虚拟机进程，同时也是一个虚拟机实例的孵化器，它通过 init 进程启动。首先会孵化出 System_Server（Android 绝大多系统服务的守护进程，它会监听 Socket 等待请求命令，当有一个应用程序启动时，就会向它发出请求，Zygote 就会 FORK 出一个新的应用程序进程）。每当系统要求执行一个 Android 应用程序时，Zygote 就会运用 Linux 的 FORK 进制产生一个子进程来执行该应用程序。

3.3.4 Android 的初始化流程

Linux 中的进程间通信方式有很多，但是 Dalvik VM 是使用信号方式来完成进程间的通信工作的。Android 的初始化流程如图 3-4 所示。

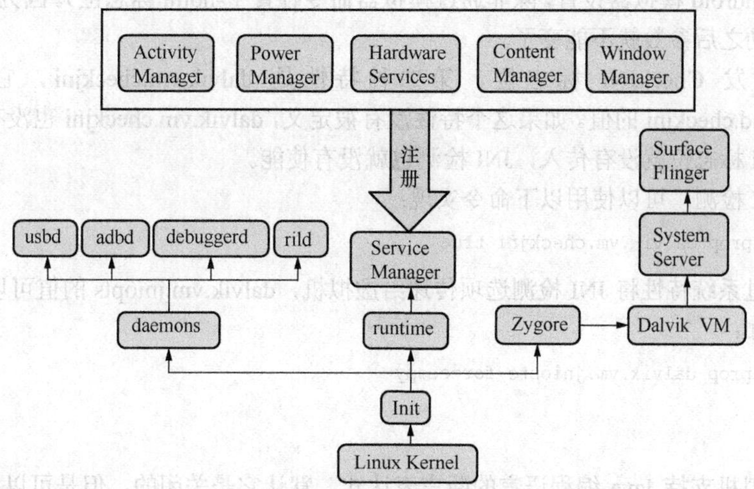

图 3-4 Android 的初始化流程

3.4 Dalvik VM 控制 VM 命令详解

Dalvik 虚拟机支持一系列的命令行参数（使用 adbshell dalvikvm –help 命令获取列表），不可能通过 Android 应用运行时来传递任意参数，但是可以通过特定的系统参数来影响虚拟机行为。在本节的内容中，将详细讲解常用的控制 VM 的命令。

3.4.1 基本命令

在 Dalvik 虚拟机中，可以通过 setprop 来设置系统特性，使用 shell 命令的语法格式如下所示：

```
adbshell setprop <name> <value>
```

在运行时必须重启 Android，从而使得改变生效（adb shell stop：adb shell start）。这是因为，这些设定在 Zygote 进程中处理，而 Zygote 是一个最早启动并且永远存活的进程。

不可以用无特权用户的身份设定 dalvik.*参数及重启系统，可以在用户调试版本的 shell 上使用 adb root 或者运行 su 命令的方式来获取 root 权限，可以通过如下命令告诉 setprop 是否发生：

```
adbshell getprop <name>
```

如果不想在设备重启之后特性随之消失，可以在/data/local.prop 中加上如下所示的命令：

```
<name>= <value>
```

重启之后这样的改变也会一直存在，但是如果 data 分区被擦除就消失了。在工作台上创建一个 local.prop，然后 adb push local.prop /data/，或者使用类似于下面格式的命令，注意这里引号很重要：

```
adb shell “echo name =value >> /data/local.prop”
```

3.4.2 扩展的 JNI 检测

JNI（Java Native Interface）意为 Java 本地接口，提供了 Java 语言程序调用本地（C/C++）代码的方法。扩展的 JNI 检测会导致系统运行更慢，但是可以发现一系列讨厌的 bug，防止它们产生问题。有两个系统参数影响这个功能，这个功能可以通过-Xcheck:jni 命令行参数来激活。第一个参数是 ro.kernel.Android.checkjni，这是通过 Android 编译系统对 development 的编译来设置的（也可以通过 Android 模拟器设置，除非通过模拟器命令行置了-nojni 标志位）。因为这是一个“ro.”特性，设备启动之后参数就不能变了。

为了能触发 CheckJNI 标志位，第二种特性是 dalvik.vm.checkjni，它的值覆盖了 ro.kernel.Android.checkjni 的值。如果这个特性没有被定义，dalvik.vm.checkjni 也没有设置成 false，那么-Xcheck:jni 标志位就没有传入，JNI 检测也就没有使能。

要打开 JNI 检测，可以使用以下命令实现：

```
adbshell setprop dalvik.vm.checkjni true
```

也可以通过系统特性将 JNI 检测选项传递给虚拟机，dalvik.vm.jniopts 的值可以通过-Xjniopts 参数传入，例如：

```
adb shellsetprop dalvik.vm.jniopts forcecopy
```

3.4.3　断言

Dalvik 虚拟机支持 Java 编程语言的断言表达式，默认它是关闭的，但是可以通过如下-ea 参数的方式设置 dalvik.vm.enableassertions 特性：

```
dalvikvm -ea..
```

在其他桌面虚拟机中这个参数同样生效，通过提供 class 名、package 名（后跟“…”），或者特殊值“all”。例如：

```
adbshell setprop dalvik.vm.enableassertion all
```

就可以在所有非系统 class 中使能断言。这个系统特性比全命令行更受限制，不可以通过-ea 入口设置更多，而且没有指定-da 入口的方法，而且未来也没有-esa/-dsa 等价的东西。

3.4.4　字节码校验和优化

Dalvik VM 会尝试预校验 DEX 文件中的所有类，从而降低 CLASS 的负担，从而可以使用一系列的优化来提升运行性能。上述功能都是通过 dexopt 命令来实现的，不论是在编译系统中还是在安装上。在开发设备上，dexopt 可能在 DEX 文件第一次被使用时运行，而不论它或者它的依赖是否更新过（Just-in-time 优化和校验，JIT）。

有两个命令行标志位控制 JIT 优化和校验：-Xverify 和-Xdexopt。Andorid 框架基于 dalvik.vm.dexopt-flags 特性来配置这俩参数，如果设定：

```
adbshell setprop dalvik.vm.dexopt-flags v=a o=v
```

那么 Android 框架会将-Xverify:all-Xdexopt:verified 传递给虚拟机，这将使能校验并且只优化校验成功的 class。这是最安全的设定，也是默认的。

另外也可以设定 dalvik.vm.dexopt-flags v=n 使得框架传输-Xverify:none －Xdexopt:verified 从而不使能校验（可以传输-Xdexopt:all 从而允许优化，但是这并不能优化更多代码，因为没有通过校验的 class 可能被优化器以同样的理由跳过）。这时 CLASS 不会被 dexopt 校验，而没被校验的代码难以执行。

使能校验会使得 dexopt 命令明显花费更多时间，因为校验过程相对较慢，一旦校验和优化过的 DEX 文件准备就绪，校验就不会占用额外的开销除非在加载预校验失败的 CLASS。

如果 DEX 文件的校验关闭了，而后来又打开了校验器，应用加载会明显变慢（大概 40%以上）因为 CLASS 会在第一次被调用的时候校验。

为了最佳效果，当特性变化时应该为 DEX 文件强制重新调用 dexopt，即：

```
adbshell "rm /data/dalvik-cache/*"
```

它删除了暂存的 DEX 文件，记住要中止再打开运行时（adb shell stop：adb shell start）。

> 注意　老的运行时版本支持布尔型的 dalvik.vm.verify-bytecode 特性，但是被 dalvik.vm.dexopt-flags 替代了。

3.4.5　Dalvik VM 的运行模式

当前 Dalvik VM 的实现包括 3 个独立的解释内核："快速"（Fast)、"可移植"（Portable)、"调试"（Debug)。快速解释器是为当前平台优化的，可能包括手动优化的汇编文件；相对的，可移植解释器是用 C 写的，可在广泛的平台上使用；调试解释器是可移植解释器的变种，包括了支持程序分析（Profiling）和单步。

VM 可能也支持 just-in-time 编译，严格地说它并不是另一个解释器，JIT 编译器也可以被同样的标志位使能/不使能。查看 dalvik－help 的输出信息来查看 JIT 编译器是否在你的虚拟机里面使能。

VM 允许人们在快速、可移植和 jit 中选择，通过使用-Xint 参数的扩展来实现，该参数的值可以通过 dalvik.vm.execution-mode 系统特性来设置。为了选择可移植解释器，此时应该用：

```
adb shell setpropdalvik.vm.execution-mode int:portable
```

如果没有指定该参数，系统会自动选择最合适的编译器，有时候机器可能允许选择其他模式，例如 JIT 编译器。

不是所有的平台都有优化的实现，有时候，快速编译器是由一系列的 C 实现的，这个结果会比可移植编译器还慢。如果对所有流行平台都有优化版本，这个命名"快速"就更准确了。

如果程序分析使能或者调试器连接了，VM 会变为调试解释器。当程序分析结束或者调试器中断连接，就会恢复原来的解释器。用调试解释器会明显变慢，这是在评估数据时要记住的。

JIT 编译器可以通过在应用程序 AndroidManifest.xml 中加入 android:vmSafeMode="rue"来不使能，当怀疑 JIT 编译器会使得应用运行不正常的时候可以使用该功能。

3.4.6　死锁预测

如果虚拟机以 WITH_DEADLOCK_PREDICTION 参数编译，那么死锁预测器会在-Xdeadlockpredict 参数中使能（dalvikvm －help 会告诉你虚拟机是否编译正确——在 Configured 中按行查找 deadlock_prediction)。这个特性会让虚拟机一直跟踪对象的锁获取的顺序，如果程序试图以与之前看到不同的顺序获取一些锁，虚拟机会 log 一个 warning 并有选择地抛出异常。

命令行参数是基于 dalvik.vm.deadlock-predict 特性设置的，正确的值是"off"，表示不使能它（默认)；"warn"表示 log 问题但是继续执行；"err"表示从 monitor-enter 指令中引发一个 dalvik.system.PotentialDeadlockError 异常；"abort"表示终止整个虚拟机。通常可以这么使用：

```
adbshell setprop dalvik.vm.deadlock-predict err
```

除非可以在 log 信息滚动的时候一直关注着。注意这个特性是死锁预测，不是死锁检测——在当前实现中，在锁被获取之后才会进行计算（这减轻了代码量，降低了互斥信息外的冗余)。在挂起的进程中执行 kill -3 时可以发现一个死锁，并且可以在 Log 信息中检测到。

这仅仅考虑了监督程序，本地的互斥量和其他资源也会引起死锁，而且不会被它检测到。

3.4.7 dump 堆栈追踪

和其他桌面虚拟机一样，Dalvik 虚拟机收到 SIGQUIT（Ctrl-\ 或者 kill -3）时，会为所有的现成 dump 所有的堆栈追踪。它默认写入 Android 的 log，但是也可以写入一个文件。

dalvik.vm.stack-trace-file 特性允许你指定要将线程堆栈追踪写入的文件名，如果不存在，将创建，新的信息将追加到文件尾，文件名通过-Xstacktracefile 参数写入虚拟机。例如：

```
adbshell setprop dalvik.vm.stack-trace-file /tmp/stack-traces.txt
```

如果这个特性没有被定义，虚拟机会在收到这个信号时将堆栈追踪信息写入 Android log。

3.4.8 dex 文件和校验

出于性能考虑，优化过的 dex 文件和校验被取消了，这通常叫安全，因为文件是在设备上产生的，并且有禁止修改的权限。

但是如果设备的存储器不可靠，就会发生数据损坏，这通常表现为重复的虚拟机崩溃。为了快速诊断这种失败，虚拟机提供了-Xcheckdexsum 参数，如果设置了，在内容被使用之前所有的 dex 文件都会进行和校验。如果 dalvik.vm.check-dex-sum 特性被使能，那么应用框架会在虚拟机创建时提供这个参数。

为了使能额外的 dex 和校验，可以用如下命令实现：

```
adbshell setprop dalvik.vm.check-dex-sum true
```

不正确的和校验会组织 dex 数据的使用，产生错误并写入 log 文件，如果设备曾经有过这样的问题，那么将这个特性写入/data/local.prop 很有用。

注意　dexdump 工具每次都会进行 dex 和校验，它也可以用于检测大量的文件。

3.4.9 产生标志位

在“Honeycomb”版本中引入了一系列的汇编，它们通过标志位写入虚拟机：

```
adb shell setprop dalvik.vm.extra-opts "flag1flag2 … flagN"
```

在这些标志位之间需要用空格隔开。可以指定任意多的标志位，只要它们在系统特性值的长度范围内，目前是 92 个字符。

这些额外的标志位会被加到命令行的底端，意味着它们会覆盖之前的设定。这些可以用于例如测试不同的-Xmx 的值，即使 Android 框架层已经设定过了。

3.5 ART 机制基础

随着 Android 4.4 和 Android L 版本的推出，Dalvik VM 系统逐渐被谷歌抛弃，取而代之的是 ART 运行机制。ART 模式英文全称为：Android runtime，是从 Android 4.4 系统开始新增的一种应用运行模式，与传统的 Dalvik 模式不同，ART 模式可以实现更为流畅的安卓系统体验，对于大家来说，只要明白 ART 模式可让系统体验更加流畅，不过只有在 Android 4.4 以上系统中采用此功能。在本章的内容中，将详细讲解 ART 技术的基础知识，为读者步入本书后面知识的学习打下基础。

3.5.1 什么是 ART 模式

ART 模式的完整名称是 Androidruntime，事实上谷歌的这次优化源于前不久其收购的一家名为 Flexycore 的公司，该公司一直致力于 Android 系统的优化，而 ART 模式也是在该公司的优化方案上演进而来。ART 的机制与 Dalvik 不同。在 Dalvik 下，应用每次运行的时候，字节码都需要通过即时编译器转换为机器码，这会拖慢应用的运行效率；而在 ART 环境中，应用在第一次安装的时候，字节码就会预先编译成机器码，使其成为真正的本地应用。这个过程称为预编译（AOT, Ahead-Of-Time）。这样，应用的启动和执行都会变得更加快速。

ART 模式与 Dalvik 模式最大的不同在于，在启用 ART 模式后，系统在安装应用的时候会进行一次预编译，在安装应用程序时会先将代码转换为机器语言存储在本地，这样在运行程序时就不会每次都进行一次编译了，执行效率也大大提升。从这方面来看，ART 模式确实能够改善 Android 平台一直以来在兼容性方面的妥协，但另一方面，应用经过预编译后的容量，以及应用是否兼容该模式也是需要重点考虑的问题。因此接下来笔者也准备通过一些测试来看看 ART 模式目前的表现是否能令人满意。

（1）产生背景。

与 iOS 相比，Android 的用户体验总是差强人意。随着 Google 的全力推动和硬件厂商的响应，Android 还是跨越各种阻碍，逐渐壮大起来了。在此过程中，Google 也在经历着重大的变化。它逐渐从一个只重视数据的公司，转变为一个重视设计和用户体验的公司。从 Android 4.0 开始，Android 拥有了自己的设计语言和应用设计指导。与此同时，Google 也在着手解决卡顿问题，Android 4.1 使系统和应用运行都更加顺畅，而 Android 4.2 提升了内存管理，使得系统能够顺利运行在硬件配置低端的设备上。

但是，所有这些都没有解决核心问题，那就是应用运行环境：Dalvik VM 效率并不是最高的。从 Android 4.4 开始，Google 开发者引进了新的 Android 运行环境 ART。Android 官方将其作为其新的虚拟机，以替代旧的 Dalvik VM。

根据一些基准测试，新的运行环境能够使大多数应用的执行时间减半。这意味着，CPU 消耗大、运行时间长的应用能够更加快速地完成，而一般的应用也能更加流畅，比如动画效果更顺畅，触控反馈更加即时。在多核处理器的设备上，多数情况下只需激活少量的核心，或者能够更好地利用 ARM 的 big.LITTLE 架构。另外，它将会显著提升电池的续航能力以及系统的性能。

预编译也会带来一些缺点。一方面，机器码占用的存储空间更大。字节码变为机器码之后，可能会增加 10%～20%，不过在应用包中，可执行的代码常常只是一部分。比如最新的 Google+ APK 是 28.3 MB，但是代码只有 6.9 MB。另一方面，应用的安装时间会变长。至于延长多少时间，取决于应用本身，一些复杂的应用，如 Facebook 和 Google+，会让你等待更长时间。

（2）Android 系统性能提升之路。

Dalvik 虚拟机作为 Android 平台的核心组成部分之一，允许在有限的内存资源中同时运行多个虚拟机实例。Dalvik 虚拟机通过以下方式提升性能。

- DEX 代码安装时或第一次动态加载时 odex 化处理。
- Android 2.2 版本提供了 JIT 机制提升性能，号称性能提升 3~5 倍。
- 提升硬件配置，如更多核 CPU、更高频率 CPU、更大的 RAM 等。

但是 Android 的系统流畅度与 iOS 系统还是有一定差距。Android 代码必须运行在 Dalvik 虚拟机上，而 iOS 直接是本地代码，性能差距也在情理之中。如果 Android 系统想拥有与 iOS 系统相同的系统性能，Dalvik 虚拟机运行机制就成为 Android 系统性能提升唯一的障碍。

从 Android 4.4 开始提供了一种与 Dalvik 截然不同的运行环境——ART（Android Runtime）的支持。

（3）ART 架构。

ART 完全兼容 Dalvik 的字节码格式 dex，因此，开发者编写软件不会受到影响，也无需担心兼容性问题。ART 的一大变化是，它不仅支持即时编译（JIT），而且支持预先编译（AOT）。在 Dalvik 上，每次软件运行，都需从字节码编译为原生代码，ART 可以只编译一次。然后，软件每次运行时，执行编译好的原生代码。预先编译也为新的优化带来了可能性。同时，这也会明显改善电池续航，因为软件运行时不用编译了，从而减少了 CPU 的使用频率，降低了能耗。

诞生 ART 也有一些缺点。其中之一是：设备首次启动以及应用的首次启动时间会变长。不过，Google 宣称说，这种差别不是很大，而且他们会在这方面努力，使其接近甚至超过 Dalvik。另一个缺点是原生代码占用空间更大，不过，现在设备的空间应该都足够。

（4）垃圾回收。

Android 虚拟机是自动内存管理，这种方式的优点是开发者无需担心内存管理，缺点是开发者失去了控制权，依赖于系统本身的机制。Dalvik 的垃圾回收机制是造成系统卡顿的原因之一。在 Dalvik 虚拟机下，启动垃圾回收机制会造成两次暂停（一次在遍历阶段，一次在标记阶段）。所谓暂停，就是应用的所有线程都不再执行。如果暂停时间过长，应用渲染中就会出现掉帧。用户体验上来说，就是应用运行的时候出现卡顿。

Google 宣称，Neuxs 5 的平均暂停时间是 54 ms，结果就是，每次垃圾回收启动，平均掉帧是 4 帧。如果应用编写得不好，情况会更加糟糕。Anandtech 测试了 FIFA 游戏。Dalvik 环境下，启动应用的几秒内，垃圾回收启动 9 次，应用暂停时间总和为 603 ms，总共掉帧为 214 帧。在 ART 下，情况有了极大改善。同样时间里，应用暂停时间总和为 12.364 ms（4 次前台垃圾回收，2 次后台垃圾回收），总共掉帧是 63 帧。

ART 能够做到这一点，是因为应用本身做了垃圾回收的一些工作。垃圾回收启动后，不再是两次暂停，而是一次暂停。在遍历阶段，应用不需要暂停，而标记阶段的暂停时间也大大缩短，因为 Google 使用了一种新技术（packard pre-cleaning），在暂停前就做了许多事情，减轻了暂停时的工作量。Google 承诺，他们已经把平均暂停时间降到了 3 ms，远远超过 Dalvik 的垃圾回收。

与此同时，Google 还改进了内存分配系统，使分配速度加快了 10 倍。垃圾回收算法也进行了修改，以增强用户体验，避免应用被打断。

（5）64 位支持。

ART 支持 64 位系统，这会带来性能上的提升，加密能力的大幅改进，同时保持与现有 32 位应用的兼容性。与苹果不同的是，Google 使用了指针压缩，以避免转换到 64 位后，空间占用大幅增加，其虚拟机仍然是 32 位指针。

Google 宣称，现有 Play Store 上应用中，85%都可以转移到 64 位，剩下的 15%有原生代码，需要重新编译。总地来说，转移到 64 位应该会在短时间内完成。

综上所述，Google 兑现了其提升性能的承诺，解决了困扰 Android 性能的诸多问题。原来 Android 的一些致命弱点是因为非原生应用和自动内存管理系统，ART 在这些方面做出了大量改进。总之，在应用的流畅度和性能方面，Android 终于可以与 iOS 一决高下了。

3.5.2　ART 优化机制基础

在安装应用程序时，Dalvik Runtime 采用的代码优化方式是：

```
dex2opt（http://124.16.139.131:24080/lxr/source/dalvik/dexopt/OptMain.cpp? v= android-4.0.4#f_OptMain.cpp）
```

而 ART Runtime 采用的代码优化方式是：

```
dex2oat (https://android.googlesource.com/platform/art/+/kitkat-release/dex2oat/dex2oat.cc)
```

Dalvik 运行环境和 ART 运行环境产生的优化代码路径及文件名都为：

```
/data/dalvik-cache/app/data@app@{package name}.apk@classes.dex
```

ART 环境产生的优化代码文件大小明显比 Dalvik 环境产生大。

在 Android L 系统中，ART 模块的实现源码被保存在 Android L 源码的根目录“art”中，如图 3-5 所示。

名称	修改日期	类型	大小
.git	2014/7/4 8:21	文件夹	
build	2014/7/4 8:21	文件夹	
compiler	2014/7/4 8:21	文件夹	
dalvikvm	2014/7/4 8:21	文件夹	
dex2oat	2014/7/4 8:21	文件夹	
disassembler	2014/7/4 8:21	文件夹	
oatdump	2014/7/4 8:21	文件夹	
runtime	2014/7/4 8:21	文件夹	
sigchainlib	2014/7/4 8:21	文件夹	
test	2014/7/4 8:21	文件夹	
tools	2014/7/4 8:21	文件夹	
.gitignore	2014/7/4 8:21	文本文档	1 KB
Android.mk	2014/7/4 8:21	VisualStudio.m...	21 KB
MODULE_LICENSE_APACHE2	2014/7/4 8:21	文件	0 KB
NOTICE	2014/7/4 8:21	文件	11 KB

图 3-5 ART 模块的源码

从源代码文件目录名称可以很清楚地了解各文件夹中相关文件的功能，其中最为主要的有 compiler、dex2oa 和 runtime 这 3 个文件夹，具体说明如下所示。

- ❑ compiler：主要负责 Dalvik 字节码到本地代码的转换，编译为 libart-compiler.so。
- ❑ dex2oat：完成 DEX 文件到 ELF 文件转换，编译为 dex2oat。
- ❑ runtime：Android ART 运行时源代码，编译为 libart.so。

第4章 分析JNI

JNI是Java Native Interface的缩写，译为Java本地接口。JNI标准是Java平台的一部分，允许Java代码和其他语言写的代码进行交互。JNI是本地编程接口，能够在Java虚拟机（VM）内部运行的Java代码能够与用其他编程语言（如C、C++和汇编语言）编写的应用程序和库进行交互操作。另外，为了讲解JNI在Java和C/C++之间的作用，特意讲解了JNI在media框架中的应用。

4.1 JNI的本质

在Android系统中，JNI是连接Java部分和C/C++部分的纽带。要想完整地使用JNI，需要仔细分析Java代码和C/C++代码。在Android中通过提供JNI的方式，让Java程序可以调用C语言程序。Android中的很多Java类都具有Native（本地）接口，这些接口由本地实现，然后注册到系统中。

JNI调用的层次非常清晰，主要分为3层，在Android系统中这3层从上到下依次为：Java→JNI→C/C++（SO库），Java可以访问C/C++中的方法，同样C/C++可以修改Java对象，图4-1清晰地描述了这三者之间的调用关系。

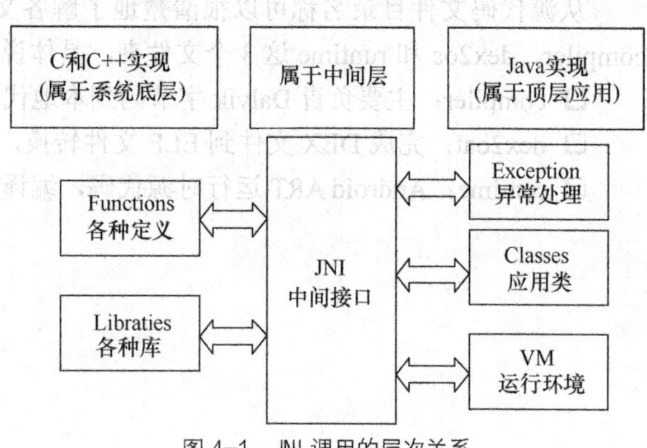

图4-1 JNI调用的层次关系

由图4-1可知，JNI的调用关系为：

Java------------------------JNI------------------------Native

在Android 4.4的源码中，主要的JNI代码放在以下的路径中：

```
frameworks/base/core/jni/
```

上述路径中的内容被编译成库libandroid_runtime.so，这是一个普通的动态库，被放置在目标系统的“/system/lib”目录下。另外，Android中还存在其他的JNI库，其实JNI中的各个文件就是普通的C++源文件；在Android中实现的JNI库，需要连接动态库libnativehelper.so。

要想弄明白JNI的本质，还要从Java的本质说起。从本质上来说，Java语言的运行完全依赖于脚本引擎对Java的代码进行解释和执行。因为现代的Java可以从源代码编译成.class之类的中间格式的二进制文件，所以这种处理会加快Java脚本的运行速度。尽管如此，基本的执行方式仍然不变，由脚本引擎（被称之为JVM）来执行。与Python、Perl之类的纯脚本相比，只是把脚本变成了二进制格式而已。另外，Java本身就是一门对面向对象语言可以调用完善的功能库。当把这个脚本引擎移植到所有平台上之后，那么这个脚本就很自然地实现“跨平台”了。绝大多数的脚本引擎都支持一个很显著的特性，就是可以通过C/C++编写模块，并在脚本中调用这些模块。

Java 也同样如此，Java 一定要提供一种在脚本中调用 C/C++编写的模块的机制，才能称得上是一个相对完善的脚本引擎。

从本质上来看，Android 平台是由 arm-linux 操作系统和一个叫做 Dalvik 的 Java 虚拟机组成的。所有在 Android 模拟器上看到的界面效果都是用 Java 语言编写的，具体请看源代码中的“frameworks/base”目录。由此可见，Dalvik 只是提供了一个标准的支持 JNI 调用的 Java 虚拟机环境。在 Android 平台中，使用 JNI 技术封装了所有的和硬件相关的操作，通过 Java 去调用 JNI 模块，而 JNI 模块使用 C/C++调用 Android 本身的 arm-linux 底层驱动，这样便实现了对硬件的调用。

在 Android 4.4 的源码中，和 JNI 相关的文件如下所示：

```
./frameworks/base/media/java/android/media/MediaScanner.java
./frameworks/base/media/jni/android_media_MediaScanner.cpp
./frameworks/base/media/jni/android_media_MediaPlayer.cpp
./frameworks/base/media/jni/AndroidRuntime.cpp
./libnativehelper/JNIHelp.cpp
```

由此可见，和 JNI 密切相关的是 Media 系统，而 Media 系统的架构基础是 MediaScanner。在启动 Android 系统之初，就会扫描出系统中的 Media 文件供后续应用使用，既有新加入的媒体，也有几微秒种前删除的媒体文件，并且还需要自动更新相应的媒体库。在 Android 系统中，和用户体验密切相关的 Music、Gallery 播放等应用，也是基于 MediaScanner 的扫描媒体文件功能的。MediaScanner 位于 Android 4.4y 源码的如下路径中：

```
packages/providers/MediaProvider
```

一个 MediaScanner 包含了 3 个主要部分：MediaScannerReceiver、MediaScannerService 和 MediaProvider。在“MediaProvider”目录下的 AndroidManifest 中可以查看 MediaProvider 的基本架构，如图 4-2 所示。

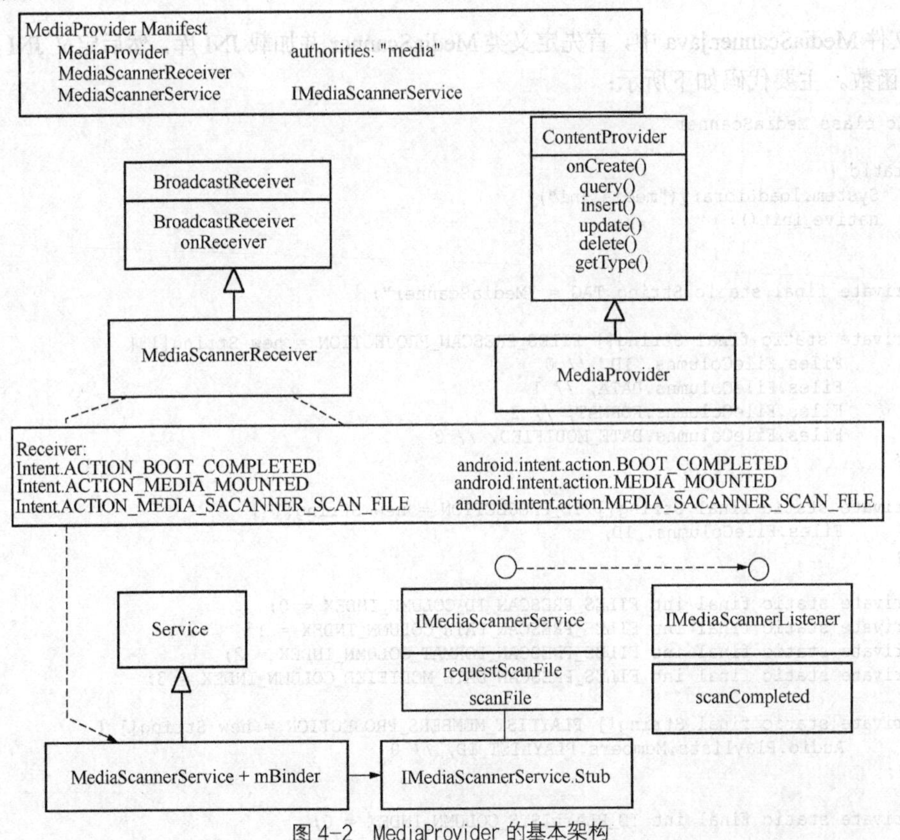

图 4-2 MediaProvider 的基本架构

❑ MediaScannerReceiver：是一个 BroadcastReceiver（接收广播），功能是进行媒体扫描，这也是 MediaScanner 提供给外界的接口之一。收到广播之后启动 MediaScannerService 具体执行扫描工作。

❑ MediaScannerService：是一个 Service，负责媒体扫描，它还要用到 Framework 中的 MediaScanner 来共同完成具体扫描工作，扫描的结果在 MediaProvider 提供的数据库中。

❑ MediaProvider：是一个 ContentProvider，媒体库（Images/Audio/Video/Playlist 等）的数据提供者。负责操作数据库，并提供给别的程序 insert、query、delete、update 等操作。

在本章接下来的内容中，以 MediaScanner 源码分析作为基础，将详细分析 JNI 在 Android 系统中的作用。

4.2 分析 Java 层

在 MediaScanner 系统中，JNI 的调用关系为：

MediaScanner -------------libmedia_jni.so -------------libmedia.so

在 Android 系统中，MediaScanner 的功能是扫描媒体文件，得到诸如歌曲时长、歌曲作者等信息，并将这些信息存放到媒体数据库中，以供其他应用程序使用。

在 JNI 应用中，Java 层 MediaScanner 的实现文件为：

```
/frameworks/base/media/java/android/media/MediaScanner.java
```

在本节的内容中，将详细讲解 MediaScanner 系统中 Java 层的具体实现过程。

4.2.1 加载 JNI 库

在文件 MediaScanner.java 中，首先定义类 MediaScanner 并加载 JNI 库，然后定义 JNI 的 Native（本地）函数。主要代码如下所示：

```
public class MediaScanner
{
    static {
        System.loadLibrary("media_jni");
        native_init();
    }

    private final static String TAG = "MediaScanner";

    private static final String[] FILES_PRESCAN_PROJECTION = new String[] {
            Files.FileColumns._ID, // 0
            Files.FileColumns.DATA, // 1
            Files.FileColumns.FORMAT, // 2
            Files.FileColumns.DATE_MODIFIED, // 3
    };

    private static final String[] ID_PROJECTION = new String[] {
            Files.FileColumns._ID,
    };

    private static final int FILES_PRESCAN_ID_COLUMN_INDEX = 0;
    private static final int FILES_PRESCAN_PATH_COLUMN_INDEX = 1;
    private static final int FILES_PRESCAN_FORMAT_COLUMN_INDEX = 2;
    private static final int FILES_PRESCAN_DATE_MODIFIED_COLUMN_INDEX = 3;

    private static final String[] PLAYLIST_MEMBERS_PROJECTION = new String[] {
            Audio.Playlists.Members.PLAYLIST_ID, // 0
     };

    private static final int ID_PLAYLISTS_COLUMN_INDEX = 0;
```

```
    private static final int PATH_PLAYLISTS_COLUMN_INDEX = 1;
    private static final int DATE_MODIFIED_PLAYLISTS_COLUMN_INDEX = 2;

    private static final String RINGTONES_DIR = "/ringtones/";
    private static final String NOTIFICATIONS_DIR = "/notifications/";
    private static final String ALARMS_DIR = "/alarms/";
    private static final String MUSIC_DIR = "/music/";
    private static final String PODCAST_DIR = "/podcasts/";
    ……
    private static native final void native_init();//声明一个 native 函数，native 为关键字
    private native final void native_setup();
    ……
}
```

函数 native_init 位于包 android.media 中，其完整路径名为：

```
android.media.MediaScanner.nantive_init
```

根据规则其对应的 JNI 层函数名称为：

```
android_media_MediaScanner_native_init
```

在调用函数 native 之前需要先加载 JNI 库，一般在类的 static 中加载调用函数 System.loadLibrary()。在加载了相应的 JNI 库之后，如果要使用相应的 native 函数，只需使用 native 声明需要被调用的函数即可：

```
private native void processDirectory(String path, String extensions, MediaScannerClient client);
private native void processFile(String path, String mimeType, MediaScannerClient client);
public native void setLocale(String locale);
```

4.2.2 实现扫描工作

在文件 MediaScanner.java 中，通过函数 scanDirectories 实现扫描工作，具体实现代码如下所示：

```
public void scanDirectories(String[] directories, String volumeName) {
    try {
        long start = System.currentTimeMillis();
        initialize(volumeName);//初始化
        prescan(null, true);//扫描前的预处理
        long prescan = System.currentTimeMillis();

        if (ENABLE_BULK_INSERTS) {
            // create MediaInserter for bulk inserts
            mMediaInserter = new MediaInserter(mMediaProvider, 500);
        }
        //函数 processDirectory 是一个 Native 函数，功能是对目标文件夹进行扫描
        for (int i = 0; i < directories.length; i++) {
            processDirectory(directories[i], mClient);
        }

        if (ENABLE_BULK_INSERTS) {
            // flush remaining inserts
            mMediaInserter.flushAll();
            mMediaInserter = null;
        }

        long scan = System.currentTimeMillis();
        postscan(directories);//扫描后的处理
        long end = System.currentTimeMillis();

        if (false) {
            Log.d(TAG, " prescan time: " + (prescan - start) + "ms\n");
            Log.d(TAG, "    scan time: " + (scan - prescan) + "ms\n");
            Log.d(TAG, "postscan time: " + (end - scan) + "ms\n");
            Log.d(TAG, "   total time: " + (end - start) + "ms\n");
        }
    } catch (SQLException e) {
        // this might happen if the SD card is removed while the media scanner is running
```

```
            Log.e(TAG, "SQLException in MediaScanner.scan()", e);
        } catch (UnsupportedOperationException e) {
            // this might happen if the SD card is removed while the media scanner is running
            Log.e(TAG, "UnsupportedOperationException in MediaScanner.scan()", e);
        } catch (RemoteException e) {
            Log.e(TAG, "RemoteException in MediaScanner.scan()", e);
        }
    }
```

在上述代码中用到了函数 initialize，此函数的功能是实现初始化操作，具体实现代码如下所示：

```
private void initialize(String volumeName) {
    //打开MediaProvider，获得它的一个实例
    mMediaProvider = mContext.getContentResolver().acquireProvider("media");
    //得到一些 uri
    mAudioUri = Audio.Media.getContentUri(volumeName);
    mVideoUri = Video.Media.getContentUri(volumeName);
    mImagesUri = Images.Media.getContentUri(volumeName);
    mThumbsUri = Images.Thumbnails.getContentUri(volumeName);
    mFilesUri = Files.getContentUri(volumeName);
    //如果需要外部存储的话，则可以支持播放列表，用缓存池实现，例如 mGenreCache 等
    if (!volumeName.equals("internal")) {
        // we only support playlists on external media
        mProcessPlaylists = true;
        mProcessGenres = true;
        mPlaylistsUri = Playlists.getContentUri(volumeName);
        mCaseInsensitivePaths = true;
    }
}
```

4.2.3　读取并保存信息

在文件 MediaScanner.java 中，函数 prescan 的功能是读取之前扫描的数据库中和文件相关的信息并保存起来。此函数创建了一个 FileCache，用来缓存扫描文件的一些信息，例如 last_modified 等。这个 FileCache 是从 MediaProvider 中已有信息构建出来的，也就是历史信息。后面根据扫描得到的新信息来对应更新历史信息。函数 prescan 的具体实现代码如下所示：

```
private void prescan(String filePath, boolean prescanFiles) throws RemoteException {
    Cursor c = null;
    String where = null;
    String[] selectionArgs = null;
    // mPlayLists 保存从数据库中获取的信息
    if (mPlayLists == null) {
        mPlayLists = new ArrayList<FileEntry>();
    } else {
        mPlayLists.clear();
    }

    if (filePath != null) {
        //只有一个文件查询
        where = MediaStore.Files.FileColumns._ID + ">?" +
            " AND " + Files.FileColumns.DATA + "=?";
        selectionArgs = new String[] { "", filePath };
    } else {
        where = MediaStore.Files.FileColumns._ID + ">?";
        selectionArgs = new String[] { "" };
    }

    //告诉提供者不删除文件.
    //如果不需要删除文件，则需要避免意外删除这个文件的机制
    //这可能在系统未被安装和未安装在扫描仪之前发生
    Uri.Builder builder = mFilesUri.buildUpon();
    builder.appendQueryParameter(MediaStore.PARAM_DELETE_DATA, "false");
    MediaBulkDeleter deleter = new MediaBulkDeleter(mMediaProvider, builder.build());

    //根据内容提供者建立文件列表
    try {
```

```
    if (prescanFiles) {
        //首先从文件表读到现有文件
        // 因为可能存在删除不存在文件的情况,所以要小批量的实现数据库查询以避免这个问题
        long lastId = Long.MIN_VALUE;
        Uri limitUri = mFilesUri.buildUpon().appendQueryParameter("limit", "1000").build();
        mWasEmptyPriorToScan = true;

        while (true) {
            selectionArgs[0] = "" + lastId;
            if (c != null) {
                c.close();
                c = null;
            }
            c = mMediaProvider.query(limitUri, FILES_PRESCAN_PROJECTION,
                    where, selectionArgs, MediaStore.Files.FileColumns._ID, null);
            if (c == null) {
                break;
            }

            int num = c.getCount();

            if (num == 0) {
                break;
            }
            mWasEmptyPriorToScan = false;
            while (c.moveToNext()) {
                long rowId = c.getLong(FILES_PRESCAN_ID_COLUMN_INDEX);
                String path = c.getString(FILES_PRESCAN_PATH_COLUMN_INDEX);
                int format = c.getInt(FILES_PRESCAN_FORMAT_COLUMN_INDEX);
                long lastModified = c.getLong(FILES_PRESCAN_DATE_MODIFIED_COLUMN_INDEX);
                lastId = rowId;

                // Only consider entries with absolute path names.
                // This allows storing URIs in the database without the
                // media scanner removing them.
                if (path != null && path.startsWith("/")) {
                    boolean exists = false;
                    try {
                        exists = Libcore.os.access(path, libcore.io.OsConstants.F_OK);
                    } catch (ErrnoException e1) {
                    }
                    if (!exists && !MtpConstants.isAbstractObject(format)) {
                        // do not delete missing playlists, since they may have been
                        // modified by the user.
                        // The user can delete them in the media player instead.
                        // instead, clear the path and lastModified fields in the row
                        MediaFile.MediaFileType mediaFileType = MediaFile.getFileType(path);
                        int fileType = (mediaFileType == null ? 0 : mediaFileType.fileType);

                        if (!MediaFile.isPlayListFileType(fileType)) {
                            deleter.delete(rowId);
                            if (path.toLowerCase(Locale.US).endsWith("/.nomedia")) {
                                deleter.flush();
                                String parent = new File(path).getParent();
                                mMediaProvider.call(MediaStore.UNHIDE_CALL, parent, null);
                            }
                        }
                    }
                }
            }
        }
    }
}
finally {
    if (c != null) {
        c.close();
    }
    deleter.flush();
}
```

```
    //计算图像的原始尺寸
    mOriginalCount = 0;
    c = mMediaProvider.query(mImagesUri, ID_PROJECTION, null, null, null, null);
    if (c != null) {
        mOriginalCount = c.getCount();
        c.close();
    }
}
```

4.2.4 删除 SD 卡外的信息

在文件 MediaScanner.java 中，函数 postscan 的功能是删除不存在于 SD 卡中的文件信息。函数 postscan 的具体实现代码如下所示：

```
private void postscan(String[] directories) throws RemoteException {

    //触发播放列表后能够知道对应存储的媒体文件
    if (mProcessPlaylists) {
        processPlayLists();
    }

    if (mOriginalCount == 0 && mImagesUri.equals(Images.Media.getContentUri("external")))
        pruneDeadThumbnailFiles();

    //允许 GC 清理
    mPlayLists = null;
    mMediaProvider = null;
}
```

4.2.5 直接转向 JNI

在文件 MediaScanner.java 中，processDirectory 是一个本地方法，能够直接转向 JNI。具体实现代码如下所示：

```
static void android_media_MediaScanner_processDirectory(JNIEnv *env, jobject thiz, jstring path,
jstring extensions, jobject client)
{   //获取 MediaScanner
    MediaScanner *mp = (MediaScanner *)env->GetIntField(thiz, fields.context);
    //参数判断，并抛出异常
    if (path == NULL) {
        jniThrowException(env, "java/lang/IllegalArgumentException", NULL);
        return;
    }
    if (extensions == NULL) {
        jniThrowException(env, "java/lang/IllegalArgumentException", NULL);
        return;
    }

    const char *pathStr = env->GetStringUTFChars(path, NULL);
    if (pathStr == NULL) {  // Out of memory
        jniThrowException(env, "java/lang/RuntimeException", "Out of memory");
        return;
    }
    const char *extensionsStr = env->GetStringUTFChars(extensions, NULL);
    if (extensionsStr == NULL) {  // Out of memory
        env->ReleaseStringUTFChars(path, pathStr);
        jniThrowException(env, "java/lang/RuntimeException", "Out of memory");
        return;
    }
    //初始化 client 实例
    MyMediaScannerClient myClient(env, client);
    //mp 调用 processDirectory
    mp->processDirectory(pathStr, extensionsStr, myClient, ExceptionCheck, env);
    //gc
    env->ReleaseStringUTFChars(path, pathStr);
    env->ReleaseStringUTFChars(extensions, extensionsStr);
}
```

4.2.6 扫描函数 scanFile

在此讲解 Java 层的函数 scanFile，功能是调用函数 doScanFile 对指定的文件进行扫描，具体实现代码如下所示：

```
public void scanFile(String path, long lastModified, long fileSize,
        boolean isDirectory, boolean noMedia) {
    //这是来自本地代码的回调函数
    // Log.v(TAG, "scanFile: "+path);
    doScanFile(path, null, lastModified, fileSize, isDirectory, false, noMedia);
}
```

4.2.7 JNI 中的异常处理

为了处理 Java 实现的方法中或者 C/C++实现的方法中抛出的 Java 异常，JNI 也提供了一套异常处理机制函数集，专门用于检查、分析和处理异常情况。例如在文件 jni.h 中，定义了主要的异常函数，具体代码如下所示：

```
//抛出异常
jint        (*Throw)(JNIEnv*, jthrowable);
//抛出新的异常
jint        (*ThrowNew)(JNIEnv *, jclass, const char *);
//异常产生
jthrowable  (*ExceptionOccurred)(JNIEnv*);
void        (*ExceptionDescribe)(JNIEnv*);
//清除异常
void        (*ExceptionClear)(JNIEnv*);
void        (*FatalError)(JNIEnv*, const char*);
```

例如在 Camera 模块中也用到了异常处理，在文件 android_hardware_Camera.cpp 中，也涉及到了异常操作，具体代码例如下所示：

```
void JNICameraContext::copyAndPost(JNIEnv* env, const sp<IMemory>& dataPtr, int
msgType)
{
  ......
  if (obj == NULL) {
            LOGE("Couldn't allocate byte array for JPEG data");
            env->ExceptionClear();
        } else {
            env->SetByteArrayRegion(obj, 0, size, data);
        }
    } else {
        LOGE("image heap is NULL");
    }
  }
  ......
}
```

在文件 android_hardware_Camera.cpp 中，函数 android_hardware_Camera_startPreview()也同样用到了异常处理机制，具体代码如下所示：

```
static void android_hardware_Camera_startPreview(JNIEnv *env, jobject thiz)
{
    LOGV("startPreview");
    sp<Camera> camera = get_native_camera(env, thiz, NULL);
    if (camera == 0) return;
    if (camera->startPreview() != NO_ERROR) {
        jniThrowRuntimeException(env, "startPreview failed");
        return;
    }
}
```

在上述代码中，android_hardware_Camera_startPreview()如果发现 startPreview()函数返回错误，

则会抛出异常并返回。这里的异常与 Java 中的异常机制很相似，读者可以对比分析它们的原理。

4.3 分析 JNI 层

由于 Android 的应用层的类都是以 Java 写的，这些 Java 类编译为 Dex 型式的 Bytecode 之后，必须借助 Dalvik 虚拟机（Vitual Machine，VM）来执行并实现。VM 在 Android 系统中扮演了一个很重要的角色，并且在执行 Java 类的过程中，如果 Java 类需要与 C 组件沟通时，VM 就会去载入 C 组件，然后让 Java 的函数顺利地调用到 C 组件的函数。此时，VM 扮演着桥梁的角色，让 Java 与 C 组件能通过标准的 JNI 介面而相互沟通。

应用层的 Java 类是在虚拟机上执行的，而 C 组件不是在 VM 上执行。如果 Java 程序又要求 VM 载入（Load）所指定的 C 组件，可以使用如下所示的指令实现这个功能：

```
System.loadLibrary(*.so 的档案名);
```

例如，在 Android 框架里所提供的 MediaPlayer.java 类中包含了下面的指令：

```
public class MediaPlayer{
    static {
            System.loadLibrary("media_jni");
    }
}
```

这要求 VM 去载入 Android 的/system/lib/libmedia_jni.so 库。载入*.so 后，Java 类与*.so 档案就汇合起来一起执行。

在 JNI 层中，MediaScanner 的对应文件是：

```
./frameworks/base/media/jni/android_media_MediaScanner.cpp
```

在本节的内容中，将详细讲解 JNI 层的基本源码。

4.3.1 将 Native 对象的指针保存到 Java 对象

在文件 android_media_MediaScanner.cpp 中，函数 android_media_MediaScanner_native_init 的功能是将 Native 对象的指针保存到 Java 对象中。函数 android_media_MediaScanner_native_init 的具体实现代码如下所示：

```
static const char* const kClassMediaScanner =
        "android/media/MediaScanner";
    ........

/*native_init 函数的 JNI 层实现*/
static void android_media_MediaScanner_native_init(JNIEnv *env)
{
    ALOGV("native_init");
    jclass clazz = env->FindClass(kClassMediaScanner);
    if (clazz == NULL) {
        return;
    }
    fields.context = env->GetFieldID(clazz, "mNativeContext", "I");
    if (fields.context == NULL) {
        return;
    }
}
```

4.3.2 创建 Native 层的 MediaScanner 对象

在文件 android_media_MediaScanner.cpp 中，函数 android_media_MediaScanner_native_setup 的功能是创建一个 Native 层的 MediaScanner 对象，但是此函数使用的是 Opencore 提供的

PVMediaScanner。函数 android_media_MediaScanner_native_setup 的具体实现代码如下所示：

```
static void android_media_MediaScanner_native_setup(JNIEnv *env, jobject thiz)
{
    ALOGV("native_setup");
    MediaScanner *mp = new StagefrightMediaScanner;
    if (mp == NULL) {
        jniThrowException(env, kRunTimeException, "Out of memory");
        return;
    }
    env->SetIntField(thiz, fields.context, (int)mp);
}
```

4.4 Native（本地）层

Java 的 Native 函数与 JNI 函数是一一对应的关系，在 Android 中使用 JNI NativeMethod 的结构体来记录这种对应关系。在本节的内容中，将详细分析 Mediascanner 系统中的 Native 层的实现源码。

4.4.1 注册 JNI 函数

在 Andoird 系统中，使用了一种“特定”的方式来定义其 Native 函数，这与传统定义 Java JNI 的方式有所差别。其中很重要的区别是在 Andorid 使用了一种 Java 和 C 函数的映射表数组，并在其中描述了函数的参数和返回值。这个数组的类型是 JNINativeMethod，具体定义如下所示：

```
typedef struct {
    const char* name;          /*Java 中函数的名字*/
    const char* signature;     /*描述了函数的参数和返回值*/
    void* fnPtr;               /*函数指针，指向 C 函数*/
} JNINativeMethod;
```

在上述代码中，比较难以理解的是第二个参数，例如：

```
"()V"
"(II)V"
"(Ljava/lang/String;Ljava/lang/String;)V"
```

实际上这些字符是与函数的参数类型一一对应的，具体说明如下所示。

- “()”中的字符表示参数，后面的则代表返回值。例如“()V”就表示 void Func()。
- “(II)V"” 表示 void Func(int, int)。

具体的每一个字符的对应关系如下所示：

字符	Java 类型	C 类型
V	void	void
Z	jboolean	boolean
I	jint	int
J	jlong	long
D	jdouble	double
F	jfloat	float
B	jbyte	byte
C	jchar	char
S	jshort	short

而数组则以“[”开始，用两个字符表示：

[I	jintArray	int[]
[F	jfloatArray	float[]

[B　jbyteArray　byte[]
[C　jcharArray　char[]
[S　jshortArray　short[]
[D　jdoubleArray　double[]
[J　jlongArray　long[]
[Z　jbooleanArray　boolean[]

上面的都是基本类型，如果 Java 函数的参数是 class，则以“L”开头，以“;”结尾，中间部分是用"/"隔开的包及类名。而其对应的 C 函数名的参数则为 jobject. 一个例外是 String 类，其对应的类为 jstring，即：

- ❑ Ljava/lang/String 中的 String jstring;
- ❑ Ljava/net/Socket 中的 Socket jobject。

如果 Java 函数位于一个嵌入类，则使用$作为类名间的分隔符。例如：

```
(Ljava/lang/String;Landroid/os/FileUtils$FileStatus;)Z"
```

定义并注册 JNINativeMethod 数组，对应的实现代码如下所示：

```
/*定义一个 JNINativeMethod 数组*/
static JNINativeMethod gMethods[] = {
    {
        "processDirectory",
        "(Ljava/lang/String;Landroid/media/MediaScannerClient;)V",
        (void *)android_media_MediaScanner_processDirectory
    },

    {
        "processFile",
        "(Ljava/lang/String;Ljava/lang/String;Landroid/media/MediaScannerClient;)V",
        (void *)android_media_MediaScanner_processFile
    },

    {
        "setLocale",
        "(Ljava/lang/String;)V",
        (void *)android_media_MediaScanner_setLocale
    },

    {
        "extractAlbumArt",
        "(Ljava/io/FileDescriptor;)[B",
        (void *)android_media_MediaScanner_extractAlbumArt
    },

    {
        "native_init",
        "()V",
        (void *)android_media_MediaScanner_native_init
    },

    {
        "native_setup",
        "()V",
        (void *)android_media_MediaScanner_native_setup
    },

    {
        "native_finalize",
        "()V",
        (void *)android_media_MediaScanner_native_finalize
    },
};
/*注册 JNINativeMethod 数组*/
```

```
int register_android_media_MediaScanner(JNIEnv *env)
{
    return AndroidRuntime::registerNativeMethods(env,
            kClassMediaScanner, gMethods, NELEM(gMethods));
}
```

4.4.2 完成注册工作

定义并注册数组 JNINativeMethod 后，接着需要调用函数 registerNativeMethods 来完成调用工作。函数 registerNativeMethods 在文件 AndroidRuntime.cpp 中实现，具体实现代码如下所示：

```
int AndroidRuntime::registerNativeMethods(JNIEnv* env,
    const char* className, const JNINativeMethod* gMethods, int numMethods)
{
    return jniRegisterNativeMethods(env, className, gMethods, numMethods);
}
```

在上述代码中，jniRegisterNativeMethods 是 Android 为方便 JNI 使用而提供的一个帮助函数，此函数在文件 JNIHelp.cpp 中实现，具体实现代码如下所示：

```
extern "C" int jniRegisterNativeMethods(C_JNIEnv* env, const char* className,
    const JNINativeMethod* gMethods, int numMethods)
{
    JNIEnv* e = reinterpret_cast<JNIEnv*>(env);

    ALOGV("Registering %s natives", className);
    scoped_local_ref<jclass> c(env, findClass(env, className));
    if (c.get() == NULL) {
        ALOGE("Native registration unable to find class '%s', aborting", className);
        abort();
    }
    if ((*env)->RegisterNatives(e, c.get(), gMethods, numMethods) < 0) {
        ALOGE("RegisterNatives failed for '%s', aborting", className);
        abort();
    }
    return 0;
}
```

通过上述代码可以了解函数 registerNativeMethods 的作用。应用层级的 Java 类别穿过 VM 呼叫到本地函数，这个过程通常是通过 VM 去寻找“*.so”格式库文件中的本地函数。如果需要连续呼叫很多次，则需要每次都寻找一遍，这会多花费很多时间。此时，组件开发人员可以自行向 VM 登记本地函数。例如，在 Android 的/system/lib/libmedia_jni.so 档案里的代码片段如下所示：

```
//#define LOG_NDEBUG 0
#define LOG_TAG "MediaPlayer-JNI"
static JNINativeMethod gMethods[] = {
    {"setDataSource",    "(Ljava/lang/String;)V",
     (void *)android_media_MediaPlayer_setDataSource},
    {"setDataSource",     "(Ljava/io/FileDescriptor;JJ)V",
(void *)android_media_MediaPlayer_setDataSourceFD},

    {"prepare",    "()V",     (void *)android_media_MediaPlayer_prepare},
    {"prepareAsync",    "()V",   (void *)android_media_MediaPlayer_prepareAsync},
    {"_start",    "()V",    (void *)android_media_MediaPlayer_start},
    {"_stop",    "()V",    (void *)android_media_MediaPlayer_stop},
    {"getVideoWidth",    "()I",     (void *)android_media_MediaPlayer_getVideoWidth},
    {"getVideoHeight",    "()I",     (void *)android_media_MediaPlayer_getVideoHeight},
    {"seekTo",        "(I)V",     (void *)android_media_MediaPlayer_seekTo},
    {"_pause",        "()V",     (void *)android_media_MediaPlayer_pause},
    {"isPlaying",        "()Z",      (void *)android_media_MediaPlayer_isPlaying},
    {"getCurrentPosition", "()I",   (void *)android_media_MediaPlayer_getCurrentPosition},
    {"getDuration",       "()I",    (void *)android_media_MediaPlayer_getDuration},
    {"_release",          "()V",    (void *)android_media_MediaPlayer_release},
    {"_reset",            "()V",     (void *)android_media_MediaPlayer_reset},
    {"setAudioStreamType","(I)V",  (void *)android_media_MediaPlayer_setAudioStreamType},
    {"setLooping",    "(Z)V",      (void *)android_media_MediaPlayer_setLooping},
```

```
    {"setVolume",     "(FF)V",     (void *)android_media_MediaPlayer_setVolume},
    {"getFrameAt",    "(I)Landroid/graphics/Bitmap;",
        (void *)android_media_MediaPlayer_getFrameAt},
    {"native_setup",   "(Ljava/lang/Object;)V",
         (void *)android_media_MediaPlayer_native_setup},
    {"native_finalize",    "()V",    (void *)android_media_MediaPlayer_native_finalize},
};
static int register_android_media_MediaPlayer(JNIEnv *env){
    return AndroidRuntime::registerNativeMethods(env,
        "android/media/MediaPlayer", gMethods, NELEM(gMethods));
}
jint JNI_OnLoad(JavaVM* vm, void* reserved){
    if (register_android_media_MediaPlayer(env) < 0) {
        LOGE("ERROR: MediaPlayer native registration failed\n");
        goto bail;
    }
}
```

这样当 VM 载入 libmedia_jni.so 档案时，就会呼叫函数 JNI_OnLoad()，然后 JNI_OnLoad()呼叫函数 register_android_media_MediaPlayer()。此时，就呼叫到函数 AndroidRuntime::registerNativeMethods()，并向 VM(即 AndroidRuntime)登记数组 gMethods[]表格所含的本地函数。由此可见，函数 registerNativeMethods 具备如下所示的两个功能。

- 更有效率地找到函数。
- 可以在执行期间进行抽换。因为 gMethods[]是一个“<名称，函数指针>”格式的对照表，所以在执行程序时可以通过多次呼叫函数 registerNativeMethods()的方式来更换本地函数之指针。

4.4.3　动态注册

当 Java 层通过 System.loadLibrary 加载完 JNI 动态库后，接着会查找函数 JNI_OnLoad，通过调用函数 JNI_OnLoad 完成动态注册工作。函数 JNI_OnLoad 在文件 android_media_MediaPlayer.cpp 中实现，具体代码如下所示：

```
jint JNI_OnLoad(JavaVM* vm, void* reserved)
{
    JNIEnv* env = NULL;
    jint result = -1;

    if (vm->GetEnv((void**) &env, JNI_VERSION_1_4) != JNI_OK) {
        ALOGE("ERROR: GetEnv failed\n");
        goto bail;
    }
    assert(env != NULL);
     ......
    if (register_android_media_MediaScanner(env) < 0) {
        ALOGE("ERROR: MediaScanner native registration failed\n");
        goto bail;
    }
   .......
    /*成功——则返回有效的版本号*/
    result = JNI_VERSION_1_4;
bail:
    return result;
}
```

函数 JNI_OnLoad 会回传 JNI_VERSION_1_4 的值给 VM，这让 VM 能够知道所使用 JNI 的版本是什么。此外，它也做了一些初期的动作（可呼叫任何本地函数），例如下面的指令：

```
if (register_android_media_MediaPlayer(env) < 0) {
     LOGE("ERROR: MediaPlayer native registration failed\n");
     goto bail;
 }
```

这样就将此组件提供的各个本地函数（Native Function）登记到 VM 里，以便能加快后续呼叫本地函数的效率。

函数 JNI_OnUnload()与 JNI_OnLoad()是相对应的。在载入 C 组件时会立即呼叫 JNI_OnLoad()进行组件内的初期动作；而当 VM 释放该 C 组件时，则会呼叫 JNI_OnUnload()函数来进行善后清除动作。当 VM 呼叫 JNI_OnLoad()或 JNI_Unload()函数时，都会将 VM 的指针（Pointer）传递给它们，其参数如下所示：

```
jint JNI_OnLoad(JavaVM* vm, void* reserved) {     }
jint JNI_OnUnload(JavaVM* vm, void* reserved){     }
```

在 JNI_OnLoad()函数里，就透过 VM 的指标而取得 JNIEnv 的指标值，并存入 env 指标变数里，如下述指令：

```
jint JNI_OnLoad(JavaVM* vm, void* reserved){
    JNIEnv* env = NULL;
    jint result = -1;
   if (vm->GetEnv((void**) &env, JNI_VERSION_1_4) != JNI_OK) {
        LOGE("ERROR: GetEnv failed\n");
        goto bail;
   }
}
```

由于 VM 通常是多执行绪(Multi-threading)的执行环境。每一个执行绪在呼叫 JNI_OnLoad()时，所传递进来的 JNIEnv 指标值都是不同的。为了配合这种多执行绪的环境，C 组件开发者在撰写本地函数时，可借用由 JNIEnv 指标值的不同而避免执行绪的资料冲突问题，才能确保所写的本地函数能安全地在 Android 的多执行绪 VM 里安全地执行。基于这个理由，当在呼叫 C 组件的函数时，都会将 JNIEnv 指标值传递给它，对应代码如下所示：

```
jint JNI_OnLoad(JavaVM* vm, void* reserved)
{
JNIEnv* env = NULL;
   if (register_android_media_MediaPlayer(env) < 0) {
   }
}
```

这样当 JNI_OnLoad()呼叫函数 register_android_media_MediaPlayer(env)时，就将 env 指标值传递过去。这样函数 register_android_media_MediaPlayer()就能借用该标识值来区别不同的执行，以便解决资料冲突的问题。

例如，在 register_android_media_MediaPlayer()函数里，可以编写如下所示的指令：

```
if ((*env)->MonitorEnter(env, obj) != JNI_OK) {
}
```

此时可以查看是否有其他执行程序进入此物件，如果没有，则此执行就进入该物件里执行了。并且也可以编写如下所示的指令：

```
if ((*env)->MonitorExit(env, obj) != JNI_OK) {
}
```

这样便可以查看是否此执行正在此物件内执行，如果是，此执行就会立即离开。

4.4.4 处理路径参数

在文件 frameworks/base/media/libmedia/MediaScanner.cpp 中，函数 processDirectory 的功能是对路径参数进行了一些处理，调用 doProcessDirectory。里面的参数“@extensions”可能包含多个扩展名，在扩展名之间用“,”分隔开。具体实现代码如下所示：

```
status_t MediaScanner::processDirectory(
        const char *path, const char *extensions,
        MediaScannerClient &client,
        ExceptionCheck exceptionCheck, void *exceptionEnv) {
```

```
    int pathLength = strlen(path);
    if (pathLength >= PATH_MAX) {
        return UNKNOWN_ERROR;
    }
    char* pathBuffer = (char *)malloc(PATH_MAX + 1);
    if (!pathBuffer) {
        return UNKNOWN_ERROR;
    }

    int pathRemaining = PATH_MAX - pathLength;
    strcpy(pathBuffer, path);
    if (pathLength > 0 && pathBuffer[pathLength - 1] != '/') {
        pathBuffer[pathLength] = '/';
        pathBuffer[pathLength + 1] = 0;
        --pathRemaining;
    }

    client.setLocale(locale());

    status_t result =
        doProcessDirectory(
            pathBuffer, pathRemaining, extensions, client,
            exceptionCheck, exceptionEnv);

    free(pathBuffer);

    return result;
}
```

4.4.5　扫描文件

当收到扫描某个文件的请求时，会调用函数 scanFile 来扫描这个文件。函数 scanFile 的具体实现代码如下所示：

```
virtual bool scanFile(const char* path, long long lastModified, long long fileSize)
 {
    jstring pathStr;
    if ((pathStr = mEnv->NewStringUTF(path)) == NULL) return false;
    //调用 Java 里面 mClient 中的 scanFile 方法
    mEnv->CallVoidMethod(mClient, mScanFileMethodID, pathStr, lastModified, fileSize);
    mEnv->DeleteLocalRef(pathStr);
    return (!mEnv->ExceptionCheck());
 }
```

4.4.6　添加 TAG 信息

在文件\frameworks\av\media\libmedia\MediaScannerClient.cpp 中，通过函数 addStringTag 添加 TAG 信息。这个 MediaScannerClient 是在 opencore 中的文件 MediaScanner.cpp 实现的，而文件 android_media_MediaScanner.cpp 中的 MyMediaScannerClient 是从 MediaScannerClient 派生下来的。函数 addStringTag 的具体实现代码如下所示：

```
status_t MediaScannerClient::addStringTag(const char* name, const char* value)
{
    if (mLocaleEncoding != kEncodingNone) {
        //不要缓存都是 ASCII 字符串
        //呼叫 handlestringtag 直接代替
        //查看值中是否有非 ASCII 字符，应该是 UTF8)
        bool nonAscii = false;
        const char* chp = value;
        char ch;
        while ((ch = *chp++)) {
            if (ch & 0x80) {
                nonAscii = true;
                break;
            }
        }
```

```
//判断 name 和 value 的编码是不是 ASCII，不是则保存到 mNames 和 mValues 中
            // save the strings for later so they can be used for native encoding detection
            mNames->push_back(name);
            mValues->push_back(value);
            return OK;
        }
        //其他的失败情形
    }
  //如果字符编码是 ASCII，则调用函数 handleStringTag
    return handleStringTag(name, value);
}
```

4.4.7 总结函数 JNI_OnLoad()与函数 JNI_OnUnload()的用途

当 Android 的 VM(Virtual Machine)执行到 System.loadLibrary()函数时，首先会去执行 C 组件里的 JNI_OnLoad()函数，其用途有如下两点。

（1）告诉 VM 此 C 组件使用哪一个 JNI 版本。如果你的*.so 档没有提供 JNI_OnLoad()函数，VM 会默认该*.so 档是使用最老的 JNI 1.1 版本。由于新版的 JNI 做了许多扩充，如果需要使用 JNI 的新版功能，例如 JNI 1.4 的 java.nio.ByteBuffer，就必须藉由 JNI_OnLoad()函数来告知 VM。

（2）由于 VM 执行到 System.loadLibrary()函数时，就会立即先呼叫 JNI_OnLoad()，所以 C 组件的开发者可以藉由 JNI_OnLoad()来进行 C 组件内的初期值的设定(Initialization) 。

例如，在 Android 的/system/lib/libmedia_jni.so 档案里，就提供了 JNI_OnLoad()函数，其代码片段如下所示：

```
//#define LOG_NDEBUG 0
#define LOG_TAG "MediaPlayer-JNI"
jint JNI_OnLoad(JavaVM* vm, void* reserved)
{
    JNIEnv* env = NULL;
    jint result = -1;
    if (vm->GetEnv((void**) &env, JNI_VERSION_1_4) != JNI_OK) {
        LOGE("ERROR: GetEnv failed\n");
        goto bail;
    }
assert(env != NULL);

    if (register_android_media_MediaPlayer(env) < 0) {
        LOGE("ERROR: MediaPlayer native registration failed\n");
        goto bail;
    }
    if (register_android_media_MediaRecorder(env) < 0) {
        LOGE("ERROR: MediaRecorder native registration failed\n");
        goto bail;
    }
    if (register_android_media_MediaScanner(env) < 0) {
        LOGE("ERROR: MediaScanner native registration failed\n");
        goto bail;
}
    if (register_android_media_MediaMetadataRetriever(env) < 0) {
        LOGE("ERROR: MediaMetadataRetriever native registration failed\n");
        goto bail;
    }
    /* success -- return valid version number */
    result = JNI_VERSION_1_4;
bail:
    return result;
}
```

此函数回传 JNI_VERSION_1_4 值给 VM，于是 VM 便知道了其所使用的 JNI 的版本。

4.4.8 Java 与 JNI 基本数据类型转换

Java 与 JNI 基本数据类型转换信息如表 4-1 所示。

表 4-1　　基本数据类型的转换关系

Java	Native 类型	本地 C 类型	字　长
boolean	jboolean	无符号	8 位
byte	jbyte	无符号	8 位
char	jchar	无符号	16 位
short	jshort	有符号	16 位
int	jint	有符号	32 位
long	jlong	有符号	64 位
float	jfloat	有符号	32 位
double	jdouble	有符号	64 位

数组类型的对应关系如表 4-2 所示。

表 4-2　　数组数据类型的对应关系

字　符	Java 类型	C 类型
[I	jintArray	int[]
[F	jfloatArray	float[]
[B	jbyteArray	byte[]
[C	jshortArray	short[]
[D	jdoubleArray	double[]
[J	jlongArray	long[]
[S	jshortArray	short[]
[Z	jbooleanArray	booean[]

对象数据类型的对应关系如表 4-3 所示。

表 4-3　　对象数据类型的对应关系

对　象	Java 类型	C 类型
Ljava/lang/String	String	jstring
Ljava/net/Socket	Socket	jobject

引用数据类型的转换关系如图 4-3 所示。

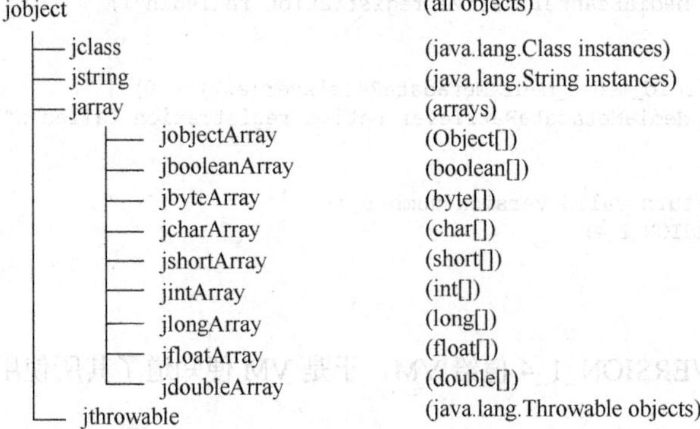

图 4-3　引用数据类型的转换关系

4.4.9 JNIEnv 接口

JNIEnv 是跟线程相关的，在 Native Method（本地方法）中，JNIEnv 作为第一个参数传入。JNIEnv 的内部结构如图 4-4 所示。

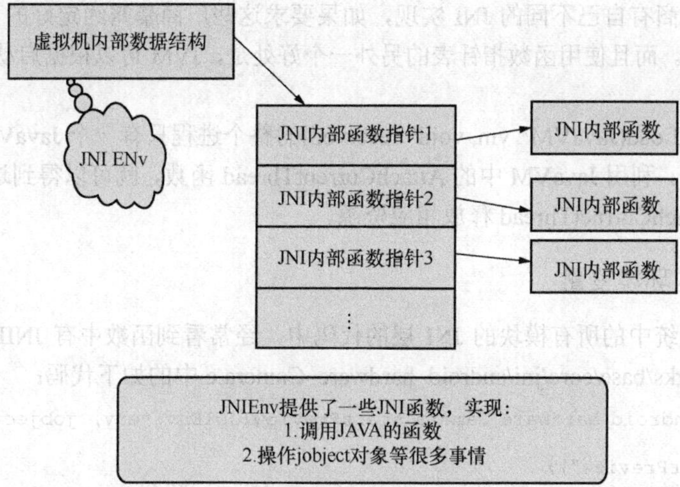

图 4-4　JNIEnv 的内部结构

在 JNIEnv 不作为参数传入的时候，JNI 提供了如下所示的两个函数以获得它。

- (*jvm)->AttachCurrentThread(jvm, (void**)&env, NULL)。
- (*jvm) ->GetEnv(jvm, (void**)&env, JNI_VERSION_1_2)。

上述两个函数都利用 JavaVM 接口获得 JNIEnv 接口，并且 JNI 可以将获得的 JNIEnv 封装成一个函数。即：

```
JNIEnv* JNU_GetEnv()
{
   JNIEnv* env;
   (*g_jvm)->GetEnv(g_jvm, (void**)&env, JNI_VERSION_1_2);
   return env;
}
```

Java 通过 JNI 机制调用 C/C++写的 native 程序，C/C++开发的 Native 程序需要遵循一定的 JNI 规范。例如下面的例子就是一个 JNI 函数声明：

```
JNIEXPORT jint JNICALL Java_jnitest_MyTest_test
 (JNIEnv * env, jobject obj, jint arg0);
```

JVM 负责从 Java Stack 转入 C/C++ Native Stack。当 Java 进入 JNI 调用，除了函数本身的参数(arg0)外会多出两个参数：JNIEnv 指针和 jobject 指针。其中 JNIEnv 指针是 JVM 创建的，用于 Native 的 C/C++方法操纵 Java 执行栈中的数据，比如 Java Class, Java Method 等。

首先，JNI 对于 JNIEnv 的使用提供了两种语法，分别是 C 语法以及 C++语法。其中 C 语法是：

```
jsize len = (*env)->GetArrayLength(env,array);
```

C++语法是：

```
jsize len =env->GetArrayLength(array);
```

因为 C 语言并不支持对象的概念，所以 C 语法中需要把 env 作为第一个参数传入，这类似于 C++的隐式参数 this 指针。

另外，在使用 JNIEnv 接口时，需要遵循如下所示的两个设计原则。

（1）JNIEnv 指针被设计成了 Thread Local Storage(TLS)变量，也就是说每一个 Thread, JNIEnv 变量都有独立的 Copy。你不能把 Thead#1 使用的 JNIEnv 传给 Thread#2 使用。

（2）在 JNIEnv 中定义了一组函数指针，C/C++ Native 程序是通过这些函数指针操纵 Java 数据。这样设计的好处是，C/C++ 程序不需要依赖任何函数库或者 DLL。由于 JVM 可能由不同的厂商实现，不同厂商有自己不同的 JNI 实现，如果要求这些厂商暴露约定好的一些头文件和库，这不是灵活的设计。而且使用函数指针表的另外一个好处是，JVM 可以根据启动参数动态替换 JNI 实现。

在 jint JNI_OnLoad(JavaVM* vm, void* reserved)的整个进程只有一个 JavaVM 对象，可以保存并在任何地方使用。利用 JavaVM 中的 AttachCurrentThread 函数，就可以得到这个线程的 JNIEnv 结构体，利用 DetachCurrnetThread 释放相应资源。

4.4.10　JNI 中的环境变量

在 Android 系统中的所有模块的 JNI 层的代码中，经常看到函数中有 JNIEnv*类型的参数，例如文件 frameworks/base/core/jni/android_hardware_Camera.c 中的如下代码：

```
static void android_hardware_Camera_startPreview(JNIEnv *env, jobject thiz)
{
    LOGV("startPreview");
    sp<Camera> camera = get_native_camera(env, thiz, NULL);
    if (camera == 0) return;
    if (camera->startPreview() != NO_ERROR) {
        jniThrowRuntimeException(env, "startPreview failed");
        return;
    }
}
```

在上述函数中，第一个参数为 JNIEnv *Env，此处的 JNIEnv *类型是一个指向 JNI 环境的指针，JNIEnv *类型在文件 Jni.h 中定义，在此结构体中包含了一些 JNI 中常用到的函数和一组函数指针，C/C++正是通过这些函数指针来操作 Java 函数的。JNIEnv 结构体在文件 jni.h 中定义，具体实现代码如下所示：

```
struct _JNIEnv {
   ......
   jint GetVersion()
   { return functions->GetVersion(this); }
   jclass FindClass(const char* name)
   { return functions->FindClass(this, name); }
   ......
   void CallVoidMethodA(jobject obj, jmethodID methodID, jvalue* args)
    { functions->CallVoidMethodA(this, obj, methodID, args); }
   ......
   jmethodID GetStaticMethodID(jclass clazz, const char* name, const char* sig)
    { return functions->GetStaticMethodID(this, clazz, name, sig); }
   ......
}
```

通过上述代码可以发现，正是通过使用这个 JNIEnv 指针才能够调用一些 JNI 环境中的方法。

第5章 分析内存系统

众所周知，内存是计算机的重要部件之一，是用户任务与 CPU 处理进行沟通的桥梁。内存（Memory）也被称为内存储器，其功能是暂时保存 CPU 中的运算数据，以及与硬盘等外部存储器交换的数据。只要是运行中的计算机，CPU 就会把需要运算的数据调到内存中进行运算处理，当运算完成后 CPU 再将结果传送出来。在本章的内容中，将和大家一起探讨 Android 内存系统的基本知识，为读者步入本书后面知识的学习打下基础。

5.1 分析 Android 的进程通信机制

要想实现对 Android 系统内存的优化，需要首先了解 Android 的内存系统，了解内存控制进程运行的机制。在本节的内容中，将带领大家一起探讨分析 Android 的进程通信机制。

5.1.1 Android 的进程间通信（IPC）机制 Binder

在 Android 系统中，每一个应用程序都是由一些 Activity 和 Service 组成的，一般 Service 运行在独立的进程中，而 Activity 可能运行在同一个进程中，也有可能运行在不同的进程中。众所周知，Android 系统是基于 Linux 内核的，而 Linux 内核继承和兼容了丰富的 Unix 系统进程间通信（IPC）机制。有传统的管道（Pipe）、信号（Signal）和跟踪（Trace），这三项通信手段只能用于父进程和子进程之间，或者只用于兄弟进程之间。随着技术的发展，后来又增加了命令管道（Named Pipe），这样使得进程之间的通信不再局限于父子进程或者兄弟进程之间。为了更好地支持商业应用中的事务处理，在 AT&T 的 Unix 系统 V 中，又增加了如下 3 种称为"System V IPC"的进程间通信机制。

- ❑ 报文队列（Message）。
- ❑ 共享内存（Share Memory）。
- ❑ 信号量（Semaphore）。

Android 系统没有采用上述提到的各种进程间通信机制，而是采用 Binder 机制。其实 Binder 并不是 Android 提出来的一套新的进程间通信机制，它是基于 OpenBinder 来实现的。Binder 是一种进程间通信机制，其类似于 COM 和 CORBA 分布式组件架构。具体来说，其实是提供了远程过程调用（RPC）功能。

在 Android 系统中，Binder 机制由 Client、Server、Service Manager 和 Binder 驱动程序等一系统组件组成。其中 Client、Server 和 Service Manager 运行在用户空间，Binder 驱动程序运行在内核空间，Binder 就是一种把这 4 个组件黏合在一起的黏结剂。在上述 Binder 组件中，核心组件是 Binder 驱动程序。Service Manager 提供了辅助管理的功能，Client 和 Server 正是在 Binder 驱动和 Service Manager 提供的基础设施上实现 Client/Server 之间通信功能的。Service Manager 和 Binder 驱动已经在 Android 平台中实现完毕，开发者只要按照规范实现自己的 Client 和

Server 组件即可。

对于初学者来说，Android 系统的 Binder 机制是最难理解的了，而 Binder 机制无论从系统开发还是应用开发的角度来看，都是 Android 系统中最重要的组成，所以很有必要深入了解 Binder 的工作方式。要深入了解 Binder 的工作方式，最好的方式是阅读 Binder 相关的源代码了。

要想深入理解 Binder 机制，必须了解 Binder 在用户空间的 3 个组件 Client、Server 和 Service Manager 之间的相互关系，并了解内核空间中 Binder 驱动程序的数据结构和设计原理。具体来说，Android 系统 Binder 机制中的 4 个组件 Client、Server、Service Manager 和 Binder 驱动程序的关系如图 5-1 所示。

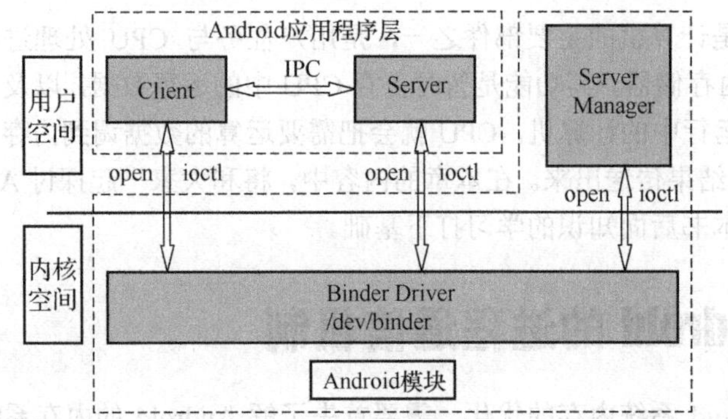

图 5-1　组件 Client、Server、Service Manager 和 Binder 驱动程序的关系

图 5-1 中所示关系的具体说明如下。

（1）Client、Server 和 Service Manager 实现在用户空间中，Binder 驱动程序实现在内核空间中。

（2）Binder 驱动程序和 Service Manager 在 Android 平台中已经实现，开发者只需要在用户空间实现自己的 Client 和 Server。

（3）Binder 驱动程序提供设备文件"/dev/binder"与用户空间交互，Client、Server 和 Service Manager 通过文件操作函数 open()和 ioctl()与 Binder 驱动程序进行通信。

（4）Client 和 Server 之间的进程间通信通过 Binder 驱动程序间接实现。

（5）Service Manager 是一个保护进程，用来管理 Server，并向 Client 提供查询 Server 接口的能力。

5.1.2　Service Manager 是 Binder 机制的上下文管理者

在 Android 系统中，Service Manager 负责告知 Binder 驱动程序它是 Binder 机制的上下文管理者。Service Manager 是整个 Binder 机制的保护进程，用来管理开发者创建的各种 Server，并且向 Client 提供查询 Server 远程接口的功能。

因为 Service Manager 组件是用来管理 Server 并且向 Client 提供查询 Server 远程接口的功能，所以 Service Manager 必然要和 Server 以及 Client 进行通信。Service Manger、Client 和 Server 三者分别是运行在独立的进程当中的，这样它们之间的通信也属于进程间的通信，而且也是采用 Binder 机制进行进程间通信。因此，Service Manager 在充当 Binder 机制的保护进程的角色的同时也在充当 Server 的角色，也是一种特殊的 Server。

Service Manager 在用户空间的源代码位于"frameworks/base/cmds/servicemanager"目录下，主要是由文件 binder.h、binder.c 和 service_manager.c 组成。Service Manager 在 Binder 机制中的基本执行流程如图 5-2 所示。

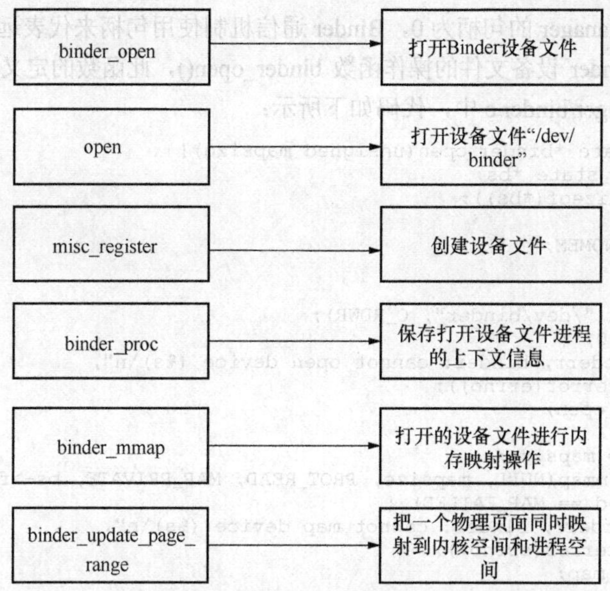

图 5-2 Service Manager 在 Binder 机制中的基本执行流程

在 Service Manager 的入口于文件 service_manager.c 中，主函数 main 的实现代码如下所示：

```
int main(int argc, char **argv){
    struct binder_state *bs;
    void *svcmgr = BINDER_SERVICE_MANAGER;
    bs = binder_open(128*1024);
    if (binder_become_context_manager(bs)) {
        LOGE("cannot become context manager (%s)\n", strerror(errno));
        return -1;
    }
    svcmgr_handle = svcmgr;
    binder_loop(bs, svcmgr_handler);
    return 0;
}
```

上述函数 main()主要有以下 3 个功能。

- 打开 Binder 设备文件。
- 告诉 Binder 驱动程序自己是 Binder 上下文管理者，即前面所说的保护进程。
- 进入一个无穷循环，充当 Server 的角色，等待 Client 的请求。

在分析上述 3 个功能之前，先来看一下这里用到的结构体 binder_state、宏 BINDER_SERVICE_MANAGER 的定义。结构体 binder_state 在文件 frameworks/base/cmds/servicemanager/binder.c 中定义，代码如下所示：

```
struct binder_state {
    int fd;
    void *mapped;
    unsigned mapsize;
};
```

其中 fd 表示文件描述符，即表示打开的“/dev/binder”设备文件描述符；mapped 表示把设备文件“/dev/binder”映射到进程空间的起始地址；mapsize 表示上述内存映射空间的大小。

宏 BINDER_SERVICE_MANAGER 在文件 frameworks/base/cmds/servicemanager/binder.h 中定义，代码如下所示：

```
/* the one magic object */
#define BINDER_SERVICE_MANAGER ((void*) 0)
```

这表示 Service Manager 的句柄为 0，Binder 通信机制使用句柄来代表远程接口。

函数首先打开 Binder 设备文件的操作函数 binder_open()，此函数的定义位于文件 frameworks/base/cmds/servicemanager/binder.c 中，代码如下所示：

```
struct binder_state *binder_open(unsigned mapsize){
    struct binder_state *bs;
    bs = malloc(sizeof(*bs));
    if (!bs) {
        errno = ENOMEM;
        return 0;
    }
    bs->fd = open("/dev/binder", O_RDWR);
    if (bs->fd < 0) {
        fprintf(stderr,"binder: cannot open device (%s)\n",
                strerror(errno));
        goto fail_open;
    }
    bs->mapsize = mapsize;
    bs->mapped = mmap(NULL, mapsize, PROT_READ, MAP_PRIVATE, bs->fd, 0);
    if (bs->mapped == MAP_FAILED) {
        fprintf(stderr,"binder: cannot map device (%s)\n",
                strerror(errno));
        goto fail_map;
    }
        /* TODO: check version */
    return bs;
fail_map:
    close(bs->fd);
fail_open:
    free(bs);
    return 0;
}
```

通过文件操作函数 open()打开设备文件“/dev/binder”，此设备文件是在 Binder 驱动程序模块初始化的时候创建的。接下来先看一下这个设备文件的创建过程，来到 kernel/common/drivers/staging/android 目录，打开文件 binder.c，可以看到如下模块初始化入口 binder_init：

```
static struct file_operations binder_fops = {
    .owner = THIS_MODULE,
    .poll = binder_poll,
    .unlocked_ioctl = binder_ioctl,
    .mmap = binder_mmap,
    .open = binder_open,
    .flush = binder_flush,
    .release = binder_release,
};

static struct miscdevice binder_miscdev = {
    .minor = MISC_DYNAMIC_MINOR,
    .name = "binder",
    .fops = &binder_fops
};

static int __init binder_init(void)
{
    int ret;

    binder_proc_dir_entry_root = proc_mkdir("binder", NULL);
    if (binder_proc_dir_entry_root)
        binder_proc_dir_entry_proc = proc_mkdir("proc", binder_proc_dir_entry_root);
    ret = misc_register(&binder_miscdev);
    if (binder_proc_dir_entry_root) {
        create_proc_read_entry("state", S_IRUGO, binder_proc_dir_entry_root, binder_read_proc_state,
NULL);
        create_proc_read_entry("stats", S_IRUGO, binder_proc_dir_entry_root, binder_read_proc_stats,
NULL);
        create_proc_read_entry("transactions", S_IRUGO, binder_proc_dir_entry_root,
binder_read_proc_transactions, NULL);
```

```
        create_proc_read_entry("transaction_log", S_IRUGO, binder_proc_dir_entry_root,
binder_read_proc_transaction_log, &binder_transaction_log);
        create_proc_read_entry("failed_transaction_log", S_IRUGO, binder_proc_dir_entry_root,
binder_read_proc_transaction_log, &binder_transaction_log_failed);
    }
    return ret;
}
device_initcall(binder_init);
```

在函数 misc_register()中实现了创建设备文件的功能，并实现了 misc 设备的注册工作，在“/proc”目录中创建了各种 Binder 相关的文件供用户访问。通过如下函数 binder_open 的执行语句即可进入到 Binder 驱动程序的 binder_open()函数：

```
bs->fd = open("/dev/binder", O_RDWR);
```

函数 binder_open()的实现代码如下所示：

```
static int binder_open(struct inode *nodp, struct file *filp)
{
    struct binder_proc *proc;

    if (binder_debug_mask & BINDER_DEBUG_OPEN_CLOSE)
        printk(KERN_INFO "binder_open: %d:%d\n", current->group_leader->pid, current->pid);

    proc = kzalloc(sizeof(*proc), GFP_KERNEL);
    if (proc == NULL)
        return -ENOMEM;
    get_task_struct(current);
    proc->tsk = current;
    INIT_LIST_HEAD(&proc->todo);
    init_waitqueue_head(&proc->wait);
    proc->default_priority = task_nice(current);
    mutex_lock(&binder_lock);
    binder_stats.obj_created[BINDER_STAT_PROC]++;
    hlist_add_head(&proc->proc_node, &binder_procs);
    proc->pid = current->group_leader->pid;
    INIT_LIST_HEAD(&proc->delivered_death);
    filp->private_data = proc;
    mutex_unlock(&binder_lock);

    if (binder_proc_dir_entry_proc) {
        char strbuf[11];
        snprintf(strbuf, sizeof(strbuf), "%u", proc->pid);
        remove_proc_entry(strbuf, binder_proc_dir_entry_proc);
        create_proc_read_entry(strbuf, S_IRUGO, binder_proc_dir_entry_proc, binder_read_proc_proc,
proc);
    }
    return 0;
}
```

函数 binder_open()的主要功能是创建一个名为 binder_proc 的数据结构，使用此数据结构可以保存打开设备文件“/dev/binder”的进程的上下文信息，并且将这个进程上下文信息保存在打开文件结构 file 的私有数据成员变量 private_data 中。

而结构体 struct binder_proc 也被定义在文件“kernel/common/drivers/staging/android/binder.c”中，具体代码如下所示：

```
struct binder_proc {
    struct hlist_node proc_node;
    struct rb_root threads;
    struct rb_root nodes;
    struct rb_root refs_by_desc;
    struct rb_root refs_by_node;
    int pid;
    struct vm_area_struct *vma;
```

```
    struct task_struct *tsk;
    struct files_struct *files;
    struct hlist_node deferred_work_node;
    int deferred_work;
    void *buffer;
    ptrdiff_t user_buffer_offset;
    struct list_head buffers;
    struct rb_root free_buffers;
    struct rb_root allocated_buffers;
    size_t free_async_space;
    struct page **pages;
    size_t buffer_size;
    uint32_t buffer_free;
    struct list_head todo;
    wait_queue_head_t wait;
    struct binder_stats stats;
    struct list_head delivered_death;
    int max_threads;
    int requested_threads;
    int requested_threads_started;
    int ready_threads;
    long default_priority;
};
```

上述结构体中的成员比较多，其中最为重要的是如下 4 个成员变量。

- Threads。
- Nodes。
- refs_by_desc。
- refs_by_node。

上述 4 个成员变量都是表示红黑树的节点，即 binder_proc 分别挂在 4 个红黑树下，具体说明如下所示。

- threads 树：用来保存 binder_proc 进程内用于处理用户请求的线程，它的最大数量由 max_threads 来决定。
- node 树：用来保存 binder_proc 进程内的 Binder 实体。
- refs_by_desc 树和 refs_by_node 树：用来保存 binder_proc 进程内的 Binder 引用，即引用的其他进程的 Binder 实体，它分别用两种方式来组织红黑树，一种是以句柄作来 key 值来组织，一种是以引用的实体节点的地址值作来 key 值来组织，它们都是表示同一样东西，只不过是为了内部查找方便而用两个红黑树来表示。

这样就完成了打开设备文件/dev/binder 的工作，接下来需要对打开的设备文件进行内存映射操作 mmap：

```
bs->mapped = mmap(NULL, mapsize, PROT_READ, MAP_PRIVATE, bs->fd, 0);
```

对应 Binder 驱动程序的是函数 binder_mmap()，实现代码如下所示：

```
static int binder_mmap(struct file *filp, struct vm_area_struct *vma)
{
    int ret;
    struct vm_struct *area;
    struct binder_proc *proc = filp->private_data;
    const char *failure_string;
    struct binder_buffer *buffer;
    if ((vma->vm_end - vma->vm_start) > SZ_4M)
        vma->vm_end = vma->vm_start + SZ_4M;
    if (binder_debug_mask & BINDER_DEBUG_OPEN_CLOSE)
        printk(KERN_INFO
            "binder_mmap: %d %lx-%lx (%ld K) vma %lx pagep %lx\n",
            proc->pid, vma->vm_start, vma->vm_end,
            (vma->vm_end - vma->vm_start) / SZ_1K, vma->vm_flags,
            (unsigned long)pgprot_val(vma->vm_page_prot));
```

```
    if (vma->vm_flags & FORBIDDEN_MMAP_FLAGS) {
        ret = -EPERM;
        failure_string = "bad vm_flags";
        goto err_bad_arg;
    }
    vma->vm_flags = (vma->vm_flags | VM_DONTCOPY) & ~VM_MAYWRITE;

    if (proc->buffer) {
        ret = -EBUSY;
        failure_string = "already mapped";
        goto err_already_mapped;
    }

    area = get_vm_area(vma->vm_end - vma->vm_start, VM_IOREMAP);
    if (area == NULL) {
        ret = -ENOMEM;
        failure_string = "get_vm_area";
        goto err_get_vm_area_failed;
    }
    proc->buffer = area->addr;
    proc->user_buffer_offset = vma->vm_start - (uintptr_t)proc->buffer;
#ifdef CONFIG_CPU_CACHE_VIPT
    if (cache_is_vipt_aliasing())
        while (CACHE_COLOUR((vma->vm_start ^ (uint32_t)proc->buffer))) {
            printk(KERN_INFO "binder_mmap: %d %lx-%lx maps %p bad alignment\n", proc->pid,
            vma->vm_start, vma->vm_end, proc->buffer);
            vma->vm_start += PAGE_SIZE;
        }
    }
#endif
    proc->pages = kzalloc(sizeof(proc->pages[0]) * ((vma->vm_end - vma->vm_start) /
    PAGE_SIZE), GFP_KERNEL);
    if (proc->pages == NULL) {
        ret = -ENOMEM;
        failure_string = "alloc page array";
        goto err_alloc_pages_failed;
    }
    proc->buffer_size = vma->vm_end - vma->vm_start;

    vma->vm_ops = &binder_vm_ops;
    vma->vm_private_data = proc;

    if (binder_update_page_range(proc, 1, proc->buffer, proc->buffer + PAGE_SIZE, vma))
{
        ret = -ENOMEM;
        failure_string = "alloc small buf";
        goto err_alloc_small_buf_failed;
    }
    buffer = proc->buffer;
    INIT_LIST_HEAD(&proc->buffers);
    list_add(&buffer->entry, &proc->buffers);
    buffer->free = 1;
    binder_insert_free_buffer(proc, buffer);
    proc->free_async_space = proc->buffer_size / 2;
    barrier();
    proc->files = get_files_struct(current);
    proc->vma = vma;

    /*printk(KERN_INFO "binder_mmap: %d %lx-%lx maps %p\n", proc->pid, vma->vm_start,
    vma->vm_end, proc->buffer);*/
    return 0;

    err_alloc_small_buf_failed:
    kfree(proc->pages);
    proc->pages = NULL;
    err_alloc_pages_failed:
    vfree(proc->buffer);
    proc->buffer = NULL;
```

```
err_get_vm_area_failed:
err_already_mapped:
err_bad_arg:
   printk(KERN_ERR "binder_mmap: %d %lx-%lx %s failed %d\n", proc->pid, vma->vm_start,
vma->vm_end, failure_string, ret);
   return ret;
}
```

在上述函数 binder_mmap()中，首先通过 filp->private_data 得到在打开设备文件“/dev/binder”时创建的结构 binder_proc，在 vma 参数中保存内存映射信息。此处 vma 的数据类型是结构 vm_area_struct，它表示的是一块连续的虚拟地址空间区域。另外，结构体 vm_struct 表示一块连续的虚拟地址空间区域。

接下来分析结构体 binder_proc 中的如下成员变量。

- buffer：是一个 void*指针，它表示要映射的物理内存在内核空间中的起始位置。
- buffer_size：是一个 size_t 类型的变量，表示要映射的内存的大小。
- pages：是一个 struct page*类型的数组，struct page 是用来描述物理页面的数据结构。
- user_buffer_offset：是一个 ptrdiff_t 类型的变量，它表示的是内核使用的虚拟地址与进程使用的虚拟地址之间的差值，即如果某个物理页面在内核空间中对应的虚拟地址为 addr，则这个物理页面在进程空间对应的虚拟地址就为如下格式。

```
addr + user_buffer_offset
```

接下来还需要看一下 Binder 驱动程序管理内存映射地址空间的方法，即如何管理 buffer～(buffer + buffer_size)这段地址空间，这个地址空间被划分为一段一段来管理，每一段是用结构体 binder_buffer 来描述的，具体代码如下所示：

```
struct binder_buffer {
   struct list_head entry; /* free and allocated entries by addesss */
   struct rb_node rb_node; /* free entry by size or allocated entry */
            /* by address */
   unsigned free : 1;
   unsigned allow_user_free : 1;
   unsigned async_transaction : 1;
   unsigned debug_id : 29;
   struct binder_transaction *transaction;
   struct binder_node *target_node;
   size_t data_size;
   size_t offsets_size;
   uint8_t data[0];
};
```

每一个 binder_buffer 通过其成员 entry 按从低地址到高地址连入到 struct binder_proc 中的 buffers 表示的链表中去，并且每一个 binder_buffer 又分为正在使用的和空闲的，通过 free 成员变量来区分。空闲的 binder_buffer 借助变量 rb_node 来到 struct binder_proc 中的 free_buffers 表示的红黑树中去；而那些正在使用的 binder_buffer，通过成员变量 rb_node 连入到 binder_proc 中的 allocated_buffers 表示的红黑树中去。这样做的目的是，方便查询和维护这块地址空间。

继续分析函数 binder_update_page_range()，查看 Binder 驱动程序把一个物理页面同时映射到内核空间和进程空间的方法。具体实现代码如下所示：

```
static int binder_update_page_range(struct binder_proc *proc, int allocate,
   void *start, void *end, struct vm_area_struct *vma)
{
   void *page_addr;
   unsigned long user_page_addr;
   struct vm_struct tmp_area;
   struct page **page;
   struct mm_struct *mm;
   if (binder_debug_mask & BINDER_DEBUG_BUFFER_ALLOC)
```

```
        printk(KERN_INFO "binder: %d: %s pages %p-%p\n",
             proc->pid, allocate ? "allocate" : "free", start, end);
    if (end <= start)
        return 0;
    if (vma)
        mm = NULL;
    else
        mm = get_task_mm(proc->tsk);
    if (mm) {
        down_write(&mm->mmap_sem);
        vma = proc->vma;
    }
    if (allocate == 0)
        goto free_range;
    if (vma == NULL) {
        printk(KERN_ERR "binder: %d: binder_alloc_buf failed to "
             "map pages in userspace, no vma\n", proc->pid);
        goto err_no_vma;
    }
    for (page_addr = start; page_addr < end; page_addr += PAGE_SIZE) {
        int ret;
        struct page **page_array_ptr;
        page = &proc->pages[(page_addr - proc->buffer) / PAGE_SIZE];
        BUG_ON(*page);
        *page = alloc_page(GFP_KERNEL | __GFP_ZERO);
        if (*page == NULL) {
            printk(KERN_ERR "binder: %d: binder_alloc_buf failed "
                 "for page at %p\n", proc->pid, page_addr);
            goto err_alloc_page_failed;
        }
        tmp_area.addr = page_addr;
        tmp_area.size = PAGE_SIZE + PAGE_SIZE /* guard page? */;
        page_array_ptr = page;
        ret = map_vm_area(&tmp_area, PAGE_KERNEL, &page_array_ptr);
        if (ret) {
            printk(KERN_ERR "binder: %d: binder_alloc_buf failed "
                 "to map page at %p in kernel\n",
                 proc->pid, page_addr);
            goto err_map_kernel_failed;
        }
        user_page_addr =
            (uintptr_t)page_addr + proc->user_buffer_offset;
        ret = vm_insert_page(vma, user_page_addr, page[0]);
        if (ret) {
            printk(KERN_ERR "binder: %d: binder_alloc_buf failed "
                 "to map page at %lx in userspace\n",
                 proc->pid, user_page_addr);
            goto err_vm_insert_page_failed;
        }
        /* vm_insert_page does not seem to increment the refcount */
    }
    if (mm) {
        up_write(&mm->mmap_sem);
        mmput(mm);
    }
    return 0;
free_range:
    for (page_addr = end - PAGE_SIZE; page_addr >= start;
         page_addr -= PAGE_SIZE) {
        page = &proc->pages[(page_addr - proc->buffer) / PAGE_SIZE];
        if (vma)
            zap_page_range(vma, (uintptr_t)page_addr +
                 proc->user_buffer_offset, PAGE_SIZE, NULL);
err_vm_insert_page_failed:
        unmap_kernel_range((unsigned long)page_addr, PAGE_SIZE);
err_map_kernel_failed:
        __free_page(*page);
        *page = NULL;
err_alloc_page_failed:
```

```
        ;
    }
err_no_vma:
    if (mm) {
        up_write(&mm->mmap_sem);
        mmput(mm);
    }
    return -ENOMEM;
}
```

通过上述代码不但可以分配物理页面，而且可以用来释放物理页面，这可以通过参数 allocate 来区别，在此只需关注分配物理页面的情况。要分配物理页面的虚拟地址空间范围为（start～end），函数前面的一些检查逻辑就不看了，只需直接查看中间的 for 循环代码，具体代码如下所示：

```
for (page_addr = start; page_addr < end; page_addr += PAGE_SIZE) {
    int ret;
    struct page **page_array_ptr;
    page = &proc->pages[(page_addr - proc->buffer) / PAGE_SIZE];
    BUG_ON(*page);
    *page = alloc_page(GFP_KERNEL | __GFP_ZERO);
    if (*page == NULL) {
        printk(KERN_ERR "binder: %d: binder_alloc_buf failed "
               "for page at %p\n", proc->pid, page_addr);
        goto err_alloc_page_failed;
    }
    tmp_area.addr = page_addr;
    tmp_area.size = PAGE_SIZE + PAGE_SIZE /* guard page? */;
    page_array_ptr = page;
    ret = map_vm_area(&tmp_area, PAGE_KERNEL, &page_array_ptr);
    if (ret) {
        printk(KERN_ERR "binder: %d: binder_alloc_buf failed "
               "to map page at %p in kernel\n",
               proc->pid, page_addr);
        goto err_map_kernel_failed;
    }
    user_page_addr =
        (uintptr_t)page_addr + proc->user_buffer_offset;
    ret = vm_insert_page(vma, user_page_addr, page[0]);
    if (ret) {
               printk(KERN_ERR "binder: %d: binder_alloc_buf failed "
               "to map page at %lx in userspace\n",
               proc->pid, user_page_addr);
               goto err_vm_insert_page_failed;
    }
    /* vm_insert_page does not seem to increment the refcount */
}
```

上述代码的具体实现流程如下所示。

（1）调用 alloc_page()分配一个物理页面，此函数返回一个结构体 page 物理页面描述符，根据这个描述的内容初始化好结构体 vm_struct tmp_area。

（2）通过 map_vm_area 将这个物理页面插入到 tmp_area 描述的内核空间中。

（3）通过 page_addr + proc->user_buffer_offset 获得进程虚拟空间地址。

（4）通过函数 vm_insert_page()将这个物理页面插入到进程地址空间去，参数 vma 表示要插入的进程的地址空间。

再次回到文件“frameworks/base/cmds/servicemanager/service_manager.c”中的 main()函数，接下来需要调用 binder_become_context_manager 来通知 Binder 驱动程序自己是 Binder 机制的上下文管理者，即保护进程。函数 binder_become_context_manager()在文件“frameworks/base/cmds/servicemanager/binder.c”中定义，具体代码如下所示：

```
int binder_become_context_manager(struct binder_state *bs){
    return ioctl(bs->fd, BINDER_SET_CONTEXT_MGR, 0);
}
```

在此通过调用 ioctl 文件操作函数通知 Binder 驱动程序自己是保护进程，命令号是 BINDER_SET_CONTEXT_MGR，并没有任何参数。BINDER_SET_CONTEXT_MGR 定义为：

```
#define    BINDER_SET_CONTEXT_MGR _IOW('b', 7, int)
```

这样就进入到 Binder 驱动程序的函数 binder_ioctl()，在此只关注如下 BINDER_SET_CONTEXT_MGR 命令即可，具体代码如下所示：

```
static long binder_ioctl(struct file *filp, unsigned int cmd, unsigned long arg)
{
    int ret;
    struct binder_proc *proc = filp->private_data;
    struct binder_thread *thread;
    unsigned int size = _IOC_SIZE(cmd);
    void __user *ubuf = (void __user *)arg;
    /*printk(KERN_INFO "binder_ioctl: %d:%d %x %lx\n", proc->pid, current->pid, cmd, arg);*/
    ret = wait_event_interruptible(binder_user_error_wait, binder_stop_on_user_error < 2);
    if (ret)
        return ret;
    mutex_lock(&binder_lock);
    thread = binder_get_thread(proc);
    if (thread == NULL) {
        ret = -ENOMEM;
        goto err;
    }
    switch (cmd) {
        ......
    case BINDER_SET_CONTEXT_MGR:
        if (binder_context_mgr_node != NULL) {
            printk(KERN_ERR "binder: BINDER_SET_CONTEXT_MGR already set\n");
            ret = -EBUSY;
            goto err;
        }
        if (binder_context_mgr_uid != -1) {
            if (binder_context_mgr_uid != current->cred->euid) {
                printk(KERN_ERR "binder: BINDER_SET_"
                    "CONTEXT_MGR bad uid %d != %d\n",
                    current->cred->euid,
                    binder_context_mgr_uid);
                ret = -EPERM;
                goto err;
            }
        } else
            binder_context_mgr_uid = current->cred->euid;
        binder_context_mgr_node = binder_new_node(proc, NULL, NULL);
        if (binder_context_mgr_node == NULL) {
            ret = -ENOMEM;
            goto err;
        }
        binder_context_mgr_node->local_weak_refs++;
        binder_context_mgr_node->local_strong_refs++;
        binder_context_mgr_node->has_strong_ref = 1;
        binder_context_mgr_node->has_weak_ref = 1;
        break;
        ......
    default:
        ret = -EINVAL;
        goto err;
    }
    ret = 0;
err:
    if (thread)
        thread->looper &= ~BINDER_LOOPER_STATE_NEED_RETURN;
    mutex_unlock(&binder_lock);
    wait_event_interruptible(binder_user_error_wait, binder_stop_on_user_error < 2);
    if (ret && ret != -ERESTARTSYS)
        printk(KERN_INFO "binder: %d:%d ioctl %x %lx returned %d\n", proc->pid, current->pid,
        cmd, arg, ret);
```

```
    return ret;
}
```

在分析函数 binder_ioctl()之前，需要先弄明白如下两个数据结构的含义。

（1）结构体 binder_thread：表示一个线程，这里就是执行 binder_become_context_manager()函数的线程。具体代码如下所示：

```
struct binder_thread {
    struct binder_proc *proc;
    struct rb_node rb_node;
    int pid;
    int looper;
    struct binder_transaction *transaction_stack;
    struct list_head todo;
    uint32_t return_error; /* Write failed, return error code in read buf */
    uint32_t return_error2; /* Write failed, return error code in read */
        /* buffer. Used when sending a reply to a dead process that */
        /* we are also waiting on */
    wait_queue_head_t wait;
    struct binder_stats stats;
};
```

在上述结构体中，proc 表示是这个线程所属的进程。结构体 binder_proc 中成员变量 thread 的类型是 rb_root，它表示一棵红黑树，把属于这个进程的所有线程都组织起来，结构体 binder_thread 的成员变量 rb_node 就是用来链入这棵红黑树的节点了。looper 成员变量表示线程的状态，可以取如下所示的值：

```
enum {
    BINDER_LOOPER_STATE_REGISTERED  = 0x01,
    BINDER_LOOPER_STATE_ENTERED     = 0x02,
    BINDER_LOOPER_STATE_EXITED      = 0x04,
    BINDER_LOOPER_STATE_INVALID     = 0x08,
    BINDER_LOOPER_STATE_WAITING     = 0x10,
    BINDER_LOOPER_STATE_NEED_RETURN = 0x20
};
```

另外，transaction_stack 表示线程正在处理的事务，todo 表示发往该线程的数据列表，return_error 和 return_error2 表示操作结果返回码，wait 用来阻塞线程等待某个事件的发生，stats 用来保存一些统计信息。这些成员变量遇到的时候再分析它们的作用。

（2）数据结构 binder_node：表示一个 binder 实体，具体代码如下所示：

```
struct binder_node {
    int debug_id;
    struct binder_work work;
    union {
        struct rb_node rb_node;
        struct hlist_node dead_node;
    };
    struct binder_proc *proc;
    struct hlist_head refs;
    int internal_strong_refs;
    int local_weak_refs;
    int local_strong_refs;
    void __user *ptr;
    void __user *cookie;
    unsigned has_strong_ref : 1;
    unsigned pending_strong_ref : 1;
    unsigned has_weak_ref : 1;
    unsigned pending_weak_ref : 1;
    unsigned has_async_transaction : 1;
    unsigned accept_fds : 1;
    int min_priority : 8;
    struct list_head async_todo;
};
```

由此可见，rb_node和dead_node组成了一个联合体，具体来说分为如下两种情形。

- ❑ 如果这个Binder实体还在正常使用，则使用rb_node来连入“proc->nodes”所表示的红黑树的节点，这棵红黑树用来组织属于这个进程的所有Binder实体。
- ❑ 如果这个Binder实体所属的进程已经销毁，而这个Binder实体又被其他进程所引用，则这个Binder实体通过dead_node进入到一个哈希表中去存放。proc成员变量就是表示这个Binder实例所属于进程了。

在上述数据结构binder_node中，主要成员的具体说明如下所示。

- ❑ refs：把所有引用了该Binder实体的Binder引用连接起来构成一个链表。
- ❑ internal_strong_refs、local_weak_refs和local_strong_refs：表示这个Binder实体的引用计数。
- ❑ ptr和cookie：分别表示这个Binder实体在用户空间的地址以及附加数据。

接下来回到函数binder_ioctl()中，首先是通过“filp->private_data”获得proc变量，此处的函数binder_mmap()是一样的，然后通过函数binder_get_thread()获得线程信息，此函数的代码如下所示：

```
static struct binder_thread *binder_get_thread(struct binder_proc *proc)
{
    struct binder_thread *thread = NULL;
    struct rb_node *parent = NULL;
    struct rb_node **p = &proc->threads.rb_node;

    while (*p) {
        parent = *p;
        thread = rb_entry(parent, struct binder_thread, rb_node);

        if (current->pid < thread->pid)
            p = &(*p)->rb_left;
        else if (current->pid > thread->pid)
            p = &(*p)->rb_right;
        else
            break;
    }
    if (*p == NULL) {
        thread = kzalloc(sizeof(*thread), GFP_KERNEL);
        if (thread == NULL)
            return NULL;
        binder_stats.obj_created[BINDER_STAT_THREAD]++;
        thread->proc = proc;
        thread->pid = current->pid;
        init_waitqueue_head(&thread->wait);
        INIT_LIST_HEAD(&thread->todo);
        rb_link_node(&thread->rb_node, parent,p);
        rb_insert_color(&thread->rb_node, &proc->threads);
        thread->looper |= BINDER_LOOPER_STATE_NEED_RETURN;
        thread->return_error = BR_OK;
        thread->return_error2 = BR_OK;
    }
    return thread;
}
```

在上述代码中，把当前线程current的pid作为键值，在进程proc->threads表示的红黑树中进行查找，看是否已经为当前线程创建过了binder_thread信息。在这个场景下，由于当前线程是第一次进到这里，所以肯定找不到，即*p == NULL成立，于是，就为当前线程创建一个线程上下文信息结构体binder_thread，并初始化相应成员变量，并插入到proc->threads所表示的红黑树中去，下次要使用时就可以从proc中找到了。注意，这里的thread->looper = BINDER_LOOPER_STATE_NEED_RETURN。

再回到函数binder_ioctl()中，接下来会有binder_context_mgr_node和binder_context_mgr_uid两个全局变量，定义如下所示：

```
static struct binder_node *binder_context_mgr_node;
static uid_t binder_context_mgr_uid = -1;
```

其中 binder_context_mgr_node 用来表示 Service Manager 实体，binder_context_mgr_uid 表示 Service Manager 保护进程的 uid。在这个场景下，由于当前线程是第一次进到这里，所以 binder_context_mgr_node 为 NULL，binder_context_mgr_uid 为–1，于是初始化 binder_context_mgr_uid 为 current->cred->euid，这样当前线程就成为 Binder 机制的保护进程了，并且通过 binder_new_node 为 Service Manager 创建 Binder 实体：

```
static struct binder_node *
binder_new_node(struct binder_proc *proc, void __user *ptr, void __user *cookie)
{
    struct rb_node **p = &proc->nodes.rb_node;
    struct rb_node *parent = NULL;
    struct binder_node *node;
    while (*p) {
        parent = *p;
        node = rb_entry(parent, struct binder_node, rb_node);
        if (ptr < node->ptr)
            p = &(*p)->rb_left;
        else if (ptr > node->ptr)
            p = &(*p)->rb_right;
        else
            return NULL;
    }
    node = kzalloc(sizeof(*node), GFP_KERNEL);
    if (node == NULL)
        return NULL;
    binder_stats.obj_created[BINDER_STAT_NODE]++;
    rb_link_node(&node->rb_node, parent, p);
    rb_insert_color(&node->rb_node, &proc->nodes);
    node->debug_id = ++binder_last_id;
    node->proc = proc;
    node->ptr = ptr;
    node->cookie = cookie;
    node->work.type = BINDER_WORK_NODE;
    INIT_LIST_HEAD(&node->work.entry);
    INIT_LIST_HEAD(&node->async_todo);
    if (binder_debug_mask & BINDER_DEBUG_INTERNAL_REFS)
        printk(KERN_INFO "binder: %d:%d node %d u%p c%p created\n",
               proc->pid, current->pid, node->debug_id,
               node->ptr, node->cookie);
    return node;
}
```

在这里传进来的 ptr 和 cookie 都为 NULL。上述函数会首先检查 proc->nodes 红黑树中是否已经存在以 ptr 为键值的 node，如果已经存在则返回 NULL。在这个场景下，由于当前线程是第一次进入到这里，所以肯定不存在，于是就新建了一个 ptr 为 NULL 的 binder_node，并且初始化其他成员变量，并插入到 proc->nodes 红黑树中去。

当 binder_new_node 返回到函数 binder_ioctl()后，会把新建的 binder_node 指针保存在 binder_context_mgr_node 中，然后又初始化 binder_context_mgr_node 的引用计数值。这样执行 BINDER_SET_CONTEXT_MGR 命令完毕，在函数 binder_ioctl()返回之前执行下面的语句：

```
if (thread)
    thread->looper &= ~BINDER_LOOPER_STATE_NEED_RETURN;
```

再次回到文件“frameworks/base/cmds/servicemanager/service_manager.c”中的 main()函数，接下来需要调用函数 binder_loop()进入循环，等待 Client 发送请求。函数 binder_loop()定义在文件“frameworks/base/cmds/servicemanager/binder.c”中：

```
void binder_loop(struct binder_state *bs, binder_handler func)
{
```

```
    int res;
    struct binder_write_read bwr;
    unsigned readbuf[32];
    bwr.write_size = 0;
    bwr.write_consumed = 0;
    bwr.write_buffer = 0;

    readbuf[0] = BC_ENTER_LOOPER;
    binder_write(bs, readbuf, sizeof(unsigned));
    for (;;) {
        bwr.read_size = sizeof(readbuf);
        bwr.read_consumed = 0;
        bwr.read_buffer = (unsigned) readbuf;
        res = ioctl(bs->fd, BINDER_WRITE_READ, &bwr);
        if (res < 0) {
            LOGE("binder_loop: ioctl failed (%s)\n", strerror(errno));
            break;
        }
        res = binder_parse(bs, 0, readbuf, bwr.read_consumed, func);
        if (res == 0) {
            LOGE("binder_loop: unexpected reply?!\n");
            break;
        }
        if (res < 0) {
            LOGE("binder_loop: io error %d %s\n", res, strerror(errno));
            break;
        }
    }
}
```

在上述代码中，首先通过函数 binder_write()执行 BC_ENTER_LOOPER 命令以告诉 Binder 驱动程序，Service Manager 马上要进入循环。在此还需要理解设备文件“/dev/binder”操作函数 ioctl 的操作码 BINDER_WRITE_READ，首先看其定义：

```
#define BINDER_WRITE_READ _IOWR('b', 1, struct binder_write_read)
```

此 IO 操作码有一个形式为 struct binder_write_read 的参数，具体代码如下所示：

```
struct binder_write_read {
    signed long    write_size;    /* bytes to write */
    signed long    write_consumed;    /* bytes consumed by driver */
    unsigned long    write_buffer;
    signed long    read_size;    /* bytes to read */
    signed long    read_consumed;    /* bytes consumed by driver */
    unsigned long    read_buffer;
};
```

用户空间程序和 Binder 驱动程序交互时，大多数是通过 BINDER_WRITE_READ 命令实现的，write_bufffer 和 read_buffer 所指向的数据结构还指定了具体要执行的操作，write_bufffer 和 read_buffer 所指向的结构体是 binder_transaction_data，定义此结构体的具体代码如下所示：

```
struct binder_transaction_data {
    /* The first two are only used for bcTRANSACTION and brTRANSACTION,
     * identifying the target and contents of the transaction.
     */
    union {
        size_t    handle;    /* target descriptor of command transaction */
        void    *ptr;    /* target descriptor of return transaction */
    } target;
    void    *cookie;    /* target object cookie */
    unsigned int    code;        /* transaction command */

    /* General information about the transaction. */
    unsigned int    flags;
    pid_t        sender_pid;
    uid_t        sender_euid;
    size_t        data_size;    /* number of bytes of data */
```

```
    size_t        offsets_size;    /* number of bytes of offsets */

    /* If this transaction is inline, the data immediately
     * follows here; otherwise, it ends with a pointer to
     * the data buffer.
     */
    union {
        struct {
            /* transaction data */
            const void    *buffer;
            /* offsets from buffer to flat_binder_object structs */
            const void    *offsets;
        } ptr;
        uint8_t    buf[8];
    } data;
}
```

到此为止，已经从源代码一步一步地分析完 Service Manager 是如何成为 Android 进程间通信（IPC）机制 Binder 保护进程的。在接下来的内容中，简要总结 Service Manager 成为 Android 进程间通信（IPC）机制 Binder 保护进程的过程。

（1）打开/dev/binder 文件：

```
open("/dev/binder", O_RDWR);
```

（2）建立 128K 内存映射：

```
mmap(NULL, mapsize, PROT_READ, MAP_PRIVATE, bs->fd, 0);
```

（3）通知 Binder 驱动程序它是保护进程：

```
binder_become_context_manager(bs);
```

（4）进入循环等待请求的到来：

```
binder_loop(bs, svcmgr_handler);
```

在这个过程中，在 Binder 驱动程序中建立了一个 struct binder_proc 结构、一个 struct binder_thread 结构和一个 struct binder_node 结构，这样，Service Manager 就在 Android 系统的进程间通信机制 Binder 担负起保护进程的职责了。

5.1.3　Service Manager 服务

众所周知，Service Manager 在 Binder 机制中既充当保护进程的角色，同时也充当着 Server 角色，但是它又与一般的 Server 不一样。对于普通的 Server 来说，Client 如果想要获得 Server 的远程接口，必须通过 Service Manager 远程接口提供的 getService 接口来获得，这本身就是一个使用 Binder 机制来进行进程间通信的过程。而对于 Service Manager 这个 Server 来说，Client 如果想要获得 Service Manager 远程接口，却不必通过进程间通信机制来获得，因为 Service Manager 远程接口是一个特殊的 Binder 引用，它的引用句柄一定是 0。

获取 Service Manager 远程接口的函数是 defaultServiceManager()，此函数在文件“frameworks/base/include/binder/IServiceManager.h”中声明，具体代码如下所示：

```
sp<IServiceManager> defaultServiceManager();
```

函数 defaultServiceManager()在文件“frameworks/base/libs/binder/IServiceManager.cpp”中实现，具体代码如下所示：

```
sp<IServiceManager> defaultServiceManager()
{
    if (gDefaultServiceManager != NULL) return gDefaultServiceManager;
    {
        AutoMutex _l(gDefaultServiceManagerLock);
```

```
        if (gDefaultServiceManager == NULL) {
            gDefaultServiceManager = interface_cast<IServiceManager>(
                ProcessState::self()->getContextObject(NULL));
        }
    }
    return gDefaultServiceManager;
}
```

其中 gDefaultServiceManagerLock 和 gDefaultServiceManager 是全局变量，在文件“frameworks/base/libs/binder/Static.cpp”中定义，具体代码如下所示：

```
Mutex gDefaultServiceManagerLock;
sp<IServiceManager> gDefaultServiceManager;
```

从上述函数可以看出，gDefaultServiceManager 是单例模式，在调用函数 defaultServiceManager()时，如果已经创建 gDefaultServiceManager 了则直接返回，否则通过 interface_cast<IServiceManager>(ProcessState::self()->getContextObject(NULL))创建一个，并保存在全局变量 gDefaultServiceManager 中。

在 Binder 机制中，类 BpServiceManager 继承了类 BpInterface<IServiceManager>，BpInterface 是一个模板类，在文件“frameworks/base/include/binder/IInterface.h”中定义，具体代码如下所示：

```
template<typename INTERFACE>
class BpInterface : public INTERFACE, public BpRefBase {
public:
    BpInterface(const sp<IBinder>& remote);
protected:
    virtual IBinder* onAsBinder();
};
```

类 IServiceManager 继承了类 IInterface，而类 IInterface 和类 BpRefBase 又分别继承了类 RefBase。

下面是创建 Service Manager 远程接口的主要代码：

```
gDefaultServiceManager = interface_cast<IServiceManager>(
  ProcessState::self()->getContextObject(NULL));
```

在上述代码中，首先调用了 ProcessState 的静态成员函数 ProcessState::self，此函数的功能是返回一个全局唯一的 ProcessState 实例变量，其实这就是单例模式，此变量名为 gProcess。如果未创建 gProcess 则执行创建操作。在 ProcessState 的构造函数中，通过文件操作函数 open()打开设备文件“/dev/binder”，并且将返回来的设备文件描述符保存在成员变量 mDriverFD 中。

接着调用函数 gProcess->getContextObject()获得一个句柄值为 0 的 Binder 引用 BpBinder。再来看函数 interface_cast<IServiceManager>的具体实现，此模板函数在文件“framework/base/include/binder/IInterface.h”中定义，具体实现代码如下所示：

```
template<typename INTERFACE>
inline sp<INTERFACE> interface_cast(const sp<IBinder>& obj) {
    return INTERFACE::asInterface(obj);
}
```

在上述代码中，INTERFACE 是 IServiceManager，调用了函数 IServiceManager::asInterface()。函数 IServiceManager::asInterface()是通过 DECLARE_META_INTERFACE(ServiceManager)宏在类 IServiceManager 中声明的，它位于文件“framework/base/include/binder/IServiceManager.h”中，展开后的代码如下所示：

```
#define DECLARE_META_INTERFACE(ServiceManager)
    static const android::String16 descriptor;
    static android::sp<IServiceManager> asInterface(
    const android::sp<android::IBinder>& obj);
    virtual const android::String16& getInterfaceDescriptor() const;
    IServiceManager();
    virtual ~IServiceManager();
```

IServiceManager::asInterface 是通过宏 IMPLEMENT_META_INTERFACE(ServiceManager, "android.os.IServiceManager")定义的，它位于文件"framework/base/libs/binder/IServiceManager.cpp"中，展开后的代码如下所示：

```
#define IMPLEMENT_META_INTERFACE(ServiceManager, "android.os.IServiceManager")
  const android::String16 IServiceManager::descriptor("android.os.IServiceManager");
  const android::String16&
   IServiceManager::getInterfaceDescriptor() const {
   return IServiceManager::descriptor;
   }
   android::sp<IServiceManager> IServiceManager::asInterface(
   const android::sp<android::IBinder>& obj)
   {
   android::sp<IServiceManager> intr;
   if (obj != NULL) {
   intr = static_cast<IServiceManager*>(
   obj->queryLocalInterface(
   IServiceManager::descriptor).get());
   if (intr == NULL) {
   intr = new BpServiceManager(obj);
   }
   }
   return intr;
   }
   IServiceManager::IServiceManager() { }
   IServiceManager::~IServiceManager() { }
```

IServiceManager::asInterface 的具体实现代码如下所示：

```
android::sp<IServiceManager> IServiceManager::asInterface(const android::sp<android::IBinder>&
obj)
{
   android::sp<IServiceManager> intr;

   if (obj != NULL) {
      intr = static_cast<IServiceManager*>(
                obj->queryLocalInterface(IServiceManager::descriptor).get());

      if (intr == NULL) {
         intr = new BpServiceManager(obj);
      }
   }
   return intr;
}
```

此处传进来的参数 obj 就是刚才创建的 new BpBinder(0)，类 BpBinder 中的成员函数 queryLocalInterface()继承自基类 IBinder，函数 IBinder::queryLocalInterface()位于文件"framework/base/libs/binder/Binder.cpp"中，具体实现代码如下所示：

```
sp<IInterface>  IBinder::queryLocalInterface(const String16& descriptor)
{
   return NULL;
}
```

由此可见，在函数 IServiceManager::asInterface()中会调用以下语句：

```
intr = new BpServiceManager(obj);
```

即：

```
intr = new BpServiceManager(new BpBinder(0));
```

创建的 Service Manager 远程接口本质上是一个 BpServiceManager，包含了一个句柄值为 0 的 Binder 引用。

5.2 匿名共享内存子系统详解

Android 系统中提供了独特的匿名共享内存子系统 Ashmem（Anonymous Shared Memory），它以驱动程序的形式实现在内核空间中。Ashmem 有如下两个特点。

- 能够辅助内存管理系统来有效地管理不再使用的内存块。
- 通过 Binder 进程间通信机制来实现进程间的内存共享。

对于 Android 系统的匿名共享内存子系统来说，其主体是以驱动程序的形式实现在内核空间的，同时，在系统运行时库层和应用程序框架层提供了访问接口。其中在系统运行时库层提供了 C/C++调用接口，而在应用程序框架层提供了 Java 调用接口。在此将直接通过应用程序框架层提供的 Java 调用接口来说明匿名共享内存子系统 Ashmem 的使用方法，毕竟在 Android 开发应用程序时，是基于 Java 语言的。其实对于 Android 系统中的应用程序框架层的 Java 调用接口来说，是通过 JNI 方法来调用系统运行时库层的 C/C++调用接口，最后需要进入到内核空间的 Ashmem 驱动程序中去。

Android 系统中的匿名共享内存 Ashmem 驱动程序，利用 Linux 的共享内存子系统导出的接口来实现自己的功能。在 Android 系统匿名共享内存系统中，其核心功能是实现创建（open）、映射（mmap）、读写（read/write）以及锁定和解锁（pin/unpin）。在本节的内容中，将详细讲解 Android 匿名共享内存子系统的基本知识。

5.2.1　基础数据结构

在 Ashmem 驱动程序中需要用到 3 个结构体，分别是 ashmem_area、ashmem_range 和 ashmem_range。其中前两个结构体在文件“kernel/goldfish/mm/ashmem.c”中定义，具体实现代码如下所示：

```
struct ashmem_area {
        char name[ASHMEM_FULL_NAME_LEN];          /* 匿名共享内存的名称 */
        struct list_head unpinned_list;           /* 解锁内存列表 */
        struct file *file;                        /* 指向临时文件系统 tmpfs 中的一个文件 */
        size_t size;                              /* 文件大小 */
        unsigned long prot_mask;                  /* 匿名共享内存的访问保护位 */
}
struct ashmem_range {
        struct list_head lru;                     /* 最近最少使用的列表 */
        struct list_head unpinned;                /* entry in its area's unpinned list */
        struct ashmem_area *asma;                 /* associated area */
        size_t pgstart;                           /* 处于解锁状态内存的开始地址 */
        size_t pgend;                             /*处于解锁状态内存的结束地址*/
        unsigned int purged;                      /* 解锁内存是否被收回 */
}
```

结构体 ashmem_area 用于表示一块匿名共享内存单元，结构体 ashmem_range 用于表示处于解锁状态的内存。

结构体 ashmem_range 用于表示被锁定或被解锁的内存，在文件“kernel/goldfish/include/linux/ashmem.h”中定义，具体实现代码如下所示：

```
struct ashmem_pin {
        __u32 offset;        /* 这块内存的偏移值*/
        __u32 len;           /* 这块内存的大小 */
}
```

结构体 ashmem_fops 定义了“dev/ashmem”的操作方法列表，具体实现代码如下所示：

```
static struct file_operations ashmem_fops = {
        .owner = THIS_MODULE,
```

```
        .open = ashmem_open,
        .release = ashmem_release,
        .mmap = ashmem_mmap,
        .unlocked_ioctl = ashmem_ioctl,
        .compat_ioctl = ashmem_ioctl,
}
```

5.2.2　初始化处理

通过 Ashmem 驱动的初始化函数，可以获取如下两点信息。

- ❑ Ashmem 给用户空间暴露了什么接口，即创建了什么样的设备文件。
- ❑ Ashmem 提供了什么函数来操作这个设备文件。

Ashmem 驱动程序在文件“kernel/common/mm/ashmem.c”中实现，其中函数 ashmem_init 用于实现模块初始化处理，主要实现代码如下所示：

```
static struct miscdevice ashmem_misc = {
    .minor = MISC_DYNAMIC_MINOR,
    .name = "ashmem",
    .fops = &ashmem_fops,
};
static int __init ashmem_init(void)
{
    int ret;
    ......
    ret = misc_register(&ashmem_misc);
    if (unlikely(ret)) {
        printk(KERN_ERR "ashmem: failed to register misc device!\n");
        return ret;
    }
    ......
    return 0;
}
```

在上述代码中，在加载 Ahshmem 驱动程序时会创建一个设备文件“/dev/ashmem”，这是一个 misc 类型的设备。通过函数 misc_register 来注册 misc 设备，调用这个函数后会在“/dev”目录下生成一个 ashmem 设备文件。在设备文件中一共提供了 open、mmap、release 和 ioctl 4 种操作，此处并没有 read 和 write 操作，原因是读写共享内存的方法是通过内存映射地址来进行的，通过 mmap 系统调用将这个设备文件映射到进程地址空间中。与此同时，直接对内存进行了读写操作，所以不需要通过 read 和 write 方式进行文件操作。

匿名共享内存创建功能是在文件“frameworks/base/core/java/android/os/MemoryFile.java”中实现的，此文件调用了类 MemoryFile 的构造函数，MemoryFile 的构造函数调用了 JNI 函数 native_open，这样便创建了匿名内存共享文件。JNI 方法 native_open 在文件“frameworks/base/core/jni/adroid_os_MemoryFile.cpp”中实现，具体代码如下所示：

```
static jobject android_os_MemoryFile_open(JNIEnv* env, jobject clazz, jstring name, jint length)
{
    const char* namestr = (name ? env->GetStringUTFChars(name, NULL) : NULL);

    int result = ashmem_create_region(namestr, length);

    if (name)
        env->ReleaseStringUTFChars(name, namestr);

    if (result < 0) {
        jniThrowException(env, "java/io/IOException", "ashmem_create_region failed");
        return NULL;
    }

    return jniCreateFileDescriptor(env, result);
}
```

函数 native_open 通过运行时库提供的接口 ashmem_create_region 创建匿名共享内存，这个接口在文件 system/core/libcutils/ashmem-dev.c 中实现，具体代码如下所示：

```
int ashmem_create_region(const char *name, size_t size)
{
    int fd, ret;

    fd = open(ASHMEM_DEVICE, O_RDWR);
    if (fd < 0)
        return fd;

    if (name) {
        char buf[ASHMEM_NAME_LEN];

        strlcpy(buf, name, sizeof(buf));
        ret = ioctl(fd, ASHMEM_SET_NAME, buf);
        if (ret < 0)
            goto error;
    }

    ret = ioctl(fd, ASHMEM_SET_SIZE, size);
    if (ret < 0)
        goto error;

    return fd;

error:
    close(fd);
    return ret;
}
```

在上述代码中，通过执行 3 个文件操作系统调用的方式与 Ashmem 驱动程序进行交互。通过 open 操作打开设备文件“ASHMEM_DEVICE”，通过 ioctl 操作设置匿名共享内存的名称和大小。

5.2.3 打开匿名共享内存设备文件

当 open 进入内核后，会调用函数 ashmem_open 打开匿名共享内存设备文件，此函数能够为程序创建一个 ashmem_area 结构体，具体实现代码如下所示：

```
static int ashmem_open(struct inode *inode, struct file *file)
{
    struct ashmem_area *asma;
    int ret;
    ret = nonseekable_open(inode, file);
    if (unlikely(ret))
        return ret;
    asma = kmem_cache_zalloc(ashmem_area_cachep, GFP_KERNEL);
    if (unlikely(!asma))
        return -ENOMEM;
    INIT_LIST_HEAD(&asma->unpinned_list);
    memcpy(asma->name, ASHMEM_NAME_PREFIX, ASHMEM_NAME_PREFIX_LEN);
    asma->prot_mask = PROT_MASK;
    file->private_data = asma;
    return 0;
}
```

上述代码的执行流程如下所示。

- 通过函数 nonseekable_open 设置这个文件不可以执行定位操作，即不可执行 seek 文件操作。
- 通过函数 kmem_cache_zalloc 在刚创建的 slab 缓冲区 ashmem_area_cachep 中创建一个 ashmem_area 结构体，并将创建的结构体保存在本地变量 asma 中。
- 初始化变量 asma 的其他域，其中域 name 初始为宏 ASHMEM_NAME_PREFIX，宏 ASHMEM_NAME_PREFIX 的定义代码为：

```
#define ASHMEM_NAME_PREFIX "dev/ashmem/"
#define ASHMEM_NAME_PREFIX_LEN (sizeof(ASHMEM_NAME_PREFIX) - 1)
```

❑ 将结构 ashmem_area 保存在打开文件结构体的 private_data 域中，此时通过使用 Ashmem 驱动程序，可以在其他模块通过 private_data 域来取回这个 ashmem_area 结构。

在函数 ashmem_create_region 中调用了两次 ioctl 文件操作，功能是设置新建匿名共享内存的名字和大小。在文件“kernel/comon/mm/include/ashmem.h”中，ASHMEM_SET_NAME 和 ASHMEM_SET_SIZE 分别表示新建内存的名字和大小，具体定义代码如下所示：

```
#define ASHMEM_NAME_LEN        256
#define __ASHMEMIOC            0x77
#define ASHMEM_SET_NAME        _IOW(__ASHMEMIOC, 1, char[ASHMEM_NAME_LEN])
#define ASHMEM_SET_SIZE        _IOW(__ASHMEMIOC, 3, size_t)
```

其中 ASHMEM_SET_NAME 的 ioctl 调用会进入到 Ashmem 驱动程序函数 ashmem_ioctl 中，此函数能够将从用户空间传进来的匿名共享内存的大小值保存在对应的 asma->size 域中。函数 ashmem_ioctl 的实现代码如下所示：

```
static long ashmem_ioctl(struct file *file, unsigned int cmd, unsigned long arg)
{
        struct ashmem_area *asma = file->private_data;
        long ret = -ENOTTY;
        switch (cmd) {
        case ASHMEM_SET_NAME:
                ret = set_name(asma, (void __user *) arg);
                break;
        case ASHMEM_GET_NAME:
                ret = get_name(asma, (void __user *) arg);
                break;
        case ASHMEM_SET_SIZE:
                ret = -EINVAL;
                if (!asma->file) {
                        ret = 0;
                        asma->size = (size_t) arg;
                }
                break;
        case ASHMEM_GET_SIZE:
                ret = asma->size;
                break;
        case ASHMEM_SET_PROT_MASK:
                ret = set_prot_mask(asma, arg);
                break;
        case ASHMEM_GET_PROT_MASK:
                ret = asma->prot_mask;
                break;
        case ASHMEM_PIN:
        case ASHMEM_UNPIN:
        case ASHMEM_GET_PIN_STATUS:
                ret = ashmem_pin_unpin(asma, cmd, (void __user *) arg);
                break;
        case ASHMEM_PURGE_ALL_CACHES:
                ret = -EPERM;
                if (capable(CAP_SYS_ADMIN)) {
                        ret = ashmem_shrink(0, GFP_KERNEL);
                        ashmem_shrink(ret, GFP_KERNEL);
                }
                break;
        }
        return ret;
}
```

上述代码主要完成如下两个功能。

❑ struct ashmem_area *asma = file->private_data：获取描述将要改名的匿名共享内存 asma。

❑ ret = set_name(asma, (void __user *) arg)：调用函数 set_name 修改匿名共享内存的名称。

函数 set_name 也是在文件“kernel/goldfish/mm/ashmem.c”中实现的，功能是把用户空间传进来的匿名共享内存的名字设置到 asma->name 域中。函数 set_name 的具体实现代码如下所示：

```
static int set_name(struct ashmem_area *asma, void __user *name)
{
	int ret = 0;
	mutex_lock(&ashmem_mutex);
	/* cannot change an existing mapping's name */
	if (unlikely(asma->file)) {
		ret = -EINVAL;
		goto out;
	}
	if (unlikely(copy_from_user(asma->name + ASHMEM_NAME_PREFIX_LEN,
				    name, ASHMEM_NAME_LEN)))
		ret = -EFAULT;
	asma->name[ASHMEM_FULL_NAME_LEN-1] = '\0';
out:
	mutex_unlock(&ashmem_mutex);
	return ret;
}
```

到此为止，创建匿名共享内存的过程就全部介绍完毕了。

5.2.4 内存映射

Ashmem 驱动程序并不提供文件的 read 操作和 write 操作，如果进程要访问这个共享内存，则必须将这个设备文件映射到自己的进程空间中，然后才能进行内存访问。在类 MemoryFile 的构造函数中，创建匿名共享内存后需要把匿名共享内存设备文件映射到进程空间。映射功能是通过调用 JNI 方法 native_mmap 实现的，此 JNI 方法在文件“frameworks/base/core/jni/adroid_os_MemoryFile.cpp”中实现，具体代码如下所示：

```
static jint android_os_MemoryFile_mmap(JNIEnv* env, jobject clazz, jobject fileDescriptor,
	jint length, jint prot)
{
    int fd = jniGetFDFromFileDescriptor(env, fileDescriptor);
    jint result = (jint)mmap(NULL, length, prot, MAP_SHARED, fd, 0);
    if (!result)
        jniThrowException(env, "java/io/IOException", "mmap failed");
    return result;
}
```

在上述代码中，在 open 匿名设备文件“/dev/ashmem”获得文件描述符 fd。有了这个文件描述符后，就可以直接通过函数 mmap 执行内存映射操作了。当调用函数 mmap 打开映射到进程的地址空间时，会立即执行 ashmem 中的函数 ashmem_mmap。函数 ashmem_mmap 的功能是，调用 Linux 内核中的函数 shmem_file_setup 在临时文件系统 tmpfs 中创建一个临时文件，这个临时文件与 Ashmem 驱动程序创建的匿名共享内存对应。函数 ashmem_mmap 在文件“kernel/goldfish/mm/ashmem.c”中定义，具体实现代码如下所示：

```
static int ashmem_mmap(struct file *file, struct vm_area_struct *vma)
{
	struct ashmem_area *asma = file->private_data;
	int ret = 0;
	mutex_lock(&ashmem_mutex);
	/* user needs to SET_SIZE before mapping */
	if (unlikely(!asma->size)) {
		ret = -EINVAL;
		goto out;
	}
	/* requested protection bits must match our allowed protection mask */
	if (unlikely((vma->vm_flags & ~asma->prot_mask) & PROT_MASK)) {
		ret = -EPERM;
```

```
            goto out;
        }
        if (!asma->file) {
            char *name = ASHMEM_NAME_DEF;
            struct file *vmfile;
            if (asma->name[ASHMEM_NAME_PREFIX_LEN] != '\0')
                name = asma->name;
            /* ... and allocate the backing shmem file */
            vmfile = shmem_file_setup(name, asma->size, vma->vm_flags);
            if (unlikely(IS_ERR(vmfile))) {
                ret = PTR_ERR(vmfile);
                goto out;
            }
            asma->file = vmfile;
        }
        get_file(asma->file);
        if (vma->vm_flags & VM_SHARED)
            shmem_set_file(vma, asma->file);
        else {
            if (vma->vm_file)
                fput(vma->vm_file);
            vma->vm_file = asma->file;
        }
        vma->vm_flags |= VM_CAN_NONLINEAR;
out:
        mutex_unlock(&ashmem_mutex);
        return ret;
}
```

在上述代码中，检查了虚拟内存 vma 是否允许在不同进程之间实现共享。如果允许，则调用函数 shmem_set_file 来设置它的映射文件和内存操作方法表。

5.2.5 读写操作

从类 MemoryFile 中可以获得读写操作的过程，对应的实现代码如下所示：

```
private static native int native_read(FileDescriptor fd, int address, byte[] buffer,
int srcOffset, int destOffset, int count, boolean isUnpinned) throws IOException;
private static native void native_write(FileDescriptor fd, int address, byte[] buffer,
int srcOffset, int destOffset, int count, boolean isUnpinned) throws IOException;
private FileDescriptor mFD;    // ashmem file descriptor
private int mAddress;   // address of ashmem memory
private int mLength;   // total length of our ashmem region
private boolean mAllowPurging = false;  // true if our ashmem region is unpinned
public int readBytes(byte[] buffer, int srcOffset, int destOffset, int count)
    throws IOException {
        if (isDeactivated()) {
            throw new IOException("Can't read from deactivated memory file.");
        }
        if (destOffset < 0 || destOffset > buffer.length || count < 0
            || count > buffer.length - destOffset
            || srcOffset < 0 || srcOffset > mLength
            || count > mLength - srcOffset) {
             throw new IndexOutOfBoundsException();
        }
        return native_read(mFD, mAddress, buffer, srcOffset, destOffset, count, mAllowPurging);
    }
    public void writeBytes(byte[] buffer, int srcOffset, int destOffset, int count)
        throws IOException {
            if (isDeactivated()) {
                throw new IOException("Can't write to deactivated memory file.");
            }
            if (srcOffset < 0 || srcOffset > buffer.length || count < 0
                || count > buffer.length - srcOffset
                || destOffset < 0 || destOffset > mLength
                || count > mLength - destOffset) {
                    throw new IndexOutOfBoundsException();
            }
```

```
        native_write(mFD, mAddress, buffer, srcOffset, destOffset, count, mAllowPurging);
}
```

通过对上述代码的分析可知，是通过调用 JNI 方法实现读写匿名共享内存操作功能。读操作的 JNI 方法是 native_read，写操作的 NI 方法是 native_write，这两个方法都在文件“frameworks/base/core/jni/adroid_os_MemoryFile.cpp”中定义，具体实现代码如下所示：

```
static jint android_os_MemoryFile_read(JNIEnv* env, jobject clazz,
        jobject fileDescriptor, jint address, jbyteArray buffer, jint srcOffset, jint destOffset,
        jint count, jboolean unpinned)
{
    int fd = jniGetFDFromFileDescriptor(env, fileDescriptor);
    if (unpinned && ashmem_pin_region(fd, 0, 0) == ASHMEM_WAS_PURGED) {
        ashmem_unpin_region(fd, 0, 0);
        jniThrowException(env, "java/io/IOException", "ashmem region was purged");
        return -1;
    }

    env->SetByteArrayRegion(buffer, destOffset, count, (const jbyte *)address + srcOffset);

    if (unpinned) {
        ashmem_unpin_region(fd, 0, 0);
    }
    return count;
}

static jint android_os_MemoryFile_write(JNIEnv* env, jobject clazz,
        jobject fileDescriptor, jint address, jbyteArray buffer, jint srcOffset, jint destOffset,
        jint count, jboolean unpinned)
{
    int fd = jniGetFDFromFileDescriptor(env, fileDescriptor);
    if (unpinned && ashmem_pin_region(fd, 0, 0) == ASHMEM_WAS_PURGED) {
        ashmem_unpin_region(fd, 0, 0);
        jniThrowException(env, "java/io/IOException", "ashmem region was purged");
        return -1;
    }
    env->GetByteArrayRegion(buffer, srcOffset, count, (jbyte *)address + destOffset);
    if (unpinned) {
        ashmem_unpin_region(fd, 0, 0);
    }
    return count;
```

在上述代码中，函数 ashmem_pin_region 和 ashmem_unpin_region 用于为系统运行时的库提供接口，功能是执行匿名共享内存的锁定和解锁操作。这样便能够通知 Ashmem 驱动程序哪些内存块是正在使用的，哪些需要锁定，哪些不需要使用，哪些可以解锁。这两个函数在文件“system/core/libcutils/ashmem-dev.c”中定义，具体实现代码如下所示：

```
int ashmem_pin_region(int fd, size_t offset, size_t len)
{
    struct ashmem_pin pin = { offset, len };
    return ioctl(fd, ASHMEM_PIN, &pin);
}
int ashmem_unpin_region(int fd, size_t offset, size_t len)
{
    struct ashmem_pin pin = { offset, len };
    return ioctl(fd, ASHMEM_UNPIN, &pin);
}
```

经过上述操作之后，Ashmem 驱动程序就可以在整个内存管理系统中管理内存了。

5.2.6 锁定和解锁

在 Android 系统中，通过如下两个 ioctl 操作实现匿名共享内存的锁定和解锁操作。

- ASHMEM_PIN。

❑ ASHMEM_UNPIN。

ASHMEM_PIN 和 ASHMEM_UNPIN 在文件"kernel/common/include/linux/ashmem.h"中定义，对应代码如下所示：

```
#define __ASHMEMIOC         0x77
#define ASHMEM_PIN         _IOW(__ASHMEMIOC, 7, struct ashmem_pin)
#define ASHMEM_UNPIN         _IOW(__ASHMEMIOC, 8, struct ashmem_pin)
struct ashmem_pin {
    __u32 offset;       /* offset into region, in bytes, page-aligned */
    __u32 len;          /* length forward from offset, in bytes, page-aligned */
}
```

再看函数 ashmem_ioctl，在其实现代码中与 ASHMEM_PIN 和 ASHMEM_UNPIN 这两个操作相关的代码如下所示：

```
static long ashmem_ioctl(struct file *file, unsigned int cmd, unsigned long arg)
{
    struct ashmem_area *asma = file->private_data;
    long ret = -ENOTTY;
    switch (cmd) {
    ......
    case ASHMEM_PIN:
    case ASHMEM_UNPIN:
        ret = ashmem_pin_unpin(asma, cmd, (void __user *) arg);
        break;
    ......
    }
    return ret;
}
```

在上述代码中，调用函数 ashmem_pin_unpin 处理控制命令 ASHMEM_PIN 和 ASHMEM_UNPIN。函数 ashmem_pin_unpin 的实现流程如下所示。

❑ 获取传递到用户空间的参数，并将获取值保存在本地变量 pin 中。这是一个 struct ashmem_pin 类型的结构体类型，在里面包括了要 pin/unpin 内存块的起始地址和大小。

❑ 因为起始地址和大小的单位都是字节，所以通过转换处理为以页面为单位的、并保存在本地变量 pgstart 和 pgend 中。

❑ 不但对参数进行安全性检查外，并且确保只要从用户空间传进来的内存块的大小值为 0，就认为是要 pin/unpin 整个匿名共享内存。

❑ 判断当前要执行操作的类别，根据 ASHMEM_PIN 操作和 ASHMEM_UNPIN 操作分别执行 ashmem_pin 和 ashmem_unpin。

❑ 当创建匿名共享内存时，所有默认的内存都是 pinned 状态的，只有用户告诉 Ashmem 驱动程序要 unpin 某一块内存时，Ashmem 驱动程序才会把这块内存 unpin。

❑ 用户告知 Ashmem 驱动程序重新 pin 某一块前面被 unpin 过的内块，这样能够将此内存从 unpinned 状态改转换 pinned 状态。

函数 ashmem_pin_unpin 在文件"kernel/goldfish/ashmem.c"中定义，具体的实现代码如下所示：

```
static int ashmem_pin_unpin(struct ashmem_area *asma, unsigned long cmd,
            void __user *p)
{
    struct ashmem_pin pin;
    size_t pgstart, pgend;
    int ret = -EINVAL;

    if (unlikely(!asma->file))
        return -EINVAL;

    if (unlikely(copy_from_user(&pin, p, sizeof(pin))))
        return -EFAULT;
```

```
    /* per custom, you can pass zero for len to mean "everything onward" */
    if (!pin.len)
        pin.len = PAGE_ALIGN(asma->size) - pin.offset;

    if (unlikely((pin.offset | pin.len) & ~PAGE_MASK))
        return -EINVAL;

    if (unlikely(((__u32) -1) - pin.offset < pin.len))
        return -EINVAL;

    if (unlikely(PAGE_ALIGN(asma->size) < pin.offset + pin.len))
        return -EINVAL;

    pgstart = pin.offset / PAGE_SIZE;
    pgend = pgstart + (pin.len / PAGE_SIZE) - 1;

    mutex_lock(&ashmem_mutex);

    switch (cmd) {
    case ASHMEM_PIN:
        ret = ashmem_pin(asma, pgstart, pgend);
        break;
    case ASHMEM_UNPIN:
        ret = ashmem_unpin(asma, pgstart, pgend);
        break;
    case ASHMEM_GET_PIN_STATUS:
        ret = ashmem_get_pin_status(asma, pgstart, pgend);
        break;
    }

    mutex_unlock(&ashmem_mutex);

    return ret;
}
```

由此可见，执行 ASHMEM_PIN 操作的目标对象必须是一块处于 unpinned 状态的内存块。

函数 ashmem_unpin 的功能是对某一块匿名共享内存进行解锁操作，具体处理流程如下所示。

- 在遍历asma->unpinned_list列表时，查找当前处于unpinned状态的内存块是否与将要unpin的内存块[pgstart, pgend]相交，如果相交则通过执行合并操作调整 pgstart 和 pgend 的大小。
- 调用函数 range_del 删除原来的已经被 unpinned 过的内存块。
- 调用函数 range_alloc 重新 unpinned 调整过后的内存块[pgstart, pgend]，此时新的内存块[pgstart, pgend]已经包含了刚才所有被删掉的 unpinned 状态的内存。
- 如果找到相交的内存块，并且调整了 pgstart 和 pgend 的大小之后，需要重新扫描asma->unpinned_list 列表。原因是新的内存块[pgstart, pgend]可能与前后的处于 unpinned 状态的内存块发生相交。

函数 ashmem_unpin 在文件“kernel/goldfish/ashmem.c”中定义，具体的实现代码如下所示：

```
static int ashmem_unpin(struct ashmem_area *asma, size_t pgstart, size_t pgend)
{
    struct ashmem_range *range, *next;
    unsigned int purged = ASHMEM_NOT_PURGED;

restart:
    list_for_each_entry_safe(range, next, &asma->unpinned_list, unpinned) {
        /* short circuit: this is our insertion point */
        if (range_before_page(range, pgstart))
            break;

        /*
         * The user can ask us to unpin pages that are already entirely
         * or partially pinned. We handle those two cases here.
         */
```

```
        if (page_range_subsumed_by_range(range, pgstart, pgend))
            return 0;
        if (page_range_in_range(range, pgstart, pgend)) {
            pgstart = min_t(size_t, range->pgstart, pgstart),
            pgend = max_t(size_t, range->pgend, pgend);
            purged |= range->purged;
            range_del(range);
            goto restart;
        }
    }
    return range_alloc(asma, range, purged, pgstart, pgend);
}
```

range_before_page 的操作是一个宏定义，功能是判断 range 描述的内存块是否在 page 页面之前，如果是则表示结束整个描述。asma->unpinned_list 列表是按照页面号从大到小进行排列的，并且每一块被 unpin 的内存都是不相交的。range_before_pag 的定义代码如下所示：

```
#define range_before_page(range, page)
  ((range)->pgend < (page))
```

page_range_subsumed_by_range 的操作也是一个宏定义，功能是判断内存块是否包含[start, end]这个内存块，如果包含则说明当前要 unpin 的内存块已经处于 unpinned 状态。如果什么也不用操作则直接返回。page_range_subsumed_by_range 的定义代码如下所示：

```
#define page_range_subsumed_by_range(range, start, end) \
  (((range)->pgstart <= (start)) && ((range)->pgend >= (end)))
```

page_range_in_range 的操作也是一个宏定义，功能是判断内存块 [start, end]是否互相包或者相交。page_range_in_range 的定义代码如下所示：

```
#define page_range_in_range(range, start, end) \
  (page_in_range(range, start) || page_in_range(range, end) || \
   page_range_subsumes_range(range, start, end))
```

page_range_subsumed_by_range 的操作也是一个宏定义，功能是判断内存块 range 是否包含内存块 [start, end]。page_range_subsumed_by_range 的定义代码如下所示：

```
#define page_range_subsumed_by_range(range, start, end) \
  (((range)->pgstart <= (start)) && ((range)->pgend >= (end)))
```

range_in_range 的操作也是一个宏定义，功能是判断内存块地址 page 是否包含在内存块 range 中。range_in_range 的定义代码如下所示：

```
#define page_in_range(range, page)  \
(((range)->pgstart <= (page)) && ((range)->pgend >= (page)))
```

再看函数 range_del，功能是从 asma->unpinned_list 中删掉内存块，并判断它是否在 lru 列表中。函数 range_del 的具体实现代码如下所示：

```
static void range_del(struct ashmem_range *range)
{
    list_del(&range->unpinned);
    if (range_on_lru(range))
        lru_del(range);
    kmem_cache_free(ashmem_range_cachep, range);
}
```

再看函数 lru_del，内存块的状态 purged 值为 ASHMEM_NOT_PURGED，表示现在没有收回对应的物理页面，那么内存块就位于 lru 列表中，则使用函数 lru_del 删除这个内存块。函数 lru_del 的具体实现代码如下所示：

```
static inline void lru_del(struct ashmem_range *range)
{
    list_del(&range->lru);
```

```
    lru_count -= range_size(range);
}
```

再看在函数 ashmem_unpin 中调用的 range_alloc 函数，其功能是从 slab 缓冲区中 ashmem_range_cachep 分配一个 ashmem_range，并进行相应的初始化处理。然后放在对应的列表 ashmem_area->unpinned_list 中，并判断这个 range 的 purged 是否处于 ASHMEM_NOT_PURGED 状态，如果是则要把它放在 lru 列表中。函数 range_alloc 在文件“kernel/goldfish/ashmem.c”中实现，具体的实现代码如下所示：

```
static int range_alloc(struct ashmem_area *asma,
                struct ashmem_range *prev_range, unsigned int purged,
                size_t start, size_t end)
{
    struct ashmem_range *range;
    range = kmem_cache_zalloc(ashmem_range_cachep, GFP_KERNEL);
    if (unlikely(!range))
        return -ENOMEM;
    range->asma = asma;
    range->pgstart = start;
    range->pgend = end;
    range->purged = purged;
    list_add_tail(&range->unpinned, &prev_range->unpinned);
    if (range_on_lru(range))
        lru_add(range);
    return 0;
}
```

再看函数 lru_add，其功能是将未被回收的已解锁内存块添加到全局列表 ashmem_lru_list 中。函数 lru_add 在文件“kernel/goldfish/ashmem.c”中实现，具体的实现代码如下所示：

```
static inline void lru_add(struct ashmem_range *range)
{
    list_add_tail(&range->lru, &ashmem_lru_list);
    lru_count += range_size(range);
}
```

再看函数 ashmem_pin，功能是锁定一块匿名共享内存区域。被 pin 的内存块肯定被保存在 unpinned_list 列表中，如果不在则什么都不用做。要想判断在 unpinned_list 列表中是否存在 pin 的内存块，需要通过遍历 asma->unpinned_list 列表的方式找出与之相交的内存块。函数 ashmem_pin 在文件“kernel/goldfish/ashmem.c”中实现，具体的实现代码如下所示：

```
static int ashmem_pin(struct ashmem_area *asma, size_t pgstart, size_t pgend)
{
    struct ashmem_range *range, *next;
    int ret = ASHMEM_NOT_PURGED;

    list_for_each_entry_safe(range, next, &asma->unpinned_list, unpinned) {
        /* moved past last applicable page; we can short circuit */
        if (range_before_page(range, pgstart))
            break;
        if (page_range_in_range(range, pgstart, pgend)) {
            ret |= range->purged;

            /* Case #1: Easy. Just nuke the whole thing. */
            if (page_range_subsumes_range(range, pgstart, pgend)) {
                range_del(range);
                continue;
            }

            /* Case #2: We overlap from the start, so adjust it */
            if (range->pgstart >= pgstart) {
                range_shrink(range, pgend + 1, range->pgend);
                continue;
            }
```

```
            /* Case #3: We overlap from the rear, so adjust it */
            if (range->pgend <= pgend) {
               range_shrink(range, range->pgstart, pgstart-1);
               continue;
            }
            /*
             * Case #4: We eat a chunk out of the middle. A bit
             * more complicated, we allocate a new range for the
             * second half and adjust the first chunk's endpoint.
             */
            range_alloc(asma, range, range->purged,
                  pgend + 1, range->pgend);
            range_shrink(range, range->pgstart, pgstart - 1);
            break;
         }
      }
      return ret;
   }
```

在上述代码中对重新锁定内存块操作实现了判断，通过 if 语句处理了如下所示的 4 种情形。

- 指定要锁定的内存块[start, end]包含了解锁状态的内存块 range，此时只要将解锁状态的内存块 range 从其宿主匿名共享内存的解锁内存块列表 unpinned_list 中删除即可。
- 合并要锁定内存块[pgstart,pgend]后部分和解锁状态内存块 range 的前半部分，此时将解锁状态内存块 range 的开始地址设置为要锁定内存块的末尾地址的下一个页面地址。
- 合并要锁定内存块[pgstart,pgend]前部分和解锁状态内存块 range 的后半部分，此时将解锁状态内存块 range 的末尾地址设置为要锁定内存块的开始地址的下一个页面地址。
- 设置要锁定内存块[pgstart,pgend]包含在解锁状态内存块 range 中。

再看函数 range_shrink，功能是设置 range 描述的内存块的起始页面号，如果还存在于 lru 列表中，则需要调整在 lru 列表中的总页面数大小。函数 range_shrink 在文件“kernel/goldfish/ashmem.c”中定义，具体的实现代码如下所示：

```
static inline void range_shrink(struct ashmem_range *range,
            size_t start, size_t end)
{
   size_t pre = range_size(range);

   range->pgstart = start;
   range->pgend = end;

   if (range_on_lru(range))
      lru_count -= pre - range_size(range);
}
```

5.2.7　回收内存块

接下来开始看最后一步：回收匿名共享内存块。再次回到前面介绍的初始化步骤，先看 Ashmem 驱动初始化函数 ashmem_init，此函数会调用函数 register_shrinker 向内存管理系统注册一个内存回收算法函数，具体实现代码如下所示：

```
static struct shrinker ashmem_shrinker = {
   .shrink = ashmem_shrink,
   .seeks = DEFAULT_SEEKS * 4,
};
static int __init ashmem_init(void)
{
   ......
   register_shrinker(&ashmem_shrinker);
   printk(KERN_INFO "ashmem: initialized\n");
   return 0;
}
```

其实在 Linux 内核程序中，当系统内存不够用时，内存管理系统就会通过调用内存回收算法的方式删除最近没有用过的内存，将这些内存从物理内存中清除，这样可以增加物理内存的容量。所以在 Android 系统中也借用了这种机制，当内存管理系统回收内存时会调用函数 ashmem_shrink 以执行内存回收操作。函数 ashmem_shrink 在文件“kernel/goldfish/ashmem.c”中实现，具体的实现代码如下所示：

```
static int ashmem_shrink(struct shrinker *s, struct shrink_control *sc)
{
    struct ashmem_range *range, *next;
    /* We might recurse into filesystem code, so bail out if necessary */
    if (sc->nr_to_scan && !(sc->gfp_mask & __GFP_FS))
        return -1;
    if (!sc->nr_to_scan)
        return lru_count;
    mutex_lock(&ashmem_mutex);
    list_for_each_entry_safe(range, next, &ashmem_lru_list, lru) {
        loff_t start = range->pgstart * PAGE_SIZE;
        loff_t end = (range->pgend + 1) * PAGE_SIZE;
        do_fallocate(range->asma->file,
                FALLOC_FL_PUNCH_HOLE | FALLOC_FL_KEEP_SIZE,
                start, end - start);
        range->purged = ASHMEM_WAS_PURGED;
        lru_del(range);
        sc->nr_to_scan -= range_size(range);
        if (sc->nr_to_scan <= 0)
            break;
    }
    mutex_unlock(&ashmem_mutex);
    return lru_count;
}
```

5.3 C++访问接口层详解

如果想在 Android 进程之间共享一个完整的匿名共享内存块，可以通过调用接口 MemoryHeapBase 来实现；如果只是想在进程之间共享匿名共享内存块中的一部分时，可以通过调用接口 MemoryBase 来实现。在本节的内容中，将详细讲解 C++访问接口层的基本知识。

5.3.1 接口 MemoryBase

接口 MemoryBase 以接口 MemoryHeapBase 为基础，这两个接口都可以作为一个 Binder 对象在进程之间进行传输。因为接口 MemoryHeapBase 是一个 Binder 对象，所以拥有 Server 端对象（必须实现一个 BnInterface 接口）和 Client 端引用（必须要实现一个 BpInterface 接口）的概念。

1. 服务器端实现

在 Server 端的实现过程中，接口 MemoryHeapBase 可以将所有涉及到的类分为以下 3 种类型。

- 业务相关类：即跟匿名共享内存操作相关的类，包括 MemoryHeapBase、BnMemoryHeap、IMemoryHeap。
- Binder 进程通信类：即和 Binder 进程通信机制相关的类，包括 IInterface、BnInterface、IBinder、BBinder、ProcessState 和 IPCThreadState。
- 智能指针类：RefBase。

在上述 3 种类型中，Binder 进程通信类和智能指针类将在本书后面的章节中进行讲解。在接口 IMemoryBase 中定义了和操作匿名共享内存的几个方法，此接口在文件“frameworks\native\include\binder\IMemory.h”中定义，具体实现代码如下所示：

```
class IMemoryHeap : public IInterface
{
public:
    DECLARE_META_INTERFACE(MemoryHeap);

    // flags returned by getFlags()
    enum {
        READ_ONLY   = 0x00000001
    };

    virtual int         getHeapID() const = 0;
    virtual void*       getBase() const = 0;
    virtual size_t      getSize() const = 0;
    virtual uint32_t    getFlags() const = 0;
    virtual uint32_t    getOffset() const = 0;

    // these are there just for backward source compatibility
    int32_t heapID() const { return getHeapID(); }
    void*   base() const  { return getBase(); }
    size_t  virtualSize() const { return getSize(); }
}
```

在上述定义代码中，有如下 3 个重要的成员函数。

- getHeapID：功能是获得匿名共享内存块的打开文件描述符。
- getBase：功能是获得匿名共享内存块的基地址，通过这个地址可以在程序中直接访问这块共享内存。
- getSize：功能是获得匿名共享内存块的大小。

类 BnMemoryHeap 是一个本地对象类，当 Client 端引用请求 Server 端对象执行命令时，Binder 系统就会调用类 BnMemoryHeap 的成员函数 onTransact 执行具体的命令。函数 onTransact 在文件“frameworks\native\libs\binder\IMemory.cpp”中定义，具体实现代码如下所示：

```
status_t BnMemory::onTransact(
    uint32_t code, const Parcel& data, Parcel* reply, uint32_t flags)
{
    switch(code) {
        case GET_MEMORY: {
            CHECK_INTERFACE(IMemory, data, reply);
            ssize_t offset;
            size_t size;
            reply->writeStrongBinder( getMemory(&offset, &size)->asBinder() );
            reply->writeInt32(offset);
            reply->writeInt32(size);
            return NO_ERROR;
        } break;
        default:
            return BBinder::onTransact(code, data, reply, flags);
    }
}
```

类 MemoryHeapBase 继承了类 BnMemoryHeap，作为 Binder 机制中的 Server 角色需要实现 IMemoryBase 接口，主要功能是实现类 IMemoryBase 中列出的成员函数，描述了一块匿名共享内存服务。类在文件“frameworks\native\include\binder\MemoryHeapBase.h”中定义，具体实现代码如下所示：

```
class MemoryHeapBase : public virtual BnMemoryHeap
{
public:
    enum {
        READ_ONLY = IMemoryHeap::READ_ONLY,
        // memory won't be mapped locally, but will be mapped in the remote
        // process.
        DONT_MAP_LOCALLY = 0x00000100,
        NO_CACHING = 0x00000200
```

```
    };

    /*
     * maps the memory referenced by fd. but DOESN'T take ownership
     * of the filedescriptor (it makes a copy with dup()
     */
    MemoryHeapBase(int fd, size_t size, uint32_t flags = 0, uint32_t offset = 0);

    /*
     * maps memory from the given device
     */
    MemoryHeapBase(const char* device, size_t size = 0, uint32_t flags = 0);

    /*
     * maps memory from ashmem, with the given name for debugging
     */
    MemoryHeapBase(size_t size, uint32_t flags = 0, char const* name = NULL);
    virtual ~MemoryHeapBase();

    /* implement IMemoryHeap interface */
    virtual int         getHeapID() const;

    /* virtual address of the heap. returns MAP_FAILED in case of error */
    virtual void*       getBase() const;

    virtual size_t      getSize() const;
    virtual uint32_t    getFlags() const;
    virtual uint32_t    getOffset() const;

    const char*         getDevice() const;

    /* this closes this heap -- use carefully */
    void dispose();

    /* this is only needed as a workaround, use only if you know
     * what you are doing */
    status_t setDevice(const char* device) {
        if (mDevice == 0)
            mDevice = device;
        return mDevice ? NO_ERROR : ALREADY_EXISTS;
    }

protected:
        MemoryHeapBase();
    // init() takes ownership of fd
    status_t init(int fd, void *base, int size,
            int flags = 0, const char* device = NULL);

private:
    status_t mapfd(int fd, size_t size, uint32_t offset = 0);

    int         mFD;//是一个文件描述符，是在打开设备文件/dev/ashmem后得到的，能够描述一个匿名共享内存块
    size_t      mSize;//内存块的大小
    void*       mBase; //内存块的映射地址
    uint32_t     mFlags;//内存块的访问保护位
    const char* mDevice;
    bool        mNeedUnmap;
    uint32_t     mOffset;
};
```

类MemoryHeapBase在文件“frameworks\native\libs\binder\MemoryHeapBase.cpp”中实现，其核心功能是包含了一块匿名共享内存。具体实现代码如下所示：

```
MemoryHeapBase::MemoryHeapBase(size_t size, uint32_t flags, char const * name)
    : mFD(-1), mSize(0), mBase(MAP_FAILED), mFlags(flags),
      mDevice(0), mNeedUnmap(false), mOffset(0)
{
    const size_t pagesize = getpagesize();
```

```
        size = ((size + pagesize-1) & ~(pagesize-1));
        int fd = ashmem_create_region(name == NULL ? "MemoryHeapBase" : name, size);
        ALOGE_IF(fd<0, "error creating ashmem region: %s", strerror(errno));
        if (fd >= 0) {
            if (mapfd(fd, size) == NO_ERROR) {
                if (flags & READ_ONLY) {
                    ashmem_set_prot_region(fd, PROT_READ);
                }
            }
        }
}
```

在上述代码中，各个参数的具体说明如下所示。

- size：表示要创建的匿名共享内存的大小。
- flags：设置这块匿名共享内存的属性，例如可读写、只读等。
- name：此参数只是作为调试信息使用的，用于标识匿名共享内存的名字，可以是空值。

接下来看 MemoryHeapBase 的成员函数 mapfd，其功能是将得到的匿名共享内存的文件描述符映射到进程地址空间。函数 mapfd 在文件"frameworks\native\libs\binder\MemoryHeapBase.cpp"中定义，具体实现代码如下所示：

```
status_t MemoryHeapBase::mapfd(int fd, size_t size, uint32_t offset)
{
    if (size == 0) {
        // try to figure out the size automatically
#ifdef HAVE_ANDROID_OS
        // first try the PMEM ioctl
        pmem_region reg;
        int err = ioctl(fd, PMEM_GET_TOTAL_SIZE, &reg);
        if (err == 0)
            size = reg.len;
#endif
        if (size == 0) { // try fstat
            struct stat sb;
            if (fstat(fd, &sb) == 0)
                size = sb.st_size;
        }
        // if it didn't work, let mmap() fail.
    }

    if ((mFlags & DONT_MAP_LOCALLY) == 0) {//条件为 true 时执行系统调用 mmap 来执行内存映射的操作
        void* base = (uint8_t*)mmap(
0,// 表示由内核来决定这个匿名共享内存文件在进程地址空间的起始位置
 size,// 表示要映射的匿名共享内文件的大小
PROT_READ|PROT_WRITE,// 表示这个匿名共享内存是可读写的
 MAP_SHARED,
 fd, //指定要映射的匿名共享内存的文件描述符
 offset//表示要从这个文件的哪个偏移位置开始映射
);
        if (base == MAP_FAILED) {
            ALOGE("mmap(fd=%d, size=%u) failed (%s)",
                    fd, uint32_t(size), strerror(errno));
            close(fd);
            return -errno;
        }
        //ALOGD("mmap(fd=%d, base=%p, size=%lu)", fd, base, size);
        mBase = base;
        mNeedUnmap = true;
    } else  {
        mBase = 0; // not MAP_FAILED
        mNeedUnmap = false;
    }
    mFD = fd;
    mSize = size;
    mOffset = offset;
    return NO_ERROR;
}
```

这样在调用个函数 mapfd 后，会进入到内核空间的 ashmem 驱动程序模块中执行函数 ashmem_map。最后看成员函数 getHeapID、getBase 和 getSize 的具体实现，具体实现代码如下所示：

```
int MemoryHeapBase::getHeapID() const {
    return mFD;
}
void* MemoryHeapBase::getBase() const {
    return mBase;
}
size_t MemoryHeapBase::getSize() const {
    return mSize;
}
```

2. 客户端实现

在客户端的实现过程中，接口 MemoryHeapBase 可以将所有涉及到的类分为如下所示的 3 种类型。

- 业务相关类：即跟匿名共享内存操作相关的类，包括 BpMemoryHeap 和 IMemoryHeap。
- Binder 进程通信类：即和 Binder 进程通信机制相关的类，包括 IInterface、BpInterface、IBinder、BpBinder、ProcessState、BpRefBase 和 IPCThreadState。
- 智能指针类：RefBase。

在上述 3 种类型中，Binder 进程通信类和智能指针类将在本书后面的章节中进行讲解，在本章将重点介绍业务相关类。

类 BpMemoryHeap 是类 MemoryHeapBase 在 Client 端进程的远接接口类，当 Client 端进程从 Service Manager 获得了一个 MemoryHeapBase 对象的引用后，会在本地创建一个 BpMemoryHeap 对象来表示这个引用。类 BpMemoryHeap 是从 RefBase 类继承下来的，也要实现 IMemoryHeap 接口，可以和智能指针来结合使用。

类 BpMemoryHeap 在文件"frameworks\native\libs\binder\IMemory.cpp"中定义，具体实现代码如下所示：

```
class BpMemoryHeap : public BpInterface<IMemoryHeap>
{
public:
    BpMemoryHeap(const sp<IBinder>& impl);
    virtual ~BpMemoryHeap();

    virtual int getHeapID() const;
    virtual void* getBase() const;
    virtual size_t getSize() const;
    virtual uint32_t getFlags() const;
    virtual uint32_t getOffset() const;

private:
    friend class IMemory;
    friend class HeapCache;

    // for debugging in this module
    static inline sp<IMemoryHeap> find_heap(const sp<IBinder>& binder) {
        return gHeapCache->find_heap(binder);
    }
    static inline void free_heap(const sp<IBinder>& binder) {
        gHeapCache->free_heap(binder);
    }
    static inline sp<IMemoryHeap> get_heap(const sp<IBinder>& binder) {
        return gHeapCache->get_heap(binder);
    }
    static inline void dump_heaps() {
        gHeapCache->dump_heaps();
```

```
    }

    void assertMapped() const;
    void assertReallyMapped() const;

    mutable volatile int32_t mHeapId;
    mutable void*       mBase;
    mutable size_t      mSize;
    mutable uint32_t    mFlags;
    mutable uint32_t    mOffset;
    mutable bool        mRealHeap;
    mutable Mutex       mLock;
```

类BpMemoryHeap对应的构造函数是BpMemoryHeap，具体实现代码如下所示：

```
BpMemoryHeap::BpMemoryHeap(const sp<IBinder>& impl)
    : BpInterface<IMemoryHeap>(impl),
        mHeapId(-1), mBase(MAP_FAILED), mSize(0), mFlags(0), mRealHeap(false)
{
}
```

成员函数getHeapID、getBase和getSize的实现代码如下所示：

```
int BpMemoryHeap::getHeapID() const {
    assertMapped();
    return mHeapId;
}

void* BpMemoryHeap::getBase() const {
    assertMapped();
    return mBase;
}

size_t BpMemoryHeap::getSize() const {
    assertMapped();
    return mSize;
}
```

在使用上述成员函数之前，通过调用函数assertMapped来确保在Client端已经准备好了匿名共享内存。函数assertMapped在文件“frameworks\native\libs\binder\IMemory.cpp”中定义，具体实现代码如下所示：

```
void BpMemoryHeap::assertMapped() const
{
    if (mHeapId == -1) {
        sp<IBinder> binder(const_cast<BpMemoryHeap*>(this)->asBinder());
        sp<BpMemoryHeap> heap(static_cast<BpMemoryHeap*>(find_heap(binder).get()));
        heap->assertReallyMapped();
        if (heap->mBase != MAP_FAILED) {
            Mutex::Autolock _l(mLock);
            if (mHeapId == -1) {
                mBase   = heap->mBase;
                mSize   = heap->mSize;
                android_atomic_write( dup( heap->mHeapId ), &mHeapId );
            }
        } else {
            // something went wrong
            free_heap(binder);
        }
    }
}
```

类HeapCache在文件“frameworks\native\libs\binder\IMemory.cpp”中定义，具体实现代码如下所示：

```
class HeapCache : public IBinder::DeathRecipient
{
public:
```

```
    HeapCache();
    virtual ~HeapCache();

    virtual void binderDied(const wp<IBinder>& who);

    sp<IMemoryHeap> find_heap(const sp<IBinder>& binder);
    void free_heap(const sp<IBinder>& binder);
    sp<IMemoryHeap> get_heap(const sp<IBinder>& binder);
    void dump_heaps();

private:
    // For IMemory.cpp
    struct heap_info_t {
        sp<IMemoryHeap> heap;
        int32_t         count;
    };

    void free_heap(const wp<IBinder>& binder);

    Mutex mHeapCacheLock;
    KeyedVector< wp<IBinder>, heap_info_t > mHeapCache;
}
```

在上述代码中定义了成员变量mHeapCache，功能是维护进程内的所有BpMemoryHeap对象。另外还提供了函数find_heap和函数get_heap来查找内部所维护的BpMemoryHeap对象，这两个函数的具体说明如下所示。

- 函数find_heap：如果在mHeapCache找不到相应的BpMemoryHeap对象，则把BpMemoryHeap对象加入到mHeapCache中。
- 函数get_heap：不会自动把BpMemoryHeap对象加入到mHeapCache中。

接下来看函数find_heap，首先以传进来的参数binder作为关键字在mHeapCache中查找，查找是否存在对应的heap_info对象info。

- 如果有：增加引用计数info.count的值，表示此BpBinder对象多了一个使用者。
- 如果没有：创建一个放到mHeapCache中的heap_info对象info。

函数find_heap在文件“frameworks\native\libs\binder\IMemory.cpp”中定义，具体实现代码如下所示：

```
sp<IMemoryHeap> HeapCache::find_heap(const sp<IBinder>& binder)
{
    Mutex::Autolock _l(mHeapCacheLock);
    ssize_t i = mHeapCache.indexOfKey(binder);
    if (i>=0) {
        heap_info_t& info = mHeapCache.editValueAt(i);
        ALOGD_IF(VERBOSE,
                "found binder=%p, heap=%p, size=%d, fd=%d, count=%d",
                binder.get(), info.heap.get(),
                static_cast<BpMemoryHeap*>(info.heap.get())->mSize,
                static_cast<BpMemoryHeap*>(info.heap.get())->mHeapId,
                info.count);
        android_atomic_inc(&info.count);
        return info.heap;
    } else {
        heap_info_t info;
        info.heap = interface_cast<IMemoryHeap>(binder);
        info.count = 1;
        //ALOGD("adding binder=%p, heap=%p, count=%d",
        //      binder.get(), info.heap.get(), info.count);
        mHeapCache.add(binder, info);
        return info.heap;
    }
}
```

由上述实现代码可知，函数 find_heap 是 BpMemoryHeap 的成员函数，能够调用全局变量 gHeapCache 执行查找的操作。对应的实现代码如下所示：

```
class BpMemoryHeap : public BpInterface<IMemoryHeap>
{
......
private:
    static inline sp<IMemoryHeap> find_heap(const sp<IBinder>& binder) {
        return gHeapCache->find_heap(binder);
    }
```

通过调用函数 find_heap 得到 BpMemoryHeap 对象中的函数 assertReallyMapped，这样可以确认它内部的匿名共享内存是否已经映射到进程空间。函数 assertReallyMapped 在文件“frameworks\native\libs\binder\IMemory.cpp”中定义，具体实现代码如下所示：

```
void BpMemoryHeap::assertReallyMapped() const
{
    if (mHeapId == -1) {

        // remote call without mLock held, worse case scenario, we end up
        // calling transact() from multiple threads, but that's not a problem,
        // only mmap below must be in the critical section.

        Parcel data, reply;
        data.writeInterfaceToken(IMemoryHeap::getInterfaceDescriptor());
        status_t err = remote()->transact(HEAP_ID, data, &reply);
        int parcel_fd = reply.readFileDescriptor();
        ssize_t size = reply.readInt32();
        uint32_t flags = reply.readInt32();
        uint32_t offset = reply.readInt32();

        ALOGE_IF(err, "binder=%p transaction failed fd=%d, size=%ld, err=%d (%s)",
                asBinder().get(), parcel_fd, size, err, strerror(-err));

        int fd = dup( parcel_fd );
        ALOGE_IF(fd==-1, "cannot dup fd=%d, size=%ld, err=%d (%s)",
                parcel_fd, size, err, strerror(errno));

        int access = PROT_READ;
        if (!(flags & READ_ONLY)) {
            access |= PROT_WRITE;
        }

        Mutex::Autolock _l(mLock);
        if (mHeapId == -1) {
            mRealHeap = true;
            mBase = mmap(0, size, access, MAP_SHARED, fd, offset);
            if (mBase == MAP_FAILED) {
                ALOGE("cannot map BpMemoryHeap (binder=%p), size=%ld, fd=%d (%s)",
                        asBinder().get(), size, fd, strerror(errno));
                close(fd);
            } else {
                mSize = size;
                mFlags = flags;
                mOffset = offset;
                android_atomic_write(fd, &mHeapId);
            }
        }
    }
}
```

5.3.2　接口 MemoryBase

接口 MemoryBase 是建立在接口 MemoryHeapBase 基础上的，两者都可以作为一个 Binder 对象在进程之间实现数据共享。

1. 在Server端的实现

首先分析类MemoryBase在Server端的实现，MemoryBase在Server端只是简单地封装了MemoryHeapBase的实现。类MemoryBase在Server端的实现跟类MemoryHeapBase在Server端的实现类似，只需在整个类图接结构中实现如下转换即可。

- 把类IMemory换成类ImemoryHeap。
- 把类BnMemory换成类BnMemoryHeap。
- 把类MemoryBase换成类MemoryHeapBase。

类IMemory在文件"frameworks\native\include\binder\IMemory.h"中实现，功能是定义类MemoryBase所需要的实现接口。类IMemory的实现代码如下所示：

```
class IMemory : public IInterface
{
public:
    DECLARE_META_INTERFACE(Memory);
    virtual sp<IMemoryHeap> getMemory(ssize_t* offset=0, size_t* size=0) const = 0;
    // helpers
    void* fastPointer(const sp<IBinder>& heap, ssize_t offset) const;
    void* pointer() const;
    size_t size() const;
    ssize_t offset() const;
}
```

在类IMemory中定义了如下所示的成员函数。

- getMemory：功能是获取内部MemoryHeapBase对象的IMemoryHeap接口。
- pointer()：功能是获取内部所维护的匿名共享内存的基地址。
- size()：功能是获取内部所维护的匿名共享内存的大小。
- offset()：功能是获取内部所维护的匿名共享内存在整个匿名共享内存中的偏移量。

类IMemory在本身定义过程中实现了3个成员函数：pointer、size和offset，其子类MemoryBase只需实现成员函数getMemory即可。类IMemory的具体实现在文件"frameworks\native\libs\binder\IMemory.cpp"中定义，具体实现代码如下所示：

```
void* IMemory::pointer() const {
    ssize_t offset;
    sp<IMemoryHeap> heap = getMemory(&offset);
    void* const base = heap!=0 ? heap->base() : MAP_FAILED;
    if (base == MAP_FAILED)
        return 0;
    return static_cast<char*>(base) + offset;
}

size_t IMemory::size() const {
    size_t size;
    getMemory(NULL, &size);
    return size;
}
ssize_t IMemory::offset() const {
    ssize_t offset;
    getMemory(&offset);
    return offset;
}
```

类MemoryBase是一个本地Binder对象类，在文件"frameworks\native\include\binder\MemoryBase.h"中声明，具体实现代码如下所示：

```
class MemoryBase : public BnMemory
{
```

```
public:
    MemoryBase(const sp<IMemoryHeap>& heap, ssize_t offset, size_t size);
    virtual ~MemoryBase();
    virtual sp<IMemoryHeap> getMemory(ssize_t* offset, size_t* size) const;
protected:
    size_t getSize() const { return mSize; }
    ssize_t getOffset() const { return mOffset; }
    const sp<IMemoryHeap>& getHeap() const { return mHeap; }
private:
    size_t          mSize;
    ssize_t         mOffset;
    sp<IMemoryHeap> mHeap;
}
} // namespace android
#endif // ANDROID_MEMORY_BASE_H
```

类 MemoryBase 的具体实现在文件“frameworks\native\libs\binder\MemoryBase.cpp”中定义，具体实现代码如下所示：

```
MemoryBase::MemoryBase(const
 sp<IMemoryHeap>& heap,//指向 MemoryHeapBase 对象，真正的匿名共享内存就是由它来维护的
 ssize_t offset,//表示这个 MemoryBase 对象所要维护的这部分匿名共享内存在整个匿名共享内存块中的起始位置
 size_t size//表示这个 MemoryBase 对象所要维护的这部分匿名共享内存的大小
)
    : mSize(size), mOffset(offset), mHeap(heap)
{
}
//功能是返回内部的 MemoryHeapBase 对象的 IMemoryHeap 接口
//如果传进来的参数 offset 和 size 不为 NULL
//会把其内部维护的这部分匿名共享内存，在整个匿名共享内存块中的偏移位置
//以及这部分匿名共享内存的大小返回给调用者
sp<IMemoryHeap> MemoryBase::getMemory(ssize_t* offset, size_t* size) const
{
    if (offset) *offset = mOffset;
    if (size)   *size = mSize;
    return mHeap;
}
```

2. MemoryBase 类在 Client 端的实现

再来看 MemoryBase 类在 Client 端的实现，类 MemoryBase 在 Client 端的实现与类 MemoryHeapBase 在 Client 端的实现类似，只需要进行如下所示的类转换即可成为 MemoryHeapBase 在 Client 端的实现。

- 把类 IMemory 换成类 ImemoryHeap。
- 把类 BpMemory 换成类 BpMemoryHeap。

类 BpMemory 用于描述类 MemoryBase 服务的代理对象，在文件“frameworks\native\libs\binder\IMemory.cpp”中定义，具体实现代码如下所示：

```
class BpMemory : public BpInterface<IMemory>
{
public:
    BpMemory(const sp<IBinder>& impl);
    virtual ~BpMemory();
    virtual sp<IMemoryHeap> getMemory(ssize_t* offset=0, size_t* size=0) const;

private:
    mutable sp<IMemoryHeap> mHeap;// 类型为 IMemoryHeap，它指向的是一个 BpMemoryHeap 对象
    mutable ssize_t mOffset;// 表示 BpMemory 对象所要维护的匿名共享内存在整个匿名共享内存块中的起始位置
    mutable size_t mSize;// 表示这个 BpMemory 对象所要维护的这部分匿名共享内存的大小
}
```

类 BpMemory 中的成员函数 getMemory 在文件“frameworks\native\libs\binder\IMemory.cpp”

中定义，具体实现代码如下所示：

```
sp<IMemoryHeap> BpMemory::getMemory(ssize_t* offset, size_t* size) const
{
    if (mHeap == 0) {
        Parcel data, reply;
        data.writeInterfaceToken(IMemory::getInterfaceDescriptor());
        if (remote()->transact(GET_MEMORY, data, &reply) == NO_ERROR) {
            sp<IBinder> heap = reply.readStrongBinder();
            ssize_t o = reply.readInt32();
            size_t s = reply.readInt32();
            if (heap != 0) {
                mHeap = interface_cast<IMemoryHeap>(heap);
                if (mHeap != 0) {
                    mOffset = o;
                    mSize = s;
                }
            }
        }
    }
    if (offset) *offset = mOffset;
    if (size) *size = mSize;
    return mHeap;
}
```

如果成员变量 mHeap 的值为 NULL，表示此 BpMemory 对象还没有建立好匿名共享内存，此时会调用一个 Binder 进程去 Server 端请求匿名共享内存信息。通过引用信息中的 Server 端的 MemoryHeapBase 对象的引用 heap，可以在 Client 端进程中创建一个 BpMemoryHeap 远程接口，最后将这个 BpMemoryHeap 远程接口保存在成员变量 mHeap 中，同时从 Server 端获得的信息还包括这块匿名共享内存在整个匿名共享内存中的偏移位置以及大小。

5.4 Java 访问接口层详解

分析完匿名共享内存的 C++访问接口层后，在本节开始分析其 Java 访问接口层的实现过程。在 Android 应用程序框架层中，通过使用接口 MemoryFile 来封装匿名共享内存文件的创建和使用。接口 MemoryFile 在文件“frameworks/base/core/java/android/os/MemoryFile.java”中定义，具体实现代码如下所示：

```
public class MemoryFile
{
    private static String TAG = "MemoryFile";

    // mmap(2) protection flags from <sys/mman.h>
    private static final int PROT_READ = 0x1;
    private static final int PROT_WRITE = 0x2;

    private static native FileDescriptor native_open(String name, int length) throws IOException;
    // returns memory address for ashmem region
    private static native int native_mmap(FileDescriptor fd, int length, int mode)
            throws IOException;
    private static native void native_munmap(int addr, int length) throws IOException;
    private static native void native_close(FileDescriptor fd);
    private static native int native_read(FileDescriptor fd, int address, byte[] buffer,
            int srcOffset, int destOffset, int count, boolean isUnpinned) throws IOException;
    private static native void native_write(FileDescriptor fd, int address, byte[] buffer,
            int srcOffset, int destOffset, int count, boolean isUnpinned) throws IOException;
    private static native void native_pin(FileDescriptor fd, boolean pin) throws IOException;
    private static native int native_get_size(FileDescriptor fd) throws IOException;

    private FileDescriptor mFD;        // ashmem file descriptor
    private int mAddress;   // address of ashmem memory
```

```
    private int mLength;    // total length of our ashmem region
    private boolean mAllowPurging = false;  // true if our ashmem region is unpinned

    /**
     * Allocates a new ashmem region. The region is initially not purgable.
     *
     * @param name optional name for the file (can be null).
     * @param length of the memory file in bytes.
     * @throws IOException if the memory file could not be created.
     */
    public MemoryFile(String name, int length) throws IOException {
        mLength = length;
        mFD = native_open(name, length);
        if (length > 0) {
            mAddress = native_mmap(mFD, length, PROT_READ | PROT_WRITE);
        } else {
            mAddress = 0;
        }
    }
```

在上述代码中，构造方法 MemoryFile 以指定的字符串调用了 JNI 方法 native_open，目的是建立一个匿名共享内存文件，这样可以得到一个文件描述符。然后使用这个文件描述符为参数调用 JNI 方法 natvie_mmap，并把匿名共享内存文件映射到进程空间中，这样就可以通过映射得到地址空间的方式直接访问内存数据。

再看 JNI 函数 android_os_MemoryFile_get_size，此函数在文件"frameworks\base\core\jni\android_os_MemoryFile.cpp"中定义，具体实现代码如下所示：

```
static jint android_os_MemoryFile_get_size(JNIEnv* env, jobject clazz,
        jobject fileDescriptor) {
    int fd = jniGetFDFromFileDescriptor(env, fileDescriptor);
    // Use ASHMEM_GET_SIZE to find out if the fd refers to an ashmem region.
    // ASHMEM_GET_SIZE should succeed for all ashmem regions, and the kernel
    // should return ENOTTY for all other valid file descriptors
    int result = ashmem_get_size_region(fd);
    if (result < 0) {
        if (errno == ENOTTY) {
            // ENOTTY means that the ioctl does not apply to this object,
            // i.e., it is not an ashmem region.
            return (jint) -1;
        }
        // Some other error, throw exception
        jniThrowIOException(env, errno);
        return (jint) -1;
    }
    return (jint) result;
}
```

再看 JNI 函数 android_os_MemoryFile_open，此函数在文件"frameworks\base\core\jni\android_os_MemoryFile.cpp"中定义，具体实现代码如下所示：

```
static jobject android_os_MemoryFile_open(JNIEnv* env, jobject clazz, jstring name, jint length)
{
    const char* namestr = (name ? env->GetStringUTFChars(name, NULL) : NULL);

    int result = ashmem_create_region(namestr, length);

    if (name)
        env->ReleaseStringUTFChars(name, namestr);

    if (result < 0) {
        jniThrowException(env, "java/io/IOException", "ashmem_create_region failed");
        return NULL;
    }
```

```
    return jniCreateFileDescriptor(env, result);
}
```

再看 JNI 函数 android_os_MemoryFile_mmap，此函数在文件“frameworks\base\core\jni\android_os_MemoryFile.cpp”中定义，具体实现代码如下所示：

```
static jint android_os_MemoryFile_mmap(JNIEnv* env, jobject clazz, jobject fileDescriptor,
        jint length, jint prot)
{
    int fd = jniGetFDFromFileDescriptor(env, fileDescriptor);
    jint result = (jint)mmap(NULL, length, prot, MAP_SHARED, fd, 0);
    if (!result)
        jniThrowException(env, "java/io/IOException", "mmap failed");
    return result;
}
```

在文件“frameworks/base/core/java/android/os/MemoryFile.java”中，再看类 MemoryFile 的成员函数 readBytes，功能是读取某一块匿名共享内存的内容。具体实现代码如下所示：

```
    public int readBytes(byte[] buffer, int srcOffset, int destOffset, int count)
            throws IOException {
        if (isDeactivated()) {
            throw new IOException("Can't read from deactivated memory file.");
        }
        if (destOffset < 0 || destOffset > buffer.length || count < 0
                || count > buffer.length - destOffset
                || srcOffset < 0 || srcOffset > mLength
                || count > mLength - srcOffset) {
            throw new IndexOutOfBoundsException();
        }
        return native_read(mFD, mAddress, buffer, srcOffset, destOffset, count, mAllowPurging);
    }
```

在文件“frameworks/base/core/java/android/os/MemoryFile.java”中，再看类 MemoryFile 的成员函数 writeBytes，功能是写入某一块匿名共享内存的内容。具体实现代码如下所示：

```
public void writeBytes(byte[] buffer, int srcOffset, int destOffset, int count)
        throws IOException {
    if (isDeactivated()) {
        throw new IOException("Can't write to deactivated memory file.");
    }
    if (srcOffset < 0 || srcOffset > buffer.length || count < 0
            || count > buffer.length - srcOffset
            || destOffset < 0 || destOffset > mLength
            || count > mLength - destOffset) {
        throw new IndexOutOfBoundsException();
    }
    native_write(mFD, mAddress, buffer, srcOffset, destOffset, count, mAllowPurging);
}
```

在文件“frameworks/base/core/java/android/os/MemoryFile.java”中，再看类 MemoryFile 的成员函数 isDeactivated，功能是保证匿名共享内存已经被映射到进程的地址空间中。具体实现代码如下所示：

```
void deactivate() {
        if (!isDeactivated()) {
            try {
                native_munmap(mAddress, mLength);
                mAddress = 0;
            } catch (IOException ex) {
                Log.e(TAG, ex.toString());
            }
    }
}
private boolean isDeactivated() {
```

```
        return mAddress == 0;
    }
```

JNI 函数 native_read 和 native_write 分别由位于 C++层的函数 android_os_MemoryFile_read 和 android_os_MemoryFile_write 实现，这两个 C++的函数在文件 frameworks\base\core\jni\android_os_MemoryFile.cpp 中定义，具体实现代码如下所示：

```
static jint android_os_MemoryFile_read(JNIEnv* env, jobject clazz,
        jobject fileDescriptor, jint address, jbyteArray buffer, jint srcOffset, jint destOffset,
        jint count, jboolean unpinned)
{
    int fd = jniGetFDFromFileDescriptor(env, fileDescriptor);
    if (unpinned && ashmem_pin_region(fd, 0, 0) == ASHMEM_WAS_PURGED) {
        ashmem_unpin_region(fd, 0, 0);
        jniThrowException(env, "java/io/IOException", "ashmem region was purged");
        return -1;
    }
    env->SetByteArrayRegion(buffer, destOffset, count, (const jbyte *)address + srcOffset);
    if (unpinned) {
        ashmem_unpin_region(fd, 0, 0);
    }
    return count;
}
static jint android_os_MemoryFile_write(JNIEnv* env, jobject clazz,
        jobject fileDescriptor, jint address, jbyteArray buffer, jint srcOffset, jint destOffset,
        jint count, jboolean unpinned)
{
    int fd = jniGetFDFromFileDescriptor(env, fileDescriptor);
    if (unpinned && ashmem_pin_region(fd, 0, 0) == ASHMEM_WAS_PURGED) {
        ashmem_unpin_region(fd, 0, 0);
        jniThrowException(env, "java/io/IOException", "ashmem region was purged");
        return -1;
    }
    env->GetByteArrayRegion(buffer, srcOffset, count, (jbyte *)address + destOffset);
    if (unpinned) {
        ashmem_unpin_region(fd, 0, 0);
    }
    return count;
}
```

第 6 章　Android 程序的生命周期管理

Android 程序如同自然界的生物一样，有自己的生命周期。应用程序的生命周期就是程序的存活时间，即在什么时间内有效。在本章的内容中，将详细讲解 Android 生命周期管理的基本知识。

6.1 Android 程序的生命周期

Android 是一构建在 Linux 之上的开源移动开发平台，里面的每个程序都各自独立运行在 Linux 进程中。当一个程序或其某些部分被请求时，它的进程就“出生”了。当这个程序没有必要再运行下去且系统需要回收这个进程的内存用于其他程序时，这个进程就“死亡”了。由此可以看出，Android 程序的生命周期是由系统控制而非程序自身直接控制。要想编写好某种类型的程序，或者完成某种平台下的程序开发工作，最关键的就是要弄清楚这种类型的程序或整个平台下的程序的一般工作模式，并且要熟记在心。在 Android 系统中，程序的生命周期控制就是属于这个范畴。在本节的内容中，将详细讲解 Android 程序的生命周期的基本知识。

6.1.1　进程和线程

当第一次运行某个组件的时候，Android 就启动了一个进程。在默认情况下，所有的组件和程序运行在这个进程和线程中。当然，也可以设置组件在其他的进程或者线程中运行。

在 Android 系统中，由 manifest file 控制组件运行的进程。在组件的节点<ctivity>，<service>，<receiver>和 <provider>中都包含一个 process 属性，通过这个属性可以设置组件运行的进程。

- 可以配置组件在一个独立进程运行，或者多个组件在同一个进程运行。
- 如果这些程序共享一个 User ID 并给定同样的权限，可以让多个程序在一个进程中运行。

另外，在<application>节点中也包含了 process 属性，用来设置程序中所有组件的默认进程。所有的组件在此进程的主线程中实例化，系统对这些组件的调用从主线程中分离。并非每个对象都会从主线程中分离。一般来说，响应例如 View.onKeyDown()用户操作的方法和通知的方法也在主线程中运行。这说明当组件被系统调用时，不应该长时间运行或者阻塞操作（如网络操作或者计算大量数据），因为这样会阻塞进程中的其他组件。此时可以把这类操作从主线程中分离出来。当更加常用的进程无法获取足够内存时，Android 可能会关闭这些不常用的进程。当下次启动程序时，会重新启动这些进程。

当决定哪个进程需要被关闭的时候，Android 会考虑哪个对用户更加有用。如 Android 会倾向于关闭一个长期不显示在界面的进程来支持一个经常显示在界面的进程。是否关闭一个进程决定于组件在进程中的状态。

即使为组件分配了不同的进程，有时候也需要再重新分配线程。比如用户界面需要很快对用户进行响应，因此某些费时的操作，如网络连接、下载或者非常占用服务器时间的操作应该放到其他线程。

在 Android 中提供了很多方便的管理线程的方法，例如通过 Looper 在线程中运行一个消息循环，使用 Handler 传递一个消息，使用 HandlerThread 创建一个带有消息循环的线程。

6.1.2　进程的类型

开发者必须理解不同的应用程序组件，尤其是 Activity、Service 和 Intent Receiver。了解这些组件是如何影响应用程序的生命周期的，这非常重要。如果不正确地使用这些组件，可能会导致系统终止正在执行重要任务的应用程序进程。

在 Android 系统中，根据进程中当前处于活动状态组件的重要程度为依据对进程进行分类。进程的优先级可能也会根据该进程与其他进程的依赖关系而增长。例如，如果进程 A 通过在进程 B 中设置 Context.BIND_AUTO_CREATE 标记或使用 ContentProvider 被绑定到一个服务（Service），那么进程 B 在分类时至少要被看成与进程 A 同等重要。Android 系统中的进程的类型被分为如下 5 种。

（1）前台进程（Foreground）。

与用户当前正在做的事情密切相关。不同的应用程序组件能够通过不同的方法将它的宿主进程移到前台。在如下任何一个条件下：进程正在屏幕的最前端运行一个与用户交互的活动（Activity），它的 onResume()方法被调用；或进程有一正在运行的 Intent Receiver（它的 Intent Receiver.onReceive()方法正在执行）；或进程有一个服务（Service），并且在服务的某个回调函数（Service.onCreate()、Service.onStart()或 Service.onDestroy()）内有正在执行的代码，系统将把进程移动到前台。

（2）可见进程（Visible）。

它有一个可以被用户从屏幕上看到的活动，但不在前台（它的 onPause()方法被调用）。例如，如果前台的活动是一个对话框，以前的活动就隐藏在对话框之后，就会现这种进程。可见进程非常重要，一般不允许被终止，除非是为了保证前台进程的运行而不得不终止它。

（3）服务进程（Service）。

拥有一个已经用 startService()方法启动的服务。虽然用户无法直接看到这些进程，但它们做的事情却是用户所关心的（如后台 MP3 回放或后台网络数据的上传下载）。因此，系统将一直运行这些进程，除非内存不足以维持所有的前台进程和可见进程。

（4）后台进程（Background）。

拥有一个当前用户看不到的活动（它的 onStop()方法被调用）。这些进程对用户体验没有直接的影响。如果它们正确执行了活动生命周期，系统可以在任意时刻终止该进程以回收内存，并提供给前面 3 种类型的进程使用。系统中通常有很多这样的进程在运行，因此要将这些进程保存在 LRU 列表中，以确保当内存不足时用户最近看到的进程最后一个被终止。

（5）空进程（Empty）。

不拥有任何活动的应用程序组件的进程。保留这种进程的唯一原因是在下次应用程序的某个组件需要运行时，不需要重新创建进程，这样可以提高启动速度。

6.2　Activity 的生命周期

在 Android 中，一般用系统管理来决定进程的生命周期。有时因为手机所具有的一些特殊性，所以需要更多地去关注各个 Android 程序部分的运行时生命周期模型。所谓手机的特殊性，主要是指如下两点。

（1）在使用手机应用时，大多数情况下只能在手机上看到一个程序的一个界面，用户除了通过程序界面上的功能按钮来在不同的窗体间切换，还可以通过 Back（返回）键和 Home（主）键来返回上一个窗口，而用户使用 Back 或者 Home 的时机是非常不确定的，任何时候用户都可以使用 Home 或 Back 来强行切换当前的界面。

（2）通常手机上一些特殊的事件发生也会强制地改变当前用户所处的操作状态，例如无论任何情况，在手机来电时，系统都会优先显示电话接听界面。

了解了手机应用的上述特殊性之后，接下来将详细介绍 Activity 在不同阶段的生命周期。

6.2.1 Activity 的几种状态

要想了解 Activity 在不同阶段的生命周期，首先需要了解 Activity 的几种状态。当 Activity 被创建或销毁时，会存在如下 4 种状态。

（1）Active（活动）。

当 Activity 在栈的顶端时是可见的，这些有焦点的前台 Activity 用来响应用户的输入。Android 会不惜一切代价来尝试保证它的活跃性，需要的话它会杀死栈中更靠下的 Activity，这样可以保证 Active Activity 需要的资源。当另一个 Activity 变成 Active 状态时，这个就会变成 Paused。

（2）Paused（暂停）。

在一些情况下，程序 Activity 可见但是不拥有焦点，这时它就是暂停的。当最前面的 Activity 是全透明或非全屏的 Activity 时，下面的 Activity 就会到达这个状态。当暂停时，这个 Activity 还是被看作是 Active 的，但不接受用户的输入事件。在极端的情况下，Android 会“杀死”一个 Paused 的 Activity 来恢复资源给 Active Activity。当一个 Activity 完全不可见时，它就变成 Stopped。

（3）Stopped（停止）。

当一个 Activity 不可见，它就“停止”了。这个 Activity 仍然留在内存里来保存所有的状态和成员信息；但是，在什么地方当系统需要内存时，它就是“罪犯”拉出去“枪毙”了。当一个 Activity 停止时，保存数据和当前 UI 状态是很重要的。一旦 Activity 退出或关闭，它就变成 Inactive。

（4）Inactive（销毁）。

当一个曾经被启动过的 Activity 被“杀死”时，它就变成 Inactive。Inactive Activity 会从 Activity 栈中移除，当它重新显示和使用时需要再次启动。

Activity 状态转换图如图 6-1 所示。

图 6-1 所示的状态的变化是由 Android 内存管理器决定的，Android 会首先关闭那些包含 Inactive Activity 的应用程序，然后关闭 Stopped 状态的程序。在极端情况下，会移除 Paused 状态下的程序。

6.2.2 分解剖析 Activity

（1）void onCreate(Bundle savedInstanceState)。

当 Activity 被第一次加载时执行 onCreate()，当启动一个新程序的时候，其主窗体的 onCreate 事件就会被执行。如果 Activity 被 onDestroy（销毁）后，再重新加载 Task（任务）时，其 onCreate() 的事件也会被重新执行。

（2）void onStart()。

在 onCreate 事件之后执行 onStart()。或者当前窗体被交换到后台后，在用户重新查看窗体前已经过去了一段时间，窗体已经执行了 onStop()事件，但是窗体和其所在进程并没有被销毁，用户再次重新查看窗体时会执行 onRestart()事件，之后会跳过 onCreate()事件，直接执行窗体的

onStart 事件。

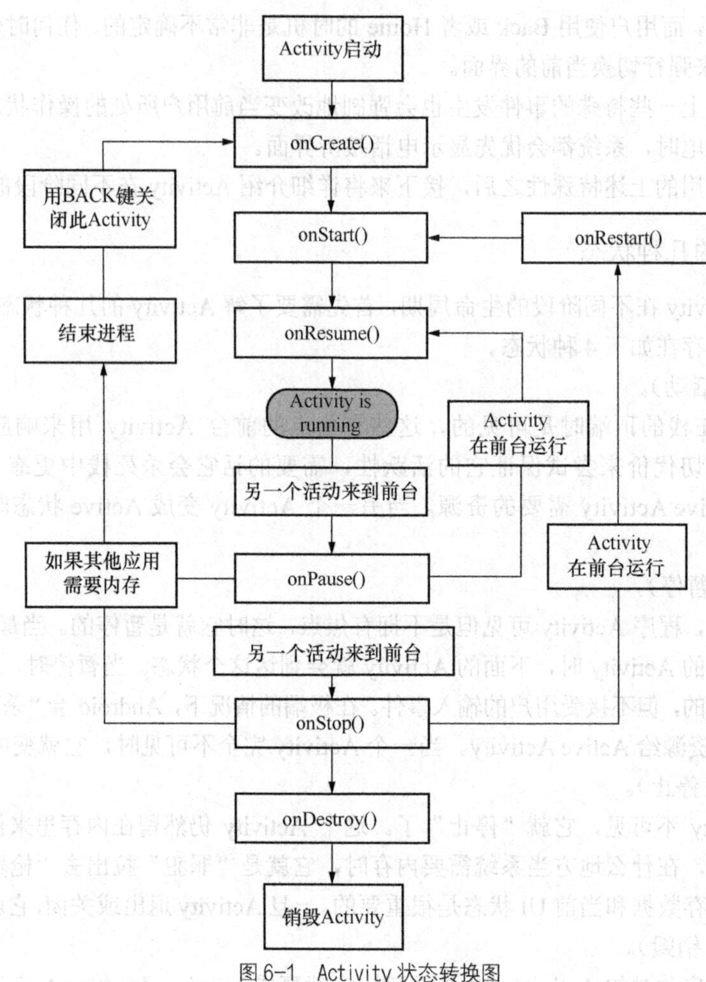

图 6-1　Activity 状态转换图

（3）void onResume()。

在 onStart 事件之后执行。或者当前窗体被交换到后台后，在用户重新查看窗体时，窗体还没有被销毁，也没有执行过 onStop()事件（窗体还继续存在于 Task 中），则会跳过窗体的 onCreate()和 onStart()事件，直接执行 onResume()事件。

（4）void onPause()。

窗体被交换到后台时执行 onPause()。

（5）void onStop()。

onPause()事件之后执行 onStop()。如果一段时间内用户还没有重新查看该窗体，则该窗体的 onStop()事件将会被执行；或者用户直接按了 Back 键，将该窗体从当前 Task 中移除，也会执行该窗体的 onStop()事件。

（6）void onRestart()。

onStop 事件执行后执行 onRestart()，如果窗体和其所在的进程没有被系统销毁，此时用户又重新查看该窗体，则会执行窗体的 onRestart 事件，onRestart 事件后会跳过窗体的 onCreate()事件直接执行 onStart()事件。

（7）void onDestroy()。

Activity 被销毁的时候执行 onDestroy()。在窗体的 onStop()事件之后，如果没有再次查看该窗体，Activity 则会被销毁。

6.2.3 几个典型的场景

根据前面讲解的 Activity 生命周期的基本知识，可以总结出如下几个典型的应用场景。

（1）Activity 从被装载到运行，执行顺序为：

```
onCreate() -> onStart()-> onResume();
```

这是一个典型的过程，发生在 Activity 被系统装载运行时。

（2）Activity 从运行到暂停，再到继续回到运行。执行顺序为：

```
onPause() -> onResume();
```

这个过程发生在 Activity 被别的 Activity 遮住了部分 UI，失去了用户焦点，另外那个 Activity 退出之后，这个 Activity 再次重新获得运行。在这个过程中，该 Activity 的实例是一直存在。

（3）Activity 从运行到停止，执行顺序为：

```
onPause() -> onStop();
```

这个过程发生在 Activity 的 UI 完全被别的 Activity 遮住了，当然也失去了用户焦点。这个过程中 Activity 的实例仍然存在。比如，当 Activity 正在运行时，按 HOME 键，该 Activity 就会被执行这个过程。

（4）Activity 从停止到运行，执行顺序为：

```
onRestart()-> onStart()-> onResume();
```

处于 STOPPED 状态并且实例仍然存在的 Activity，再次被系统运行时，执行这个过程。这个过程是（3）的逆过程，只是要先执行 onRestart()而重新获得执行。

（5）Activity 从运行到销毁，执行顺序为：

```
onPause() -> onStop() -> onDestroy();
```

这个过程发生在 Activity 完全停掉并被销毁了，所以该 Activity 的实例也就不存在了。比如，当 Activity 正在运行时，按 BACK 键，该 Activity 就会被执行这个过程。这个过程可看作是（1）的逆过程。

（6）被清除出内存的 Activity 重新运行，执行顺序为：

```
onCreate() -> onStart()-> onResume();
```

这个过程对用户是透明的，用户并不会知道这个过程的发生，看起来如同（1）的执行顺序，不同的是如果保存有系统被清除存内存时的信息，会在调用 onCreate()时，系统以参数的形式给出，而（1）中 onCreate()的参数为 null。

6.2.4 管理 Activity 的生命周期

此处说的管理 Activity 的生命周期，更确切地说应该是参与生命周期的管理，因为 Android 系统框架已经很好地管理了这其中的绝大部分，应用开发者要做的就是在 Android 的框架下，在 Activity 状态转换的各个时间点上，做出自己的实现，而实现这些要做的只是在你的 Activity 子类中 Override 这些 Activity 的方法即可。

图 6-2 列出了 Activity 生命周期相关的方法。

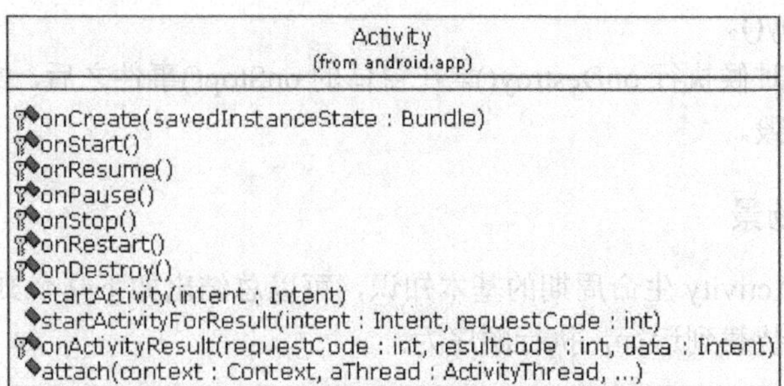

图 6-2 Activity 生命周期相关的方法

6.2.5 Activity 的实例化与启动

Activity 的实例化工作是由 Android 系统完成的，当用户点击执行一个 Activity 时，或另一个 Activity 需要执行这个 Activity 的时候，如果该 Activity 的实例不存在，Android 系统都会将其实例化，并在该 Activity 所在进程的主线程中调用该 Activity 的 onCreate()方法，实现 Activity 实例化时的工作。

由此可见，当 Activity::onCreate()是系统实例化的 Activity 时，Activity 可以被当作自身初始化的时机。在这里可以实例化变量，并可以调用 Activity::setContentView()设置 UI 显示的内容。

一般来说，在 Activity 实例化之后就要启动该 Activity，这样会在该 Activity 所在进程的主线程中顺序调用 Activity 的 onStart()、onResume()。在 Activity 的生命周期的典型时序中，一般 onStart()在所有的时序中都不是很特别的过程，所以一般不怎么实现。

图 6-3 演示了 Activity 实例化与启动的时序。

在 Activity 存续期内，只会调用 onCreate()一次。如果另外又启动了一个 Activity 的实例，并通过 onCreate()的参数传递进先前杀掉的 Activtiy 里保留的信息。因为 onStart()可因为已经停止了，再次执行而被调用多次。onResume()可以因为 Activity 的 PAUSED/RESUMED 的不停转换，而被频繁调用。

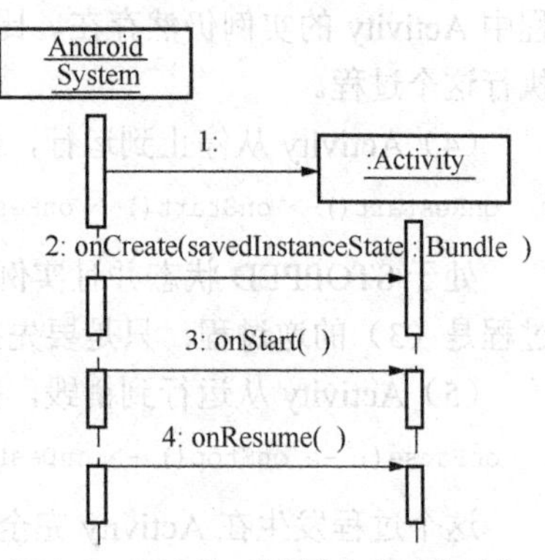

图 6-3 Activity 实例化与启动的时序

6.2.6 Activity 的暂停与继续

Activity 因为被别的 Activity 遮住部分 UI 界面，并因此失去了焦点而被中段暂停，这种情况通常发生在系统进入睡眠时或被一个对话框打断时。在被暂停之前，系统会通过 onPause()让 Activity 保留被暂停前状态的时机。Activity 可以在 onPause()中保存所做的修改到永久存储区，以停止动画显示等常见的操作。因为在 onPause()返回之前不会调入其他的 Activity 运行，所以在 onPause()中的操作必须简短并快速返回。此时 Activity 因为还会显示部分 UI，并与 Window Manager 的链接依旧存在，所以一般不需保留对 UI 的修改。即便是在极端的情况下，当 PAUSED 的 Activity 所在的进程被“杀死”时，也不可能使 Activity 的 UI 显示完整一致。

系统在被唤醒或者打断它的对话框消失之后，会继续运行，此时系统会调用 Activity 的 onResume()方法。在 onResume()方法中可以做与 onPause()中相对应的事情。

在图 6-4 中，演示了一个 Activity 启动同一个进程内另外一个 Activity 的时序图。

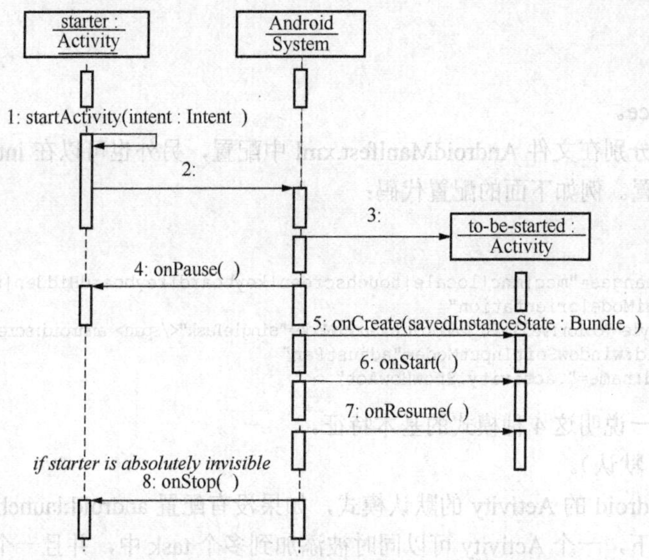

图 6-4 一个 Activity 启动另一个 Activity 的时序图

图 6-4 可以很好地说明一个 Activity 启动时，与另外一个 Activity 之间的各种 PAUSE/RESUME 交互过程。在此需要注意，PAUSE/RESUME 是众多 Activity 状态转换中的一个子集，很多其他的场景也需要经过这个过程。

6.2.7 Activity 的关闭/销毁与重新运行

当 Activity 被 Stop 时，可能是完全被别的 Activity 覆盖掉了，也可能是用户显式地按了 BACK 或 HOME 键。Activity 被 Stop 之前，它的 onStop()方法会被提前调用，来做一些 Stop 前的处理。如果再次运行处于 STOPPED 的 Activity，它的 onRestart()方法会被调用，这是区分其他调用场景比较合适的实现处理的地方。

因为处于 Paused 状态的 Activity 在内存极端不足的情况下，它所在的进程也可能被“杀掉”，这样 onStop()在被“杀掉”前，不一定会被调用，所以 onPause()是比 onStop()更合适的保留信息到永久存储区的时机。

Activity 被销毁可能是显式地按了 BACK 键，也可能是处于 Paused 或 Stopped 状态，因为内存不足而被“杀掉”的。还有种情况是配置信息改变（比如屏的方向改变）之后，根据设置需要杀掉所有的 Activity（是否关闭还要看 Activity 自己的配置），再重新运行它们。

被系统隐式“杀死”的 Activity，在被“杀死”（onStop()调用）之前，一般的会调用 onSaveInstanceState()保留该 Activity 此时的状态信息。该方法中传入 Bundle 参数，可在此方法中把此时的状态信息写入，系统保留这些信息。而当该 Activty 再次被实例化运行时，系统会把保留在 Bundler 的信息再次以参数形式，通过 onCreate()方法传入。

通常在 onSaveInstanceState()中保留 UI 的信息，永久存储的信息最好还是在 onPause()中保存。Activity 的 onSaveInstanceState()已经缺省实现来保留通用 View 的 UI 信息，所以不管是否保留当前 Activity 的信息，通常都要在 onSaveInstanceState()中先调用 super.onSaveInstanceState()来保留通用的 UI 信息。

6.2.8 Activity 的启动模式

在 Android 系统中，有以下 4 种启动 Activity 的模式。

- ❑ standard（默认）。

- singleTop。
- singleTask。
- singleInstance。

上述 4 种模式分别在文件 AndroidManifest.xml 中配置，另外也可以在 intent 启动 Activity 时添加必要参数来设置。例如下面的配置代码：

```
<activity

android:configChanges="mcc|mnc|locale|touchscreen|keyboard|keyboardHidden|navigation|screenLa
yout|fontScale|uiMode|orientation"
        <span style="color:#e53333;">android:launchMode="singleTask"</span> android:screenOrientation="portrait"
         android:windowSoftInputMode="adjustPan"
        android:name=".activity.ShowHowAct" >
```

接下来开始一一说明这 4 种模式的基本特征。

（1）standard（默认）。

standard 是 Android 的 Activity 的默认模式，如果没有配置 android:launchMode，则默认是这个模式。在该模式下，一个 Activity 可以同时被添加到多个 task 中，并且一个 task 可以有多个实例，且每次通过 intent 启动时，都会生成一个新的实例。

（2）singleTop。

属性 singleTop 和 standard 较类似，不同的地方就是，当当前 Activity 的实例在当前 task 的栈顶时，intent 启动时，则不生成新的实例，会重用（不生成新的实例）原有的实例，如果显式指定 intent 的参数 FLAG_ACTIVITY_NEW_TASK，如果提供了 FLAG_ACTIVITY_NEW_TASK 参数，会启动到别的 task 里。

（3）singleTask。

在 singleTask 模式下，Activity 只会有一个实例。如果某一个 task 中已有该 Activity 的一个实例存在，则不再启动新的，每次都会被重用（重用就是如果该 Activity 在 task 的栈底，则会被调到栈顶），且可以和其他的 Activity 共存于一个 task 中。

（4）singleInstance。

singleInstance 模式和 singleTask 模式一样，唯一的区别就是，在该模式下的 Activity 会独自拥有一个 task，不会和其他 Activity 公用，每次 Activity 都会被重用，且全局只能有一个实例。

6.3 进程与线程

当某个组件第一次运行的时候，Android 启动了一个进程。在默认情形下，所有的组件和程序运行在这个进程和线程中。另外，也可以安排组件在其他的进程或者线程中运行。在本节的内容中，将简要介绍 Android 进程与线程的基本知识。

6.3.1 进程

组件运行的进程由 Manifest File 控制，组件中的节点<activity>、<service>、<receiver>和<provider>都包含了一个 Process 属性。通过 Process 属性，可以设置组件运行的进程，既可以配置组件在一个独立进程中运行，也可以配置多个组件在同一个进程运行，甚至可以配置多个程序在同一个进程中运行（前提是这些程序共享一个 User ID 并给定同样的权限）。另外，在<application>节点中也包含了 Process 属性，可以用来设置程序中所有组件的默认进程。

所有的组件在此进程的主线程中实例化，系统对这些组件的调用从主线程中分离。并非每个对象都会从主线程中分离。一般来说，响应例如 View.onKeyDown()用户操作的方法和通知的方法也在主线程中运行。这就表示，组件被系统调用的时候不应该长时间运行或者阻塞操作（如网络操作或者计算大量数据），因为这样会阻塞进程中的其他组件。可以把这类操作从主线程中分离。

当更加常用的进程无法获取足够内存，Android 可能会关闭不常用的进程。下次启动程序的时候会重新启动进程。

当决定哪个进程需要被关闭的时候，Android 会考虑哪个对用户更加有用。如 Android 会倾向于关闭一个长期不显示在界面的进程来支持一个经常显示在界面的进程，是否关闭一个进程决定于组件在进程中的状态。

6.3.2 线程

即使为组件分配了不同的进程，有时候也需要再分配线程。比如用户界面需要很快对用户进行响应，因此某些费时的操作，如网络连接、下载或者非常占用服务器时间的操作应该放到其他线程。

线程通过 Java 的标准对象 Thread 创建，在 Android 中提供了如下管理线程的方法。

- ❑ Looper：在线程中运行一个消息循环。
- ❑ Handler：传递一个消息。
- ❑ HandlerThread：创建一个带有消息循环的线程。

Android 会让一个应用程序在单独的线程中，指导它创建自己的线程。除了上述方法外，通过使用应用程序组件，例如 Activity、service、broadcast receiver，可以在主线程中实现实例化操作。

6.3.3 线程安全的方法

了解了进程和线程的基本知识后，很有必要了解线程安全方面的知识。在某些情况下，方法可能调用不止一个线程，因此需要注意方法的线程安全。例如当一个调用在 IBinder（是一个接口，是对跨进程的对象的抽象）对象中的方法的程序启动了和 IBinder 对象相同的进程，方法就在 IBinder 的进程中执行。但是，如果调用者发起另外一个进程，方法在另外一个线程中运行，这个线程与 IBinder 对象在一个线程池中，它不会在进程的主线程中运行。如果一个 Service 从主线程被调用 onBind()方法，onBind()返回的对象中的方法会被从线程池中调用。因为一个服务可能有多个客户端请求，不止一个线程池会在同一时间调用 IBinder 的方法，所以此时 IBinder 必须保证线程安全。

6.3.4 Android 的线程模型

Android 包括一个应用程序框架、几个应用程序库和一个基于 Dalvik 虚拟机的运行时，所有这些都运行在 Linux 内核之上。通过利用 Linux 内核的优势，Android 得到了大量操作系统服务，包括进程和内存管理、网络堆栈、驱动程序、硬件抽象层、安全性等相关的服务。

在安装 Android 应用程序的时候，Android 会为每个程序分配一个 Linux 用户 ID，并设置相应的权限，这样其他应用程序就不能访问此应用程序所拥有的数据和资源了。

在 Linux 中，一个用户 ID 识别一个给定用户。在 Android 上，一个用户 ID 识别一个应用程序。应用程序在安装时被分配用户 ID，应用程序在设备上的存续期间内，用户 ID 保持不变。

在默认情况下，每个 APK 运行在它自己的 Linux 进程中。当需要执行应用程序中的代码时，Android 会启动一个 JVM，即一个新的进程来执行，因此不同的 apk 运行在相互隔离的环境中。

不同的 Android 应用程序可以运行在相同的进程中，要想实现这个功能，首先必须使用相同的私钥签署这些应用程序，然后使用 manifest 文件给它们分配相同的 Linux 用户 ID，这通过用相同的“值/名”定义 manifest 属性 android:sharedUserId 来实现。

（1）Android 的单线程模型。

当第一次启动一个程序时，Android 会同时启动一个对应的主线程（Main Thread），主线程主要负责处理与 UI 相关的事件，如用户的按键事件、用户接触屏幕的事件以及屏幕绘图事件，并把相关的事件分发到对应的组件进行处理，因此主线程通常又被称为 UI 线程。

在开发 Android 应用时必须遵守单线程模型的原则：Android UI 操作并不是线程安全的，并且这些操作必须在 UI 线程中执行。

如果在非 UI 线程中直接操作 UI 线程，则会抛出如下异常：

```
android.view.ViewRoot$CalledFromWrongThreadException: Only the original thread that created a view hierarchy can touch its views
```

这与普通的 Java 程序是不同的。

由于 UI 线程负责事件的监听和绘图处理，因此必须保证 UI 线程能够随时响应用户的需求，UI 线程里的操作应该向中断事件那样短小，费时的操作（如网络连接）需要另开线程；否则，如果 UI 线程超过 5 s 没有响应用户请求，会弹出对话框提醒用户终止应用程序。

如果在新开的线程中需要对 UI 进行设定，就可能违反单线程模型，因此 Android 采用一种复杂的 Message Queue 机制保证线程间通信。

（2）Message Queue。

Message Queue 是一个消息队列，用来存放通过 Handler 发布的消息。Android 在第一次启动程序时默认会为 UI thread 创建一个关联的消息队列，可以通过 Looper.myQueue()得到当前线程的消息队列，用来管理程序的一些上层组件，例如 activities 和 broadcast receivers 等。可以在自己的子线程中创建 Handler 与 UI thread 通信。

通过 Handler 可以发布或者处理一个消息或者是一个 Runnable 的实例，每个 Handler 都会与唯一的一个线程以及该线程的消息队列管理。Looper 扮演着一个 Handler 和消息队列之间通信桥梁的角色。程序组件首先通过 Handler 把消息传递给 Looper，Looper 把消息放入队列。Looper 也把消息队列里的消息广播给所有的 Handler，Handler 接收到消息后调用 handleMessage 进行处理。例如下面的演示代码：

```
public void onCreate(Bundle savedInstanceState) {
   super.onCreate(savedInstanceState);
   setContentView(R.layout.main);
   editText = (EditText) findViewById(R.id.weather_city_edit);
   Button button = (Button) findViewById(R.id.goQuery);
   button.setOnClickListener(this);
   Looper looper = Looper.myLooper();  //得到当前线程的 Looper 实例，由于当前线程是 UI 线程也可以通过
Looper.getMainLooper()//得到
    messageHandler = new MessageHandler(looper);  //此处甚至可以不需要设置 Looper，因为 Handler 默
//认就使用当前线程的 Looper
}
public void onClick(View v) {
   new Thread() {
      public void run() {
         Message message = Message.obtain();
         message.obj = "abc";
         messageHandler.sendMessage(message);  //发送消息
      }
   }.start();
}
Handler messageHandler = new Handler {
   public MessageHandler(Looper looper) {
      super(looper);
   }
   public void handleMessage(Message msg) {
      setTitle((String) msg.obj);
   }
}
```

对于上述演示代码，当这个 activity 执行完 oncreate、onstart 和 onresume 后，就监听 UI 的各种事件和消息。当点击一个按钮后，启动一个线程，线程执行结束后，通过 handler 发送一个消息，由于这个 handler 属于 UI 线程，因此这个消息也发送给 UI 线程，然后 UI 线程又把这个消息给 handler 处理，而这个 handler 是 UI 线程创造的，它可以访问 UI 组件，因此就更新了页面。

由于通过 handler 需要自己管理线程类，如果业务稍微复杂，代码看起来就比较混乱，因此 Android 提供了类 AsyncTask 来解决这个问题。

（3）AsyncTask。

首先继承类 publishProgress，实现如下所示的方法。

- onPreExecute()：该方法将在执行实际的后台操作前被 UI thread 调用。可以在该方法中做一些准备工作，如在界面上显示一个进度条。
- doInBackground(Params...)：在方法 onPreExecute 执行后马上执行，该方法运行在后台线程中。这里将主要负责执行那些很耗时的后台计算工作。
- publishProgress()：更新实时的任务进度。该方法是抽象方法，必须实现子类。
- onProgressUpdate(Progress...)：在 publishProgress 方法被调用后，UI thread 将调用这个方法从而在界面上展示任务的进展情况，例如通过一个进度条进行展示。
- onPostExecute(Result)：在 doInBackground 执行完成后，onPostExecute 方法将被 UI thread 调用，后台的计算结果将通过该方法传递到 UI thread。

使用 publishProgress 类时需要遵循以下规则。

- Task 的实例必须在 UI thread 中创建。
- execute 方法必须在 UI thread 中调用。
- 不要手动调用这些方法，只调用 execute 即可。
- 该 task 只能被执行一次，否则多次调用时将会出现异常。

例如下面的演示代码：

```
public void onCreate(Bundle savedInstanceState) {
        super.onCreate(savedInstanceState);
        setContentView(R.layout.main);
        editText = (EditText) findViewById(R.id.weather_city_edit);
        Button button = (Button) findViewById(R.id.goQuery);
        button.setOnClickListener(this);
}
public void onClick(View v) {
        new GetWeatherTask().execute("aaa");
}
class GetWeatherTask extends AsyncTask<String, Integer, String> {
    protected String doInBackground(String... params) {
        return getWetherByCity(params[0]);
    }
    protected void onPostExecute(String result) {
        setTitle(result);
    }
```

6.4 测试生命周期

经过本书前面内容的学习，相信读者已经了解了 Android 生命周期的基本知识。在本节中，将通过几段应用代码来测试 Android 的生命周期。

（1）首先看 MainActivity 的代码，这是软件启动时默认打开的 Activity。

```
package cn.itcast.life;

import android.app.Activity;
```

```
import android.content.Intent;
import android.os.Bundle;
import android.util.Log;
import android.view.View;
import android.widget.Button;

public class MainActivity extends Activity {
    private static final String TAG = "MainActivity";

    @Override
    public void onCreate(Bundle savedInstanceState) {
        super.onCreate(savedInstanceState);
        setContentView(R.layout.main);
        Log.i(TAG, "onCreate()");

        Button button = (Button) this.findViewById(R.id.button);
        button.setOnClickListener(new View.OnClickListener() {
            @Override
            public void onClick(View v) {
                Intent intent = new Intent(MainActivity.this, OtherActivity.class);
                startActivity(intent);
            }
        });

        Button threebutton = (Button) this.findViewById(R.id.threebutton);
        threebutton.setOnClickListener(new View.OnClickListener() {

            @Override
            public void onClick(View v) {
                Intent intent = new Intent(MainActivity.this, ThreeActivity.class);
                startActivity(intent);
            }
        });
    }

    @Override
    protected void onDestroy() {
        Log.i(TAG, "onDestroy()");
        super.onDestroy();
    }

    @Override
    protected void onPause() {
        Log.i(TAG, "onPause()");
        super.onPause();
    }

    @Override
    protected void onRestart() {
        Log.i(TAG, "onRestart()");
        super.onRestart();
    }

    @Override
    protected void onResume() {
        Log.i(TAG, "onResume()");
        super.onResume();
    }

    @Override
    protected void onStart() {
        Log.i(TAG, "onStart()");
        super.onStart();
    }

    @Override
    protected void onStop() {
        Log.i(TAG, "onStop()");
```

```
        super.onStop();
    }

}
```

（2）以下是 MainActivity 匹配的 XML 布局代码：

```
<?xml version="1.0" encoding="utf-8"?>
<LinearLayout xmlns:android="http://schemas.android.com/apk/res/android"
    android:orientation="vertical"
    android:layout_width="fill_parent"
    android:layout_height="fill_parent"
    >
<TextView
    android:layout_width="fill_parent"
    android:layout_height="wrap_content"
    android:text="@string/hello"
    />

    <Button
        android:layout_width="wrap_content"
        android:layout_height="wrap_content"
         android:text="打开 OtherActivity"
        android:id="@+id/button"
        />

     <Button
        android:layout_width="wrap_content"
        android:layout_height="wrap_content"
         android:text="打开 ThreeActivity"
        android:id="@+id/threebutton"
        />
</LinearLayout>
```

（3）以下是一个新的 Activity，为了验证“onstop”方法，使用下面的 OtherActivity 将前面的 MainActivity 覆盖掉：

```
package cn.itcast.life;

import android.app.Activity;
import android.os.Bundle;

public class OtherActivity extends Activity {

    @Override
    protected void onCreate(Bundle savedInstanceState) {
        // TODO Auto-generated method stub
        super.onCreate(savedInstanceState);
        setContentView(R.layout.other);
    }
}
```

以下是 OtherActivity 匹配的 XML 布局代码：

```
<?xml version="1.0" encoding="utf-8"?>
<LinearLayout
  xmlns:android="http://schemas.android.com/apk/res/android"
  android:orientation="vertical"
  android:layout_width="fill_parent"
  android:layout_height="fill_parent">
    <TextView
    android:layout_width="fill_parent"
    android:layout_height="wrap_content"
    android:text="这是 OtherActivity"
    />
</LinearLayout>
```

（4）以下的 ThreeActivity 用于测试 onpause 方法，使用半透明或者提示框的形式，覆盖掉前面的 MainActivity：

```
package cn.itcast.life;

import android.app.Activity;
import android.os.Bundle;

public class ThreeActivity extends Activity {
    @Override
    protected void onCreate(Bundle savedInstanceState) {
        // TODO Auto-generated method stub
        super.onCreate(savedInstanceState);
        setContentView(R.layout.three);
    }
}
```

以下是 ThreeActivity 匹配的 XML 布局代码：

```
<?xml version="1.0" encoding="utf-8"?>
<LinearLayout
  xmlns:android="http://schemas.android.com/apk/res/android"
  android:layout_width="wrap_content"
  android:layout_height="wrap_content">

  <TextView
    android:layout_width="fill_parent"
    android:layout_height="wrap_content"
    android:text="第三个 Activity"
    />
</LinearLayout>
```

（5）以下是项目清单文件，在此使用了 android:theme="@android:style/Theme.Dialog"来设置 Activity 的样式风格的弹出框：

```
<?xml version="1.0" encoding="utf-8"?>
<manifest xmlns:android="http://schemas.android.com/apk/res/android"
      package="cn.itcast.life"
      android:versionCode="1"
      android:versionName="1.0">
    <application android:icon="@drawable/icon" android:label="@string/app_name">
        <activity android:name=".MainActivity"
                  android:label="@string/app_name">
            <intent-filter>
                <action android:name="android.intent.action.MAIN" />
                <category android:name="android.intent.category.LAUNCHER" />
            </intent-filter>
        </activity>
        <activity android:name=".OtherActivity" android:theme="@android:style/Theme.Dialog"/>
        <activity android:name=".ThreeActivity"/>
    </application>
    <uses-sdk android:minSdkVersion="8" />
</manifest>
```

运行上述程序，如果在 Debug 状态时切换到 DDMS 界面，可以马上看到所打印出来的 Log 信息，这样就可以很清楚地分析程序的运行过程。

Activity 的 onSaveInstanceState()和 onRestoreInstanceState()并不是生命周期方法，它们不同于 onCreate()和 onPause()等生命周期方法，它们并不一定会被触发。当应用遇到意外情况时，例如内存不足、用户直接按 Home 键等操作，当系统销毁一个 Activity 时，onSaveInstanceState()才会被调用。但是当用户主动去销毁一个 Activity 时，例如在应用中按返回键，onSaveInstanceState()就不会被调用。因为在这种情况下，用户的行为决定了不需要保存 Activity 的状态。通常 onSaveInstanceState()只适合用于保存一些临时性的状态，而 onPause()适合用于数据的持久化保存。

6.5 Service 的生命周期

在本章前面的内容中，已经讲解了 Android 中 Activity 的生命周期。在本节的内容中，将详细讲解 Android 中 Service 的生命周期知识，为读者步入本书后面知识的学习打下基础。

6.5.1 Service 的基本概念和用途

Android 中的服务，它与 Activity 不同，它是不能与用户交互的，不能自己启动的，运行在后台的程序，如果退出应用时，Service 进程并没有结束，它仍然在后台运行，那什么时候会用到 Service 呢？比如播放音乐的时候，有可能想边听音乐边干些其他事情，当退出播放音乐的应用，如果不用 Service，就听不到歌了，所以这时候就得用到 Service 了。又比如当一个应用的数据是通过网络获取的，不同时间（一段时间）的数据是不同的这时候可以用 Service 在后台定时更新，而无需每打开应用的时候在去获取。

6.5.2 Service 的生命周期详解

Android Service 的生命周期并不像 Activity 那么复杂，它只继承了 onCreate()、onStart()、onDestroy()3 个方法，当第一次启动 Service 时，先后调用了 onCreate()和 onStart()这两个方法，当停止 Service 时，则执行 onDestroy()方法，这里需要注意的是，如果 Service 已经启动了，当再次启动 Service 时，不会再执行 onCreate()方法，而是直接执行 onStart()方法，具体的可以看下面的实例。

6.5.3 Service 与 Activity 通信

Service 后端的数据最终还是要呈现在前端 Activity 之上的，因为启动 Service 时，系统会重新开启一个新的进程，这就涉及到不同进程间通信的问题了（AIDL）这一节不作过多描述，当想获取启动的 Service 实例时，可以用到 bindService 和 onBindService 方法，它们分别执行了 Service 中 IBinder()和 onUnbind()方法。

为了让大家更容易理解 Service 与 Activity 通信过程，接下来用一段演示代码来详细讲解。

（1）新建一个 Android 工程，在此命名为 ServiceDemo。

（2）修改文件 main.xml，在此增加了 4 个按钮，具体实现代码如下所示：

```
<?xml version="1.0" encoding="utf-8"?>
<LinearLayout xmlns:android="http://schemas.android.com/apk/res/android"
    android:orientation="vertical"
    android:layout_width="fill_parent"
    android:layout_height="fill_parent"
    >
    <TextView
        android:id="@+id/text"
        android:layout_width="fill_parent"
        android:layout_height="wrap_content"
        android:text="@string/hello"
        />
    <Button
        android:id="@+id/startservice"
        android:layout_width="fill_parent"
        android:layout_height="wrap_content"
        android:text="startService"
    />
    <Button
        android:id="@+id/stopservice"
        android:layout_width="fill_parent"
```

```
        android:layout_height="wrap_content"
        android:text="stopService"
    />
    <Button
        android:id="@+id/bindservice"
        android:layout_width="fill_parent"
        android:layout_height="wrap_content"
        android:text="bindService"
    />
    <Button
        android:id="@+id/unbindservice"
        android:layout_width="fill_parent"
        android:layout_height="wrap_content"
        android:text="unbindService"
    />
</LinearLayout>
```

（3）新建一个 Service，命名为 MyService.java，具体实现代码如下所示：

```
public class MyService extends Service {
    //定义个一个 Tag 标签
    private static final String TAG = "MyService";
    //这里定义吧一个 Binder 类，用在 onBind()有方法里，这样 Activity 那边可以获取到
    private MyBinder mBinder = new MyBinder();
    @Override
    public IBinder onBind(Intent intent) {
        Log.e(TAG, "start IBinder~~~");
        return mBinder;
    }
    @Override
    public void onCreate() {
        Log.e(TAG, "start onCreate~~~");
        super.onCreate();
    }

    @Override
    public void onStart(Intent intent, int startId) {
        Log.e(TAG, "start onStart~~~");
        super.onStart(intent, startId);
    }
    @Override
    public void onDestroy() {
        Log.e(TAG, "start onDestroy~~~");
        super.onDestroy();
    }


    @Override
    public boolean onUnbind(Intent intent) {
        Log.e(TAG, "start onUnbind~~~");
        return super.onUnbind(intent);
    }

    //这里我写了一个获取当前时间的函数，不过没有格式化就先这么着吧
    public String getSystemTime(){

        Time t = new Time();
        t.setToNow();
        return t.toString();
    }

    public class MyBinder extends Binder{
        MyService getService()
        {
            return MyService.this;
        }
    }
}
```

（4）修改文件ServiceDemo.java，具体实现代码如下所示：

```
public class ServiceDemo extends Activity implements OnClickListener{

    private MyService  mMyService;
    private TextView mTextView;
    private Button startServiceButton;
    private Button stopServiceButton;
    private Button bindServiceButton;
    private Button unbindServiceButton;
    private Context mContext;

    //这里需要用到 ServiceConnection 在 Context.bindService 和 context.unBindService()里用到
    private ServiceConnection mServiceConnection = new ServiceConnection() {
        //当我 bindService 时，让 TextView 显示 MyService 里 getSystemTime()方法的返回值
        public void onServiceConnected(ComponentName name, IBinder service) {
            // TODO Auto-generated method stub
            mMyService = ((MyService.MyBinder)service).getService();
            mTextView.setText("I am frome Service :" + mMyService.getSystemTime());
        }

        public void onServiceDisconnected(ComponentName name) {
            // TODO Auto-generated method stub

        }
    };
    public void onCreate(Bundle savedInstanceState) {
        super.onCreate(savedInstanceState);
        setContentView(R.layout.main);
        setupViews();
    }

    public void setupViews(){

        mContext = ServiceDemo.this;
        mTextView = (TextView)findViewById(R.id.text);



        startServiceButton = (Button)findViewById(R.id.startservice);
        stopServiceButton = (Button)findViewById(R.id.stopservice);
        bindServiceButton = (Button)findViewById(R.id.bindservice);
        unbindServiceButton = (Button)findViewById(R.id.unbindservice);

        startServiceButton.setOnClickListener(this);
        stopServiceButton.setOnClickListener(this);
        bindServiceButton.setOnClickListener(this);
        unbindServiceButton.setOnClickListener(this);
    }

    public void onClick(View v) {
        // TODO Auto-generated method stub
        if(v == startServiceButton){
            Intent i  = new Intent();
            i.setClass(ServiceDemo.this, MyService.class);
            mContext.startService(i);
        }else if(v == stopServiceButton){
            Intent i  = new Intent();
            i.setClass(ServiceDemo.this, MyService.class);
            mContext.stopService(i);
        }else if(v == bindServiceButton){
            Intent i  = new Intent();
            i.setClass(ServiceDemo.this, MyService.class);
            mContext.bindService(i, mServiceConnection, BIND_AUTO_CREATE);
        }else{
            mContext.unbindService(mServiceConnection);
        }
    }
}
```

（5）修改文件 AndroidManifest.xml 的代码，在此注册新建的 MyService。

```
<?xml version="1.0" encoding="utf-8"?>
<manifest xmlns:android="http://schemas.android.com/apk/res/android"
     package="com.tutor.servicedemo"
     android:versionCode="1"
     android:versionName="1.0">
   <application android:icon="@drawable/icon" android:label="@string/app_name">
      <activity android:name=".ServiceDemo"
              android:label="@string/app_name">
         <intent-filter>
            <action android:name="android.intent.action.MAIN" />
            <category android:name="android.intent.category.LAUNCHER" />
         </intent-filter>
      </activity>
      <service android:name=".MyService" android:exported="true"></service>
   </application>
   <uses-sdk android:minSdkVersion="7" />
</manifest>
```

执行上述代码后的效果如图 6-5 所示。

当单击“startServie”按钮时，先后执行了 Service 中 onCreate()→onStart()这两个方法，如果打开 DDMS 的 Logcat 窗口，会看到如图 6-6 所示的界面。

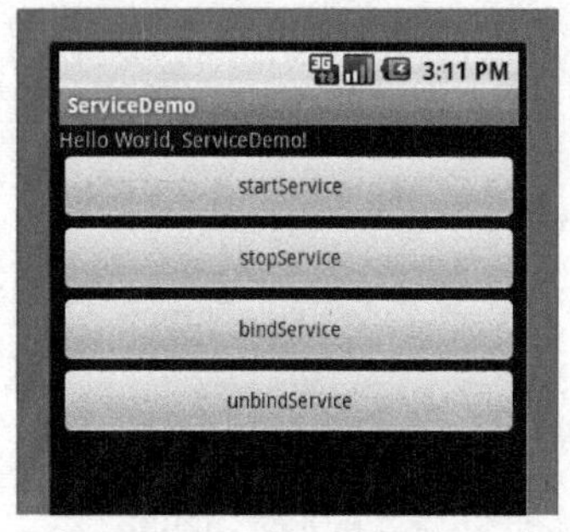

图 6-5　执行效果

pid	tag	Message
254	MyService	start onCreate~~~~
254	MyService	start onStart~~~~

图 6-6　Logcat 窗口

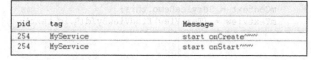

这时可以按 HOME 键进入 Settings（设置）→Applications（应用）→Running Services（正在运行的服务）看一下新启动了一个服务，效果如图 6-7 所示。

当单击 stopService 按钮时，Service 执行了 onDestroy()方法，效果如图 6-8 所示。

图 6-7　新启动了一个服务

pid	tag	Message
254	MyService	start onDestroy~~~~

图 6-8　Logcat 窗口

如果此时再次单击“startService”按钮，然后单击“bindService”按钮(通常 bindService 都是 bind 已经启动的 Service)，看一下 Service 执行了 IBinder()方法，以及 TextView 的值也有所变化了，如图 6-9 和图 6-10 所示。

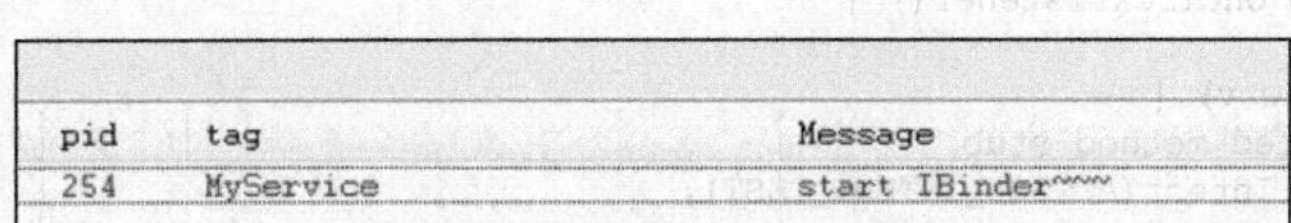

pid	tag	Message
254	MyService	start IBinder~~~

图 6-9 Logcat 窗口

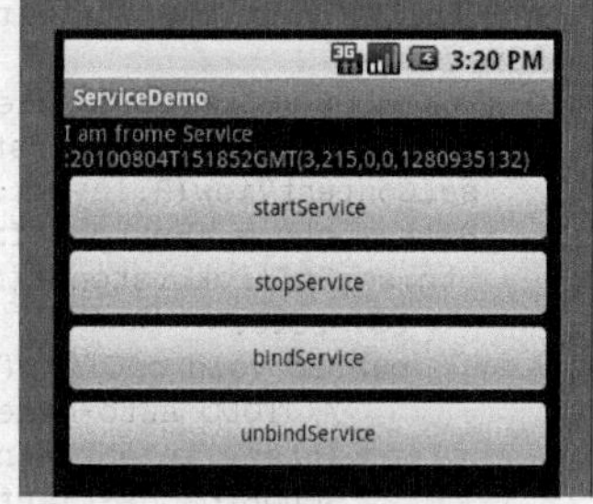

图 6-10 执行效果

最后单击“unbindService”按钮，则 Service 执行了 onUnbind()方法，如图 6-11 所示。

pid	tag	Message
254	MyService	start onUnbind~~~

图 6-11 Logcat 窗口

6.6 Android 广播的生命周期

收听收音机就是一种广播，在收音机中有很多个广播电台，每个广播电台播放的内容都不相同。接收广播时广播（发送方）并不在意接收方接收到广播时如何处理。好比收听交通电台的广播，电台中告诉人们现在在交通状况如何，但它并不关心人们接收到广播时如何做出处理，这不是广播应该关心的问题。那么 Android 中的广播是如何操作的呢？这个问题将在本节的内容中进行解答。

6.6.1 Android 的广播机制

在 Android 系统中有各种各样的广播，比如电池的使用状态、电话的接收和短信的接收都会产生一个广播，应用程序开发者也可以监听这些广播并做出程序逻辑的处理。图 6-12 演示了广播的运行机制。

在 Android 系统中有各式各样的广播，具体运作流程如下所示。

（1）当“系统/应用”程序运行时会向 Android 注册各种广播。

（2）当 Android 接收到广播时，会判断哪种广播需要哪种事件。

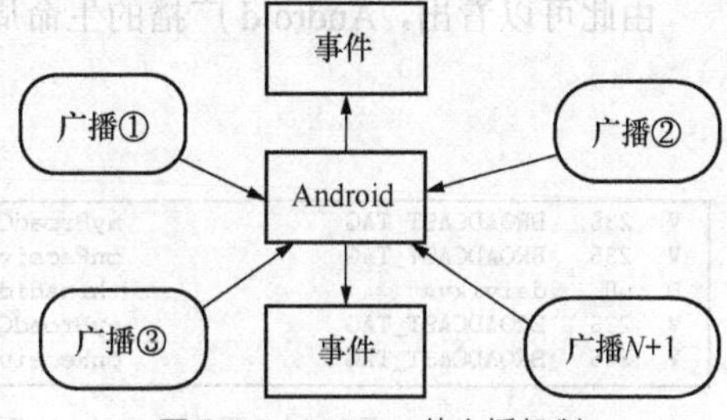

图 6-12 Android 的广播机制

（3）向不同需要事件的应用程序注册事件，不同的广播可能处理不同的事件，也可能处理相同的广播事件，这时就需要 Android 系统做筛选。例如在一个经典的电话黑名单应用程序中，首先通过将黑名单号码保存在数据库中，当来电时，我们接收到来电广播并将黑名单号码与数据库中的某个数据做匹配。如果匹配则做出相应的处理，例如挂掉电话和静音等。

6.6.2 编写广播程序

下面通过演示代码来讲解在 Android 中如何编写广播程序，在代码中设置了一个按钮，为按钮设置单击监听，通过单击发送广播，在后台中接收到广播并打印 LOG 信息：

```
public class BroadCastActivity extends Activity {
    public static final String ACTION_INTENT_TEST = "com.terry.broadcast.test";
```

```
    /** Called when the activity is first created. */
    @Override
    public void onCreate(Bundle savedInstanceState) {
        super.onCreate(savedInstanceState);
        setContentView(R.layout.main);
        Button btn = (Button) findViewById(R.id.Button01);
        btn.setOnClickListener(new OnClickListener() {
            @Override
            public void onClick(View v) {
                // TODO Auto-generated method stub
                Intent intent = new Intent(ACTION_INTENT_TEST);
                sendBroadcast(intent);
            }
        });
    }
}
```

接收器的代码如下所示：

```
public class myBroadCast extends BroadcastReceiver {
    public myBroadCast() {
        Log.v("BROADCAST_TAG", "myBroadCast");
    }
    @Override
    public void onReceive(Context context, Intent intent) {
        // TODO Auto-generated method stub
        Log.v("BROADCAST_TAG", "onReceive");
    }
}
```

在上面的接收器中，继承了 BroadcastReceiver，并重写了它的 onReceive，并构造了一个函数。当点击一下按钮，它向 Android 发送了一个广播，如图 6-13 所示。

V	235	BROADCAST_TAG	myBroadCast
V	235	BROADCAST_TAG	onReceive

图 6-13　向 Android 发送了一个广播

如果此时再单击一下按钮，还是会再向 Android 系统发送广播，此时日志信息如图 6-14 所示。

由此可以看出，Android 广播的生命周期并不像 Activity 一样复杂，基本过程如图 6-15 所示。

V	235	BROADCAST_TAG	myBroadCast
V	235	BROADCAST_TAG	onReceive
D	60	dalvikvm	threadid=17: bogus mon 1+0>0; adjusting
V	235	BROADCAST_TAG	myBroadCast
V	235	BROADCAST_TAG	onReceive

图 6-14　再次向 Android 系统发送广播

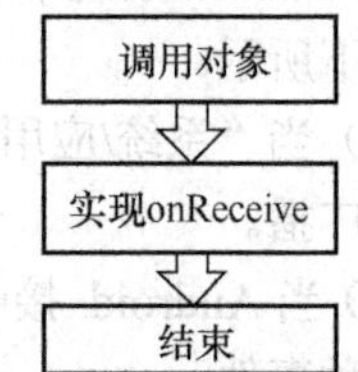

图 6-15　Android 广播生命周期的过程

前面说过 Android 的广播各式各样，那么 Android 系统是如何帮人们处理需要哪种广播并提供相应的广播服务呢？读者需要注意，每当实现一个广播接收类时，必须在应用程序中的 manifest 中显式注明需要广播哪一个类，并为其设置过滤器，如图 6-16 所示。

其中 action 代表一个要执行的动作，在 Andriod 中有很多种 action，例如 ACTION_VIEW 和 ACTION_EDIT。

也可能有读者会问：如果在一个广播接收器中要处理多个动作呢？那要如何去处理？在 Android 的接收器中 onReceive 已经为我们想到的。同理，必须在 Intent-filter 中注册该动作，可以是系统的广播动作也可以是自己需要的广播，之后需要在 onReceive 方法中，通过 intent.getAction()判断传进来的动作，这样即可做出不同的处理和不同的动作。

```
    package="com.terry"
    android:versionCode="1"
    android:versionName="1.0">
  <application android:icon="@drawable/icon" android:label="@string/app_name">
      <activity android:name=".BroadCastActivity"
               android:label="@string/app_name">
          <intent-filter>
              <action android:name="android.intent.action.MAIN" />
              <category android:name="android.intent.category.LAUNCHER" />
          </intent-filter>
      </activity>

  <receiver android:name=".myBroadCast">
      <intent-filter >
          <action android:name="com.terry.broadcast.test"></action>
      </intent-filter>
  </receiver>
/application>
```

对应的广播接收类

在该节点下的**action**都是可以通过的动作

表示接受广播注册的广播动作，这里是自定义的一个动作

图 6-16　需要广播的类

6.7 ART 进程管理

ART 虚拟机的实现离不开进程管理，其进程管理特别依赖于 Linux 的进程体系。例如要为应用程序创建一个进程，会使用 Linux 的 FORK 机制复制一个进程，因为复制进程的过程比创建进程的效率要高。并且在 Linux 中进程之间的通信方式有很多，例如管道、信号、报文和共享内存等，这样对进程管理将更加方便。

在 Android 系统中，所有的应用程序进程以及系统服务进程 SystemServer 都是由 Zygote 进程孕育（fork）出来的。当 ActivityManagerService 启动一个应用程序的时候，就会通过 Socket 与 Zygote 进程进行通信，请求它 fork 一个子进程出来作为这个即将要启动的应用程序的进程。在系统中有两个重要服务：PackageManagerService 和 ActivityManagerService，都是由 SystemServer 进程来负责启动的，而 SystemServer 进程本身是 Zygote 进程在启动的过程中 fork 出来的。

Android 系统是基于 Linux 内核的，而在 Linux 系统中，所有的进程都是 init 进程的子孙进程，也就是说，所有的进程都是直接或者间接地由 init 进程 fork 出来的。Zygote 进程也不例外，它是在系统启动的过程，由 init 进程创建的。在系统启动脚本文件“system/core/rootdir/init.rc”中，可以看到如下启动 Zygote 进程的脚本命令。

Zygote 本身是一个应用层的程序，和驱动、内核模块没有任何关系。Zygote 的启动由 Linux 的祖先 init 启动。启动后看到的进程名叫 zygote，其最初的名字是 app_process，通过直接调用 pctrl 把名字给改成了“zygote”。

（1）app_process.main 函数。

此函数定义在文件“frameworks/base/cmds/app_process/App_main.cpp”中，具体实现代码如下所示：

```
int main(int argc, char* const argv[])
{
#ifdef __arm__
    /*
     * b/7188322 - Temporarily revert to the compat memory layout
     * to avoid breaking third party apps.
     *
     * THIS WILL GO AWAY IN A FUTURE ANDROID RELEASE.
     *
     * http://git.kernel.org/?p=linux/kernel/git/torvalds/linux-2.6.git;a=commitdiff;h=7dbaa466
     * changes the kernel mapping from bottom up to top-down.
     * This breaks some programs which improperly embed
     * an out of date copy of Android's linker.
     */
    char value[PROPERTY_VALUE_MAX];
```

```
    property_get("ro.kernel.qemu", value, "");
    bool is_qemu = (strcmp(value, "1") == 0);
    if ((getenv("NO_ADDR_COMPAT_LAYOUT_FIXUP") == NULL) && !is_qemu) {
        int current = personality(0xFFFFFFFF);
        if ((current & ADDR_COMPAT_LAYOUT) == 0) {
            personality(current | ADDR_COMPAT_LAYOUT);
            setenv("NO_ADDR_COMPAT_LAYOUT_FIXUP", "1", 1);
            execv("/system/bin/app_process", argv);
            return -1;
        }
    }
    unsetenv("NO_ADDR_COMPAT_LAYOUT_FIXUP");
#endif

    // These are global variables in ProcessState.cpp
    mArgC = argc;
    mArgV = argv;

    mArgLen = 0;
    for (int i=0; i<argc; i++) {
        mArgLen += strlen(argv[i]) + 1;
    }
    mArgLen--;

    AppRuntime runtime;
    const char* argv0 = argv[0];
    // Process command line arguments
    // ignore argv[0]
    argc--;
    argv++;

    // Everything up to '--' or first non '-' arg goes to the vm

    int i = runtime.addVmArguments(argc, argv);

    // Parse runtime arguments. Stop at first unrecognized option.
    bool zygote = false;
    bool startSystemServer = false;
    bool application = false;
    const char* parentDir = NULL;
    const char* niceName = NULL;
    const char* className = NULL;
    while (i < argc) {
        const char* arg = argv[i++];
        if (!parentDir) {
            parentDir = arg;
        } else if (strcmp(arg, "--zygote") == 0) {
            zygote = true;
            niceName = "zygote";
        } else if (strcmp(arg, "--start-system-server") == 0) {
            startSystemServer = true;
        } else if (strcmp(arg, "--application") == 0) {
            application = true;
        } else if (strncmp(arg, "--nice-name=", 12) == 0) {
            niceName = arg + 12;
        } else {
            className = arg;
            break;
        }
    }

    if (niceName && *niceName) {
        setArgv0(argv0, niceName);
        set_process_name(niceName);
    }

    runtime.mParentDir = parentDir;
```

```
    if (zygote) {
        runtime.start("com.android.internal.os.ZygoteInit",
                startSystemServer ? "start-system-server" : "");
    } else if (className) {
        // Remainder of args get passed to startup class main()
        runtime.mClassName = className;
        runtime.mArgC = argc - i;
        runtime.mArgV = argv + i;
        runtime.start("com.android.internal.os.RuntimeInit",
                application ? "application" : "tool");
    } else {
        fprintf(stderr, "Error: no class name or --zygote supplied.\n");
        app_usage();
        LOG_ALWAYS_FATAL("app_process: no class name or --zygote supplied.");
        return 10;
    }
}
```

此函数的主要作用就是创建一个 AppRuntime 变量，然后调用它的 start 成员函数。runtime 是 AppRuntime 的实例，AppRuntime 继承自 AndroidRuntime，进入 AndroidRuntime 类的 start 函数。函数 start 在文件“frameworks/base/core/jni/AndroidRuntime.cpp”中定义，具体代码如下所示：

```
void AndroidRuntime::start(const char* className, const char* options)
{
    ALOGD("\n>>>>>> AndroidRuntime START %s <<<<<<\n",
            className != NULL ? className : "(unknown)");

    /*
     * 'startSystemServer == true' means runtime is obsolete and not run from
     * init.rc anymore, so we print out the boot start event here.
     */
    if (strcmp(options, "start-system-server") == 0) {
        /* track our progress through the boot sequence */
        const int LOG_BOOT_PROGRESS_START = 3000;
        LOG_EVENT_LONG(LOG_BOOT_PROGRESS_START,
                       ns2ms(systemTime(SYSTEM_TIME_MONOTONIC)));
    }

    const char* rootDir = getenv("ANDROID_ROOT");
    if (rootDir == NULL) {
        rootDir = "/system";
        if (!hasDir("/system")) {
            LOG_FATAL("No root directory specified, and /android does not exist.");
            return;
        }
        setenv("ANDROID_ROOT", rootDir, 1);
    }

    //const char* kernelHack = getenv("LD_ASSUME_KERNEL");
    //ALOGD("Found LD_ASSUME_KERNEL='%s'\n", kernelHack);

    /* start the virtual machine */
    JniInvocation jni_invocation;
    jni_invocation.Init(NULL);
    JNIEnv* env;
    if (startVm(&mJavaVM, &env) != 0) {
        return;
    }
    onVmCreated(env);

    /*
     * Register android functions.
     */
    if (startReg(env) < 0) {
        ALOGE("Unable to register all android natives\n");
        return;
    }
```

```
    /*
     * We want to call main() with a String array with arguments in it.
     * At present we have two arguments, the class name and an option string.
     * Create an array to hold them.
     */
    jclass stringClass;
    jobjectArray strArray;
    jstring classNameStr;
    jstring optionsStr;

    stringClass = env->FindClass("java/lang/String");
    assert(stringClass != NULL);
    strArray = env->NewObjectArray(2, stringClass, NULL);
    assert(strArray != NULL);
    classNameStr = env->NewStringUTF(className);
    assert(classNameStr != NULL);
    env->SetObjectArrayElement(strArray, 0, classNameStr);
    optionsStr = env->NewStringUTF(options);
    env->SetObjectArrayElement(strArray, 1, optionsStr);
    /*
     * Start VM. This thread becomes the main thread of the VM, and will
     * not return until the VM exits.
     */
    char* slashClassName = toSlashClassName(className);
    jclass startClass = env->FindClass(slashClassName);
    if (startClass == NULL) {
        ALOGE("JavaVM unable to locate class '%s'\n", slashClassName);
        /* keep going */
    } else {
        jmethodID startMeth = env->GetStaticMethodID(startClass, "main",
            "([Ljava/lang/String;)V");
        if (startMeth == NULL) {
            ALOGE("JavaVM unable to find main() in '%s'\n", className);
            /* keep going */
        } else {
            env->CallStaticVoidMethod(startClass, startMeth, strArray);

#if 0
            if (env->ExceptionCheck())
                threadExitUncaughtException(env);
#endif
        }
    }
    free(slashClassName);

    ALOGD("Shutting down VM\n");
    if (mJavaVM->DetachCurrentThread() != JNI_OK)
        ALOGW("Warning: unable to detach main thread\n");
    if (mJavaVM->DestroyJavaVM() != 0)
        ALOGW("Warning: VM did not shut down cleanly\n");
}
```

此函数的作用是启动 Android 系统运行时库，它主要做了 3 件事情：一是调用函数 startVM 启动虚拟机，二是调用函数 startReg 注册 JNI 方法，三是调用了 com.android.internal.os.ZygoteInit 类的 main 函数。在上述代码中，还调用了类 JniInvocation 的 Init 函数，该函数在文件“libnativehelper/JniInvocation.cpp”中定义，具体实现代码如下所示：

```
bool JniInvocation::Init(const char* library) {
#ifdef HAVE_ANDROID_OS
  char default_library[PROPERTY_VALUE_MAX];
  property_get("persist.sys.dalvik.vm.lib", default_library, "libdvm.so");
#else
  const char* default_library = "libdvm.so";
#endif
  if (library == NULL) {
    library = default_library;
  }
```

```
    handle_ = dlopen(library, RTLD_NOW);
    if (handle_ == NULL) {
      ALOGE("Failed to dlopen %s: %s", library, dlerror());
      return false;
    }
    if (!FindSymbol(reinterpret_cast<void**>(&JNI_GetDefaultJavaVMInitArgs_),
                    "JNI_GetDefaultJavaVMInitArgs")) {
      return false;
    }
    if (!FindSymbol(reinterpret_cast<void**>(&JNI_CreateJavaVM_),
                    "JNI_CreateJavaVM")) {
      return false;
    }
    if (!FindSymbol(reinterpret_cast<void**>(&JNI_GetCreatedJavaVMs_),
                    "JNI_GetCreatedJavaVMs")) {
      return false;
    }
    return true;
  }
```

在上述代码中，首先定义了 default_library 字符数组，然后从属性系统中获得名为 persist.sys.dalvik.vm.lib 的属性值，默认值为 libdvm.so。对于 ART 环境来说，该值为 libart.so。由于参数 library 为 NULL，因此将 default_library 赋值给 library。接下来调用 dlopen 打开 libart.so 文件，分别调用 FindSymbol 函数从打开的 libart.so 文件中搜索到对应的导出符号，例如 JNI_CreateJavaVM 对应的是文件"/art/runtime/jni_internal.cc"中的 JNI_CreateJavaVM 函数，而不是"/dalvik/vm/Jni.cpp"的 JNI_CreateJavaVM 函数，因为现在打开的是 libart.so，这是需要注意的地方。

而在前面的函数 AndroidRuntime::start 中，调用 startVm 函数启动虚拟机，该函数最终会调用 JNI_CreateJavaVM 函数，此处的 JNI_ CreateJavaVM 函数在文件"/art/runtime/jni_internal.cc"中定义，具体实现代码如下所示：

```
extern "C" jint JNI_CreateJavaVM(JavaVM** p_vm, JNIEnv** p_env, void* vm_args) {
  const JavaVMInitArgs* args = static_cast<JavaVMInitArgs*>(vm_args);
  if (IsBadJniVersion(args->version)) {
    LOG(ERROR) << "Bad JNI version passed to CreateJavaVM: " << args->version;
    return JNI_EVERSION;
  }
  Runtime::Options options;
  for (int i = 0; i < args->nOptions; ++i) {
    JavaVMOption* option = &args->options[i];
    options.push_back(std::make_pair(std::string(option->optionString), option->extraInfo));
  }
  bool ignore_unrecognized = args->ignoreUnrecognized;
  if (!Runtime::Create(options, ignore_unrecognized)) {
    return JNI_ERR;
  }
  Runtime* runtime = Runtime::Current();
  bool started = runtime->Start();
  if (!started) {
    delete Thread::Current()->GetJniEnv();
    delete runtime->GetJavaVM();
    LOG(WARNING) << "CreateJavaVM failed";
    return JNI_ERR;
  }
  *p_env = Thread::Current()->GetJniEnv();
  *p_vm = runtime->GetJavaVM();
  return JNI_OK;
}
```

在上述代码中，首先解析虚拟机启动参数并存入到 Runtime::Options 实例中。然后根据解析的参数信息调用函数 Create 创建 Runtime 的实例，该函数的具体实现代码如下所示：

```
1. bool Runtime::Create(const Options& options, bool ignore_unrecognized) {
2. ......
3. InitLogging(NULL);
4. instance_ = new Runtime;
5. if (!instance_->Init(options, ignore_unrecognized)) {
6. .....
7. return false;
8. }
9. return true;
10. }
```

在上述代码中，第 3 行初始化 Log 系统，第 4 行创建 Runtime 实例，第 7 行初始化 Runtime。

再次回到 JNI_ Create JavaVM 函数中，获得 Runtime 当前实例后，Runtime 使用单例模式实现，并调用 Start 函数，该函数的实现代码如下：

```
1. bool Runtime::Start() {
2. ......
3. Thread* self = Thread::Current();
4. self->TransitionFromRunnableToSuspended(kNative);
5. started_ = true;
6. InitNativeMethods();
7. ......
8. if (is_zygote_) {
9. if (!InitZygote()) {
10. return false;
11. }
12. } else {
13. DidForkFromZygote();
14. }
15. StartDaemonThreads();
16. ......
17. return true;
18. }
```

在上述代码中，第 3 行获得当前运行线程；第 4 行将该线程状态从 Runnable 切换到 Suspend；第 6 行完成 Native 函数的初始化工作，函数 InitNativeMethods 的实现代码如下所示：

```
1. void Runtime::InitNativeMethods() {
2. .....
3. JNIEnv* env = self->GetJniEnv();
4. ......
5. RegisterRuntimeNativeMethods(env);
6. ......
7. }
```

在上述代码中，第 3 行获取 JNI 环境；第 5 行调用 RegisterRuntimeNativeMethods 函数完成 Native 函数的注册，至于注册了哪些 Native 函数，有兴趣的可以继续跟踪源码。

继续分析 Runtime:: Start 函数，调用 InitZygote 完成一些文件文件系统的 mount 后，会最终通过调用 java.lang.Daemons.start()函数启动守护进程。

再看函数 ZygoteInit.main，此函数在文件“frameworks/base/core/java/com/android/internal/os/ZygoteInit.java”中定义，具体实现代码如下所示：

```
public class ZygoteInit {
  ......
  public static void main(String argv[]) {
  try {
  ......
  registerZygoteSocket();
  ......
  ......
  if (argv[1].equals("true")) {
  startSystemServer();
  } else if (!argv[1].equals("false")) {
  ......
  }
```

```
......
if (ZYGOTE_FORK_MODE) {
......
} else {
runSelectLoopMode();
}
......
} catch (MethodAndArgsCaller caller) {
......
} catch (RuntimeException ex) {
......
}
}
......
}
```

上述函数主要做了 3 件事情：一是调用 registerZygoteSocket 函数创建了一个 socket 接口，用来和 ActivityManagerService 通信；二是调用 startSystemServer 函数来启动 SystemServer 组件；三是调用 runSelectLoopMode 函数进入一个无限循环，在前面创建的 socket 接口上等待 Activity ManagerService 请求创建新的应用程序进程。

再看 ZygoteInit.registerZygoteSocket，此函数在文件"frameworks/base/core/java/com/android/internal/os/ZygoteInit.java"中定义，具体实现代码如下所示：

```
public class ZygoteInit {
    ......
    /**
    * Registers a server socket for zygote command connections
    *
    * @throws RuntimeException when open fails
    */
    private static void registerZygoteSocket() {
    if (sServerSocket == null) {
    int fileDesc;
    try {
    String env = System.getenv(ANDROID_SOCKET_ENV);
    fileDesc = Integer.parseInt(env);
    } catch (RuntimeException ex) {
    ......
    }
    try {
    sServerSocket = new LocalServerSocket(
    createFileDescriptor(fileDesc));
    } catch (IOException ex) {
    .......
    }
    }
    ......
```

这个接口 socket 是通过文件描述符来创建的，这个文件描符代表的就是前面说的"/dev/socket/zygote"文件了。这个文件描述符是通过环境变量 ANDROID_SOCKET_ENV 得到的，它定义如下：

```
public class ZygoteInit {
......
private static final String ANDROID_SOCKET_ENV = "ANDROID_SOCKET_zygote";
......
}
```

那么这个环境变量的值又是由谁来设置的呢？我们知道，系统启动脚本文件"system/core/rootdir/init.rc"是由 init 进程来解释执行的，而 init 进程的源代码位于"system/core/init"目录中，在 init.c 文件中，是由函数 service_start 来解释文件 init.rc 中的 service 命令的，具体实现代码如下所示：

```
void service_start(struct service *svc, const char *dynamic_args)
{
......
pid_t pid;
......
pid = fork();
if (pid == 0) {
struct socketinfo *si;
......
for (si = svc->sockets; si; si = si->next) {
int socket_type = (
!strcmp(si->type, "stream") ? SOCK_STREAM :
(!strcmp(si->type, "dgram") ? SOCK_DGRAM : SOCK_SEQPACKET));
int s = create_socket(si->name, socket_type,
si->perm, si->uid, si->gid);
if (s >= 0) {
publish_socket(si->name, s);
}
}
......
}
......
}
```

每一个 service 命令都会促使 init 进程调用 fork 函数来创建一个新的进程，在新的进程里面，会分析里面的 socket 选项，对于每一个 socket 选项，都会通过 create_socket 函数来在“/dev/socket”目录下创建一个文件，在这个场景中，这个文件便是 zygote 了，然后得到的文件描述符通过 publish_socket 函数写入到环境变量中去：

```
static void publish_socket(const char *name, int fd)
{
char key[64] = ANDROID_SOCKET_ENV_PREFIX;
char val[64];
strlcpy(key + sizeof(ANDROID_SOCKET_ENV_PREFIX) - 1,
name,
sizeof(key) - sizeof(ANDROID_SOCKET_ENV_PREFIX));
snprintf(val, sizeof(val), "%d", fd);
add_environment(key, val);
/* make sure we don't close-on-exec */
fcntl(fd, F_SETFD, 0);
}
```

这里传进来的参数 name 的值为"zygote"，而 ANDROID_SOCKET_ENV_PREFIX 在文件“system/core/include/cutils/sockets.h”定义为：

```
view plain#define ANDROID_SOCKET_ENV_PREFIX "ANDROID_SOCKET_"
```

因此，这里就把上面得到的文件描述符写入到以"ANDROID_SOCKET_zygote"为 key 值的环境变量中。又因为上面的 ZygoteInit.registerZygoteSocket 函数与这里创建 socket 文件的 create_socket 函数是运行在同一个进程中，因此，上面的 ZygoteInit.registerZygoteSocket 函数可以直接使用这个文件描述符来创建一个 Java 层的 LocalServerSocket 对象。如果其他进程也需要打开这个“/dev/socket/zygote”文件来和 Zygote 进程进行通信，那就必须要通过文件名来连接这个 LocalServerSocket 了。ActivityManagerService 是通过 Process.start 函数来创建一个新的进程的，而 Process.start 函数会首先通过 Socket 连接到 Zygote 进程中，最终由 Zygote 进程来完成创建新的应用程序进程，而类 Process 是通过函数 openZygoteSocketIfNeeded 来连接到 Zygote 进程中的 Socket 的，具体实现代码如下所示：

```
public class Process {
......
private static void openZygoteSocketIfNeeded()
throws ZygoteStartFailedEx {
......
```

```
for (int retry = 0
; (sZygoteSocket == null) && (retry < (retryCount + 1))
; retry++ ) {
......
try {
sZygoteSocket = new LocalSocket();
sZygoteSocket.connect(new LocalSocketAddress(ZYGOTE_SOCKET,
LocalSocketAddress.Namespace.RESERVED));
sZygoteInputStream
= new DataInputStream(sZygoteSocket.getInputStream());
sZygoteWriter =
new BufferedWriter(
new OutputStreamWriter(
sZygoteSocket.getOutputStream()),
256);
......
} catch (IOException ex) {
......
}
}
......
}
......
}
```

这里的 ZYGOTE_SOCKET 定义如下。

```
public class Process {
  ......
  private static final String ZYGOTE_SOCKET = "zygote";
  ......
  }
```

它刚好就是对应"/dev/socket"目录下的 zygote 文件了。

Android 系统中的 socket 机制和 binder 机制一样，都是可以用来进行进程间通信。当 Socket 对象创建完成之后，回到第三步中的 ZygoteInit.main 函数中，startSystemServer 函数来启动 SystemServer 组件。

（2）ZygoteInit.startSystemServer 函数。

此函数在文件"frameworks/base/core/java/com/android/internal/os/ZygoteInit.java"中定义，具体实现代码如下所示：

```
public class ZygoteInit {
  ......
  private static boolean startSystemServer()
  throws MethodAndArgsCaller, RuntimeException {
  /* Hardcoded command line to start the system server */
  String args[] = {
  "--setuid=1000",
  "--setgid=1000",
  "--setgroups=1001,1002,1003,1004,1005,1006,1007,1008,1009,1010,1018,3001,3002,3003",
  "--capabilities=130104352,130104352",
  "--runtime-init",
  "--nice-name=system_server",
  "com.android.server.SystemServer",
  };
  ZygoteConnection.Arguments parsedArgs = null;
  int pid;
  try {
  parsedArgs = new ZygoteConnection.Arguments(args);
  ......
  /* Request to fork the system server process */
  pid = Zygote.forkSystemServer(
  parsedArgs.uid, parsedArgs.gid,
  parsedArgs.gids, debugFlags, null,
  parsedArgs.permittedCapabilities,
  parsedArgs.effectiveCapabilities);
```

```
} catch (IllegalArgumentException ex) {
......
}
/* For child process */
if (pid == 0) {
handleSystemServerProcess(parsedArgs);
}
return true;
}
......
}
```

这里可以看到，Zygote 进程通过 Zygote.forkSystemServer 函数来创建一个新的进程来启动 SystemServer 组件，返回值 pid 等于 0 的地方就是新的进程要执行的路径，即新创建的进程会执行 handleSystemServerProcess 函数。

（3）ZygoteInit.handleSystemServerProcess 函数。

此函数在文件“frameworks/base/core/java/com/android/internal/os/ZygoteInit.java”中定义，具体实现代码如下所示：

```
public class ZygoteInit {
  ......
  private static void handleSystemServerProcess(
  ZygoteConnection.Arguments parsedArgs)
  throws ZygoteInit.MethodAndArgsCaller {
  closeServerSocket();
  /*
  * Pass the remaining arguments to SystemServer.
  * "--nice-name=system_server com.android.server.SystemServer"
  */
  RuntimeInit.zygoteInit(parsedArgs.remainingArgs);
  /* should never reach here */
  }
  ......
}
```

由于由 Zygote 进程创建的子进程会继承 Zygote 进程在前面第（4）步中创建的 Socket 文件描述符，而这里的子进程又不会用到它，因此，这里就调用 closeServerSocket 函数来关闭它。此函数接着调用 RuntimeInit.zygoteInit 函数来进一步执行启动 SystemServer 组件的操作。

（4）RuntimeInit.zygoteInit 函数。

此函数在文件“frameworks/base/core/java/com/android/internal/os/RuntimeInit.java”中定义，具体实现代码如下所示：

```
public class RuntimeInit {
  ......
  public static final void zygoteInit(String[] argv)
  throws ZygoteInit.MethodAndArgsCaller {
  ......
  zygoteInitNative();
  ......
  // Remaining arguments are passed to the start class's static main
  String startClass = argv[curArg++];
  String[] startArgs = new String[argv.length - curArg];
  System.arraycopy(argv, curArg, startArgs, 0, startArgs.length);
  invokeStaticMain(startClass, startArgs);
  }
  ......
}
```

此函数会执行两个操作：一个是调用 zygoteInitNative 函数来执行一个 Binder 进程间通信机制的初始化工作，这个工作完成之后，这个进程中的 Binder 对象就可以方便地进行进程间通信了；另一个是调用第（5）步传进来的 com.android.server.SystemServer 类的 main 函数。

（5）RuntimeInit.zygoteInitNative 函数。

此函数定义在文件“frameworks/base/core/java/com/android/internal/os/RuntimeInit.java”中，具体实现代码如下所示：

```
public class RuntimeInit {
  ......
  public static final native void zygoteInitNative();
  ......
  }
```

从这里可以看出，函数 zygoteInitNative 是一个 Native 函数，实现在文件“frameworks/base/core/jni/AndroidRuntime.cpp”中。完成这一步后，这个进程的 Binder 进程间通信机制基础设施就准备好了。

回到步骤（7）中的 RuntimeInit.zygoteInitNative 函数，下一步它就要执行 com.android.server.System Server 类的 main 函数了。

（6）SystemServer.main 函数。

此函数在文件“frameworks/base/services/java/com/android/server/SystemServer.java”中定义，具体实现代码如下所示：

```
public class SystemServer
  {
  ......
  native public static void init1(String[] args);
  ......
  public static void main(String[] args) {
  init1(args);
  ......
  }
  public static final void init2() {
  Slog.i(TAG, "Entered the Android system server!");
  Thread thr = new ServerThread();
  thr.setName("android.server.ServerThread");
  thr.start();
  }
  ......
  }
```

这里的 main 函数首先会执行 JNI 方法 init1，然后 init1 会调用这里的 init2 函数，在 init2 函数里面，会创建一个 ServerThread 线程对象来执行一些系统关键服务的启动操作。

执行完成这一步骤后，层层返回，最后回到第（3）步中的 ZygoteInit.main 函数中，接下来它就要调用函数 runSelectLoopMode 进入一个无限循环，在前面第（4）步中创建的 socket 接口上等待 ActivityManagerService 请求创建新的应用程序进程了。

（7）ZygoteInit.runSelectLoopMode 函数。

此函数在文件“frameworks/base/core/java/com/android/internal/os/ZygoteInit.java”中定义，具体实现代码如下所示：

```
public class ZygoteInit {
  ......
  private static void runSelectLoopMode() throws MethodAndArgsCaller {
  ArrayList fds = new ArrayList();
  ArrayList peers = new ArrayList();
  FileDescriptor[] fdArray = new FileDescriptor[4];
  fds.add(sServerSocket.getFileDescriptor());
  peers.add(null);
  int loopCount = GC_LOOP_COUNT;
  while (true) {
  int index;
  ......
```

```
try {
fdArray = fds.toArray(fdArray);
index = selectReadable(fdArray);
} catch (IOException ex) {
throw new RuntimeException("Error in select()", ex);
}
if (index < 0) {
throw new RuntimeException("Error in select()");
} else if (index == 0) {
ZygoteConnection newPeer = acceptCommandPeer();
peers.add(newPeer);
fds.add(newPeer.getFileDesciptor());
} else {
boolean done;
done = peers.get(index).runOnce();
if (done) {
peers.remove(index);
fds.remove(index);
}
}
}
}
......
}
```

此函数用于等待 ActivityManagerService 来连接这个 Socket，然后调用 ZygoteConnection.runOnce 函数创建新的应用程序。

这样，Zygote 进程就启动完成了，到此为止，终于都对 Android 系统中的进程有了一个深刻的认识，在此总结如下 3 点。

- 系统启动时 init 进程会创建 Zygote 进程，Zygote 进程负责后续 Android 应用程序框架层其他进程的创建和启动工作。
- Zygote 进程会首先创建一个 SystemServer 进程，SystemServer 进程负责启动系统的关键服务，如包管理服务 PackageManagerService 和应用程序组件管理服务 ActivityManagerService。
- 当需要启动一个 Android 应用程序时，ActivityManagerService 会通过 Socket 进程间通信机制，通知 Zygote 进程为这个应用程序创建一个新的进程。

第7章　IPC进程通信机制

要想了解Dalvik VM线程管理的知识，除了需要了解Android的内存系统的基本知识外，还需要了解内存控制进程运行的机制（IPC）。在本章的内容中，将详细讲解Android系统中IPC进程通信机制的基本知识，为读者步入本书后面知识的学习打下基础。

7.1 Binder机制概述

Binder是Android系统提供的一种IPC（进程间通信）机制。由于Android是基于Linux内核的，所以除了Binder以外，还存在其他的IPC机制，例如管道和socket等。Binder相对于其他IPC机制来说更加灵活和方便，其驱动代码保存在文件“kernel/drivers/staing/android/binder.c”中，另外该目录下还有一个binder.h头文件。Binder是一个虚拟设备，所以它的代码相比而言还算简单，读者只要有基本的Linux驱动开发方面的知识就能读懂。在“/proc/binder”目录下的内容，可以用来查看Binder设备的运行状况。

可以将Android系统看作是一个基于Binder通信的C/S架构，Binder就像网络一样把系统的各个部分连接在了一起。在基于Binder通信的C/S架构体系中，除了C/S架构所包括的Client端和Server端外，Android还有一个全局的Service Manager端，它的作用是管理系统中的各种服务（Service）。

在Android系统的Binder机制中，由Client、Server、Service Manager、Binder Driver 4个组件组成，具体关系如图7-1所示。

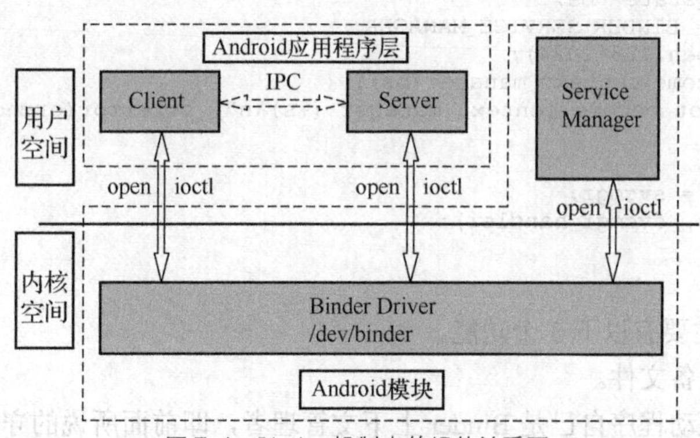

图7-1　Binder机制中的组件关系图

在Android系统中，Client、Server和ServiceManager三者之间的交互关系如下所示。

- Client、Server和Service Manager在用户空间中实现，Binder驱动程序在内核空间中实现。
- Server进程要先注册一些Service到Service Manager中，所以Server是Service Manager的客户端，而Service Manager就是服务端了。Service Manager是一个守护进程，能够管理Server并向Client提供查询Server接口。

- 如果某个 Client 进程要使用某个 Service，必须先到 Service Manager 中获取该 Service 的相关信息，所以 Client 是 Service Manager 的客户端。另外，Client 根据得到的 Service 信息与 Service 所在的 Server 进程建立通信的通路，然后就可以直接与 Service 交互了，所以 Client 也是 Server 的客户端。
- Binder 驱动程序提供设备文件“/dev/binder”与用户空间交互，Client、Server 和 Service Manager 通过 open 和 ioctl 文件操作函数与 Binder 驱动程序进行通信。

在 Android 平台中已经实现了 Binder 驱动程序和 Service Manager，开发者只需要在用户空间实现自己的 Client 和 Server 即可。三者的交互都是基于 Binder 通信的，所以通过任意两者之间的关系，都可以获取 Binder 的奥秘。

7.2 Service Manager 是 Binder 机制的上下文管理者

在分析 Binder 源代码时，需要先弄清楚 Service Manager 是如何告知 Binder 驱动程序它是 Binder 机制的上下文管理者。Service Manager 是整个 Binder 机制的守护进程，用来管理开发者创建的各种 Server，并且向 Client 提供查询 Server 远程接口的功能。因为 Service Manager 组件是用来管理 Server 并且向 Client 提供查询 Server 远程接口的功能，所以 Service Manager 必然要和 Server 以及 Client 进行通信。我们知道，Service Manger、Client 和 Server 三者分别是运行在独立的进程当中的，这样它们之间的通信也属于进程间的通信，而且也是采用 Binder 机制进行进程间通信。因此，Service Manager 在充当 Binder 机制守护进程的角色的同时，也在充当 Server 的角色，但是它是一种特殊的 Server，要想了解具体的特殊之处请看本节下面的内容。

7.2.1　入口函数

Service Manager 在用户空间的源代码位于“frameworks/base/cmds/servicemanager”目录下，主要是由文件 binder.h、binder.c 和 service_manager.c 组成。Service Manager 的入口位于文件“service_manager.c”中的函数 main()中，代码如下所示：

```
int main(int argc, char **argv){
    struct binder_state *bs;
    void *svcmgr = BINDER_SERVICE_MANAGER;
    bs = binder_open(128*1024);
    if (binder_become_context_manager(bs)) {
        LOGE("cannot become context manager (%s)\n", strerror(errno));
        return -1;
    }
    svcmgr_handle = svcmgr;
    binder_loop(bs, svcmgr_handler);
    return 0;
}
```

上述函数 main()主要有以下 3 个功能。

- 打开 Binder 设备文件。
- 告诉 Binder 驱动程序自己是 Binder 上下文管理者，即前面所说的守护进程。
- 进入一个无穷循环，充当 Server 的角色，等待 Client 的请求。

在进入上述 3 个功能之前，先来看一下这里用到的结构体 binder_state、宏 BINDER_SERVICE_MANAGER 的定义。结构体 binder_state 定义在文件“frameworks/base/cmds/servicemanager/binder.c”中，代码如下：

```
struct binder_state {
    int fd;
    void *mapped;
```

```
    unsigned mapsize;
};
```

其中 fd 表示文件描述符，即表示打开的“/dev/binder”设备文件描述符；mapped 表示把设备文件“/dev/binder”映射到进程空间的起始地址；mapsize 表示上述内存映射空间的大小。

宏 BINDER_SERVICE_MANAGER 在文件“frameworks/base/cmds/servicemanager/binder.h”中定义，代码如下。

```
/* the one magic object */
#define BINDER_SERVICE_MANAGER ((void*) 0)
```

这表示 Service Manager 的句柄为 0。Binder 通信机制使用句柄来代表远程接口，此句柄的意义和 Windows 编程中用到的句柄差不多。前面说到，Service Manager 在充当守护进程的同时，也充当 Server 的角色，当它作为远程接口使用时，它的句柄值便为 0，这就是它的特殊之处，其余 Server 的远程接口句柄值都大于 0 而且由 Binder 驱动程序自动进行分配。

7.2.2 打开 Binder 设备文件

函数首先打开 Binder 设备文件的操作函数 binder_open()，此函数的定义位于文件“frameworks/base/cmds/servicemanager/binder.c”中，代码如下：

```
struct binder_state *binder_open(unsigned mapsize){
    struct binder_state *bs;
    bs = malloc(sizeof(*bs));
    if (!bs) {
        errno = ENOMEM;
        return 0;
    }
    bs->fd = open("/dev/binder", O_RDWR);
    if (bs->fd < 0) {
        fprintf(stderr,"binder: cannot open device (%s)\n",
                strerror(errno));
        goto fail_open;
    }
    bs->mapsize = mapsize;
    bs->mapped = mmap(NULL, mapsize, PROT_READ, MAP_PRIVATE, bs->fd, 0);
    if (bs->mapped == MAP_FAILED) {
        fprintf(stderr,"binder: cannot map device (%s)\n",
                strerror(errno));
        goto fail_map;
    }
        /* TODO: check version */
    return bs;
fail_map:
    close(bs->fd);
fail_open:
    free(bs);
    return 0;
}
```

通过文件操作函数 open()打开设备文件“/dev/binder”，此设备文件是在 Binder 驱动程序模块初始化的时候创建的。

7.2.3 创建设备文件

接下来先看一下这个设备文件的创建过程，来到“kernel/common/drivers/staging/android”目录，打开文件“binder.c”，可以看到如下模块初始化入口 binder_init 的代码：

```
static struct file_operations binder_fops = {
    .owner = THIS_MODULE,
    .poll = binder_poll,
    .unlocked_ioctl = binder_ioctl,
```

```
    .mmap = binder_mmap,
    .open = binder_open,
    .flush = binder_flush,
    .release = binder_release,
};
static struct miscdevice binder_miscdev = {
    .minor = MISC_DYNAMIC_MINOR,
    .name = "binder",
    .fops = &binder_fops
};

static int __init binder_init(void)
{
    int ret;

    binder_proc_dir_entry_root = proc_mkdir("binder", NULL);
    if (binder_proc_dir_entry_root)
        binder_proc_dir_entry_proc = proc_mkdir("proc", binder_proc_dir_entry_root);
       ret = misc_register(&binder_miscdev);
    if (binder_proc_dir_entry_root) {
        create_proc_read_entry("state", S_IRUGO, binder_proc_dir_entry_root,
 binder_read_proc_state, NULL);
        create_proc_read_entry("stats",  S_IRUGO,  binder_proc_dir_entry_root,  binder_read_proc_stats,
NULL);
        create_proc_read_entry("transactions", S_IRUGO, binder_proc_dir_entry_root, binder_read_proc_
transactions, NULL);
        create_proc_read_entry("transaction_log",  S_IRUGO,  binder_proc_dir_entry_root,  binder_
read_proc_transaction_log, &binder_transaction_log);
        create_proc_read_entry("failed_transaction_log", S_IRUGO, binder_proc_dir_entry_root, binder_
read_proc_transaction_log, &binder_transaction_log_failed);
    }
    return ret;
}
device_initcall(binder_init);
```

在函数 misc_register()中实现了创建设备文件的功能，并实现了 misc 设备的注册工作，在“/proc”目录中创建了各种 Binder 相关的文件供用户访问。从设备文件的操作方法 binder_fops 可以看出，通过如下函数 binder_open 的执行语句：

```
bs->fd = open("/dev/binder", O_RDWR);
```

即可进入到 Binder 驱动程序的 binder_open()函数，具体实现代码如下所示：

```
static int binder_open(struct inode *nodp, struct file *filp)
{
    struct binder_proc *proc;

    if (binder_debug_mask & BINDER_DEBUG_OPEN_CLOSE)
        printk(KERN_INFO "binder_open: %d:%d\n", current->group_leader->pid, current->pid);

    proc = kzalloc(sizeof(*proc), GFP_KERNEL);
    if (proc == NULL)
        return -ENOMEM;
    get_task_struct(current);
    proc->tsk = current;
    INIT_LIST_HEAD(&proc->todo);
    init_waitqueue_head(&proc->wait);
    proc->default_priority = task_nice(current);
    mutex_lock(&binder_lock);
    binder_stats.obj_created[BINDER_STAT_PROC]++;
    hlist_add_head(&proc->proc_node, &binder_procs);
    proc->pid = current->group_leader->pid;
    INIT_LIST_HEAD(&proc->delivered_death);
    filp->private_data = proc;
    mutex_unlock(&binder_lock);

    if (binder_proc_dir_entry_proc) {
        char strbuf[11];
```

```
        snprintf(strbuf, sizeof(strbuf), "%u", proc->pid);
        remove_proc_entry(strbuf, binder_proc_dir_entry_proc);
        create_proc_read_entry(strbuf, S_IRUGO, binder_proc_dir_entry_proc, binder_read_
        proc_proc, proc);
    }
    return 0;
}
```

上述函数的主要作用是创建一个名为 binder_proc 的数据结构，用此数据结构来保存打开设备文件“/dev/binder”的进程的上下文信息，并且将这个进程上下文信息保存在打开文件结构 file 的私有数据成员变量 private_data 中。这样当在执行其他文件操作时，就通过打开文件结构 file 来取回这个进程上下文信息了。这个进程上下文信息同时还会保存在一个全局哈希表 binder_procs 中，供驱动程序内部使用。哈希表 binder_procs 定义在文件的开头：

```
static HLIST_HEAD(binder_procs);
```

而结构体 struct binder_proc 也被定义在文件“kernel/common/drivers/staging/android/binder.c”中，具体实现代码如下所示：

```
struct binder_proc {
    struct hlist_node proc_node;
    struct rb_root threads;
    struct rb_root nodes;
    struct rb_root refs_by_desc;
    struct rb_root refs_by_node;
    int pid;
    struct vm_area_struct *vma;
    struct task_struct *tsk;
    struct files_struct *files;
    struct hlist_node deferred_work_node;
    int deferred_work;
    void *buffer;
    ptrdiff_t user_buffer_offset;
    struct list_head buffers;
    struct rb_root free_buffers;
    struct rb_root allocated_buffers;
    size_t free_async_space;
    struct page **pages;
    size_t buffer_size;
    uint32_t buffer_free;
    struct list_head todo;
    wait_queue_head_t wait;
    struct binder_stats stats;
    struct list_head delivered_death;
    int max_threads;
    int requested_threads;
    int requested_threads_started;
    int ready_threads;
    long default_priority;
};
```

上述结构体的成员比较多，其中最终重要的有 4 个成员变量：

- Threads;
- Nodes;
- refs_by_desc;
- refs_by_node。

上述 4 个成员变量都是表示红黑树的节点，也就是说，binder_proc 分别挂在 4 个红黑树下，具体说明如下所示。

- threads 树：用来保存 binder_proc 进程内用于处理用户请求的线程，它的最大数量由 max_threads 来决定。

❑ node 树：用来保存 binder_proc 进程内的 Binder 实体。

❑ refs_by_desc 树和 refs_by_node 树：用来保存 binder_proc 进程内的 Binder 引用，即引用的其他进程的 Binder 实体，它分别用两种方式来组织红黑树，一种是以句柄作来 key 值来组织，一种是以引用的实体节点的地址值作来 key 值来组织。它们都是表示同一样东西，只不过是为了内部查找方便而用两个红黑树来表示。

这样，打开设备文件“/dev/binder”的操作就完成了，接下来需要对打开的设备文件进行内存映射操作 mmap：

```
bs->mapped = mmap(NULL, mapsize, PROT_READ, MAP_PRIVATE, bs->fd, 0);
```

对应 Binder 驱动程序的是函数 binder_mmap()，具体实现代码如下所示：

```
static int binder_mmap(struct file *filp, struct vm_area_struct *vma)
{
    int ret;
    struct vm_struct *area;
    struct binder_proc *proc = filp->private_data;
    const char *failure_string;
    struct binder_buffer *buffer;
    if ((vma->vm_end - vma->vm_start) > SZ_4M)
        vma->vm_end = vma->vm_start + SZ_4M;
    if (binder_debug_mask & BINDER_DEBUG_OPEN_CLOSE)
        printk(KERN_INFO
            "binder_mmap: %d %lx-%lx (%ld K) vma %lx pagep %lx\n",
            proc->pid, vma->vm_start, vma->vm_end,
            (vma->vm_end - vma->vm_start) / SZ_1K, vma->vm_flags,
            (unsigned long)pgprot_val(vma->vm_page_prot));
    if (vma->vm_flags & FORBIDDEN_MMAP_FLAGS) {
        ret = -EPERM;
        failure_string = "bad vm_flags";
        goto err_bad_arg;
    }
    vma->vm_flags = (vma->vm_flags | VM_DONTCOPY) & ~VM_MAYWRITE;

    if (proc->buffer) {
        ret = -EBUSY;
        failure_string = "already mapped";
        goto err_already_mapped;
    }

    area = get_vm_area(vma->vm_end - vma->vm_start, VM_IOREMAP);
    if (area == NULL) {
        ret = -ENOMEM;
        failure_string = "get_vm_area";
        goto err_get_vm_area_failed;
    }
    proc->buffer = area->addr;
    proc->user_buffer_offset = vma->vm_start - (uintptr_t)proc->buffer;

#ifdef CONFIG_CPU_CACHE_VIPT
    if (cache_is_vipt_aliasing()) {
        while (CACHE_COLOUR((vma->vm_start ^ (uint32_t)proc->buffer))) {
            printk(KERN_INFO "binder_mmap: %d %lx-%lx maps %p bad alignment\n", proc->pid,
            vma->vm_start, vma->vm_end, proc->buffer);
            vma->vm_start += PAGE_SIZE;
        }
    }
#endif
    proc->pages = kzalloc(sizeof(proc->pages[0]) * ((vma->vm_end - vma->vm_start) /
PAGE_SIZE), GFP_KERNEL);
    if (proc->pages == NULL) {
        ret = -ENOMEM;
        failure_string = "alloc page array";
        goto err_alloc_pages_failed;
    }
```

```
    proc->buffer_size = vma->vm_end - vma->vm_start;

    vma->vm_ops = &binder_vm_ops;
    vma->vm_private_data = proc;

    if (binder_update_page_range(proc, 1, proc->buffer, proc->buffer + PAGE_SIZE, vma))
    {
        ret = -ENOMEM;
        failure_string = "alloc small buf";
        goto err_alloc_small_buf_failed;
    }
    buffer = proc->buffer;
    INIT_LIST_HEAD(&proc->buffers);
    list_add(&buffer->entry, &proc->buffers);
    buffer->free = 1;
    binder_insert_free_buffer(proc, buffer);
    proc->free_async_space = proc->buffer_size / 2;
    barrier();
    proc->files = get_files_struct(current);
    proc->vma = vma;

    /*printk(KERN_INFO "binder_mmap: %d %lx-%lx maps %p\n", proc->pid, vma->vm_start,
vma->vm_end, proc->buffer);*/
    return 0;

err_alloc_small_buf_failed:
    kfree(proc->pages);
    proc->pages = NULL;
err_alloc_pages_failed:
    vfree(proc->buffer);
    proc->buffer = NULL;
err_get_vm_area_failed:
err_already_mapped:
err_bad_arg:
    printk(KERN_ERR "binder_mmap: %d %lx-%lx %s failed %d\n", proc->pid, vma->vm_start,
vma->vm_end, failure_string, ret);
    return ret;
}
```

在上述函数 binder_mmap()中，首先通过 filp->private_data 得到在打开设备文件“/dev/binder”时创建的结构 binder_proc。内存映射信息放在 vma 参数中。读者需要注意，这里的 vma 的数据类型是结构 vm_area_struct，它表示的是一块连续的虚拟地址空间区域。在函数变量声明的地方，还看到有一个类似的结构体 vm_struct，这个数据结构也是表示一块连续的虚拟地址空间区域。那么，这两者的区别是什么呢？在 Linux 系统中，结构体 vm_area_struct 表示的虚拟地址是给进程使用的，而结构体 vm_struct 表示的虚拟地址是给内核使用的，它们对应的物理页面都可以是不连续的。结构体 vm_area_struct 表示的地址空间范围是 0～3 G，而结构体 vm_struct 表示的地址空间范围是（3 G + 896 M + 8 M）～4 G。为什么结构体 vm_struct 表示的地址空间范围不是 3 G～4 G 呢？因为 3 G～（3 G + 896 M）范围的地址是用来映射连续的物理页面的，这个范围的虚拟地址和对应的实际物理地址有着简单的对应关系，即对应 0～896 M 的物理地址空间，而（3 G + 896 M）～（3 G + 896 M + 8 M）是安全保护区域。例如所有指向这 8 M 地址空间的指针都是非法的，所以结构体 vm_struct 使用（3 G + 896 M + 8 M）～4 G 地址空间来映射非连续的物理页面。

此处为什么会同时使用进程虚拟地址空间和内核虚拟地址空间来映射同一个物理页面呢？这就是 Binder 进程间通信机制的精髓所在了。在同一个物理页面，一方面映射到进程虚拟地址空间，一方面映射到内核虚拟地址空间，这样进程和内核之间就可以减少一次内存拷贝工作，提高了进程之间的通信效率。

讲解了 binder_mmap 的原理之后，整个函数的逻辑就很好理解了。但是在此还是先要解释一下 binder_proc 结构体中的如下成员变量。

- buffer：是一个 void*指针，它表示要映射的物理内存在内核空间中的起始位置。
- buffer_size：是一个 size_t 类型的变量，表示要映射的内存的大小。
- pages：是一个 struct page*类型的数组，struct page 是用来描述物理页面的数据结构；
- user_buffer_offset：是一个 ptrdiff_t 类型的变量，它表示的是内核使用的虚拟地址与进程使用的虚拟地址之间的差值，即如果某个物理页面在内核空间中对应的虚拟地址是 addr，则这个物理页面在进程空间对应的虚拟地址就为“addr + user_buffer_offset”格式。

7.2.4　管理内存映射地址空间

接下来还需要看一下 Binder 驱动程序管理内存映射地址空间的方法，即如何管理 buffer ～ (buffer + buffer_size)这段地址空间的，这个地址空间被划分为一段一段来管理，每一段是用结构体 binder_buffer 来描述的，具体实现代码如下所示：

```
struct binder_buffer {
    struct list_head entry; /* free and allocated entries by addesss */
    struct rb_node rb_node; /* free entry by size or allocated entry */
            /* by address */
    unsigned free : 1;
    unsigned allow_user_free : 1;
    unsigned async_transaction : 1;
    unsigned debug_id : 29;
    struct binder_transaction *transaction;
    struct binder_node *target_node;
    size_t data_size;
    size_t offsets_size;
    uint8_t data[0];
};
```

每一个 binder_buffer 通过其成员 entry 按从低地址到高地址连入到 struct binder_proc 中的 buffers 表示的链表中去，同时，每一个 binder_buffer 又分为正在使用的和空闲的，通过 free 成员变量来区分。空闲的 binder_buffer 通过成员变量 rb_node 的帮助，连入到 struct binder_proc 中的 free_buffers 表示的红黑树中去。而那些正在使用的 binder_buffer，通过成员变量 rb_node 连入到 binder_proc 中的 allocated_buffers 表示的红黑树中去。这样做当然是为了方便查询和维护这块地址空间了。

然后回到函数 binder_mmap()，首先是对参数做一些检查，例如要映射的内存大小不能超过 SIZE_4 M，即 4 MB。在来到文件 service_manager.c 中的 main()函数，这里传进来的值是 128×1 024 个字节，即 128 KB，这个检查没有问题。通过检查之后，调用函数 get_vm_area()获得一个空闲的 vm_struct 区间，并初始化“proc”结构体的 buffer、user_buffer_offset、pages 和 buffer_size 等成员变量，接着调用 binder_update_page_range 为虚拟地址空间 proc->buffer ～ proc->buffer + PAGE_SIZE 分配一个空闲的物理页面，同时这段地址空间使用一个 binder_buffer 来描述，分别插入到 proc->buffers 链表和 proc->free_buffers 红黑树中去，最后还初始化了 proc 结构体的 free_async_space、files 和 vma 3 个成员变量。

然后继续分析函数 binder_update_page_range()，看一下 Binder 驱动程序是如何实现把一个物理页面同时映射到内核空间和进程空间去的。具体实现代码如下所示：

```
static int binder_update_page_range(struct binder_proc *proc, int allocate,
    void *start, void *end, struct vm_area_struct *vma)
{
    void *page_addr;
    unsigned long user_page_addr;
    struct vm_struct tmp_area;
    struct page **page;
    struct mm_struct *mm;
    if (binder_debug_mask & BINDER_DEBUG_BUFFER_ALLOC)
```

```
        printk(KERN_INFO "binder: %d: %s pages %p-%p\n",
            proc->pid, allocate ? "allocate" : "free", start, end);
    if (end <= start)
        return 0;
    if (vma)
        mm = NULL;
    else
        mm = get_task_mm(proc->tsk);
    if (mm) {
        down_write(&mm->mmap_sem);
        vma = proc->vma;
    }
    if (allocate == 0)
        goto free_range;
    if (vma == NULL) {
        printk(KERN_ERR "binder: %d: binder_alloc_buf failed to "
            "map pages in userspace, no vma\n", proc->pid);
        goto err_no_vma;
    }
    for (page_addr = start; page_addr < end; page_addr += PAGE_SIZE) {
        int ret;
        struct page **page_array_ptr;
        page = &proc->pages[(page_addr - proc->buffer) / PAGE_SIZE];
        BUG_ON(*page);
        *page = alloc_page(GFP_KERNEL | __GFP_ZERO);
        if (*page == NULL) {
            printk(KERN_ERR "binder: %d: binder_alloc_buf failed "
                "for page at %p\n", proc->pid, page_addr);
            goto err_alloc_page_failed;
        }
        tmp_area.addr = page_addr;
        tmp_area.size = PAGE_SIZE + PAGE_SIZE /* guard page? */;
        page_array_ptr = page;
        ret = map_vm_area(&tmp_area, PAGE_KERNEL, &page_array_ptr);
        if (ret) {
            printk(KERN_ERR "binder: %d: binder_alloc_buf failed "
                "to map page at %p in kernel\n",
                proc->pid, page_addr);
            goto err_map_kernel_failed;
        }
        user_page_addr =
            (uintptr_t)page_addr + proc->user_buffer_offset;
        ret = vm_insert_page(vma, user_page_addr, page[0]);
        if (ret) {
            printk(KERN_ERR "binder: %d: binder_alloc_buf failed "
                "to map page at %lx in userspace\n",
                proc->pid, user_page_addr);
            goto err_vm_insert_page_failed;
        }
        /* vm_insert_page does not seem to increment the refcount */
    }
    if (mm) {
        up_write(&mm->mmap_sem);
        mmput(mm);
    }
    return 0;
free_range:
    for (page_addr = end - PAGE_SIZE; page_addr >= start;
         page_addr -= PAGE_SIZE) {
        page = &proc->pages[(page_addr - proc->buffer) / PAGE_SIZE];
        if (vma)
            zap_page_range(vma, (uintptr_t)page_addr +
                proc->user_buffer_offset, PAGE_SIZE, NULL);
err_vm_insert_page_failed:
        unmap_kernel_range((unsigned long)page_addr, PAGE_SIZE);
err_map_kernel_failed:
        __free_page(*page);
        *page = NULL;
err_alloc_page_failed:
```

```
            ;
        }
    err_no_vma:
        if (mm) {
            up_write(&mm->mmap_sem);
            mmput(mm);
        }
        return -ENOMEM;
    }
```

通过上述函数不但可以分配物理页面，而且可以用来释放物理页面，这可以通过参数 allocate 来区别，在此只需关注分配物理页面的情况。要分配物理页面的虚拟地址空间范围为(start ～ end)，只需直接看中间的 for 循环部分，具体实现代码如下所示：

```
for (page_addr = start; page_addr < end; page_addr += PAGE_SIZE) {
    int ret;
    struct page **page_array_ptr;
    page = &proc->pages[(page_addr - proc->buffer) / PAGE_SIZE];
    BUG_ON(*page);
    *page = alloc_page(GFP_KERNEL | __GFP_ZERO);
    if (*page == NULL) {
        printk(KERN_ERR "binder: %d: binder_alloc_buf failed "
               "for page at %p\n", proc->pid, page_addr);
        goto err_alloc_page_failed;
    }
    tmp_area.addr = page_addr;
    tmp_area.size = PAGE_SIZE + PAGE_SIZE /* guard page? */;
    page_array_ptr = page;
    ret = map_vm_area(&tmp_area, PAGE_KERNEL, &page_array_ptr);
    if (ret) {
        printk(KERN_ERR "binder: %d: binder_alloc_buf failed "
               "to map page at %p in kernel\n",
               proc->pid, page_addr);
        goto err_map_kernel_failed;
    }
    user_page_addr =
        (uintptr_t)page_addr + proc->user_buffer_offset;
    ret = vm_insert_page(vma, user_page_addr, page[0]);
    if (ret) {
        printk(KERN_ERR "binder: %d: binder_alloc_buf failed "
               "to map page at %lx in userspace\n",
               proc->pid, user_page_addr);
        goto err_vm_insert_page_failed;
    }
    /* vm_insert_page does not seem to increment the refcount */
}
```

在上述代码中，首先调用 alloc_page()分配一个物理页面，此函数返回一个结构体 page 物理页面描述符，根据这个描述的内容初始化好结构体 vm_struct tmp_area，然后通过 map_vm_area 将这个物理页面插入到 tmp_area 描述的内核空间去，接着通过 page_addr + proc->user_buffer_offset 获得进程虚拟空间地址，并通过函数 vm_insert_page()将这个物理页面插入到进程地址空间去，参数 vma 表示要插入的进程的地址空间。

7.2.5 发生通知

此时，文件“frameworks/base/cmds/servicemanager/binder.c”中的函数 binder_open()讲解完毕。再次回到文件“frameworks/base/cmds/servicemanager/service_manager.c”中的 main()函数，接下来需要调用 binder_become_context_manager 来通知 Binder 驱动程序自己是 Binder 机制的上下文管理者，即守护进程。函数 binder_become_context_manager()位于文件“frameworks/base/cmds/servicemanager/binder.c”中，具体代码如下所示：

```
int binder_become_context_manager(struct binder_state *bs){
    return ioctl(bs->fd, BINDER_SET_CONTEXT_MGR, 0);
}
```

在此通过调用 ioctl 文件操作函数通知 Binder 驱动程序自己是守护进程，命令号是 BINDER_SET_CONTEXT_MGR，并没有任何参数。BINDER_SET_CONTEXT_MGR 定义为：

```
#define    BINDER_SET_CONTEXT_MGR         _IOW('b', 7, int)
```

这样就进入到 Binder 驱动程序的函数 binder_ioctl()，在此只关注如下 BINDER_SET_CONTEXT_MGR 命令即可：

```
static long binder_ioctl(struct file *filp, unsigned int cmd, unsigned long arg)
{
    int ret;
    struct binder_proc *proc = filp->private_data;
    struct binder_thread *thread;
    unsigned int size = _IOC_SIZE(cmd);
    void __user *ubuf = (void __user *)arg;
    /*printk(KERN_INFO "binder_ioctl: %d:%d %x %lx\n", proc->pid, current->pid, cmd,
    arg);*/
    ret = wait_event_interruptible(binder_user_error_wait, binder_stop_on_user_error < 2);
    if (ret)
        return ret;
    mutex_lock(&binder_lock);
    thread = binder_get_thread(proc);
    if (thread == NULL) {
        ret = -ENOMEM;
        goto err;
    }
    switch (cmd) {
        ......
    case BINDER_SET_CONTEXT_MGR:
        if (binder_context_mgr_node != NULL) {
            printk(KERN_ERR "binder: BINDER_SET_CONTEXT_MGR already set\n");
            ret = -EBUSY;
            goto err;
        }
        if (binder_context_mgr_uid != -1) {
            if (binder_context_mgr_uid != current->cred->euid) {
                printk(KERN_ERR "binder: BINDER_SET_"
                    "CONTEXT_MGR bad uid %d != %d\n",
                    current->cred->euid,
                    binder_context_mgr_uid);
                ret = -EPERM;
                goto err;
            }
        } else
            binder_context_mgr_uid = current->cred->euid;
        binder_context_mgr_node = binder_new_node(proc, NULL, NULL);
        if (binder_context_mgr_node == NULL) {
            ret = -ENOMEM;
            goto err;
        }
        binder_context_mgr_node->local_weak_refs++;
        binder_context_mgr_node->local_strong_refs++;
        binder_context_mgr_node->has_strong_ref = 1;
        binder_context_mgr_node->has_weak_ref = 1;
        break;
        ......
    default:
        ret = -EINVAL;
        goto err;
    }
    ret = 0;
err:
    if (thread)
        thread->looper &= ~BINDER_LOOPER_STATE_NEED_RETURN;
```

```
    mutex_unlock(&binder_lock);
    wait_event_interruptible(binder_user_error_wait, binder_stop_on_user_error < 2);
    if (ret && ret != -ERESTARTSYS)
        printk(KERN_INFO "binder: %d:%d ioctl %x %lx returned %d\n", proc->pid,
        current->pid, cmd, arg, ret);
    return ret;
}
```

在分析函数 binder_ioctl()之前，需要先弄明白如下两个数据结构。

❑ 结构体 binder_thread：表示一个线程，这里就是执行 binder_become_context_manager()函数的线程，具体实现代码如下所示：

```
struct binder_thread {
    struct binder_proc *proc;
    struct rb_node rb_node;
    int pid;
    int looper;
    struct binder_transaction *transaction_stack;
    struct list_head todo;
    uint32_t return_error; /* Write failed, return error code in read buf */
    uint32_t return_error2; /* Write failed, return error code in read */
        /* buffer. Used when sending a reply to a dead process that */
        /* we are also waiting on */
    wait_queue_head_t wait;
    struct binder_stats stats;
};
```

在上述结构体中，proc 表示是这个线程所属的进程。结构体 binder_proc 中成员变量 thread 的类型是 rb_root，它表示一查红黑树，把属于这个进程的所有线程都组织起来，结构体 binder_thread 的成员变量 rb_node 就是用来链入这棵红黑树的节点了。looper 成员变量表示线程的状态，它可以取以下值：

```
enum {
    BINDER_LOOPER_STATE_REGISTERED  = 0x01,
    BINDER_LOOPER_STATE_ENTERED     = 0x02,
    BINDER_LOOPER_STATE_EXITED      = 0x04,
    BINDER_LOOPER_STATE_INVALID     = 0x08,
    BINDER_LOOPER_STATE_WAITING     = 0x10,
    BINDER_LOOPER_STATE_NEED_RETURN = 0x20
};
```

至于其余的成员变量，transaction_stack 表示线程正在处理的事务，todo 表示发往该线程的数据列表，return_error 和 return_error2 表示操作结果返回码，wait 用来阻塞线程等待某个事件的发生，stats 用来保存一些统计信息。这些成员变量遇到的时候再分析它们的作用。

❑ 数据结构 binder_node：表示一个 binder 实体，具体实现代码如下所示：

```
struct binder_node {
    int debug_id;
    struct binder_work work;
    union {
        struct rb_node rb_node;
        struct hlist_node dead_node;
    };
    struct binder_proc *proc;
    struct hlist_head refs;
    int internal_strong_refs;
    int local_weak_refs;
    int local_strong_refs;
    void __user *ptr;
    void __user *cookie;
    unsigned has_strong_ref : 1;
    unsigned pending_strong_ref : 1;
    unsigned has_weak_ref : 1;
    unsigned pending_weak_ref : 1;
    unsigned has_async_transaction : 1;
```

```
    unsigned accept_fds : 1;
    int min_priority : 8;
    struct list_head async_todo;
};
```

由此可见，rb_node 和 dead_node 组成了一个联合体，具体来说分为如下两种情形。

- 如果这个 Binder 实体还在正常使用，则使用 rb_node 来连入“proc->nodes”所表示的红黑树的节点，这棵红黑树用来组织属于这个进程的所有 Binder 实体。
- 如果这个 Binder 实体所属的进程已经销毁，而这个 Binder 实体又被其他进程所引用，则这个 Binder 实体通过 dead_node 进入到一个哈希表中去存放。proc 成员变量就是表示这个 Binder 实例所属的进程了。

refs 成员变量把所有引用了该 Binder 实体的 Binder 引用连接起来构成一个链表。internal_strong_refs、local_weak_refs 和 local_strong_refs 表示这个 Binder 实体的引用计数。ptr 和 cookie 成员变量分别表示这个 Binder 实体在用户空间的地址以及附加数据。其余的成员变量就不描述了，遇到的时候再分析。

接下来回到函数 binder_ioctl()中，首先是通过“filp->private_data”获得 proc 变量，此处的函数 binder_mmap()是一样的，然后通过函数 binder_get_thread()获得线程信息，具体实现代码如下所示：

```
static struct binder_thread *binder_get_thread(struct binder_proc *proc)
{
    struct binder_thread *thread = NULL;
    struct rb_node *parent = NULL;
    struct rb_node **p = &proc->threads.rb_node;

    while (*p) {
        parent = *p;
        thread = rb_entry(parent, struct binder_thread, rb_node);

        if (current->pid < thread->pid)
            p = &(*p)->rb_left;
        else if (current->pid > thread->pid)
            p = &(*p)->rb_right;
        else
            break;
    }
    if (*p == NULL) {
        thread = kzalloc(sizeof(*thread), GFP_KERNEL);
        if (thread == NULL)
            return NULL;
        binder_stats.obj_created[BINDER_STAT_THREAD]++;
        thread->proc = proc;
        thread->pid = current->pid;
        init_waitqueue_head(&thread->wait);
        INIT_LIST_HEAD(&thread->todo);
        rb_link_node(&thread->rb_node, parent, p);
        rb_insert_color(&thread->rb_node, &proc->threads);
        thread->looper |= BINDER_LOOPER_STATE_NEED_RETURN;
        thread->return_error = BR_OK;
        thread->return_error2 = BR_OK;
    }
    return thread;
}
```

在上述代码中，把当前线程 current 的 pid 作为键值，在进程“proc->threads”表示的红黑树中进行查找，看是否已经为当前线程创建过了 binder_thread 信息。在这个场景下，由于当前线程是第一次进到这里，所以肯定找不到，即*p == NULL 成立，于是，就为当前线程创建一个线程上下文信息结构体 binder_thread，并初始化相应成员变量，并插入到“proc->threads”所表示的红黑树中去，下次要使用时就可以从 proc 中找到了。注意，这里的 thread->looper = BINDER_

LOOPER_STATE_NEED_RETURN。

再回到函数 binder_ioctl()中，接下来会有两个全局变量 binder_context_mgr_node 和 binder_context_mgr_uid，定义如下：

```
static struct binder_node *binder_context_mgr_node;
static uid_t binder_context_mgr_uid = -1;
```

其中 binder_context_mgr_node 用来表示 Service Manager 实体，binder_context_mgr_uid 表示 Service Manager 守护进程的 uid。在这个场景下，由于当前线程是第一次进到这里，所以 binder_context_mgr_node 为 NULL，binder_context_mgr_uid 为-1，于是初始化 binder_context_mgr_uid 为 current->cred->euid，这样当前线程就成为 Binder 机制的守护进程了，并且通过 binder_new_node 为 Service Manager 创建 Binder 实体，具体实现代码如下所示：

```
static struct binder_node *
binder_new_node(struct binder_proc *proc, void __user *ptr, void __user *cookie)
{
    struct rb_node **p = &proc->nodes.rb_node;
    struct rb_node *parent = NULL;
    struct binder_node *node;
    while (*p) {
        parent = *p;
        node = rb_entry(parent, struct binder_node, rb_node);
        if (ptr < node->ptr)
            p = &(*p)->rb_left;
        else if (ptr > node->ptr)
            p = &(*p)->rb_right;
        else
            return NULL;
    }
    node = kzalloc(sizeof(*node), GFP_KERNEL);
    if (node == NULL)
        return NULL;
    binder_stats.obj_created[BINDER_STAT_NODE]++;
    rb_link_node(&node->rb_node, parent, p);
    rb_insert_color(&node->rb_node, &proc->nodes);
    node->debug_id = ++binder_last_id;
    node->proc = proc;
    node->ptr = ptr;
    node->cookie = cookie;
    node->work.type = BINDER_WORK_NODE;
    INIT_LIST_HEAD(&node->work.entry);
    INIT_LIST_HEAD(&node->async_todo);
    if (binder_debug_mask & BINDER_DEBUG_INTERNAL_REFS)
        printk(KERN_INFO "binder: %d:%d node %d u%p c%p created\n",
        proc->pid, current->pid, node->debug_id,
        node->ptr, node->cookie);
        return node;
}
```

在这里传进来的 ptr 和 cookie 都为 NULL。上述函数会首先检查“proc->nodes”红黑树中是否已经存在以 ptr 为键值的 node，如果已经存在则返回 NULL。在这个场景下，由于当前线程是第一次进入到这里，所以肯定不存在，于是就新建了一个 ptr 为 NULL 的 binder_node，并且初始化其他成员变量，并插入到“proc->nodes”红黑树中去。

当 binder_new_node 返回到函数 binder_ioctl()后，会把新建的 binder_node 指针保存在 binder_context_mgr_node 中，然后又初始化 binder_context_mgr_node 的引用计数值。这样执行 BINDER_SET_CONTEXT_MGR 命令完毕，在函数 binder_ioctl()返回之前执行下面的语句。

```
if (thread)
    thread->looper &= ~BINDER_LOOPER_STATE_NEED_RETURN;
```

在执行 binder_get_thread 时，thread->looper = BINDER_LOOPER_STATE_NEED_RETURN，执行了这条语句后，thread->looper = 0。

7.2.6 循环等待

再次回到文件“frameworks/base/cmds/servicemanager/service_manager.c”中的 main()函数，接下来需要调用函数 binder_loop()进入循环，等待“Client”发送请求。函数 binder_loop()在文件“frameworks/base/cmds/servicemanager/binder.c”中定义，具体实现代码如下所示：

```
void binder_loop(struct binder_state *bs, binder_handler func)
{
    int res;
    struct binder_write_read bwr;
    unsigned readbuf[32];
    bwr.write_size = 0;
    bwr.write_consumed = 0;
    bwr.write_buffer = 0;

    readbuf[0] = BC_ENTER_LOOPER;
    binder_write(bs, readbuf, sizeof(unsigned));
    for (;;) {
        bwr.read_size = sizeof(readbuf);
        bwr.read_consumed = 0;
        bwr.read_buffer = (unsigned) readbuf;
        res = ioctl(bs->fd, BINDER_WRITE_READ, &bwr);
        if (res < 0) {
            LOGE("binder_loop: ioctl failed (%s)\n", strerror(errno));
            break;
        }
        res = binder_parse(bs, 0, readbuf, bwr.read_consumed, func);
        if (res == 0) {
            LOGE("binder_loop: unexpected reply?!\n");
            break;
        }
        if (res < 0) {
            LOGE("binder_loop: io error %d %s\n", res, strerror(errno));
            break;
        }
    }
}
```

在上述代码中，首先通过函数 binder_write()执行 BC_ENTER_LOOPER 命令以告诉 Binder 驱动程序，Service Manager 马上要进入循环。在此还需要理解设备文件“/dev/binder”操作函数 ioctl 的操作码 BINDER_WRITE_READ，首先看其定义：

```
#define BINDER_WRITE_READ           _IOWR('b', 1, struct binder_write_read)
```

此 io 操作码有一个形式为 struct binder_write_read 的参数：

```
struct binder_write_read {
    signed long   write_size;       /* bytes to write */
    signed long   write_consumed;   /* bytes consumed by driver */
    unsigned long   write_buffer;
    signed long   read_size;        /* bytes to read */
    signed long   read_consumed;    /* bytes consumed by driver */
    unsigned long   read_buffer;
};
```

用户空间程序和 Binder 驱动程序交互时，大多数是通过 BINDER_WRITE_READ 命令实现的，write_bufffer 和 read_buffer 所指向的数据结构还指定了具体要执行的操作，write_bufffer 和 read_buffer 所指向的结构体是 binder_transaction_data，定义此结构体的代码如下所示：

```
struct binder_transaction_data {
    /* The first two are only used for bcTRANSACTION and brTRANSACTION,
     * identifying the target and contents of the transaction.
     */
    union {
        size_t    handle;          /* target descriptor of command transaction */
        void     *ptr;             /* target descriptor of return transaction */
    } target;
```

```
    void        *cookie;          /* target object cookie */
    unsigned int    code;         /* transaction command */

    /* General information about the transaction. */
    unsigned int    flags;
    pid_t       sender_pid;
    uid_t       sender_euid;
    size_t       data_size;    /* number of bytes of data */
    size_t        offsets_size;   /* number of bytes of offsets */

    /* If this transaction is inline, the data immediately
     * follows here; otherwise, it ends with a pointer to
     * the data buffer.
     */
    union {
        struct {
            /* transaction data */
            const void    *buffer;
            /* offsets from buffer to flat_binder_object structs */
            const void    *offsets;
        } ptr;
        uint8_t    buf[8];
    } data;
};
```

到此为止，已经从源代码一步一步地分析完 Service Manager 是如何成为 Android 进程间通信（IPC）机制 Binder 守护进程的了。在接下来的内容中，简要总结 Service Manager 成为 Android 进程间通信（IPC）机制 Binder 守护进程的过程。

（1）打开/dev/binder 文件：

```
open("/dev/binder", O_RDWR);
```

（2）建立 128 KB 内存映射：

```
mmap(NULL, mapsize, PROT_READ, MAP_PRIVATE, bs->fd, 0);
```

（3）通知 Binder 驱动程序它是守护进程：

```
binder_become_context_manager(bs);
```

（4）进入循环等待请求的到来：

```
binder_loop(bs, svcmgr_handler);
```

在这个过程中，在 Binder 驱动程序中建立了一个 struct binder_proc 结构、一个 struct binder_thread 结构和一个 struct binder_node 结构，这样，Service Manager 就在 Android 系统的进程间通信机制 Binder 担负起守护进程的职责了。

7.3 内存映射

在 Android 系统中，当打开 Binder 设备文件“/dev/binder”后，需要调用函数 mmap 把设备内存映射到用户进程地址空间中，这样就可以像操作用户内存那样操作设备内存。在 Binder 设备中，对内存的映射操作是有限制的，比如 Binder 不能映射具有写权限的内存区域，最大能映射 4 MB 的内存区域等。在 Android 系统中，大多数设备本身具有设备映射的设备内存，或者是在驱动初始化时由 vmalloc 或 kmalloc 等内核内存函数分配的，在 mmap 操作时分配 Binder 的设备内存。在本节的内容中，将详细讲解内存映射的基本知识。

7.3.1 实现内存分配功能

在 Android 系统中，函数 mmap 实现分配功能的实现流程如下所示：

① 在内核虚拟映射表上获取一个可以使用的区域。

② 分配物理页，并把物理页映射到获取的虚拟空间上。

③ 每个进程/线程只能执行一次映射操作，后面的操作都会返回错误。

函数 mmap 的具体实现流程如下所示。

① 检查内存映射条件，包括映射内存大小（4 MB）、flags、是否是第一次 mmap 等。

② 获得地址空间，并把此空间的地址记录在进程信息（buffer）中。

③ 分配物理页面（pages）并记录下来。

④ 将 buffer 插入到进程信息的 buffer 列表中。

⑤ 调用函数 binder_update_page_range 将分配的物理页面和 vm 空间对应起来。

⑥ 调用函数 binder_insert_free_buffer 把进程中的 buffer 插入到进程信息中。

函数 mmap 的具体实现代码如下所示：

```
static int binder_mmap(struct file *filp, struct vm_area_struct *vma)
{
	int ret;
	struct vm_struct *area;
	struct binder_proc *proc = filp->private_data;
	const char *failure_string;
	struct binder_buffer *buffer;

	if ((vma->vm_end - vma->vm_start) > SZ_4M)
		vma->vm_end = vma->vm_start + SZ_4M;

	binder_debug(BINDER_DEBUG_OPEN_CLOSE,
		     "binder_mmap: %d %lx-%lx (%ld K) vma %lx pagep %lx\n",
		     proc->pid, vma->vm_start, vma->vm_end,
		     (vma->vm_end - vma->vm_start) / SZ_1K, vma->vm_flags,
		     (unsigned long)pgprot_val(vma->vm_page_prot));

	if (vma->vm_flags & FORBIDDEN_MMAP_FLAGS) {
		ret = -EPERM;
		failure_string = "bad vm_flags";
		goto err_bad_arg;
	}
	vma->vm_flags = (vma->vm_flags | VM_DONTCOPY) & ~VM_MAYWRITE;

	mutex_lock(&binder_mmap_lock);
	if (proc->buffer) {
		ret = -EBUSY;
		failure_string = "already mapped";
		goto err_already_mapped;
	}

	area = get_vm_area(vma->vm_end - vma->vm_start, VM_IOREMAP);
	if (area == NULL) {
		ret = -ENOMEM;
		failure_string = "get_vm_area";
		goto err_get_vm_area_failed;
	}
	proc->buffer = area->addr;
	proc->user_buffer_offset = vma->vm_start - (uintptr_t)proc->buffer;
	mutex_unlock(&binder_mmap_lock);

#ifdef CONFIG_CPU_CACHE_VIPT
	if (cache_is_vipt_aliasing()) {
		while (CACHE_COLOUR((vma->vm_start ^ (uint32_t)proc->buffer))) {
			printk(KERN_INFO "binder_mmap: %d %lx-%lx maps %p bad alignment\n", proc->pid,
			vma->vm_start, vma->vm_end, proc->buffer);
			vma->vm_start += PAGE_SIZE;
		}
	}
#endif
	proc->pages = kzalloc(sizeof(proc->pages[0]) * ((vma->vm_end - vma->vm_start) / PAGE_SIZE), GFP_KERNEL);
```

```
        if (proc->pages == NULL) {
                ret = -ENOMEM;
                failure_string = "alloc page array";
                goto err_alloc_pages_failed;
        }
        proc->buffer_size = vma->vm_end - vma->vm_start;

        vma->vm_ops = &binder_vm_ops;
        vma->vm_private_data = proc;

        if (binder_update_page_range(proc, 1, proc->buffer, proc->buffer + PAGE_SIZE, vma)) {
                ret = -ENOMEM;
                failure_string = "alloc small buf";
                goto err_alloc_small_buf_failed;
        }
        buffer = proc->buffer;
        INIT_LIST_HEAD(&proc->buffers);
        list_add(&buffer->entry, &proc->buffers);
        buffer->free = 1;
        binder_insert_free_buffer(proc, buffer);
        proc->free_async_space = proc->buffer_size / 2;
        barrier();
        proc->files = get_files_struct(proc->tsk);
        proc->vma = vma;
        proc->vma_vm_mm = vma->vm_mm;

        /*printk(KERN_INFO "binder_mmap: %d %lx-%lx maps %p\n",
                 proc->pid, vma->vm_start, vma->vm_end, proc->buffer);*/
        return 0;

err_alloc_small_buf_failed:
        kfree(proc->pages);
        proc->pages = NULL;
err_alloc_pages_failed:
        mutex_lock(&binder_mmap_lock);
        vfree(proc->buffer);
        proc->buffer = NULL;
err_get_vm_area_failed:
err_already_mapped:
        mutex_unlock(&binder_mmap_lock);
err_bad_arg:
        printk(KERN_ERR "binder_mmap: %d %lx-%lx %s failed %d\n",
               proc->pid, vma->vm_start, vma->vm_end, failure_string, ret);
        return ret;
}
```

在上述代码中，参数 vm_area_struct 是一个结构体，在 mmp 的具体实现中会非常有用。为了优化查找方法，内核专门维护了 VMA 的链表和树形结构。在结构 vm_area_struct 中，很多成员函数都是用来维护这个树形结构的。VMA 的功能是管理进程地址空间中不同区域的数据结构。该函数首先对内存映射进行检查，主要包括映射内存的大小、flags 以及是否已经映射过了，并判断其映射条件是否合法；然后，通过内核函数 get_vm_area 从系统中申请可用的虚拟内存空间，在内核中申请并保留一块连续的内核虚拟内存空间区域；接着，将 binder_proc 的用户地址偏移（即用户进程的 VMA 地址与 Binder 申请的 VMA 地址的偏差）存放到 proc->user_buffer_offset 中；再接着，使用 kzalloc 函数根据请求映射的内存空间大小，分配 Binder 的核心数据结构 binder_proc 的 pages 成员，它主要用来保存指向申请的物理页的指针；最后，为 VMA 指定了 vm_operations_struct 操作，并且将 vma->vm_private_data 指向了核心数据 proc。

7.3.2 分配物理内存

到目前为止，就可以真正地开始分配物理内存（page）了。物理内存的分配工作是通过函数 binder_update_page_range 实现的，该函数主要完成以下工作。

- alloc_page：分配页面。

- map_vm_area：为分配的内存做映射关系。
- vm_insert_page：把分配的物理页插入到用户 VMA 区域。

函数 binder_update_page_range 的具体实现代码如下所示：

```
static int binder_update_page_range(struct binder_proc *proc, int allocate,
                    void *start, void *end,
                    struct vm_area_struct *vma)
{
    void *page_addr;
    unsigned long user_page_addr;
    struct vm_struct tmp_area;
    struct page **page;
    struct mm_struct *mm;

    binder_debug(BINDER_DEBUG_BUFFER_ALLOC,
          "binder: %d: %s pages %p-%p\n", proc->pid,
          allocate ? "allocate" : "free", start, end);

    if (end <= start)
        return 0;

    trace_binder_update_page_range(proc, allocate, start, end);

    if (vma)
        mm = NULL;
    else
        mm = get_task_mm(proc->tsk);

    if (mm) {
        down_write(&mm->mmap_sem);
        vma = proc->vma;
        if (vma && mm != proc->vma_vm_mm) {
            pr_err("binder: %d: vma mm and task mm mismatch\n",
                proc->pid);
            vma = NULL;
        }
    }

    if (allocate == 0)
        goto free_range;

    if (vma == NULL) {
        printk(KERN_ERR "binder: %d: binder_alloc_buf failed to "
            "map pages in userspace, no vma\n", proc->pid);
        goto err_no_vma;
    }

    for (page_addr = start; page_addr < end; page_addr += PAGE_SIZE) {
        int ret;
        struct page **page_array_ptr;
        page = &proc->pages[(page_addr - proc->buffer) / PAGE_SIZE];

        BUG_ON(*page);
        *page = alloc_page(GFP_KERNEL | __GFP_HIGHMEM | __GFP_ZERO);
        if (*page == NULL) {
            printk(KERN_ERR "binder: %d: binder_alloc_buf failed "
                "for page at %p\n", proc->pid, page_addr);
            goto err_alloc_page_failed;
        }
        tmp_area.addr = page_addr;
        tmp_area.size = PAGE_SIZE + PAGE_SIZE /* guard page? */;
        page_array_ptr = page;
        ret = map_vm_area(&tmp_area, PAGE_KERNEL, &page_array_ptr);
        if (ret) {
            printk(KERN_ERR "binder: %d: binder_alloc_buf failed "
            "to map page at %p in kernel\n",
            proc->pid, page_addr);
            goto err_map_kernel_failed;
        }
        user_page_addr =
```

```
                (uintptr_t)page_addr + proc->user_buffer_offset;
        ret = vm_insert_page(vma, user_page_addr, page[0]);
        if (ret) {
                printk(KERN_ERR "binder: %d: binder_alloc_buf failed "
                        "to map page at %lx in userspace\n",
                        proc->pid, user_page_addr);
                goto err_vm_insert_page_failed;
        }
        /* vm_insert_page does not seem to increment the refcount */
    }
    if (mm) {
        up_write(&mm->mmap_sem);
        mmput(mm);
    }
    return 0;

free_range:
    for (page_addr = end - PAGE_SIZE; page_addr >= start;
        page_addr -= PAGE_SIZE) {
        page = &proc->pages[(page_addr - proc->buffer) / PAGE_SIZE];
        if (vma)
                zap_page_range(vma, (uintptr_t)page_addr +
                        proc->user_buffer_offset, PAGE_SIZE, NULL);
                err_vm_insert_page_failed:
                        unmap_kernel_range((unsigned long)page_addr, PAGE_SIZE);
                err_map_kernel_failed:
                        __free_page(*page);
        *page = NULL;
        err_alloc_page_failed:
                ;
    }
err_no_vma:
    if (mm) {
        up_write(&mm->mmap_sem);
        mmput(mm);
    }
    return -ENOMEM;
}
```

其中 vm_operations_struct 只包括了一个打开操作和一个关闭操作，具体的定义代码如下所示：

```
static struct vm_operations_struct binder_vm_ops = {
    .open = binder_vma_open,
    .close = binder_vma_close,
};
```

7.3.3　释放物理页面

在 Android 系统的 Binder 机制中，函数 binder_insert_free_buffer 的功能是把进程中的 buffer 插入到进程信息中。也就是说，通过此函数能够将一个空闲内核缓冲区加入到进程中的空闲内核缓冲区的红黑树中。函数 binder_insert_free_buffer 的具体实现代码如下所示：

```
static void binder_insert_free_buffer(struct binder_proc *proc,
                    struct binder_buffer *new_buffer)
{
    struct rb_node **p = &proc->free_buffers.rb_node;
    struct rb_node *parent = NULL;
    struct binder_buffer *buffer;
    size_t buffer_size;
    size_t new_buffer_size;
    BUG_ON(!new_buffer->free);
    new_buffer_size = binder_buffer_size(proc, new_buffer);
    binder_debug(BINDER_DEBUG_BUFFER_ALLOC,
            "binder: %d: add free buffer, size %zd, "
            "at %p\n", proc->pid, new_buffer_size, new_buffer);
    while (*p) {
        parent = *p;
        buffer = rb_entry(parent, struct binder_buffer, rb_node);
        BUG_ON(!buffer->free);
        buffer_size = binder_buffer_size(proc, buffer);
```

```
        if (new_buffer_size < buffer_size)
            p = &parent->rb_left;
        else
            p = &parent->rb_right;
    }
    rb_link_node(&new_buffer->rb_node, parent, p);
    rb_insert_color(&new_buffer->rb_node, &proc->free_buffers);
}
```

7.3.4 分配内核缓冲区

在 Android 系统中，Binder 在使用 buffer 的时候一次声明一个 proc（对应一个进程）的 buffer 总大小，然后分配一页并做好映射。当在使用时如果发现空间不足，会接着映射并把这个 buffer 拆成两个，并把剩余的继续放到 free_buffers 里面。在 Binder 驱动程序中，函数*binder_alloc_buf 的功能是分配内核缓冲区，具体代码如下所示：

```
static struct binder_buffer *binder_alloc_buf(struct binder_proc *proc,
                                              size_t data_size,
                                              size_t offsets_size, int is_async)
{
    struct rb_node *n = proc->free_buffers.rb_node;
    struct binder_buffer *buffer;
    size_t buffer_size;
    struct rb_node *best_fit = NULL;
    void *has_page_addr;
    void *end_page_addr;
    size_t size;
    if (proc->vma == NULL) {
        printk(KERN_ERR "binder: %d: binder_alloc_buf, no vma\n",
               proc->pid);
        return NULL;
    }
    size = ALIGN(data_size, sizeof(void *)) +
        ALIGN(offsets_size, sizeof(void *));
    if (size < data_size || size < offsets_size) {
        binder_user_error("binder: %d: got transaction with invalid "
            "size %zd-%zd\n", proc->pid, data_size, offsets_size);
        return NULL;
    }
    if (is_async &&
        proc->free_async_space < size + sizeof(struct binder_buffer)) {
        binder_debug(BINDER_DEBUG_BUFFER_ALLOC,
                     "binder: %d: binder_alloc_buf size %zd"
                     "failed, no async space left\n", proc->pid, size);
        return NULL;
    }
    while (n) {
        buffer = rb_entry(n, struct binder_buffer, rb_node);
        BUG_ON(!buffer->free);
        buffer_size = binder_buffer_size(proc, buffer);
        if (size < buffer_size) {
            best_fit = n;
            n = n->rb_left;
        } else if (size > buffer_size)
            n = n->rb_right;
        else {
            best_fit = n;
            break;
        }
    }
    if (best_fit == NULL) {
        printk(KERN_ERR "binder: %d: binder_alloc_buf size %zd failed, "
               "no address space\n", proc->pid, size);
        return NULL;
    }
    if (n == NULL) {
        buffer = rb_entry(best_fit, struct binder_buffer, rb_node);
        buffer_size = binder_buffer_size(proc, buffer);
    }
    binder_debug(BINDER_DEBUG_BUFFER_ALLOC,
```

```
			"binder: %d: binder_alloc_buf size %zd got buff"
			"er %p size %zd\n", proc->pid, size, buffer, buffer_size);
	has_page_addr =
		(void *)(((uintptr_t)buffer->data + buffer_size) & PAGE_MASK);
	if (n == NULL) {
		if (size + sizeof(struct binder_buffer) + 4 >= buffer_size)
			buffer_size = size; /* no room for other buffers */
		else
			buffer_size = size + sizeof(struct binder_buffer);
	}
	end_page_addr =
		(void *)PAGE_ALIGN((uintptr_t)buffer->data + buffer_size);
	if (end_page_addr > has_page_addr)
		end_page_addr = has_page_addr;
	if (binder_update_page_range(proc, 1,
	    (void *)PAGE_ALIGN((uintptr_t)buffer->data), end_page_addr, NULL))
		return NULL;
	rb_erase(best_fit, &proc->free_buffers);
	buffer->free = 0;
	binder_insert_allocated_buffer(proc, buffer);
	if (buffer_size != size) {
		struct binder_buffer *new_buffer = (void *)buffer->data + size;
		list_add(&new_buffer->entry, &buffer->entry);
		new_buffer->free = 1;
		binder_insert_free_buffer(proc, new_buffer);
	}
	binder_debug(BINDER_DEBUG_BUFFER_ALLOC,
			"binder: %d: binder_alloc_buf size %zd got "
			"%p\n", proc->pid, size, buffer);
	buffer->data_size = data_size;
	buffer->offsets_size = offsets_size;
	buffer->async_transaction = is_async;
	if (is_async) {
		proc->free_async_space -= size + sizeof(struct binder_buffer);
		binder_debug(BINDER_DEBUG_BUFFER_ALLOC_ASYNC,
				"binder: %d: binder_alloc_buf size %zd "
				"async free %zd\n", proc->pid, size,
				proc->free_async_space);
	}
	return buffer;
}
```

再看函数 binder_insert_allocated_buffer，功能是将分配的内核缓冲区添加到目标进程的已分配物理页面的内核缓冲区红黑树中。函数 binder_insert_allocated_buffer 的具体实现代码如下所示：

```
static void binder_insert_allocated_buffer(struct binder_proc *proc,
					   struct binder_buffer *new_buffer)
{
	struct rb_node **p = &proc->allocated_buffers.rb_node;
	struct rb_node *parent = NULL;
	struct binder_buffer *buffer;
	BUG_ON(new_buffer->free);
	while (*p) {
		parent = *p;
		buffer = rb_entry(parent, struct binder_buffer, rb_node);
		BUG_ON(buffer->free);
		if (new_buffer < buffer)
			p = &parent->rb_left;
		else if (new_buffer > buffer)
			p = &parent->rb_right;
		else
			BUG();
	}
	rb_link_node(&new_buffer->rb_node, parent, p);
	rb_insert_color(&new_buffer->rb_node, &proc->allocated_buffers);
}
```

7.3.5 释放内核缓冲区

在 Android 系统中，函数 binder_free_buf 的功能是释放内核缓冲区的操作，具体实现代码如下所示：

```
static void binder_free_buf(struct binder_proc *proc,
                    struct binder_buffer *buffer)
{
      size_t size, buffer_size;
      //计算要释放的内核缓冲区 buffer 的大小，保存在 buffer_size 中
      buffer_size = binder_buffer_size(proc, buffer);
      //计算数据缓冲区和偏移数组缓冲区的大小，并保存在 size 中
      size = ALIGN(buffer->data_size, sizeof(void *)) +
            ALIGN(buffer->offsets_size, sizeof(void *));

      binder_debug(BINDER_DEBUG_BUFFER_ALLOC,
                  "binder: %d: binder_free_buf %p size %zd buffer"
                  "_size %zd\n", proc->pid, buffer, size, buffer_size);

      BUG_ON(buffer->free);
      BUG_ON(size > buffer_size);
      BUG_ON(buffer->transaction != NULL);
      BUG_ON((void *)buffer < proc->buffer);
      BUG_ON((void *)buffer > proc->buffer + proc->buffer_size);
      //检查要释放的内核缓冲区 buffer 是否用于异步事物
      if (buffer->async_transaction) {
            proc->free_async_space += size + sizeof(struct binder_buffer);

            binder_debug(BINDER_DEBUG_BUFFER_ALLOC_ASYNC,
                        "binder: %d: binder_free_buf size %zd "
                        "async free %zd\n", proc->pid, size,
                        proc->free_async_space);
      }
//释放内核缓冲区
      binder_update_page_range(proc, 0,
            (void *)PAGE_ALIGN((uintptr_t)buffer->data),
            (void *)(((uintptr_t)buffer->data + buffer_size) & PAGE_MASK),
            NULL);
      rb_erase(&buffer->rb_node, &proc->allocated_buffers);
      buffer->free = 1;
      if (!list_is_last(&buffer->entry, &proc->buffers)) {
            struct binder_buffer *next = list_entry(buffer->entry.next,
                                    struct binder_buffer, entry);
            if (next->free) {
                  rb_erase(&next->rb_node, &proc->free_buffers);
                  binder_delete_free_buffer(proc, next);
            }
      }
      if (proc->buffers.next != &buffer->entry) {
            struct binder_buffer *prev = list_entry(buffer->entry.prev,
                                    struct binder_buffer, entry);
            if (prev->free) {
                  binder_delete_free_buffer(proc, buffer);
                  rb_erase(&prev->rb_node, &proc->free_buffers);
                  buffer = prev;
            }
      }
      binder_insert_free_buffer(proc, buffer);
}
```

函数*buffer_start_page 和*buffer_end_page 用于计算结构体 binder_buffer 所占用的虚拟页面的地址，具体实现代码如下所示：

```
static void *buffer_start_page(struct binder_buffer *buffer)
{
      return (void *)((uintptr_t)buffer & PAGE_MASK);
}
static void *buffer_end_page(struct binder_buffer *buffer)
{
      return (void *)(((uintptr_t)(buffer + 1) - 1) & PAGE_MASK);
}
```

函数 binder_delete_free_buffer 的功能是删除结构体 binder_buffer，具体实现代码如下所示：

```
static void binder_delete_free_buffer(struct binder_proc *proc,
                              struct binder_buffer *buffer)
{
      struct binder_buffer *prev, *next = NULL;
```

```
        int free_page_end = 1;
        int free_page_start = 1;

        BUG_ON(proc->buffers.next == &buffer->entry);
        prev = list_entry(buffer->entry.prev, struct binder_buffer, entry);
        BUG_ON(!prev->free);
        if (buffer_end_page(prev) == buffer_start_page(buffer)) {
                free_page_start = 0;
                if (buffer_end_page(prev) == buffer_end_page(buffer))
                        free_page_end = 0;
                binder_debug(BINDER_DEBUG_BUFFER_ALLOC,
                          "binder: %d: merge free, buffer %p "
                          "share page with %p\n", proc->pid, buffer, prev);
        }

        if (!list_is_last(&buffer->entry, &proc->buffers)) {
                next = list_entry(buffer->entry.next,
                               struct binder_buffer, entry);
                if (buffer_start_page(next) == buffer_end_page(buffer)) {
                        free_page_end = 0;
                        if (buffer_start_page(next) ==
                            buffer_start_page(buffer))
                                free_page_start = 0;
                        binder_debug(BINDER_DEBUG_BUFFER_ALLOC,
                                  "binder: %d: merge free, buffer"
                                  " %p share page with %p\n", proc->pid,
                                  buffer, prev);
                }
        }
        list_del(&buffer->entry);
        if (free_page_start || free_page_end) {
                binder_debug(BINDER_DEBUG_BUFFER_ALLOC,
                          "binder: %d: merge free, buffer %p do "
                          "not share page%s%s with with %p or %p\n",
                          proc->pid, buffer, free_page_start ? "" : " end",
                          free_page_end ? "" : " start", prev, next);
                binder_update_page_range(proc, 0, free_page_start ?
                        buffer_start_page(buffer) : buffer_end_page(buffer),
                        (free_page_end ? buffer_end_page(buffer) :
                        buffer_start_page(buffer)) + PAGE_SIZE, NULL);
        }
}
```

7.3.6 查询内核缓冲区

在 Android 系统中，函数*binder_buffer_lookup 的功能是根据一个用户空间地址查询一个内核缓冲区，具体实现代码如下所示：

```
static struct binder_buffer *binder_buffer_lookup(struct binder_proc *proc,
                                          void __user *user_ptr)
{
        struct rb_node *n = proc->allocated_buffers.rb_node;
//对于已经分配的buffer空间，以内存地址为索引加入红黑树allocated_buffers中
        struct binder_buffer *buffer;
        struct binder_buffer *kern_ptr;

        kern_ptr = user_ptr - proc->user_buffer_offset
                - offsetof(struct binder_buffer, data);
/* 进程ioctl传下来的指针并不是binder_buffer的地址，而直接是binder_buffer.data的地址。user_buffer_offset 用户空间和内核空间，被映射区域起始地址之间的偏移*/
        while (n) {
                buffer = rb_entry(n, struct binder_buffer, rb_node);
                BUG_ON(buffer->free);

                if (kern_ptr < buffer)
                        n = n->rb_left;
                else if (kern_ptr > buffer)
                        n = n->rb_right;
                else
                        return buffer;
        }
        return NULL;
}
```

第 8 章　init 进程详解

在运行 Android 程序之后，首先会启动 init 进程，这个进程是 Linux 系统中用户空间的第一个进程，其进程号为 1。在本章的内容中，将详细讲解 Android 4.3 中 init 进程的运行过程，希望通过本章内容的学习，为读者步入本书后面高级知识的学习打下基础。

8.1 init 基础

我们知道，Android 本质上就是一个基于 Linux 内核的操作系统。与 Linux 的最大的区别是，Android 在应用层专门为移动设备添加了一些特有的支持。目前 Linux 有很多通信机制可以在用户空间和内核空间之间交互，例如设备驱动文件（位于"/dev"目录中）、内存文件（"/proc"、"/sys"目录等）。Android 在加载 Linux 基本内核后，就开始运行一个初始化进程 init。从 Android 加载 Linux 内核时设置了如下所示的参数。

```
Kernelcommand line: noinitrd root=/dev/nfs console=ttySAC0 init=/initnfsroot=192.168.1.103:/nfsbootip=
192.168.1.20:192.168.1.103:192.168.1.1:255.255.255.0::eth0:on
```

在上述命令中，告诉 Linux 内核初始化完成后开始运行 init 进程，由于 init 进程就是放在系统根目录下面。而 init 进程的代码位于源码的目录"system/core/init"下面。在分析 init 的核心代码之前，还需要做如下所示的工作。

- ❑ 初始化属性。
- ❑ 处理配置文件的命令（主要是 init.rc 文件），包括处理各种 Action。
- ❑ 性能分析（使用 bootchart 工具）。
- ❑ 无限循环执行 command（启动其他的进程）。

init 程序并不是由一个源代码文件组成的，而是由一组源代码文件的目标文件链接而成的。这些文件位于如下所示的目录中：

```
/system/core/init
```

主要的 JNI 代码放在以下的路径中：

```
frameworks/base/core/jni/
```

另外，还涉及了其他目录中的以下文件：

```
\bionic\libc\bionic\libc_init_common.h
\bionic\libc\bionic\libc_init_common.c
\bionic\libc\bionic\libc_init_dynamic.c
\bionic\libc\bionic\libc_init_static.c
\system\core\libcutils\properties.c
```

在本章的内容中，将详细分析 init 进程的启动过程，了解 Android 系统是如何启动起来的。

8.2 分析入口函数

进程 init 入口函数是 main，具体实现文件的路径是：

```
system\core\init\init.c
```

函数 main 的实现非常复杂，从这个 main 函数可以看出，init 实际上就分为如下两个部分。

（1）初始化。

初始化主要包括建立“/dev、”“/proc”等目录，初始化属性，执行 init.rc 等初始化文件中的 action 等。

（2）使用 for 循环、无限循环建立子进程。

这两项工作是 init 中的核心。在 Linux 系统中 init 是一切应用空间进程的父进程，因此平常在 Linux 终端执行命令，并建立进程，实际上都是在这个无限的 for 循环中完成的。也就是说，在 Linux 终端执行 ps-e 命令后，看到的所有除了 init 外的其他进程，都是由 init 负责创建的，而且 init 也会常驻内容。当然，如果 init 崩溃了，那么 Linux 系统就会基本上崩溃了。

函数 main 的具体实现代码如下所示：

```
int main(int argc, char **argv)
{
    int fd_count = 0;
    struct pollfd ufds[4];
    char *tmpdev;
    char* debuggable;
    char tmp[32];
    int property_set_fd_init = 0;
    int signal_fd_init = 0;
    int keychord_fd_init = 0;
    bool is_charger = false;

    if (!strcmp(basename(argv[0]), "ueventd"))
        return ueventd_main(argc, argv);

    if (!strcmp(basename(argv[0]), "watchdogd"))
        return watchdogd_main(argc, argv);

    /* clear the umask */
    umask(0);
    //  下面的代码开始建立各种用户空间的目录，如/dev、/proc、/sys 等
    mkdir("/dev", 0755);
    mkdir("/proc", 0755);
    mkdir("/sys", 0755);

    mount("tmpfs", "/dev", "tmpfs", MS_NOSUID, "mode=0755");
    mkdir("/dev/pts", 0755);
    mkdir("/dev/socket", 0755);
    mount("devpts", "/dev/pts", "devpts", 0, NULL);
    mount("proc", "/proc", "proc", 0, NULL);
    mount("sysfs", "/sys", "sysfs", 0, NULL);

        /* 检测/dev/.booting 文件是否可读写和创建*/
    close(open("/dev/.booting", O_WRONLY | O_CREAT, 0000));

    open_devnull_stdio();
    klog_init();
    //  初始化属性
    property_init();

    get_hardware_name(hardware, &revision);
    //  处理内核命令行
    process_kernel_cmdline();
    … …
```

```
    is_charger = !strcmp(bootmode, "charger");

    INFO("property init\n");
    if (!is_charger)
        property_load_boot_defaults();

    INFO("reading config file\n");
    // 分析/init.rc文件的内容
    init_parse_config_file("/init.rc");
    ……// 执行初始化文件中的动作
    action_for_each_trigger("init", action_add_queue_tail);
    // 在charger模式下略过mount文件系统的工作
    if (!is_charger) {
        action_for_each_trigger("early-fs", action_add_queue_tail);
        action_for_each_trigger("fs", action_add_queue_tail);
        action_for_each_trigger("post-fs", action_add_queue_tail);
        action_for_each_trigger("post-fs-data", action_add_queue_tail);
    }

    queue_builtin_action(property_service_init_action, "property_service_init");
    queue_builtin_action(signal_init_action, "signal_init");
    queue_builtin_action(check_startup_action, "check_startup");

    if (is_charger) {
        action_for_each_trigger("charger", action_add_queue_tail);
    } else {
        action_for_each_trigger("early-boot", action_add_queue_tail);
        action_for_each_trigger("boot", action_add_queue_tail);
    }

        /* run all property triggers based on current state of the properties */
    queue_builtin_action(queue_property_triggers_action, "queue_property_triggers");


#if BOOTCHART
    queue_builtin_action(bootchart_init_action, "bootchart_init");
#endif
    // 进入无限循环，建立init的子进程（init是所有进程的父进程）
    for(;;) {
        int nr, i, timeout = -1;
        // 执行命令（子进程对应的命令）
        execute_one_command();
        restart_processes();

        if (!property_set_fd_init && get_property_set_fd() > 0) {
            ufds[fd_count].fd = get_property_set_fd();
            ufds[fd_count].events = POLLIN;
            ufds[fd_count].revents = 0;
            fd_count++;
            property_set_fd_init = 1;
        }
        if (!signal_fd_init && get_signal_fd() > 0) {
            ufds[fd_count].fd = get_signal_fd();
            ufds[fd_count].events = POLLIN;
            ufds[fd_count].revents = 0;
            fd_count++;
            signal_fd_init = 1;
        }
        if (!keychord_fd_init && get_keychord_fd() > 0) {
            ufds[fd_count].fd = get_keychord_fd();
            ufds[fd_count].events = POLLIN;
            ufds[fd_count].revents = 0;
            fd_count++;
            keychord_fd_init = 1;
        }

        if (process_needs_restart) {
            timeout = (process_needs_restart - gettime()) * 1000;
```

```
        if (timeout < 0)
            timeout = 0;
    }

    if (!action_queue_empty() || cur_action)
        timeout = 0;
//  bootchart是一个性能统计工具，用于搜集硬件和系统的信息，并将其写入磁盘，以便其
//  他程序使用
#if BOOTCHART
    if (bootchart_count > 0) {
        if (timeout < 0 || timeout > BOOTCHART_POLLING_MS)
            timeout = BOOTCHART_POLLING_MS;
        if (bootchart_step() < 0 || --bootchart_count == 0) {
            bootchart_finish();
            bootchart_count = 0;
        }
    }
#endif
    //  等待下一个命令的提交
    nr = poll(ufds, fd_count, timeout);
    if (nr <= 0)
        continue;

    for (i = 0; i < fd_count; i++) {
        if (ufds[i].revents == POLLIN) {
            if (ufds[i].fd == get_property_set_fd())
                handle_property_set_fd();
            else if (ufds[i].fd == get_keychord_fd())
                handle_keychord();
            else if (ufds[i].fd == get_signal_fd())
                handle_signal();
        }
    }
  }

  return 0;
}
```

8.3　配置文件详解

在 init 进程中，配置文件是指文件 init.rc，其路径是：

```
\system\core\rootdir\init.rc
```

在本节的内容中，将详细剖析讲解配置文件 init.rc 的基本知识。

8.3.1　init.rc 简介

文件 init.rc 是一个可配置的初始化文件，在里面定制了厂商可以配置的额外的初始化配置信息，具体格式为：

```
init.%PRODUCT%.rc
```

文件 init.rc 是在文件“/system/core/init/init.c”中读取的，它基于“行”，包含一些用空格隔开的关键字（它属于特殊字符）。如果在关键字中含有空格，则使用“/“表示转义，使用“”防止关键字被断开，如果“/”在末尾则表示换行，以“#”开头的表示注释。

文件 init.rc 包含 4 种状态类别，分别是 Actions、Commands、Services 和 Options，当声明一个 Service 或者 Action 的时候，它将隐式声明一个 section，它之后跟随的 Command 或者 Option 都将属于这个 section。另外，Action 和 Service 不能重名，否则忽略为 error。

（1）Actions。

Actions 就是在某种条件下触发一系列的命令，通常有一个 trigger，形式如下所示：

```
on <trigger>
    <command>
    <command>
```

（2）Service。

Service 的结构如下所示：

```
service <name> <pathname> [ <argument> ]*
  <option>
  <option>
```

（3）Option。

Option 是 Service 的修饰词，主要包括如下所示的选项。

- critical：表示如果服务在 4 分钟内存在多于 4 次，则系统重启到 recovery mode。
- disabled：表示服务不会自动启动，需要手动调用名字启动。
- setEnv <name> <value>：设置启动环境变量。
- socket <name> <type> <permission> [<user> [<group>]]：开启一个 unix 域的 socket，名字为/dev/socket/<name>，<type>只能是 dgram 或者 stream，<user>和<group>默认为 0。
- user <username>：表示将用户切换为<username>，用户名已经定义好了，只能是 system/root。
- group <groupname>：表示将组切换为<groupname>。
- oneshot：表示这个 Service 只启动一次。
- class <name>：指定一个要启动的类，这个类中如果有多个 service，将会被同时启动。默认的 class 将会是“default”。
- onrestart：在重启时执行一条命令。

（4）trigger。

主要包括如下所示的选项。

- boot：当/init.conf 加载完毕时。
- <name>=<value>：当<name>被设置为<value>时。
- device-added-<path>：设备<path>被添加时。
- device-removed-<path>：设备<path>被移除时。
- service-exited-<name>：服务<name>退出时。

（5）主要包含的操作命令及具体说明。

exec <path> [<argument>]*：执行一个<path>指定的程序。

- export <name> <value>：设置一个全局变量。
- ifup <interface>：使网络接口<interface>连接。
- import <filename>：引入其他的配置文件。
- hostname <name>：设置主机名。
- chdir <directory>：切换工作目录。
- chmod <octal-mode> <path>：设置访问权限。
- chown <owner> <group> <path>：设置用户和组。
- chroot <directory>：设置根目录
- class_start <serviceclass>：启动类中的 service。
- class_stop <serviceclass>：停止类中的 service。
- domainname <name>：设置域名。
- insmod <path>：安装模块。
- mkdir <path> [mode] [owner] [group] ：创建一个目录，并可以指定权限，用户和组。

- mount <type> <device> <dir> [<mountoption>]* ：加载指定设备到目录下。
- <mountoption>：包括"ro", "rw", "remount", "noatime"。
- setprop <name> <value>：设置系统属性。
- setrlimit <resource> <cur> <max>：设置资源访问权限。
- start <service>：开启服务。
- stop <service>：停止服务。
- symlink <target> <path>：创建一个动态链接。
- sysclktz <mins_west_of_gmt>：设置系统时钟。
- trigger <event>：触发事件。
- write <path> <string> [<string>]*：向<path>路径的文件写入多个<string>。

读者可以进到 Android 的 shell，会看到根目录有一个 init.rc 文件。启动 Android 后，会将文件 init.rc 装载到内存。而修改文件 init.rc 的内容实际上只是修改内存中的 init.rc 文件的内容。一旦重启 Android，文件 init.rc 的内容又会恢复到最初的装载。想彻底修改文件 init.rc 内容，唯一方法是修改 Android 的 ROM 中的内核镜像（boot.img）。

8.3.2 分析 init.rc 的过程

文件 init.rc 是一个配置文件，内部有许多的语言规则，所有语言会在函数 init_parse_config_file 中进行解析。前面的主函数 main 读取完配置文件 init.rc 后，会调用函数 parse_config 进行解析。整个实现流程如下所示：

```
init_parse_config_file->read_file->parse_config
```

（1）函数 parse_config 和 init_parse_config_file 在如下文件中实现：

```
\system\core\init\init_parser.c
```

函数 parse_config 和 init_parse_config_file 的具体实现代码如下所示：

```
static void parse_config(const char *fn, char *s)//s 为 init.rc 中字符串的内容
{
    struct parse_state state;
    char *args[INIT_PARSER_MAXARGS];
    int nargs;

    nargs = 0;
    state.filename = fn;
    state.line = 1;
    state.ptr = s;
    state.nexttoken = 0;
    state.parse_line = parse_line_no_op;
    for (;;) {
        switch (next_token(&state)) {
        case T_EOF:       //文件的结尾
            state.parse_line(&state, 0, 0);
            return;
        case T_NEWLINE://新的一行
            if (nargs) {
                int kw = lookup_keyword(args[0]);//读取 init.rc 返回关键字例如 service,返回 K_service
                if (kw_is(kw, SECTION)) {        //查看关键字是否为 SECTION,只有 service 和 on 满足
                    state.parse_line(&state, 0, 0);
                    parse_new_section(&state, kw, nargs, args);
                } else {
                    state.parse_line(&state, nargs, args);//on 和 service 两个段下面的内容
                }
                nargs = 0;
            }
            break;
        case T_TEXT://文本内容
```

```
            if (nargs < INIT_PARSER_MAXARGS) {
                args[nargs++] = state.text;
            }
            break;
        }
    }
}

int init_parse_config_file(const char *fn)
{
    char *data;
    data = read_file(fn, 0);
    if (!data) return -1;

    parse_config(fn, data);
    DUMP();
    return 0;
}
```

通过上述代码可以看出，在 for 的无限循环中对文件 init.rc 的内容进行了解析，以一行一行的形式进行了读取。

（2）每读取完一行内容换行时到下一行时，使用函数 lookup_keyword 分析已经读取的一行的第一个参数。函数 lookup_keyword 的具体实现代码如下所示：

```
int lookup_keyword(const char *s)
{
    switch (*s++) {
    case 'c':
    if (!strcmp(s, "opy")) return K_copy;
        if (!strcmp(s, "apability")) return K_capability;
        if (!strcmp(s, "hdir")) return K_chdir;
        if (!strcmp(s, "hroot")) return K_chroot;
        if (!strcmp(s, "lass")) return K_class;
        if (!strcmp(s, "lass_start")) return K_class_start;
        if (!strcmp(s, "lass_stop")) return K_class_stop;
        if (!strcmp(s, "lass_reset")) return K_class_reset;
        if (!strcmp(s, "onsole")) return K_console;
        if (!strcmp(s, "hown")) return K_chown;
        if (!strcmp(s, "hmod")) return K_chmod;
        if (!strcmp(s, "ritical")) return K_critical;
        break;
    case 'd':
        if (!strcmp(s, "isabled")) return K_disabled;
        if (!strcmp(s, "omainname")) return K_domainname;
        break;
    case 'e':
        if (!strcmp(s, "xec")) return K_exec;
        if (!strcmp(s, "xport")) return K_export;
        break;
    case 'g':
        if (!strcmp(s, "roup")) return K_group;
        break;
    case 'h':
        if (!strcmp(s, "ostname")) return K_hostname;
        break;
    case 'i':
        if (!strcmp(s, "oprio")) return K_ioprio;
        if (!strcmp(s, "fup")) return K_ifup;
        if (!strcmp(s, "nsmod")) return K_insmod;
        if (!strcmp(s, "mport")) return K_import;
        break;
    case 'k':
        if (!strcmp(s, "eycodes")) return K_keycodes;
        break;
    case 'l':
        if (!strcmp(s, "oglevel")) return K_loglevel;
        if (!strcmp(s, "oad_persist_props")) return K_load_persist_props;
        break;
```

```
    case 'm':
        if (!strcmp(s, "kdir")) return K_mkdir;
        if (!strcmp(s, "ount_all")) return K_mount_all;
        if (!strcmp(s, "ount")) return K_mount;
        break;
    case 'o':
        if (!strcmp(s, "n")) return K_on;
        if (!strcmp(s, "neshot")) return K_oneshot;
        if (!strcmp(s, "nrestart")) return K_onrestart;
        break;
    case 'r':
        if (!strcmp(s, "estart")) return K_restart;
        if (!strcmp(s, "estorecon")) return K_restorecon;
        if (!strcmp(s, "mdir")) return K_rmdir;
        if (!strcmp(s, "m")) return K_rm;
        break;
    case 's':
        if (!strcmp(s, "eclabel")) return K_seclabel;
        if (!strcmp(s, "ervice")) return K_service;
        if (!strcmp(s, "etcon")) return K_setcon;
        if (!strcmp(s, "etenforce")) return K_setenforce;
        if (!strcmp(s, "etenv")) return K_setenv;
        if (!strcmp(s, "etkey")) return K_setkey;
        if (!strcmp(s, "etprop")) return K_setprop;
        if (!strcmp(s, "etrlimit")) return K_setrlimit;
        if (!strcmp(s, "etsebool")) return K_setsebool;
        if (!strcmp(s, "ocket")) return K_socket;
        if (!strcmp(s, "tart")) return K_start;
        if (!strcmp(s, "top")) return K_stop;
        if (!strcmp(s, "ymlink")) return K_symlink;
        if (!strcmp(s, "ysclktz")) return K_sysclktz;
        break;
    case 't':
        if (!strcmp(s, "rigger")) return K_trigger;
        break;
    case 'u':
        if (!strcmp(s, "ser")) return K_user;
        break;
    case 'w':
        if (!strcmp(s, "rite")) return K_write;
        if (!strcmp(s, "ait")) return K_wait;
        break;
    }
    return K_UNKNOWN;
}
```

由此可见，函数 lookup_keyword 主要对每一行的第一个字符做 case 判断，然后在 if 语句中调用 strcmp 命令，这些命令都是按照文件 init.rc 的格式要求进行的。比如常用的 service 和 on 等经过 lookup_keyword 后返回 K_servcie 和 K_on。随后使用 kw_is（kw, SECTION）判断返回的 kw 是否属于 SECTION 类型。在文件 init.rc 中，只有 service 和 on 满足该类型。

（3）定义关键字。

在文件 keywords.h 中定义了 init 使用的关键字，在此文件中定义了如 do_class_start、do_class_stop 之类的函数，并且还定义了枚举。文件 keywords.h 的路径如下所示：

```
\system\core\init\
```

文件 keywords.h 的具体实现代码如下所示：

```
#ifndef KEYWORD
int do_chroot(int nargs, char **args);
int do_chdir(int nargs, char **args);
int do_class_start(int nargs, char **args);
int do_class_stop(int nargs, char **args);
int do_class_reset(int nargs, char **args);
......
#define __MAKE_KEYWORD_ENUM__
```

```
#define KEYWORD(symbol, flags, nargs, func) K_##symbol,
enum {
    K_UNKNOWN,
#endif
    KEYWORD(capability,  OPTION,  0, 0)
    KEYWORD(chdir,      COMMAND, 1, do_chdir)
    KEYWORD(chroot,     COMMAND, 1, do_chroot)
    KEYWORD(class,      OPTION,  0, 0)
    KEYWORD(class_start, COMMAND, 1, do_class_start)
    KEYWORD(class_stop,  COMMAND, 1, do_class_stop)
    KEYWORD(class_reset, COMMAND, 1, do_class_reset)
    KEYWORD(console,     OPTION,  0, 0)
    KEYWORD(critical,    OPTION,  0, 0)
    KEYWORD(disabled,    OPTION,  0, 0)
    KEYWORD(domainname,  COMMAND, 1, do_domainname)
    KEYWORD(exec,       COMMAND, 1, do_exec)
    KEYWORD(export,     COMMAND, 2, do_export)
……
    KEYWORD(load_persist_props,    COMMAND, 0, do_load_persist_props)
    KEYWORD(ioprio,     OPTION,  0, 0)
#ifdef __MAKE_KEYWORD_ENUM__
    KEYWORD_COUNT,
};
#undef __MAKE_KEYWORD_ENUM__
#undef KEYWORD
#endif
```

文件 keywords.h 在文件 init_parser.c 中被用到了两次，具体引用代码如下所示：

```
#define SECTION 0x01
#define COMMAND 0x02
#define OPTION  0x04

#include "keywords.h"

#define KEYWORD(symbol, flags, nargs, func) \
    [ K_##symbol ] = { #symbol, func, nargs + 1, flags, },

struct {
    const char *name;
    int (*func)(int nargs, char **args);
    unsigned char nargs;
    unsigned char flags;
} keyword_info[KEYWORD_COUNT] = {
    [ K_UNKNOWN ] = { "unknown", 0, 0, 0 },
#include "keywords.h"
};
#undef KEYWORD

#define kw_is(kw, type) (keyword_info[kw].flags & (type))
#define kw_name(kw) (keyword_info[kw].name)
#define kw_func(kw) (keyword_info[kw].func)
#define kw_nargs(kw) (keyword_info[kw].nargs)
```

8.4 解析 service

由前面的函数 lookup_keyword 可知，在调用过程中会对 on 和 service 所在的段进行解析，这里首先分析 service，在分析时以文件 init.rc 中的 service zygote 为例。

8.4.1 Zygote 对应的 service action

在文件 init.rc 中，zygote 对应的 service action 的代码如下所示：

```
service zygote /system/bin/app_process -Xzygote /system/bin --zygote --start-system-server
    class main
    socket zygote stream 660 root system
```

```
    onrestart write /sys/android_power/request_state wake
    onrestart write /sys/power/state on
    onrestart restart media
    onrestart restart netd
```

解析 action 的入口函数是 parse_new_section，在此函数中再分别对 service 或者 on 关键字开头的内容进行解析。函数 parse_new_section 的具体实现代码如下所示：

```
void parse_new_section(struct parse_state *state, int kw,
                       int nargs, char **args)
{
    printf("[ %s %s ]\n", args[0],
        nargs > 1 ? args[1] : "");
    switch(kw) {
    case K_service:
        state->context = parse_service(state, nargs, args);
        if (state->context) {
            state->parse_line = parse_line_service;
            return;
        }
        break;
    case K_on:
        state->context = parse_action(state, nargs, args);
        if (state->context) {
            state->parse_line = parse_line_action;
            return;
        }
        break;
    case K_import:
        parse_import(state, nargs, args);
        break;
    }
    state->parse_line = parse_line_no_op;
}
```

8.4.2 init 组织 service

在 init 进程中，使用了一个名为 service 的结构体保存和 service action 有关的信息。此结构体是在如下文件中中定义的：

```
\system\core\init\init.h
```

结构体 service 的具体实现代码如下所示：

```
struct service {
        /* list of all services */
    struct listnode slist;

    const char *name;
    const char *classname;

    unsigned flags;
    pid_t pid;
    time_t time_started;    /* time of last start */
    time_t time_crashed;    /* first crash within inspection window */
    int nr_crashed;        /* number of times crashed within window */

    uid_t uid;
    gid_t gid;
    gid_t supp_gids[NR_SVC_SUPP_GIDS];
    size_t nr_supp_gids;

#ifdef HAVE_SELINUX
    char *seclabel;
#endif

    struct socketinfo *sockets;
    struct svcenvinfo *envvars;
```

```
    struct action onrestart;  /* Actions to execute on restart. */

    /* keycodes for triggering this service via /dev/keychord */
    int *keycodes;
    int nkeycodes;
    int keychord_id;

    int ioprio_class;
    int ioprio_pri;

    int nargs;
    /* "MUST BE AT THE END OF THE STRUCT" */
    char *args[1];
}; /*      ^-------'args' MUST be at the end of this struct! */
```

另外，在文件 init.h 中还定义了结构体 action，具体实现代码如下所示：

```
struct action {
        /* node in list of all actions */
    struct listnode alist;
        /* node in the queue of pending actions */
    struct listnode qlist;
        /* node in list of actions for a trigger */
    struct listnode tlist;

    unsigned hash;
    const char *name;

    struct listnode commands;
    struct command *current;
};
```

这样便通过上述两个结构体对 service 进行了组织。

8.4.3 函数 parse_service 和 parse_line_service

在解析 Service 会用到两个函数，分别是 parse_service 和 parse_line_service。

（1）当解析文件 init.rc 中的 service zygote 时会执行函数 parse_service，此函数的功能是构建 service 的骨架，对 service 关键字开头的内容进行解析。函数 parse_service 的具体实现代码如下所示：

```
static void *parse_service(struct parse_state *state, int nargs, char **args)
{
    struct service *svc;
    if (nargs < 3) {
        parse_error(state, "services must have a name and a program\n");
        return 0;
    }
    if (!valid_name(args[1])) {
        parse_error(state, "invalid service name '%s'\n", args[1]);
        return 0;
    }

    svc = service_find_by_name(args[1]);//查找服务是否已经存在
    if (svc) {
        parse_error(state, "ignored duplicate definition of service '%s'\n", args[1]);
        return 0;
    }

    nargs -= 2;
    svc = calloc(1, sizeof(*svc) + sizeof(char*) * nargs);
    if (!svc) {
        parse_error(state, "out of memory\n");
        return 0;
    }
    svc->name = args[1];            //sevice 的名字
    svc->classname = "default"; //svc 的类名默认是 default
    memcpy(svc->args, args + 2, sizeof(char*) * nargs);//首个参数放的是可执行文件
```

```
    svc->args[nargs] = 0;
    svc->nargs = nargs;//参数个数
    svc->onrestart.name = "onrestart";
    list_init(&svc->onrestart.commands);
    list_add_tail(&service_list, &svc->slist);
    return svc;
}
```

在上述代码中，agrs[1]就是 zygote，系统会先查找是否已经存在该服务，然后构建一个 service svc 并进行相关的填充，包括服务名、服务所属的类别名字和已经服务启动带入的参数个数（要减去 service 和服务名 zygote），最后将这个 svc 加入到 service_list 全局链表中。

（2）函数 parse_line_service 的功能是根据配置文件的内容填充 service 结构体，并解析 Service 中剩余行中的 Option，比如 class、socket、onrestart 等。函数 parse_line_service 的具体实现代码如下所示：

```
static void parse_line_service(struct parse_state *state, int nargs, char **args)
{
    struct service *svc = state->context;
    struct command *cmd;
    int i, kw, kw_nargs;

    if (nargs == 0) {
        return;
    }

    svc->ioprio_class = IoSchedClass_NONE;

    kw = lookup_keyword(args[0]);
    switch (kw) {
    case K_capability:
        break;
    case K_class:
        if (nargs != 2) {
            parse_error(state, "class option requires a classname\n");
        } else {
            svc->classname = args[1];
        }
        break;
    case K_console:
        svc->flags |= SVC_CONSOLE;
        break;
    case K_disabled:
        svc->flags |= SVC_DISABLED;
        svc->flags |= SVC_RC_DISABLED;
        break;
    case K_ioprio:
        if (nargs != 3) {
            parse_error(state, "ioprio optin usage: ioprio <rt|be|idle> <ioprio 0-7>\n");
        } else {
            svc->ioprio_pri = strtoul(args[2], 0, 8);

            if (svc->ioprio_pri < 0 || svc->ioprio_pri > 7) {
                parse_error(state, "priority value must be range 0 - 7\n");
                break;
            }

            if (!strcmp(args[1], "rt")) {
                svc->ioprio_class = IoSchedClass_RT;
            } else if (!strcmp(args[1], "be")) {
                svc->ioprio_class = IoSchedClass_BE;
            } else if (!strcmp(args[1], "idle")) {
                svc->ioprio_class = IoSchedClass_IDLE;
            } else {
                parse_error(state, "ioprio option usage: ioprio <rt|be|idle> <0-7>\n");
            }
        }
```

```
        break;
    case K_group:
        if (nargs < 2) {
            parse_error(state, "group option requires a group id\n");
        } else if (nargs > NR_SVC_SUPP_GIDS + 2) {
            parse_error(state, "group option accepts at most %d supp. groups\n",
                        NR_SVC_SUPP_GIDS);
        } else {
            int n;
            svc->gid = decode_uid(args[1]);
            for (n = 2; n < nargs; n++) {
                svc->supp_gids[n-2] = decode_uid(args[n]);
            }
            svc->nr_supp_gids = n - 2;
        }
        break;
    case K_keycodes:
        if (nargs < 2) {
            parse_error(state, "keycodes option requires atleast one keycode\n");
        } else {
            svc->keycodes = malloc((nargs - 1) * sizeof(svc->keycodes[0]));
            if (!svc->keycodes) {
                parse_error(state, "could not allocate keycodes\n");
            } else {
                svc->nkeycodes = nargs - 1;
                for (i = 1; i < nargs; i++) {
                    svc->keycodes[i - 1] = atoi(args[i]);
                }
            }
        }
        break;
    case K_oneshot:
        svc->flags |= SVC_ONESHOT;
        break;
    case K_onrestart:
        nargs--;
        args++;
        kw = lookup_keyword(args[0]);
        if (!kw_is(kw, COMMAND)) {
            parse_error(state, "invalid command '%s'\n", args[0]);
            break;
        }
        kw_nargs = kw_nargs(kw);
        if (nargs < kw_nargs) {
            parse_error(state, "%s requires %d %s\n", args[0], kw_nargs - 1,
                kw_nargs > 2 ? "arguments" : "argument");
            break;
        }

        cmd = malloc(sizeof(*cmd) + sizeof(char*) * nargs);
        cmd->func = kw_func(kw);
        cmd->nargs = nargs;
        memcpy(cmd->args, args, sizeof(char*) * nargs);
        list_add_tail(&svc->onrestart.commands, &cmd->clist);
        break;
    case K_critical:
        svc->flags |= SVC_CRITICAL;
        break;
    case K_setenv: { /* name value */
        struct svcenvinfo *ei;
        if (nargs < 2) {
            parse_error(state, "setenv option requires name and value arguments\n");
            break;
        }
        ei = calloc(1, sizeof(*ei));
        if (!ei) {
            parse_error(state, "out of memory\n");
            break;
        }
```

```
        ei->name = args[1];
        ei->value = args[2];
        ei->next = svc->envvars;
        svc->envvars = ei;
        break;
    }
    case K_socket: {/* name type perm [ uid gid ] */
        struct socketinfo *si;
        if (nargs < 4) {
            parse_error(state, "socket option requires name, type, perm arguments\n");
            break;
        }
        if (strcmp(args[2],"dgram") && strcmp(args[2],"stream")
                && strcmp(args[2],"seqpacket")) {
            parse_error(state, "socket type must be 'dgram', 'stream' or 'seqpacket'\n");
            break;
        }
        si = calloc(1, sizeof(*si));
        if (!si) {
            parse_error(state, "out of memory\n");
            break;
        }
        si->name = args[1];
        si->type = args[2];
        si->perm = strtoul(args[3], 0, 8);
        if (nargs > 4)
            si->uid = decode_uid(args[4]);
        if (nargs > 5)
            si->gid = decode_uid(args[5]);
        si->next = svc->sockets;
        svc->sockets = si;
        break;
    }
    case K_user:
        if (nargs != 2) {
            parse_error(state, "user option requires a user id\n");
        } else {
            svc->uid = decode_uid(args[1]);
        }
        break;
    case K_seclabel:
#ifdef HAVE_SELINUX
        if (nargs != 2) {
            parse_error(state, "seclabel option requires a label string\n");
        } else {
            svc->seclabel = args[1];
        }
#endif
        break;

    default:
        parse_error(state, "invalid option '%s'\n", args[0]);
    }
}
```

8.5 字段 on

在本节的内容中，将详细剖析 on 字段的内容，并以 on boot 这个 section 作为例子进行分析。希望读者认真学习，为步入本书后面知识的学习打下基础。

8.5.1 Zygote 对应的 on action

字段 on 的内容比较复杂，此处将以 on boot 这个 section 作为列子进行分析。在文件 init.rc 中，zygote 对应的 on boot action 的代码如下所示：

```
on boot
# basic network init
    ifup lo
    hostname localhost
    domainname localdomain

# set RLIMIT_NICE to allow priorities from 19 to -20
    setrlimit 13 40 40

# Memory management.  Basic kernel parameters, and allow the high
# level system server to be able to adjust the kernel OOM driver
# parameters to match how it is managing things.
    write /proc/sys/vm/overcommit_memory 1
    write /proc/sys/vm/min_free_order_shift 4
    chown root system /sys/module/lowmemorykiller/parameters/adj
    chmod 0664 /sys/module/lowmemorykiller/parameters/adj
    chown root system /sys/module/lowmemorykiller/parameters/minfree
    chmod 0664 /sys/module/lowmemorykiller/parameters/minfree

    # Tweak background writeout
    write /proc/sys/vm/dirty_expire_centisecs 200
    write /proc/sys/vm/dirty_background_ratio  5

    # Permissions for System Server and daemons.
    chown radio system /sys/android_power/state
    chown radio system /sys/android_power/request_state
    chown radio system /sys/android_power/acquire_full_wake_lock
    chown radio system /sys/android_power/acquire_partial_wake_lock
    chown radio system /sys/android_power/release_wake_lock
    chown system system /sys/power/autosleep
    chown system system /sys/power/state
    chown system system /sys/power/wakeup_count
    chown radio system /sys/power/wake_lock
    chown radio system /sys/power/wake_unlock
    chmod 0660 /sys/power/state
    chmod 0660 /sys/power/wake_lock
    chmod 0660 /sys/power/wake_unlock
……
# Set this property so surfaceflinger is not started by system_init
    setprop system_init.startsurfaceflinger 0

    class_start core
    class_start main
```

和前面对 service 的分析类似，case 中进入 K_on 选项执行函数 parse_action。函数 parse_action 的具体实现代码如下所示：

```
static void *parse_action(struct parse_state *state, int nargs, char **args)
{
    struct action *act;
    if (nargs < 2) {
        parse_error(state, "actions must have a trigger\n");
        return 0;
    }
    if (nargs > 2) {
        parse_error(state, "actions may not have extra parameters\n");
        return 0;
    }
    act = calloc(1, sizeof(*act));
    act->name = args[1];
    list_init(&act->commands);
    list_add_tail(&action_list, &act->alist);
        /* XXX add to hash */
    return act;
}
```

8.5.2　init 组织 on

在 init 进程中可以看到一个名为 action 结构体类似于 service，这个 action 的名字为 boot，最后会将这个 action 加入到全局链表 action_list 中。结构体 action 的具体实现代码如下所示：

```
struct action {
        /* node in list of all actions */
    struct listnode alist;
        /* node in the queue of pending actions */
    struct listnode qlist;
        /* node in list of actions for a trigger */
    struct listnode tlist;

    unsigned hash;
    const char *name;

    struct listnode commands;
    struct command *current;
};
```

8.5.3 解析 on 用到的函数

在解析 on 时会用到函数 parse_line_action，功能是对 on 字段所在的 option 进行解析。函数 parse_line_action 的具体实现代码如下所示：

```
static void parse_line_action(struct parse_state* state, int nargs, char **args)
//action 所在的行
{
    struct command *cmd;
    struct action *act = state->context;//on boot 启动
    int (*func)(int nargs, char **args);
    int kw, n;

    if (nargs == 0) {
        return;
    }

    kw = lookup_keyword(args[0]);//命令的参数个数
    if (!kw_is(kw, COMMAND)) {
        parse_error(state, "invalid command '%s'\n", args[0]);
        return;
    }

    n = kw_nargs(kw);
    if (nargs < n) {
        parse_error(state, "%s requires %d %s\n", args[0], n - 1,
            n > 2 ? "arguments" : "argument");
        return;
    }
    cmd = malloc(sizeof(*cmd) + sizeof(char*) * nargs);
    cmd->func = kw_func(kw);
    cmd->nargs = nargs;
    memcpy(cmd->args, args, sizeof(char*) * nargs);
    list_add_tail(&act->commands, &cmd->clist);  //
}
```

到此为止，on 和 service 两个 section 的分析工作全部完成。

8.6 在 init 控制 service

在 Android 系统中，当进行 init 进程初始化的时候，除了对系统做一些必要的初始化外操作，还需要启动 Service。而 Service 是在 init 脚本中定义的，所以很有必要了解一下在 init 中对 service 进行控制的知识，这部分知识将在本节的内容中呈现出来。

8.6.1 启动 Zygote

Android 系统是基于 Linux 内核的，而在 Linux 系统中的所有进程都是 init 进程的子进程或孙进程。也就是说，所有的进程都是直接或者间接地由 init 进程 fork 出来的。Zygote 进程也不例外，它是在系统启动的过程，由 init 进程创建的。在系统启动脚本文件“system/core /rootdir/init.rc”中，可以看到如下启动 Zygote 进程的脚本命令：

```
service zygote /system/bin/app_process -Xzygote /system/bin --zygote --start-system-server
    socket zygote stream 666
    onrestart write /sys/android_power/request_state wake
    onrestart write /sys/power/state on
    onrestart restart media
    onrestart restart netd
```

在上述代码中，各个关键字的具体说明如下所示。

- service：用于通知 init 进程创建一个名为“Zygote”的进程，这个 Zygote 进程要执行的程序是“/system/bin/app_process”，后面是要传给 app_process 的参数。
- Socket:表示这个 Zygote 进程需要一个名称为“Zygote”的 Socket 资源，这样启动系统后，就可以在“/dev/socket”目录下看到有一个名为 Zygote 的文件。这里定义的 Socket 的类型为 Unix Domain Socket，它是用来作本地进程间通信用的。

8.6.2 启动 service

首先看函数 do_class_start，此函数的功能是启动 service，此函数在如下文件中定义：

```
\system\core\init\builtins.c
```

函数 do_class_start 的具体实现代码如下所示：

```
int do_class_start(int nargs, char **args)
{
      /* Starting a class does not start services
       * which are explicitly disabled.  They must
       * be started individually.
       */
    service_for_each_class(args[1], service_start_if_not_disabled);
    return 0;
}
```

在上述代码中，调用了函数 service_start_if_not_disabled 实现启动功能，此函数也在文件 builtins.c 中实现，具体实现代码如下所示：

```
static void service_start_if_not_disabled(struct service *svc)
{
    if (!(svc->flags & SVC_DISABLED)) {
        service_start(svc, NULL);
    }
}
```

在上述代码中，调用了函数 service_start 实现启动功能。函数 service_start 是整个启动功能的核心，在文件 init.c 中定义，具体实现代码如下所示：

```
void service_start(struct service *svc, const char *dynamic_args)
{
    struct stat s;
    pid_t pid;
    int needs_console;
    int n;
#ifdef HAVE_SELINUX
    char *scon = NULL;
    int rc;
#endif
        /* starting a service removes it from the disabled or reset
         * state and immediately takes it out of the restarting
         * state if it was in there
         */
```

```
    svc->flags &= (~(SVC_DISABLED|SVC_RESTARTING|SVC_RESET));
    svc->time_started = 0;
        /* running processes require no additional work -- if
         * they're in the process of exiting, we've ensured
         * that they will immediately restart on exit, unless
         * they are ONESHOT
         */
    if (svc->flags & SVC_RUNNING) {
        return;
    }

    needs_console = (svc->flags & SVC_CONSOLE) ? 1 : 0;
    if (needs_console && (!have_console)) {
        ERROR("service '%s' requires console\n", svc->name);
        svc->flags |= SVC_DISABLED;
        return;
    }

    if (stat(svc->args[0], &s) != 0) {
        ERROR("cannot find '%s', disabling '%s'\n", svc->args[0], svc->name);
        svc->flags |= SVC_DISABLED;
        return;
    }

    if ((!(svc->flags & SVC_ONESHOT)) && dynamic_args) {
        ERROR("service '%s' must be one-shot to use dynamic args, disabling\n",
              svc->args[0]);
        svc->flags |= SVC_DISABLED;
        return;
    }

#ifdef HAVE_SELINUX
    if (is_selinux_enabled() > 0) {
        char *mycon = NULL, *fcon = NULL;

        INFO("computing context for service '%s'\n", svc->args[0]);
        rc = getcon(&mycon);
        if (rc < 0) {
            ERROR("could not get context while starting '%s'\n", svc->name);
            return;
        }

        rc = getfilecon(svc->args[0], &fcon);
        if (rc < 0) {
            ERROR("could not get context while starting '%s'\n", svc->name);
            freecon(mycon);
            return;
        }

        rc = security_compute_create(mycon, fcon, string_to_security_class("process"), &scon);
        freecon(mycon);
        freecon(fcon);
        if (rc < 0) {
            ERROR("could not get context while starting '%s'\n", svc->name);
            return;
        }
    }
#endif

    NOTICE("starting '%s'\n", svc->name);

    pid = fork();

    if (pid == 0) {
        struct socketinfo *si;
        struct svcenvinfo *ei;
        char tmp[32];
        int fd, sz;
```

```
    umask(077);
#ifdef __arm__
    /*
     * b/7188322 - Temporarily revert to the compat memory layout
     * to avoid breaking third party apps.
     *
     * THIS WILL GO AWAY IN A FUTURE ANDROID RELEASE.
     *
     *
http://git.kernel.org/?p=linux/kernel/git/torvalds/linux-2.6.git;a=commitdiff;h=7dbaa466
     * changes the kernel mapping from bottom up to top-down.
     * This breaks some programs which improperly embed
     * an out of date copy of Android's linker.
     */
    int current = personality(0xffffFFFF);
    personality(current | ADDR_COMPAT_LAYOUT);
#endif
    if (properties_inited()) {
        get_property_workspace(&fd, &sz);
        sprintf(tmp, "%d,%d", dup(fd), sz);
        add_environment("ANDROID_PROPERTY_WORKSPACE", tmp);
    }

    for (ei = svc->envvars; ei; ei = ei->next)
        add_environment(ei->name, ei->value);

#ifdef HAVE_SELINUX
    setsockcreatecon(scon);
#endif

    for (si = svc->sockets; si; si = si->next) {
        int socket_type = (
                !strcmp(si->type, "stream") ? SOCK_STREAM :
                    (!strcmp(si->type, "dgram") ? SOCK_DGRAM : SOCK_SEQPACKET));
        int s = create_socket(si->name, socket_type,
                              si->perm, si->uid, si->gid);
        if (s >= 0) {
            publish_socket(si->name, s);
        }
    }

#ifdef HAVE_SELINUX
    freecon(scon);
    scon = NULL;
    setsockcreatecon(NULL);
#endif

    if (svc->ioprio_class != IoSchedClass_NONE) {
        if (android_set_ioprio(getpid(), svc->ioprio_class, svc->ioprio_pri)) {
            ERROR("Failed to set pid %d ioprio = %d,%d: %s\n",
                  getpid(), svc->ioprio_class, svc->ioprio_pri, strerror(errno));
        }
    }

    if (needs_console) {
        setsid();
        open_console();
    } else {
        zap_stdio();
    }

#if 0
    for (n = 0; svc->args[n]; n++) {
        INFO("args[%d] = '%s'\n", n, svc->args[n]);
    }
    for (n = 0; ENV[n]; n++) {
        INFO("env[%d] = '%s'\n", n, ENV[n]);
    }
```

```
#endif

        setpgid(0, getpid());

    /* as requested, set our gid, supplemental gids, and uid */
        if (svc->gid) {
            if (setgid(svc->gid) != 0) {
                ERROR("setgid failed: %s\n", strerror(errno));
                _exit(127);
            }
        }
        if (svc->nr_supp_gids) {
            if (setgroups(svc->nr_supp_gids, svc->supp_gids) != 0) {
                ERROR("setgroups failed: %s\n", strerror(errno));
                _exit(127);
            }
        }
        if (svc->uid) {
            if (setuid(svc->uid) != 0) {
                ERROR("setuid failed: %s\n", strerror(errno));
                _exit(127);
            }
        }

#ifdef HAVE_SELINUX
        if (svc->seclabel) {
            if (is_selinux_enabled() > 0 && setexeccon(svc->seclabel) < 0) {
                ERROR("cannot setexeccon('%s'): %s\n", svc->seclabel, strerror(errno));
                _exit(127);
            }
        }
#endif

        if (!dynamic_args) {
            if (execve(svc->args[0], (char**) svc->args, (char**) ENV) < 0) {
                ERROR("cannot execve('%s'): %s\n", svc->args[0], strerror(errno));
            }
        } else {
            char *arg_ptrs[INIT_PARSER_MAXARGS+1];
            int arg_idx = svc->nargs;
            char *tmp = strdup(dynamic_args);
            char *next = tmp;
            char *bword;

            /* Copy the static arguments */
            memcpy(arg_ptrs, svc->args, (svc->nargs * sizeof(char *)));

            while((bword = strsep(&next, " "))) {
                arg_ptrs[arg_idx++] = bword;
                if (arg_idx == INIT_PARSER_MAXARGS)
                    break;
            }
            arg_ptrs[arg_idx] = '\0';
            execve(svc->args[0], (char**) arg_ptrs, (char**) ENV);
        }
        _exit(127);
    }

#ifdef HAVE_SELINUX
    freecon(scon);
#endif

    if (pid < 0) {
        ERROR("failed to start '%s'\n", svc->name);
        svc->pid = 0;
        return;
    }

    svc->time_started = gettime();
```

```
    svc->pid = pid;
    svc->flags |= SVC_RUNNING;

    if (properties_inited())
        notify_service_state(svc->name, "running");
}
```

在函数 service_start 中，参数 dynamic_args 只有当 service 的 option 中有 oneshot 时才会用到，此时会通过替换掉启动服务的命令参数启动服务。因为 service 的 option 会记录在 struct service 中，所以在启动 service 时只需考虑到这些选项即可。同时，会记录下 service 的 pid 和状态等信息。

8.6.3 4 种启动 service 的方式

其实在 init 进程中，可以使用如下所示的方式启动 service。

（1）在 action 下面添加和启动服务相关的 command，在 action 中和操作服务相关的命令如下。

- class_start <serviceclass> #：启动所有指定 class 的服务。
- class_stop <serviceclass> #：停止所有指定 class 的服务，后续没法通过 class_start 启动。
- class_reset <serviceclass> #：停止服务，后续可以通过 class_start 启动。
- restart <servicename> #：重启指定名称的服务，先 stop，再 start。
- start <servicename> #：启动指定名称的服务。
- stop <servicename> #：停止指定名称的服务。

在启动 command 时，在文件 init.c 的主函数 main 中，通过使用“for(;;)”循环执行函数 execute_one_command 的方式实现，此函数的具体实现代码如下所示：

```
void execute_one_command(void)
{
    int ret;

    if (!cur_action || !cur_command || is_last_command(cur_action, cur_command)) {
        cur_action = action_remove_queue_head();
        cur_command = NULL;
        if (!cur_action)
            return;
        INFO("processing action %p (%s)\n", cur_action, cur_action->name);
        cur_command = get_first_command(cur_action);
    } else {
        cur_command = get_next_command(cur_action, cur_command);
    }

    if (!cur_command)
        return;

    ret = cur_command->func(cur_command->nargs, cur_command->args);//执行class_start等
    INFO("command '%s' r=%d\n", cur_command->args[0], ret);
}
```

（2）使用函数 restart_processes 和 restart_service_if_needed 重启 Service，该函数位于 init 的主线程循环中，功能是查看是否有需要重新启动的 Service。在文件 init.c 中，函数 restart_processes 和 restart_service_if_needed 的具体实现代码如下所示：

```
static void restart_service_if_needed(struct service *svc)
{
    time_t next_start_time = svc->time_started + 5;

    if (next_start_time <= gettime()) {
        svc->flags &= (~SVC_RESTARTING);
        service_start(svc, NULL);
        return;
    }
```

```
    if ((next_start_time < process_needs_restart) ||
        (process_needs_restart == 0)) {
        process_needs_restart = next_start_time;
    }
}

static void restart_processes()
{
    process_needs_restart = 0;
    service_for_each_flags(SVC_RESTARTING,
                          restart_service_if_needed);
}
```

在重启过程中，会重启 flag 为 SVC_RESTARTING 的服务。这部分进程的重启其实在 init 由 handle_signal 来管理，一旦出现 Service 崩溃，函数 poll 会接收到相关文件变化的信息，并执行 handle_signal 中的函数 wait_for_one_process。函数 wait_for_one_process 的具体实现代码如下所示：

```
static int wait_for_one_process(int block)
{
    pid_t pid;
    int status;
    struct service *svc;
    struct socketinfo *si;
    time_t now;
    struct listnode *node;
    struct command *cmd;

    while ( (pid = waitpid(-1, &status, block ? 0 : WNOHANG)) == -1 && errno == EINTR );
    if (pid <= 0) return -1;
    INFO("waitpid returned pid %d, status = %08x\n", pid, status);

    svc = service_find_by_pid(pid);
    if (!svc) {
        ERROR("untracked pid %d exited\n", pid);
        return 0;
    }
 .....
  svc->flags |= SVC_RESTARTING;

    /* Execute all onrestart commands for this service. */
    list_for_each(node, &svc->onrestart.commands) {
        cmd = node_to_item(node, struct command, clist);
        cmd->func(cmd->nargs, cmd->args);
    }
    notify_service_state(svc->name, "restarting");
    return 0;
}
```

在上述代码中，使用 waitpid 找到子进程退出的进程号 pid，然后查找到该 Service，对 Service 中的 onrestart 这个 commands 进行操作。同时将 Service 的 flag 设置为 SVC_RESTARTING，这样就结合前面讲到的 restart_processes 便重新启动了该服务进程。

（3）在文件“\system\core\init\property_service.c”中，使用函数 handle_property_set_f 向 socket 中名称为 property_service 的属性服务发送控制的消息，这样便可以进入到该函数中。函数 handle_property_set_f 的具体实现代码如下所示：

```
void handle_property_set_fd()
{
    prop_msg msg;
    int s;
    int r;
    int res;
    struct ucred cr;
    struct sockaddr_un addr;
    socklen_t addr_size = sizeof(addr);
    socklen_t cr_size = sizeof(cr);
```

```
    char * source_ctx = NULL;

    if ((s = accept(property_set_fd, (struct sockaddr *) &addr, &addr_size)) < 0) {
        return;
    }

    /* Check socket options here */
    if (getsockopt(s, SOL_SOCKET, SO_PEERCRED, &cr, &cr_size) < 0) {
        close(s);
        ERROR("Unable to recieve socket options\n");
        return;
    }

    r = TEMP_FAILURE_RETRY(recv(s, &msg, sizeof(msg), 0));
    if(r != sizeof(prop_msg)) {
        ERROR("sys_prop: mis-match msg size recieved: %d expected: %d errno: %d\n",
              r, sizeof(prop_msg), errno);
        close(s);
        return;
    }

    switch(msg.cmd) {
    case PROP_MSG_SETPROP:
        msg.name[PROP_NAME_MAX-1] = 0;
        msg.value[PROP_VALUE_MAX-1] = 0;

#ifdef HAVE_SELINUX
        getpeercon(s, &source_ctx);
#endif

        if(memcmp(msg.name,"ctl.",4) == 0) {
            // Keep the old close-socket-early behavior when handling
            // ctl.* properties.
            close(s);
            if (check_control_perms(msg.value, cr.uid, cr.gid, source_ctx)) {
                handle_control_message((char*) msg.name + 4, (char*) msg.value);
            } else {
                ERROR("sys_prop: Unable to %s service ctl [%s] uid:%d gid:%d pid:%d\n",
                        msg.name + 4, msg.value, cr.uid, cr.gid, cr.pid);
            }
        } else {
            if (check_perms(msg.name, cr.uid, cr.gid, source_ctx)) {
                property_set((char*) msg.name, (char*) msg.value);
            } else {
                ERROR("sys_prop: permission denied uid:%d  name:%s\n",
                      cr.uid, msg.name);
            }

            // Note: bionic's property client code assumes that the
            // property server will not close the socket until *AFTER*
            // the property is written to memory.
            close(s);
        }
#ifdef HAVE_SELINUX
        freecon(source_ctx);
#endif

        break;

    default:
        close(s);
        break;
    }
}
```

（4）使用函数 handle_keychord 启动，该函数和 chorded keyboard 有关，功能是处理注册在 Service structure 上的 keychord，通常是启动 Service。函数 handle_keychord 在文件“system/core/init/keychords.c”中定义，具体实现代码如下所示：

```
void handle_keychord()
{
    struct service *svc;
    const char* debuggable;
    const char* adb_enabled;
    int ret;
    __u16 id;

    // only handle keychords if ro.debuggable is set or adb is enabled.
    // the logic here is that bugreports should be enabled in userdebug or eng builds
    // and on user builds for users that are developers.
    debuggable = property_get("ro.debuggable");
    adb_enabled = property_get("init.svc.adbd");
    ret = read(keychord_fd, &id, sizeof(id));
    if (ret != sizeof(id)) {
        ERROR("could not read keychord id\n");
        return;
    }

    if ((debuggable && !strcmp(debuggable, "1")) ||
        (adb_enabled && !strcmp(adb_enabled, "running"))) {
        svc = service_find_by_keychord(id);
        if (svc) {
            INFO("starting service %s from keychord\n", svc->name);
            service_start(svc, NULL);
        } else {
            ERROR("service for keychord %d not found\n", id);
        }
    }
}
```

8.7 控制属性服务

编写过 Windows 本地应用的读者应该知道，在 Windows 中的注册表机制中提供了大量的属性，其实在 Linux 中也有类似的机制：属性服务。init 在启动的过程中会启动属性服务（Socket 服务），并且在内存中建立一块存储区域，用来存储这些属性。当读取这些属性时，直接从这一内存区域读取，如果修改属性值，需要通过 Socket 连接属性服务完成。在文件 init.c 中，函数 action 调用了函数 start_property_service 来启动属性服务，action 是 init.rc 及其类似文件中的一种执行机制。

在文件 init.c 中，可以看到和属性操作相关的代码情景，例如：

```
property_init()
property_set_fd
```

在本节的内容中，将详细讲解在 Android 4.3 系统中 init 控制属性服务的基本知识。

8.7.1　引入属性

在文件 init.c 的主函数 main 中，调用函数 property_init 为属性分配了一些存储空间。如果查看文件 init.rc，会发现该文件开始部分用一些 import 语句导入了其他的配置文件，例如，/init.usb.rc。大多数配置文件都直接使用了确定的文件名，只有如下的代码使用了一个变量（${ro.hardware}）执行了配置文件名的一部分。

```
import /init.${ro.hardware}.rc
```

要想了解上述变量的获得过程，首先需要了解配置文件 init.${ro.hardware}.rc 的内容，这些通常与当前的硬件有关，其中函数 get_hardware_name 用于获取硬件的名称信息，具体代码如下所示：

```
void get_hardware_name(char *hardware, unsigned int *revision)
{
    char data[1024];
```

```
    int fd, n;
    char *x, *hw, *rev;

    /* 如果 hardware 已经有值了，说明 hardware 通过内核命令行提供，直接返回 */
    if (hardware[0])
        return;
    //  打开/proc/cpuinfo 文件
    fd = open("/proc/cpuinfo", O_RDONLY);
    if (fd < 0) return;
    //  读取/proc/cpuinfo 文件的内容
    n = read(fd, data, 1023);
    close(fd);
    if (n < 0) return;

    data[n] = 0;
    //  从/proc/cpuinfo 文件中获取 Hardware 字段的值
    hw = strstr(data, "\nHardware");
    rev = strstr(data, "\nRevision");
    //  成功获取 Hardware 字段的值
    if (hw) {
        x = strstr(hw, ": ");
        if (x) {
            x += 2;
            n = 0;
            while (*x && *x != '\n') {
                if (!isspace(*x))
                    //  将 Hardware 字段的值都转换为小写，并更新 hardware 参数的值
                    //  hardware 也就是在 init.c 文件中定义的 hardware 数组
                    hardware[n++] = tolower(*x);
                x++;
                if (n == 31) break;
            }
            hardware[n] = 0;
        }
    }

    if (rev) {
        x = strstr(rev, ": ");
        if (x) {
            *revision = strtoul(x + 2, 0, 16);
        }
    }
}
```

从上述代码可以得知，该函数主要用于确定 hardware 和 revision 的变量的值。获取 hardware 的来源是从 Linux 内核命令行或文件“/proc/cpuinfo”中的内容，文件“/proc/cpuinfo”是虚拟文件（内存文件），执行 cat /proc/cpuinfo 命令会看到该文件中的内容，如图 8-1 所示。

```
Processor       : ARMv7 Processor rev 9 (v7l)
processor       : 0
BogoMIPS        : 1993.93

processor       : 1
BogoMIPS        : 1993.93

processor       : 2
BogoMIPS        : 1993.93

processor       : 3
BogoMIPS        : 1993.93

Features        : swp half thumb fastmult vfp edsp neon vfpv3 tls
CPU implementer : 0x41
CPU architecture: 7
CPU variant     : 0x2
CPU part        : 0xc09
CPU revision    : 9

Hardware        : grouper
Revision        : 0000
Serial          : 0f410a0001440200
root@android:/ #
```

图 8-1　显示文件内容

在图 8-1 中，白框中的内容就是 Hardware 字段的值。由于该设备是 Nexus 7，所以值为 grouper。如果程序就到此位置，那么与硬件有关的配置文件名是 init.grouper.rc。有 Nexus 7 的读者会看到在根目录下确实有一个 init.grouper.rc 文件。说明 Nexus 7 的原生 ROM 并没有在其他的地方设置配置文件名，所以配置文件名就是从“/proc/cpuinfo”文件的 Hardware 字段中取的值。

接下来看在函数 get_hardware_name 后面调用的函数 process_kernel_cmdline，具体实现代码如下所示：

```
static void process_kernel_cmdline(void)
{
    /* don't expose the raw commandline to nonpriv processes */
    chmod("/proc/cmdline", 0440);

    // 导入内核命令行参数
    import_kernel_cmdline(0, import_kernel_nv);
    if (qemu[0])
        import_kernel_cmdline(1, import_kernel_nv);

    // 用属性值设置内核变量
    export_kernel_boot_props();
}
```

在上述代码中，除了使用函数 import_kernel_cmdline 导入内核变量外，其主要功能是调用函数 export_kernel_boot_props 通过属性设置内核变量。例如，通过属性 ro.boot.hardware 设置 hardware 变量。也就是说，可以通过属性值 ro.boot.hardware 修改函数 get_hardware_name 中从文件“/proc/cpuinfo”得到的 hardware 字段值。函数 export_kernel_boot_props 的具体实现代码如下所示：

```
static void export_kernel_boot_props(void)
{
    char tmp[PROP_VALUE_MAX];
    const char *pval;
    unsigned i;
    struct {
        const char *src_prop;
        const char *dest_prop;
        const char *def_val;
    } prop_map[] = {
        { "ro.boot.serialno", "ro.serialno", "", },
        { "ro.boot.mode", "ro.bootmode", "unknown", },
        { "ro.boot.baseband", "ro.baseband", "unknown", },
        { "ro.boot.bootloader", "ro.bootloader", "unknown", },
    };
    // 通过内核的属性设置应用层配置文件的属性
    for (i = 0; i < ARRAY_SIZE(prop_map); i++) {
        pval = property_get(prop_map[i].src_prop);
        property_set(prop_map[i].dest_prop, pval ?: prop_map[i].def_val);
    }
    // 根据 ro.boot.console 属性的值设置 console 变量
    pval = property_get("ro.boot.console");
    if (pval)
        strlcpy(console, pval, sizeof(console));

    /* save a copy for init's usage during boot */
    strlcpy(bootmode, property_get("ro.bootmode"), sizeof(bootmode));

    /* if this was given on kernel command line, override what we read
     * before (e.g. from /proc/cpuinfo), if anything */
    // 获取 ro.boot.hardware 属性的值
    pval = property_get("ro.boot.hardware");
    if (pval)
        // 这里通过 ro.boot.hardware 属性再次改变 hardware 变量的值
        strlcpy(hardware, pval, sizeof(hardware));
    // 利用 hardware 变量的值设置设置 ro.hardware 属性
    // 这个属性就是前面提到的设置初始化文件名的属性，实际上是通过 hardware 变量设置的
    property_set("ro.hardware", hardware);
```

```
    snprintf(tmp, PROP_VALUE_MAX, "%d", revision);
    property_set("ro.revision", tmp);

    /* TODO: these are obsolete. We should delete them */
    if (!strcmp(bootmode,"factory"))
       property_set("ro.factorytest", "1");
    else if (!strcmp(bootmode,"factory2"))
       property_set("ro.factorytest", "2");
    else
       property_set("ro.factorytest", "0");
}
```

由上述代码可以看出，该函数实际上就是来回设置一些属性值，并且利用某些属性值修改console、hardware 等变量。其中 hardware 变量（就是一个长度为 32 的字符数组）在函数 get_ hardware_name 中已经从文件“/proc/cpuinfo”中获得过一次值了，在函数 export_kernel_ boot_props 中又通过属性 ro.boot.hardware 设置了一次值，不过在 Nexus 7 中并没有设置该属性，所以 hardware 的值仍为 grouper。最后用变量 hardware 设置属性 ro.hardware，所以最后的初始化文件名为 init.grouper.rc。

8.7.2 初始化属性服务

在文件“\system\core\init\property_service.c”中，使用函数 property_init 初始化属性存储区域，具体实现代码如下所示：

```
void property_init(void)
{
    init_property_area();
}
```

在上述代码中，函数 init_property_area 也是在文件 property_service.c 中实现的，该函数用于初始化属性内存区域，也就是__system_property_area__变量。函数 init_property_area 的具体实现代码如下所示：

```
static int init_property_area(void)
{
    prop_area *pa;

    if(pa_info_array)
       return -1;

    if(init_workspace(&pa_workspace, PA_SIZE))
       return -1;

    fcntl(pa_workspace.fd, F_SETFD, FD_CLOEXEC);

    pa_info_array = (void*) (((char*) pa_workspace.data) + PA_INFO_START);

    pa = pa_workspace.data;
    memset(pa, 0, PA_SIZE);
    pa->magic = PROP_AREA_MAGIC;
    pa->version = PROP_AREA_VERSION;

       /* plug into the lib property services */
    __system_property_area__ = pa;
    property_area_inited = 1;
    return 0;
}
```

8.7.3 启动属性服务

在文件“\system\core\init\property_service.c”中，使用函数 start_property_service 启动一个属性服务器，具体实现代码如下所示：

```
void start_property_service(void)
{
    int fd;
    //  装载不同的属性文件
    load_properties_from_file(PROP_PATH_SYSTEM_BUILD);
    load_properties_from_file(PROP_PATH_SYSTEM_DEFAULT);
    load_override_properties();
    /* Read persistent properties after all default values have been loaded. */
    load_persistent_properties();
    //  创建 socket 服务（属性服务）
    fd = create_socket(PROP_SERVICE_NAME, SOCK_STREAM, 0666, 0, 0);
    if(fd < 0) return;
    fcntl(fd, F_SETFD, FD_CLOEXEC);
    fcntl(fd, F_SETFL, O_NONBLOCK);
    //  开始服务监听
    listen(fd, 8);
    property_set_fd = fd;
}
```

通过上述代码可以知道属性服务的启动方式，另外在函数 start_property_service 中还涉及如下所示的两个宏。

- PROP_PATH_SYSTEM_BUILD。
- PROP_PATH_SYSTEM_DEFAULT。

上述两个宏都是系统预定义的属性文件名的路径，为了获取这些宏的定义，需要先分析函数 property_get，该函数也是在 Property_service.c 中实现，具体实现代码如下所示：

```
const char* property_get(const char *name)
{
    prop_info *pi;
    if(strlen(name) >= PROP_NAME_MAX) return 0;
    pi = (prop_info*) __system_property_find(name);
    if(pi != 0) {
        return pi->value;
    } else {
        return 0;
    }
}
```

通过上述代码可以看到，在函数 property_get 中调用了核心函数__system_property_find，该核心函数真正实现了获取属性值的功能。函数__system_property_find 属于 bionic 的一个 library，在文件 system_properties.c 中实现，可以在如下的目录找到该文件。

```
/bionic/libc/bionic
```

函数__system_property_find 的具体实现代码如下所示：

```
const prop_info *__system_property_find(const char *name)
{
    //  获取属性存储内存区域的首地址
    prop_area *pa = __system_property_area__;
    unsigned count = pa->count;
    unsigned *toc = pa->toc;
    unsigned len = strlen(name);
    prop_info *pi;

    while(count--) {
        unsigned entry = *toc++;
        if(TOC_NAME_LEN(entry) != len) continue;

        pi = TOC_TO_INFO(pa, entry);
        if(memcmp(name, pi->name, len)) continue;
        return pi;
    }
    return 0;
}
```

从上述函数__system_property_find的实现代码可以看出，第一行使用了一个__system_property_area__变量，该变量是全局的。

在文件system_properties.c对应的头文件system_properties.h中，定义了前面提到的两个表示属性文件路径的宏，其实还有另外两个表示路径的宏，共4个属性文件。文件system_properties.h可以在如下所示的目录中找到

```
/bionic/libc/include/sys
```

这4个宏的具体定义如下所示：

```
#define PROP_PATH_RAMDISK_DEFAULT  "/default.prop"
#define PROP_PATH_SYSTEM_BUILD     "/system/build.prop"
#define PROP_PATH_SYSTEM_DEFAULT   "/system/default.prop"
#define PROP_PATH_LOCAL_OVERRIDE   "/data/local.prop"
```

此时可以进入Android设备的相应目录找到上述4个文件，一般会被保存在根目录中，通常在文件default.prop和catdefault.prop中会看到该文件的内容。而属性服务就是装载所有这4个属性文件中的所有属性以及使用 property_set 设置的属性。在 Android 设备的终端可以直接使用getprop命令从属性服务获取所有的属性值，如图8-2所示。另外，getprop命令还可以直接根属性名还获取具体的属性值，例如：

```
getprop ro.build.product
```

```
2|root@android:/dev/socket # getprop
[camera.flash_off]: [0]
[dalvik.vm.dexopt-flags]: [m=y]
[dalvik.vm.heapgrowthlimit]: [64m]
[dalvik.vm.heapmaxfree]: [8m]
[dalvik.vm.heapminfree]: [512k]
[dalvik.vm.heapsize]: [384m]
[dalvik.vm.heapstartsize]: [8m]
[dalvik.vm.heaptargetutilization]: [0.75]
[dalvik.vm.stack-trace-file]: [/data/anr/traces.txt]
[debug.nfc.fw_download]: [false]
[debug.nfc.se]: [false]
[dev.bootcomplete]: [1]
[dhcp.wlan0.dns1]: [192.168.17.254]
[dhcp.wlan0.dns2]: []
[dhcp.wlan0.dns3]: []
[dhcp.wlan0.dns4]: []
[dhcp.wlan0.gateway]: [192.168.17.254]
[dhcp.wlan0.ipaddress]: [192.168.17.104]
[dhcp.wlan0.leasetime]: [7200]
[dhcp.wlan0.mask]: [255.255.255.0]
[dhcp.wlan0.pid]: [10740]
[dhcp.wlan0.reason]: [RENEW]
[dhcp.wlan0.result]: [ok]
[dhcp.wlan0.server]: [192.168.17.254]
```

图8-2 从属性服务获取所有的属性值

在Android 4.3源码中，getprop命令的源代码文件是getprop.c。读者可以在如下所示的目录中找到该文件：

```
/system/core/toolbox/
```

其实getprop获取属性值也是通过函数property_get完成的，此函数实际上调用了函数__system_property_find，从__system_property_area__变量指定的内存区域获取相应的属性值。另外，在文件system_properties.c中还有如下两个函数用于通过属性服务修改或添加某个属性的值。

```
static int send_prop_msg(prop_msg *msg)
{
    struct pollfd pollfds[1];
    struct sockaddr_un addr;
    socklen_t alen;
    size_t namelen;
    int s;
    int r;
    int result = -1;
    // 创建用于连接属性服务的socket
    s = socket(AF_LOCAL, SOCK_STREAM, 0);
    if(s < 0) {
        return result;
```

```
    }

    memset(&addr, 0, sizeof(addr));
    // property_service_socket 是 Socket 设备文件名称
    namelen = strlen(property_service_socket);
    strlcpy(addr.sun_path, property_service_socket, sizeof addr.sun_path);
    addr.sun_family = AF_LOCAL;
    alen = namelen + offsetof(struct sockaddr_un, sun_path) + 1;

    if(TEMP_FAILURE_RETRY(connect(s, (struct sockaddr *) &addr, alen)) < 0) {
        close(s);
        return result;
    }

    r = TEMP_FAILURE_RETRY(send(s, msg, sizeof(prop_msg), 0));

    if(r == sizeof(prop_msg)) {
        pollfds[0].fd = s;
        pollfds[0].events = 0;
        r = TEMP_FAILURE_RETRY(poll(pollfds, 1, 250 /* ms */));
        if (r == 1 && (pollfds[0].revents & POLLHUP) != 0) {
            result = 0;
        } else {

            result = 0;
        }
    }

    close(s);
    return result;
}
// 用户可以直接调用该函数设置属性值
int __system_property_set(const char *key, const char *value)
{
    int err;
    int tries = 0;
    int update_seen = 0;
    prop_msg msg;

    if(key == 0) return -1;
    if(value == 0) value = "";
    if(strlen(key) >= PROP_NAME_MAX) return -1;
    if(strlen(value) >= PROP_VALUE_MAX) return -1;

    memset(&msg, 0, sizeof msg);
    msg.cmd = PROP_MSG_SETPROP;
    strlcpy(msg.name, key, sizeof msg.name);
    strlcpy(msg.value, value, sizeof msg.value);
    //  设置属性值
    err = send_prop_msg(&msg);
    if(err < 0) {
        return err;
    }
    return 0;
}
```

在函数 send_prop_msg 中，涉及了重要变量 property_service_socket，具体定义如下所示：

```
static const char property_service_socket[] = "/dev/socket/" PROP_SERVICE_NAME;
```

实际上，send_prop_msg 通过这个设备文件与属性服务通信。读者可以在 Android 设备的终端进入“/dev/socket”目录，通常会看到一个名为 property_service 的文件，该文件就是属性服务映射的设备文件。

8.7.4　处理设置属性的请求

当属性服务器收到客户端的请求时，init 会调用函数 handle_property_set_fd 进行处理。当客户端的权限满足要求时，init 就调用函数 property_set 进行相关的处理。

```
int property_set(const char *name, const char *value)
{
    prop_area *pa;
    prop_info *pi;

    int namelen = strlen(name);
    int valuelen = strlen(value);

    if(namelen >= PROP_NAME_MAX) return -1;
    if(valuelen >= PROP_VALUE_MAX) return -1;
    if(namelen < 1) return -1;

    pi = (prop_info*) __system_property_find(name);

    if(pi != 0) {
        /* ro.* properties may NEVER be modified once set */
        if(!strncmp(name, "ro.", 3)) return -1;

        pa = __system_property_area__;
        update_prop_info(pi, value, valuelen);
        pa->serial++;
        __futex_wake(&pa->serial, INT32_MAX);
    } else {
        pa = __system_property_area__;
        if(pa->count == PA_COUNT_MAX) return -1;

        pi = pa_info_array + pa->count;
        pi->serial = (valuelen << 24);
        memcpy(pi->name, name, namelen + 1);
        memcpy(pi->value, value, valuelen + 1);

        pa->toc[pa->count] =
            (namelen << 24) | (((unsigned) pi) - ((unsigned) pa));

        pa->count++;
        pa->serial++;
        __futex_wake(&pa->serial, INT32_MAX);
    }
    /* If name starts with "net." treat as a DNS property. */
    if (strncmp("net.", name, strlen("net.")) == 0) {
        if (strcmp("net.change", name) == 0) {
            return 0;
        }
       /*
        * The 'net.change' property is a special property used track when any
        * 'net.*' property name is updated. It is _ONLY_ updated here. Its value
        * contains the last updated 'net.*' property.
        */
        property_set("net.change", name);
    } else if (persistent_properties_loaded &&
            strncmp("persist.", name, strlen("persist.")) == 0) {
        /*
         * Don't write properties to disk until after we have read all default properties
         * to prevent them from being overwritten by default values.
         */
        write_persistent_property(name, value);
#ifdef HAVE_SELINUX
    } else if (strcmp("selinux.reload_policy", name) == 0 &&
               strcmp("1", value) == 0) {
        selinux_reload_policy();
#endif
    }
    property_changed(name, value);
    return 0;
}
```

到此为止，整个属性服务器的源码知识就介绍完毕了。

第9章　Dalvik VM的进程系统

在Android系统中，存在如下3个十分重要的进程系统。

- Zygote进程：被称为“孵化”进程或“孕育”进程，功能和Linux系统的fork类似，用于“孕育”产生出不同的子进程。
- System进程：是系统进程，是整个Android Framework所在的进程，用于启动Android系统。其核心进程是system_server，它的父进程就是Zygote。
- 应用程序进程：每个Android应用程序运行后都会拥有自己的进程，这和Windows资源管理器中的体现的进程是同一个含义。

在本章的内容中，将详细分析Android L中上述3大进程系统的基本知识，彻底深入了解Android进程系统的方方面面。

9.1 Zygote（孕育）进程详解

在本书前面的内容中，已经了解过Zygote进程的一些知识。在Android系统中，如果查看进程列表，会发现进程Zygote的父进程是init，而且它是所有应用的父进程；还有一个进程是system_server，它的父进程是Zygote。Zygote服务实际上是一种Select服务模型，是为启动Java代码而生，完成了一次androidRuntime的打开和关闭操作。

9.1.1 Zygote基础

Android系统是基于Linux内核的，在Linux系统中的所有进程都是直接或者间接地由init进程fork（孕育）出来的，Zygote进程也不例外。Zygote进程是在系统启动的过程由init进程创建的，是Android系统的核心进程之一，被认为是Android ramework大家族的祖先。事实上，Zygote正是平常所说的Java运行环境（JVM）。从总体架构上看，Zygote是一个简单的典型C/S结构。其他进程作为一个客服端向Zygote发出“孕育”请求，当Zygote接收到命令后就“孕育”出一个Activity进程。具体“孕育”过程如图9-1所示。

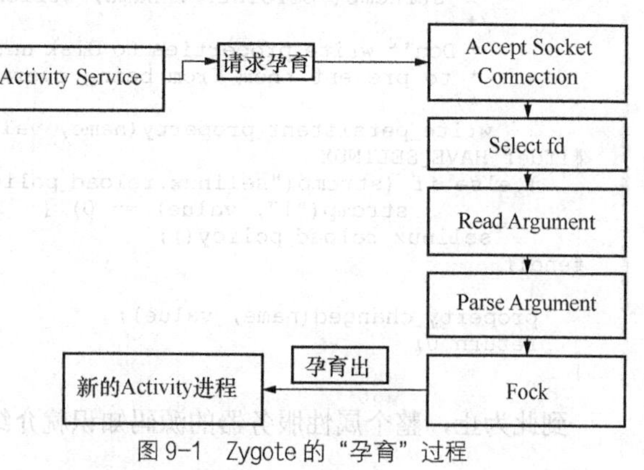

图9-1　Zygote的“孕育”过程

在Android系统中，Zygote本身是一个应用层的程序，和驱动、内核模块等没有任何关系。Zygote系统源码的组成及其调用结构如下所示。

（1）Zygote.java。

提供访问Dalvik的“zygote”接口，主要是包装Linux系统的Fork（孕育），以建立一个新的 VM 实例进程。

（2）ZygoteConnection.java。

Zygote 的套接口连接管理及其参数解析。其他 Actvitiy 建立进程请求是通过套接口发送命令参数给 Zygote。

（3）ZygoteInit.java。

Zygote 系统的 main 函数入口。

9.1.2 分析 Zygote 的启动过程

在 Android L 源码中，在文件“system\core\rootdir\init.rc”中可以看到启动 Zygote 进程的脚本命令，具体实现代码如下所示：

```
service zygote /system/bin/app_process -Xzygote /system/bin --zygote --start-system-server
    socket zygote stream 666
```

通过上述代码，系统启动后会在“/dev/socket”目录下看到有一个名为“zygote”的文件。在上述代码中，相关关键字的具体说明如下所示。

- service：通知 init 进程创建一个名为“zygote”的进程，此 zygote 进程要执行的程序是 /system/bin/app_process，后面部分需要传给 app_proces。
- socket：表示这个 zygote 进程需要一个名称为“zygote”的 socket 资源。

Zygote 最初的名字是 app_process，通过直接调用 pctrl 后把名字给改成了“zygote”。Zygote 进程执行的程序是 system/bin/app_process，其对应的源代码在如下文件中定义：

```
frameworks/base/cmds /app_process/app_main.cpp
```

文件 app_main.cpp 的入口函数是 main，接下来的内容中，将详细讲解启动 Zygote 进程的过程。

（1）分析启动脚本。

在文件“system\core\init\init.c”中，以服务的形式来启动 Zygote 进程。在启动初始化进程 init 时，会调用函数 service_start 来启动 Zygote。函数 service_start 的具体实现代码如下所示：

```
void service_start(struct service *svc, const char *dynamic_args)
{
    struct stat s;
    pid_t pid;
    int needs_console;
    int n;
    char *scon = NULL;
    int rc;

        /* starting a service removes it from the disabled or reset
         * state and immediately takes it out of the restarting
         * state if it was in there
         */
    svc->flags &= (~(SVC_DISABLED|SVC_RESTARTING|SVC_RESET));
    svc->time_started = 0;

        /* running processes require no additional work -- if
         * they're in the process of exiting, we've ensured
         * that they will immediately restart on exit, unless
         * they are ONESHOT
         */
    if (svc->flags & SVC_RUNNING) {
        return;
    }

    needs_console = (svc->flags & SVC_CONSOLE) ? 1 : 0;
    if (needs_console && (!have_console)) {
        ERROR("service '%s' requires console\n", svc->name);
        svc->flags |= SVC_DISABLED;
```

```
        return;
    }

    if (stat(svc->args[0], &s) != 0) {
        ERROR("cannot find '%s', disabling '%s'\n", svc->args[0], svc->name);
        svc->flags |= SVC_DISABLED;
        return;
    }

    if ((!(svc->flags & SVC_ONESHOT)) && dynamic_args) {
        ERROR("service '%s' must be one-shot to use dynamic args, disabling\n",
               svc->args[0]);
        svc->flags |= SVC_DISABLED;
        return;
    }

    if (is_selinux_enabled() > 0) {
        if (svc->seclabel) {
            scon = strdup(svc->seclabel);
            if (!scon) {
                ERROR("Out of memory while starting '%s'\n", svc->name);
                return;
            }
        } else {
            char *mycon = NULL, *fcon = NULL;

            INFO("computing context for service '%s'\n", svc->args[0]);
            rc = getcon(&mycon);
            if (rc < 0) {
                ERROR("could not get context while starting '%s'\n", svc->name);
                return;
            }

            rc = getfilecon(svc->args[0], &fcon);
            if (rc < 0) {
                ERROR("could not get context while starting '%s'\n", svc->name);
                freecon(mycon);
                return;
            }

            rc = security_compute_create(mycon, fcon, string_to_security_class("process"), &scon);
            freecon(mycon);
            freecon(fcon);
            if (rc < 0) {
                ERROR("could not get context while starting '%s'\n", svc->name);
                return;
            }
        }
    }

    NOTICE("starting '%s'\n", svc->name);

    pid = fork();

    if (pid == 0) {
        struct socketinfo *si;
        struct svcenvinfo *ei;
        char tmp[32];
        int fd, sz;

        umask(077);
        if (properties_inited()) {
            get_property_workspace(&fd, &sz);
            sprintf(tmp, "%d,%d", dup(fd), sz);
            add_environment("ANDROID_PROPERTY_WORKSPACE", tmp);
        }

        for (ei = svc->envvars; ei; ei = ei->next)
            add_environment(ei->name, ei->value);
```

```
    setsockcreatecon(scon);
    for (si = svc->sockets; si; si = si->next) {
        int socket_type = (
                !strcmp(si->type, "stream") ? SOCK_STREAM :
                    (!strcmp(si->type, "dgram") ? SOCK_DGRAM : SOCK_SEQPACKET));
        int s = create_socket(si->name, socket_type,
                              si->perm, si->uid, si->gid);
        if (s >= 0) {
            publish_socket(si->name, s);
        }
    }

    freecon(scon);
    scon = NULL;
    setsockcreatecon(NULL);

    if (svc->ioprio_class != IoSchedClass_NONE) {
        if (android_set_ioprio(getpid(), svc->ioprio_class, svc->ioprio_pri)) {
            ERROR("Failed to set pid %d ioprio = %d,%d: %s\n",
                  getpid(), svc->ioprio_class, svc->ioprio_pri, strerror(errno));
        }
    }

    if (needs_console) {
        setsid();
        open_console();
    } else {
        zap_stdio();
    }

#if 0
    for (n = 0; svc->args[n]; n++) {
        INFO("args[%d] = '%s'\n", n, svc->args[n]);
    }
    for (n = 0; ENV[n]; n++) {
        INFO("env[%d] = '%s'\n", n, ENV[n]);
    }
#endif

    setpgid(0, getpid());

  /* as requested, set our gid, supplemental gids, and uid */
    if (svc->gid) {
        if (setgid(svc->gid) != 0) {
            ERROR("setgid failed: %s\n", strerror(errno));
            _exit(127);
        }
    }
    if (svc->nr_supp_gids) {
        if (setgroups(svc->nr_supp_gids, svc->supp_gids) != 0) {
            ERROR("setgroups failed: %s\n", strerror(errno));
            _exit(127);
        }
    }
    if (svc->uid) {
        if (setuid(svc->uid) != 0) {
            ERROR("setuid failed: %s\n", strerror(errno));
            _exit(127);
        }
    }
    if (svc->seclabel) {
        if (is_selinux_enabled() > 0 && setexeccon(svc->seclabel) < 0) {
            ERROR("cannot setexeccon('%s'): %s\n", svc->seclabel, strerror(errno));
            _exit(127);
        }
    }
```

```
        if (!dynamic_args) {
            if (execve(svc->args[0], (char**) svc->args, (char**) ENV) < 0) {
                ERROR("cannot execve('%s'): %s\n", svc->args[0], strerror(errno));
            }
        } else {
            char *arg_ptrs[INIT_PARSER_MAXARGS+1];
            int arg_idx = svc->nargs;
            char *tmp = strdup(dynamic_args);
            char *next = tmp;
            char *bword;

            /* Copy the static arguments */
            memcpy(arg_ptrs, svc->args, (svc->nargs * sizeof(char *)));

            while((bword = strsep(&next, " "))) {
                arg_ptrs[arg_idx++] = bword;
                if (arg_idx == INIT_PARSER_MAXARGS)
                    break;
            }
            arg_ptrs[arg_idx] = '\0';
            execve(svc->args[0], (char**) arg_ptrs, (char**) ENV);
        }
        _exit(127);
    }

    freecon(scon);

    if (pid < 0) {
        ERROR("failed to start '%s'\n", svc->name);
        svc->pid = 0;
        return;
    }

    svc->time_started = gettime();
    svc->pid = pid;
    svc->flags |= SVC_RUNNING;

    if (properties_inited())
        notify_service_state(svc->name, "running");
}
```

在上述代码中，每一个 service 命令都会促使 init 进程调用 fork 函数来创建一个新的进程，在新的进程中会分析里面的 socket 选项。对于每一个 socket 选项来说，都会通过函数 create_socket 来在“/dev/socket”目录下创建一个文件。由此可见，函数 service_start 起了一个解释文件 init.rc 中 service 命令的作用。

再看函数 create_socket，其功能是调用函数 socket 创建一个 socket，使用文件描述符 fd 来描述此 socket。函数 create_socket 的具体实现代码如下所示：

```
int create_socket(const char *name, int type, mode_t perm, uid_t uid, gid_t gid)
{
    struct sockaddr_un addr;
    int fd, ret;
    char *secon;
//调用函数 socket 创建一个 socket,使用文件描述符 fd 来描述此 socket。
    fd = socket(PF_UNIX, type, 0);
    if (fd < 0) {
        ERROR("Failed to open socket '%s': %s\n", name, strerror(errno));
        return -1;
    }
//为 socket 创建一个类型为 AF_UNIX 的 socket 地址 addr
    memset(&addr, 0 , sizeof(addr));
    addr.sun_family = AF_UNIX;
    snprintf(addr.sun_path, sizeof(addr.sun_path), ANDROID_SOCKET_DIR"/%s",
            name);
    ret = unlink(addr.sun_path);
    if (ret != 0 && errno != ENOENT) {
```

```
        ERROR("Failed to unlink old socket '%s': %s\n", name, strerror(errno));
        goto out_close;
    }
    secon = NULL;
    if (sehandle) {
        ret = selabel_lookup(sehandle, &secon, addr.sun_path, S_IFSOCK);
        if (ret == 0)
            setfscreatecon(secon);
    }
    ret = bind(fd, (struct sockaddr *) &addr, sizeof (addr));
    if (ret) {
        ERROR("Failed to bind socket '%s': %s\n", name, strerror(errno));
        goto out_unlink;
    }
    setfscreatecon(NULL);
    freecon(secon);
    //设置设备文件的/dev/socket/zygote的用户id、用户组id和用户权限
    chown(addr.sun_path, uid, gid);
    chmod(addr.sun_path, perm);
    INFO("Created socket '%s' with mode '%o', user '%d', group '%d'\n",
         addr.sun_path, perm, uid, gid);
    return fd;
out_unlink:
    unlink(addr.sun_path);
out_close:
    close(fd);
    return -1;
}
```

再看函数 publish_socket，具体实现代码如下所示：

```
//参数fd是文件描述符，指向函数create_socket创建的socket
static void publish_socket(const char *name, int fd)
{
    char key[64] = ANDROID_SOCKET_ENV_PREFIX;
    char val[64];
//将宏ANDROID_SOCKET_ENV_PREFIX和参数name描述的字符串连接在一起，并保存在字符串key中
    strlcpy(key + sizeof(ANDROID_SOCKET_ENV_PREFIX) - 1,
            name,
            sizeof(key) - sizeof(ANDROID_SOCKET_ENV_PREFIX));
    snprintf(val, sizeof(val), "%d", fd);
    add_environment(key, val);

    /* make sure we don't close-on-exec */
    fcntl(fd, F_SETFD, 0);
}
```

（2）分析入口函数。

Zygote 的入口函数是 main，功能是创建 AppRuntime 变量，然后调用成员函数 start 启动进程。函数是 main 在文件"frameworks\base\cmds\app_process\app_main.cpp"中定义的，具体实现代码如下所示：

```
int main(int argc, char* const argv[])
{
#ifdef __arm__
    /*
     * b/7188322 - Temporarily revert to the compat memory layout
     * to avoid breaking third party apps.
     *
     * THIS WILL GO AWAY IN A FUTURE ANDROID RELEASE.
     *
     * http://git.kernel.org/?p=linux/kernel/git/torvalds/linux-2.6.git;a=commitdiff;h=7dbaa466
     * changes the kernel mapping from bottom up to top-down.
     * This breaks some programs which improperly embed
     * an out of date copy of Android's linker.
     */
    char value[PROPERTY_VALUE_MAX];
    property_get("ro.kernel.qemu", value, "");
```

```
    bool is_qemu = (strcmp(value, "1") == 0);
    if ((getenv("NO_ADDR_COMPAT_LAYOUT_FIXUP") == NULL) && !is_qemu) {
        int current = personality(0xFFFFFFFF);
        if ((current & ADDR_COMPAT_LAYOUT) == 0) {
            personality(current | ADDR_COMPAT_LAYOUT);
            setenv("NO_ADDR_COMPAT_LAYOUT_FIXUP", "1", 1);
            execv("/system/bin/app_process", argv);
            return -1;
        }
    }
    unsetenv("NO_ADDR_COMPAT_LAYOUT_FIXUP");
#endif

    // These are global variables in ProcessState.cpp
    mArgC = argc;
    mArgV = argv;

    mArgLen = 0;
    for (int i=0; i<argc; i++) {
        mArgLen += strlen(argv[i]) + 1;
    }
    mArgLen--;

    AppRuntime runtime;
    const char* argv0 = argv[0];

    // Process command line arguments
    // ignore argv[0]
    argc--;
    argv++;

    // Everything up to '--' or first non '-' arg goes to the vm

    int i = runtime.addVmArguments(argc, argv);

    // Parse runtime arguments.  Stop at first unrecognized option.
    bool zygote = false;
    bool startSystemServer = false;
    bool application = false;
    const char* parentDir = NULL;
    const char* niceName = NULL;
    const char* className = NULL;
    while (i < argc) {
        const char* arg = argv[i++];
        if (!parentDir) {
            parentDir = arg;
        } else if (strcmp(arg, "--zygote") == 0) {
            zygote = true;
            niceName = "zygote";
        } else if (strcmp(arg, "--start-system-server") == 0) {
            startSystemServer = true;
        } else if (strcmp(arg, "--application") == 0) {
            application = true;
        } else if (strncmp(arg, "--nice-name=", 12) == 0) {
            niceName = arg + 12;
        } else {
            className = arg;
            break;
        }
    }

    if (niceName && *niceName) {
        setArgv0(argv0, niceName);
        set_process_name(niceName);
    }

    runtime.mParentDir = parentDir;

    if (zygote) {
```

```
        runtime.start("com.android.internal.os.ZygoteInit",
                startSystemServer ? "start-system-server" : "");
    } else if (className) {
        // Remainder of args get passed to startup class main()
        runtime.mClassName = className;
        runtime.mArgC = argc - i;
        runtime.mArgV = argv + i;
        runtime.start("com.android.internal.os.RuntimeInit",
                application ? "application" : "tool");
    } else {
        fprintf(stderr, "Error: no class name or --zygote supplied.\n");
        app_usage();
        LOG_ALWAYS_FATAL("app_process: no class name or --zygote supplied.");
        return 10;
    }
}
```

（3）分析启动函数。

Zygote 的启动函数是 start，功能是调用函数 startVm 在 Zygote 中创建一个虚拟机实例。函数是 start 在文件“frameworks\base\core\jni\AndroidRuntime.cpp”中定义的，具体实现代码如下所示：

```
void AndroidRuntime::start(const char* className, const char* options)
{
    ALOGD("\n>>>>>> AndroidRuntime START %s <<<<<<\n",
            className != NULL ? className : "(unknown)");

    blockSigpipe();

    /*
     * 'startSystemServer == true' means runtime is obsolete and not run from
     * init.rc anymore, so we print out the boot start event here.
     */
    if (strcmp(options, "start-system-server") == 0) {
        /* track our progress through the boot sequence */
        const int LOG_BOOT_PROGRESS_START = 3000;
        LOG_EVENT_LONG(LOG_BOOT_PROGRESS_START,
                       ns2ms(systemTime(SYSTEM_TIME_MONOTONIC)));
    }

    const char* rootDir = getenv("ANDROID_ROOT");
    if (rootDir == NULL) {
        rootDir = "/system";
        if (!hasDir("/system")) {
            LOG_FATAL("No root directory specified, and /android does not exist.");
            return;
        }
        setenv("ANDROID_ROOT", rootDir, 1);
    }

    //const char* kernelHack = getenv("LD_ASSUME_KERNEL");
    //ALOGD("Found LD_ASSUME_KERNEL='%s'\n", kernelHack);

    /*创建一个虚拟机实例 */
    JNIEnv* env;
    if (startVm(&mJavaVM, &env) != 0) {
        return;
    }
    onVmCreated(env);

    /*
     * 调用函数 startReg 在虚拟机实例中注册 JNI 方法
     */
    if (startReg(env) < 0) {
        ALOGE("Unable to register all android natives\n");
        return;
    }

    /*
```

```
     * We want to call main() with a String array with arguments in it.
     * At present we have two arguments, the class name and an option string.
     * Create an array to hold them.
     */
    jclass stringClass;
    jobjectArray strArray;
    jstring classNameStr;
    jstring optionsStr;

    stringClass = env->FindClass("java/lang/String");
    assert(stringClass != NULL);
    strArray = env->NewObjectArray(2, stringClass, NULL);
    assert(strArray != NULL);
    classNameStr = env->NewStringUTF(className);
    assert(classNameStr != NULL);
    env->SetObjectArrayElement(strArray, 0, classNameStr);
    optionsStr = env->NewStringUTF(options);
    env->SetObjectArrayElement(strArray, 1, optionsStr);

    /*
     * Start VM.  This thread becomes the main thread of the VM, and will
     * not return until the VM exits.
     */
    char* slashClassName = toSlashClassName(className);
    jclass startClass = env->FindClass(slashClassName);
    if (startClass == NULL) {
        ALOGE("JavaVM unable to locate class '%s'\n", slashClassName);
        /* keep going */
    } else {
        jmethodID startMeth = env->GetStaticMethodID(startClass, "main",
            "([Ljava/lang/String;)V");
        if (startMeth == NULL) {
            ALOGE("JavaVM unable to find main() in '%s'\n", className);
            /* keep going */
        } else {
//调用类 com.android.internal.os.ZygoteInit 的静态成员函数 main 来启动 Zygote 进程
            env->CallStaticVoidMethod(startClass, startMeth, strArray);

#if 0
            if (env->ExceptionCheck())
                threadExitUncaughtException(env);
#endif
        }
    }
    free(slashClassName);
    ALOGD("Shutting down VM\n");
    if (mJavaVM->DetachCurrentThread() != JNI_OK)
        ALOGW("Warning: unable to detach main thread\n");
    if (mJavaVM->DestroyJavaVM() != 0)
        ALOGW("Warning: VM did not shut down cleanly\n");
}
```

在上述代码中，通过调用类 com.android.internal.os.ZygoteInit 中的函数 main 启动了 Zygote 进程。成员函数 main 在文件“frameworks\base\core\java\com\android\internal\os\ZygoteInit.java”中定义，具体实现代码如下所示：

```
    public static void main(String argv[]) {
        try {
            // Start profiling the zygote initialization.
            SamplingProfilerIntegration.start();
            //调用函数 registerZygoteSocket 创建一个 socket 接口
            registerZygoteSocket();
            EventLog.writeEvent(LOG_BOOT_PROGRESS_PRELOAD_START,
                SystemClock.uptimeMillis());
            preload();
            EventLog.writeEvent(LOG_BOOT_PROGRESS_PRELOAD_END,
                SystemClock.uptimeMillis());
```

```
        // Finish profiling the zygote initialization.
        SamplingProfilerIntegration.writeZygoteSnapshot();

        // Do an initial gc to clean up after startup
        gc();

        // Disable tracing so that forked processes do not inherit stale tracing tags from
        // Zygote.
        Trace.setTracingEnabled(false);

        // If requested, start system server directly from Zygote
        if (argv.length != 2) {
            throw new RuntimeException(argv[0] + USAGE_STRING);
        }
        //调用函数 startSystemServer 启动 SystemServer 组件
        if (argv[1].equals("start-system-server")) {
            startSystemServer();
        } else if (!argv[1].equals("")) {
            throw new RuntimeException(argv[0] + USAGE_STRING);
        }

        Log.i(TAG, "Accepting command socket connections");
        //调用函数 runSelectLoop 进入一个无限循环
        //在前面创建的 socket 接口中等待 ActivityManagerService 请求，以创建新的应用程序进程
        runSelectLoop();

        closeServerSocket();
    } catch (MethodAndArgsCaller caller) {
        caller.run();
    } catch (RuntimeException ex) {
        Log.e(TAG, "Zygote died with exception", ex);
        closeServerSocket();
        throw ex;
    }
}
```

（4）和 Zygote 进程中的 Socket 实现连接。

在 Android 系统中，ActivityManagerService 通过函数 Process.start 创建一个新的进程。函数 Process.start 会先通过 Socket 连接到 Zygote 进程，并由 Zygote 进程实现创建新应用程序进程的功能。另外，类 Process 是通过函数 openZygoteSocketIfNeeded 来连接到 Zygote 进程中的 Socket。函数 openZygoteSocketIfNeeded 在文件“frameworks\base\core\java\android\os\Process.java”中定义，具体实现代码如下所示：

```
private static void openZygoteSocketIfNeeded()
        throws ZygoteStartFailedEx {
    int retryCount;
    if (sPreviousZygoteOpenFailed) {
        /*
         * If we've failed before, expect that we'll fail again and
         * don't pause for retries.
         */
        retryCount = 0;
    } else {
        retryCount = 10;
    }
    /*
     * See bug #811181: Sometimes runtime can make it up before zygote.
     * Really, we'd like to do something better to avoid this condition,
     * but for now just wait a bit...
     */
    for (int retry = 0
        ; (sZygoteSocket == null) && (retry < (retryCount + 1))
        ; retry++ ) {
        if (retry > 0) {
            try {
                Log.i("Zygote", "Zygote not up yet, sleeping...");
```

```
                Thread.sleep(ZYGOTE_RETRY_MILLIS);
            } catch (InterruptedException ex) {
                // should never happen
            }
        }
        try {
            sZygoteSocket = new LocalSocket();
            sZygoteSocket.connect(new LocalSocketAddress(ZYGOTE_SOCKET,
                    LocalSocketAddress.Namespace.RESERVED));
            sZygoteInputStream
                    = new DataInputStream(sZygoteSocket.getInputStream());
            sZygoteWriter =
                new BufferedWriter(
                        new OutputStreamWriter(
                                sZygoteSocket.getOutputStream()),
                        256);
            Log.i("Zygote", "Process: zygote socket opened");
            sPreviousZygoteOpenFailed = false;
            break;
        } catch (IOException ex) {
            if (sZygoteSocket != null) {
                try {
                    sZygoteSocket.close();
                } catch (IOException ex2) {
                    Log.e(LOG_TAG,"I/O exception on close after exception",
                            ex2);
                }
            }
            sZygoteSocket = null;
        }
    }
    if (sZygoteSocket == null) {
        sPreviousZygoteOpenFailed = true;
        throw new ZygoteStartFailedEx("connect failed");
    }
}
```

在文件 ZygoteInit.java 中的函数 main 的实现代码中，用到了函数 registerZygoteSocket，此函数在文件“frameworks\base\core\java\com\android\internal\os\ZygoteInit.java”中定义，具体实现代码如下所示：

```
private static void registerZygoteSocket() {
    if (sServerSocket == null) {
        int fileDesc;
        try {
            String env = System.getenv(ANDROID_SOCKET_ENV);
            fileDesc = Integer.parseInt(env);
        } catch (RuntimeException ex) {
            throw new RuntimeException(
                    ANDROID_SOCKET_ENV + " unset or invalid", ex);
        }

        try {
            sServerSocket = new LocalServerSocket(
                    createFileDescriptor(fileDesc));
        } catch (IOException ex) {
            throw new RuntimeException(
                    "Error binding to local socket '" + fileDesc + "'", ex);
        }
    }
}
```

在上述代码中，通过文件描述符创建了 socket 接口，此文件描符就是本书前面所讲的文件“/dev/socket/zygote”，而此文件描述符通过环境变量 ANDROID_SOCKET_ENV 获得。另外，由 init 进程负责解释执行系统启动脚本文件“system\core\rootdir\init.rc”，而 init 进程的源代码位于文件“system/core/init/init.c”中，由函数 service_start 负责解释文件 init.rc 中的 service 命令。在

service_start 函数中，每一个 service 命令都会促使进程 init 调用函数 fork 创建一个新的进程，在新进程中会解析里面的 socket 选项。对于每一个 socket 选项来说，全部会通过 t 函数 create_socke 在“/dev/socket”目录下创建一个 zygote 文件，然后通过函数 publish_socket 将得到的文件描述符写入到环境变量中。函数 publish_socket 的具体实现代码如下所示：

```
static void publish_socket(const char *name, int fd)
{
    char key[64] = ANDROID_SOCKET_ENV_PREFIX;
    char val[64];

    strlcpy(key + sizeof(ANDROID_SOCKET_ENV_PREFIX) - 1, name,
        sizeof(key) - sizeof(ANDROID_SOCKET_ENV_PREFIX));
    snprintf(val, sizeof(val), "%d", fd);
    add_environment(key, val);

    /* make sure we don't close-on-exec */
    fcntl(fd, F_SETFD, 0);
}
```

在上述代码中，传进的参数 name 值为"zygote"，而 ANDROID_SOCKET_ENV_PREFIX 在文件“system\core\include\cutils\sockets.h”中的定义代码为：

```
#define ANDROID_SOCKET_ENV_PREFIX    "ANDROID_SOCKET_"
```

这样就将得到的文件描述符写入到“ANDROID_SOCKET_zygote”的环境变量，这个环境变量值为 key 值。因为函数 ZygoteInit.registerZygoteSocket 和函数 create_socket 都是运行在同一个进程中，所以函数 ZygoteInit.registerZygoteSocket 可以直接使用文件描述符来创建一个 Java 层的 LocalServerSocket 对象。如果其它进程也需要打开“/dev/socket/zygote”文件以和 Zygote 进程进行通信，则必须通过文件名作为中介来连接 LocalServerSocket。

在文件 ZygoteInit.java 中的函数 main 的实现代码中，用到了函数 startSystemServer，此函数也是在文件“frameworks\base\core\java\com\android\internal\os\ZygoteInit.java”中定义，具体实现代码如下所示：

```
    private static boolean startSystemServer()
            throws MethodAndArgsCaller, RuntimeException {
        /* Hardcoded command line to start the system server */
        String args[] = {
            "--setuid=1000",
            "--setgid=1000",
"--setgroups=1001,1002,1003,1004,1005,1006,1007,1008,1009,1010,1018,3001,3002,3003,3006,3007",
            "--capabilities=130104352,130104352",
            "--runtime-init",
            "--nice-name=system_server",
            "com.android.server.SystemServer",
        };
        ZygoteConnection.Arguments parsedArgs = null;

        int pid;

        try {
            parsedArgs = new ZygoteConnection.Arguments(args);
            ZygoteConnection.applyDebuggerSystemProperty(parsedArgs);
            ZygoteConnection.applyInvokeWithSystemProperty(parsedArgs);

            /* Request to fork the system server process */
            pid = Zygote.forkSystemServer(
                    parsedArgs.uid, parsedArgs.gid,
                    parsedArgs.gids,
                    parsedArgs.debugFlags,
                    null,
```

```
            parsedArgs.permittedCapabilities,
            parsedArgs.effectiveCapabilities);
    } catch (IllegalArgumentException ex) {
        throw new RuntimeException(ex);
    }
    /* For child process */
    if (pid == 0) {
        handleSystemServerProcess(parsedArgs);
    }
    return true;
}
```

在文件 ZygoteInit.java 中的函数 main 的实现代码中，用到了函数 startSystemServer，此函数在文件“frameworks\base\core\java\com\android\internal\os\ZygoteInit.java”中定义，具体实现代码如下所示：

```
private static boolean startSystemServer()
        throws MethodAndArgsCaller, RuntimeException {
    /* Hardcoded command line to start the system server */
    String args[] = {
        "--setuid=1000",
        "--setgid=1000",
"--setgroups=1001,1002,1003,1004,1005,1006,1007,1008,1009,1010,1018,3001,3002,3003,3006,3007",
        "--capabilities=130104352,130104352",
        "--runtime-init",
        "--nice-name=system_server",
        "com.android.server.SystemServer",
    };
    ZygoteConnection.Arguments parsedArgs = null;
    int pid;

    try {
        parsedArgs = new ZygoteConnection.Arguments(args);
        ZygoteConnection.applyDebuggerSystemProperty(parsedArgs);
        ZygoteConnection.applyInvokeWithSystemProperty(parsedArgs);
        /* Request to fork the system server process */
        pid = Zygote.forkSystemServer(
                parsedArgs.uid, parsedArgs.gid,
                parsedArgs.gids,
                parsedArgs.debugFlags,
                null,
                parsedArgs.permittedCapabilities,
                parsedArgs.effectiveCapabilities);
    } catch (IllegalArgumentException ex) {
        throw new RuntimeException(ex);
    }
    /* For child process */
    if (pid == 0) {
        handleSystemServerProcess(parsedArgs);
    }
    return true;
}
```

在上述代码中，Zygote 进程通过函数 forkSystemServer“孕育”了一个新的进程来启动 SystemServer 组件，返回值 pid 为 0 的位置标示新进程的执行路径，即新建进程会执行函数 handleSystemServerProcess。函数 handleSystemServerProcess 在文件“frameworks\base\core\java\com\android\internal\os\ZygoteInit.java”中定义，具体实现代码如下所示：

```
private static void handleSystemServerProcess(
        ZygoteConnection.Arguments parsedArgs)
        throws ZygoteInit.MethodAndArgsCaller {
    closeServerSocket();
    // set umask to 0077 so new files and directories will default to owner-only permissions.
    Libcore.os.umask(S_IRWXG | S_IRWXO);
```

```
        if (parsedArgs.niceName != null) {
            Process.setArgV0(parsedArgs.niceName);
        }
        if (parsedArgs.invokeWith != null) {
            WrapperInit.execApplication(parsedArgs.invokeWith,
                    parsedArgs.niceName, parsedArgs.targetSdkVersion,
                    null, parsedArgs.remainingArgs);
        } else {
            /*
             * Pass the remaining arguments to SystemServer.
             */
            RuntimeInit.zygoteInit(parsedArgs.targetSdkVersion, parsedArgs.remainingArgs);
        }
        /* should never reach here */
    }
```

在上述代码中，调用函数closeServerSocket关闭了子进程，然后调用函数RuntimeInit.zygoteInit进一步启动SystemServer组件。函数RuntimeInit.zygoteInit在文件“frameworks/base/core/java/com/android/internal/os/RuntimeInit.java”中定义，具体实现代码如下所示：

```
    public static final void zygoteInit(int targetSdkVersion, String[] argv)
            throws ZygoteInit.MethodAndArgsCaller {
        if (DEBUG) Slog.d(TAG, "RuntimeInit: Starting application from zygote");

        redirectLogStreams();
        commonInit();
//调用函数 zygoteInitNative 来执行一个 Binder 进程间通信机制的初始化工作
//当完成这个工作后，这个进程中的 Binder 对象就可以方便地进行进程间通信
        nativeZygoteInit();

        applicationInit(targetSdkVersion, argv);
    }
```

在文件ZygoteInit.java中的函数main的实现代码中，用到了函数runSelectLoop，功能是进入一个无限循环在前面创建的socket接口中，并等待ActivityManagerService请求创建新的应用程序进程。函数runSelectLoop在文件“frameworks\base\core\java\com\android\internal\os\ZygoteInit.java”中定义，具体实现代码如下所示：

```
    private static void runSelectLoop() throws MethodAndArgsCaller {
        ArrayList<FileDescriptor> fds = new ArrayList<FileDescriptor>();
        ArrayList<ZygoteConnection> peers = new ArrayList<ZygoteConnection>();
        FileDescriptor[] fdArray = new FileDescriptor[4];

        fds.add(sServerSocket.getFileDescriptor());
        peers.add(null);

        int loopCount = GC_LOOP_COUNT;
        while (true) {
            int index;

            /*
             * Call gc() before we block in select().
             * It's work that has to be done anyway, and it's better
             * to avoid making every child do it.  It will also
             * madvise() any free memory as a side-effect.
             *
             * Don't call it every time, because walking the entire
             * heap is a lot of overhead to free a few hundred bytes.
             */
            if (loopCount <= 0) {
                gc();
                loopCount = GC_LOOP_COUNT;
            } else {
                loopCount--;
            }
```

```
            try {
               fdArray = fds.toArray(fdArray);
               index = selectReadable(fdArray);
            } catch (IOException ex) {
               throw new RuntimeException("Error in select()", ex);
            }

            if (index < 0) {
               throw new RuntimeException("Error in select()");
            } else if (index == 0) {
               ZygoteConnection newPeer = acceptCommandPeer();
               peers.add(newPeer);
               fds.add(newPeer.getFileDesciptor());
            } else {
               boolean done;
  //将数据通过 Socket 接口发送出去后会执行下面的语句
  // peers.get(index)得到的是一个 ZygoteConnection 对象，表示一个 Socket 连接
  //调用 ZygoteConnection.runOnce 函数进一步处理
      done = peers.get(index).runOnce();

               if (done) {
                  peers.remove(index);
                  fds.remove(index);
               }
            }
         }
      }
```

通过上述代码，可以等待 ActivityManagerService 连接这个 Socket，然后调用函数 ZygoteConnection.runOnce 创建新的应用程序。

9.2 System 进程详解

在 Android 系统中，System 进程和系统服务有着重要的关系。几乎所有的 Android 核心服务都在这个进程中，例如 ActivityManagerService、PowerManagerService 和 WindowManagerService 等。在本节的内容中，将详细分析 Android L 中 System 进程的基本知识，为读者步入本书后面知识的学习打下基础。

9.2.1 启动 System 进程前的准备工作

在 Android 系统中，通过静态类 ZygoteInit 的成员函数 handleSystemServerProcess 来启动 System 进程。具体启动过程如图 9-2 所示。

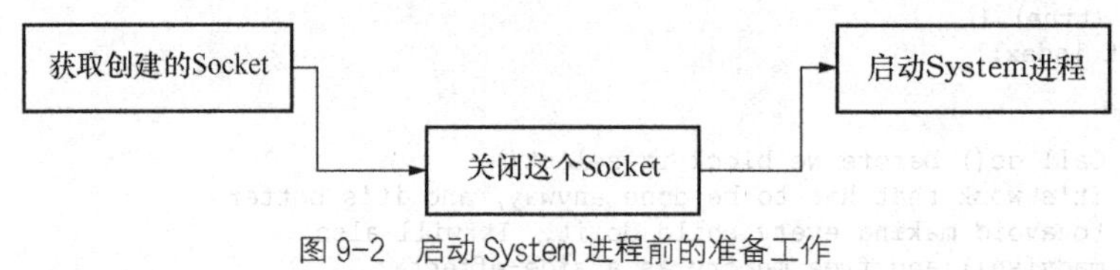

图 9-2　启动 System 进程前的准备工作

（1）首先在文件“frameworks\base\core\java\com\android\internal\os\ZygoteInit.java”中，获取 Zygote 进程在启动过程中创建的 Socket。其实 System 进程不需要这个 Socket，所以会调用类 ZygoteInit 的成员函数 closeServerSocket 关闭这个 Socket。对应的代码如下所示：

```
    private static void handleSystemServerProcess(
           ZygoteConnection.Arguments parsedArgs)
           throws ZygoteInit.MethodAndArgsCaller {
  //关闭这个 Socket
       closeServerSocket();
```

```
        // set umask to 0077 so new files and directories will default to owner-only permissions.
        Libcore.os.umask(S_IRWXG | S_IRWXO);
        if (parsedArgs.niceName != null) {
            Process.setArgV0(parsedArgs.niceName);
        }
        if (parsedArgs.invokeWith != null) {
            WrapperInit.execApplication(parsedArgs.invokeWith,
                    parsedArgs.niceName, parsedArgs.targetSdkVersion,
                    null, parsedArgs.remainingArgs);
        } else {
            /*
             * Pass the remaining arguments to SystemServer.
             */
//调用类 RuntimeInit 的静态函数 zygoteInit 启动 System 进程
            RuntimeInit.zygoteInit(parsedArgs.targetSdkVersion, parsedArgs.remainingArgs);
        }
        /* should never reach here */
    }
```

（2）接下来调用类 RuntimeInit 的静态函数 zygoteInit 启动 System 进程，此函数在文件“frameworks\ base\core\java\com\android\internal\os\ZygoteInit.java”中定义，具体实现代码如下所示：

```
    public static final void zygoteInit(int targetSdkVersion, String[] argv)
            throws ZygoteInit.MethodAndArgsCaller {
        if (DEBUG) Slog.d(TAG, "RuntimeInit: Starting application from zygote");
        redirectLogStreams();
//调用函数 commonInit 设置 Sysrem 进程的时区和键盘布局等信息
        commonInit ();
//调用函数 nativeZygoteInit 启动一个 Binder 线程池
        nativeZygoteInit();
        applicationInit(targetSdkVersion, argv);
    }
```

9.2.2 分析 SystemServer

SystemServer 是 Android Java 的两大支柱进程之一，另一个是专门负责孵化 Java 进程的 Zygote。如果这两大支柱中的任何一个崩溃了，都会导致 Android 中 Java 层的崩溃。如果 Java 层真的崩溃了，则 Linux 系统中的进程 init 会重新启动 SystemServer 和 Zygote，以重新建立 Android 的 Java 层。在本节的内容中，将首先纵览分析 SystemServer 的源码。

（1）分析主函数 main。

SystemServer 是由 Zygote 孵化而来的一个进程，通过 ps 命令，可知其进程名为 system_server。在 DDMS 中可以看到，进程 system_server 的进程名为 system_process。SystemServer 核心逻辑的入口是函数 main，此入口函数在如下所示的文件中实现：

```
\frameworks\base\services\java\com\android\server\SystemServer.java
```

文件 SystemServer.java 的入口函数是 main，具体实现代码如下所示：

```
    public static void main(String[] args) {
        if (System.currentTimeMillis() < EARLIEST_SUPPORTED_TIME) {
            //如果系统时钟早于 1970 年，则设置系统时钟从 1970 年开始
            Slog.w(TAG, "System clock is before 1970; setting to 1970.");
            SystemClock.setCurrentTimeMillis(EARLIEST_SUPPORTED_TIME);
        }

        if (SamplingProfilerIntegration.isEnabled()) {
            SamplingProfilerIntegration.start();
            timer = new Timer();
            timer.schedule(new TimerTask() {
                @Override
                public void run() {
                //SystemServer 性能统计，每小时统计一次，统计结果输出为文件
```

```
                SamplingProfilerIntegration.writeSnapshot("system_server", null);
            } // SNAPSHOT_INTERVAL 定义为 1 小时
        }, SNAPSHOT_INTERVAL, SNAPSHOT_INTERVAL);
    }

    // //和 Dalvik 虚拟机相关的设置，主要是内存使用方面的控制
    dalvik.system.VMRuntime.getRuntime().clearGrowthLimit();

    // The system server has to run all of the time, so it needs to be
    // as efficient as possible with its memory usage.
    VMRuntime.getRuntime().setTargetHeapUtilization(0.8f);
    //加载动态库 libandroid_servers.so
    System.loadLibrary("android_servers");
    init1(args); //调用 native 的 init1 函数
}

public static final void init2() {
    Slog.i(TAG, "Entered the Android system server!");
    Thread thr = new ServerThread();
    thr.setName("android.server.ServerThread");
    thr.start();
}
}
```

由此可见，函数 main 首先做一些初始化工作，然后加载动态库 libandroid_servers.so，最后调用 native 的函数 init1。该函数在 libandroid_servers.so 库中实现，在如下所示的文件中定义。

```
\frameworks\base\services\jni\com_android_server_SystemServer.cpp
```

函数 init1 的具体实现代码如下所示：

```
extern "C" int system_init();
static void android_server_SystemServer_init1(JNIEnv* env, jobject clazz)
{
  system_init();          //调用上面那个用 extern 声明的 system_init 函数
}
```

而函数 system_init 在另外一个库 libsystem_server.so 中实现，在如下所示的文件中定义。

```
\frameworks\base\cmds\system_server\library\System_init.cpp
```

函数 system_init 的具体实现代码如下所示：

```
extern "C" status_t system_init()
{
    ALOGI("Entered system_init()");

    sp<ProcessState> proc(ProcessState::self());

    sp<IServiceManager> sm = defaultServiceManager();
    ALOGI("ServiceManager: %p\n", sm.get());

    sp<GrimReaper> grim = new GrimReaper();
    sm->asBinder()->linkToDeath(grim, grim.get(), 0);

    char propBuf[PROPERTY_VALUE_MAX];
    property_get("system_init.startsurfaceflinger", propBuf, "1");
    if (strcmp(propBuf, "1") == 0) {
        // Start the SurfaceFlinger
        SurfaceFlinger::instantiate();
    }

    property_get("system_init.startsensorservice", propBuf, "1");
    if (strcmp(propBuf, "1") == 0) {
        // Start the sensor service
        SensorService::instantiate();
    }

    // And now start the Android runtime.  We have to do this bit
    // of nastiness because the Android runtime initialization requires
```

```
        // some of the core system services to already be started.
        // All other servers should just start the Android runtime at
        // the beginning of their processes's main(), before calling
        // the init function.
        ALOGI("System server: starting Android runtime.\n");
        AndroidRuntime* runtime = AndroidRuntime::getRuntime();

        ALOGI("System server: starting Android services.\n");
        JNIEnv* env = runtime->getJNIEnv();
        if (env == NULL) {
            return UNKNOWN_ERROR;
        }
        jclass clazz = env->FindClass("com/android/server/SystemServer");
        if (clazz == NULL) {
            return UNKNOWN_ERROR;
        }
        jmethodID methodId = env->GetStaticMethodID(clazz, "init2", "()V");
        if (methodId == NULL) {
            return UNKNOWN_ERROR;
        }
        env->CallStaticVoidMethod(clazz, methodId);

        ALOGI("System server: entering thread pool.\n");
        ProcessState::self()->startThreadPool();
        IPCThreadState::self()->joinThreadPool();
        ALOGI("System server: exiting thread pool.\n");

        return NO_ERROR;
    }
```

通过上述代码可知，SystemServer 中的函数 main 通过函数 init1，从 Java 层穿越到 Native 层，实现了一些初始化工作后，又通过 JNI 从 Native 层穿越到 Java 层去调用函数 init2。函数 init2 返回后，最终又回归到 Native 层。

（2）分析函数 init2。

在文件 SystemServer.java 中，函数 init1 较简单，其实重点内容都在函数 init2 中。函数 init2 的具体实现代码如下所示：

```
public static final void init2() {
    Thread thr = new ServerThread();
    thr.setName("android.server.ServerThread");
    thr.start();//启动一个线程，这个线程就像英雄大会一样，聚集了各路英雄
}
```

通过上述代码将创建一个新的线程 ServerThread，该线程的 run 函数的实现代码有 600 多行，如此之长的原因是，Android 平台中众多 Service 都汇集于此。

在 Android 平台中，共有 7 大类 43 个 Service（包括 Watchdog）。实际上，还有一些 Service 并没有在 ServerThread 的 run 函数中露面。这 7 大类服务主要如下。

- 第一大类：是 Android 的核心服务，如 ActivityManagerService、WindowManager-Service 等。
- 第二大类：是和通信相关的服务，如 Wi-Fi 相关服务、Telephone 相关服务。
- 第三大类：是和系统功能相关的服务，如 AudioService、MountService、Usb-Service 等。
- 第四大类：是 BatteryService、VibratorService 等服务。
- 第五大类：是 EntropyService、DiskStatsService、Watchdog 等相对独立的服务。
- 第六大类：是蓝牙服务。
- 第七大类：是和 UI 紧密相关的服务，如状态栏服务、通知管理服务等。

在本章后面的内容中，将详细分析其中的第五类服务。该类中的 Service 之间关系简单，而且功能相对独立。第五大类服务包括如下所示的服务。

- EntropyService：熵服务，它和随机数的生成有关。

- ClipboardService：剪贴板服务。
- DropBoxManagerService：该服务和系统运行时日志的存储与管理有关。
- DiskStatsService 和 DeviceStorageMonitorService：这两个服务用于查看和监测系统存储空间。
- SamplingProfilerService：这个服务是从 Android 4.0 新增的，功能非常简单。
- Watchdog：即看门狗，是 Android 的“老员工”了。在卷 I 第 4 章“深入理解 Zygote”中曾分析过它。Android 2.3 以后其内存检测功能被去掉，所以与 Android 2.2 相比，显得更简单了。

9.2.3 分析 EntropyService

EntropyService 是 SystemServer 启动的第一个 Service，它以 3 个小时为单位周期性进行加载和保存熵池（/dev/urandom）的工作。但是因为/dev/urandom 本身的安全性比/dev/random 差一些，所以每隔 3 小时，Android 系统在 kernel 的熵池中增加了一些附加信息，这些信息用于提高随机数的质量。在 Android 会添加如下所示的额外信息。

- out.println("Copyright (C) 2009 The Android Open Source Project")。
- out.println("All Your Randomness Are Belong To Us")。
- out.println(START_TIME)。
- out.println(START_NANOTIME)。
- out.println(SystemProperties.get("ro.serialno")。
- out.println(SystemProperties.get("ro.bootmode"))。
- out.println(SystemProperties.get("ro.baseband"))。
- out.println(SystemProperties.get("ro.carrier"))。
- out.println(SystemProperties.get("ro.bootloader"))。
- out.println(SystemProperties.get("ro.hardware"))。
- out.println(SystemProperties.get("ro.revision"))。
- out.println(System.currentTimeMillis())。
- out.println(System.nanoTime())。

物理学存在一个基本原理：如果一个系统的熵越大，该系统就越不稳定。在 Android 中，目前也只有随机数常处于这种不稳定的系统中了。在 Android 系统中，SystemServer 中添加该服务的代码如下所示：

```
ServiceManager.addService("entropy", new EntropyService());
```

上述代码非常简单，从中可直接分析 EntropyService 的构造函数，此函数在文件 EntropyService.java 中定义，具体实现代码如下所示：

```
public EntropyService() {
    //调用另外一个构造函数，getSystemDir 函数返回的是/data/system 目录
    this(getSystemDir() + "/entropy.dat", "/dev/urandom");
}
public EntropyService(String entropyFile, String randomDevice) {
 this.randomDevice = randomDevice;//urandom 是 Linux 系统中产生随机数的设备
 // /data/system/entropy.dat 文件保存了系统此前的熵信息
 this.entropyFile = entropyFile;
//下面有 4 个关键函数
 loadInitialEntropy();
 addDeviceSpecificEntropy();
 writeEntropy();
 scheduleEntropyWriter();
}
```

从以上代码中可以看出，EntropyService 构造函数中依次调用了 4 个关键函数，这 4 个函数比

较简单，具体功能如下所示。

（1）函数 loadInitialEntropy。

函数 loadInitialEntropy 的功能是将文件 entropy.dat 的中内容写到 urandom 设备，这样可以增加系统的随机性。在系统中有一个 entropy pool，pool 在刚启动系统时的内容为空，这样会导致早期生成的随机数变得可以预测。函数 loadInitialEntropy 的具体实现代码如下所示：

```
private void loadInitialEntropy() {
    try {
        RandomBlock.fromFile(entropyFile).toFile(randomDevice);
    } catch (IOException e) {
        Slog.w(TAG, "unable to load initial entropy (first boot?)", e);
    }
}
```

（2）函数 addDeviceSpecificEntropy。

函数 addDeviceSpecificEntropy 的功能是，将一些和设备相关的信息写入 urandom 设备，具体实现代码如下所示：

```
private void addDeviceSpecificEntropy() {
    PrintWriter out = null;
    try {
        out = new PrintWriter(new FileOutputStream(randomDevice));
        out.println("Copyright (C) 2009 The Android Open Source Project");
        out.println("All Your Randomness Are Belong To Us");
        out.println(START_TIME);
        out.println(START_NANOTIME);
        out.println(SystemProperties.get("ro.serialno"));
        out.println(SystemProperties.get("ro.bootmode"));
        out.println(SystemProperties.get("ro.baseband"));
        out.println(SystemProperties.get("ro.carrier"));
        out.println(SystemProperties.get("ro.bootloader"));
        out.println(SystemProperties.get("ro.hardware"));
        out.println(SystemProperties.get("ro.revision"));
        out.println(System.currentTimeMillis());
        out.println(System.nanoTime());
    } catch (IOException e) {
        Slog.w(TAG, "Unable to add device specific data to the entropy pool", e);
    } finally {
        if (out != null) {
            out.close();
        }
    }
}
```

由上述代码可知，即使向 urandom 的 entropy pool 中写入了固定信息，也能增加随机数生成的随机性。从熵的角度考虑，系统的质量越大（即 pool 中的内容越多），该系统就越不稳定。

（3）函数 writeEntropy。

函数 writeEntropy 的功能是读取 urandom 设备的内容到 entropy.dat 文件。具体实现代码如下所示：

```
private void writeEntropy() {
    try {
        RandomBlock.fromFile(randomDevice).toFile(entropyFile);
    } catch (IOException e) {
        Slog.w(TAG, "unable to write entropy", e);
    }
}
```

（4）函数 scheduleEntropyWriter。

函数 scheduleEntropyWriter 的功能是，向 EntropyService 内部的 Handler 发送一个 ENTROPY_WHAT 消息。该消息每 3 小时发送一次。收到该消息后，EntropyService 会再次调用 writeEntropy

函数，将 urandom 设备的内容写到 entropy.dat 中。具体实现代码如下所示：

```
private void scheduleEntropyWriter() {
    mHandler.removeMessages(ENTROPY_WHAT);
    mHandler.sendEmptyMessageDelayed(ENTROPY_WHAT, ENTROPY_WRITE_PERIOD);
}
```

通过上面的分析可知，文件 entropy.dat 保存了 urandom 设备内容的快照（每 3 小时更新一次）。当重新系统启动时，EntropyService 又利用这个文件来增加系统的熵，通过这种方式使随机数的生成更加不可预测。

9.2.4 分析 DropBoxManagerService

在 Android 应用中，DropBoxManagerService（DBMS）用于生成和管理系统运行时的一些日志文件。这些日志文件大多记录的是系统或某个应用程序出错时的信息。其中向 SystemServer 添加 DBMS 的代码如下所示：

```
ServiceManager.addService(Context.DROPBOX_SERVICE,              //服务名为"dropbox"
  new DropBoxManagerService(context,
  new File("/data/system/dropbox")));
```

（1）分析 DBMS 构造函数。

DBMS 构造函数在如下所示的文件中实现。

```
\frameworks\base\services\java\com\android\server\DropBoxManagerService.java
```

DBMS 构造函数 DropBoxManagerService 的具体实现代码如下所示：

```
public DropBoxManagerService(final Context context, File path) {
    mDropBoxDir = path; //path指定dropbox目录为/data/system/dropbox

    // Set up intent receivers
    mContext = context;
    mContentResolver = context.getContentResolver();

    IntentFilter filter = new IntentFilter();
    filter.addAction(Intent.ACTION_DEVICE_STORAGE_LOW);
    filter.addAction(Intent.ACTION_BOOT_COMPLETED);
//注册一个Broadcast监听对象，当系统启动完毕或者设备存储空间不足时，会收到广播
    context.registerReceiver(mReceiver, filter);
//当Settings数据库相应项发生变化时候，也需要告知DBMS进行相应处理
    mContentResolver.registerContentObserver(
        Settings.Global.CONTENT_URI, true,
        new ContentObserver(new Handler()) {
            @Override
            public void onChange(boolean selfChange) {
                //当Settings数据库发生变化时候， BroadcastReceiver的onReceive函数
                //将被调用。注意第二个参数为null
                mReceiver.onReceive(context, (Intent) null);
            }
        });

    mHandler = new Handler() {
        @Override
        public void handleMessage(Message msg) {
            if (msg.what == MSG_SEND_BROADCAST) {
                mContext.sendBroadcastAsUser((Intent)msg.obj, UserHandle.OWNER,
                        android.Manifest.permission.READ_LOGS);
            }
        }
    };

    // The real work gets done lazily in init() -- that way service creation always
    // succeeds, and things like disk problems cause individual method failures.
}
```

```
    /** Unregisters broadcast receivers and any other hooks -- for test instances */
    public void stop() {
        mContext.unregisterReceiver(mReceiver);
    }
```

通过上述代码可知，DBMS 注册一个 BroadcastReceiver 对象，同时会监听 Settings 数据库的变动。其核心逻辑都在此 BroadcastReceiver 的 onReceive 函数中。函数 onReceive 的主要功能是，存储空间不足时需要删除一些旧的日志文件以节省存储空间。函数 onReceive 的具体实现代码如下所示：

```
        public void onReceive(Context context, Intent intent) {
            if (intent != null && Intent.ACTION_BOOT_COMPLETED.equals(intent.getAction())) {
                mBooted = true;
                return;
            }

            // Else, for ACTION_DEVICE_STORAGE_LOW:
            mCachedQuotaUptimeMillis = 0;  // Force a re-check of quota size

            // Run the initialization in the background (not this main thread).
            // The init() and trimToFit() methods are synchronized, so they still
            // block other users -- but at least the onReceive() call can finish.
            new Thread() {
                public void run() {
                    try {
                        init();
                        trimToFit();
                    } catch (IOException e) {
                        Slog.e(TAG, "Can't init", e);
                    }
                }
            }.start();
        }
    };
```

函数 onReceive 会在以下 3 种情况发生时被调用。

- 当系统启动完毕时，由 BOOT_COMPLETED 广播触发。
- 当设备存储空间不足时，由 DEVICE_STORAGE_LOW 广播触发。
- 当 Settings 数据库相应项发生变化时候，该函数也会被触发。

（2）添加 dropbox 日志文件。

在 Android L 系统中，要想理清一个 Service，最好从它提供的服务开始进行分析。当某个应用程序因为发生异常而崩溃（crash）时，会调用 ActivityManagerService（AMS）的函数 handleApplicationCrash，此函数在如下所示 id 文件中定义。

```
\frameworks\base\services\java\com\android\server\am\ActivityManagerService.java
```

函数 handleApplicationCrash 的具体实现代码如下所示：

```
    public void handleApplicationCrash(IBinder app, ApplicationErrorReport.CrashInfo crashInfo)
{
        ProcessRecord r = findAppProcess(app, "Crash");
        final String processName = app == null ? "system_server"
                : (r == null ? "unknown" : r.processName);

        EventLog.writeEvent(EventLogTags.AM_CRASH, Binder.getCallingPid(),
                UserHandle.getUserId(Binder.getCallingUid()), processName,
                r == null ? -1 : r.info.flags,
                crashInfo.exceptionClassName,
                crashInfo.exceptionMessage,
                crashInfo.throwFileName,
                crashInfo.throwLineNumber);
        //调用 addErrorToDropBox 函数，第一个参数是一个字符串，为“crash”
```

```
        addErrorToDropBox("crash", r, processName, null, null, null, null, null, crashInfo);

        crashApplication(r, crashInfo);
    }
```

下面来看函数 addErrorToDropBox，此函数也在文件 ActivityManagerService.java 中实现，具体实现代码如下所示：

```
    public void addErrorToDropBox(String eventType,
            ProcessRecord process, String processName, ActivityRecord activity,
            ActivityRecord parent, String subject,
            final String report, final File logFile,
            final ApplicationErrorReport.CrashInfo crashInfo) {
        // NOTE -- this must never acquire the ActivityManagerService lock,
        // otherwise the watchdog may be prevented from resetting the system.

        final String dropboxTag = processClass(process) + "_" + eventType;
        final DropBoxManager dbox = (DropBoxManager)
                mContext.getSystemService(Context.DROPBOX_SERVICE);

        // Exit early if the dropbox isn't configured to accept this report type.
        if (dbox == null || !dbox.isTagEnabled(dropboxTag)) return;

        final StringBuilder sb = new StringBuilder(1024);
        appendDropBoxProcessHeaders(process, processName, sb);
        if (activity != null) {
            sb.append("Activity: ").append(activity.shortComponentName).append("\n");
        }
        if (parent != null && parent.app != null && parent.app.pid != process.pid) {
            sb.append("Parent-Process: ").append(parent.app.processName).append("\n");
        }
        if (parent != null && parent != activity) {
            sb.append("Parent-Activity: ").append(parent.shortComponentName).append("\n");
        }
        if (subject != null) {
            sb.append("Subject: ").append(subject).append("\n");
        }
        sb.append("Build: ").append(Build.FINGERPRINT).append("\n");
        if (Debug.isDebuggerConnected()) {
            sb.append("Debugger: Connected\n");
        }
        sb.append("\n");

        // Do the rest in a worker thread to avoid blocking the caller on I/O
        // (After this point, we shouldn't access AMS internal data structures.)
        Thread worker = new Thread("Error dump: " + dropboxTag) {
            @Override
            public void run() {
                if (report != null) {
                    sb.append(report);
                }
                if (logFile != null) {
                    try {
                        sb.append(FileUtils.readTextFile(logFile, 128 * 1024, "\n\n[[TRUNCATED]]"));
                    } catch (IOException e) {
                        Slog.e(TAG, "Error reading " + logFile, e);
                    }
                }
                if (crashInfo != null && crashInfo.stackTrace != null) {
                    sb.append(crashInfo.stackTrace);
                }

                String setting = Settings.Global.ERROR_LOGCAT_PREFIX + dropboxTag;
                int lines = Settings.Global.getInt(mContext.getContentResolver(), setting, 0);
                if (lines > 0) {
                    sb.append("\n");

                    // Merge several logcat streams, and take the last N lines
                    InputStreamReader input = null;
```

```
            try {
                java.lang.Process logcat = new ProcessBuilder("/system/bin/logcat",
                    "-v", "time", "-b", "events", "-b", "system", "-b", "main",
                        "-t", String.valueOf(lines)).redirectErrorStream(true).start();

                try { logcat.getOutputStream().close(); } catch (IOException e) {}
                try { logcat.getErrorStream().close(); } catch (IOException e) {}
                input = new InputStreamReader(logcat.getInputStream());

                int num;
                char[] buf = new char[8192];
                while ((num = input.read(buf)) > 0) sb.append(buf, 0, num);
            } catch (IOException e) {
                Slog.e(TAG, "Error running logcat", e);
            } finally {
                if (input != null) try { input.close(); } catch (IOException e) {}
            }
        }

        dbox.addText(dropboxTag, sb.toString());
    }
};

if (process == null) {
    // If process is null, we are being called from some internal code
    // and may be about to die -- run this synchronously.
    worker.run();
} else {
    worker.start();
}
}
```

由上述代码可知，函数 addErrorToDropBox 的核心功能是生成日志内容，并调用函数 addText 将内容传给 DBMS 的功能。函数 addText 定义在如下所示的文件中。

```
\frameworks\base\core\java\android\os\DropBoxManager.java
```

在 DropBoxManager 类中，函数 addText 的实现代码如下所示：

```
public void addText(String tag, String data) {
    try { mService.add(new Entry(tag, 0, data)); } catch (RemoteException e) {}
}
```

在上述代码中实现了 mService 和 DBMS 的交互。DBMS 对外只提供一个 add 函数实现日志添加工作，而 DBM 提供了 3 个函数，分别是 addText、addData、addFile，这样以方便使用。

DBM 向 DBMS 传递的数据被封装在一个 Entry 中，DBMS 中的函数 add 在文件 DropBoxManagerService.java 中定义，具体实现代码如下所示：

```
public void add(DropBoxManager.Entry entry) {
    File temp = null;
    OutputStream output = null;
    final String tag = entry.getTag();
    try {
        int flags = entry.getFlags();
        if ((flags & DropBoxManager.IS_EMPTY) != 0) throw new IllegalArgumentException();

        init();
        if (!isTagEnabled(tag)) return;
        long max = trimToFit();
        long lastTrim = System.currentTimeMillis();

        byte[] buffer = new byte[mBlockSize];
        InputStream input = entry.getInputStream();

        // First, accumulate up to one block worth of data in memory before
        // deciding whether to compress the data or not.
```

```
            int read = 0;
            while (read < buffer.length) {
                int n = input.read(buffer, read, buffer.length - read);
                if (n <= 0) break;
                read += n;
            }

            // If we have at least one block, compress it -- otherwise, just write
            // the data in uncompressed form.

            temp = new File(mDropBoxDir, "drop" + Thread.currentThread().getId() + ".tmp");
            int bufferSize = mBlockSize;
            if (bufferSize > 4096) bufferSize = 4096;
            if (bufferSize < 512) bufferSize = 512;
            FileOutputStream foutput = new FileOutputStream(temp);
            output = new BufferedOutputStream(foutput, bufferSize);
            if (read == buffer.length && ((flags & DropBoxManager.IS_GZIPPED) == 0)) {
                output = new GZIPOutputStream(output);
                flags = flags | DropBoxManager.IS_GZIPPED;
            }

            do {
                output.write(buffer, 0, read);

                long now = System.currentTimeMillis();
                if (now - lastTrim > 30 * 1000) {
                    max = trimToFit();  // In case data dribbles in slowly
                    lastTrim = now;
                }

                read = input.read(buffer);
                if (read <= 0) {
                    FileUtils.sync(foutput);
                    output.close();  // Get a final size measurement
                    output = null;
                } else {
                    output.flush();  // So the size measurement is pseudo-reasonable
                }

                long len = temp.length();
                if (len > max) {
                    Slog.w(TAG, "Dropping: " + tag + " (" + temp.length() + " > " + max + " bytes)");
                    temp.delete();
                    temp = null;  // Pass temp = null to createEntry() to leave a tombstone
                    break;
                }
            } while (read > 0);

            long time = createEntry(temp, tag, flags);
            temp = null;

            final Intent dropboxIntent = new Intent(DropBoxManager.ACTION_DROPBOX_ENTRY_ADDED);
            dropboxIntent.putExtra(DropBoxManager.EXTRA_TAG, tag);
            dropboxIntent.putExtra(DropBoxManager.EXTRA_TIME, time);
            if (!mBooted) {
                dropboxIntent.addFlags(Intent.FLAG_RECEIVER_REGISTERED_ONLY);
            }
            // Call sendBroadcast after returning from this call to avoid deadlock. In particular
            // the caller may be holding the WindowManagerService lock but sendBroadcast requires a
            // lock in ActivityManagerService. ActivityManagerService has been caught holding that
            // very lock while waiting for the WindowManagerService lock.
            mHandler.sendMessage(mHandler.obtainMessage(MSG_SEND_BROADCAST, dropboxIntent));
        } catch (IOException e) {
            Slog.e(TAG, "Can't write: " + tag, e);
        } finally {
            try { if (output != null) output.close(); } catch (IOException e) {}
            entry.close();
            if (temp != null) temp.delete();
        }
    }
```

从上述代码可知，DBMS 非常爱惜“/data”分区的空间，需要考虑每一个日志文件的压缩以节省存储空间。

（3）DBMS 和 settings 数据库。

DBMS 的运行依赖一些配置项。其实除了 DBMS，SystemServer 中很多服务都依赖相关的配置项。这些配置项都是通过 SettingsProvider 操作 Settings 数据库来设置和查询的。SettingsProvider 是系统中很重要的一个 APK，如果将其删除后系统就不能正常启动了。

和系统相关的配置项都在 Settings 数据库的 Secure 表内保存，具体说明如下所示：

```
//用来判断是否允许记录该 tag 类型的日志文件。默认是允许生成任何 tag 类型的文件
Secure.DROPBOX_TAG_PREFIX+tag: "dropbox:"+tag
//用于控制每个日志文件的存活时间，默认是 3 天。大于 3 天的日志文件就会被删除以节省空间
Secure.DROPBOX_AGE_SECONDS: "dropbox_age_seconds"
//用于控制系统保存的日志文件个数，默认是 1000 个文件
Secure.DROPBOX_MAX_FILES: "dropbox_max_files"
//用于控制 dropbox 目录最多占存储空间容量的比例，默认是 10%
Secure.DROPBOX_QUOTA_PERCENT:"dropbox_quota_percent"
//不允许 dropbox 使用的存储空间的比例，默认是 10%，也就是 dropbox 最多只能使用 90%的空间
Secure.DROPBOX_RESERVE_PERCENT:"dropbox_reserve_percent"
//dropbox 最大能使用的空间大小，默认是 5MB
Secure.DROPBOX_QUOTA_KB: "dropbox_quota_kb"
```

读者可以利用 adb shell 进入“/data/data/com.android.providers.settings/databases/”目录，然后利用 sqlite3 命令操作 settings.db，通过里面的表 Secure 可以了解相关内容。不过系统中的很多选项在该表中都没有相关设置，因此实际运行时都会使用代码中设置的默认值。

9.2.5 分析 DiskStatsService

在 Android L 中，DiskStatsService 在如下所示的文件中实现：

```
\frameworks\base\services\java\com\android\server\DiskStatsService.java
```

文件 DiskStatsService.java 的具体实现代码如下所示：

```
import java.io.File;
import java.io.FileDescriptor;
import java.io.FileOutputStream;
import java.io.IOException;
import java.io.PrintWriter;

/**
 * This service exists only as a "dumpsys" target which reports
 * statistics about the status of the disk.
 */
public class DiskStatsService extends Binder {
    private static final String TAG = "DiskStatsService";

    private final Context mContext;

    public DiskStatsService(Context context) {
        mContext = context;
    }

    @Override
    protected void dump(FileDescriptor fd, PrintWriter pw, String[] args) {
        mContext.enforceCallingOrSelfPermission(android.Manifest.permission.DUMP, TAG);

        // Run a quick-and-dirty performance test: write 512 bytes
        byte[] junk = new byte[512];
        for (int i = 0; i < junk.length; i++) junk[i] = (byte) i;  // Write nonzero bytes

        File tmp = new File(Environment.getDataDirectory(), "system/perftest.tmp");
        FileOutputStream fos = null;
        IOException error = null;
```

```
        long before = SystemClock.uptimeMillis();
        try {
            fos = new FileOutputStream(tmp);
            fos.write(junk);
        } catch (IOException e) {
            error = e;
        } finally {
            try { if (fos != null) fos.close(); } catch (IOException e) {}
        }

        long after = SystemClock.uptimeMillis();
        if (tmp.exists()) tmp.delete();

        if (error != null) {
            pw.print("Test-Error: ");
            pw.println(error.toString());
        } else {
            pw.print("Latency: ");
            pw.print(after - before);
            pw.println("ms [512B Data Write]");
        }

        reportFreeSpace(Environment.getDataDirectory(), "Data", pw);
        reportFreeSpace(Environment.getDownloadCacheDirectory(), "Cache", pw);
        reportFreeSpace(new File("/system"), "System", pw);

        // TODO: Read /proc/yaffs and report interesting values;
        // add configurable (through args) performance test parameters.
    }

    private void reportFreeSpace(File path, String name, PrintWriter pw) {
        try {
            StatFs statfs = new StatFs(path.getPath());
            long bsize = statfs.getBlockSize();
            long avail = statfs.getAvailableBlocks();
            long total = statfs.getBlockCount();
            if (bsize <= 0 || total <= 0) {
                throw new IllegalArgumentException(
                        "Invalid stat: bsize=" + bsize + " avail=" + avail + " total=" + total);
            }

            pw.print(name);
            pw.print("-Free: ");
            pw.print(avail * bsize / 1024);
            pw.print("K / ");
            pw.print(total * bsize / 1024);
            pw.print("K total = ");
            pw.print(avail * 100 / total);
            pw.println("% free");
        } catch (IllegalArgumentException e) {
            pw.print(name);
            pw.print("-Error: ");
            pw.println(e.toString());
            return;
        }
    }
}
```

从上述代码可以看出：虽然 DiskStatsService 从 Binder 中派生，但是并没有实现任何接口，也就是说 DiskStatsService 没有任何可调用的业务函数。但是在系统中为什么会存在这样的服务呢？要想解决这个问题，需要先了解介绍系统中的命令——dumpsys，此命令用于打印系统中指定服务的信息，在如下所示的文件中定义。

```
\frameworks\native\cmds\dumpsys\dumpsys.cpp
```

文件 dumpsys.cpp 的具体实现代码如下所示：

```
#define LOG_TAG "dumpsys"

#include <utils/Log.h>
#include <binder/Parcel.h>
#include <binder/ProcessState.h>
#include <binder/IServiceManager.h>
#include <utils/TextOutput.h>
#include <utils/Vector.h>

#include <getopt.h>
#include <stdlib.h>
#include <stdio.h>
#include <string.h>
#include <unistd.h>
#include <sys/time.h>

using namespace android;

static int sort_func(const String16* lhs, const String16* rhs)
{
    return lhs->compare(*rhs);
}

int main(int argc, char* const argv[])
{
    signal(SIGPIPE, SIG_IGN);
    sp<IServiceManager> sm = defaultServiceManager();
    fflush(stdout);
    if (sm == NULL) {
        ALOGE("Unable to get default service manager!");
        aerr << "dumpsys: Unable to get default service manager!" << endl;
        return 20;
    }

    Vector<String16> services;
    Vector<String16> args;
    if (argc == 1) {
        services = sm->listServices();
        services.sort(sort_func);
        args.add(String16("-a"));
    } else {
        services.add(String16(argv[1]));
        for (int i=2; i<argc; i++) {
            args.add(String16(argv[i]));
        }
    }

    const size_t N = services.size();

    if (N > 1) {
        // first print a list of the current services
        aout << "Currently running services:" << endl;

        for (size_t i=0; i<N; i++) {
            sp<IBinder> service = sm->checkService(services[i]);
            if (service != NULL) {
                aout << "  " << services[i] << endl;
            }
        }
    }

    for (size_t i=0; i<N; i++) {
        sp<IBinder> service = sm->checkService(services[i]);
        if (service != NULL) {
            if (N > 1) {
                aout << "------------------------------------------------------------"
                        "-------------------" << endl;
                aout << "DUMP OF SERVICE " << services[i] << ":" << endl;
            }
```

```
            int err = service->dump(STDOUT_FILENO, args);
            if (err != 0) {
                aerr << "Error dumping service info: (" << strerror(err)
                        << ") " << services[i] << endl;
            }
        } else {
            aerr << "Can't find service: " << services[i] << endl;
        }
    }

    return 0;
}
```

通过上述代码可知，dumpsys 通过 Binder 调用某个 Service 的 dump 函数。上述代码的具体实现流程如下所示。

（1）先获取与 ServiceManager 进程通信的 BpServiceManager 对象。

（2）如果输入参数个数为 1，则先查询在 SM 中注册的所有 Service。

（3）将 Service 排序。

（4）指定查询某个 Service。

（5）保存剩余参数，以后可以传给 Service 的 dump 函数。

（6）通过 Binder 调用该 Service 的 dump 函数，将 args 也传给 dump 函数。

接下来看文件 DiskStatsService.java 中的函数 dump，具体实现代码如下所示：

```
protected void dump(FileDescriptor fd, PrintWriter pw, String[] args) {
    mContext.enforceCallingOrSelfPermission(android.Manifest.permission.DUMP, TAG);

    // Run a quick-and-dirty performance test: write 512 bytes
    byte[] junk = new byte[512];
    for (int i = 0; i < junk.length; i++) junk[i] = (byte) i;  // Write nonzero bytes

    File tmp = new File(Environment.getDataDirectory(), "system/perftest.tmp");
    FileOutputStream fos = null;
    IOException error = null;

    long before = SystemClock.uptimeMillis();
    try {
        fos = new FileOutputStream(tmp);
        fos.write(junk);
    } catch (IOException e) {
        error = e;
    } finally {
        try { if (fos != null) fos.close(); } catch (IOException e) {}
    }

    long after = SystemClock.uptimeMillis();
    if (tmp.exists()) tmp.delete();

    if (error != null) {
        pw.print("Test-Error: ");
        pw.println(error.toString());
    } else {
        pw.print("Latency: ");
        pw.print(after - before);
        pw.println("ms [512B Data Write]");
    }

    reportFreeSpace(Environment.getDataDirectory(), "Data", pw);
    reportFreeSpace(Environment.getDownloadCacheDirectory(), "Cache", pw);
    reportFreeSpace(new File("/system"), "System", pw);

    // TODO: Read /proc/yaffs and report interesting values;
    // add configurable (through args) performance test parameters.
}
```

从上述代码可知，DiskStatsService 没有实现任何业务接口，只是为了调试而存在。

9.2.6 分析 DeviceStorageManagerService

在 Android L 中，DeviceStorageManagerService（DSMS）用于监测系统内部存储空间的状态，添加该服务的代码如下所示：

```
//DSMS 的服务名为 devicestoragemonitor
ServiceManager.addService(DeviceStorageMonitorService.SERVICE,
new DeviceStorageMonitorService(context));
```

DSMS 的构造函数在如下所示的文件中实现。

```
\frameworks\base\services\java\com\android\server\DeviceStorageMonitorService.java
```

函数 DeviceStorageMonitorService 的具体实现代码如下所示：

```
public DeviceStorageMonitorService(Context context) {
      mLastReportedFreeMemTime = 0;
      mContext = context;
      mContentResolver = mContext.getContentResolver();
      mDataFileStats = new StatFs(DATA_PATH);//获取 data 分区的信息
      mSystemFileStats = new StatFs(SYSTEM_PATH);// 获取 system 分区的信息
      mCacheFileStats = new StatFs(CACHE_PATH);// 获取 cache 分区的信息
      //获得 data 分区的总大小
      mTotalMemory = ((long)mDataFileStats.getBlockCount() *
                mDataFileStats.getBlockSize())/100L;
      /*
      创建 3 个 Intent，分别用于通知存储空间不足、存储空间恢复正常和存储空间满。
      由于设置了 REGISTERED_ONLY_BEFORE_BOOT 标志，这 3 个 Intent 广播只能由
      系统服务接收
      */
     mStorageLowIntent = new Intent(Intent.ACTION_DEVICE_STORAGE_LOW);
     mStorageLowIntent.addFlags(
              Intent.FLAG_RECEIVER_REGISTERED_ONLY_BEFORE_BOOT);
      mStorageOkIntent = new Intent(Intent.ACTION_DEVICE_STORAGE_OK);
      mStorageOkIntent.addFlags(
            Intent.FLAG_RECEIVER_REGISTERED_ONLY_BEFORE_BOOT);
      mStorageFullIntent = new Intent(Intent.ACTION_DEVICE_STORAGE_FULL);
      mStorageFullIntent.addFlags(
            Intent.FLAG_RECEIVER_REGISTERED_ONLY_BEFORE_BOOT);
      mStorageNotFullIntent = new
            Intent(Intent.ACTION_DEVICE_STORAGE_NOT_FULL);
      mStorageNotFullIntent.addFlags(
            Intent.FLAG_RECEIVER_REGISTERED_ONLY_BEFORE_BOOT);
      //查询 Settings 数据库中 sys_storage_threshold_percentage 的值，默认是 10，
      //即当/data 空间只剩下 10%的时候，认为空间不足
      mMemLowThreshold = getMemThreshold();
     //查询 Settings 数据库中的 sys_storage_full_threshold_bytes 的值，默认是 1MB，
     //即当 data 分区只剩 1MB 时，就认为空间已满，剩下的这 1MB 空间保留给系统自用
      mMemFullThreshold = getMemFullThreshold();
      //检查内存
      checkMemory(true);
}
```

再来看函数 checkMemory，此函数也是在文件 DeviceStorageMonitorService.java 中定义的，具体实现代码如下所示：

```
private final void checkMemory(boolean checkCache) {
    //if the thread that was started to clear cache is still running do nothing till its
    //finished clearing cache. Ideally this flag could be modified by clearCache
    // and should be accessed via a lock but even if it does this test will fail now and
    //hopefully the next time this flag will be set to the correct value.
    if(mClearingCache) {
       if(localLOGV) Slog.i(TAG, "Thread already running just skip");
       //make sure the thread is not hung for too long
       long diffTime = System.currentTimeMillis() - mThreadStartTime;
```

```
            if(diffTime > (10*60*1000)) {
                Slog.w(TAG, "Thread that clears cache file seems to run for ever");
            }
        } else {
            restatDataDir();
            if (localLOGV) Slog.v(TAG, "freeMemory="+mFreeMem);

            //post intent to NotificationManager to display icon if necessary
            if (mFreeMem < mMemLowThreshold) {
                if (checkCache) {
                    // We are allowed to clear cache files at this point to
                    // try to get down below the limit, because this is not
                    // the initial call after a cache clear has been attempted.
                    // In this case we will try a cache clear if our free
                    // space has gone below the cache clear limit.
                    if (mFreeMem < mMemCacheStartTrimThreshold) {
                        // We only clear the cache if the free storage has changed
                        // a significant amount since the last time.
                        if ((mFreeMemAfterLastCacheClear-mFreeMem)
                                >= ((mMemLowThreshold-mMemCacheStartTrimThreshold)/4)) {
                            // See if clearing cache helps
                            // Note that clearing cache is asynchronous and so we do a
                            // memory check again once the cache has been cleared.
                            mThreadStartTime = System.currentTimeMillis();
                            mClearSucceeded = false;
                            clearCache();
                        }
                    }
                } else {
                    // This is a call from after clearing the cache.  Note
                    // the amount of free storage at this point.
                    mFreeMemAfterLastCacheClear = mFreeMem;
                    if (!mLowMemFlag) {
                        // We tried to clear the cache, but that didn't get us
                        // below the low storage limit.  Tell the user.
                        Slog.i(TAG, "Running low on memory. Sending notification");
                        sendNotification();
                        mLowMemFlag = true;
                    } else {
                        if (localLOGV) Slog.v(TAG, "Running low on memory " +
                                "notification already sent. do nothing");
                    }
                }
            } else {
                mFreeMemAfterLastCacheClear = mFreeMem;
                if (mLowMemFlag) {
                    Slog.i(TAG, "Memory available. Cancelling notification");
                    cancelNotification();
                    mLowMemFlag = false;
                }
            }
            if (mFreeMem < mMemFullThreshold) {
                if (!mMemFullFlag) {
                    sendFullNotification();
                    mMemFullFlag = true;
                }
            } else {
                if (mMemFullFlag) {
                    cancelFullNotification();
                    mMemFullFlag = false;
                }
            }
        }
        if(localLOGV) Slog.i(TAG, "Posting Message again");
        //keep posting messages to itself periodically
        postCheckMemoryMsg(true, DEFAULT_CHECK_INTERVAL);
    }
```

当空间不足时，DSMS 会先使用函数 clearCache 进行处理，在此函数内部会与 PackageManager-

Service（简称 PKMS）进行交互。函数 clearCache 在文件 DeviceStorageManagerService.java 中定义，具体实现代码如下所示：

```
private final void clearCache() {
    if (mClearCacheObserver == null) {
        // Lazy instantiation
        mClearCacheObserver = new CachePackageDataObserver();
    }
    mClearingCache = true;
    try {
        if (localLOGV) Slog.i(TAG, "Clearing cache");
        IPackageManager.Stub.asInterface(ServiceManager.getService("package")).
                freeStorageAndNotify(mMemCacheTrimToThreshold, mClearCacheObserver);
    } catch (RemoteException e) {
        Slog.w(TAG, "Failed to get handle for PackageManger Exception: "+e);
        mClearingCache = false;
        mClearSucceeded = false;
    }
}
```

CachePackageDataObserver 是 DSMS 定义的内部类，其中的函数 onRemoveCompleted 用于重新发送消息，让 DSMS 再检测一次存储空间。函数 DeviceStorageManagerService 并没有重载 dump 函数。

9.2.7 分析 SamplingProfilerService

在 Android L 的源码中，添加 SamplingProfilerService 服务的代码如下所示：

```
ServiceManager.addService("samplingprofiler",//服务名
                    new SamplingProfilerService(context));
```

在本节的内容中，将详细分析 Android L 中 SamplingProfilerService 的源码。

（1）分析 SamplingProfilerService 构造函数。

SamplingProfilerService 的构造函数在如下所示的文件中实现：

```
\frameworks\base\services\java\com\android\server\SamplingProfilerService.java
```

在文件 SamplingProfilerService.java 中，函数 SamplingProfilerService 的具体实现代码如下所示：

```
public SamplingProfilerService(Context context) {
    //注册一个 CotentObserver，用于监测 Settings 数据库的变化
    registerSettingObserver(context);
    startWorking(context);//① startWorking 函数，见下文的分析
}
```

上述代码的核心是函数 startWorking，此函数在文件 SamplingProfilerService.java 中定义，具体实现代码如下所示：

```
private void startWorking(Context context) {
    if (LOCAL_LOGV) Slog.v(TAG, "starting SamplingProfilerService!");

    final DropBoxManager dropbox =
            (DropBoxManager) context.getSystemService(Context.DROPBOX_SERVICE);

    // before FileObserver is ready, there could have already been some snapshots
    // in the directory, we don't want to miss them
    File[] snapshotFiles = new File(SNAPSHOT_DIR).listFiles();
    for (int i = 0; snapshotFiles != null && i < snapshotFiles.length; i++) {
        handleSnapshotFile(snapshotFiles[i], dropbox);
    }

    // detect new snapshot and put it in dropbox
    // delete it afterwards no matter what happened before
    // Note: needs listening at event ATTRIB rather than CLOSE_WRITE, because we set the
    // readability of snapshot files after writing them!
```

```
        snapshotObserver = new FileObserver(SNAPSHOT_DIR, FileObserver.ATTRIB) {
            @Override
            public void onEvent(int event, String path) {
                handleSnapshotFile(new File(SNAPSHOT_DIR, path), dropbox);
            }
        };
        snapshotObserver.startWatching();

        if (LOCAL_LOGV) Slog.v(TAG, "SamplingProfilerService activated");
    }
```

通过上述代码可知，SamplingProfilerService 本身并不提供性能统计的功能。统计功能是通过类 SamplingProfilerIntegration 实现的，这个类封装了一个 SamplingProfiler（由 dalvik 虚拟机提供）对象，并提供了方便利用的函数进行性能统计。

（2）分析 SamplingProfilerIntegration。

通过使用 SamplingProfilerIntegration 可以进行性能统计。在 Andorid 系统中有很多重要进程都需要对性能进行分析，比如 Zygote，其相关代码在如下所示的文件中实现：

```
\frameworks\base\core\java\com\android\internal\os\ZygoteInit.java
```

在文件 ZygoteInit.java 中，和性能分析相关的代码如下所示：

```
    public static void main(String argv[]) {
        try {
            // Start profiling the zygote initialization.
            SamplingProfilerIntegration.start();

            registerZygoteSocket();
            EventLog.writeEvent(LOG_BOOT_PROGRESS_PRELOAD_START,
                SystemClock.uptimeMillis());
            preload();
            EventLog.writeEvent(LOG_BOOT_PROGRESS_PRELOAD_END,
                SystemClock.uptimeMillis());

            // Finish profiling the zygote initialization.
            SamplingProfilerIntegration.writeZygoteSnapshot();

            // Do an initial gc to clean up after startup
            gc();

            // If requested, start system server directly from Zygote
            if (argv.length != 2) {
                throw new RuntimeException(argv[0] + USAGE_STRING);
            }

            if (argv[1].equals("start-system-server")) {
                startSystemServer();
            } else if (!argv[1].equals("")) {
                throw new RuntimeException(argv[0] + USAGE_STRING);
            }

            Log.i(TAG, "Accepting command socket connections");

            if (ZYGOTE_FORK_MODE) {
                runForkMode();
            } else {
                runSelectLoopMode();
            }

            closeServerSocket();
        } catch (MethodAndArgsCaller caller) {
            caller.run();
        } catch (RuntimeException ex) {
            Log.e(TAG, "Zygote died with exception", ex);
            closeServerSocket();
            throw ex;
        }
    }
```

在上述代码中，函数 start 在如下所示的文件中实现。

```
\frameworks\base\core\java\com\android\internal\os\SamplingProfilerIntegration.java
```

函数 start 的具体实现代码如下所示：

```
public static void start() {
    if (!enabled) {       //判断是否开启性能统计
        return;
    }
    if (samplingProfiler != null) {
        Log.e(TAG, "SamplingProfilerIntegration already started at " + new Date(startMillis));
        return;
    }

    ThreadGroup group = Thread.currentThread().getThreadGroup();
    //创建一个 dalvik 的 SamplingProfiler
    SamplingProfiler.ThreadSet threadSet = SamplingProfiler.newThreadGroupTheadSet(group);
    samplingProfiler = new SamplingProfiler(samplingProfilerDepth, threadSet);
    //启动统计
    samplingProfiler.start(samplingProfilerMilliseconds);
    startMillis = System.currentTimeMillis();
}
```

在上述代码中，使用该类的 static 语句来判断是否启动性能统计的 enable 变量由谁控制。在文件 SamplingProfilerIntegration.java 中，static 语句的实现代码如下所示：

```
static {
    samplingProfilerMilliseconds = SystemProperties.getInt("persist.sys.profiler_ms", 0);
    samplingProfilerDepth = SystemProperties.getInt("persist.sys.profiler_depth", 4);
    if (samplingProfilerMilliseconds > 0) {
        File dir = new File(SNAPSHOT_DIR);
        dir.mkdirs();
        // the directory needs to be writable to anybody to allow file writing
        dir.setWritable(true, false);
        // the directory needs to be executable to anybody to allow file creation
        dir.setExecutable(true, false);
        if (dir.isDirectory()) {
            snapshotWriter = Executors.newSingleThreadExecutor(new ThreadFactory() {
                    public Thread newThread(Runnable r) {
                        return new Thread(r, TAG);
                    }
                });
            enabled = true;
            Log.i(TAG, "Profiling enabled. Sampling interval ms: "
                    + samplingProfilerMilliseconds);
        } else {
            snapshotWriter = null;
            enabled = true;
            Log.w(TAG, "Profiling setup failed. Could not create " + SNAPSHOT_DIR);
        }
    } else {
        snapshotWriter = null;
        enabled = false;
        Log.i(TAG, "Profiling disabled.");
    }
}
```

由上述代码可知，enable 的控制在 static 语句中实现，这表明要使用性能统计，就必须重新启动要统计的进程。

当启动性能统计后，需要输出统计文件，此功能由函数 writeZygoteSnapshot 实现。在文件 SamplingProfilerIntegration.java 中，函数 writeZygoteSnapshot 的具体实现代码如下所示：

```
public static void writeZygoteSnapshot() {
    if (!enabled) {
        return;
    }
```

```
        writeSnapshotFile("zygote", null);
        samplingProfiler.shutdown();
        samplingProfiler = null;
        startMillis = 0;
    }
```

在上述代码中，调用了 writeSnapshotFile 函数，其第一个参数为 zygote 表示进程名。函数 writeSnapshotFile 的功能是在 shots 目录下生成一个统计文件，这个统计文件的名称由两部分组成，其格式是“进程名_开始性能统计的时刻.snapshot”。另外，在 writeSnapshotfile 内部会调用函数 generateSnapshotHeader 在该统计文件头部写一些固定信息，例如版本号、编译信息等。在文件 SamplingProfilerIntegration.java 中，函数 writeSnapshotFile 的具体实现代码如下所示：

```
    private static void writeSnapshotFile(String processName, PackageInfo packageInfo) {
        if (!enabled) {
            return;
        }
        samplingProfiler.stop();

        /*
         * We use the global start time combined with the process name
         * as a unique ID. We can't use a counter because processes
         * restart. This could result in some overlap if we capture
         * two snapshots in rapid succession.
         */
        String name = processName.replaceAll(":", ".");
        String path = SNAPSHOT_DIR + "/" + name + "-" + startMillis + ".snapshot";
        long start = System.currentTimeMillis();
        OutputStream outputStream = null;
        try {
            outputStream = new BufferedOutputStream(new FileOutputStream(path));
            PrintStream out = new PrintStream(outputStream);
            generateSnapshotHeader(name, packageInfo, out);
            if (out.checkError()) {
                throw new IOException();
            }
            BinaryHprofWriter.write(samplingProfiler.getHprofData(), outputStream);
        } catch (IOException e) {
            Log.e(TAG, "Error writing snapshot to " + path, e);
            return;
        } finally {
            IoUtils.closeQuietly(outputStream);
        }
        // set file readable to the world so that SamplingProfilerService
        // can put it to dropbox
        new File(path).setReadable(true, false);

        long elapsed = System.currentTimeMillis() - start;
        Log.i(TAG, "Wrote snapshot " + path + " in " + elapsed + "ms.");
        samplingProfiler.start(samplingProfilerMilliseconds);
    }
```

SamplingProfilerIntegration 的核心是类 SamplingProfiler，这个类定义在如下所示 id 文件中：

```
libcore/dalvik /src/main/java/dalvik/system/profiler/SamplingProfiler.java
```

文件 SamplingProfiler.java 的具体实现代码如下所示：

```
public final class SamplingProfiler {
    private final Map<HprofData.StackTrace, int[]> stackTraces
            = new HashMap<HprofData.StackTrace, int[]>();
    private final HprofData hprofData = new HprofData(stackTraces);
    private final Timer timer = new Timer("SamplingProfiler", true);
    private Sampler sampler;
    private final int depth;
    private final ThreadSet threadSet;
    private int nextThreadId = 200001;
    private int nextStackTraceId = 300001;
```

```
private int nextObjectId = 1;
private Thread[] currentThreads = new Thread[0];
private final Map<Thread, Integer> threadIds = new HashMap<Thread, Integer>();
private final HprofData.StackTrace mutableStackTrace = new HprofData.StackTrace();
private final ThreadSampler threadSampler;
public SamplingProfiler(int depth, ThreadSet threadSet) {
    this.depth = depth;
    this.threadSet = threadSet;
    this.threadSampler = findDefaultThreadSampler();
    threadSampler.setDepth(depth);
    hprofData.setFlags(BinaryHprof.ControlSettings.CPU_SAMPLING.bitmask);
    hprofData.setDepth(depth);
}

private static ThreadSampler findDefaultThreadSampler() {
    if ("Dalvik Core Library".equals(System.getProperty("java.specification.name"))) {
        String className = "dalvik.system.profiler.DalvikThreadSampler";
        try {
            return (ThreadSampler) Class.forName(className).newInstance();
        } catch (Exception e) {
            System.out.println("Problem creating " + className + ": " + e);
        }
    }
    return new PortableThreadSampler();
}

/**
 * A ThreadSet specifies the set of threads to sample.
 */
public static interface ThreadSet {
    /**
     * Returns an array containing the threads to be sampled. The
     * array may be longer than the number of threads to be
     * sampled, in which case the extra elements must be null.
     */
    public Thread[] threads();
}

/**
 * Returns a ThreadSet for a fixed set of threads that will not
 * vary at runtime. This has less overhead than a dynamically
 * calculated set, such as {@link #newThreadGroupTheadSet}, which has
 * to enumerate the threads each time profiler wants to collect
 * samples.
 */
public static ThreadSet newArrayThreadSet(Thread... threads) {
    return new ArrayThreadSet(threads);
}

/**
 * An ArrayThreadSet samples a fixed set of threads that does not
 * vary over the life of the profiler.
 */
private static class ArrayThreadSet implements ThreadSet {
    private final Thread[] threads;
    public ArrayThreadSet(Thread... threads) {
        if (threads == null) {
            throw new NullPointerException("threads == null");
        }
        this.threads = threads;
    }
    public Thread[] threads() {
        return threads;
    }
}

/**
 * Returns a ThreadSet that is dynamically computed based on the
 * threads found in the specified ThreadGroup and that
```

```
     * ThreadGroup's children.
     */
    public static ThreadSet newThreadGroupTheadSet(ThreadGroup threadGroup) {
        return new ThreadGroupThreadSet(threadGroup);
    }

    /**
     * An ThreadGroupThreadSet sample the threads from the specified
     * ThreadGroup and the ThreadGroup's children
     */
    private static class ThreadGroupThreadSet implements ThreadSet {
        private final ThreadGroup threadGroup;
        private Thread[] threads;
        private int lastThread;

        public ThreadGroupThreadSet(ThreadGroup threadGroup) {
            if (threadGroup == null) {
                throw new NullPointerException("threadGroup == null");
            }
            this.threadGroup = threadGroup;
            resize();
        }

        private void resize() {
            int count = threadGroup.activeCount();
            // we can only tell if we had enough room for all active
            // threads if we actually are larger than the the number of
            // active threads. making it larger also leaves us room to
            // tolerate additional threads without resizing.
            threads = new Thread[count*2];
            lastThread = 0;
        }

        public Thread[] threads() {
            int threadCount;
            while (true) {
                threadCount = threadGroup.enumerate(threads);
                if (threadCount == threads.length) {
                    resize();
                } else {
                    break;
                }
            }
            if (threadCount < lastThread) {
                // avoid retaining pointers to threads that have ended
                Arrays.fill(threads, threadCount, lastThread, null);
            }
            lastThread = threadCount;
            return threads;
        }
    }

    /**
     * Starts profiler sampling at the specified rate.
     *
     * @param interval The number of milliseconds between samples
     */
    public void start(int interval) {
        if (interval < 1) {
            throw new IllegalArgumentException("interval < 1");
        }
        if (sampler != null) {
            throw new IllegalStateException("profiling already started");
        }
        sampler = new Sampler();
        hprofData.setStartMillis(System.currentTimeMillis());
        timer.scheduleAtFixedRate(sampler, 0, interval);
    }
```

```java
    /**
     * Stops profiler sampling. It can be restarted with {@link
     * #start(int)} to continue sampling.
     */
    public void stop() {
        if (sampler == null) {
            return;
        }
        synchronized(sampler) {
            sampler.stop = true;
            while (!sampler.stopped) {
                try {
                    sampler.wait();
                } catch (InterruptedException ignored) {
                }
            }
        }
        sampler = null;
    }

    /**
     * Shuts down profiling after which it can not be restarted. It is
     * important to shut down profiling when done to free resources
     * used by the profiler. Shutting down the profiler also stops the
     * profiling if that has not already been done.
     */
    public void shutdown() {
        stop();
        timer.cancel();
    }

    /**
     * Returns the hprof data accumulated by the profiler since it was
     * created. The profiler needs to be stopped, but not necessarily
     * shut down, in order to access the data. If the profiler is
     * restarted, there is no thread safe way to access the data.
     */
    public HprofData getHprofData() {
        if (sampler != null) {
            throw new IllegalStateException("cannot access hprof data while sampling");
        }
        return hprofData;
    }

    /**
     * The Sampler does the real work of the profiler.
     *
     * At every sample time, it asks the thread set for the set
     * of threads to sample. It maintains a history of thread creation
     * and death events based on changes observed to the threads
     * returned by the {@code ThreadSet}.
     *
     * For each thread to be sampled, a stack is collected and used to
     * update the set of collected samples. Stacks are truncated to a
     * maximum depth. There is no way to tell if a stack has been truncated.
     */
    private class Sampler extends TimerTask {

        private boolean stop;
        private boolean stopped;

        private Thread timerThread;

        public void run() {
            synchronized(this) {
                if (stop) {
                    cancel();
                    stopped = true;
                    notifyAll();
```

```
                return;
            }
        }

        if (timerThread == null) {
            timerThread = Thread.currentThread();
        }

        // process thread creation and death first so that we
        // assign thread ids to any new threads before allocating
        // new stacks for them
        Thread[] newThreads = threadSet.threads();
        if (!Arrays.equals(currentThreads, newThreads)) {
            updateThreadHistory(currentThreads, newThreads);
            currentThreads = newThreads.clone();
        }

        for (Thread thread : currentThreads) {
            if (thread == null) {
                break;
            }
            if (thread == timerThread) {
                continue;
            }

            StackTraceElement[] stackFrames = threadSampler.getStackTrace(thread);
            if (stackFrames == null) {
                continue;
            }
            recordStackTrace(thread, stackFrames);
        }
    }

    /**
     * Record a new stack trace. The thread should have been
     * previously registered with addStartThread.
     */
    private void recordStackTrace(Thread thread, StackTraceElement[] stackFrames) {
        Integer threadId = threadIds.get(thread);
        if (threadId == null) {
            throw new IllegalArgumentException("Unknown thread " + thread);
        }
        mutableStackTrace.threadId = threadId;
        mutableStackTrace.stackFrames = stackFrames;

        int[] countCell = stackTraces.get(mutableStackTrace);
        if (countCell == null) {
            countCell = new int[1];
            // cloned because the ThreadSampler may reuse the array
            StackTraceElement[] stackFramesCopy = stackFrames.clone();
            HprofData.StackTrace stackTrace
                    = new HprofData.StackTrace(nextStackTraceId++, threadId, stackFramesCopy);
            hprofData.addStackTrace(stackTrace, countCell);
        }
        countCell[0]++;
    }

    private void updateThreadHistory(Thread[] oldThreads, Thread[] newThreads) {
        // thread start/stop shouldn't happen too often and
        // these aren't too big, so hopefully this approach
        // won't be too slow...
        Set<Thread> n = new HashSet<Thread>(Arrays.asList(newThreads));
        Set<Thread> o = new HashSet<Thread>(Arrays.asList(oldThreads));

        // added = new-old
        Set<Thread> added = new HashSet<Thread>(n);
        added.removeAll(o);

        // removed = old-new
```

```
        Set<Thread> removed = new HashSet<Thread>(o);
        removed.removeAll(n);
        for (Thread thread : added) {
            if (thread == null) {
                continue;
            }
            if (thread == timerThread) {
                continue;
            }
            addStartThread(thread);
        }
        for (Thread thread : removed) {
            if (thread == null) {
                continue;
            }
            if (thread == timerThread) {
                continue;
            }
            addEndThread(thread);
        }
    }
    /**
     * Record that a newly noticed thread.
     */
    private void addStartThread(Thread thread) {
        if (thread == null) {
            throw new NullPointerException("thread == null");
        }
        int threadId = nextThreadId++;
        Integer old = threadIds.put(thread, threadId);
        if (old != null) {
            throw new IllegalArgumentException("Thread already registered as " + old);
        }

        String threadName = thread.getName();
        // group will become null when thread is terminated
        ThreadGroup group = thread.getThreadGroup();
        String groupName = group == null ? null : group.getName();
        ThreadGroup parentGroup = group == null ? null : group.getParent();
        String parentGroupName = parentGroup == null ? null : parentGroup.getName();

        HprofData.ThreadEvent event
            = HprofData.ThreadEvent.start(nextObjectId++, threadId,
                                          threadName, groupName, parentGroupName);
        hprofData.addThreadEvent(event);
    }

    /**
     * Record that a thread has disappeared.
     */
    private void addEndThread(Thread thread) {
        if (thread == null) {
            throw new NullPointerException("thread == null");
        }
        Integer threadId = threadIds.remove(thread);
        if (threadId == null) {
            throw new IllegalArgumentException("Unknown thread " + thread);
        }
        HprofData.ThreadEvent event = HprofData.ThreadEvent.end(threadId);
        hprofData.addThreadEvent(event);
    }
}
```

9.2.8 分析 ClipboardService

在 Android L 的源码中，类 content.ClipboardManager 继承自类 text.ClipboardManager，早期的

剪贴功能只支持文本。ClipboardManager 由剪贴板服务的客户端使用，在 SDK 中有相应文档说明。目前，Android 系统中的剪贴板支持 3 种类型的数据（Text、Intent，以及 URL 列表）。

下面通过一个例子来分析 CBS，该例子来源于 Android SDK 提供的一段示例代码，路径如下所示：

```
\sdk\samples\android-17\
```

1. 复制数据到剪贴板

在 Android SDK 的实例源码中，截取如下所示的与复制操作相关的代码：

```
//获取能与 CBS 交互的 ClipboardManager 对象
ClipboardManager clipboard = (ClipboardManager)
                getSystemService(Context.CLIPBOARD_SERVICE);
//调用 setPrimaryClip 函数，参数是 ClipData.newUri 函数的返回值
clipboard.setPrimaryClip(ClipData.newUri(
                        getContentResolver(),"Note",noteUri));
```

在上述代码中，ClipData 中的 newUri 是一个 static 函数，用于返回一个存储 URI 数据类型的 ClipData，代码如下。ClipData 对象装载的就是可保存在剪贴板中的数据。函数 newUri 在如下所示的文件中实现：

```
\frameworks\base\core\java\android\content\ClipData.java
```

函数 newUri 的具体实现代码如下所示：

```
static public ClipData newUri(ContentResolver resolver, CharSequence label,
        Uri uri) {
    Item item = new Item(uri);
    String[] mimeTypes = null;
    if ("content".equals(uri.getScheme())) {
        String realType = resolver.getType(uri);
        mimeTypes = resolver.getStreamTypes(uri, "*/*");
        if (mimeTypes == null) {
            if (realType != null) {
                mimeTypes = new String[] { realType, ClipDescription.MIMETYPE_TEXT_URILIST };
            }
        } else {
            String[] tmp = new String[mimeTypes.length + (realType != null ? 2 : 1)];
            int i = 0;
            if (realType != null) {
                tmp[0] = realType;
                i++;
            }
            System.arraycopy(mimeTypes, 0, tmp, i, mimeTypes.length);
            tmp[i + mimeTypes.length] = ClipDescription.MIMETYPE_TEXT_URILIST;
            mimeTypes = tmp;
        }
    }
    if (mimeTypes == null) {
        mimeTypes = MIMETYPES_TEXT_URILIST;
    }
    return new ClipData(label, mimeTypes, item);
}
```

函数 newUri 的主要功能是获得 uri 所指向的数据的数据类型。对于使用剪切板服务的程序来说，了解剪切板中数据的数据类型相当重要，因为这样可以判断自己能否处理这种类型的数据。在上述代码中，uri 指向数据的位置，这和 PC 上文件的存储位置类似，例如 c:/dfp。MIME 则表示该数据的数据类型。在 Windows 平台上是采用后缀名来表示文件类型的，前面提到的 C 盘下的 DFP 文件，后缀是.wav，表示该文件是一个 WAV 格式音频。对于剪切板来说，数据源由 uri 指定，数据类型由 MIME 表示，两者缺一不可。

当获得一个 ClipData 后，会调用函数 setPrimaryClip，功能是将数据传递到 CBS。函数

setPrimaryClip 在如下所示的文件中实现：

```
\frameworks\base\core\java\android\content\ClipboardManager.java
```

函数 setPrimaryClip 的具体实现代码如下所示：

```
public void setPrimaryClip(ClipData clip) {
try {
        //跨 Binder 调用，先要把参数打包。有兴趣的读者可以查看 writToParcel 函数
        getService().setPrimaryClip(clip);
    } catch (RemoteException e) {
  }
}
```

通过 Binder 发送 setPrimaryClip 请求后，由 CBS 完成实际功能。在文件“\frameworks\base\services\java\com\android\server\ClipboardService.java”中，函数 setPrimaryClip 的具体实现代码如下所示：

```
public void setPrimaryClip(ClipData clip) {
    synchronized (this) {
        if (clip != null && clip.getItemCount() <= 0) {
            throw new IllegalArgumentException("No items");
        }
        checkDataOwnerLocked(clip, Binder.getCallingUid());
        clearActiveOwnersLocked();
        PerUserClipboard clipboard = getClipboard();
        clipboard.primaryClip = clip;
        final int n = clipboard.primaryClipListeners.beginBroadcast();
        for (int i = 0; i < n; i++) {
            try {
clipboard.primaryClipListeners.getBroadcastItem(i).dispatchPrimaryClipChanged();
            } catch (RemoteException e) {
                // The RemoteCallbackList will take care of removing
                // the dead object for us.
            }
        }
        clipboard.primaryClipListeners.finishBroadcast();
    }
}
```

2. 从剪切板粘贴数据

请读者继续看 SDK 安装包中的实例，看如下所示的演示代码：

```
final void performPaste() {
  //获取 ClipboardManager 对象
  ClipboardManager clipboard = (ClipboardManager)
      getSystemService(Context.CLIPBOARD_SERVICE);
  //获取 ContentResolver 对象
  ContentResolver cr = getContentResolver();
  //从剪贴板中取出 ClipData
  ClipData clip = clipboard.getPrimaryClip();
  if (clip != null) {
     String text=null;
     String title=null;
     //取剪切板 ClipData 中的第一项 Item
     ClipData.Item item = clip.getItemAt(0);
     /*
     取出 Item 中所包含的 uri。
     */
     Uri uri = item.getUri();
     Cursor orig = cr.query(uri,PROJECTION, null, null,null);
       ....../查询数据库并获取信息
       orig.close();
       }
```

```
        }
        if (text == null) {
            //如果paste方不了解ClipData中的数据类型，可调用coerceToText函数，
          //强制得到文本类型的数据
            text = item.coerceToText(this).toString();//强制为文本
        }
```

在上述代码中用到了函数 getPrimaryClip，此函数在文件 ClipboardManager.java 中定义，具体实现代码如下所示：

```
    public ClipData getPrimaryClip() {
        try {
            return getService().getPrimaryClip(mContext.getPackageName());
        } catch (RemoteException e) {
            return null;
        }
    }
```

在文件 ClipboardManagerService.java 中，函数的具体实现代码如下所示：

```
public ClipData getPrimaryClip(String pkg) {
    synchronized (this) {
        //赋予该pkg相应的权限，后文再作分析
        addActiveOwnerLocked(Binder.getCallingUid(), pkg);
        return mPrimaryClip;//返回ClipData给客户端
    }
}
```

在上述代码中，函数 coerceToText 在 paste 方不了解 ClipData 中数据类型的情况下，可以强制得到文本类型的数据。

再看文件 ClipData.java，在里面定义了 coerceToText，具体实现代码如下所示：

```
        public CharSequence coerceToText(Context context) {
            // If this Item has an explicit textual value, simply return that.
            CharSequence text = getText();
            if (text != null) {
                return text;
            }

            // If this Item has a URI value, try using that.
            Uri uri = getUri();
            if (uri != null) {

                // First see if the URI can be opened as a plain text stream
                // (of any sub-type).  If so, this is the best textual
                // representation for it.
                FileInputStream stream = null;
                try {
                    // Ask for a stream of the desired type.
                    AssetFileDescriptor descr = context.getContentResolver()
                            .openTypedAssetFileDescriptor(uri, "text/*", null);
                    stream = descr.createInputStream();
                    InputStreamReader reader = new InputStreamReader(stream, "UTF-8");

                    // Got it...  copy the stream into a local string and return it.
                    StringBuilder builder = new StringBuilder(128);
                    char[] buffer = new char[8192];
                    int len;
                    while ((len=reader.read(buffer)) > 0) {
                        builder.append(buffer, 0, len);
                    }
                    return builder.toString();

                } catch (FileNotFoundException e) {
                    // Unable to open content URI as text...  not really an
                    // error, just something to ignore.

                } catch (IOException e) {
```

```
                // Something bad has happened.
                Log.w("ClippedData", "Failure loading text", e);
                return e.toString();

            } finally {
                if (stream != null) {
                    try {
                        stream.close();
                    } catch (IOException e) {
                    }
                }
            }

            // If we couldn't open the URI as a stream, then the URI itself
            // probably serves fairly well as a textual representation.
            return uri.toString();
        }

        // Finally, if all we have is an Intent, then we can just turn that
        // into text.  Not the most user-friendly thing, but it's something.
        Intent intent = getIntent();
        if (intent != null) {
            return intent.toUri(Intent.URI_INTENT_SCHEME);
        }

        // Shouldn't get here, but just in case...
        return "";
    }
```

由上述实现代码可知，针对 URI 类型的数据，函数 coerceToText 实现了处理功能。当然，还需要提供该 URI 的 ContentProvider 实现相应的函数。

3. 管理 CBS 中的权限

在 Android L 源码中，CBS 和权限管理相关的函数调用如下所示：

```
//copy 方设置 ClipData 在 CBS 的 setPrimaryClip 函数中进行:
checkDataOwnerLocked(clip, Binder.getCallingUid());
clearActiveOwnersLocked();
//paste 方获取 ClipData 在 CBS 的 getPrimaryClip 函数中进行:
addActiveOwnerLocked(Binder.getCallingUid(), pkg);
```

（1）URI 权限管理介绍。

Android 系统的权限管理中有一类是专门针对 URI 的，先来看一个示例，该例来自“package/providers/ContactsProvider”，在其对应的文件 AndroidManifest.xml 中有如下所示的声明代码：

```
<prvider android:name="ContactsProvider2"
        ......
        android:readPermission="android.permission.READ_CONTACTS"
        android:writePermission="android.permission.WRITE_CONTACTS">
        ......
        <grant-uri-permission android:pathPattern=".*" />
</provider>
```

在上述代码中声明了一个名为 ContactsProvider2 的 ContentProvider，并定义了几个权限声明，具体说明如下所示。

- readPermission：要求调用 query 函数的客户端必须声明一个 use-permission 为 READ_CONTACTS 的权限。
- writePermission：要求调用 update 或 insert 函数的客户端必须声明一个 use-permission 为 WRITE_CONTACTS 的权限。
- grant-uri-permission：和授权有关。

Contacts 和 ContactProvider 这两个 APP 都是由系统提供的程序，而且两者的关系十分紧密，所以 Contacts 一定会声明 use_Permission 为 READ_CONTACTS 和 WRITE_CONTACT 的权限。这样，Contacts 就可以通过 ContactsProvider 来查询或更新数据库了。

（2）分析函数 checkDataOwnerLocked。

函数 checkDataOwnerLocked 在文件 ClipboardService.java 中定义，具体实现代码如下所示：

```
    private final void checkDataOwnerLocked(ClipData data, int uid) {
        final int N = data.getItemCount();
        for (int i=0; i<N; i++) {
            checkItemOwnerLocked(data.getItemAt(i), uid);
        }
    }
private final void checkItemOwnerLocked(ClipData.Item item, int uid) {
      if (item.getUri() != null) {//检查 uri
         checkUriOwnerLocked(item.getUri(), uid);
      }
      Intent intent = item.getIntent();
      //getData 函数返回的也是一个 uri，因此这里实际上检查的也是 uri
      if (intent != null && intent.getData() != null) {
         checkUriOwnerLocked(intent.getData(), uid);
      }
}
```

由此可知，权限检查就是针对 uri 进行的，因为 uri 所指向的数据可能是系统内部使用或私密的。接下来分析文件 ClipboardService.java 中的函数 checkUriOwnerLocked，具体实现代码如下所示：

```
    private final void checkUriOwnerLocked(Uri uri, int uid) {
        if (!"content".equals(uri.getScheme())) {
            return;
        }
        long ident = Binder.clearCallingIdentity();
        try {
            // This will throw SecurityException for us.
            mAm.checkGrantUriPermission(uid, null, uri, Intent.FLAG_GRANT_READ_URI_PERMISSION);
        } catch (RemoteException e) {
        } finally {
            Binder.restoreCallingIdentity(ident);
        }
    }
```

通过上述代码，检查 copy 方是否有读取 uri 的权限。

（3）分析函数 clearActiveOwnersLocked。

函数 clearActiveOwnersLocked 在文件 ClipboardService.java 中定义，具体实现代码如下所示：

```
    private final void addActiveOwnerLocked(int uid, String pkg) {
        final IPackageManager pm = AppGlobals.getPackageManager();
        final int targetUserHandle = UserHandle.getCallingUserId();
        final long oldIdentity = Binder.clearCallingIdentity();
        try {
            PackageInfo pi = pm.getPackageInfo(pkg, 0, targetUserHandle);
            if (pi == null) {
                throw new IllegalArgumentException("Unknown package " + pkg);
            }
            if (!UserHandle.isSameApp(pi.applicationInfo.uid, uid)) {
                throw new SecurityException("Calling uid " + uid
                        + " does not own package " + pkg);
            }
        } catch (RemoteException e) {
            // Can't happen; the package manager is in the same process
        } finally {
            Binder.restoreCallingIdentity(oldIdentity);
        }
        PerUserClipboard clipboard = getClipboard();
        if (clipboard.primaryClip != null && !clipboard.activePermissionOwners.contains(pkg)) {
            final int N = clipboard.primaryClip.getItemCount();
            for (int i=0; i<N; i++) {
                grantItemLocked(clipboard.primaryClip.getItemAt(i), pkg);
```

```
        }
        clipboard.activePermissionOwners.add(pkg);
    }
}
```

再看文件 ClipboardService.java 中的函数 grantUriLocked，具体实现代码如下所示：

```
private final void grantUriLocked(Uri uri, String pkg) {
    long ident = Binder.clearCallingIdentity();
    try {
        mAm.grantUriPermissionFromOwner(mPermissionOwner, Process.myUid(), pkg, uri,
                Intent.FLAG_GRANT_READ_URI_PERMISSION);
    } catch (RemoteException e) {
    } finally {
        Binder.restoreCallingIdentity(ident);
    }
}
```

当客户端使用完毕后就需要撤销授权，这个工作是在函数 setPrimaryClip 的 clearActiveOwnersLocked 中完成的。当为剪切板设置新的 ClipData 时，自然需要将与旧 ClipData 相关的权限撤销。函数 clearActiveOwnersLocked 在文件 ClipboardService.java 中定义，具体实现代码如下所示：

```
    private final void clearActiveOwnersLocked() {
        PerUserClipboard clipboard = getClipboard();
        clipboard.activePermissionOwners.clear();
        if (clipboard.primaryClip == null) {
            return;
        }
        final int N = clipboard.primaryClip.getItemCount();
        for (int i=0; i<N; i++) {
            revokeItemLocked(clipboard.primaryClip.getItemAt(i));
        }
    }
}
```

9.3 应用程序进程详解

在启动 Android 应用程序过程中，除了可以获得虚拟机实例外，还可以获得一个消息循环和一个 Binder 线程池。这样在应用程序中运行的组件，可以使用系统的信息处理机制和 Binder 通信机制实现自己的业务逻辑。在本节将详细分析创建应用程序的实现源码，为读者步入本书后面知识的学习打下基础。

9.3.1 创建应用程序

在 Android 系统中，当 ActivityManagerService 创建新进程来启动某个应用程序组件时，会调用类 ActivityManagerService 中的函数 startProcessLocked 向孵化进程 Zygote 发送创建应用程序进程的请求。函数 startProcessLocked 在文件“frameworks\base\services\java\com\android\server\am\ActivityManagerService.java”中定义，具体实现代码如下所示：

```
private final void startProcessLocked(ProcessRecord app,
        String hostingType, String hostingNameStr) {
    if (app.pid > 0 && app.pid != MY_PID) {
        synchronized (mPidsSelfLocked) {
            mPidsSelfLocked.remove(app.pid);
            mHandler.removeMessages(PROC_START_TIMEOUT_MSG, app);
        }
        app.setPid(0);
    }

    if (DEBUG_PROCESSES && mProcessesOnHold.contains(app)) Slog.v(TAG,
            "startProcessLocked removing on hold: " + app);
```

```
        mProcessesOnHold.remove(app);

        updateCpuStats();

        System.arraycopy(mProcDeaths, 0, mProcDeaths, 1, mProcDeaths.length-1);
        mProcDeaths[0] = 0;
        //获取创建应用程序进程的用户 ID 和用户组 ID
        try {
            int uid = app.uid;

            int[] gids = null;
            int mountExternal = Zygote.MOUNT_EXTERNAL_NONE;
            if (!app.isolated) {
                int[] permGids = null;
                try {
                    final PackageManager pm = mContext.getPackageManager();
                    permGids = pm.getPackageGids(app.info.packageName);

                    if (Environment.isExternalStorageEmulated()) {
                        if (pm.checkPermission(
                                android.Manifest.permission.ACCESS_ALL_EXTERNAL_STORAGE,
                                app.info.packageName) == PERMISSION_GRANTED) {
                            mountExternal = Zygote.MOUNT_EXTERNAL_MULTIUSER_ALL;
                        } else {
                            mountExternal = Zygote.MOUNT_EXTERNAL_MULTIUSER;
                        }
                    }
                } catch (PackageManager.NameNotFoundException e) {
                    Slog.w(TAG, "Unable to retrieve gids", e);
                }

                /*
                 * Add shared application GID so applications can share some
                 * resources like shared libraries
                 */
                if (permGids == null) {
                    gids = new int[1];
                } else {
                    gids = new int[permGids.length + 1];
                    System.arraycopy(permGids, 0, gids, 1, permGids.length);
                }
                gids[0] = UserHandle.getSharedAppGid(UserHandle.getAppId(uid));
            }
            if (mFactoryTest != SystemServer.FACTORY_TEST_OFF) {
                if (mFactoryTest == SystemServer.FACTORY_TEST_LOW_LEVEL
                        && mTopComponent != null
                        && app.processName.equals(mTopComponent.getPackageName())) {
                    uid = 0;
                }
                if (mFactoryTest == SystemServer.FACTORY_TEST_HIGH_LEVEL
                        && (app.info.flags&ApplicationInfo.FLAG_FACTORY_TEST) != 0) {
                    uid = 0;
                }
            }
            int debugFlags = 0;
            if ((app.info.flags & ApplicationInfo.FLAG_DEBUGGABLE) != 0) {
                debugFlags |= Zygote.DEBUG_ENABLE_DEBUGGER;
                // Also turn on CheckJNI for debuggable apps. It's quite
                // awkward to turn on otherwise.
                debugFlags |= Zygote.DEBUG_ENABLE_CHECKJNI;
            }
            // Run the app in safe mode if its manifest requests so or the
            // system is booted in safe mode.
            if ((app.info.flags & ApplicationInfo.FLAG_VM_SAFE_MODE) != 0 ||
                Zygote.systemInSafeMode == true) {
                debugFlags |= Zygote.DEBUG_ENABLE_SAFEMODE;
            }
            if ("1".equals(SystemProperties.get("debug.checkjni"))) {
                debugFlags |= Zygote.DEBUG_ENABLE_CHECKJNI;
```

```
            }
            if ("1".equals(SystemProperties.get("debug.jni.logging"))) {
                debugFlags |= Zygote.DEBUG_ENABLE_JNI_LOGGING;
            }
            if ("1".equals(SystemProperties.get("debug.assert"))) {
                debugFlags |= Zygote.DEBUG_ENABLE_ASSERT;
            }
            //调用函数start创建应用程序进程
            Process.ProcessStartResult startResult =
            Process.start("android.app.ActivityThread",
                    app.processName, uid, uid, gids, debugFlags, mountExternal,
                    app.info.targetSdkVersion, app.info.seinfo, null);
            BatteryStatsImpl bs = app.batteryStats.getBatteryStats();
            synchronized (bs) {
                if (bs.isOnBattery()) {
                    app.batteryStats.incStartsLocked();
                }
            }

            EventLog.writeEvent(EventLogTags.AM_PROC_START,
                    UserHandle.getUserId(uid), startResult.pid, uid,
                    app.processName, hostingType,
                    hostingNameStr != null ? hostingNameStr : "");
            if (app.persistent) {
                Watchdog.getInstance().processStarted(app.processName, startResult.pid);
            }
            StringBuilder buf = mStringBuilder;
            buf.setLength(0);
            buf.append("Start proc ");
            buf.append(app.processName);
            buf.append(" for ");
            buf.append(hostingType);
            if (hostingNameStr != null) {
                buf.append(" ");
                buf.append(hostingNameStr);
            }
            buf.append(": pid=");
            buf.append(startResult.pid);
            buf.append(" uid=");
            buf.append(uid);
            buf.append(" gids={");
            if (gids != null) {
                for (int gi=0; gi<gids.length; gi++) {
                    if (gi != 0) buf.append(", ");
                    buf.append(gids[gi]);
                }
            }
            buf.append("}");
            Slog.i(TAG, buf.toString());
            app.setPid(startResult.pid);
            app.usingWrapper = startResult.usingWrapper;
            app.removed = false;
            synchronized (mPidsSelfLocked) {
                this.mPidsSelfLocked.put(startResult.pid, app);
                Message msg = mHandler.obtainMessage(PROC_START_TIMEOUT_MSG);
                msg.obj = app;
                mHandler.sendMessageDelayed(msg, startResult.usingWrapper
                        ? PROC_START_TIMEOUT_WITH_WRAPPER : PROC_START_TIMEOUT);
            }
        } catch (RuntimeException e) {
            // XXX do better error recovery.
            app.setPid(0);
            Slog.e(TAG, "Failure starting process " + app.processName, e);
        }
    }
```

类Process中的函数start在文件“frameworks\base\core\java\android\os\Process.java”中定义，具体实现代码如下所示：

```
    public static final ProcessStartResult start(final String processClass,
                                  final String niceName,
                                  int uid, int gid, int[] gids,
                                  int debugFlags, int mountExternal,
                                  int targetSdkVersion,
                                  String seInfo,
                                  String[] zygoteArgs) {
        try {
//调用函数 startViaZygote 让 Zygote 进程创建一个应用程序进程。
            return startViaZygote(processClass, niceName, uid, gid, gids,
                    debugFlags, mountExternal, targetSdkVersion, seInfo, zygoteArgs);
        } catch (ZygoteStartFailedEx ex) {
            Log.e(LOG_TAG,
                    "Starting VM process through Zygote failed");
            throw new RuntimeException(
                    "Starting VM process through Zygote failed", ex);
        }
    }
```

在上述代码中用到了函数 startViaZygote，功能是将要创建的应用程序进程的启动参数保存在字符串列表 argsForZygote 中，并调用函数 zygoteSendArgsAndGetResult 请求进程 Zygote 创建应用程序。函数 startViaZygote 在文件“frameworks\base\core\java\android\os\Process.java”中定义，具体实现代码如下所示：

```
private static ProcessStartResult startViaZygote(final String processClass,
                                  final String niceName,
                                  final int uid, final int gid,
                                  final int[] gids,
                                  int debugFlags, int mountExternal,
                                  int targetSdkVersion,
                                  String seInfo,
                                  String[] extraArgs)
                                  throws ZygoteStartFailedEx {
    synchronized(Process.class) {
        ArrayList<String> argsForZygote = new ArrayList<String>();
        // --runtime-init, --setuid=, --setgid=,
        // and --setgroups= must go first
        argsForZygote.add("--runtime-init");
        argsForZygote.add("--setuid=" + uid);
        argsForZygote.add("--setgid=" + gid);
        if ((debugFlags & Zygote.DEBUG_ENABLE_JNI_LOGGING) != 0) {
            argsForZygote.add("--enable-jni-logging");
        }
        if ((debugFlags & Zygote.DEBUG_ENABLE_SAFEMODE) != 0) {
            argsForZygote.add("--enable-safemode");
        }
        if ((debugFlags & Zygote.DEBUG_ENABLE_DEBUGGER) != 0) {
            argsForZygote.add("--enable-debugger");
        }
        if ((debugFlags & Zygote.DEBUG_ENABLE_CHECKJNI) != 0) {
            argsForZygote.add("--enable-checkjni");
        }
        if ((debugFlags & Zygote.DEBUG_ENABLE_ASSERT) != 0) {
            argsForZygote.add("--enable-assert");
        }
        if (mountExternal == Zygote.MOUNT_EXTERNAL_MULTIUSER) {
            argsForZygote.add("--mount-external-multiuser");
        } else if (mountExternal == Zygote.MOUNT_EXTERNAL_MULTIUSER_ALL) {
            argsForZygote.add("--mount-external-multiuser-all");
        }
        argsForZygote.add("--target-sdk-version=" + targetSdkVersion);
        //TODO optionally enable debuger
        //argsForZygote.add("--enable-debugger");
        // --setgroups is a comma-separated list
        if (gids != null && gids.length > 0) {
            StringBuilder sb = new StringBuilder();
            sb.append("--setgroups=");
            int sz = gids.length;
```

```
            for (int i = 0; i < sz; i++) {
                if (i != 0) {
                    sb.append(',');
                }
                sb.append(gids[i]);
            }
            argsForZygote.add(sb.toString());
        }
        if (niceName != null) {
            argsForZygote.add("--nice-name=" + niceName);
        }
        if (seInfo != null) {
            argsForZygote.add("--seinfo=" + seInfo);
        }
        argsForZygote.add(processClass);
        if (extraArgs != null) {
            for (String arg : extraArgs) {
                argsForZygote.add(arg);
            }
        }
        //请求进程 Zygote 创建应用程序
        return zygoteSendArgsAndGetResult(argsForZygote);
    }
}
```

在上述代码中，通过函数 zygoteSendArgsAndGetResult 调用 Zygote 进程创建了一个指定的应用程序。函数 zygoteSendArgsAndGetResult 在文件“frameworks\base\core\java\android\os\Process.java”中定义，具体实现代码如下所示：

```
private static ProcessStartResult zygoteSendArgsAndGetResult(ArrayList<String> args)
        throws ZygoteStartFailedEx {
//调用函数 openZygoteSocketIfNeeded 创建一个连接到 Zygote 进程的本地对象 LocalSocket
    openZygoteSocketIfNeeded();
    try {
        /**
         * 将要创建的应用程序进程启动参数列表写入到本地对象 LocalSocket 中
         * Zygote 进程接收到数据之后会创建一个新的应用程序进程
         *将创建的进程 pid 返回给 ActivityManagerService
         */
        sZygoteWriter.write(Integer.toString(args.size()));
        sZygoteWriter.newLine();
        int sz = args.size();
        for (int i = 0; i < sz; i++) {
            String arg = args.get(i);
            if (arg.indexOf('\n') >= 0) {
                throw new ZygoteStartFailedEx(
                        "embedded newlines not allowed");
            }
            sZygoteWriter.write(arg);
            sZygoteWriter.newLine();
        }
        sZygoteWriter.flush();
        // Should there be a timeout on this?
        ProcessStartResult result = new ProcessStartResult();
        result.pid = sZygoteInputStream.readInt();
        if (result.pid < 0) {
            throw new ZygoteStartFailedEx("fork() failed");
        }
        result.usingWrapper = sZygoteInputStream.readBoolean();
        return result;
    } catch (IOException ex) {
        try {
            if (sZygoteSocket != null) {
                sZygoteSocket.close();
            }
        } catch (IOException ex2) {
            // we're going to fail anyway
            Log.e(LOG_TAG,"I/O exception on routine close", ex2);
```

```
            }
            sZygoteSocket = null;
            throw new ZygoteStartFailedEx(ex);
        }
    }
```

在上述代码中用到了函数 openZygoteSocketIfNeeded，功能是创建一个连接到 Zygote 进程的本地对象 LocalSocket。函数 openZygoteSocketIfNeeded 在文件“frameworks\base\core\java\android\os\Process.java”中定义，具体实现代码如下所示：

```
    private static void openZygoteSocketIfNeeded()
            throws ZygoteStartFailedEx {
        int retryCount;
        if (sPreviousZygoteOpenFailed) {
            /*
             * If we've failed before, expect that we'll fail again and
             * don't pause for retries.
             */
            retryCount = 0;
        } else {
            retryCount = 10;
        }
        /*
         * See bug #811181: Sometimes runtime can make it up before zygote.
         * Really, we'd like to do something better to avoid this condition,
         * but for now just wait a bit...
         */
        for (int retry = 0
                ; (sZygoteSocket == null) && (retry < (retryCount + 1))
                ; retry++ ) {
            if (retry > 0) {
                try {
                    Log.i("Zygote", "Zygote not up yet, sleeping...");
                    Thread.sleep(ZYGOTE_RETRY_MILLIS);
                } catch (InterruptedException ex) {
                    // should never happen
                }
            }
            try {
//创建一个保存在 sZygoteSocket 中的 LocalSocket 对象
                sZygoteSocket = new LocalSocket();
//将创建的的 LocalSocket 对象和名为 ZYGOTE_SOCKET 的 zygote 进程建立连接
                sZygoteSocket.connect(new LocalSocketAddress(ZYGOTE_SOCKET,
                        LocalSocketAddress.Namespace.RESERVED));
//将获得的 LocalSocket 对象 sZygoteSocket 的输入流保存在变量 sZygoteInputStream 中
                sZygoteInputStream
                        = new DataInputStream(sZygoteSocket.getInputStream());
//将获得的 LocalSocket 对象 sZygoteSocket 的输出流保存在变量 sZygoteWriter 中
                sZygoteWriter =
                    new BufferedWriter(
                            new OutputStreamWriter(
                                    sZygoteSocket.getOutputStream()),
                            256);
                Log.i("Zygote", "Process: zygote socket opened");
                sPreviousZygoteOpenFailed = false;
                break;
            } catch (IOException ex) {
                if (sZygoteSocket != null) {
                    try {
                        sZygoteSocket.close();
                    } catch (IOException ex2) {
                        Log.e(LOG_TAG,"I/O exception on close after exception",
                                ex2);
                    }
                }
                sZygoteSocket = null;
            }
        }
```

```
    if (sZygoteSocket == null) {
        sPreviousZygoteOpenFailed = true;
        throw new ZygoteStartFailedEx("connect failed");
    }
}
```

在上述代码中，sZygoteSocket是一个LocalSocket类型的成员变量，能够连接Zygote进程中的名为“zygote”的Socket，这个Socket和设备文件“/dev/socket/zygote”相对应。

接下来Zygote进程会在函数runSelectLoop中接收一个创建新应用程序的要求。函数runSelectLoop在文件“frameworks\base\core\java\com\android\internal\os\ZygoteInit.java”中定义，具体实现代码如下所示：

```
private static void runSelectLoop() throws MethodAndArgsCaller {
    ArrayList<FileDescriptor> fds = new ArrayList<FileDescriptor>();
    ArrayList<ZygoteConnection> peers = new ArrayList<ZygoteConnection>();
    FileDescriptor[] fdArray = new FileDescriptor[4];

    fds.add(sServerSocket.getFileDescriptor());
    peers.add(null);

    int loopCount = GC_LOOP_COUNT;
    while (true) {
        int index;

        /*
         * Call gc() before we block in select().
         * It's work that has to be done anyway, and it's better
         * to avoid making every child do it.  It will also
         * madvise() any free memory as a side-effect.
         *
         * Don't call it every time, because walking the entire
         * heap is a lot of overhead to free a few hundred bytes.
         */
        if (loopCount <= 0) {
            gc();
            loopCount = GC_LOOP_COUNT;
        } else {
            loopCount--;
        }

        try {
            fdArray = fds.toArray(fdArray);
            index = selectReadable(fdArray);
        } catch (IOException ex) {
            throw new RuntimeException("Error in select()", ex);
        }

        if (index < 0) {
            throw new RuntimeException("Error in select()");
        } else if (index == 0) {
            ZygoteConnection newPeer = acceptCommandPeer();
            peers.add(newPeer);
            fds.add(newPeer.getFileDesciptor());
        } else {
            boolean done;
            done = peers.get(index).runOnce();
            if (done) {
                peers.remove(index);
                fds.remove(index);
            }
        }
    }
}
```

在上述代码中，会调用函数runOnce处理接收到创建新应用程序的要求。函数runOnce在文件“frameworks\base\core\java\com\android\internal\os\ZygoteConnection.java”中定义，具体实现代码如下所示：

```
boolean runOnce() throws ZygoteInit.MethodAndArgsCaller {

    String args[];
    Arguments parsedArgs = null;
    FileDescriptor[] descriptors;
    try {
      args = readArgumentList();//获得启动要创建应用程序进程的参数
      descriptors = mSocket.getAncillaryFileDescriptors();
    } catch (IOException ex) {
      Log.w(TAG, "IOException on command socket " + ex.getMessage());
      closeSocket();
      return true;
    }

    if (args == null) {
      // EOF reached.
      closeSocket();
      return true;
    }

    /** the stderr of the most recent request, if avail */
    PrintStream newStderr = null;

    if (descriptors != null && descriptors.length >= 3) {
      newStderr = new PrintStream(
            new FileOutputStream(descriptors[2]));
    }

    int pid = -1;
    FileDescriptor childPipeFd = null;
    FileDescriptor serverPipeFd = null;

    try {
      parsedArgs = new Arguments(args);

      applyUidSecurityPolicy(parsedArgs, peer, peerSecurityContext);
      applyRlimitSecurityPolicy(parsedArgs, peer, peerSecurityContext);
      applyCapabilitiesSecurityPolicy(parsedArgs, peer, peerSecurityContext);
      applyInvokeWithSecurityPolicy(parsedArgs, peer, peerSecurityContext);
      applyseInfoSecurityPolicy(parsedArgs, peer, peerSecurityContext);

      applyDebuggerSystemProperty(parsedArgs);
      applyInvokeWithSystemProperty(parsedArgs);

      int[][] rlimits = null;

      if (parsedArgs.rlimits != null) {
        rlimits = parsedArgs.rlimits.toArray(intArray2d);
      }

      if (parsedArgs.runtimeInit && parsedArgs.invokeWith != null) {
        FileDescriptor[] pipeFds = Libcore.os.pipe();
        childPipeFd = pipeFds[1];
        serverPipeFd = pipeFds[0];
        ZygoteInit.setCloseOnExec(serverPipeFd, true);
      }
     //调用函数 forkAndSpecialize 创建应用程序进程
      pid = Zygote.forkAndSpecialize(parsedArgs.uid, parsedArgs.gid, parsedArgs.gids,
            parsedArgs.debugFlags, rlimits, parsedArgs.mountExternal, parsedArgs.seInfo,
            parsedArgs.niceName);
    } catch (IOException ex) {
      logAndPrintError(newStderr, "Exception creating pipe", ex);
    } catch (ErrnoException ex) {
      logAndPrintError(newStderr, "Exception creating pipe", ex);
    } catch (IllegalArgumentException ex) {
```

```
            logAndPrintError(newStderr, "Invalid zygote arguments", ex);
        } catch (ZygoteSecurityException ex) {
            logAndPrintError(newStderr,
                    "Zygote security policy prevents request: ", ex);
        }

        try {
            if (pid == 0) {
                // in child
                IoUtils.closeQuietly(serverPipeFd);
                serverPipeFd = null;
                handleChildProc(parsedArgs, descriptors, childPipeFd, newStderr);

                // should never get here, the child is expected to either
                // throw ZygoteInit.MethodAndArgsCaller or exec().
                return true;
            } else {
                // in parent...pid of < 0 means failure
                IoUtils.closeQuietly(childPipeFd);
                childPipeFd = null;
                return handleParentProc(pid, descriptors, serverPipeFd, parsedArgs);
            }
        } finally {
            IoUtils.closeQuietly(childPipeFd);
            IoUtils.closeQuietly(serverPipeFd);
        }
    }
```

在上述代码中，通过函数 readArgumentList 获得启动要创建应用程序进程的参数，并通过函数 forkAndSpecialize 创建了这个要启动应用程序的进程。其中函数 readArgumentList 在文件"frameworks\ base\core\java\com\android\internal\os\ZygoteConnection.java"中定义，具体实现代码如下所示：

```
    private String[] readArgumentList()
            throws IOException {

        /**
         * See android.os.Process.zygoteSendArgsAndGetPid()
         * Presently the wire format to the zygote process is:
         * a) a count of arguments (argc, in essence)
         * b) a number of newline-separated argument strings equal to count
         *
         * After the zygote process reads these it will write the pid of
         * the child or -1 on failure.
         */

        int argc;

        try {
            String s = mSocketReader.readLine();
            if (s == null) {
                // EOF reached.
                return null;
            }
            argc = Integer.parseInt(s);
        } catch (NumberFormatException ex) {
            Log.e(TAG, "invalid Zygote wire format: non-int at argc");
            throw new IOException("invalid wire format");
        }
        // See bug 1092107: large argc can be used for a DOS attack
        if (argc > MAX_ZYGOTE_ARGC) {
            throw new IOException("max arg count exceeded");
        }

        String[] result = new String[argc];
        for (int i = 0; i < argc; i++) {
```

```
            result[i] = mSocketReader.readLine();
            if (result[i] == null) {
                // We got an unexpected EOF.
                throw new IOException("truncated request");
            }
        }

        return result;
    }
```

函数 forkAndSpecialize 在文件“libcore\dalvik\src\main\java\dalvik\system\Zygote.java”中定义，具体实现代码如下所示：

```
public static int forkAndSpecialize(int uid, int gid, int[] gids, int debugFlags,
        int[][] rlimits, int mountExternal, String seInfo, String niceName) {
    preFork();
    int pid = nativeForkAndSpecialize(
            uid, gid, gids, debugFlags, rlimits, mountExternal, seInfo, niceName);
    postFork();
    return pid;
}
```

在上述代码中，当创建一个进程的子进程时，如果返回值为 0，则表示在新创建的进程中执行。此时需要调用函数 handleChildProc 来启动这个子进程，并在 handleChildProc 中调用函数 zygoteInit 在新创建的应用程序进程中初始化运行库，这样便可以启动一个 Binder 线程池。

9.3.2 启动线程池

在创建新应用程序完毕之前，需要调用类 RuntimeInit 中的函数 nativeZygoteInit 启动一个新的 Binder 线程池，具体启动流程如下所示：

调用类 RuntimeInit 中的函数 nativeZygoteInit，此函数在文件“frameworks\base\core\java\com\android\internal\os\RuntimeInit.java”中定义，对应的实现代码如下所示：

```
public class RuntimeInit {
    private final static String TAG = "AndroidRuntime";
    private final static boolean DEBUG = false;

    /** true if commonInit() has been called */
    private static boolean initialized;

    private static IBinder mApplicationObject;

    private static volatile boolean mCrashing = false;

    private static final native void nativeZygoteInit();
    private static final native void nativeFinishInit();
```

函数 nativeZygoteInit 是一个 JNI 函数，在文件“frameworks\base\core\jni\AndroidRuntime.cpp”中定义实现，对应代码如下所示：

```
static void com_android_internal_os_RuntimeInit_nativeZygoteInit(JNIEnv* env, jobject clazz)
{
    gCurRuntime->onZygoteInit();
}
```

在上述实现代码中，gCurRuntime 是一个全局变量，上述代码用到了 gCurRuntime 的成员函数 onZygoteInit 启动了一个 Binder 线程池。函数 onZygoteInit 在文件“frameworks\base\cmds\app_process\app_main.cpp”中定义，具体实现代码如下所示：

```
virtual void onZygoteInit()
{
    // Re-enable tracing now that we're no longer in Zygote.
    atrace_set_tracing_enabled(true);
```

```
        sp<ProcessState> proc = ProcessState::self();
        ALOGV("App process: starting thread pool.\n");
//调用函数 startThreadPool 启动一个 Binder 线程池
        proc->startThreadPool();
    }
```

在上述代码中，当调用函数 startThreadPool 启动一个 Binder 线程池后，当前应用程序进程就可以通过 Binder 机制和其他进程实现通信。函数 startThreadPool 在文件“frameworks\native\libs\binder\ProcessState.cpp”中定义，具体实现代码如下所示：

```
void ProcessState::startThreadPool()
{
    AutoMutex _l(mLock);
    if (!mThreadPoolStarted) {
        mThreadPoolStarted = true;//默认值为 false
        spawnPooledThread(true);
    }
}
```

在上述代码中，mThreadPoolStarted 的默认值为 false。当第一次调用函数 startThreadPool 时，会在当前进程中启动 Binder 线程池，并将 mThreadPoolStarted 设置为 true，这样做的目的是防止在以后重复启动 Binder 线程池。

9.3.3 创建信息循环

当创建新应用程序进程完毕以后，会调用函数 invokeStaticMain 将类 ActivityThread 的函数 main 设置为新程序的入口函数。当使用函数 main 时，会在当前程序的进程中建立一个信息循环。

接下来首先看函数 invokeStaticMain 的具体实现，此函数在文件“frameworks\base\core\java\com\android\internal\os\RuntimeInit.java”中定义，具体实现代码如下所示：

```
    private static void invokeStaticMain(String className, String[] argv)
            throws ZygoteInit.MethodAndArgsCaller {
        Class<?> cl;

        try {
            cl = Class.forName(className);
        } catch (ClassNotFoundException ex) {
            throw new RuntimeException(
                    "Missing class when invoking static main " + className,
                    ex);
        }

        Method m;
        try {
        //获得静态成员函数 main，并保存在 Method 对象中
            m = cl.getMethod("main", new Class[] { String[].class });
        } catch (NoSuchMethodException ex) {
            throw new RuntimeException(
                    "Missing static main on " + className, ex);
        } catch (SecurityException ex) {
            throw new RuntimeException(
                    "Problem getting static main on " + className, ex);
        }

        int modifiers = m.getModifiers();
        if (! (Modifier.isStatic(modifiers) && Modifier.isPublic(modifiers))) {
            throw new RuntimeException(
                    "Main method is not public and static on " + className);
        }

        /*
         *将 method 对象封装在静态成员函数 main 中，并保存在一个 Method 对象中
         * 将 MethodAndArgsCaller 对象作为异常抛给当前程序进程来处理
```

```
    */
    throw new ZygoteInit.MethodAndArgsCaller(m, argv);
}
```

静态成员函数 main 在文件“frameworks\base\core\java\com\android\internal\os\RuntimeInit.java”中定义，具体实现代码如下所示：

```
public static final void main(String[] argv) {
    if (argv.length == 2 && argv[1].equals("application")) {
        if (DEBUG) Slog.d(TAG, "RuntimeInit: Starting application");
        redirectLogStreams();
    } else {
        if (DEBUG) Slog.d(TAG, "RuntimeInit: Starting tool");
    }

    commonInit();

    /*
     * Now that we're running in interpreted code, call back into native code
     * to run the system.
     */
    nativeFinishInit();

    if (DEBUG) Slog.d(TAG, "Leaving RuntimeInit!");
}
```

在上述代码中，当函数 main 捕获到 MethodAndArgsCaller 异常后，会调用 MethodAndArgsCaller 成员函数 run 进行后面处理。接下来看函数 MethodAndArgsCaller 和 run，这两个函数都是在文件“frameworks\base\core\java\com\android\internal\os\ZygoteInit.java”中定义，具体实现代码如下所示：

```
public static class MethodAndArgsCaller extends Exception
        implements Runnable {
    /** method to call */
    private final Method mMethod;

    /** argument array */
    private final String[] mArgs;

    public MethodAndArgsCaller(Method method, String[] args) {
        mMethod = method;
        mArgs = args;
    }

    public void run() {
        try {
       //执行函数invoke，这样就执行了类android.app.ActivityThread中的函数main
            mMethod.invoke(null, new Object[] { mArgs });
        } catch (IllegalAccessException ex) {
            throw new RuntimeException(ex);
        } catch (InvocationTargetException ex) {
            Throwable cause = ex.getCause();
            if (cause instanceof RuntimeException) {
                throw (RuntimeException) cause;
            } else if (cause instanceof Error) {
                throw (Error) cause;
            }
            throw new RuntimeException(ex);
        }
    }
  }
}
```

在上述代码中，变量 mMethod 和 mArgs 是在构造异常对象时传递进来的，其中变量 mMethod 和类 android.app.ActivityThread 中的函数 main 相对应。

第 10 章　Dalvik VM 运作流程详解

经过本书前面内容的学习，已经了解了 Android 虚拟机 Dalvik 的基本知识和具体结构。在本章的内容中，将详细讲解 Dalvik VM 运作流程的基本知识，详细讲解每一个步骤的具体实现过程，为读者步入本书后面知识的学习打下基础。

10.1 Dalvik VM 相关的可执行程序

在 Android 源码中，会发现好几处和 Dalvik 这个概念相关的可执行程序，正确区分这些可执行程序将有助于理解 Framework 内部结构。这些可执行程序的名称和源码路径如表 10-1 所示。

表 10-1　和虚拟机相关的源码

名　称	源 码 路 径
dalvikvm	dalvik/dalvikvm
dvz	dalvik/dvz
app_process	frameworks/base/cmds/app_process

在本节的内容中，将分别介绍这些可执行程序的作用。

10.1.1　dalvikvm、dvz 和 app_process 简介

（1）dalvikvm。

当运行 Java 程序时，都是由一个虚拟机来解释 Java 的字节码，虚拟机将这些字节码翻译成本地 CPU 的指令码，然后才被执行。对于 Java 程序来说，由虚拟机负责解释并执行工作。而对于 Linux 系统而言，这个进程只是一个普通的进程，与一个只有一行代码的“Hello World”可执行程序没有什么本质区别。所以说，启动一个虚拟机的方法与启动一个可执行程序的方法是相同的。都是先在命令行下输入可执行程序的名称，然后在参数中指定要执行的 Java 类。执行 dalvikvm 的命令格式如下所示：

```
dalvikvm -cp 类路径 类名
```

由此可以看到，dalvikvm 的作用就像在 PC 上执行 Java 程序一样，提供了一个运行环境。

（2）dvz。

在 Dalvik 虚拟机中，dvz 的作用是从 Zygote 进程中孕育出一个新的进程，新的进程也是一个 Dalvik 虚拟机。dvz 进程与 dalvikvm 启动的虚拟机进程相比，区别在于该进程中已经预装了 Framework 的大部分类和资源。使用 dvz 的语法格式如下：

```
dvz  -classpath 包名称 类名
```

一个 APK 的入口类是 ActivityThread 类。因为类 Activity 仅仅是被回调的类，所以不可以通过类 Activity 来启动一个 APK，dvz 工具仅仅用于 Framework 开发过程的调试。

（3）app_process。

前面讲解的 dalvikvm 和 dvz 是通用的两个工具，然而 Framework 在启动时需要加载并运行如

下两个特定 Java 类：

- ZygoteInit.java；
- SystemServer.java。

为了便于使用，系统才提供了一个 app_process 进程，该进程会自动运行这两个类，从这个角度来讲，app_process 的本质就是使用 dalvikvm 启动 ZygoteInit.java，并在启动后加载 Framework 中的大部分类和资源。

10.1.2 对比 app_process 和 dalvikvm 的执行过程

在接下来的内容中，将详细讲解 app_process 和 dalvikvm 的主要执行过程。

（1）首先看 dalvikvm，其源码在文件“dalvik/dalvikvm/Main.c”中，该源码中的关键代码有两点：

```
    /*
     * 第 1 点：通过如下代码创建一个 vm 对象
     */
    if (JNI_CreateJavaVM(&vm, &env, &initArgs) < 0) {
        fprintf(stderr, "Dalvik VM init failed (check log file)\n");
        goto bail;
    }

    /*
     * Make sure they provided a class name.  We do this after VM init
     * so that things like "-Xrunjdwp:help" have the opportunity to emit
     * a usage statement.
     */
    if (argIdx == argc) {
        fprintf(stderr, "Dalvik VM requires a class name\n");
        goto bail;
    }

    /*
     * We want to call main() with a String array with our arguments in it.
     * Create an array and populate it.  Note argv[0] is not included.
     */
    jobjectArray strArray;
    strArray = createStringArray(env, &argv[argIdx+1], argc-argIdx-1);
    if (strArray == NULL)
        goto bail;

    /*
     * Find [class].main(String[]).
     */
    jclass startClass;
    jmethodID startMeth;
    char* cp;

    /* convert "com.android.Blah" to "com/android/Blah" */
    slashClass = strdup(argv[argIdx]);
    for (cp = slashClass; *cp != '\0'; cp++)
        if (*cp == '.')
            *cp = '/';
/*第 2 点：创建好 JavaVm 对象后，使用该对象去加载指定的类*/
    startClass = (*env)->FindClass(env, slashClass);
    if (startClass == NULL) {
        fprintf(stderr, "Dalvik VM unable to locate class '%s'\n", slashClass);
        goto bail;
    }

    startMeth = (*env)->GetStaticMethodID(env, startClass,
                    "main", "([Ljava/lang/String;)V");
    if (startMeth == NULL) {
        fprintf(stderr, "Dalvik VM unable to find static main(String[]) in '%s'\n",
            slashClass);
```

```
        goto bail;
    }

    /*
     * Make sure the method is public.  JNI doesn't prevent us from calling
     * a private method, so we have to check it explicitly.
     */
    if (!methodIsPublic(env, startClass, startMeth))
        goto bail;

    /*
     * Invoke main().
     */
    (*env)->CallStaticVoidMethod(env, startClass, startMeth, strArray);

    if (!(*env)->ExceptionCheck(env))
        result = 0;
```

在上述第一点关键代码处，该段代码通过调用 JNI_CreateJavaVM()，并同时创建了 JavaVm 对象和 JNIEnv 对象，这两个对象的定义如下。

```
JNIEnvExt* pEnv = NULL;
JavaVMExt* pVM = NULL;
```

该函数的参数是“指针的指针“类型，其原型如下所示：

```
jint JNI_CreateJavaVM(JavaVM** p_vm, JNIEnv** p_env, void* vm_args)
```

在上述第二点关键代码处，首先调用 FindClass()找到指定的 class 文件，然后调用 GetStaticMethodID()找到 main()函数，最后调用 CallStaticVoidMethod 执行该 main()函数。

（2）接下来看 app_process 中是如何创建虚拟机并执行指定的 class 文件的，这部分的源代码在文件“frameworks/base/cmds/app_process/app_main.cpp”中定义，该文件中的关键代码有如下两点。

- 第一点：先创建一个 AppRuntime 对象，具体代码如下所示：

```
virtual void onVmCreated(JNIEnv* env)
{
    if (mClassName == NULL) {
        return; // Zygote. Nothing to do here.
    }
    char* slashClassName = toSlashClassName(mClassName);
    mClass = env->FindClass(slashClassName);
    if (mClass == NULL) {
        ALOGE("ERROR: could not find class '%s'\n", mClassName);
    }
    free(slashClassName);

    mClass = reinterpret_cast<jclass>(env->NewGlobalRef(mClass));
}
```

- 第二点：调用 runtime 的 start()方法启动指定的 class，对应代码如下所示：

```
if (niceName && *niceName) {
    setArgv0(argv0, niceName);
    set_process_name(niceName);
}
runtime.mParentDir = parentDir;
if (zygote) {
    runtime.start("com.android.internal.os.ZygoteInit",
            startSystemServer ? "start-system-server" : "");
} else if (className) {
    // Remainder of args get passed to startup class main()
    runtime.mClassName = className;
    runtime.mArgC = argc - i;
    runtime.mArgV = argv + i;
    runtime.start("com.android.internal.os.RuntimeInit",
            application ? "application" : "tool");
} else {
```

```
        fprintf(stderr, "Error: no class name or --zygote supplied.\n");
        app_usage();
        LOG_ALWAYS_FATAL("app_process: no class name or --zygote supplied.");
        return 10;
    }
```

在系统中只有在文件 init.rc 中使用了 app_process，因为在使用时参数包含了“--zygote”和“start-system-server”，所以这里仅分析包含这两个参数的情况。start()方式是类 AppRuntime 的成员函数，而 AppRuntime 是在该文件中定义的一个应用类，其父类是 AndroidRuntime，该类的实现在文件“frameworkds/base/core/jni/AndroidRuntime.cpp”中。

在函数 start()中，首先调用函数 startVm()创建一个 vm 对象，然后就可以和 dalvikvm 一样先找到 Class()，再执行 class 中函数 main()，最后使用函数 startVm()创建 vm 对象。

由上述过程可以看出，app_process 和 dalvikvm 在本质上是相同的。两者唯一的区别是，app_process 可以指定一些特别的参数，这些参数有利于 Framework 启动特定的类，并进行一些特别的系统环境参数设置。

注意　有关函数 startVm()的实现过程，将在本书后面的内容中进行讲解。

10.2 初始化 Dalvik 虚拟机

在运行 Dalvik 虚拟机伊始，最先进行的是初始化工作，此工作的核心实现文件是 Init.c 和 Init.c。在本节的内容中，将详细讲解 Dalvik 虚拟机的初始化过程。

10.2.1 开始虚拟机的准备工作

在 Dalvik 虚拟机的初始化过程中，使用函数 dvmStartup()实现所有开始虚拟机的准备工作，此函数的具体实现代码如下所示：

```
int dvmStartup(int argc, const char* const argv[], bool ignoreUnrecognized,
    JNIEnv* pEnv)
{
    int i, cc;
    assert(gDvm.initializing);
    LOGV("VM init args (%d):\n", argc);
    for (i = 0; i < argc; i++)
        LOGV("  %d: '%s'\n", i, argv[i]);
    setCommandLineDefaults();
    /* prep properties storage */
    if (!dvmPropertiesStartup(argc))
        goto fail;
    /*
     * Process the option flags (if any).
     */
    cc = dvmProcessOptions(argc, argv, ignoreUnrecognized);
    if (cc != 0) {
        if (cc < 0) {
            dvmFprintf(stderr, "\n");
            dvmUsage("dalvikvm");
        }
        goto fail;
    }
```

10.2.2 初始化跟踪显示系统

在 Dalvik 虚拟机的初始化过程中，使用函数 dvmAllocTrackerStartup()初始化跟踪显示系统，跟踪系统主要用生成调试系统的数据包。此函数的具体实现代码如下所示：

```
bool dvmAllocTrackerStartup(void)
{
    /* prep locks */
    dvmInitMutex(&gDvm.allocTrackerLock);
    /* initialized when enabled by DDMS */
    assert(gDvm.allocRecords == NULL);
    return true;
}
```

上述函数的实现保存在文件 AllocTracker.c 中。

10.2.3 初始化垃圾回收器

在 Dalvik 虚拟机的初始化过程中，使用函数 dvmGcStartup()初始化垃圾回收器，此函数的具体实现代码如下所示：

```
bool dvmGcStartup(void)
{
    dvmInitMutex(&gDvm.gcHeapLock);
    return dvmHeapStartup();
}
```

10.2.4 初始化线程列表和主线程环境参数

在 Dalvik 虚拟机的初始化过程中，使用函数 dvmThreadStartup()初始化线程列表和主线程环境参数，此函数的具体实现代码如下所示：

```
bool dvmThreadStartup(void)
{
    Thread* thread;

    /* allocate a TLS slot */
    if (pthread_key_create(&gDvm.pthreadKeySelf, threadExitCheck) != 0) {
        LOGE("ERROR: pthread_key_create failed\n");
        return false;
    }

    /* test our pthread lib */
    if (pthread_getspecific(gDvm.pthreadKeySelf) != NULL)
        LOGW("WARNING: newly-created pthread TLS slot is not NULL\n");

    /* prep thread-related locks and conditions */
    dvmInitMutex(&gDvm.threadListLock);
    pthread_cond_init(&gDvm.threadStartCond, NULL);
    //dvmInitMutex(&gDvm.vmExitLock);
    pthread_cond_init(&gDvm.vmExitCond, NULL);
    dvmInitMutex(&gDvm._threadSuspendLock);
    dvmInitMutex(&gDvm.threadSuspendCountLock);
    pthread_cond_init(&gDvm.threadSuspendCountCond, NULL);
#ifdef WITH_DEADLOCK_PREDICTION
    dvmInitMutex(&gDvm.deadlockHistoryLock);
#endif

    /*
     * Dedicated monitor for Thread.sleep().
     * TODO: change this to an Object* so we don't have to expose this
     * call, and we interact better with JDWP monitor calls.  Requires
     * deferring the object creation to much later (e.g. final "main"
     * thread prep) or until first use.
     */
    gDvm.threadSleepMon = dvmCreateMonitor(NULL);

    gDvm.threadIdMap = dvmAllocBitVector(kMaxThreadId, false);

    thread = allocThread(gDvm.stackSize);
    if (thread == NULL)
        return false;
```

```
    /* switch mode for when we run initializers */
    thread->status = THREAD_RUNNING;

    /*
     * We need to assign the threadId early so we can lock/notify
     * object monitors.  We'll set the "threadObj" field later.
     */
    prepareThread(thread);
    gDvm.threadList = thread;

#ifdef COUNT_PRECISE_METHODS
    gDvm.preciseMethods = dvmPointerSetAlloc(200);
#endif
    return true;
}
```

上述函数的实现保存在文件 Thread.c 中。

10.2.5　分配内部操作方法的表格内存

在 Dalvik 虚拟机的初始化过程中，使用函数 dvmInlineNativeStartup()分配内部操作方法的表格内存，此函数的具体实现代码如下所示：

```
bool dvmInlineNativeStartup(void)
{
#ifdef WITH_PROFILER
    gDvm.inlinedMethods =
        (Method**) calloc(NELEM(gDvmInlineOpsTable), sizeof(Method*));
    if (gDvm.inlinedMethods == NULL)
        return false;
#endif
    return true;
}
```

上述函数的实现保存在文件 InlineNative.c 中。

10.2.6　初始化虚拟机的指令码相关的内容

在 Dalvik 虚拟机的初始化过程中，使用函数 dvmVerificationStartup()初始化虚拟机的指令码相关的内容，以便检查指令是否正确。此函数的具体实现代码如下所示：

```
bool dvmVerificationStartup(void)
{
    gDvm.instrWidth = dexCreateInstrWidthTable();
    gDvm.instrFormat = dexCreateInstrFormatTable();
    gDvm.instrFlags = dexCreateInstrFlagsTable();
    if (gDvm.instrWidth == NULL || gDvm.instrFormat == NULL ||
        gDvm.instrFlags == NULL)
    {
        LOGE("Unable to create instruction tables\n");
        return false;
    }
    return true;
}
```

上述函数的实现保存在文件“analysis\DexVerify.c”中。

10.2.7　分配指令寄存器状态的内存

在 Dalvik 虚拟机的初始化过程中，使用函数 dvmRegisterMapStartup()分配指令寄存器状态的内存。此函数的具体实现代码如下所示：

```
bool dvmRegisterMapStartup(void)
```

```
{
#ifdef REGISTER_MAP_STATS
    MapStats* pStats = calloc(1, sizeof(MapStats));
    gDvm.registerMapStats = pStats;
#endif
    return true;
}
```

上述函数的实现保存在文件“analysis\RegisterMap.c”中。

10.2.8 分配指令寄存器状态的内存和最基本用的 Java 库

在 Dalvik 虚拟机的初始化过程中，使用函数 dvmInstanceofStartup()分配虚拟机使用的缓存。此函数的具体实现代码如下所示：

```
bool dvmInstanceofStartup(void)
{
    gDvm.instanceofCache = dvmAllocAtomicCache(INSTANCEOF_CACHE_SIZE);
    if (gDvm.instanceofCache == NULL)
        return false;
    return true;
}
```

上述函数的实现保存在文件“oo\TypeCheck.c”中。

在 Dalvik 虚拟机的初始化过程中，使用函数 dvmClassStartup()初始化虚拟机最基本用的 Java 库。此函数的具体实现代码如下所示：

```
bool dvmClassStartup(void)
{
    ClassObject* unlinkedClass;
    /* make this a requirement -- don't currently support dirs in path */
    if (strcmp(gDvm.bootClassPathStr, ".") == 0) {
        LOGE("ERROR: must specify non-'.' bootclasspath\n");
        return false;
    }
    gDvm.loadedClasses =
        dvmHashTableCreate(256, (HashFreeFunc) dvmFreeClassInnards);
    gDvm.pBootLoaderAlloc = dvmLinearAllocCreate(NULL);
    if (gDvm.pBootLoaderAlloc == NULL)
        return false;
    if (false) {
        linearAllocTests();
        exit(0);
    }
    /*
     * Class serial number.  We start with a high value to make it distinct
     * in binary dumps (e.g. hprof).
     */
    gDvm.classSerialNumber = INITIAL_CLASS_SERIAL_NUMBER;
    /* Set up the table we'll use for tracking initiating loaders for
     * early classes.
     * If it's NULL, we just fall back to the InitiatingLoaderList in the
     * ClassObject, so it's not fatal to fail this allocation.
     */
    gDvm.initiatingLoaderList =
        calloc(ZYGOTE_CLASS_CUTOFF, sizeof(InitiatingLoaderList));
    /* This placeholder class is used while a ClassObject is
     * loading/linking so those not in the know can still say
     * "obj->clazz->...".
     */
    unlinkedClass = &gDvm.unlinkedJavaLangClassObject;
    memset(unlinkedClass, 0, sizeof(*unlinkedClass));
    /* Set obj->clazz to NULL so anyone who gets too interested
     * in the fake class will crash.
     */
    DVM_OBJECT_INIT(&unlinkedClass->obj, NULL);
    unlinkedClass->descriptor = "!unlinkedClass";
```

```
    dvmSetClassSerialNumber(unlinkedClass);
    gDvm.unlinkedJavaLangClass = unlinkedClass;
    /*
     * Process the bootstrap class path.  This means opening the specified
     * DEX or Jar files and possibly running them through the optimizer.
     */
    assert(gDvm.bootClassPath == NULL);
    processClassPath(gDvm.bootClassPathStr, true);
    if (gDvm.bootClassPath == NULL)
        return false;
    return true;
}
```

上述函数的实现保存在文件“oo\Class.c”中。

10.2.9　初始化使用的 Java 类库线程类

在 Dalvik 虚拟机的初始化过程中，使用函数 dvmThreadObjStartup()初始化虚拟机进一步使用的 Java 类库线程类。此函数的具体实现代码如下所示：

```
bool dvmThreadObjStartup(void)
{
    /*
     * Cache the locations of these classes.  It's likely that we're the
     * first to reference them, so they're being loaded now.
     */
    gDvm.classJavaLangThread =
        dvmFindSystemClassNoInit("Ljava/lang/Thread;");
    gDvm.classJavaLangVMThread =
        dvmFindSystemClassNoInit("Ljava/lang/VMThread;");
    gDvm.classJavaLangThreadGroup =
        dvmFindSystemClassNoInit("Ljava/lang/ThreadGroup;");
    if (gDvm.classJavaLangThread == NULL ||
        gDvm.classJavaLangThreadGroup == NULL ||
        gDvm.classJavaLangThreadGroup == NULL)
    {
        LOGE("Could not find one or more essential thread classes\n");
        return false;
    }

    /*
     * Cache field offsets.  This makes things a little faster, at the
     * expense of hard-coding non-public field names into the VM.
     */
    gDvm.offJavaLangThread_vmThread =
        dvmFindFieldOffset(gDvm.classJavaLangThread,
            "vmThread", "Ljava/lang/VMThread;");
    gDvm.offJavaLangThread_group =
        dvmFindFieldOffset(gDvm.classJavaLangThread,
            "group", "Ljava/lang/ThreadGroup;");
    gDvm.offJavaLangThread_daemon =
        dvmFindFieldOffset(gDvm.classJavaLangThread, "daemon", "Z");
    gDvm.offJavaLangThread_name =
        dvmFindFieldOffset(gDvm.classJavaLangThread,
            "name", "Ljava/lang/String;");
    gDvm.offJavaLangThread_priority =
        dvmFindFieldOffset(gDvm.classJavaLangThread, "priority", "I");

    if (gDvm.offJavaLangThread_vmThread < 0 ||
        gDvm.offJavaLangThread_group < 0 ||
        gDvm.offJavaLangThread_daemon < 0 ||
        gDvm.offJavaLangThread_name < 0 ||
        gDvm.offJavaLangThread_priority < 0)
    {
        LOGE("Unable to find all fields in java.lang.Thread\n");
        return false;
    }
```

```
    gDvm.offJavaLangVMThread_thread =
        dvmFindFieldOffset(gDvm.classJavaLangVMThread,
            "thread", "Ljava/lang/Thread;");
    gDvm.offJavaLangVMThread_vmData =
        dvmFindFieldOffset(gDvm.classJavaLangVMThread, "vmData", "I");
    if (gDvm.offJavaLangVMThread_thread < 0 ||
        gDvm.offJavaLangVMThread_vmData < 0)
    {
        LOGE("Unable to find all fields in java.lang.VMThread\n");
        return false;
    }

    /*
     * Cache the vtable offset for "run()".
     *
     * We don't want to keep the Method* because then we won't find see
     * methods defined in subclasses.
     */
    Method* meth;
    meth = dvmFindVirtualMethodByDescriptor(gDvm.classJavaLangThread, "run", "()V");
    if (meth == NULL) {
        LOGE("Unable to find run() in java.lang.Thread\n");
        return false;
    }
    gDvm.voffJavaLangThread_run = meth->methodIndex;

    /*
     * Cache vtable offsets for ThreadGroup methods.
     */
    meth = dvmFindVirtualMethodByDescriptor(gDvm.classJavaLangThreadGroup,
        "removeThread", "(Ljava/lang/Thread;)V");
    if (meth == NULL) {
        LOGE("Unable to find removeThread(Thread) in java.lang.ThreadGroup\n");
        return false;
    }
    gDvm.voffJavaLangThreadGroup_removeThread = meth->methodIndex;

    return true;
}
```

上述函数的实现保存在文件 Thread.c 中。

10.2.10 初始化虚拟机使用的异常 Java 类库

在 Dalvik 虚拟机的初始化过程中，使用函数 dvmExceptionStartup()初始化虚拟机使用的异常 Java 类库。此函数的具体实现代码如下所示：

```
bool dvmExceptionStartup(void)
{
    gDvm.classJavaLangThrowable =
        dvmFindSystemClassNoInit("Ljava/lang/Throwable;");
    gDvm.classJavaLangRuntimeException =
        dvmFindSystemClassNoInit("Ljava/lang/RuntimeException;");
    gDvm.classJavaLangError =
        dvmFindSystemClassNoInit("Ljava/lang/Error;");
    gDvm.classJavaLangStackTraceElement =
        dvmFindSystemClassNoInit("Ljava/lang/StackTraceElement;");
    gDvm.classJavaLangStackTraceElementArray =
        dvmFindArrayClass("[Ljava/lang/StackTraceElement;", NULL);
    if (gDvm.classJavaLangThrowable == NULL ||
        gDvm.classJavaLangStackTraceElement == NULL ||
        gDvm.classJavaLangStackTraceElementArray == NULL)
    {
        LOGE("Could not find one or more essential exception classes\n");
        return false;
    }

    /*
     * Find the constructor.  Note that, unlike other saved method lookups,
```

```
     * we're using a Method* instead of a vtable offset.  This is because
     * constructors don't have vtable offsets.  (Also, since we're creating
     * the object in question, it's impossible for anyone to sub-class it.)
     */
    Method* meth;
    meth = dvmFindDirectMethodByDescriptor(gDvm.classJavaLangStackTraceElement,
        "<init>", "(Ljava/lang/String;Ljava/lang/String;Ljava/lang/String;I)V");
    if (meth == NULL) {
        LOGE("Unable to find constructor for StackTraceElement\n");
        return false;
    }
    gDvm.methJavaLangStackTraceElement_init = meth;

    /* grab an offset for the stackData field */
    gDvm.offJavaLangThrowable_stackState =
        dvmFindFieldOffset(gDvm.classJavaLangThrowable,
            "stackState", "Ljava/lang/Object;");
    if (gDvm.offJavaLangThrowable_stackState < 0) {
        LOGE("Unable to find Throwable.stackState\n");
        return false;
    }

    /* and one for the message field, in case we want to show it */
    gDvm.offJavaLangThrowable_message =
        dvmFindFieldOffset(gDvm.classJavaLangThrowable,
            "detailMessage", "Ljava/lang/String;");
    if (gDvm.offJavaLangThrowable_message < 0) {
        LOGE("Unable to find Throwable.detailMessage\n");
        return false;
    }

    /* and one for the cause field, just 'cause */
    gDvm.offJavaLangThrowable_cause =
        dvmFindFieldOffset(gDvm.classJavaLangThrowable,
            "cause", "Ljava/lang/Throwable;");
    if (gDvm.offJavaLangThrowable_cause < 0) {
        LOGE("Unable to find Throwable.cause\n");
        return false;
    }
    return true;
}
```

上述函数的实现保存在文件 Exception.c 中。

10.2.11　初始化其他对象

除了本章前面介绍的初始化对象外，Android 虚拟机还能够初始化如下所示的对象。

（1）初始化虚拟机解释器使用的字符串哈希表。

在 Dalvik 虚拟机的初始化过程中，使用函数 dvmStringInternStartup()初始化虚拟机解释器使用的字符串哈希表。此函数的具体实现代码如下所示：

```
bool dvmStringInternStartup(void)
{
    gDvm.internedStrings = dvmHashTableCreate(256, NULL);
    if (gDvm.internedStrings == NULL)
        return false;
    return true;
}
```

上述函数的实现保存在文件 Intern.c 中。

（2）初始化本地方法库的表。

在 Dalvik 虚拟机的初始化过程中，使用函数 dvmNativeStartup()来初始化本地方法库的表。此函数的具体实现代码如下所示：

```
bool dvmNativeStartup(void)
{
```

```
    gDvm.nativeLibs = dvmHashTableCreate(4, freeSharedLibEntry);
    if (gDvm.nativeLibs == NULL)
        return false;
    return true;
}
```

上述函数的实现保存在文件 Native.c 中。

（3）初始化内部本地方法。

在 Dalvik 虚拟机的初始化过程中，使用函数 dvmInternalNativeStartup()初始化内部本地方法，建立哈希表，方便快速查找到。此函数的具体实现代码如下所示：

```
bool dvmInternalNativeStartup()
{
    DalvikNativeClass* classPtr = gDvmNativeMethodSet;
    while (classPtr->classDescriptor != NULL) {
        classPtr->classDescriptorHash =
            dvmComputeUtf8Hash(classPtr->classDescriptor);
        classPtr++;
    }
    gDvm.userDexFiles = dvmHashTableCreate(2, dvmFreeDexOrJar);
    if (gDvm.userDexFiles == NULL)
        return false;
    return true;
}
```

上述函数的实现保存在文件“native/InternalNative.cpp”中。

（4）初始化 JNI 调用表。

在 Dalvik 虚拟机的初始化过程中，使用函数 dvmJniStartup()初始化 JNI 调用表，以便快速找到本地方法调用的入口。此函数的具体实现代码如下所示：

```
bool dvmJniStartup(void)
{
#ifdef USE_INDIRECT_REF
    if (!dvmInitIndirectRefTable(&gDvm.jniGlobalRefTable,
            kGlobalRefsTableInitialSize, kGlobalRefsTableMaxSize,
            kIndirectKindGlobal))
        return false;
#else
    if (!dvmInitReferenceTable(&gDvm.jniGlobalRefTable,
            kGlobalRefsTableInitialSize, kGlobalRefsTableMaxSize))
        return false;
#endif
    dvmInitMutex(&gDvm.jniGlobalRefLock);
    gDvm.jniGlobalRefLoMark = 0;
    gDvm.jniGlobalRefHiMark = kGrefWaterInterval * 2;
    if (!dvmInitReferenceTable(&gDvm.jniPinRefTable,
            kPinTableInitialSize, kPinTableMaxSize))
        return false;
    dvmInitMutex(&gDvm.jniPinRefLock);
    /*
     * Look up and cache pointers to some direct buffer classes, fields,
     * and methods.
     */
    Method* meth;
    ClassObject* platformAddressClass =
        dvmFindSystemClassNoInit("Lorg/apache/harmony/luni/platform/PlatformAddress;");
    ClassObject* platformAddressFactoryClass =
        dvmFindSystemClassNoInit("Lorg/apache/harmony/luni/platform/PlatformAddressFactory;");
    ClassObject* directBufferClass =
        dvmFindSystemClassNoInit("Lorg/apache/harmony/nio/internal/DirectBuffer;");
    ClassObject* readWriteBufferClass =
        dvmFindSystemClassNoInit("Ljava/nio/ReadWriteDirectByteBuffer;");
    ClassObject* bufferClass =
        dvmFindSystemClassNoInit("Ljava/nio/Buffer;");
    if (platformAddressClass == NULL || platformAddressFactoryClass == NULL ||
        directBufferClass == NULL || readWriteBufferClass == NULL ||
```

```
        bufferClass == NULL)
    {
        LOGE("Unable to find internal direct buffer classes\n");
        return false;
    }
    gDvm.classJavaNioReadWriteDirectByteBuffer = readWriteBufferClass;
    gDvm.classOrgApacheHarmonyNioInternalDirectBuffer = directBufferClass;
    /* need a global reference for extended CheckJNI tests */
    gDvm.jclassOrgApacheHarmonyNioInternalDirectBuffer =
        addGlobalReference((Object*) directBufferClass);
    /*
     * We need a Method* here rather than a vtable offset, because
     * DirectBuffer is an interface class.
     */
    meth = dvmFindVirtualMethodByDescriptor(
            gDvm.classOrgApacheHarmonyNioInternalDirectBuffer,
            "getEffectiveAddress",
            "()Lorg/apache/harmony/luni/platform/PlatformAddress;");
    if (meth == NULL) {
        LOGE("Unable to find PlatformAddress.getEffectiveAddress\n");
        return false;
    }
    gDvm.methOrgApacheHarmonyNioInternalDirectBuffer_getEffectiveAddress = meth;
    meth = dvmFindVirtualMethodByDescriptor(platformAddressClass,
            "toLong", "()J");
    if (meth == NULL) {
        LOGE("Unable to find PlatformAddress.toLong\n");
        return false;
    }
    gDvm.voffOrgApacheHarmonyLuniPlatformPlatformAddress_toLong =
        meth->methodIndex;
    meth = dvmFindDirectMethodByDescriptor(platformAddressFactoryClass,
            "on",
            "(I)Lorg/apache/harmony/luni/platform/PlatformAddress;");
    if (meth == NULL) {
        LOGE("Unable to find PlatformAddressFactory.on\n");
        return false;
    }
    gDvm.methOrgApacheHarmonyLuniPlatformPlatformAddress_on = meth;
    meth = dvmFindDirectMethodByDescriptor(readWriteBufferClass,
            "<init>",
            "(Lorg/apache/harmony/luni/platform/PlatformAddress;II)V");
    if (meth == NULL) {
        LOGE("Unable to find ReadWriteDirectByteBuffer.<init>\n");
        return false;
    }
    gDvm.methJavaNioReadWriteDirectByteBuffer_init = meth;
    gDvm.offOrgApacheHarmonyLuniPlatformPlatformAddress_osaddr =
        dvmFindFieldOffset(platformAddressClass, "osaddr", "I");
    if (gDvm.offOrgApacheHarmonyLuniPlatformPlatformAddress_osaddr < 0) {
        LOGE("Unable to find PlatformAddress.osaddr\n");
        return false;
    }
    gDvm.offJavaNioBuffer_capacity =
        dvmFindFieldOffset(bufferClass, "capacity", "I");
    if (gDvm.offJavaNioBuffer_capacity < 0) {
        LOGE("Unable to find Buffer.capacity\n");
        return false;
    }
    gDvm.offJavaNioBuffer_effectiveDirectAddress =
        dvmFindFieldOffset(bufferClass, "effectiveDirectAddress", "I");
    if (gDvm.offJavaNioBuffer_effectiveDirectAddress < 0) {
        LOGE("Unable to find Buffer.effectiveDirectAddress\n");
        return false;
    }
    return true;
}
```

上述函数的实现保存在文件 Jni.c 中。

（5）缓存 Java 类库里的反射类。

在 Dalvik 虚拟机的初始化过程中，使用函数 dvmReflectStartup()缓存 Java 类库里的反射类。此函数的具体实现代码如下所示：

```
bool dvmReflectStartup(void)
{
    gDvm.classJavaLangReflectAccessibleObject =
        dvmFindSystemClassNoInit("Ljava/lang/reflect/AccessibleObject;");
    gDvm.classJavaLangReflectConstructor =
        dvmFindSystemClassNoInit("Ljava/lang/reflect/Constructor;");
    gDvm.classJavaLangReflectConstructorArray =
        dvmFindArrayClass("[Ljava/lang/reflect/Constructor;", NULL);
    gDvm.classJavaLangReflectField =
        dvmFindSystemClassNoInit("Ljava/lang/reflect/Field;");
    gDvm.classJavaLangReflectFieldArray =
        dvmFindArrayClass("[Ljava/lang/reflect/Field;", NULL);
    gDvm.classJavaLangReflectMethod =
        dvmFindSystemClassNoInit("Ljava/lang/reflect/Method;");
    gDvm.classJavaLangReflectMethodArray =
        dvmFindArrayClass("[Ljava/lang/reflect/Method;", NULL);
    gDvm.classJavaLangReflectProxy =
        dvmFindSystemClassNoInit("Ljava/lang/reflect/Proxy;");
    if (gDvm.classJavaLangReflectAccessibleObject == NULL ||
        gDvm.classJavaLangReflectConstructor == NULL ||
        gDvm.classJavaLangReflectConstructorArray == NULL ||
        gDvm.classJavaLangReflectField == NULL ||
        gDvm.classJavaLangReflectFieldArray == NULL ||
        gDvm.classJavaLangReflectMethod == NULL ||
        gDvm.classJavaLangReflectMethodArray == NULL ||
        gDvm.classJavaLangReflectProxy == NULL)
    {
        LOGE("Could not find one or more reflection classes\n");
        return false;
    }

    gDvm.methJavaLangReflectConstructor_init =
        dvmFindDirectMethodByDescriptor(gDvm.classJavaLangReflectConstructor, "<init>",
        "(Ljava/lang/Class;[Ljava/lang/Class;[Ljava/lang/Class;I)V");
    gDvm.methJavaLangReflectField_init =
        dvmFindDirectMethodByDescriptor(gDvm.classJavaLangReflectField, "<init>",
        "(Ljava/lang/Class;Ljava/lang/Class;Ljava/lang/String;I)V");
    gDvm.methJavaLangReflectMethod_init =
        dvmFindDirectMethodByDescriptor(gDvm.classJavaLangReflectMethod, "<init>",
"(Ljava/lang/Class;[Ljava/lang/Class;[Ljava/lang/Class;Ljava/lang/Class;Ljava/lang/String;I)V");
    if (gDvm.methJavaLangReflectConstructor_init == NULL ||
        gDvm.methJavaLangReflectField_init == NULL ||
        gDvm.methJavaLangReflectMethod_init == NULL)
    {
        LOGE("Could not find reflection constructors\n");
        return false;
    }

    gDvm.classJavaLangClassArray =
        dvmFindArrayClass("[Ljava/lang/Class;", NULL);
    gDvm.classJavaLangObjectArray =
        dvmFindArrayClass("[Ljava/lang/Object;", NULL);
    if (gDvm.classJavaLangClassArray == NULL ||
        gDvm.classJavaLangObjectArray == NULL)
    {
        LOGE("Could not find class-array or object-array class\n");
        return false;
    }

    gDvm.offJavaLangReflectAccessibleObject_flag =
        dvmFindFieldOffset(gDvm.classJavaLangReflectAccessibleObject, "flag",
          "Z");
```

```
    gDvm.offJavaLangReflectConstructor_slot =
        dvmFindFieldOffset(gDvm.classJavaLangReflectConstructor, "slot", "I");
    gDvm.offJavaLangReflectConstructor_declClass =
        dvmFindFieldOffset(gDvm.classJavaLangReflectConstructor,
            "declaringClass", "Ljava/lang/Class;");

    gDvm.offJavaLangReflectField_slot =
        dvmFindFieldOffset(gDvm.classJavaLangReflectField, "slot", "I");
    gDvm.offJavaLangReflectField_declClass =
        dvmFindFieldOffset(gDvm.classJavaLangReflectField,
            "declaringClass", "Ljava/lang/Class;");

    gDvm.offJavaLangReflectMethod_slot =
        dvmFindFieldOffset(gDvm.classJavaLangReflectMethod, "slot", "I");
    gDvm.offJavaLangReflectMethod_declClass =
        dvmFindFieldOffset(gDvm.classJavaLangReflectMethod,
            "declaringClass", "Ljava/lang/Class;");

    if (gDvm.offJavaLangReflectAccessibleObject_flag < 0 ||
        gDvm.offJavaLangReflectConstructor_slot < 0 ||
        gDvm.offJavaLangReflectConstructor_declClass < 0 ||
        gDvm.offJavaLangReflectField_slot < 0 ||
        gDvm.offJavaLangReflectField_declClass < 0 ||
        gDvm.offJavaLangReflectMethod_slot < 0 ||
        gDvm.offJavaLangReflectMethod_declClass < 0)
    {
        LOGE("Could not find reflection fields\n");
        return false;
    }
    if (!dvmReflectProxyStartup())
        return false;
    if (!dvmReflectAnnotationStartup())
        return false;
    return true;
}
```

上述函数的实现保存在文件“reflect\Reflect.c”中。

（6）剩余类的初始化工作。

经过前面的初始化函数处理之后，接着把下面的类先进行初始化操作：

```
staticconst char*earlyClasses[] = {
"Ljava/lang/InternalError;",
"Ljava/lang/StackOverflowError;",
"Ljava/lang/UnsatisfiedLinkError;",
"Ljava/lang/NoClassDefFoundError;",
NULL
};
```

初始化这些类，就是调用函数 dvmFindSystemClassNoInit 来初始化。接着调用函数 dvm ValidateBoxClasses()来初始化 Java 基本类型库：

```
staticconstchar*classes[] = {
"Ljava/lang/Boolean;",
"Ljava/lang/Character;",
"Ljava/lang/Float;",
"Ljava/lang/Double;",
"Ljava/lang/Byte;",
"Ljava/lang/Short;",
"Ljava/lang/Integer;",
"Ljava/lang/Long;",
NULL
};
```

这些类调用函数，不是上面使用系统函数来初始化，而是调用函数 dvmFindClassNoInit()来初始化。此函数的实现代码如下所示：

```
ClassObject* dvmFindClassNoInit(const char* descriptor,
        Object* loader)
{
    assert(descriptor != NULL);
    //assert(loader != NULL);

    LOGVV("FindClassNoInit '%s' %p\n", descriptor, loader);

    if (*descriptor == '[') {
        /*
         * Array class.  Find in table, generate if not found.
         */
        return dvmFindArrayClass(descriptor, loader);
    } else {
        /*
         * Regular class.  Find in table, load if not found.
         */
        if (loader != NULL) {
            return findClassFromLoaderNoInit(descriptor, loader);
        } else {
            return dvmFindSystemClassNoInit(descriptor);
        }
    }
}
```

调用函数 dvmPrepMainForJni()准备主线程里的解释栈可以调用 JNI 的方法，此函数的实现代码如下所示：

```
bool dvmPrepMainForJni(JNIEnv* pEnv)
{
    Thread* self;

    /* main thread is always first in list at this point */
    self = gDvm.threadList;
    assert(self->threadId == kMainThreadId);

    /* create a "fake" JNI frame at the top of the main thread interp stack */
    if (!createFakeEntryFrame(self))
        return false;

    /* fill these in, since they weren't ready at dvmCreateJNIEnv time */
    dvmSetJniEnvThreadId(pEnv, self);
    dvmSetThreadJNIEnv(self, (JNIEnv*) pEnv);

    return true;
}
```

调用函数 registerSystemNatives()来注册 Java 库里的 JNI 方法，此函数的实现代码如下所示：

```
static bool registerSystemNatives(JNIEnv* pEnv)
{
    Thread* self;

    /* main thread is always first in list */
    self = gDvm.threadList;

    /* must set this before allowing JNI-based method registration */
    self->status = THREAD_NATIVE;

    if (jniRegisterSystemMethods(pEnv) < 0) {
        LOGE("jniRegisterSystemMethods failed");
        return false;
    }

    /* back to run mode */
    self->status = THREAD_RUNNING;

    return true;
}
```

调用函数 dvmCreateStockExceptions()分配异常出错的内存，此函数的实现代码如下所示：

```
bool dvmCreateStockExceptions(void)
{
    /*
     * Pre-allocate some throwables.  These need to be explicitly added
     * to the GC's root set (see dvmHeapMarkRootSet()).
     */
    gDvm.outOfMemoryObj = createStockException("Ljava/lang/OutOfMemoryError;",
        "[memory exhausted]");
    dvmReleaseTrackedAlloc(gDvm.outOfMemoryObj, NULL);
    gDvm.internalErrorObj = createStockException("Ljava/lang/InternalError;",
        "[pre-allocated]");
    dvmReleaseTrackedAlloc(gDvm.internalErrorObj, NULL);
    gDvm.noClassDefFoundErrorObj =
        createStockException("Ljava/lang/NoClassDefFoundError;", NULL);
    dvmReleaseTrackedAlloc(gDvm.noClassDefFoundErrorObj, NULL);

    if (gDvm.outOfMemoryObj == NULL || gDvm.internalErrorObj == NULL ||
        gDvm.noClassDefFoundErrorObj == NULL)
    {
        LOGW("Unable to create stock exceptions\n");
        return false;
    }

    return true;
}
```

调用函数 dvmPrepMainThread()实现解释器主线程的初始化，此函数的实现代码如下所示：

```
bool dvmPrepMainThread(void)
{
    Thread* thread;
    Object* groupObj;
    Object* threadObj;
    Object* vmThreadObj;
    StringObject* threadNameStr;
    Method* init;
    JValue unused;
    LOGV("+++ finishing prep on main VM thread\n");
    /* main thread is always first in list at this point */
    thread = gDvm.threadList;
    assert(thread->threadId == kMainThreadId);
    /*
     * Make sure the classes are initialized.  We have to do this before
     * we create an instance of them.
     */
    if (!dvmInitClass(gDvm.classJavaLangClass)) {
        LOGE("'Class' class failed to initialize\n");
        return false;
    }
    if (!dvmInitClass(gDvm.classJavaLangThreadGroup) ||
        !dvmInitClass(gDvm.classJavaLangThread) ||
        !dvmInitClass(gDvm.classJavaLangVMThread))
    {
        LOGE("thread classes failed to initialize\n");
        return false;
    }
    groupObj = dvmGetMainThreadGroup();
    if (groupObj == NULL)
        return false;
    /*
     * Allocate and construct a Thread with the internal-creation
     * constructor.
     */
    threadObj = dvmAllocObject(gDvm.classJavaLangThread, ALLOC_DEFAULT);
    if (threadObj == NULL) {
        LOGE("unable to allocate main thread object\n");
        return false;
```

```
    }
    dvmReleaseTrackedAlloc(threadObj, NULL);

    threadNameStr = dvmCreateStringFromCstr("main", ALLOC_DEFAULT);
    if (threadNameStr == NULL)
        return false;
    dvmReleaseTrackedAlloc((Object*)threadNameStr, NULL);
    init = dvmFindDirectMethodByDescriptor(gDvm.classJavaLangThread, "<init>",
            "(Ljava/lang/ThreadGroup;Ljava/lang/String;IZ)V");
    assert(init != NULL);
    dvmCallMethod(thread, init, threadObj, &unused, groupObj, threadNameStr,
        THREAD_NORM_PRIORITY, false);
    if (dvmCheckException(thread)) {
        LOGE("exception thrown while constructing main thread object\n");
        return false;
    }
```

调用函数 dvmDebuggerStartup()进行调试器的初始化，此函数的实现代码如下所示：

```
bool dvmDebuggerStartup(void)
{
    gDvm.dbgRegistry = dvmHashTableCreate(1000, NULL);
    return (gDvm.dbgRegistry != NULL);
}
```

调用 dvmInitZygote()或者 dvmInitAfterZygote()来初始化线程的模式，此函数的实现代码如下所示：

```
static bool dvmInitZygote(void)
{
    /* zygote goes into its own process group */
    setpgid(0,0);
    return true;
}
bool dvmInitAfterZygote(void)
{
    u8 startHeap, startQuit, startJdwp;
    u8 endHeap, endQuit, endJdwp;
    startHeap = dvmGetRelativeTimeUsec();
    /*
     * Post-zygote heap initialization, including starting
     * the HeapWorker thread.
     */
    if (!dvmGcStartupAfterZygote())
        return false;
    endHeap = dvmGetRelativeTimeUsec();
    startQuit = dvmGetRelativeTimeUsec();
    /* start signal catcher thread that dumps stacks on SIGQUIT */
    if (!gDvm.reduceSignals && !gDvm.noQuitHandler) {
        if (!dvmSignalCatcherStartup())
            return false;
    }
    /* start stdout/stderr copier, if requested */
    if (gDvm.logStdio) {
        if (!dvmStdioConverterStartup())
            return false;
    }
    endQuit = dvmGetRelativeTimeUsec();
    startJdwp = dvmGetRelativeTimeUsec();
    /*
     * Start JDWP thread. If the command-line debugger flags specified
     * "suspend=y", this will pause the VM. We probably want this to
     * come last.
     */
    if (!dvmInitJDWP()) {
        LOGD("JDWP init failed; continuing anyway\n");
    }
    endJdwp = dvmGetRelativeTimeUsec();
    LOGV("thread-start heap=%d quit=%d jdwp=%d total=%d usec\n",
```

```
        (int)(endHeap-startHeap), (int)(endQuit-startQuit),
        (int)(endJdwp-startJdwp), (int)(endJdwp-startHeap));
#ifdef WITH_JIT
    if (gDvm.executionMode == kExecutionModeJit) {
        if (!dvmCompilerStartup())
            return false;
    }
#endif
    return true;
}
```

调用函数 dvmCheckException()检查是否有异常情况出现，此函数的实现代码如下所示。

10.3 启动 Zygote

在本书第 9 章的内容中，已经分析了启动孕育进程 Zygote 的基本源码。在本节的内容中，将详细分析 Zygote 进程的内部启动过程。

10.3.1　在 init.rc 中配置 Zygote 启动参数

init.rc 保存在设备的根目录下，可以使用“adb pull /init.rc～/Desktop”命令取出该文件，文件中和 zygote 相关的配置信息如下所示：

```
service zygote /system/bin/app_process -Xzygote
/system/bin --zygote --start-system-server
    socket zygote stream 666
    onrestart write /sys/android_power/request_state wake
    onrestart write /sys/power/state on
    onrestart restart media
    onrestart restart netd
```

10.3.2　启动 Socket 服务端口

在 Android 系统中，当 zygote 服务从 app_process 启动后会启动一个 Dalvik 虚拟机。因为虚拟机执行的第一个 Java 类文件是“frameworks\base\core\java\com\android\internal\os\ZygoteInit.java”，所以接下来的过程需要从类 ZygoteInit 中的函数 main()开始说起。在函数 main()的实现代码中，首先要做的工作便是启动一个 Socket 服务端口，该 Socket 端口用于接收启动新进程的命令。

- 在静态函数 registerZygoteSocket()中，完成启动 Socket 服务端口的功能。
- 当准备好 LocalServerSocket 端口后，在函数 main()中调用 runSelectLoopMode()进入非阻塞读操作，该函数会先将 ServerSocket 加入到被监测的文件描述符列表中，然后在 while(true)循环中将该文件描述符添加到 select 的列表中，并调用 ZygoteConnection 类的 runOnce()函数处理每一个 Socket 接收到的命令。
- 在 SystemServer 进程中创建一个 Socket 客户端，此功能在文件 Process.java 中实现，而调用 Process 类的工作在类 AmS 中的 startProcessLocked()函数中实现。
- 函数 start()内部又调用了静态函数 startViaZygote()，该函数的实体是使用一个本地 Socket 向 zygote 中的 Socket 发送进行启动命令，其执行流程如下所示。
  - 将 startViaZygote()的函数参数转换为一个 ArrayList<String>列表。
  - 然后再构造出一个 LocalSocket 本地 Socket 接口。
  - 通过该 LocalSocket 对象构造出一个 BufferedWriter 对象。
  - 通过该对象将 ArralyList<String>列表中的参数传递给 zygote 中的 LocalServerSocket 对象。
  - 在 zygote 端调用 Zygote.forkAndSpecialize()函数孕育出一个新的应用进程。

10.3.3 加载 preload-classes

使用类 ZygoteInit 中的函数 main 创建完 Socket 服务端后还不能接着孕育新的进程，因为这个"卵"中还没有预装的 Framework 大部分类及资源。

在 framework.jar 中有一个名为“preload-classes”的文本文件列表，这个列表就是预装的类列表。列表“preload-classes”在文本文件“frameworks/base/preload-classes”中进行原始定义，而该文件又是通过如下类生成的：

```
frameworks/base/tools/preload/WritePreloadedClassFile.java
```

在 Android 根目录下，通过执行如下命令生成 preload-classes：

```
$java -Xss512M -cp /path/to/preload.jar WritePreloadedClassFile /path/to/.compiled
1517 classses were loaded by more than one app.
Added 147 more to speed up applications.
1664 total classes will be preloaded.
Writing object model...
Done!
```

在上述命令中，“/path/to/preload.jar”是指如下文件：

```
out/host/darwin-x86/framework/preload.jar
```

上述.jar 文件是由“frameworks/base/tools/preload”子项目编译而成的。

“/path/to/.compiled/”是指如下目录下的几个.compiled 文件。

```
frameworks/base/tools/preload
```

- 参数“-Xss”：用于执行该程序所需要的 Java 虚拟机栈大小，此处使用 512 MB，默认的大小不能满足该程序的运行，会抛出 java.lang.StackOverflowError 错误信息。
- WritePreloadedClassFile：表示要执行的具体类。

当执行完以上命令后，会在“frameworks/base”目录下产生 preload-classes 文本文件。从该命令的执行情况来看，预装的 Java 类信息包含在.compiled 文件中，而这个文件却是一个二进制文件，尽管目前能够确知如何产生 preload-classes，但却无法明确这个.compiled 文件是如何产生的。

在编译 Android 源码的时候，会最终把 preload-classes 文件打包到 framework.jar 中。这样有了这个列表后，ZygoteInit 中通过调用 preloadClasses()完成装载这些类。装载的方法很简单，就是读取 preload-classes 列表中的每一行，因为每一行代表了一个具体的类，然后调用 Class.forName()装载目标类。在装载的过程中，忽略以“#”开始的目标类，并忽略换行符及空格。

10.3.4 加载 preload-resources

在 Android 系统中，preload-resources 是在如下文件中被定义的：

```
frameworks/base/core/res/res/values/arrays.xml
```

在 preload-resources 中包含了如下两类资源：

- drawable 资源；
- color 资源。

下面是定义 preload-resources 对应的代码：

```
<array name="preloaded_drawables">
    <item>@drawable/sym_def_app_icon</item>
... ...
  </array>
  <array name="preloaded_color_state_lists">
    <item>@color/hint_foreground_dark</item>
      ... ...
  </array>
```

加载这些资源功能是在函数 preloadResources()中实现的，在该函数中分别调用了如下两个函数来加载这两类资源：

- ❑ preloadDrawables();
- ❑ preloadColorStateLists()。

具体的加载原理非常简单，就是把这些资源读出来放到一个全局变量中，只要该类对象不被销毁，这些全局变量就会一直保存。

通过全局变量是 mResources 来保存 Drawable 资源，该变量的类型是 Resources 类，由于在该类内部会保存一个 Drawable 资源列表，因此实际上是在 Resources 内部缓存这些 Drawable 资源的。保存 Color 资源的全局变量的功能也是 mResources 实现的。同样，在类 Resources 内部也有一个 Color 资源列表。

10.3.5　使用 folk 启动新进程

folk 是 Linux 系统中的一个系统调用，其功能是复制当前进程并产生一个新的进程。除了进程 id 不同，新的进程将拥有和原始进程完全相同的进程信息。进程信息包括该进程所打开的文件描述符列表、所分配的内存等。当创建新进程后，两个进程将共享已经分配的内存空间，直到其中一个需要向内存中写入数据时，操作系统才负责复制一份目标地址空间，并将要写的数据写入到新的地址中，这就是“copy-on-write（仅当写的时候才复制）”机制，这种机制可以最大限度地在多个进程中共享物理内存。

在所有的操作系统中都存在一个程序装载器，程序装载器一般会作为操作系统的一部分，并由 Shell 程序调用。当内核启动后，会首先启动 Shell 程序。常见的 Shell 程序包含如下两大类：

- ❑ 命令行界面；
- ❑ 窗口界面。

Windows 系统中的 Shell 程序就是桌面程序，Ubuntu 系统中的 Shell 程序就是 GNOME 桌面程序。当启动 Shell 程序后，用户可以双击桌面图标启动指定的应用程序，而在操作系统内部，启动新的进程包含如下 3 个过程。

（1）第一个过程：在内核中创建一个进程数据结构，用于表示将要启动的进程。

（2）第二个过程：在内核中调用程序装载器函数，从指定的程序文件读取程序代码，并将这些程序代码装载到预先设定的内存地址。

（3）第三个过程：装载完毕后，内核将程序指针指向到目标程序地址的入口处开始执行指定的进程。

10.4　启动 SystemServer 进程

在 Android 系统中，Zygote 孕育出的第一个进程是 SystemServer，进程 SystemServer 是从文件 ZygoteInit.java 中的 main()函数中调用 startSystemServer()函数开始的。与启动普通进程过程相比，类 Zygote 为启动 SystemServer 提供了专门的函数 startSystemServer()，而不是使用标准的 forAndSpecilize()函数。

函数 startSystemServer()的主要功能如下所示。

（1）定义了一个 String[]数组，数组中包含了要启动的进程的相关信息，其中最后一项指定新进程启动后装载的第一个 Java 类，此处即为类 com.android.server.SystemServer。

（2）调用 forkSystemServer()从当前的 Zygote 进程孕育出新的进程。该函数是一个 native 函数，其作用与 folkAndSpecilize()相似。

（3）启动新进程后，在函数 handleSystemServerProcess()中主要完成如下两件事情。

- 关闭 Socket 服务端。
- 执行 com.android.server.SystemServer 类中的函数 main()。

除了这两个主要事情外，还做了一些额外的运行环境配置，这些配置主要在函数 commonInit()和函数 zygoteInitNative()中完成。一旦配置好 SystemServer 的进程环境后，就从类 SystemServer 中的 main()函数开始运行。

10.4.1 启动各种系统服务线程

SystemServer 进程在 Android 的运行环境中扮演了“中枢”的作用，在 APK 应用中能够直接交互的大部分系统服务都在这个进程中运行，例如 WindowManagerServer（Wms）、ActivityManagerSystemService（AmS）、PackageManagerServer（PmS）等常见的应用，这些系统服务都是以一个线程的方式存在于 SystemServer 进程中。下面就来介绍到底都有哪些服务线程，及其启动的顺序。

SystemServer 中的 main()函数首先调用的是函数 init1()，这是一个 native 函数，内部会进行一些与 Dalvik 虚拟机相关的初始化工作。该函数执行完毕后，其内部会调用 Java 端的 init2()函数，这就是为什么 Java 源码中没有引用 init2()的地方，主要的系统服务都是在 init2()函数中完成的。

该函数首先创建了一个 ServerThread 对象，该对象是一个线程，然后直接运行该线程，从 ServerThread 的 run()方法内部开始真正启动各种服务线程。基本上每个服务都有对应的 Java 类，从编码规范的角度来看，启动这些服务的模式可归类为如下 3 种。

- 模式一：是指直接使用构造函数构造一个服务，由于大多数服务都对应一个线程，因此，在构造函数内部就会创建一个线程并自动运行。
- 模式二：是指服务类会提供一个 getInstance()方法，通过该方法获取该服务对象，这样的好处是保证系统中仅包含一个该服务对象。
- 模式三：是指从服务类的 main()函数中开始执行。

无论以上何种模式，当创建了服务对象后，有时可能还需要调用该服务类的 init()函数或者 systemReady()函数来完成该对象的启动，当然这些都是服务类内部自定义的。为了区分以上启动的不同，以下采用一种新的方式描述该启动过程。

在表 10-2 中列出了 SystemServer 中所启动的所有服务，以及这些服务的启动模式。

表 10-2　　SystemServer 中启动服务列表

服务类名称	作 用 描 述	启 动 模 式
EntropyService	提供伪随机数	1.0
PowerManagerService	电源管理服务	1.2/3
ActivityManagerService	最核心的服务之一，管理 Activity	自定义
TelephonyRegistry	通过该服务注册电话模块的事件响应，比如重启、关闭、启动等	1.0
PackageManagerService	程序包管理服务	3.3
AccountManagerService	账户管理服务，是指联系人账户，而不是 Linux 系统的账户	1.0
ContentService	ContentProvider 服务，提供跨进程数据交换	3.0
BatteryService	电池管理服务	1.0
LightsService	自然光强度感应传感器服务	1.0

续表

服务类名称	作 用 描 述	启 动 模 式
VibratorService	震动器服务	1.0
AlarmManagerService	定时器管理服务，提供定时提醒服务	1.0
WindowManagerService	Framework 最核心的服务之一，负责窗口管理	3.3
BluetoothService	蓝牙服务	1.0 +
DevicePolicyManagerService	提供一些系统级别的设置及属性	1.3
StatusBarManagerService	状态栏管理服务	1.3
ClipboardService	系统剪切板服务	1.0
InputMethodManagerService	输入法管理服务	1.0
NetStatService	网络状态服务	1.0
NetworkManagementService	网络管理服务	NMS.create()
ConnectivityService	网络连接管理服务	2.3
ThrottleService	暂不清楚其作用	1.3
AccessibilityManagerService	辅助管理程序截获所有的用户输入，并根据这些输入给用户一些额外的反馈，起到辅助的效果	1.0
MountService	挂载服务，可通过该服务调用 Linux 层面的 mount 程序	1.0
NotificationManagerService	通知栏管理服务，Android 中的通知栏和状态栏在一起，只是界面上前者在左边，后者在右边	1.3
DeviceStorageMonitorService	磁盘空间状态检测服务	1.0
LocationManagerService	地理位置服务	1.3
SearchManagerService	搜索管理服务	1.0
DropBoxManagerService	通过该服务访问 Linux 层面的 Dropbox 程序	1.0
WallpaperManagerService	墙纸管理服务，墙纸不等同于桌面背景，在 View 系统内部，墙纸可以作为任何窗口的背景	1.3
AudioService	音频管理服务	1.0
BackupManagerService	系统备份服务	1.0
AppWidgetService	Widget 服务	1.3
RecognitionManagerService	身份识别服务	1.3
DiskStatsService	磁盘统计服务	1.0

AmS 的启动模式如下所示。

- 调用函数 main()返回一个 Context 对象，而不是 AmS 服务本身。
- 调用 AmS.setSystemProcess()。
- 调用 AmS.installProviders()。
- 调用 systemReady()，当 AmS 执行完 systemReady()后，会相继启动相关联服务的 systemReady()函数，完成整体初始化。

10.4.2　启动第一个 Activity

当启动以上服务线程后，ActivityManagerService（AmS）服务是以 systemReady()调用完成最后启动的，而在 AmS 的函数 systemReady()内部的最后一段代码则发出了启动任务队列中最上面一个 Activity 的消息。因为在系统刚启动时，mMainStack 队列中并没有任何 Activity 对象，所以在类 ActivityStack 中将调用函数 startHomeActivityLocked()。

在开机后，系统从哪个 Activity 开始执行这一动作，完全取决于 mMainStack 中的第一个 Activity 对象。如果在 ActivityManagerService 启动时能够构造一个 Activity 对象（并不是说构造出一个 Activity 类的对象），并将其放到 mMainStack 中，那么第一个运行的 Activity 就是这个 Activity，这一点不像其他操作系统中通过设置一个固定程序作为第一个启动程序。

在 AmS 的 startHomeActivityLocked()中，系统发出了一个 catagory 字段包含 CATEGORY_HOME 的 intent。

无论是哪个应用程序，只要声明自己能够响应该 intent，那么就可以被认为是 Home 程序，这就是为什么在 Android 领域会存在各种“Home 程序”的原因。系统并没有给任何程序赋予“Home”特权，而只是把这个权利交给了用户。当在系统中有多个程序能够响应该 intent 时，系统会弹出一个对话框，请求用户选择启动哪个程序，并允许用户记住该选择，从而使得以后每次按“Home”键后都启动相同的 Activity。这就是第一个 Activity 的启动过程。

10.5 加载 class 类文件

Java 的源代码经过编译后会生成“.class”格式的文件，即字节码文件。然后在 Android 中使用 dx 工具将其转换为后缀为“.jar”格式的 dex 类型文件。Dalvik 虚拟机负责解释并执行编译后的字节码。在解释执行字节码之前，当然要读取文件，分析文件的内容，得到字节码，然后才能解释并执行。在整个的加载过程中，最为重要的就是对 Class 的加载，Class 包含 Method，Method 又包含 code。通过对 Class 的加载，即可获得所需执行的字节码。在本节的内容中，将从 dexfile 文件分析及 Class 加载中的数据结构入手，结合主要流程，对整个加载过程进行分析。

10.5.1 DexFile 在内存中的映射

在 Android 系统中，java 源文件会被编译为“.jar”格式的 dex 类型文件，在代码中称为 dexfile。在加载 Class 之前，必先读取相应的 jar 文件。通常使用 read()函数来读取文件中的内容。但在 Dalvik 中使用 mmap()函数。和 read()不同，mmap()函数会将 dex 文件映射到内存中，这样通过普通的内存读取操作即可访问 dex file 中的内容。

Dexfile 的文件格式如图 10-1 所示，主要有 3 部分组成：头部，索引，数据。通过头部可知索引的位置和数目，可知数据区的起始位置。其中 classDefsOff 指定了 ClassDef 在文件的起始位置，dataOff 指定了数据在文件的起始位置，ClassDef 即可理解为 Class 的索引。通过读取 ClassDef 可获知 Class 的基本信息，其中 classDataOff 指定了 Class 数据在数据区的位置。

当把 dexfile 文件映射到内存后，会调用函数 dexFileParse()对其分析，并将分析的结果保存在名为 DexFile 的数据结构中，此结构中数据成员的具体说明如下所示。

- baseAddr：指向映射区的起始位置。
- pClassDefs：指向 ClassDefs（即 class 索引）的起始位置。

另外，因为在查找 class 时是根据 class 的名字进行的，所以为了加快查找速度，创建了一个 hash 表。在 hash 表中对 class 名字进行 hash，并生成 index。这些操作都是在对文件解析时所完成的，这样虽然在加载过程中比较耗时，但是在运行过程中确可节省大量查找时间。

解析完毕之后，接下来开始加载 class 文件。在此需要将加载类用 ClassObject 来保存，所以在此需要先分析和 ClassObject 相关的几个数据结构。

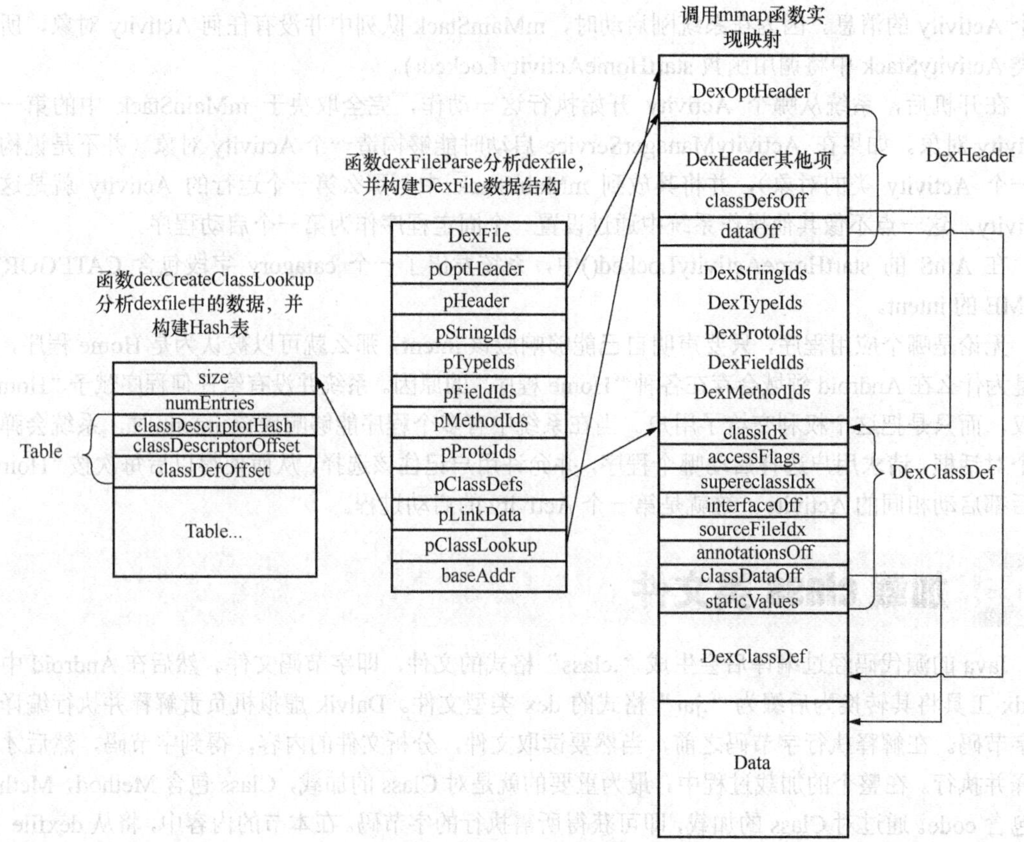

图 10-1　Dexfile 的文件格式

首先在文件 Object.h 中可以看到如下对结构体 Object 的定义：

```
typedef struct Object {
    /* ptr to class object */
    ClassObject*    clazz;
    /*
     * A word containing either a "thin" lock or a "fat" monitor.  See
     * the comments in Sync.c for a description of its layout.
     */
    u4 lock;
} Object;
```

通过结构体 Object 定义了基本类的实现，这里有如下两个变量。

- lock：对应 Obejct 对像中的锁实现，即 notify wait 的处理。
- clazz：是结构体指针，姑且不看结构体内容，这里用了指针的定义。

下面会有更多的结构体定义：

```
struct DataObject {
    Object obj;                     /* MUST be first item */

    /* variable #of u4 slots; u8 uses 2 slots */
    u4 instanceData[1];
};
struct StringObject {
    Object      obj;            /* MUST be first item */
    /* variable #of u4 slots; u8 uses 2 slots */
    u4 instanceData[1];
};
```

可以看到最熟悉的一个词 StringObject，把这个结构体展开后是下面的样子：

```
struct StringObject {
   /* ptr to class object */
   ClassObject*    clazz;
   /*
    * A word containing either a "thin" lock or a "fat" monitor. See
    * the comments in Sync.c for a description of its layout.
    */
   u4              lock;

   /* variable #of u4 slots; u8 uses 2 slots */
   u4              instanceData[1];
};
```

由此不难发现，任何对象的内存结构体中第一行都是 Object 结构体，而这个结构体第一个总是一个 ClassObejct，第二个总是 lock。按照 C++中的技巧，这些结构体可以当成 Object 结构体使用，因此所有的类在内存中都具有“对象”的功能，即可以找到一个类（ClassObject），可以有一个锁（lock）。

StringObject 是对 String 类进行管理的数据对象，ArrayObejct 是数据相关的管理。

10.5.2　ClassObject——Class 在加载后的表现形式

在 Dalvik VM 中，由数据结构 ClassObject 负责存放加载的信息。在加载过程，需要在内存中 alloc 几个区域，分别存放 directMethods、virtualMethods、sfields、ifields。这些信息是从 dex 文件的数据区中读取的，首先会读取 Class 的详细信息，从中获得 directMethod、virtualMethod、sfield、ifield 等的信息，然后再读取。在此需要注意，在 ClassObject 结构中有个名为 super 的成员，通过 super 成员可以指向它的超类。

10.5.3　加载 Class 并生成相应 ClassObject 的函数

在讲解完加载数据结构的知识后，接下来开始分析负责加载工作的函数 findClassNoInit()。在获取 Class 索引时，会分为基本类库文件和用户类文件两种情况进行处理。在文件 grund.sh 中通过如下语句指定 Dalvik VM 所需的基本库文件，如果没有此语句，Dalvik 在启动过程中就会报错退出：

```
export
BOOTCLASSPATH=$bootpath/core.jar:$bootpath/ext.jar:$bootpath/framework.jar:$bootpath/android.
police.jar
```

函数 LoadClassFromDex()会先读取 Class 的具体数据（从 ClassDataoff 处），然后分别加载 directMethod、virtualMethod、ifield 和 sfield。

为了追求效率，在加载后需要将其缓存起来，以便以后使用方便。另外，如果在查找过程中使用顺序查找会很慢，所以需要使用 gDvm.loadedClasses 这个 Hash 表来帮忙。如果一个子类需要调用超类的函数，它当然要先加载超类了，有可能的话甚至会加载超类的超类。

接下来用 GDB 调试，在函数 findClassNoInit()处设置断点（在 GDB 提示符后输入“b findClassNoInit”）。在 GDB 提示符后连续几次执行“c”和“bt”，此时可能会出现如下信息，在此可以看到在函数调用栈上可以多次看到 findClassNoInit()函数：

```
(gdb) bt
#0  findClassNoInit (descriptor=0xfef4c7f4 "??????%", loader=0x0, pDvmDex=0x0)
    at dalvik/vm/oo/Class.c:1373
#1  0xf6fc4d53 in dvmFindClassNoInit (descriptor=0xf5046a63 "Ljava/lang/Object;", loader=0x0)
    at dalvik/vm/oo/Class.c:1194
#2  0xf6fc6c0a in dvmResolveClass (referrer=0xf5837400, classIdx=290,
    fromUnverifiedConstant=false) at dalvik/vm/oo/Resolve.c:94
#3  0xf6fc3476 in dvmLinkClass (clazz=0xf5837400, classesResolved=false)
```

```
    at dalvik/vm/oo/Class.c:2537
#4  0xf6fc1b67 in findClassNoInit (descriptor=0xf6ff0df6 "Ljava/lang/Class;", loader=0x0,
    pDvmDex=0xa04c720) at dalvik/vm/oo/Class.c:1489
```

10.5.4 加载基本类库文件

在文件 class.c 的 2 575 行设置断点，然后等待程序停下为止。下面是 clazz 的内容：

```
 (gdb) p clazz->super->descriptor
$6 = 0xf5046a63 "Ljava/lang/Object;"
 (gdb) p clazz->descriptor
$7 = 0xf5046121 "Ljava/lang/Class;"
```

然后先在函数 findClassNoInit()所在位置设置断点，然后运行程序并等待程序停下为止：

```
(gdb) b findClassNoInit
Breakpoint 2 at 0xf6fc13e0: file dalvik/vm/oo/Class.c, line 1373.
(gdb) c
Continuing.
```

此时可以看出谁才是第一个加载的 Class：

```
 (gdb) bt
#0  findClassNoInit (descriptor=0x0, loader=0x0, pDvmDex=0x0) at dalvik/vm/oo/Class.c:1373
#1  0xf6fc32a1 in dvmLinkClass (clazz=0xf5837350, classesResolved=false)
    at dalvik/vm/oo/Class.c:2491
#2  0xf6fc1b67 in findClassNoInit (descriptor=0xf6ff1ded "Ljava/lang/Thread;", loader=0x0,
    pDvmDex=0xa04c720) at dalvik/vm/oo/Class.c:1489
#3  0xf6f92692 in dvmThreadObjStartup () at dalvik/vm/Thread.c:328
#4  0xf6f800e6 in dvmStartup (argc=2, argv=0xa041190, ignoreUnrecognized=false, pEnv=0xa0411a0)
    at dalvik/vm/Init.c:1155
#5  0xf6f8b8e3 in JNI_CreateJavaVM (p_vm=0xf6ff0df6, p_env=0xf6ff0df6, vm_args=0xfef4d0b0)
    at dalvik/vm/Jni.c:4198
#6  0x08048893 in main (argc=3, argv=0xfef4d168) at dalvik/dalvikvm/Main.c:212
```

由上述函数的调用顺序可得出加载基本类库文件的过程依次是：

① main；

② JNI_CreateJavaVM；

③ dvmStartup；

④ dvmThreadObjStartup；

⑤ dvmFindSystemClassNoInit；

⑥ findClassNoInit。

在上述调用栈中没有 dvmFindSystemClassNoInit，是因为编译器将其作为 inline 优化了，导致 GDB 看不到有 dvmFindSystemClassNoInit 的栈。但是不要担心，可以从回溯栈中看到 dvmFindSystemClassNoInit。

10.5.5 加载用户类文件

在加载用户类文件时，会先加载一个 Class，然后这个 Class 去负责用户类文件的加载，而这个 Class 又会通过 JNI 的方式去调用 findClassNoInit。具体加载过程和前面介绍的基本类库加载类似，读者可以参考前面的知识来理解。在此为节省本书篇幅，所以不再详细介绍。

第11章 DEX文件详解

DEX 文件是 Android 平台上一种可执行文件的类型。可以使用 Android 打包工具（aapt）将 DEX 文件、资源文件以及 AndroidManifest.xml 文件（二进制格式）组合成一个应用程序包（APK）。应用程序包可以被发布到手机上运行。在本章的内容中，将详细讲解 Android 系统中 DEX 文件的基本知识，为读者步入本书后面知识的学习打下基础。

11.1 DEX 文件介绍

在 Android 系统中，每一个应用都运行在一个 Dalvik VM 实例里，而每一个虚拟机实例都是一个独立的进程空间。虚拟机的线程机制、内存分配和管理以及 Mutex 等都是依赖底层操作系统而实现的。

在 Android 系统中，因为所有应用的线程都对应一个 Linux 线程，所以虚拟机可以更多地依赖操作系统的线程调度和管理机制。不同的应用在不同的进程空间里运行，加之对不同来源的应用都使用不同的 Linux 用户来运行，可以最大程度地保护应用的安全和独立运行。

Zygote 是一个虚拟机进程，同时也是一个虚拟机实例的孵化器，每当系统要求执行一个 Android 应用程序，Zygote 就会 FORK（孵化）出一个子进程来执行该应用程序。这样做的好处显而易见：Zygote 进程是在系统启动时产生的，它会完成虚拟机的初始化、库的加载、预置类库的加载和初始化等操作，而在系统需要一个新的虚拟机实例时。

Zygote 通过复制自身，最快速地提供一个系统。另外，对于一些只读的系统库，所有虚拟机实例都和 Zygote 共享一块内存区域，大大节省了内存开销。Android 应用开发和 Dalvik 虚拟机 Android 应用所使用的编程语言是 Java 语言，和 Java SE 一样，编译时使用 Oracle JDK 将 Java 源程序编程成标准的 Java 字节码文件（.class 文件）。

而后通过工具软件 DX 把所有的字节码文件转成 Android DEX 文件（classes.DEX）。最后使用 Android 打包工具（aapt）将 DEX 文件、资源文件以及 AndroidManifest.xml 文件（二进制格式）组合成一个应用程序包（APK）。应用程序包可以被发布到手机上运行。

11.2 DEX 文件的格式

假设有一个名为“test.java”的 Java 文件，其代码如下所示：

```
class test{
   public static void main(String[] argc){
      System.out.println("test!");
   }
}
```

通过如下命令可以得到 DEX 文件。

```
$ javac test.java
$ dx --DEX --output=test.DEX test.class
$ hexdump test.DEX
```

DEX 文件 test.DEX 的内容如下所示：

```
0000000 6564 0a78 3330 0035  5eb4 4f7a 94e6 65f0
0000010 fb3e d5f3 e185 dd62 fce7 c887 a7ec 5329
0000020 02d8 0000  0070 0000  5678 1234  0000 0000
0000030 0000 0000  0238 0000  000e 0000  0070 0000
0000040 0007 0000  00a8 0000  0003 0000  00c4 0000
0000050 0001 0000  00e8 0000  0004 0000  00f0 0000
0000060 0001 0000  0110 0000  01a8 0000  0130 0000
0000070 0176 0000  017e 0000 0195 0000 01a9 0000
0000080 01bd 0000 01d1 0000 01d9 0000 01dc 0000
0000090 01e0 0000 01f5 0000 01fb 0000 0200 0000
00000a0 0209 0000 0210 0000 0001 0000  0002 0000
00000b0 0003 0000 0004 0000 0005 0000 0006 0000
00000c0 0008 0000 0006 0000 0005 0000 0000 0000
00000d0 0007 0000 0005 0000 0168 0000 0007 0000
00000e0 0005 0000 0170 0000 0003 0000 000a 0000
00000f0 0000 0001 000b 0000  0001 0000 0000 0000
0000100 0004 0000 0000 0000 0004 0002 0009 0000
0000110 0004  0000 0000 0000 0001 0000 0000 0000
0000120 000d 0000 0000 0000 0227 0000 0000 0000
0000130 0001 0001 0001 0000 021b 0000 0004 0000
0000140 1070 0001 0000 000e 0003 0001 0002 0000
0000150 0220 0000 0008 0000 0062 0000 011a 000c
0000160 206e 0000 0010 000e 0001 0000 0002 0000
0000170 0001 0000 0006 3c06 6e69 7469 003e 4c15
0000180 616a 6176 692f 2f6f 7250 6e69 5374 7274
0000190 6165 3b6d 1200 6a4c 7661 2f61 616c 676e
00001a0 4f2f 6a62 6365 3b74 1200 6a4c 7661 2f61
00001b0 616c 676e 532f 7274 6e69 3b67 1200 6a4c
00001c0 7661 2f61 616c 676e 532f 7379 6574 3b6d
00001d0 0600 744c 7365 3b74 0100 0056 5602 004c
00001e0 5b13 6a4c 7661 2f61 616c 676e 532f 7274
00001f0 6e69 3b67 0400 616d 6e69 0300 756f 0074
0000200 7007 6972 746e 6e6c 0500 6574 7473 0021
0000210 7409 7365 2e74 616a 6176 0100 0700 000e
0000220 0103 0700 780e 0000 0200 0200 8080 b004
0000230 0102 c809 0002 0000 000d 0000  0000 0000
0000240 0001 0000 0000 0000 0001 0000 000e 0000
0000250 0070 0000 0002 0000 0007 0000 00a8 0000
0000260 0003 0000 0003 0000 00c4 0000 0004 0000
0000270 0001 0000 00e8 0000 0005 0000 0004 0000
0000280 00f0 0000 0006 0000 0001 0000 0110 0000
0000290 2001 0000 0002 0000 0130 0000 1001 0000
00002a0 0002 0000 0168 0000 2002 0000 000e 0000
00002b0 0176 0000 2003 0000 0002 0000 021b 0000
00002c0 2000 0000 0001 0000 0227 0000 1000 0000
00002d0 0001 0000 0238 0000
00002d8
```

接下来开始分析这个 DEX 文件的具体格式。

11.2.1 map_list

map_list 数据结构如表 11-1 所示。

表 11-1 map_list 数据结构

名　字	格　式
size	uint
list	map_item[size]

第一项为 map_list 的大小，其中 map_item 的结构为如表 11-2 所示。

表 11-2 map_item 的结构

名　字	格　式
type	ushort
unused	ushort
size	uint
offset	uint

type 的值如表 11-3 所示。

表 11-3 type 的值

Item 类型	常　量	值	Item 的大小/Byte
header_item	TYPE_HEADER_ITEM	0x0000	0x70
string_id_item	TYPE_STRING_ID_ITEM	0x0001	0x04
type_id_item	TYPE_TYPE_ID_ITEM	0x0002	0x04
proto_id_item	TYPE_PROTO_ID_ITEM	0x0003	0x0c
field_id_item	TYPE_FIELD_ID_ITEM	0x0004	0x08
method_id_item	TYPE_METHOD_ID_ITEM	0x0005	0x08
class_def_item	TYPE_CLASS_DEF_ITEM	0x0006	0x20
map_list	TYPE_MAP_LIST	0x1000	4 + (item.size×12)
type_list	TYPE_TYPE_LIST	0x1001	4 + (item.size×2)
annotation_set_ref_list	TYPE_ANNOTATION_SET_REF_LIST	0x1002	4 + (item.size×4)
annotation_set_item	TYPE_ANNOTATION_SET_ITEM	0x1003	4 + (item.size×4)
class_data_item	TYPE_CLASS_DATA_ITEM	0x2000	implicit; must parse
code_item	TYPE_CODE_ITEM	0x2001	implicit; must parse
string_data_item	TYPE_STRING_DATA_ITEM	0x2002	implicit; must parse
debug_info_item	TYPE_DEBUG_INFO_ITEM	0x2003	implicit; must parse
annotation_item	TYPE_ANNOTATION_ITEM	0x2004	implicit; must parse
encoded_array_item	TYPE_ENCODED_ARRAY_ITEM	0x2005	implicit; must parse
annotations_directory_item	TYPE_ANNOTATIONS_DIRECTORY_ITEM	0x2006	implicit; must parse

这个 map_list 有 13 个 map_item，具体说明如表 11-4 所示。

表 11-4 13 个 map_item

值	类　型	大小	偏　移
0x0000	header_item	0x1	0x0
0x0001	string_id_item	0xe	0x70
0x0002	type_id_item	0x7	0xa8
0x0003	proto_id_item	0x3	0xc4
0x0004	field_id_item	0x1	0xe8
0x0005	method_id_item	0x4	0xf0
0x0006	class_def_item	0x1	0x110
0x2001	code_item	0x2	0x130

续表

值	类　型	大小	偏　移
0x1001	type_list	0x2	0x168
0x2002	string_data_item	0xe	0x176
0x2003	debug_info_item	0x2	0x21b
0x2000	class_data_item	0x1	0x227
0x1000	map_list	0x1	0x238

由此可以看出，表中的 size 和 offset 和 header_item 中的值一致。

11.2.2　string_id_item

在前面的 test.DEX 文件中，通过第三行中的“0070 0000”得出 string id 列表的位置为 0x70，通过第三行的“000e 0000”得出 string id 列表中 string_id_item 的数量为 0xe。string id 的结构为 string_id_item，其格式说明如表 11-5 所示。

表 11-5　格式说明

Name	Format
string_data_off	uint

string_data_off 指向 string 的数据，string 的数据的结构为 string_data_item，其格式说明如表 11-6 所示。

表 11-6　格式说明

Name	Format
utf16_size	uleb128
data	ubyte[]

每个 LEB128 由 1 到 5 个字节组成，所有字节组合到一起代表一个 32 位值。除了最后一个字节的最高标志位为 0，其他的为 1，剩下的 7 位为有效负荷，第二个字节的 7 位接上。有符号 LEB128 的符号由最后字节的有效负荷最高位决定。如图 11-1 所示。

Bitwise diagram of a two-byte LEB128 value															
First byte								Second byte							
1	bit_6	bit_5	bit_4	bit_3	bit_2	bit_1	bit_0	0	bit_{13}	bit_{12}	bit_{11}	bit_{10}	bit_9	bit_8	bit_7

图 11-1　LEB128 数据类型

如果是有符号的 LEB128，符号位取决于 bit13。如表 11-7 所示，uleb128p1 的值加 1 表示为 uleb128。

表 11-7　LEB128 的扩展

Encoded Sequence	As sleb128	As uleb128	As uleb128p1
0	0	0	–1
1	1	1	0
7f	–1	127	126
80 7f	–128	16256	16255

从文件的 0x70 得出 string id 的列表如下，共有 14 个 string_id_item。

```
0176 0000 017e 0000 0195 0000 01a9 0000
01bd 0000 01d1 0000 01d9 0000 01dc 0000
01e0 0000 01f5 0000 01fb 0000 0200 0000
0209 0000 0210 0000
```

例如：

（1）String Data 在 0x176 处，可以从文件的 0x176 得到以下数据，以 0 结尾。

```
3c06 6e69 7469 003e
```

先读取第一个字节为 0x06，得出 String Data 的长度为 6，所以 String Data 的 ASCII 码序列为：3c 69 6e 69 74 3e 得到：<init>。

（2）String Data 在 0x17e 处，可以从文件的 0x17e 得到以下数据，以 0 结尾。

```
4c15 616a 6176 692f 2f6f 7250 6e69 5374 7274 6165 3b6d 12 00
```

先读取第一个字节为 0x15，得出 String Data 的长度为 21，所以 String Data 的 ASCII 码序列为：

4c 6a 61 76 61 2f 69 6f 2f 50 72 69 6e 74 53 74 72 65 61 6d 3b 得到：Ljava/io/PrintStream，以此类推，可以得到其他 String Data。

（3）6a4c 7661 2f61 616c 676e 4f2f 6a62 6365 3b74 得到：Ljava/lang/Object。

（4）6a4c 7661 2f61 616c 676e 532f 7274 6e69 3b67 得到：Ljava/lang/String。

（5）6a4c 7661 2f61 616c 676e 532f 7379 6574 3b6d 得到：Ljava/lang/System。

（6）744c 7365 3b74 得到：Ltest。

（7）56 得到：V。

（8）56 4c 得到：VL。

（9）5b 6a4c 7661 2f61 616c 676e 532f 7274 6e69 3b67 得到：Ljava/lang/String。

（10）616d 6e69 得到：main。

（11）756f 74 得到：out。

（12）70 6972 746e 6e6c 得到：println。

（13）6574 7473 21 得到：test!

（14）74 7365 2e74 616a 6176 得到：test.java。

具体的算法实现位于 libDEX\Leb1211.h 中，如下面的代码所示：

```
DEX_INLINE int readUnsignedLeb128(const u1** pStream) {
    const u1* ptr = *pStream;
    int result = *(ptr++);

    if (result > 0x7f) {     //如果第一个字节的最高位是 1
        int cur = *(ptr++);    //指向第二个字节
        //当前值是第一个字节的 7 位加上第二个字节的 7 位
        result = (result & 0x7f) | ((cur & 0x7f) << 7);
        if (cur > 0x7f) {   //如果第二个字节的最高位是 1
            cur = *(ptr++);    //指向第三个字节
            result |= (cur & 0x7f) << 14;//当前值加上第三个字节的 7 位
            if (cur > 0x7f) {//如果第三个字节的最高位是 1
                cur = *(ptr++);
                result |= (cur & 0x7f) << 21;//当前值加上第四个字节的 7 位
                if (cur > 0x7f) {//如果第四个字节的最高位是 1
                    /*
                     * Note: We don't check to see if cur is out of
                     * range here, meaning we tolerate garbage in the
                     * high four-order bits.
                     */
                    cur = *(ptr++);
                    result |= cur << 28;//当前值加上第五个字节的 7 位
                }
            }
        }
```

```
    }

    *pStream = ptr;
    return result;
}

/*
* 读取有符号的，符号位取决于最后字节的有效负荷最高位。>>是到符号的
 */
DEX_INLINE int readSignedLeb128(const u1** pStream) {
    const u1* ptr = *pStream;
    int result = *(ptr++);

    if (result <= 0x7f) {
        result = (result << 25) >> 25;
    } else {
        int cur = *(ptr++);
        result = (result & 0x7f) | ((cur & 0x7f) << 7);
        if (cur <= 0x7f) {
            result = (result << 18) >> 18;
        } else {
            cur = *(ptr++);
            result |= (cur & 0x7f) << 14;
            if (cur <= 0x7f) {
                result = (result << 11) >> 11;
            } else {
                cur = *(ptr++);
                result |= (cur & 0x7f) << 21;
                if (cur <= 0x7f) {
                    result = (result << 4) >> 4;
                } else {
                    /*
                     * Note: We don't check to see if cur is out of
                     * range here, meaning we tolerate garbage in the
                     * high four-order bits.
                     */
                    cur = *(ptr++);
                    result |= cur << 28;
                }
            }
        }
    }
    *pStream = ptr;
    return result;
}
/*
* 读取并验证无符号的 Leb128
 */
int readAndVerifyUnsignedLeb128(const u1** pStream, const u1* limit,
        bool* okay);
/*
* 读取并验证有符号的 Leb128
 */
int readAndVerifySignedLeb128(const u1** pStream, const u1* limit, bool* okay);
/*
* 写入无符号的 Leb128
 */
DEX_INLINE u1* writeUnsignedLeb128(u1* ptr, u4 data)
{
    while (true) {
        u1 out = data & 0x7f;
        if (out != data) {
            *ptr++ = out | 0x80;
            data >>= 7;
        } else {
            *ptr++ = out;
            break;
        }
    }
```

```
    return ptr;
}
/*
* 尺寸
 */
DEX_INLINE int unsignedLeb128Size(u4 data)
{
    int count = 0;

    do {
        data >>= 7;
        count++;
    } while (data != 0);

    return count;
}
```

找到文件"libDEX\DEXFile.h"和"libDEX\DEXFile.c"，.DEX 文件会被映射到 DEXMapList，其定义结构体的代码如下所示：

```
typedef struct DEXMapList {
    u4  size;                /* #of entries in list */
    DEXMapItem list[1];      /* entries */
} DEXMapList;
```

size 表示 map_list 的大小，即条目数；DEXMapItem 结构体表示单个条目，其定义代码如下所示：

```
typedef struct DEXMapItem {
    u2  type;              /* type code (see kDEXType* above) */
    u2  unused;
    u4  size;              /* count of items of the indicated type */
    u4  offset;            /* file offset to the start of data */
} DEXMapItem;
```

而在结构体 DEXHeader 中存储了各个数据类型的实地址和偏移量等信息，其定义代码如下所示：

```
typedef struct DEXHeader {
    u1  magic[8];          /*includes version number */
    u4  checksum;          /* adler32 checksum */
    u1  signature[kSHA1DigestLen]; /* SHA-1 hash */
    u4  fileSize;          /* length of entire file */
    u4  headerSize;        /* offset to start of next section */
    u4  endianTag;
    u4  linkSize;
    u4  linkOff;
    u4  mapOff;
    u4  stringIdsSize;
    u4  stringIdsOff;
    u4  typeIdsSize;
    u4  typeIdsOff;
    u4  protoIdsSize;
    u4  protoIdsOff;
    u4  fieldIdsSize;
    u4  fieldIdsOff;
    u4  methodIdsSize;
    u4  methodIdsOff;
    u4  classDefsSize;
    u4  classDefsOff;
    u4  dataSize;
    u4  dataOff;
} DEXHeader;
```

11.2.3 type_id_item

由第 5 行的"00a8 0000"得出 type id 列表的位置为 0xa8，由第 5 行的"0007 0000"得出 type id 列表中 type_id_item 的数量为 0x7。type id 的结构为 type_id_item，如表 11-8 所示。

表 11-8 type_id_item

名　字	格　式
descriptor_idx	uint

descriptor_idx 为 String id 列表的索引，索引为：

```
0001 0000 0002 0000 0003 0000 0004 0000 0005 0000 0006 0000 0008 0000
```

依次代表：Ljava/io/PrintStream、Ljava/lang/Object、Ljava/lang/String、Ljava/lang/System、Ltest、V [Ljava/lang/String。

Type_id_list 列表如表 11-9 所示。

表 11-9 Type_id_list 列表

0	Ljava/io/PrintStream;
1	Ljava/lang/Object;
2	Ljava/lang/String;
3	Ljava/lang/System;
4	Ltest;
5	V
6	[Ljava/lang/String;

11.2.4 proto_id_item

由第 5 行的“00c4 0000”得出 prototype id 列表的位置为 0xc4，第 5 行的“0003 0000”处 prototype id 列表中 proto_id_item 的数量为 0x3。prototype id 的结构为 proto_id_item，如表 11-10 所示。

表 11-10 proto_id_item

名　字	格　式
shorty_idx	uint
return_type_idx	uint
parameters_off	uint

shorty_idx 为 String Id 列表的索引，return_type_idx 为 Type Id 列表的索引，parameters_off 指向 type_list。type_list 结构如表 11-11 所示。

表 11-11 type_list 结构

Name	Format
size	uint
list	type_item[size]

type_item 结构如表 11-12 所示。

表 11-12 type_item 结构

Name	Format
type_idx	ushort

type_idx 为 type id 列表的索引。

从文件的 0xc4 得到 prototype id 列表如下，共有如下 3 个 proto_id_item。

（1）0006 0000 0005 0000 0000 0000

string_id_list[0x6]代表 V，返回类型 type_id_list[0x5]代表 V，没有参数。

（2）0007 0000 0005 0000 0168 0000

string_id_list[0x7]代表 VL，返回类型 type_id_list[0x5]代表 V，参数从 0x168 处的值为：0001 0000 0002，索引为 0x2，type_id_list[0x2]代表 Ljava/lang/String;。

（3）0007 0000 0005 0000 0170 0000

string_id_list[0x7]代表 VL，返回类型 type_id_list[0x5]代表 V，参数从 0x170 处的值为：0001 0000 0006，索引为 0x6，type_id_list[0x6]代表[Ljava/lang/String;。

11.2.5　ield_id_item

由第 6 行的“00e8 0000”得出 field id 列表的位置为 0xe8，由第 6 行的“0001 0000”处 field id 列表中 field_id_item 的数量为 0x1。Field id 的结构为 field_id_item，如表 11-13 所示。

表 11-13　field_id_item

Name	Format
class_idx	ushort
type_idx	ushort
name_idx	uint

- class_idx：表示类的类型，即该字段所属的类。
- type_idx：表示此字段的类型。
- name_idx：表示此字段的名字。

从文件的 0xe8 得到如下 filed id 的列表。

```
0003 0000 000a 0000
```

由此可见，共有 1 个 field_id_item。该字段所属的类为：Ljava/lang/System;，此字段的类型为：Ljava/io/PrintStream;，此字段的名字为：out。

11.2.6　method_id_item

由第 6 行的“00f0 0000”得出 method id 列表的位置为 0xf0，第 6 行“0004 000”处 method id 列表中 method_id_item 的数量为 0x4。Method id 的结构为 method_id_item，如表 11-14 所示。

表 11-14　method_id_item

名　字	格　式
class_idx	ushort
proto_idx	ushort
name_idx	uint

- class_idx：表示类的类型，即该方法所属的类。
- proto_idx：表示此方法原型。
- name_idx：表示此方法名字。

从文件的 0xf0 得到 method id 列表如下，共有 4 个 method_id_item。

- 0000 0001 000b 0000　类：Ljava/io/PrintStream;，原型：VL，名字为：println。
- 0001 0000 0000 0000　类：Ljava/lang/Object;，原型：V，名字为：<init>。

- ❑ 0004 0000 0000 0000 类：Ljava/lang/System;，原型：V，名字为：<init>。
- ❑ 0004 0002 0009 0000 类：Ljava/lang/System;，原型：VL，名字为：main。

11.2.7 class_def_item

由第 7 行的“0110 0000”得出 class definitions 列表的位置为 0x110，第 7 行的“0001 0000”处 class definitions 列表中 class_def__item 的数量为 0x1。class definitions 的结构为 class_def_item，如表 11-15 所示。

表 11-15 class_def_item

名　字	格　式
class_idx	uint
access_flags	uint
superclass_idx	uint
interfaces_off	uint
source_file_idx	uint
annotations_off	uint
class_data_off	uint
static_values_off	uint

- ❑ class_idx：表示类的类型：Ltest。
- ❑ access_flags：表示访问权限。
- ❑ superclass_idx：表示父类：Ljava/lang/Object。
- ❑ interfaces_off：表示没有接口。
- ❑ source_file_idx：表示文件名 test.java。
- ❑ annotations_off：表示没有注释。
- ❑ class_data_off：表示指向 class_data_item。
- ❑ static_values_off：表示暂时无。

class_data_item 如表 11-16 所示。

表 11-16 class_data_item

名　字	格　式
static_fields_size	uleb128
instance_fields_size	uleb128
direct_methods_size	uleb128
virtual_methods_size	uleb128
static_fields	encoded_field[static_fields_size]
instance_fields	encoded_field[instance_fields_size]
direct_methods	encoded_method[direct_methods_size]
virtual_methods	encoded_method[virtual_methods_size]

文件的 0x227 处为 class_data_item 结构。从 0x227 处得来的字节序为：

```
00 00 02 00 02 80 80 04 b0 02 01 09 c8 02 00 00 00
```

- ❑ static_fields_size 为 0;

❑ instance_fields_size 为 0；

❑ direct_methods_size 为 2；

❑ virtual_methods_size 为 0；

因为前两个为 0，所以下一个字节开始就是 direct_methods，encoded_method 和 encoded_field 机构如下，两个 direct_method 的具体说明如下所示。

（1）02 80 80 04 b0 02。

method_idx_diff：为 0x2　<init>。

access_flags：为 0x10000 (80 80 04)，代表 constructor method。

code_off：为 0x130 (b0 02)　，指向 code_item。

从 0x130 解析 code_item 如下：

```
registers_size            0x1
ins_size                  0x1
outs_size                 0x1
tries_size                0
debug_info_off            0x21b
insns_size                0x4
insns    ushort[insns_size]    1070 0001 0000 000e
```

❑ 0x70 的 opcode 为：invoke-direct。

其格式如下。

```
invoke-direct {vD, vE, vF, vG, vA}, meth@CCCC
B: argument word count (4 bits)
C: method inDEX (16 bits)
D..G, A: argument registers (4 bits each)
```

分布如下。

```
B|A|op CCCC G|F|E|D [B=5] op {vD, vE, vF, vG, vA}, meth@CCCC
[B=5] op {vD, vE, vF, vG, vA}, type@CCCC
[B=4] op {vD, vE, vF, vG}, kind@CCCC
[B=3] op {vD, vE, vF}, kind@CCCC
[B=2] op {vD, vE}, kind@CCCC
[B=1] op {vD}, kind@CCCC
[B=0] op {}, kind@CCCC
```

由于 B=1，D=0，CCCC=0x0001，对应的 method 为 Ljava/lang/Object;的<init>，即构造方法。

❑ 0x0e 的 opcode 为 return-void。

经过上述分析，得到如下指令。

```
|0000: invoke-direct {v0}, Ljava/lang/Object;.<init>:()V // method@0001
|0003: return-void
```

（2）01 09 c8 02。

method_idx_diff：为 0x1，<init>。

access_flags：为 0x9 (0x8 and 0x1)，代表 static and public

code_off：为 0x148 (c8 02)，指向 code_item。

从 0x148 解析 code_item 如下。

```
registers_size            0x3
ins_size                  0x1
outs_size                 0x2
tries_size                0
debug_info_off            0x220
insns_size                0x8
insns    ushort[insns_size]    0062 0000 011a 000c 206e 0000 0010 000e
```

❑ 0x62 的 opcode 为 sget-object。

其格式如下。

```
sget-object vAA, field@BBBB
A: value register or pair; may be source or dest (8 bits)
B: static field reference inDEX (16 bits)
```

分布如下。

```
AA|op BBBB
```

由于 AA=0，BBBB=0x0000，字段为 out。

- 0x1a 的 opcode 为 const-string。

其格式如下。

```
const-string vAA, string@BBBB
A: destination register (8 bits)
B: string inDEX
```

分布如下。

```
AA|op BBBB
```

由于 AA=1，BBBB=0x000c，对应的字符串为：test!

- 0x6e 的 opcode 为：invoke-virtual。

其格式和分布同 invoke-direct。

```
B|A|op CCCC G|F|E|D [B=5] op {vD, vE, vF, vG, vA}, meth@CCCC
[B=5] op {vD, vE, vF, vG, vA}, type@CCCC
[B=4] op {vD, vE, vF, vG}, kind@CCCC
[B=3] op {vD, vE, vF}, kind@CCCC
[B=2] op {vD, vE}, kind@CCCC
[B=1] op {vD}, kind@CCCC
[B=0] op {}, kind@CCCC
```

这里 B=2，A=0，E=1，D=0，CCCC=0x0000，方法为“Ljava/io/PrintStream;”的 println。

- 0x0e 的 opcode 为 return-void。

经过上述分析，得到指令如下。

```
sget-object v0, Ljava/lang/System;.out:Ljava/io/PrintStream; // field@0000
const-string v1, "test!" // string@000c
invoke-virtual {v0, v1}, Ljava/io/PrintStream;.println:(Ljava/lang/String;)V // method@0000
return-void
```

表 11-17　encoded_method

名　字	格　式
method_idx_diff	uleb128
access_flags	uleb128
code_off	uleb128

表 11-18　encoded_field

Name	Format
field_idx_diff	uleb128
access_flags	uleb128

表 11-19　code_item

Name	Format
registers_size	ushort
ins_size	ushort

续表

Name	Format
outs_size	ushort
tries_size	ushort
debug_info_off	uint
insns_size	uint
insns	ushort[insns_size]
padding	ushort (optional) = 0
tries	try_item[tries_size] (optional)
handlers	encoded_catch_handler_list (optional)

11.3 DEX 文件结构

在 Android 系统中，DEX 文件是可以直接在 Dalvik VM 中加载运行的文件。通过 ADT 和复杂的编译处理后，可以把 Java 源代码转换为 DEX 文件。那么这个文件的格式是什么样的呢？为什么 Android 不直接使用 class 文件，而采用这个不一样文件呢？其实它是针对嵌入式系统优化的结果，Dalvik 虚拟机的指令码并不是标准的 Java 虚拟机指令码，而是使用了自己独有的一套指令集。如果有自己的编译系统，可以不生成 class 文件， 直接生成 DEX 文件。DEX 文件中共用了很多类名称、常量字符串，使它的体积比较小，运行效率也比较高。但归根到底，Dalvik 还是基于寄存器的虚拟机的一个实现。在本节的内容中，将详细讲解 DEX 文件结构的基本知识。

11.3.1 文件头（File Header）

DEX 文件头主要包括校验和以及其他结构的偏移地址和长度信息，文件头的结构如表 11-20 所示。

表 11-20 文件头的结构

字段名称	偏移值	长度	描 述
magic	0x0	8	'Magic'值，即魔数字段，格式如"DEX/n035/0"，其中的 035 表示结构的版本
checksum	0x8	4	校验码
signature	0xC	20	SHA-1 签名
file_size	0x20	4	DEX 文件的总长度
header_size	0x24	4	文件头长度，009 版本=0x5C,035 版本=0x70
endian_tag	0x28	4	标识字节顺序的常量,根据这个常量可以判断文件是否交换了字节顺序，缺省情况下=0x78563412
link_size	0x2C	4	连接段的大小，如果为 0 就表示是静态连接
link_off	0x30	4	连接段的开始位置，从本文件头开始算起。如果连接段的大小为 0，这里也是 0
map_off	0x34	4	map 数据基地址
string_ids_size	0x38	4	字符串列表的字符串个数
string_ids_off	0x3C	4	字符串列表表基地址
type_ids_size	0x40	4	类型列表里类型个数
type_ids_off	0x44	4	类型列表基地址

续表

字段名称	偏移值	长度	描　　述
proto_ids_size	0x48	4	原型列表里原型个数
proto_ids_off	0x4C	4	原型列表基地址
field_ids_size	0x50	4	字段列表里字段个数
field_ids_off	0x54	4	字段列表基地址
method_ids_size	0x58	4	方法列表里方法个数
method_ids_off	0x5C	4	方法列表基地址
class_defs_size	0x60	4	类定义类表中类的个数
class_defs_off	0x64	4	类定义列表基地址
data_size	0x68	4	数据段的大小，必须以 4 字节对齐
data_off	0x6C	4	数据段基地址

11.3.2　魔数字段

魔数字段，主要就是 DEX 文件的标识符，它占用 4 个字节，在目前的源码里是 “DEX\n”，它的作用主要是用来标识 DEX 文件的，比如有一个文件也以 DEX 为后缀名，仅此并不会被认为是 Davlik 虚拟机运行的文件，还要判断这 4 个字节。另外 Davlik 虚拟机也有优化的 DEX，也是通过个字段来区分的，当它是优化的 DEX 文件时，它的值就变成“dey\n”了。根据这 4 个字节，就可以识别不同类型的 DEX 文件了。

紧跟在“DEX\n”后面的是版本字段，主要用来标识 DEX 文件的版本。目前支持的版本号为“035\0”，无论是否优化的版本，都是使用这个版本号。

11.3.3　检验码字段

主要用来检查从这个字段开始到文件结尾，这段数据是否完整，有没有人修改过，或者传送过程中是否有出错等等。通常用来检查数据是否完整的算法有 CRC32、SHA128 等，但这里并不是采用这两类，而采用一个比较特别的算法，叫做 Adler32，这是在开源 zlib 里常用的算法，用来检查文件的完整性。该算法由 MarkAdler 发明，其可靠程度跟 CRC32 差不多，不过还是弱一点点，但它有一个很好的优点，就是使用软件来计算检验码时比 CRC32 要快很多。可见 Android 系统，在算法上就已经为移动设备进行优化了。

下面是 Adler32 算法的 C 源码，注意在 Java 中可使用 java.util.zip.Adler32 类做校验操作：

```
#define ZLIB_INTERNAL
#include "zlib.h"
#define BASE 65521UL /* largest prime smaller than 65536 */
#define NMAX 5552
/*NMAX is the largest n such that 255n(n+1)/2 + (n+1)(BASE-1) <=2^32-1 */

#define DO1(buf,i){adler += (buf)[i]; sum2 += adler;}
#define DO2(buf,i) DO1(buf,i); DO1(buf,i+1);
#define DO4(buf,i) DO2(buf,i); DO2(buf,i+2);
#define DO8(buf,i) DO4(buf,i); DO4(buf,i+4);
#define DO16(buf) DO8(buf,0); DO8(buf,8);

/*use NO_DIVIDE if your processor does not do division in hardware */
#ifdef NO_DIVIDE
#define MOD(a) \
do{ \
if(a >= (BASE << 16)) a -= (BASE << 16); \
if(a >= (BASE << 15)) a -= (BASE << 15); \
```

```
if(a >= (BASE << 14)) a -= (BASE << 14); \
if(a >= (BASE << 13)) a -= (BASE << 13); \
if(a >= (BASE << 12)) a -= (BASE << 12); \
if(a >= (BASE << 11)) a -= (BASE << 11); \
if(a >= (BASE << 10)) a -= (BASE << 10); \
if(a >= (BASE << 9)) a -= (BASE << 9); \
if(a >= (BASE << 8)) a -= (BASE << 8); \
if(a >= (BASE << 7)) a -= (BASE << 7); \
if(a >= (BASE << 6)) a -= (BASE << 6); \
if(a >= (BASE << 5)) a -= (BASE << 5); \
if(a >= (BASE << 4)) a -= (BASE << 4); \
if(a >= (BASE << 3)) a -= (BASE << 3); \
if(a >= (BASE << 2)) a -= (BASE << 2); \
if(a >= (BASE << 1)) a -= (BASE << 1); \
if(a >= BASE) a -= BASE; \
}while (0)
# define MOD4(a) \
do{ \
if(a >= (BASE << 4)) a -= (BASE << 4); \
if(a >= (BASE << 3)) a -= (BASE << 3); \
if(a >= (BASE << 2)) a -= (BASE << 2); \
if(a >= (BASE << 1)) a -= (BASE << 1); \
if(a >= BASE) a -= BASE; \
}while (0)
#else
#define MOD(a) a %= BASE
#define MOD4(a) a %= BASE
#endif

/*========================================================================*/
uLong ZEXPORT adler32(adler, buf, len)
   uLong adler;
   const Bytef *buf;
   uInt len;
{
   unsigned long sum2;
   unsigned n;

   /*split Adler-32 into component sums */
   sum2= (adler >> 16) & 0xffff;
   adler&= 0xffff;

   /*in case user likes doing a byte at a time, keep it fast */
   if(len == 1) {
      adler+= buf[0];
      if(adler >= BASE)adler-= BASE;
      sum2+= adler;
      if(sum2 >= BASE)sum2-= BASE;
      return adler|(sum2 << 16);
   }

   /*initial Adler-32 value (deferred check for len == 1 speed) */
   if(buf == Z_NULL)return 1L;

   /*in case short lengths are provided, keep it somewhat fast */
   if(len < 16) {
      while(len--) {
         adler+= *buf++;
         sum2+= adler;
      }
      if(adler >= BASE)
         adler-= BASE;
      MOD4(sum2); /* only added so many BASE's */
      return adler|(sum2 << 16);
   }

   /*do length NMAX blocks -- requires just one modulo operation */
   while(len >= NMAX) {
      len-= NMAX;
```

```
        n= NMAX/16; /* NMAX is divisible by 16 */
        do{
            DO16(buf); /* 16 sums unrolled */
            buf+= 16;
        }while (--n);
        MOD(adler);
        MOD(sum2);
    }

    /*do remaining bytes (less than NMAX, still just one modulo) */
    if(len) {
        /* avoid modulos if none remaining */
        while(len >= 16) {
            len-= 16;
            DO16(buf);
            buf+= 16;
        }
        while(len--) {
            adler+= *buf++;
            sum2+= adler;
        }
        MOD(adler);
        MOD(sum2);
    }
    /*return recombined sums */
    return adler|(sum2 << 16);
}
```

11.3.4　SHA-1 签名字段

在 DEX 文件里，前面已经有了面有一个 4 字节的检验字段码了，为什么还有 SHA-1 签名字段呢？这样不是造成重复了吗？因为 DEX 文件一般都不是很小，简单的应用程序都有几十千字节，这么多数据使用一个 4 字节的检验码，重复的机率还是有的，也就是说当文件里的数据修改了，还是很有可能检验不出来的。这时检验码就失去了作用，需要使用更加强大的检验码，这就是 SHA-1。SHA-1 校验码有 20 个字节，比前面的检验码多了 16 个字节，几乎不同的文件计算出来的检验不会是一样的。设计两个检验码的目的，就是先使用第一个检验码进行快速检查，这样可以先把简单出错的 DEX 文件丢掉了，接着再使用第二个复杂的检验码进行复杂计算，验证文件是否完整，这样确保执行的文件完整和安全。

SHA 是 Secure Hash Algorithm 的缩写，意为安全散列算法。SHA 是美国国家安全局设计，美国国家标准与技术研究院发布的一系列密码散列函数。SHA-1 的内部比 MD5 更强，其摘要比 MD5 的 16 字节长 4 个字节，这个算法成功经受了密码分析专家的攻击，也因而受到密码学界的推崇。

11.3.5　map_off 字段

map_off 字段主要用于保存 map 开始位置，就是从文件头开始到 map 数据的长度，通过这个索引就可以找到 map 数据。map 的数据结构如表 11-21 所示。

表 11-21　map 的数据结构

名　称	大　小	说　明
size	4 字节	map 里项的个数
list	变长	每一项定义为 12 字节，项的个数由上面项大小决定

定义 map 数据排列结构的格式如下所示：

```
/*
*Direct-mapped "map_list".
*/
truct DEXMapList {
```

```
    u4 size; /* #of entries inlist */
    DEXMapItem list[1]; /* entries */
}DEXMapList;
```

每一个 map 项的结构定义格式如下：

```
/*
*Direct-mapped "map_item".
*/
typedef struct DEXMapItem {
    u2 type; /* type code (seekDEXType* above) */
    u2 unused;
    u4 size; /* count of items ofthe indicated type */
    u4 offset; /* file offset tothe start of data */
}DEXMapItem;
```

结构 DEXMapItem 的功能是定义每一项的数据意义，例如类型、类型个数、类型开始位置。其中定义类型的代码如下：

```
/*map item type codes */
enum{
    kDEXTypeHeaderItem = 0x0000,
    kDEXTypeStringIdItem = 0x0001,
    kDEXTypeTypeIdItem = 0x0002,
    kDEXTypeProtoIdItem = 0x0003,
    kDEXTypeFieldIdItem = 0x0004,
    kDEXTypeMethodIdItem = 0x0005,
    kDEXTypeClassDefItem = 0x0006,
    kDEXTypeMapList = 0x1000,
    kDEXTypeTypeList = 0x1001,
    kDEXTypeAnnotationSetRefList = 0x1002,
    kDEXTypeAnnotationSetItem = 0x1003,
    kDEXTypeClassDataItem = 0x2000,
    kDEXTypeCodeItem = 0x2001,
    kDEXTypeStringDataItem = 0x2002,
    kDEXTypeDebugInfoItem = 0x2003,
    kDEXTypeAnnotationItem = 0x2004,
    kDEXTypeEncodedArrayItem = 0x2005,
    kDEXTypeAnnotationsDirectoryItem = 0x2006,
};
```

从上面的类型可知，它包括了在 DEX 文件里可能出现的所有类型。可以看出这里的类型与文件头里定义的类型有很多是一样的，这里的类型其实就是文件头里定义的类型。其实此 map 数据就是头中类型的重复，完全是为了检验作用而存在的。当 Android 系统加载 DEX 文件时，如果比较文件头类型个数与 map 里类型不一致时，就会停止使用这个 DEX 文件。

11.3.6 string_ids_size 和 off 字段

string_ids_size 字段和 off 字段主要用来标识字符串资源。在编译源程序后，程序里用到的字符串都保存在这个数据段里，以便解释执行这个 DEX 文件使用。其中包括调用库函数里的类名称描述，用于输出显示的字符串等。

string_ids_size 标识了有多少个字符串，string_ids_off 标识字符串数据区的开始位置。字符串的存储结构如下所示：

```
/*
 * Direct-mapped "string_id_item".
 */
typedef struct DEXStringId {
    u4  stringDataOff;      /* file offset to string_data_item */
} DEXStringId;
```

由此可以看出，这个数据区保存的只是字符串表的地址索引。如果要找到字符串的实际数据，还需要通过个地址索引找到文件的相应开始位置，然后才能得到字符串数据。每一个字符串项的索引占用 4 个字节，因此这个数据区的大小就为 4×string_ids_size。实际数据区中的字符串采用 UTF8 格式保存。

例如，如果 DEX 文件使用 16 进制显示出来内容如下：

```
063c 696e 6974 3e00
```

其实际数据则是“<init>\0”。

另外这段数据中不仅包括字符串的内容和结束标志，在最开头的位置还标明了字符串的长度。上例中第一个字节 06 就是表示这个字符串有 6 个字符。

关于字符串的长度，有如下两点需要注意的地方。

（1）关于长度的编码格式。

DEX 文件里采用了变长方式表示字符串长度。一个字符串的长度可能是一个字节（小于 256）或者 4 个字节（1 G 大小以上）。字符串的长度大多数都是小于 256 个字节，因此需要使用一种编码，既可以表示一个字节的长度，也可以表示 4 个字节的长度，并且 1 个字节的长度占绝大多数。能满足这种表示的编码方式有很多，但 DEX 文件里采用的是 uleb128 方式。leb128 编码是一种变长编码，每个字节采用 7 位来表达原来的数据，最高位用来表示是否有后继字节。它的编码算法如下所示：

```
/*
 * Writes a 32-bit value in unsigned ULEB128 format.
 * Returns the updated pointer.
 */
DEX_INLINE u1* writeUnsignedLeb128(u1* ptr, u4 data)
{
    while (true) {
        u1 out = data & 0x7f;
        if (out != data) {
            *ptr++ = out | 0x80;
            data >>= 7;
        } else {
            *ptr++ = out;
            break;
        }
    }
    return ptr;
}
```

它的解码算法如下所示：

```
/*
 * Reads an unsigned LEB128 value, updating the given pointer to point
 * just past the end of the read value. This function tolerates
 * non-zero high-order bits in the fifth encoded byte.
 */
DEX_INLINE int readUnsignedLeb128(const u1** pStream) {
    const u1* ptr = *pStream;
    int result = *(ptr++);
   if (result > 0x7f) {
        int cur = *(ptr++);
        result = (result & 0x7f) | ((cur & 0x7f) << 7);
        if (cur > 0x7f) {
            cur = *(ptr++);
            result |= (cur & 0x7f) << 14;
            if (cur > 0x7f) {
                cur = *(ptr++);
                result |= (cur & 0x7f) << 21;
                if (cur > 0x7f) {
                    /*
                     * Note: We don't check to see if cur is out of
                     * range here, meaning we tolerate garbage in the
                     * high four-order bits.
                     */
```

```
                cur = *(ptr++);
                result |= cur << 28;
            }
        }
    }
}
*pStream = ptr;
return result;
}
```

根据上面的算法分析上面例子字符串，取得第一个字节是 06，最高位为 0，因此没有后继字节，那么取出这个字节里 7 位有效数据，就是 6，也就是说这个字符串是 6 个字节，但不包括结束字符“\0”。

（2）长度的意义。

由于字符串内容采用的是 UTF-8 格式编码，这表示一个字符的字节数是不定的。即有时是一个字节表示一个字符，有时是两个、三个甚至四个字节表示一个字符。而这里的长度代表的并不是整个字符串所占用的字节数，表示这个字符串包含的字符个数。所以在读取时需要注意，尤其是在包含中文字符时，往往会因为读取的长度不正确导致字符串被截断。

11.4 DEXFile 接口详解

在 Android 系统中，DEXFile 接口在文件“libcore\dalvik\src\main\java\dalvik\system\DexFile.java”中定义，各个接口与的继承关系如下所示。

- public final class DEXFile extends Object。
- java.lang.Object。
- dalvik.system.DEXFile。

操作 DEX 文件的原理上和操作 ZipFile 相似，主要是在类装载器中被使用。在实际应用中，不直接打开和读取 DEX 文件，它们被虚拟机以只读方式映射到内存了。在本节的内容中，将详细讲解 DEXFile 接口的基本知识。

11.4.1 构造函数

（1）public DEXFile (File file)。

函数 DEXFile (File file)的功能是通过指定的 File 对象打开 DEX 文件，指定的文件通常是一个“ZIP/JAR”格式的压缩文件，在里面包含一个名为“classes.DEX”的文件。虚拟机将在目录“/data/dalvik-cache”下生成对应的文件名字并打开它，如果系统权限允许的话会首先创建或更新它。不要传目录“/data/dalvik-cache”下的文件名给它，因为这个文件被认为处于初始状态（DEX 被优化之前）。

函数 DEXFile (File file)的原型如下所示：

```
public DexFile(File file) throws IOException {
    this(file.getPath());
}
```

构造函数 DEXFile (File file)的参数是 File，表示引用实际 DEX 文件的 File 对象。此函数会发生 I/O 异常，例如文件不存在或者没有权限访问。

（2）public DEXFile (String fileName)。

此函数的功能是打开指定文件名的 DEX 文件。此处指定的文件通常是一个 ZIP/JAR 文件，里面包含一个“classes.DEX”。虚拟机将在目录“/data/dalvik-cache”下生成对应的文件名字并打开它，如果系统权限允许的话会首先创建或更新它。不要传目录“/data/dalvik-cache”下的文件名给它，因为这个文件被认为处于初始状态（DEX 被优化之前）。

函数 DEXFile (String fileName)的原型如下所示：

```
public DexFile(String fileName) throws IOException {
    mCookie = openDexFile(fileName, null, 0);
    mFileName = fileName;
    guard.open("close");
    //System.out.println("DEX FILE cookie is "+ mCookie);
}
```

此构造函数的参数是 fileName，表示 DEX 文件名。此函数会发生 I/O 异常，例如文件不存在或者没有权限访问。

11.4.2　公共方法

（1）public void close ()方法。

公共方法 close 的功能是关闭 DEX 文件。在系统中有可能存在无法释放任何资源，也就是说，如果来自 DEX 文件的类还存活着，DEX 文件就不能被取消映射。方法 close 可能会在关闭文件的过程中可能发生 I/O 异常，但是一般不会发生。

方法 close 的原型如下所示：

```
public void close() throws IOException {
    guard.close();
    closeDexFile(mCookie);
    mCookie = 0;
}
```

（2）public Enumeration<String> entries ()方法。

此公共方法的功能是枚举 DEX 文件里面的类名。返回值是 DEX 文件所包含类名的枚举，类名的类型是一般内部格式（像 java/lang/String）。

Enumeration<String> entries 的原型如下所示：

```
public Enumeration<String> entries() {
    return new DFEnum(this);
}
```

（3）public String getName ()方法。

公共方法 getName 的功能是获取（已打开）DEX 文件名，返回值是文件名。方法 getName 的原型如下所示：

```
public String getName() {
    return mFileName;
}
```

（4）public static boolean isDEXOptNeeded (String fileName)方法。

公共方法 isDEXOptNeeded 的功能是，如果虚拟机认为 apk/jar 文件已经过期返回 true，并且应该再次通过“DEXopt”传递。参数 fileName 表示被检查 apk/jar 文件的绝对路径名。如果应该调用 DEXopt 处理文件返回 true；否则 false。

isDEXOptNeeded 的原型如下所示：

```
    native public static boolean isDexOptNeeded(String fileName)
            throws FileNotFoundException, IOException;
}
```

公共方法 isDEXOptNeeded 会发生如下异常。

- FileNotFounDEXception：文件不可读、不是一个文件或者文件不存在。

❑ IOException：fileName 不是有效的 apk/jar 文件，或者在解析文件时出现问题。

❑ NullPointerException：fileName 是空的。

❑ StaleDEXCacheError：优化过的 DEX 文件已过期且位于只读分区。

（5）public Class loadClass (String name, ClassLoader loader)方法。

公共方法 loadClass 的功能是装载一个类，会返回成功装载的类，如果失败则返回空值。如果在类装载器之外调用它，往往不会得到你想要的结果，这时请使用 forName(String)。loadClass 的原型如下所示：

```
public Class loadClass(String name, ClassLoader loader) {
    String slashName = name.replace('.', '/');
    return loadClassBinaryName(slashName, loader);
}
```

方法 loadClass 不会在找不到类的时候抛出 ClassNotFounDEXception 异常，因为每次第一个 DEX 文件里找不到类就粗暴地抛出异常是不合理的。

公共方法 loadClass 的参数如下所示。

❑ name：类名，应该是一个“java/lang/String”。

❑ loader：试图装载类的类装载器，大多数情况下就是该方法的调用者。

公共方法 loadClass 的返回值是类名对应的对象，当装载失败时返回空。

（6）public static DEXFile loadDEX (String sourcePathName, String outputPathName, int flags)。

公共方法 loadDEX 的功能是打开一个 DEX 文件，并提供一个文件来保存优化过的 DEX 数据。如果优化过的格式已存在并且是最新的，就直接使用它。如果不是，虚拟机将试图重新创建一个。该方法主要用于应用希望在通常的应用安装机制之外下载和执行 DEX 文件。不能在应用里直接调用该方法，而应该通过一个类装载器例如 dalvik.system.DEXClassLoader。

loadDEX 的原型如下所示：

```
static public DexFile loadDex(String sourcePathName, String outputPathName,
    int flags) throws IOException {

    /*
     * TODO: we may want to cache previously-opened DexFile objects.
     * The cache would be synchronized with close(). This would help
     * us avoid mapping the same DEX more than once when an app
     * decided to open it multiple times. In practice this may not
     * be a real issue.
     */
    return new DexFile(sourcePathName, outputPathName, flags);
}
```

公共方法 loadDEX 的参数如下所示。

❑ sourcePathName：包含“classes.DEX”的 Jar 或者 APK 文件（将来可能会扩展支持"raw DEX"）。

❑ outputPathName：保存优化过的 DEX 数据的文件。

❑ flags：打开可选功能。

方法 loadDEX 的返回值是一个新的或者先前已经打开的 DEXFile，可能发生 IOException 异常，表示无法打开输入或输出文件。

> **注意** 在 DEXFile 接口中还有一个受保护的方法 protected void finalize()，此方法的功能是在类结束时调用，确保 DEX 文件被关闭。另外，此受保护的方法在关闭文件时可能发生 I/O 异常。

11.5　DEX 和动态加载类机制

在 Android 应用开发的一般情况下，常规的开发方式和代码架构就能满足人们的普通需求。但是有些特殊问题，常常引发我们进一步的沉思。我们从沉思中产生顿悟，从而产生新的技术形式。例如应该如何开发一个可以自定义控件的 Android 应用？就像 Eclipse 一样，可以动态加载插件。如何让 Android 应用执行服务器上不可预知的代码？如何对 Android 应用加密，而只在执行时自解密，从而防止被破解？等等上述问题，可以使用下面将要讲解的类加载器来实现。

11.5.1　类加载机制

Dalvik 虚拟机如同其他 Java 虚拟机一样，在运行程序时首先需要将对应的类加载到内存中。而在 Java 标准的虚拟机中，可以从 class 文件中读取类加载，也可以是其他形式的二进制流。所以基于此，在程序运行时通过手动加载 Class 的方式达到代码动态加载执行的目的。

然而 Dalvik 虚拟机毕竟不算是标准的 Java 虚拟机，因此在类加载机制上既有很多相同的地方，也有很多不同之处，这些必须区别对待。Android 中 ClassLoader 的 defineClass 方法具体是调用 VMClassLoader 的 defineClass 本地静态方法，而这个本地方法除了抛出一个 “UnsupportedOperation Exception”之外什么都没做，甚至连返回值都为空。在文件“dalvik\vm\native\java_lang_ VMClassLoader.cpp”中，Dalvik_java_lang_VMClassLoader_defineClass 的实现代码如下所示：

```
static void Dalvik_java_lang_VMClassLoader_defineClass(const u4* args,
    JValue* pResult)
{
    Object* loader = (Object*) args[0];
    StringObject* nameObj = (StringObject*) args[1];
    const u1* data = (const u1*) args[2];
    int offset = args[3];
    int len = args[4];
    Object* pd = (Object*) args[5];
    char* name = NULL;
    name = dvmCreateCstrFromString(nameObj);
    LOGE("ERROR: defineClass(%p, %s, %p, %d, %d, %p)\n",
        loader, name, data, offset, len, pd);
    dvmThrowException("Ljava/lang/UnsupportedOperationException;",
        "can't load this type of class file");
    free(name);
    RETURN_VOID();
}
```

11.5.2　具体加载

那如果在 Dalvik 虚拟机里 ClassLoader 不太好用， Android 从 ClassLoader 派生出了两个类：DEXClassLoader 和 PathClassLoader。其中类 DEXClassLoader 在文件“libcore\dalvik\src\main\java\dalvik\system\DexClassLoader.java”中定义，具体实现代码如下所示：

```
public class DexClassLoader extends BaseDexClassLoader {
    /**
     * Creates a {@code DexClassLoader} that finds interpreted and native
     * code.  Interpreted classes are found in a set of DEX files contained
     * in Jar or APK files.
     *
     * <p>The path lists are separated using the character specified by the
     * {@code path.separator} system property, which defaults to {@code :}.
     *
     * @param dexPath the list of jar/apk files containing classes and
```

```
 *     resources, delimited by {@code File.pathSeparator}, which
 *     defaults to {@code ":"} on Android
 * @param optimizedDirectory directory where optimized dex files
 *     should be written; must not be {@code null}
 * @param libraryPath the list of directories containing native
 *     libraries, delimited by {@code File.pathSeparator}; may be
 *     {@code null}
 * @param parent the parent class loader
 */
public DexClassLoader(String dexPath, String optimizedDirectory,
        String libraryPath, ClassLoader parent) {
    super(dexPath, new File(optimizedDirectory), libraryPath, parent);
}
}
```

类 PathClassLoader 在文件“libcore\dalvik\src\main\java\dalvik\system\PathClassLoader.java”中定义，具体实现代码如下所示：

```
public class PathClassLoader extends BaseDexClassLoader {
    /**
     * Creates a {@code PathClassLoader} that operates on a given list of files
     * and directories. This method is equivalent to calling
     * {@link #PathClassLoader(String, String, ClassLoader)} with a
     * {@code null} value for the second argument (see description there).
     *
     * @param dexPath the list of jar/apk files containing classes and
     * resources, delimited by {@code File.pathSeparator}, which
     * defaults to {@code ":"} on Android
     * @param parent the parent class loader
     */
    public PathClassLoader(String dexPath, ClassLoader parent) {
        super(dexPath, null, null, parent);
    }

    /**
     * Creates a {@code PathClassLoader} that operates on two given
     * lists of files and directories. The entries of the first list
     * should be one of the following:
     *
     * <ul>
     * <li>JAR/ZIP/APK files, possibly containing a "classes.dex" file as
     * well as arbitrary resources.
     * <li>Raw ".dex" files (not inside a zip file).
     * </ul>
     *
     * The entries of the second list should be directories containing
     * native library files.
     *
     * @param dexPath the list of jar/apk files containing classes and
     * resources, delimited by {@code File.pathSeparator}, which
     * defaults to {@code ":"} on Android
     * @param libraryPath the list of directories containing native
     * libraries, delimited by {@code File.pathSeparator}; may be
     * {@code null}
     * @param parent the parent class loader
     */
    public PathClassLoader(String dexPath, String libraryPath,
            ClassLoader parent) {
        super(dexPath, null, libraryPath, parent);
    }
}
```

通过上述实现代码可知，这两个继承自 ClassLoader 的类加载器，本质上是重载了 ClassLoader 的 findClass 方法。在执行 loadClass 时，可以参看文件“libcore\luni\src\main\java\java\lang\ClassLoader.java”中对 ClassLoader 的如下定义源码：

```
protected Class<?> loadClass(String className, boolean resolve)
throws ClassNotFounDEXception {
```

```
Class<?> clazz = findLoadedClass(className);
    if (clazz == null) {
        try {
            clazz = parent.loadClass(className, false);
        } catch (ClassNotFounDEXception e) {
            // Don't want to see this.
        }
        if (clazz == null) {
            clazz = findClass(className);
        }
    }
    return clazz;
}
```

因此 DEXClassLoader 和 PathClassLoader 都属于符合双亲委派模型的类加载器（因为它们没有重载 loadClass 方法）。也就是说，它们在加载一个类之前，回去检查自己以及自己以上的类加载器是否已经加载了这个类。如果已经加载过了，就会直接将之返回，而不会重复加载。

DEXClassLoader 和 PathClassLoader 其实都是通过 DEXFile 这个类来实现类加载的。这里需要顺便提一下的是，Dalvik 虚拟机识别的是 DEX 文件，而不是 class 文件。因此，供类加载的文件也只能是 DEX 文件，或者包含有 DEX 文件的.apk 或.jar 文件。

也许有人想到，既然 DEXFile 可以直接加载类，那么为什么还要使用 ClassLoader 的子类呢？DEXFile 在加载类时，具体是调用成员方法 loadClass 或者 loadClassBinaryName。其中 loadClassBinary Name 需要将包含包名的类名中的"."转换为"/"。看一下 loadClass 代码就会明白：

```
public Class loadClass(String name, ClassLoader loader) {
        String slashName = name.replace('.', '/');
        return loadClassBinaryName(slashName, loader);
}
```

PathClassLoader 是通过构造函数 new DEXFile(path)来产生 DEXFile 对象的；而 DEXClassLoader 则是通过其静态方法 loadDEX（path, outpath, 0）得到 DEXFile 对象的。这两者的区别在于 DEXClassLoader 需要提供一个可写的 outpath 路径，用来释放.apk 包或者.jar 包中的 DEX 文件。换个说法来说，就是 PathClassLoader 不能主动从 zip 包中释放出 DEX，因此只支持直接操作 DEX 格式文件，或者已经安装的 apk（因为已经安装的 apk 在 cache 中存在缓存的 DEX 文件）。而 DEXClassLoader 可以支持.apk、.jar 和.DEX 文件，并且会在指定的 outpath 路径释放出 DEX 文件。

另外，PathClassLoader 在加载类时调用的是 DEXFile 的 loadClassBinaryName，而 DEXClassLoader 调用的是 loadClass。因此，在使用 PathClassLoader 时类全名需要用"/"替换"."。

这样具体的加载操作比较简单，可能使用到的工具有：javac、dx、eclipse 等。其中 dx 工具最好指明"--no-strict"，因为 class 文件的路径可能不匹配。

在加载类之后，通常可以通过 Java 反射机制来使用这个类。但是这样效率相对不高，而且经常用反射代码也会比较复杂凌乱。更好的做法是定义一个 interface，并将这个 interface 写进容器端。待加载的类，继承自这个 interface，并且有一个参数为空的构造函数，以能够通过 Class 的 newInstance 方法产生对象。然后将对象强制转换为 interface 对象，于是就可以直接调用成员方法了。

11.5.3 代码加密

在实现代码加密时，最初设想是将 DEX 文件加密，然后通过 JNI 将解密代码写在 Native 层。解密之后直接传上二进制流，再通过 defineClass 将类加载到内存中。其实现在也可以这样做，由于不能直接使用 defineClass，而必须传文件路径给 dalvik 虚拟机内核，因此解密后的文件需要写到磁盘上，增加了被破解的风险。

Dalvik 虚拟机内核仅支持从 DEX 文件加载类的方式是不灵活的。在 RawDEXFile 诞生之前，都只能使用这种存在一定风险的加密方式。需要注意释放的 DEX 文件路径及权限管理。另外在加载完毕类之后，除非出于其他的目的，否则应该马上删除临时的解密文件。

11.6 动态加载 jar 和 DEX

在目前的软硬件环境下，Native App 与 Web App 在用户体验上有着明显的优势。但是在实际项目应用中，有些会因为业务的频繁变更而频繁地升级客户端，造成较差的用户体验，而这也正是 Web App 的优势。在接下来的内容中，将简要讲解 Android 动态加载 jar 文件和 DEX 文件的基本过程。

在 Android 系统中可以实现动态加载，但是无法像 Java 中那样方便地动态加载 jar。这是因为 Android 虚拟机（Dalvik VM）不认识 Java 打出 jar 的 byte code，需要通过 dx 工具来优化转换成 Dalvik byte code 后才行。这一点在 Android 项目打包的 apk 中可以看出：引入其他 Jar 的内容都被打包进了 classes.DEX。

在当前的 Android 应用中，有如下两个 API 可以实现动态加载功能。

- DEXClassLoader：这个可以加载“jar/apk/DEX”，也可以从 SD 卡中加载。
- PathClassLoader：只能加载已经安装到 Android 系统中的 APK 文件。

有关类 DEXClassLoader 和类 PathClassLoader 的具体实现，已经在本章前面进行了详细讲解，在此不再阐述。

第 12 章 Dvlik VM 内存系统详解

Dalvik 虚拟机是 Google 在 Android 平台上的 Java 虚拟机的实现，内存管理是 Dalvik 虚拟机中的一个重要组件。在本章的内容中，将详细讲解 Android 系统中 Dvlik VM 内存系统的基本知识，为读者步入本书后面知识的学习打下基础。

12.1 如何分配内存

内存管理是 Dalvik VM 中的一个重要组件。从概念上来说，如下两个部分内存管理的核心。

- ❑ 内存分配。
- ❑ 内存回收。

Java 语言使用 new 操作符来分配内存，但是与 C/C++等语言不同的是，Java 语言并没有提供任何操作来释放内存，而是通过一种叫做垃圾收集的机制来回收内存。对于内存管理的实现，通过如下 3 个方面来加以分析。

- ❑ 内存分配。
- ❑ 内存回收。
- ❑ 内存管理调试。

在本节的内容中，将首先分析 Dalvik 虚拟机是如何分配内存的。

1. 对象布局

内存管理的主要操作之一是为 Java 对象分配内存，Java 对象在虚拟机中的内存布局如图 12-1 所示。

所有的对象都有一个相同的头部 clazz 和 lock。

（1）clazz：指向该对象的类对象，类对象用来描述该对象所属的类，这样可以很容易地从一个对象获取该对象所属的类的具体信息。

（2）lock：是一个无符号整数，用以实现对象的同步。

（3）data：用于存放对象数据，根据对象的不同数据区的大小是不同的。

2. 堆

堆是 Dalvik 虚拟机从操作系统分配的一块连续的虚拟内存，如图 12-2 所示。其中 heapBase 表示堆的起始地址，heapLimit 表示堆的最大地址，堆大小的最大值可以通过“-Xmx”选项或 dalvik.vm.heapsize 指定。在原生系统中，一般 dalvik.vm.heapsize 值是 32 MB，在 MIUI 中将其设为 64 MB。

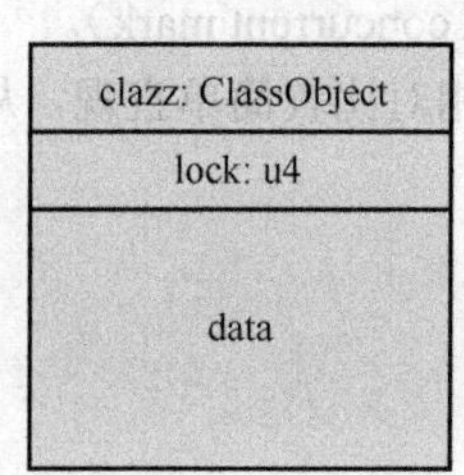

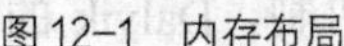
图 12-1 内存布局

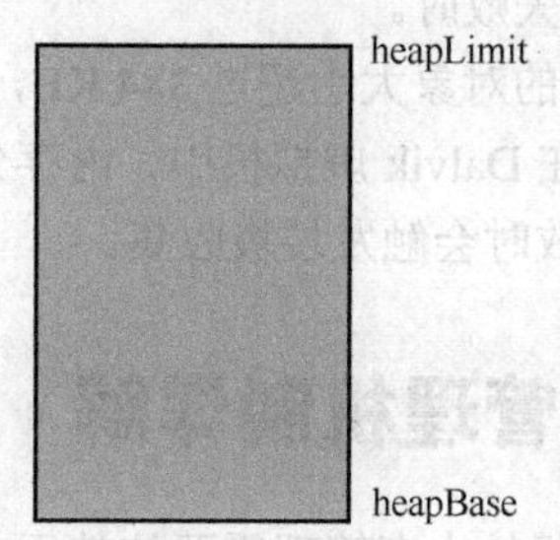

图 12-2 堆结构

3. 堆内存位图

在虚拟机中维护了两个对应于堆内存的位图，分别被称为 liveBits 和 markBits。在对象布局中会发现对象最小占用 8 个字节，当为对象分配内存时要求必须 8 字节对齐。这也就是说，对象的大小会调整为 8 字节的倍数。比如说一个对象的实际大小是 13 字节，但是在分配内存的时候分配 16 字节。因此所有对象的起始地址一定是 8 字节的倍数。堆内存位图就是用来描述堆内存的，每一个 bit 描述 8 个字节，因此堆内存位图的大小是对的 64 分之一。对于 MIUI 的实现来说，这两个位图各占 1 MB。

liveBits 的功能是跟踪堆中以分配的内存，每当分配一个对象时，对象的内存起始地址对应于位图中的位被设为 1。

4. 堆的内存管理

在 Dalvik 虚拟机的实现中，是通过底层的 bionicC 库的 malloc/free 操作来分配/释放内存的。库 bionicC 的“malloc/free”操作是基于 DougLea 的实现（dlmalloc），这是一个经典的 C 内存管理库。有关库 dlmalloc 的基本知识，读者可以参阅相关资料。

5. dvmAllocObject

在 Dalvik 虚拟机中，操作符 new 最终对应 C 函数 dvmAllocObject()。下面通过伪码的形式列出 dvmAllocObject 的实现：

```
Object*dvmAllocObject(ClassObject *clazz, int flags) {
      n = get object size form class object clazz
      first try: allocate n bytes from heap
      if first try failed {
            run garbage collector without collecting soft references
            second try: allocate n bytes from heap
      }
      if second try failed {
            third try: grow the heap and allocate n bytes from heap
/*堆是虚拟内存，一开始并未分配所有的物理内存，只要还没有达到虚拟内存的最大值，可以通过获取更多物理内存的方式来扩展堆*/
      }
      if third try failed {
            run garbage collector with collecting soft references
            fourth try: grow the hap and allocate n bytes from heap
      }
      if fourth try failed, return null pointer, dalvik vm will abort
}
```

由此可以看出，为了分配内存，虚拟机尽了最大的努力，做了 4 次尝试。其中进行了两次垃圾收集，第一次不收集 SoftReference，第二次收集 SoftReference。从中也可以看出垃圾收集的时机，实质上在 Dalvik 虚拟机实现中有 3 个时机可以触发垃圾收集的运行。

- 程序员显式的调用 System.gc()。

- 内存分配失败时。
- 如果分配的对象大小超过 384 KB，运行并发标记（concurrent mark）。

综上所述，在 Dalvik 虚拟机中，内存分配操作的流程相对比较简单直观，从一个堆中分配可用内存，分配失败时会触发垃圾收集。

12.2 内存管理机制详解

Dalvik 虚拟机的内存管理需要依赖于 Lnux 的内存管理机制，Dalvik 的内存管理的实现源码保存在“vm\alloc”目录中。在本节的内容中，将通过对源码的分析来简要讲解 Dalvik 内存管理机制的基本知识。

（1）表示堆的结构体。

在文件 HeapSource.c 中定义表示堆的结构体，其源码如下所示：

```
typedef struct {
    /*使用 dlmalloc 分配的内存.
     */
    mspace *msp;

    HeapBitmap objectBitmap;

    /* 堆可以增长的最大值.
     */
    size_t absoluteMaxSize;

    /* 已分配的字节数.
     */
    size_t bytesAllocated;
    /* 已分配的对象数.
     */
    size_t objectsAllocated;
} Heap;
```

（2）表示位图堆的结构体数据。

在文件 HeapBitmap.h 中定义表示位图堆的结构体数据，其源码如下所示：

```
typedef struct {
    /* 位图数据.
     */
    unsigned long int *bits;

    /* 位图的大小
     */
    size_t bitsLen;

    /* 位图对应的对象指针数组的首地址.
     */
    uintptr_t base;

    /* 位图使用中的最后一位被设置的对象指针地址，如果全没设置则(max < base).
     */
    uintptr_t max;
} HeapBitmap;
```

（3）HeapSource 结构体。

在 Dalvik 虚拟机中，使用结构体 HeapSource 来管理各种 Heap 数据，Heap 只是其中的一个子项，其源码在文件 HeapSource.c 中定义：

```
struct HeapSource {
    /* 堆的使用率，范围从 1 到 HEAP_UTILIZATION_MAX
```

```
     */
    size_t targetUtilization;

    /* 分配堆的最小尺寸.
     */
    size_t minimumSize;

    /* 堆分配的初始尺寸.
     */
    size_t startSize;

    /* 允许分配的堆增长到的最大尺寸.
     */
    size_t absoluteMaxSize;

    /* 理想的堆的最大尺寸.
     */
    size_t idealSize;

    /* 在垃圾收集前允许堆分配的最大尺寸.
     */
    size_t softLimit;

    /* 堆数组，最大尺寸为 3.
     */
    Heap heaps[HEAP_SOURCE_MAX_HEAP_COUNT];

    /* 当前堆的个数.
     */
    size_t numHeaps;

    /* 对外分配计数.
     */
    size_t externalBytesAllocated;

    /* 允许外部分配的最大值
     */
    size_t externalLimit;

    /* 在创建这个 HeapSource 的时候是否是 Zygote 模式，确定是否有 Zygote 进程.
     */
    bool sawZygote;
};
```

（4）和 mark bits 相关的结构体。

在文件 MarkSweep.h 中定义了和 mark bits 相关的结构体，其源码如下所示：

```
typedef struct {
    /* 允许增长到的最低地址
     */
    const Object **limit;

    /* 栈顶
     */
    const Object **top;

    /* 栈底
     */
    const Object **base;
} GcMarkStack;

typedef struct {
/* 存放位图的数组
     */
    HeapBitmap bitmaps[HEAP_SOURCE_MAX_HEAP_COUNT];
/* 位图数
     */
    size_t numBitmaps;
/* GC 标记栈
```

```
    */
    GcMarkStack stack;
/* 存放地址上限的标志
    */
    const void *finger;   // only used while scanning/recursing.
} GcMarkContext;
```

（5）结构体 GcHeap。

在文件 HeapInternal.h 中定义了 Dalvik 的垃圾回收机制需要用到结构体 GcHeap，其源码如下所示：

```
struct GcHeap {
 /* HeapSource 结构，包含了所有的堆数据      */
   HeapSource      *heapSource;

    /* 存储不能被垃圾回收对象的参考表
     */
    HeapRefTable    nonCollectableRefs;

    /*存储一些当被垃圾回收时需要执行 finalization 方法的参考表
     *
     */
    LargeHeapRefTable  *finalizableRefs;

    /*存储需要执行 finalization 方法的对象的参考表
     */
    LargeHeapRefTable  *pendingFinalizationRefs;

    /* 软引用对象列表
     */
    Object          *softReferences;
 /* 弱引用对象列表
     */
    Object          *weakReferences;
 /* 影子引用对象列表
     */
    Object          *phantomReferences;

    /* 需要被执行 clear 或 enqueue 方法的引用对象列表
     */
    LargeHeapRefTable  *referenceOperations;

    /* 如果对象不为空，则表示 HeapWorker 线程正在执行
     * executing.
     */
    Object *heapWorkerCurrentObject;
    Method *heapWorkerCurrentMethod;

    /*如果 heapWorkerCurrentObject 不为空，表示 HeapWorker 开始执行这个方法的时间
     */
    u8 heapWorkerInterpStartTime;
    /*如果 heapWorkerCurrentObject 不为空，表示 HeapWorker CPU 开始执行这个方法的时间
     */
    u8 heapWorkerInterpCpuStartTime;

    /* 下一次裁剪 Heap Source 的时间
     */
    struct timespec heapWorkerNextTrim;

    /* 标记步骤中的状态
     */
    GcMarkContext   markContext;

    /* GC 开始的时间.
     */
    u8              gcStartTime;

    /* 是否正在执行 GC.
```

```
    */
    bool          gcRunning;

    /* GC 时引用对象回收多少
     * 不回收，回收一半，全部回收
     */
    enum { SR_COLLECT_NONE, SR_COLLECT_SOME, SR_COLLECT_ALL }
                  softReferenceCollectionState;

    /* 存在多少软引用对象时开始回收引用对象
     */
    size_t        softReferenceHeapSizeThreshold;

    /*当软引用回收策略为回收一半时使用的概率值
     */
    int           softReferenceColor;

    /* 引用收集策略
     */
    bool          markAllReferents;
#if DVM_TRACK_HEAP_MARKING
    /* Every time an unmarked object becomes marked, markCount
     * is incremented and markSize increases by the size of
     * that object.
     */
    size_t        markCount;
    size_t        markSize;
#endif

    /* 下面是和调试相关的信息
     */
    int           ddmHpifWhen;
    int           ddmHpsgWhen;
    int           ddmHpsgWhat;
    int           ddmNhsgWhen;
    int           ddmNhsgWhat;

#if WITH_HPROF
    bool          hprofDumpOnGc;
    const char*    hprofFileName;
    hprof_context_t *hprofContext;
    int           hprofResult;
#endif
};
```

（6）初始化垃圾回收器。

在文件 Init.c 中，通过函数 dvmGcStartup()来初始化垃圾回收器：

```
bool dvmGcStartup(void)
{
    dvmInitMutex(&gDvm.gcHeapLock);
    return dvmHeapStartup();
}
```

（7）初始化和 Heap 相关的信息。

在文件 alloc\Heap.c 中，通过 dvmHeapStartup()函数来初始化和 Heap 相关的信息，例如常见的内存分配和内存管理等工作。其源码如下所示：

```
bool dvmHeapStartup()
{
    GcHeap *gcHeap;
#if defined(WITH_ALLOC_LIMITS)
    gDvm.checkAllocLimits = false;
    gDvm.allocationLimit = -1;
#endif
    gcHeap = dvmHeapSourceStartup(gDvm.heapSizeStart, gDvm.heapSizeMax);
    if (gcHeap == NULL) {
        return false;
```

```
    }
    gcHeap->heapWorkerCurrentObject = NULL;
    gcHeap->heapWorkerCurrentMethod = NULL;
    gcHeap->heapWorkerInterpStartTime = 0LL;
    gcHeap->softReferenceCollectionState = SR_COLLECT_NONE;
    gcHeap->softReferenceHeapSizeThreshold = gDvm.heapSizeStart;
    gcHeap->ddmHpifWhen = 0;
    gcHeap->ddmHpsgWhen = 0;
    gcHeap->ddmHpsgWhat = 0;
    gcHeap->ddmNhsgWhen = 0;
    gcHeap->ddmNhsgWhat = 0;
#if WITH_HPROF
    gcHeap->hprofDumpOnGc = false;
    gcHeap->hprofContext = NULL;
#endif
    /* This needs to be set before we call dvmHeapInitHeapRefTable().
     */
    gDvm.gcHeap = gcHeap;
    /* Set up the table we'll use for ALLOC_NO_GC.
     */
    if (!dvmHeapInitHeapRefTable(&gcHeap->nonCollectableRefs,
                    kNonCollectableRefDefault))
    {
        LOGE_HEAP("Can't allocate GC_NO_ALLOC table/n");
        goto fail;
    }
    /* Set up the lists and lock we'll use for finalizable
     * and reference objects.
     */
    dvmInitMutex(&gDvm.heapWorkerListLock);
    gcHeap->finalizableRefs = NULL;
    gcHeap->pendingFinalizationRefs = NULL;
    gcHeap->referenceOperations = NULL;
    /* Initialize the HeapWorker locks and other state
     * that the GC uses.
     */
    dvmInitializeHeapWorkerState();
    return true;
fail:
    gDvm.gcHeap = NULL;
    dvmHeapSourceShutdown(gcHeap);
    return false;
}
```

（8）创建 GcHeap。

在文件 alloc\HeapSource.c 中，通过函数 dvmHeapSourceStartup()来创建 GcHeap。其源码如下所示：

```
GcHeap *
dvmHeapSourceStartup(size_t startSize, size_t absoluteMaxSize)
{
    GcHeap *gcHeap;
    HeapSource *hs;
    Heap *heap;
    mspace msp;
    assert(gHs == NULL);

    if (startSize > absoluteMaxSize) {
        LOGE("Bad heap parameters (start=%d, max=%d)\n",
          startSize, absoluteMaxSize);
        return NULL;
    }

    /* Create an unlocked dlmalloc mspace to use as
     * the small object heap source.
     */
    msp = createMspace(startSize, absoluteMaxSize, 0);
```

```
    if (msp == NULL) {
        return false;
    }

    /* Allocate a descriptor from the heap we just created.
     */
    gcHeap = mspace_malloc(msp, sizeof(*gcHeap));
    if (gcHeap == NULL) {
        LOGE_HEAP("Can't allocate heap descriptor\n");
        goto fail;
    }
    memset(gcHeap, 0, sizeof(*gcHeap));

    hs = mspace_malloc(msp, sizeof(*hs));
    if (hs == NULL) {
        LOGE_HEAP("Can't allocate heap source\n");
        goto fail;
    }
    memset(hs, 0, sizeof(*hs));

    hs->targetUtilization = DEFAULT_HEAP_UTILIZATION;
    hs->minimumSize = 0;
    hs->startSize = startSize;
    hs->absoluteMaxSize = absoluteMaxSize;
    hs->idealSize = startSize;
    hs->softLimit = INT_MAX;    // no soft limit at first
    hs->numHeaps = 0;
    hs->sawZygote = gDvm.zygote;
    if (!addNewHeap(hs, msp, absoluteMaxSize)) {
        LOGE_HEAP("Can't add initial heap\n");
        goto fail;
    }

    gcHeap->heapSource = hs;

    countAllocation(hs2heap(hs), gcHeap, false);
    countAllocation(hs2heap(hs), hs, false);

    gHs = hs;
    return gcHeap;

fail:
    destroy_contiguous_mspace(msp);
    return NULL;
}
```

（9）追踪位置。

在文件 alloc\HeapSource.c 中，通过函数 countAllocation()在 Heap::obiect Bitmap 上进行标记，以便追踪这些区域的位置。其源码如下所示：

```
static inline void
countAllocation(Heap *heap, const void *ptr, bool isObj)
{
    assert(heap->bytesAllocated < mspace_footprint(heap->msp));

    heap->bytesAllocated += mspace_usable_size(heap->msp, ptr) +
            HEAP_SOURCE_CHUNK_OVERHEAD;
    if (isObj) {
        heap->objectsAllocated++;
//标记回收
        dvmHeapBitmapSetObjectBit(&heap->objectBitmap, ptr);
    }
    assert(heap->bytesAllocated < mspace_footprint(heap->msp));
}
HB_INLINE_PROTO(
    bool
    dvmHeapBitmapMayContainObject(const HeapBitmap *hb,
            const void *obj)
```

```
)
{
    const uintptr_t p = (const uintptr_t)obj;

    assert((p & (HB_OBJECT_ALIGNMENT - 1)) == 0);

    return p >= hb->base && p <= hb->max;
}
HB_INLINE_PROTO(
    bool
    dvmHeapBitmapCoversAddress(const HeapBitmap *hb, const void *obj)
)
{
    assert(hb != NULL);

    if (obj != NULL) {
        const uintptr_t offset = (uintptr_t)obj - hb->base;
        const size_t index = HB_OFFSET_TO_INDEX(offset);
        return index < hb->bitsLen / sizeof(*hb->bits);
    }
    return false;
}
......
```

（10）分配空间。

在文件 Heap.c 中，通过函数 dvmMalloc()实现空间的分配工作。其源码如下所示：

```
void* dvmMalloc(size_t size, int flags)
{
    GcHeap *gcHeap = gDvm.gcHeap;
    DvmHeapChunk *hc;
    void *ptr;
    bool triedGc, triedGrowing;

#if 0
    /* handy for spotting large allocations */
    if (size >= 100000) {
        LOGI("dvmMalloc(%d):\n", size);
        dvmDumpThread(dvmThreadSelf(), false);
    }
#endif

#if defined(WITH_ALLOC_LIMITS)
    /*
     * See if they've exceeded the allocation limit for this thread.
     *
     * A limit value of -1 means "no limit".
     *
     * This is enabled at compile time because it requires us to do a
     * TLS lookup for the Thread pointer. This has enough of a performance
     * impact that we don't want to do it if we don't have to. (Now that
     * we're using gDvm.checkAllocLimits we may want to reconsider this,
     * but it's probably still best to just compile the check out of
     * production code -- one less thing to hit on every allocation.)
     */
    if (gDvm.checkAllocLimits) {
        Thread* self = dvmThreadSelf();
        if (self != NULL) {
            int count = self->allocLimit;
            if (count > 0) {
                self->allocLimit--;
            } else if (count == 0) {
                /* fail! */
                assert(!gDvm.initializing);
                self->allocLimit = -1;
                dvmThrowException("Ldalvik/system/AllocationLimitError;",
                    "thread allocation limit exceeded");
                return NULL;
            }
```

```
        }
    }

    if (gDvm.allocationLimit >= 0) {
        assert(!gDvm.initializing);
        gDvm.allocationLimit = -1;
        dvmThrowException("Ldalvik/system/AllocationLimitError;",
            "global allocation limit exceeded");
        return NULL;
    }
#endif

    dvmLockHeap();

    /* Try as hard as possible to allocate some memory.
     */
    hc = tryMalloc(size);
    if (hc != NULL) {
alloc_succeeded:
        /* We've got the memory.
         */
        if ((flags & ALLOC_FINALIZABLE) != 0) {
            /* This object is an instance of a class that
             * overrides finalize().  Add it to the finalizable list.
             *
             * Note that until DVM_OBJECT_INIT() is called on this
             * object, its clazz will be NULL.  Since the object is
             * in this table, it will be scanned as part of the root
             * set.  scanObject() explicitly deals with the NULL clazz.
             */
            if (!dvmHeapAddRefToLargeTable(&gcHeap->finalizableRefs,
                                    (Object *)hc->data))
            {
                LOGE_HEAP("dvmMalloc(): no room for any more "
                        "finalizable objects\n");
                dvmAbort();
            }
        }

#if WITH_OBJECT_HEADERS
        hc->header = OBJECT_HEADER;
        hc->birthGeneration = gGeneration;
#endif
        ptr = hc->data;

        /* The caller may not want us to collect this object.
         * If not, throw it in the nonCollectableRefs table, which
         * will be added to the root set when we GC.
         *
         * Note that until DVM_OBJECT_INIT() is called on this
         * object, its clazz will be NULL.  Since the object is
         * in this table, it will be scanned as part of the root
         * set.  scanObject() explicitly deals with the NULL clazz.
         */
        if ((flags & ALLOC_NO_GC) != 0) {
            if (!dvmHeapAddToHeapRefTable(&gcHeap->nonCollectableRefs, ptr)) {
                LOGE_HEAP("dvmMalloc(): no room for any more "
                        "ALLOC_NO_GC objects: %zd\n",
                        dvmHeapNumHeapRefTableEntries(
                                &gcHeap->nonCollectableRefs));
                dvmAbort();
            }
        }

#ifdef WITH_PROFILER
        if (gDvm.allocProf.enabled) {
            Thread* self = dvmThreadSelf();
            gDvm.allocProf.allocCount++;
            gDvm.allocProf.allocSize += size;
```

```
            if (self != NULL) {
                self->allocProf.allocCount++;
                self->allocProf.allocSize += size;
            }
        }
#endif
    } else {
        /* The allocation failed.
         */
        ptr = NULL;

#ifdef WITH_PROFILER
        if (gDvm.allocProf.enabled) {
            Thread* self = dvmThreadSelf();
            gDvm.allocProf.failedAllocCount++;
            gDvm.allocProf.failedAllocSize += size;
            if (self != NULL) {
                self->allocProf.failedAllocCount++;
                self->allocProf.failedAllocSize += size;
            }
        }
#endif
    }

    dvmUnlockHeap();

    if (ptr != NULL) {
        /*
         * If this block is immediately GCable, and they haven't asked us not
         * to track it, add it to the internal tracking list.
         *
         * If there's no "self" yet, we can't track it. Calls made before
         * the Thread exists should use ALLOC_NO_GC.
         */
        if ((flags & (ALLOC_DONT_TRACK | ALLOC_NO_GC)) == 0) {
            dvmAddTrackedAlloc(ptr, NULL);
        }
    } else {
        /*
         * The allocation failed; throw an OutOfMemoryError.
         */
        throwOOME();
    }

    return ptr;
}
```

上述具体分配过程由 tryMalloc 控制，具体执行流程为：

```
tryMalloc()-gcForMalloc()-dvmCollectGarbageInternal()
```

当满足条件时，会调用 dvmCollectGarbageInternal()进行垃圾回收。上述函数在文件“/dalvik/vm/alloc/Heap.cpp”中定义，接下来将一一讲解上述函数的具体实现。

首先看函数 tryMalloc，功能是尝试尽可能的分配一些内存，具体实现代码如下所示：

```
static void *tryMalloc(size_t size)
{
    void *ptr;

//TODO: figure out better heuristics
//    There will be a lot of churn if someone allocates a bunch of
//    big objects in a row, and we hit the frag case each time.
//    A full GC for each.
//    Maybe we grow the heap in bigger leaps
//    Maybe we skip the GC if the size is large and we did one recently
//      (number of allocations ago) (watch for thread effects)
//    DeflateTest allocs a bunch of ~128k buffers w/in 0-5 allocs of each other
//      (or, at least, there are only 0-5 objects swept each time)
```

```
    ptr = dvmHeapSourceAlloc(size);
    if (ptr != NULL) {
        return ptr;
    }

    /*
     * The allocation failed.  If the GC is running, block until it
     * completes and retry.
     */
    if (gDvm.gcHeap->gcRunning) {
        /*
         * The GC is concurrently tracing the heap.  Release the heap
         * lock, wait for the GC to complete, and retrying allocating.
         */
        dvmWaitForConcurrentGcToComplete();
    } else {
        /*
         * Try a foreground GC since a concurrent GC is not currently running.
         */
        gcForMalloc(false);
    }

    ptr = dvmHeapSourceAlloc(size);
    if (ptr != NULL) {
        return ptr;
    }

    /* Even that didn't work;  this is an exceptional state.
     * Try harder, growing the heap if necessary.
     */
    ptr = dvmHeapSourceAllocAndGrow(size);
    if (ptr != NULL) {
        size_t newHeapSize;

        newHeapSize = dvmHeapSourceGetIdealFootprint();
//TODO: may want to grow a little bit more so that the amount of free
//      space is equal to the old free space + the utilization slop for
//      the new allocation.
        LOGI_HEAP("Grow heap (frag case) to "
                "%zu.%03zuMB for %zu-byte allocation",
                FRACTIONAL_MB(newHeapSize), size);
        return ptr;
    }

    /* Most allocations should have succeeded by now, so the heap
     * is really full, really fragmented, or the requested size is
     * really big.  Do another GC, collecting SoftReferences this
     * time.  The VM spec requires that all SoftReferences have
     * been collected and cleared before throwing an OOME.
     */
//TODO: wait for the finalizers from the previous GC to finish
    LOGI_HEAP("Forcing collection of SoftReferences for %zu-byte allocation",
            size);
    gcForMalloc(true);
    ptr = dvmHeapSourceAllocAndGrow(size);
    if (ptr != NULL) {
        return ptr;
    }
//TODO: maybe wait for finalizers and try one last time

    LOGE_HEAP("Out of memory on a %zd-byte allocation.", size);
//TODO: tell the HeapSource to dump its state
    dvmDumpThread(dvmThreadSelf(), false);

    return NULL;
}
```

函数 gcForMalloc 的功能是执行回收机制，其回收机制就利用对象的标记与否来判定是否要回收。函数 gcForMalloc 的具体实现代码如下所示：

```
/* Do a full garbage collection, which may grow the
 * heap as a side-effect if the live set is large.
 */
static void gcForMalloc(bool clearSoftReferences)
{
    if (gDvm.allocProf.enabled) {
        Thread* self = dvmThreadSelf();
        gDvm.allocProf.gcCount++;
        if (self != NULL) {
            self->allocProf.gcCount++;
        }
    }
    /* This may adjust the soft limit as a side-effect.
     */
    const GcSpec *spec = clearSoftReferences ? GC_BEFORE_OOM : GC_FOR_MALLOC;
    dvmCollectGarbageInternal(spec);
}
```

再看函数 dvmCollectGarbageInternal，功能是实现垃圾回收，具体实现代码如下所示：

```
void dvmCollectGarbageInternal(const GcSpec* spec)
{
    GcHeap *gcHeap = gDvm.gcHeap;
    u4 gcEnd = 0;
    u4 rootStart = 0 , rootEnd = 0;
    u4 dirtyStart = 0, dirtyEnd = 0;
    size_t numObjectsFreed, numBytesFreed;
    size_t currAllocated, currFootprint;
    size_t percentFree;
    int oldThreadPriority = INT_MAX;

    /* The heap lock must be held.
     */

    if (gcHeap->gcRunning) {
        LOGW_HEAP("Attempted recursive GC");
        return;
    }

    // Trace the beginning of the top-level GC.
    if (spec == GC_FOR_MALLOC) {
        ATRACE_BEGIN("GC (alloc)");
    } else if (spec == GC_CONCURRENT) {
        ATRACE_BEGIN("GC (concurrent)");
    } else if (spec == GC_EXPLICIT) {
        ATRACE_BEGIN("GC (explicit)");
    } else if (spec == GC_BEFORE_OOM) {
        ATRACE_BEGIN("GC (before OOM)");
    } else {
        ATRACE_BEGIN("GC (unknown)");
    }

    gcHeap->gcRunning = true;

    rootStart = dvmGetRelativeTimeMsec();
    ATRACE_BEGIN("GC: Threads Suspended"); // Suspend A
    dvmSuspendAllThreads(SUSPEND_FOR_GC);

    /*
     * If we are not marking concurrently raise the priority of the
     * thread performing the garbage collection.
     */
    if (!spec->isConcurrent) {
        oldThreadPriority = os_raiseThreadPriority();
    }
    if (gDvm.preVerify) {
        LOGV_HEAP("Verifying roots and heap before GC");
        verifyRootsAndHeap();
    }
```

```
    dvmMethodTraceGCBegin();

    /* Set up the marking context.
     */
    if (!dvmHeapBeginMarkStep(spec->isPartial)) {
        ATRACE_END(); // Suspend A
        ATRACE_END(); // Top-level GC
        LOGE_HEAP("dvmHeapBeginMarkStep failed; aborting");
        dvmAbort();
    }

    /* Mark the set of objects that are strongly reachable from the roots.
     */
    LOGD_HEAP("Marking...");
    dvmHeapMarkRootSet();

    /* dvmHeapScanMarkedObjects() will build the lists of known
     * instances of the Reference classes.
     */
    assert(gcHeap->softReferences == NULL);
    assert(gcHeap->weakReferences == NULL);
    assert(gcHeap->finalizerReferences == NULL);
    assert(gcHeap->phantomReferences == NULL);
    assert(gcHeap->clearedReferences == NULL);

    if (spec->isConcurrent) {
        /*
         * Resume threads while tracing from the roots.  We unlock the
         * heap to allow mutator threads to allocate from free space.
         */
        dvmClearCardTable();
        dvmUnlockHeap();
        dvmResumeAllThreads(SUSPEND_FOR_GC);
        ATRACE_END(); // Suspend A
        rootEnd = dvmGetRelativeTimeMsec();
    }

    /* Recursively mark any objects that marked objects point to strongly.
     * If we're not collecting soft references, soft-reachable
     * objects will also be marked.
     */
    LOGD_HEAP("Recursing...");
    dvmHeapScanMarkedObjects();

    if (spec->isConcurrent) {
        /*
         * Re-acquire the heap lock and perform the final thread
         * suspension.
         */
        dirtyStart = dvmGetRelativeTimeMsec();
        dvmLockHeap();
        ATRACE_BEGIN("GC: Threads Suspended"); // Suspend B
        dvmSuspendAllThreads(SUSPEND_FOR_GC);
        /*
         * As no barrier intercepts root updates, we conservatively
         * assume all roots may be gray and re-mark them.
         */
        dvmHeapReMarkRootSet();
        /*
         * With the exception of reference objects and weak interned
         * strings, all gray objects should now be on dirty cards.
         */
        if (gDvm.verifyCardTable) {
            dvmVerifyCardTable();
        }
        /*
         * Recursively mark gray objects pointed to by the roots or by
         * heap objects dirtied during the concurrent mark.
```

```
        */
        dvmHeapReScanMarkedObjects();
    }

    /*
     * All strongly-reachable objects have now been marked. Process
     * weakly-reachable objects discovered while tracing.
     */
    dvmHeapProcessReferences(&gcHeap->softReferences,
                             spec->doPreserve == false,
                             &gcHeap->weakReferences,
                             &gcHeap->finalizerReferences,
                             &gcHeap->phantomReferences);

#if defined(WITH_JIT)
    /*
     * Patching a chaining cell is very cheap as it only updates 4 words. It's
     * the overhead of stopping all threads and synchronizing the I/D cache
     * that makes it expensive.
     *
     * Therefore we batch those work orders in a queue and go through them
     * when threads are suspended for GC.
     */
    dvmCompilerPerformSafePointChecks();
#endif

    LOGD_HEAP("Sweeping...");

    dvmHeapSweepSystemWeaks();

    /*
     * Live objects have a bit set in the mark bitmap, swap the mark
     * and live bitmaps. The sweep can proceed concurrently viewing
     * the new live bitmap as the old mark bitmap, and vice versa.
     */
    dvmHeapSourceSwapBitmaps();

    if (gDvm.postVerify) {
        LOGV_HEAP("Verifying roots and heap after GC");
        verifyRootsAndHeap();
    }

    if (spec->isConcurrent) {
        dvmUnlockHeap();
        dvmResumeAllThreads(SUSPEND_FOR_GC);
        ATRACE_END(); // Suspend B
        dirtyEnd = dvmGetRelativeTimeMsec();
    }
    dvmHeapSweepUnmarkedObjects(spec->isPartial, spec->isConcurrent,
                                &numObjectsFreed, &numBytesFreed);
    LOGD_HEAP("Cleaning up...");
    dvmHeapFinishMarkStep();
    if (spec->isConcurrent) {
        dvmLockHeap();
    }

    LOGD_HEAP("Done.");

    /* Now's a good time to adjust the heap size, since
     * we know what our utilization is.
     *
     * This doesn't actually resize any memory;
     * it just lets the heap grow more when necessary.
     */
    dvmHeapSourceGrowForUtilization();

    currAllocated = dvmHeapSourceGetValue(HS_BYTES_ALLOCATED, NULL, 0);
    currFootprint = dvmHeapSourceGetValue(HS_FOOTPRINT, NULL, 0);
```

```
    dvmMethodTraceGCEnd();
    LOGV_HEAP("GC finished");

    gcHeap->gcRunning = false;

    LOGV_HEAP("Resuming threads");
    if (spec->isConcurrent) {
        /*
         * Wake-up any threads that blocked after a failed allocation
         * request.
         */
        dvmBroadcastCond(&gDvm.gcHeapCond);
    }
    if (!spec->isConcurrent) {
        dvmResumeAllThreads(SUSPEND_FOR_GC);
        ATRACE_END(); // Suspend A
        dirtyEnd = dvmGetRelativeTimeMsec();
        /*
         * Restore the original thread scheduling priority if it was
         * changed at the start of the current garbage collection.
         */
        if (oldThreadPriority != INT_MAX) {
            os_lowerThreadPriority(oldThreadPriority);
        }
    }

    /*
     * Move queue of pending references back into Java.
     */
    dvmEnqueueClearedReferences(&gDvm.gcHeap->clearedReferences);

    gcEnd = dvmGetRelativeTimeMsec();
    percentFree = 100 - (size_t)(100.0f * (float)currAllocated / currFootprint);
    if (!spec->isConcurrent) {
        u4 markSweepTime = dirtyEnd - rootStart;
        u4 gcTime = gcEnd - rootStart;
        bool isSmall = numBytesFreed > 0 && numBytesFreed < 1024;
        ALOGD("%s freed %s%zdK, %d%% free %zdK/%zdK, paused %ums, total %ums",
             spec->reason,
             isSmall ? "<" : "",
             numBytesFreed ? MAX(numBytesFreed / 1024, 1) : 0,
             percentFree,
             currAllocated / 1024, currFootprint / 1024,
             markSweepTime, gcTime);
    } else {
        u4 rootTime = rootEnd - rootStart;
        u4 dirtyTime = dirtyEnd - dirtyStart;
        u4 gcTime = gcEnd - rootStart;
        bool isSmall = numBytesFreed > 0 && numBytesFreed < 1024;
        ALOGD("%s freed %s%zdK, %d%% free %zdK/%zdK, paused %ums+%ums, total %ums",
             spec->reason,
             isSmall ? "<" : "",
             numBytesFreed ? MAX(numBytesFreed / 1024, 1) : 0,
             percentFree,
             currAllocated / 1024, currFootprint / 1024,
             rootTime, dirtyTime, gcTime);
    }
    if (gcHeap->ddmHpifWhen != 0) {
        LOGD_HEAP("Sending VM heap info to DDM");
        dvmDdmSendHeapInfo(gcHeap->ddmHpifWhen, false);
    }
    if (gcHeap->ddmHpsgWhen != 0) {
        LOGD_HEAP("Dumping VM heap to DDM");
        dvmDdmSendHeapSegments(false, false);
    }
    if (gcHeap->ddmNhsgWhen != 0) {
        LOGD_HEAP("Dumping native heap to DDM");
```

```
        dvmDdmSendHeapSegments(false, true);
    }

    ATRACE_END(); // Top-level GC
}
```

由此可见，函数 dvmCollectGarbageInternal 的功能十分强大，除了具备资源回收功能外，还具有如下所示的功能。

- Suspend（暂停）此程序里所有的 thread（线程）。
- 初始化一个 Mark Stack 用来作为追踪 HeapBitmap 中的 bitmark。
- 在 Heap memory 中的所有对象中，只要不是被 Soft Reference 或是 Weak Reference 所参考的，全部都做一个标记，并且在 bitmap memory 中的位做设定，并且将对象推进 Mark Stack。
- 处理 heap memory 中的所有 soft reference 和 weak reference，将这些的 reference 全部都设为 NULL，使其可及对象失去参考对象。
- 将所有在 Heap Memory 中没被标记的系统对象全部回收。
- 如果活动对象在 mark bitmap 中有一个位设定，则将 live bitmap 与 mark bitmap 互换，反之亦然。将有标记跟没标记的活动对象做分类。
- 访问 bitmap memory 检查位，将有设定的位与未设定的位取出来，以找出未标记的对象并做回收。
- 将 bitmap memory 中的所有位重新进行初始化处理，并删除 Mark Stack。
- 利用 Heap Source 重新调整 Heap memory size。
- 重启程序中的所有的 thread。

到此为止，完成了 Android 虚拟机进行内存分配的过程。此时可以了解到 Android DVM 的内存管理是如何实现资源回收机制的。在 Android 一开机的时候就开始处理 DVM 的初始化，这初始化过程中会做如下三件事。

（1）划分一块 mSpace 的内存区域。

（2）在这个 mSpace 内存区域中配置一块 Heap memory，由 Heap source 来设定此块内存的属性，并配置一块 Bitmap memory 来记录此 Heap memory 的状态。

（3）配置一块 mark stack memory 来存放所配置的对象。

当 Java 程序需要实体化一个对象时，其流程会从 dvmMalloc 函数开始。一开始便在 Heap Memory 中配置了一块内存，在 Heap memory 中的 Bitmap Memory 中去设定位。然后由 gcForMalloc 去执行回收机制，在回收机制中利用对象的标记与否来判定是否要回收。

注意　在本章的分析中没有提到强可及对象，这一概念在 Java 比较常见。在 Java 的语法中会参考指向 new 出来的对象，这个参考就称为强参考或是强引用，而相对的对象就称为强可及物件。如果一旦参考，在程序某处被指定为 NULL 或是指向别的对象，此时原本所指的对象就退化为软可及对象。强可及对象和软可及对象与被指向的参考有关。Java 中的参考有分强、软、弱、虚 4 类，具体内容读者可以参阅 Java 虚拟机的知识。

12.3　优化 Dalvik 虚拟机的堆内存分配

对于 Android 平台来说，其托管层使用的是 Dalvik Java VM。从目前的表现来看，还有很多地方可以进行优化处理，比如在开发一些大型游戏或耗资源的应用中可能考虑手动干涉 GC 处理，使用类 dalvik.system.VMRuntime 提供的方法 setTargetHeapUtilization()可以增强程序堆内存的处理效率。下面是具体使用方法：

```
private final static float TARGET_HEAP_UTILIZATION = 0.75f;
```

在程序 onCreate 时就可以调用如下方法即可：

```
VMRuntime.getRuntime().setTargetHeapUtilization(TARGET_HEAP_UTILIZATION);
```

对于一些大型 Android 项目或游戏来说，在算法处理上没有问题外，影响性能瓶颈的主要是 Android 自己内存管理机制问题。目前手机厂商对 RAM 都比较吝啬，对于软件的流畅性来说 RAM 对性能的影响十分敏感，除了优化 Dalvik 虚拟机的堆内存分配外，还可以强制定义自己软件堆内存大小。

可以使用 Dalvik 提供的类 dalvik.system.VMRuntime 来设置最小堆内存。类 VMRuntime 提供了对虚拟机全局的特定功能的接口，Android 为每个程序分配的堆内存可以通过类 Runtime 的方法 totalMemory()和方法 freeMemory()获取 VM 的一些内存信息。例如：

```
private final static int CWJ_HEAP_SIZE = 6* 1024* 1024 ;
VMRuntime.getRuntime().setMinimumHeapSize(CWJ_HEAP_SIZE); //设置最小 heap 内存为 6MB 大小
```

当然对于内存吃紧的机器来说，还可以通过手动干涉 GC 的方式去处理。比如在处理图片时，通常需要销毁 Android 上的 Bitmap 对象，可以借助方法 recycle()显式让 GC 回收一个 Bitmap 对象，通常对一个不用的 Bitmap 可以使用下面的方式，例如：

```
if(bitmapObject.isRecycled()==false) //如果没有回收
bitmapObject.recycle();
```

其中用 Max Heap Size 表示堆内存的上限值，Android 的缺省值是 16 MB（某些机型是 24 MB），对于普通应用这是不能改的。函数 setMinimumHeapSize 其实只是改变了堆的下限值，它可以防止过于频繁的堆内存分配，当设置最小堆内存大小超过上限值时仍然采用堆的上限值(16 MB)，对于内存不足没什么作用。

方法 setTargetHeapUtilization（float newTarget）可以设定内存利用率的百分比，当实际的利用率偏离这个百分比时，虚拟机会在 GC 的时候调整堆内存大小，让实际占用率向这个百分比靠拢。

第13章　Dalvik VM 垃圾收集机制

垃圾收集是 Dalvik VM 内存管理的核心，垃圾收集的性能在很大程度上影响了一个 Android 应用程序内存使用的效率。顾名思义，垃圾收集就是收集垃圾内存加以回收。在本章的内容中，将详细讲解 Android 系统中 Dalvik 垃圾收集机制的基本知识，为读者步入本书后面知识的学习打下基础。

13.1 引用计数算法

在垃圾回收算法技术中，主要有以下 3 种经典的算法。

- 引用计数。
- MarkSweep 算法。
- SemiSpaceCopy 算法。

其他算法或者混合以上 3 种法来使用，可以根据不同的场合来选择不同的算法。在本节的内容中，将首先讲解引用计数算法的基本知识。

引用计数算法非常简单，就是使用一个变量记录这块内存或者对象的使用次数。比如在 COM 技术中，就是使用引用计数来确认这个 COM 对象是什么时候删除的。当一个 COM 对象给不同线程使用时，由于不同的线程生命周期不同，因此，没有办法知道这个 COM 对象到底在哪个线程删除，只能使用引用计数来删除，否则还需要不同线程之间添加同步机制，这样是非常麻烦和复杂的，如果 COM 对象有很多，就变成基本上不能实现了。

引用计数算法的优点如下。

- 在对象变成垃圾时，可以马上进行回收，回收效率和成本都是最低。
- 内存使用率最高，基本上没有时间花费，不需要把所有访问 COM 对象线程都停下来。

引用计数算法的缺点是。

- 引用计数会影响执行效率，每引用一次都需要更新引用计数，对于 COM 对象那是人工控制的，因此次数很少，没有什么影响。但是在 Java 里是由编译程序来控制的，因此引用次数非常多。
- 不能解决交叉引用或者环形引用的问题。比如在一个环形链表里，每一个元素都引用前面的元素，这样首尾相连的链表，当所有元素都变成不需要时，就没有办法识别出来，并进行内存回收。

13.2 Mark Sweep 算法

该算法又被称为“标记－清除”算法，依赖于对所有存活对象进行一次全局遍历来确定哪些对象可以回收，遍历的过程从根出发，找到所有可到达对象，其他不可到达的对象就是垃圾对象，可被回收。正如其名称所暗示的那样，这个算法分为两大阶段：标记和清除。这种分步执行的思路构成了现代垃圾收集算法的思想基础。与引用计数算法不同的是，“标记－清除”算法不需要监测每一次内存分配和指针操作，只需要在标记阶段进行一次统计就行了。“标记－清除”算法可以非常自然地处理环形问题，另外在创建对象和销毁对象时少了操作引用计数值的开销。不过“标记－清除”算法也有一个缺点，就是需要标记和清除阶段中把所有对象停止执行。在垃圾回收器运行过程中，应用程序必须暂时停止，并等到垃圾回收器全部运行完成后，才能重新启动应用程序运行。

Dalvik 虚拟机最常用的算法便是 Mark Sweep 算法，该算法一般分 Mark 阶段（标记出活动对象）、Sweep 阶段（回收垃圾内存）和可选的 Compact 阶段（减少堆中的碎片）。Dalvik 虚拟机的实现不进行可选的 Compact 阶段。

（1）Mark 阶段。

垃圾收集的第一步是标记出活动对象，因为没有办法识别那些不可访问的对象（unreachableobjects）、因此只能标记出活动对象，这样所有未被标记的对象就是可以回收的垃圾。

❑ 根集合（RootSet）。

当进行垃圾收集时，需要停止 Dalvik 虚拟机的运行（当然，除了垃圾收集之外）。因此垃圾收集又被称作 STW（stop-the-world，整个世界因我而停止）。Dalvik 虚拟机在运行过程中要维护一些状态信息，这些信息包括：每个线程所保存的寄存器、Java 类中的静态字段、局部和全局的 JNI 引用，JVM 中的所有函数调用会对应一个相应 C 的栈帧。每一个栈帧里可能包含对对象的引用，比如包含对象引用的局部变量和参数。

所有这些引用信息被加入到一个集合中，称为根集合。然后从根集合开始，递归地查找可以从根集合出发访问的对象。因此，Mark 过程又被称为追踪，追踪所有可被访问的对象。如图 13-1 所示，假定从根集合{a}开始，可以访问的对象集合为{a,b,c,d}，这样就追踪出所有可被访问的对象集合。

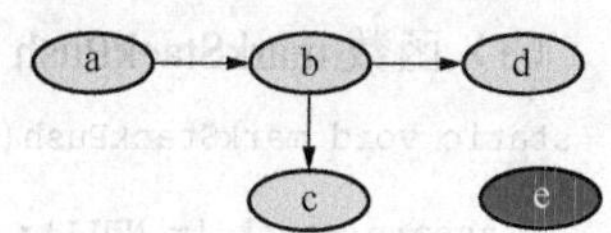

图 13-1 Mark 过程

❑ 标记栈（MarkStack）。

垃圾收集使用栈来保存根集合，然后对栈中的每一个元素，递归追踪所有可访问的对象，对于所有可访问的对象，在 markBits 位图中该将对象的内存起始地址对应的位设为 1。这样当栈为空时，markBits 位图就是所有可访问的对象集合。

（2）Sweep 阶段。

垃圾收集的第二步就是回收内存，在 Mark 阶段通过 markBits 位图可以得到所有可访问的对象集合，而 liveBits 位图表示所有已经分配的对象集合。因此通过比较这两个位图，liveBits 位图和 markBits 位图的差异就是所有可回收的对象集合。Sweep 阶段调用 free 来释放这些内存给堆。

（3）Concurrent Mark（并发标记）。

为了运行垃圾收集，需要停止虚拟机的运行，这可能会导致程序比较长时间的停顿。垃圾收集的主要工作位于 Mark 阶段，Dalvik 虚拟机使用了 Concurrent Mark 技术以缩短停顿时间。在 Concurrent Mark 中引入一个单独的 gc 线程，由该线程去跟踪自己的根集合中所有可访问的对象，同时所有其他的线程也在运行。这也是 Concurrent 一词的含义。但是为了回收内存，即运行 Sweep 阶段，必需停止虚拟机的运行。这会导入一个问题，即在 gc 线程 mark 对象的时候，其他线程的运行又引入了新的访问对象。因此在 Sweep 阶段，又重新运行 mark 阶段，但是在这个阶段对于已经 mark 的对象来说，可以不用继续递归追踪了，这样从一定程度上降低了程序的停顿时间。

13.3 和垃圾收集算法有关的函数

在源文件“alloc/MarkSweep.h”中，定义了和垃圾收集有关的函数，各个函数的具体说明如下所示。

（1）函数 createMarkStack 的功能是创建初始化堆栈顶部和建议需要的标记堆栈页。具体实现代码如下所示：

```
static bool createMarkStack(GcMarkStack *stack)
{
    assert(stack != NULL);
    size_t length = dvmHeapSourceGetIdealFootprint() * sizeof(Object*) /
        (sizeof(Object) + HEAP_SOURCE_CHUNK_OVERHEAD);
    madvise(stack->base, length, MADV_NORMAL);
    stack->top = stack->base;
    return true;
}
```

（2）函数 destroyMarkStack 的功能是销毁无效的栈顶和建议不需要的标记堆栈页。具体实现代码如下所示：

```
static void destroyMarkStack(GcMarkStack *stack)
{
    assert(stack != NULL);
    madvise(stack->base, stack->length, MADV_DONTNEED);
    stack->top = NULL;
}
```

（3）函数 markStackPush 的功能是在堆栈中做一个标记，具体实现代码如下所示：

```
static void markStackPush(GcMarkStack *stack, const Object *obj)
{
    assert(stack != NULL);
    assert(stack->base <= stack->top);
    assert(stack->limit > stack->top);
    assert(obj != NULL);
    *stack->top = obj;
    ++stack->top;
}
```

（4）函数 markStackPop 的功能是在标记栈中加入一个对象，具体实现代码如下所示：

```
static const Object *markStackPop(GcMarkStack *stack)
{
    assert(stack != NULL);
    assert(stack->base < stack->top);
    assert(stack->limit > stack->top);
    --stack->top;
    return *stack->top;
}
```

（5）调用函数 dvmHeapBeginMarkStep()创建位图，并从对象位图里拷贝一份位图出来，以便后面对这个位图进行标记。函数 dvmHeapBeginMarkStep()的具体实现代码如下所示：

```
bool dvmHeapBeginMarkStep(bool isPartial)
{
    GcMarkContext *ctx = &gDvm.gcHeap->markContext;

    if (!createMarkStack(&ctx->stack)) {
        return false;
    }
    ctx->finger = NULL;
    ctx->immuneLimit = (char*)dvmHeapSourceGetImmuneLimit(isPartial);
    return true;
}
```

（6）函数 setAndReturnMarkBit 的功能是设置并返回有标记位的对象，具体实现代码如下所示：

```
static long setAndReturnMarkBit(GcMarkContext *ctx, const void *obj)
{
    return dvmHeapBitmapSetAndReturnObjectBit(ctx->bitmap, obj);
}
```

（7）函数 markObjectNonNul 的功能是标记非空对象，具体实现代码如下所示：

```
static void markObjectNonNull(const Object *obj, GcMarkContext *ctx,
                              bool checkFinger)
{
    assert(ctx != NULL);
    assert(obj != NULL);
    assert(dvmIsValidObject(obj));
    if (obj < (Object *)ctx->immuneLimit) {
        assert(isMarked(obj, ctx));
        return;
    }
    if (!setAndReturnMarkBit(ctx, obj)) {
        /* This object was not previously marked.
         */
        if (checkFinger && (void *)obj < ctx->finger) {
            /* This object will need to go on the mark stack.
             */
            markStackPush(&ctx->stack, obj);
        }
    }
}
```

（8）函数 markObject 的功能是通过递归算法来标记对象，具体实现代码如下所示：

```
static void markObject(const Object *obj, GcMarkContext *ctx)
{
    if (obj != NULL) {
        markObjectNonNull(obj, ctx, true);
    }
}
```

任何新标记的对象的地址是低于指定访问的扫描位图，所以这些对象需要被添加到标记栈。

（9）函数 rootMarkObjectVisitor 的功能是标记根对象的访问者，具体实现代码如下所示：

```
static void rootMarkObjectVisitor(void *addr, u4 thread, RootType type,
                                  void *arg)
{
    assert(addr != NULL);
    assert(arg != NULL);
    Object *obj = *(Object **)addr;
    GcMarkContext *ctx = (GcMarkContext *)arg;
    if (obj != NULL) {
        markObjectNonNull(obj, ctx, false);
    }
}
```

（10）调用函数 dvmHeapMarkRootSet()标记所有根对象，即负责标记 heap 中没有任何引用连接的 root 对象。其实现代码如下所示：

```
void dvmHeapMarkRootSet()
{
    GcHeap *gcHeap = gDvm.gcHeap;
    dvmMarkImmuneObjects(gcHeap->markContext.immuneLimit);
    dvmVisitRoots(rootMarkObjectVisitor, &gcHeap->markContext);
}
```

（11）函数 rootReMarkObjectVisitor 的功能是重新标记根对象的访问者，具体实现代码如下所示：

```
static void rootReMarkObjectVisitor(void *addr, u4 thread, RootType type,
                                    void *arg)
{
```

```
    assert(addr != NULL);
    assert(arg != NULL);
    Object *obj = *(Object **)addr;
    GcMarkContext *ctx = (GcMarkContext *)arg;
    if (obj != NULL) {
        markObjectNonNull(obj, ctx, true);
    }
}
```

（12）函数 dvmHeapReMarkRootSet 的功能是标灰根中的所有引用，具体实现代码如下所示：

```
void dvmHeapReMarkRootSet()
{
    GcMarkContext *ctx = &gDvm.gcHeap->markContext;
    assert(ctx->finger == (void *)ULONG_MAX);
    dvmVisitRoots(rootReMarkObjectVisitor, ctx);
}
```

（13）函数 scanFields 的功能是扫描所有的实例字段，具体实现代码如下所示：

```
static void scanFields(const Object *obj, GcMarkContext *ctx)
{
    assert(obj != NULL);
    assert(obj->clazz != NULL);
    assert(ctx != NULL);
    if (obj->clazz->refOffsets != CLASS_WALK_SUPER) {
        unsigned int refOffsets = obj->clazz->refOffsets;
        while (refOffsets != 0) {
            size_t rshift = CLZ(refOffsets);
            size_t offset = CLASS_OFFSET_FROM_CLZ(rshift);
            Object *ref = dvmGetFieldObject(obj, offset);
            markObject(ref, ctx);
            refOffsets &= ~(CLASS_HIGH_BIT >> rshift);
        }
    } else {
        for (ClassObject *clazz = obj->clazz;
             clazz != NULL;
             clazz = clazz->super) {
            InstField *field = clazz->ifields;
            for (int i = 0; i < clazz->ifieldRefCount; ++i, ++field) {
                void *addr = BYTE_OFFSET(obj, field->byteOffset);
                Object *ref = ((JValue *)addr)->l;
                markObject(ref, ctx);
            }
        }
    }
}
```

（14）函数 scanStaticFields 的功能是扫描类对象中的静态域，具体实现代码如下所示：

```
static void scanStaticFields(const ClassObject *clazz, GcMarkContext *ctx)
{
    assert(clazz != NULL);
    assert(ctx != NULL);
    for (int i = 0; i < clazz->sfieldCount; ++i) {
        char ch = clazz->sfields[i].signature[0];
        if (ch == '[' || ch == 'L') {
            Object *obj = clazz->sfields[i].value.l;
            markObject(obj, ctx);
        }
    }
}
```

（15）函数 scanInterfaces 的功能是访问一个类对象的接口，具体实现代码如下所示：

```
static void scanInterfaces(const ClassObject *clazz, GcMarkContext *ctx)
{
    assert(clazz != NULL);
    assert(ctx != NULL);
    for (int i = 0; i < clazz->interfaceCount; ++i) {
```

```
            markObject((const Object *)clazz->interfaces[i], ctx);
        }
}
```

（16）函数 scanClassObject 的功能是分别扫描头、静态字段的引用和一个类对象的接口指针。具体实现代码如下所示：

```
static void scanClassObject(const Object *obj, GcMarkContext *ctx)
{
    assert(obj != NULL);
    assert(dvmIsClassObject(obj));
    assert(ctx != NULL);
    markObject((const Object *)obj->clazz, ctx);
    const ClassObject *asClass = (const ClassObject *)obj;
    if (IS_CLASS_FLAG_SET(asClass, CLASS_ISARRAY)) {
        markObject((const Object *)asClass->elementClass, ctx);
    }
    /* Do super and the interfaces contain Objects and not dex idx values? */
    if (asClass->status > CLASS_IDX) {
        markObject((const Object *)asClass->super, ctx);
    }
    markObject((const Object *)asClass->classLoader, ctx);
    scanFields(obj, ctx);
    scanStaticFields(asClass, ctx);
    if (asClass->status > CLASS_IDX) {
        scanInterfaces(asClass, ctx);
    }
}
```

（17）函数 scanArrayObject 的功能是扫描所有数组对象的标题。如果数组对象是专门为一个引用类型服务的，则需要扫描其阵列的数据和。具体实现代码如下所示：

```
static void scanArrayObject(const Object *obj, GcMarkContext *ctx)
{
    assert(obj != NULL);
    assert(obj->clazz != NULL);
    assert(ctx != NULL);
    markObject((const Object *)obj->clazz, ctx);
    if (IS_CLASS_FLAG_SET(obj->clazz, CLASS_ISOBJECTARRAY)) {
        const ArrayObject *array = (const ArrayObject *)obj;
        const Object **contents = (const Object **)(void *)array->contents;
        for (size_t i = 0; i < array->length; ++i) {
            markObject(contents[i], ctx);
        }
    }
}
```

（18）函数 referenceClassFlags 的功能是返回有关参考子类标志，具体实现代码如下所示：

```
static int referenceClassFlags(const Object *obj)
{
    int flags = CLASS_ISREFERENCE |
                CLASS_ISWEAKREFERENCE |
                CLASS_ISFINALIZERREFERENCE |
                CLASS_ISPHANTOMREFERENCE;
    return GET_CLASS_FLAG_GROUP(obj->clazz, flags);
}
```

（19）函数 isSoftReference、isWeakReference、isFinalizerReference 和 isPhantomReference 的功能是，如果对象是软引用、来自弱引用、派生自 FinalizerReference、派生自 PhantomReference，则都返回 true，具体实现代码如下所示：

```
static bool isSoftReference(const Object *obj)
{
    return referenceClassFlags(obj) == CLASS_ISREFERENCE;
}
static bool isWeakReference(const Object *obj)
{
```

```
    return referenceClassFlags(obj) & CLASS_ISWEAKREFERENCE;
}

/*
 * Returns true if the object derives from FinalizerReference.
 */
static bool isFinalizerReference(const Object *obj)
{
    return referenceClassFlags(obj) & CLASS_ISFINALIZERREFERENCE;
}

/*
 * Returns true if the object derives from PhantomReference.
 */
static bool isPhantomReference(const Object *obj)
{
    return referenceClassFlags(obj) & CLASS_ISPHANTOMREFERENCE;
}
```

（20）函数 enqueuePendingReference 的功能是处理排队等待的引用的高优先的线程，具体实现代码如下所示：

```
static void enqueuePendingReference(Object *ref, Object **list)
{
    assert(ref != NULL);
    assert(list != NULL);
    size_t offset = gDvm.offJavaLangRefReference_pendingNext;
    if (*list == NULL) {
        dvmSetFieldObject(ref, offset, ref);
        *list = ref;
    } else {
        Object *head = dvmGetFieldObject(*list, offset);
        dvmSetFieldObject(ref, offset, head);
        dvmSetFieldObject(*list, offset, ref);
    }
}
```

（21）函数 dequeuePendingReference 的功能是移除引用队列中的一个参考，具体实现代码如下所示：

```
static Object *dequeuePendingReference(Object **list)
{
    assert(list != NULL);
    assert(*list != NULL);
    size_t offset = gDvm.offJavaLangRefReference_pendingNext;
    Object *head = dvmGetFieldObject(*list, offset);
    Object *ref;
    if (*list == head) {
        ref = *list;
        *list = NULL;
    } else {
        Object *next = dvmGetFieldObject(head, offset);
        dvmSetFieldObject(*list, offset, next);
        ref = head;
    }
    dvmSetFieldObject(ref, offset, NULL);
    return ref;
}
```

（22）函数 delayReferenceReferent 的功能是根据对象的类型，将其放入相应的待释放队列中。如果对象是 fianlizeable 对象，则放入 finalizerReferences 队列中。如对象是 WeakReference 对象，则将其放入 weakReferences 队列中。函数 delayReferenceReferent 的具体实现代码如下所示：

```
static void delayReferenceReferent(Object *obj, GcMarkContext *ctx)
{
    assert(obj != NULL);
    assert(obj->clazz != NULL);
```

```
    assert(IS_CLASS_FLAG_SET(obj->clazz, CLASS_ISREFERENCE));
    assert(ctx != NULL);
    GcHeap *gcHeap = gDvm.gcHeap;
    size_t pendingNextOffset = gDvm.offJavaLangRefReference_pendingNext;
    size_t referentOffset = gDvm.offJavaLangRefReference_referent;
    Object *pending = dvmGetFieldObject(obj, pendingNextOffset);
    Object *referent = dvmGetFieldObject(obj, referentOffset);
    if (pending == NULL && referent != NULL && !isMarked(referent, ctx)) {
        Object **list = NULL;
        if (isSoftReference(obj)) {
            list = &gcHeap->softReferences;
        } else if (isWeakReference(obj)) {
            list = &gcHeap->weakReferences;
        } else if (isFinalizerReference(obj)) {
            list = &gcHeap->finalizerReferences;
        } else if (isPhantomReference(obj)) {
            list = &gcHeap->phantomReferences;
        }
        assert(list != NULL);
        enqueuePendingReference(obj, list);
    }
}
```

（23）函数 scanDataObject 的功能是扫描对象的各个成员，并标记其所有引用到的对象。具体实现代码如下所示：

```
static void scanDataObject(const Object *obj, GcMarkContext *ctx)
{
    assert(obj != NULL);
    assert(obj->clazz != NULL);
    assert(ctx != NULL);
    markObject((const Object *)obj->clazz, ctx);
    scanFields(obj, ctx);
    if (IS_CLASS_FLAG_SET(obj->clazz, CLASS_ISREFERENCE)) {
        delayReferenceReferent((Object *)obj, ctx);
    }
}
```

（24）函数 scanObject 的功能是处理"mark stack"中的每个对象，首先判断对象是保存 Java 类型信息的类型对象，还是数组对象，还是普通的 Java 对象，针对这 3 种对象进行不同的处理。由于 finalize 对象是普通的 Java 对象，因此这里只看相应的 scanDataObject 函数。函数 scanObject 的具体实现代码如下所示：

```
static void scanObject(const Object *obj, GcMarkContext *ctx)
{
    assert(obj != NULL);
    assert(obj->clazz != NULL);
    if (obj->clazz == gDvm.classJavaLangClass) {
        scanClassObject(obj, ctx);
    } else if (IS_CLASS_FLAG_SET(obj->clazz, CLASS_ISARRAY)) {
        scanArrayObject(obj, ctx);
    } else {
        scanDataObject(obj, ctx);
    }
}
```

（25）函数 processMarkStack 的功能是处理需要特殊处理的对象，调用 scanObject 函数处理"mark stack"中的每个对象。具体实现代码如下所示：

```
static void processMarkStack(GcMarkContext *ctx)
{
    assert(ctx != NULL);
    assert(ctx->finger == (void *)ULONG_MAX);
    assert(ctx->stack.top >= ctx->stack.base);
    GcMarkStack *stack = &ctx->stack;
    while (stack->top > stack->base) {
        const Object *obj = markStackPop(stack);
```

```
        scanObject(obj, ctx);
    }
}
```

（26）函数 objectSize 的功能是计算对象的大小，具体实现代码如下所示：

```
static size_t objectSize(const Object *obj)
{
    assert(dvmIsValidObject(obj));
    assert(dvmIsValidObject((Object *)obj->clazz));
    if (IS_CLASS_FLAG_SET(obj->clazz, CLASS_ISARRAY)) {
        return dvmArrayObjectSize((ArrayObject *)obj);
    } else if (obj->clazz == gDvm.classJavaLangClass) {
        return dvmClassObjectSize((ClassObject *)obj);
    } else {
        return obj->clazz->objectSize;
    }
}
```

（27）函数 nextGrayObject 的功能是向前扫描标记对象，如果没有标记对象，在这一区域返回 null。函数 nextGrayObject 的具体实现代码如下所示：

```
static Object *nextGrayObject(const u1 *base, const u1 *limit,
                              const HeapBitmap *markBits)
{
    const u1 *ptr;

    assert(base < limit);
    assert(limit - base <= GC_CARD_SIZE);
    for (ptr = base; ptr < limit; ptr += HB_OBJECT_ALIGNMENT) {
        if (dvmHeapBitmapIsObjectBitSet(markBits, ptr))
            return (Object *)ptr;
    }
    return NULL;
}
```

（28）函数 scanDirtyCards 的功能是在开始和结束范围之间扫描“脏卡片”。在垃圾收集时，需要对与老一代中卡片相关联的标记位进行检查，对“脏的卡片”扫描以寻找对年轻代有引用的对象。函数 scanDirtyCards 的具体实现代码如下所示：

```
const u1 *scanDirtyCards(const u1 *start, const u1 *end,
                         GcMarkContext *ctx)
{
    const HeapBitmap *markBits = ctx->bitmap;
    const u1 *card = start, *prevAddr = NULL;
    while (card < end) {
        if (*card != GC_CARD_DIRTY) {
            return card;
        }
        const u1 *ptr = prevAddr ? prevAddr : (u1*)dvmAddrFromCard(card);
        const u1 *limit = ptr + GC_CARD_SIZE;
        while (ptr < limit) {
            Object *obj = nextGrayObject(ptr, limit, markBits);
            if (obj == NULL) {
                break;
            }
            scanObject(obj, ctx);
            ptr = (u1*)obj + ALIGN_UP(objectSize(obj), HB_OBJECT_ALIGNMENT);
        }
        if (ptr < limit) {
            /* Ended within the current card, advance to the next card. */
            ++card;
            prevAddr = NULL;
        } else {
            /* Ended past the current card, skip ahead. */
            card = dvmCardFromAddr(ptr);
            prevAddr = ptr;
        }
```

```
    }
    return NULL;
}
```

而函数 scanGrayObjects 的功能是实现具体扫描工作，具体实现代码如下所示：

```
static void scanGrayObjects(GcMarkContext *ctx)
{
    GcHeap *h = gDvm.gcHeap;
    const u1 *base, *limit, *ptr, *dirty;

    base = &h->cardTableBase[0];
    limit = dvmCardFromAddr((u1 *)dvmHeapSourceGetLimit());
    assert(limit <= &h->cardTableBase[h->cardTableLength]);

    ptr = base;
    for (;;) {
        dirty = (const u1 *)memchr(ptr, GC_CARD_DIRTY, limit - ptr);
        if (dirty == NULL) {
            break;
        }
        assert((dirty > ptr) && (dirty < limit));
        ptr = scanDirtyCards(dirty, limit, ctx);
        if (ptr == NULL) {
            break;
        }
        assert((ptr > dirty) && (ptr < limit));
    }
}
```

（29）函数 scanBitmapCallback 的功能是扫描每个对象位图中的回调，当前的被设置为对应于位图中的位的下一个最低地址的地址。函数 scanBitmapCallback 的具体实现代码如下所示：

```
static void scanBitmapCallback(Object *obj, void *finger, void *arg)
{
    GcMarkContext *ctx = (GcMarkContext *)arg;
    ctx->finger = (void *)finger;
    scanObject(obj, ctx);
}
```

（30）调用函数 dvmHeapScanMarkedObjects()根据上一个函数给出的根对象位图，对每一个根相关的位图进行计算，如果这个根对象有被引用，就标记为使用。这个过程是递归调用的过程，从根开始不断重复地对子树进行标记的过程。函数 dvmHeapScanMarkedObjects()的具体实现代码如下所示：

```
void dvmHeapScanMarkedObjects(void)
{
    GcMarkContext *ctx = &gDvm.gcHeap->markContext;

    assert(ctx->finger == NULL);

    /* The bitmaps currently have bits set for the root set.
     * Walk across the bitmaps and scan each object.
     */
    dvmHeapBitmapScanWalk(ctx->bitmap, scanBitmapCallback, ctx);

    ctx->finger = (void *)ULONG_MAX;

    /* We've walked the mark bitmaps.  Scan anything that's
     * left on the mark stack.
     */
    processMarkStack(ctx);
}
```

而函数 dvmHeapReScanMarkedObjects 的功能是重复扫描标记的对象，具体实现代码如下所示：

```
void dvmHeapReScanMarkedObjects()
{
    GcMarkContext *ctx = &gDvm.gcHeap->markContext;

    /*
     * The finger must have been set to the maximum value to ensure
     * that gray objects will be pushed onto the mark stack.
     */
    assert(ctx->finger == (void *)ULONG_MAX);
    scanGrayObjects(ctx);
    processMarkStack(ctx);
}
```

（31）函数 clearReference 的功能是清除参考区域，具体实现代码如下所示：

```
static void clearReference(Object *reference)
{
    size_t offset = gDvm.offJavaLangRefReference_referent;
    dvmSetFieldObject(reference, offset, NULL);
}
```

（32）函数 preserveSomeSoftReferences 的功能是保留一些软引用，具体实现代码如下所示：

```
static void preserveSomeSoftReferences(Object **list)
{
    assert(list != NULL);
    GcMarkContext *ctx = &gDvm.gcHeap->markContext;
    size_t referentOffset = gDvm.offJavaLangRefReference_referent;
    Object *clear = NULL;
    size_t counter = 0;
    while (*list != NULL) {
        Object *ref = dequeuePendingReference(list);
        Object *referent = dvmGetFieldObject(ref, referentOffset);
        if (referent == NULL) {
            /* Referent was cleared by the user during marking. */
            continue;
        }
        bool marked = isMarked(referent, ctx);
        if (!marked && ((++counter) & 1)) {
            /* Referent is white and biased toward saving, mark it. */
            markObject(referent, ctx);
            marked = true;
        }
        if (!marked) {
            /* Referent is white, queue it for clearing. */
            enqueuePendingReference(ref, &clear);
        }
    }
    *list = clear;
    /*
     * Restart the mark with the newly black references added to the
     * root set.
     */
    processMarkStack(ctx);
}
```

（33）函数 clearWhiteReferences 的功能是清除白色参考引用，具体实现代码如下所示：

```
static void clearWhiteReferences(Object **list)
{
    assert(list != NULL);
    GcMarkContext *ctx = &gDvm.gcHeap->markContext;
    size_t referentOffset = gDvm.offJavaLangRefReference_referent;
    while (*list != NULL) {
        Object *ref = dequeuePendingReference(list);
        Object *referent = dvmGetFieldObject(ref, referentOffset);
        if (referent != NULL && !isMarked(referent, ctx)) {
            /* Referent is white, clear it. */
            clearReference(ref);
            if (isEnqueueable(ref)) {
```

```
                enqueueReference(ref);
            }
        }
    }
    assert(*list == NULL);
}
```

（34）函数 enqueueFinalizerReferences 的功能是通过 JNI 方式将 finalize 对象的引用传递到 Java 端的一个 java.lang.ref.ReferenceQueue 中，具体实现代码如下所示：

```
static void enqueueFinalizerReferences(Object **list)
{
    assert(list != NULL);
    GcMarkContext *ctx = &gDvm.gcHeap->markContext;
    size_t referentOffset = gDvm.offJavaLangRefReference_referent;
    size_t zombieOffset = gDvm.offJavaLangRefFinalizerReference_zombie;
    bool hasEnqueued = false;
    while (*list != NULL) {
        Object *ref = dequeuePendingReference(list);
        Object *referent = dvmGetFieldObject(ref, referentOffset);
        if (referent != NULL && !isMarked(referent, ctx)) {
            markObject(referent, ctx);
            /* If the referent is non-null the reference must queuable. */
            assert(isEnqueuable(ref));
            dvmSetFieldObject(ref, zombieOffset, referent);
            clearReference(ref);
            enqueueReference(ref);
            hasEnqueued = true;
        }
    }
    if (hasEnqueued) {
        processMarkStack(ctx);
    }
    assert(*list == NULL);
}
```

（35）函数 dvmHeapProcessReferences 的功能是在垃圾对象收集完毕后，将 finalize 队列从虚拟机的 Native（本地）端传递到 Java 端。函数 dvmHeapProcessReferences 的具体实现代码如下所示：

```
void dvmHeapProcessReferences(Object **softReferences, bool clearSoftRefs,
                              Object **weakReferences,
                              Object **finalizerReferences,
                              Object **phantomReferences)
{
    assert(softReferences != NULL);
    assert(weakReferences != NULL);
    assert(finalizerReferences != NULL);
    assert(phantomReferences != NULL);
    /*
     * Unless we are in the zygote or required to clear soft
     * references with white references, preserve some white
     * referents.
     */
    if (!gDvm.zygote && !clearSoftRefs) {
        preserveSomeSoftReferences(softReferences);
    }
    /*
     * Clear all remaining soft and weak references with white
     * referents.
     */
    clearWhiteReferences(softReferences);
    clearWhiteReferences(weakReferences);
    /*
     * Preserve all white objects with finalize methods and schedule
     * them for finalization.
     */
    enqueueFinalizerReferences(finalizerReferences);
```

```
    /*
     * Clear all f-reachable soft and weak references with white
     * referents.
     */
    clearWhiteReferences(softReferences);
    clearWhiteReferences(weakReferences);
    /*
     * Clear all phantom references with white referents.
     */
    clearWhiteReferences(phantomReferences);
    /*
     * At this point all reference lists should be empty.
     */
    assert(*softReferences == NULL);
    assert(*weakReferences == NULL);
    assert(*finalizerReferences == NULL);
    assert(*phantomReferences == NULL);
}
```

（36）函数 dvmEnqueueClearedReferences 的功能是将一个清除列表引用托管堆，具体实现代码如下所示：

```
void dvmEnqueueClearedReferences(Object **cleared)
{
    assert(cleared != NULL);
    if (*cleared != NULL) {
        Thread *self = dvmThreadSelf();
        assert(self != NULL);
        Method *meth = gDvm.methJavaLangRefReferenceQueueAdd;
        assert(meth != NULL);
        JValue unused;
        Object *reference = *cleared;
        dvmCallMethod(self, meth, NULL, &unused, reference);
        *cleared = NULL;
    }
}
```

（37）调用函数 dvmHeapFinishMarkStep()回收已经删除对象的内存，可以调用堆管理函数改变目前堆使用的内存，并整理内存，这样就可以得到更多空闲的内存了。函数 dvmHeapFinishMarkStep()的具体实现代码如下所示：

```
void dvmHeapFinishMarkStep()
{
    HeapBitmap *markBitmap;
    HeapBitmap objectBitmap;
    GcMarkContext *markContext;

    markContext = &gDvm.gcHeap->markContext;

    /* The sweep step freed every object that appeared in the
     * HeapSource bitmaps that didn't appear in the mark bitmaps.
     * The new state of the HeapSource is exactly the final
     * mark bitmaps, so swap them in.
     *
     * The old bitmaps will be swapped into the context so that
     * we can clean them up.
     */
    dvmHeapSourceReplaceObjectBitmaps(markContext->bitmaps,
            markContext->numBitmaps);
    /* Clean up the old HeapSource bitmaps and anything else associated
     * with the marking process.
     */
    dvmHeapBitmapDeleteList(markContext->bitmaps, markContext->numBitmaps);
    destroyMarkStack(&markContext->stack);
    memset(markContext, 0, sizeof(*markContext));
}
```

（38）函数 sweepBitmapCallback 的功能是扫描位图回调，具体实现代码如下所示：

```
static void sweepBitmapCallback(size_t numPtrs, void **ptrs, void *arg)
{
    assert(arg != NULL);
    SweepContext *ctx = (SweepContext *)arg;
    if (ctx->isConcurrent) {
        dvmLockHeap();
    }
    ctx->numBytes += dvmHeapSourceFreeList(numPtrs, ptrs);
    ctx->numObjects += numPtrs;
    if (ctx->isConcurrent) {
        dvmUnlockHeap();
    }
}
```

（39）函数 sweepWeakJniGlobals 的功能是扫描弱全局 JNI，具体实现代码如下所示：

```
static void sweepWeakJniGlobals()
{
    IndirectRefTable* table = &gDvm.jniWeakGlobalRefTable;
    GcMarkContext* ctx = &gDvm.gcHeap->markContext;
    typedef IndirectRefTable::iterator It; // TODO: C++0x auto
    for (It it = table->begin(), end = table->end(); it != end; ++it) {
        Object** entry = *it;
        if (!isMarked(*entry, ctx)) {
            *entry = kClearedJniWeakGlobal;
        }
    }
}
```

（40）调用函数 dvmHeapSweepUnmarkedObjects()清除未曾标记的对象，也就是删除没有再使用的对象。函数 dvmHeapSweepUnmarkedObjects()的具体实现代码如下所示：

```
dvmHeapSweepUnmarkedObjects(int *numFreed, size_t *sizeFreed)
{
    const HeapBitmap *markBitmaps;
    const GcMarkContext *markContext;
    HeapBitmap objectBitmaps[HEAP_SOURCE_MAX_HEAP_COUNT];
    size_t origObjectsAllocated;
    size_t origBytesAllocated;
    size_t numBitmaps;

    /* All reachable objects have been marked.
     * Detach any unreachable interned strings before
     * we sweep.
     */
    dvmGcDetachDeadInternedStrings(isUnmarkedObject);

    /* Free any known objects that are not marked.
     */
    origObjectsAllocated = dvmHeapSourceGetValue(HS_OBJECTS_ALLOCATED, NULL, 0);
    origBytesAllocated = dvmHeapSourceGetValue(HS_BYTES_ALLOCATED, NULL, 0);

    markContext = &gDvm.gcHeap->markContext;
    markBitmaps = markContext->bitmaps;
    numBitmaps = dvmHeapSourceGetObjectBitmaps(objectBitmaps,
        HEAP_SOURCE_MAX_HEAP_COUNT);
#ifndef NDEBUG
    if (numBitmaps != markContext->numBitmaps) {
        LOGE("heap bitmap count mismatch: %zd != %zd\n",
                numBitmaps, markContext->numBitmaps);
        dvmAbort();
    }
```

（41）调用函数 dvmHeapHandleReferences()处理 Java 类对象的引用类型。在此主要处理如下 3 个直接子类。

- SoftReference：对象封装了对引用目标的“软引用”。
- WeakReference：封装了对引用目标的“弱引用”。

❑ PhantomReference：封装了对引用目标的“影子引用”。强引用禁止引用目标被垃圾收集，而软引用、弱引用和影子引用不禁止。

函数 dvmHeapHandleReferences()的具体实现代码如下所示：

```
void dvmHeapHandleReferences(Object *refListHead, enum RefType refType)
{
    Object *reference;
    GcMarkContext *markContext = &gDvm.gcHeap->markContext;
    const int offVmData = gDvm.offJavaLangRefReference_vmData;
    const int offReferent = gDvm.offJavaLangRefReference_referent;
    bool workRequired = false;

size_t numCleared = 0;
size_t numEnqueued = 0;
    reference = refListHead;
    while (reference != NULL) {
        Object *next;
        Object *referent;

        /* Pull the interesting fields out of the Reference object.
         */
        next = dvmGetFieldObject(reference, offVmData);
        referent = dvmGetFieldObject(reference, offReferent);

        //TODO: when handling REF_PHANTOM, unlink any references
        //      that fail this initial if().  We need to re-walk
        //      the list, and it would be nice to avoid the extra
        //      work.
        if (referent != NULL && !isMarked(ptr2chunk(referent), markContext)) {
            bool schedClear, schedEnqueue;

            /* This is the strongest reference that refers to referent.
             * Do the right thing.
             */
            switch (refType) {
            case REF_SOFT:
            case REF_WEAK:
                schedClear = clearReference(reference);
                schedEnqueue = enqueueReference(reference);
                break;
            case REF_PHANTOM:
                /* PhantomReferences are not cleared automatically.
                 * Until someone clears it (or the reference itself
                 * is collected), the referent must remain alive.
                 *
                 * It's necessary to fully mark the referent because
                 * it will still be present during the next GC, and
                 * all objects that it points to must be valid.
                 * (The referent will be marked outside of this loop,
                 * after handing all references of this strength, in
                 * case multiple references point to the same object.)
                 */
                schedClear = false;

                /* A PhantomReference is only useful with a
                 * queue, but since it's possible to create one
                 * without a queue, we need to check.
                 */
                schedEnqueue = enqueueReference(reference);
                break;
            default:
                assert(!"Bad reference type");
                schedClear = false;
                schedEnqueue = false;
                break;
            }
numCleared += schedClear ? 1 : 0;
numEnqueued += schedEnqueue ? 1 : 0;
```

```
        if (schedClear || schedEnqueue) {
            uintptr_t workBits;

            /* Stuff the clear/enqueue bits in the bottom of
             * the pointer. Assumes that objects are 8-byte
             * aligned.
             *
             * Note that we are adding the *Reference* (which
             * is by definition already marked at this point) to
             * this list; we're not adding the referent (which
             * has already been cleared).
             */
            assert(((intptr_t)reference & 3) == 0);
            assert(((WORKER_CLEAR | WORKER_ENQUEUE) & ~3) == 0);
            workBits = (schedClear ? WORKER_CLEAR : 0) |
                    (schedEnqueue ? WORKER_ENQUEUE : 0);
            if (!dvmHeapAddRefToLargeTable(
                    &gDvm.gcHeap->referenceOperations,
                    (Object *)((uintptr_t)reference | workBits)))
            {
                LOGE_HEAP("dvmMalloc(): no room for any more "
                        "reference operations\n");
                dvmAbort();
            }
            workRequired = true;
        }

        if (refType != REF_PHANTOM) {
            /* Let later GCs know not to reschedule this reference.
             */
            dvmSetFieldObject(reference, offVmData,
                    SCHEDULED_REFERENCE_MAGIC);
        } // else this is handled later for REF_PHANTOM

    } // else there was a stronger reference to the referent.

    reference = next;
  }
#define refType2str(r) \
   ((r) == REF_SOFT ? "soft" : ( \
    (r) == REF_WEAK ? "weak" : ( \
    (r) == REF_PHANTOM ? "phantom" : "UNKNOWN" )))
LOGD_HEAP("dvmHeapHandleReferences(): cleared %zd, enqueued %zd %s references\n", numCleared,
numEnqueued, refType2str(refType));

   /* Walk though the reference list again, and mark any non-clear/marked
    * referents.  Only PhantomReferences can have non-clear referents
    * at this point.
    */
   if (refType == REF_PHANTOM) {
       bool scanRequired = false;

       HPROF_SET_GC_SCAN_STATE(HPROF_ROOT_REFERENCE_CLEANUP, 0);
       reference = refListHead;
       while (reference != NULL) {
           Object *next;
           Object *referent;

           /* Pull the interesting fields out of the Reference object.
            */
           next = dvmGetFieldObject(reference, offVmData);
           referent = dvmGetFieldObject(reference, offReferent);

           if (referent != NULL && !isMarked(ptr2chunk(referent), markContext)) {
               markObjectNonNull(referent, markContext);
               scanRequired = true;

               /* Let later GCs know not to reschedule this reference.
```

```
            */
            dvmSetFieldObject(reference, offVmData,
                    SCHEDULED_REFERENCE_MAGIC);
        }

        reference = next;
    }
    HPROF_CLEAR_GC_SCAN_STATE();

    if (scanRequired) {
        processMarkStack(markContext);
    }
}

if (workRequired) {
    dvmSignalHeapWorker(false);
}
}
```

（42）调用函数 dvmHeapScheduleFinalizations()调用未曾标记的对象，让每一个对象最后删除动作可以运行，以便后面从内存里把对象删除，相当于对象的析构作用。函数 dvmHeapScheduleFinalizations()的具体实现代码如下所示：

```
void dvmHeapScheduleFinalizations()
{
    HeapRefTable newPendingRefs;
    LargeHeapRefTable *finRefs = gDvm.gcHeap->finalizableRefs;
    Object **ref;
    Object **lastRef;
    size_t totalPendCount;
    GcMarkContext *markContext = &gDvm.gcHeap->markContext;

    /*
     * All reachable objects have been marked.
     * Any unmarked finalizable objects need to be finalized.
     */

    /* Create a table that the new pending refs will
     * be added to.
     */
    if (!dvmHeapInitHeapRefTable(&newPendingRefs, 128)) {
        //TODO: mark all finalizable refs and hope that
        //      we can schedule them next time.  Watch out,
        //      because we may be expecting to free up space
        //      by calling finalizers.
        LOGE_GC("dvmHeapScheduleFinalizations(): no room for "
                "pending finalizations\n");
        dvmAbort();
    }

    /* Walk through finalizableRefs and move any unmarked references
     * to the list of new pending refs.
     */
    totalPendCount = 0;
    while (finRefs != NULL) {
        Object **gapRef;
        size_t newPendCount = 0;

        gapRef = ref = finRefs->refs.table;
        lastRef = finRefs->refs.nextEntry;
        while (ref < lastRef) {
            DvmHeapChunk *hc;

            hc = ptr2chunk(*ref);
            if (!isMarked(hc, markContext)) {
                if (!dvmHeapAddToHeapRefTable(&newPendingRefs, *ref)) {
                    //TODO: add the current table and allocate
                    //      a new, smaller one.
```

```
                LOGE_GC("dvmHeapScheduleFinalizations(): "
                        "no room for any more pending finalizations: %zd\n",
                        dvmHeapNumHeapRefTableEntries(&newPendingRefs));
                dvmAbort();
            }
            newPendCount++;
        } else {
            /* This ref is marked, so will remain on finalizableRefs.
             */
            if (newPendCount > 0) {
                /* Copy it up to fill the holes.
                 */
                *gapRef++ = *ref;
            } else {
                /* No holes yet; don't bother copying.
                 */
                gapRef++;
            }
        }
        ref++;
    }
    finRefs->refs.nextEntry = gapRef;
    //TODO: if the table is empty when we're done, free it.
    totalPendCount += newPendCount;
    finRefs = finRefs->next;
}
LOGD_GC("dvmHeapScheduleFinalizations(): %zd finalizers triggered.\n",
        totalPendCount);
if (totalPendCount == 0) {
    /* No objects required finalization.
     * Free the empty temporary table.
     */
    dvmClearReferenceTable(&newPendingRefs);
    return;
}

/* Add the new pending refs to the main list.
 */
if (!dvmHeapAddTableToLargeTable(&gDvm.gcHeap->pendingFinalizationRefs,
            &newPendingRefs))
{
    LOGE_GC("dvmHeapScheduleFinalizations(): can't insert new "
            "pending finalizations\n");
    dvmAbort();
}

//TODO: try compacting the main list with a memcpy loop

/* Mark the refs we just moved;  we don't want them or their
 * children to get swept yet.
 */
ref = newPendingRefs.table;
lastRef = newPendingRefs.nextEntry;
assert(ref < lastRef);
HPROF_SET_GC_SCAN_STATE(HPROF_ROOT_FINALIZING, 0);
while (ref < lastRef) {
    markObjectNonNull(*ref, markContext);
    ref++;
}
HPROF_CLEAR_GC_SCAN_STATE();

/* Set markAllReferents so that we don't collect referents whose
 * only references are in final-reachable objects.
 * TODO: eventually provide normal reference behavior by properly
 *       marking these references.
 */
gDvm.gcHeap->markAllReferents = true;
processMarkStack(markContext);
gDvm.gcHeap->markAllReferents = false;
```

```
    dvmSignalHeapWorker(false);
}
```

上述算法函数的整个过程，就是 Dalvik 虚拟机的整个标记和删除的算法过程，实际的代码会相当复杂，算法上是很清楚的，就是细节、时间方面要求相当严格，否则乱删除还在使用的对象，就会导致整个虚拟机运行出错。

13.4 垃圾回收的时机

通过本章前面内容的学习，了解了垃圾回收的原理和过程。那么 Dalvik 虚拟机是什么时候进行垃圾回收呢？要回答这个问题，就得继续分析代码，继续进入下面的学习。其实，垃圾回收主要有两种方式，一种是虚拟机线程自动进行的，一种是手动进行的。现在先来学习自动进行的方式，所谓自动方式，就是虚拟机创建一个线程，这个线程定时进行。虚拟机在初始化时就创建这个线程，例如下面的代码。

```
if(gDvm.zygote){
if(!dvmInitZygote())
gotofail;
} else{
if(!dvmInitAfterZygote())
gotofail;
}
```

在上述代码中调用了函数 dvmInitAfterZygote()，此函数中会调用函数 dvmSignalCatcherStartup()来创建垃圾回收线程。函数 dvmInitAfterZygote()的具体实现代码如下所示：

```
bool dvmInitAfterZygote(void)
{
    u8 startHeap, startQuit, startJdwp;
    u8 endHeap, endQuit, endJdwp;
    startHeap = dvmGetRelativeTimeUsec();
    /*
     * Post-zygote heap initialization, including starting
     * the HeapWorker thread.
     */
    if (!dvmGcStartupAfterZygote())
        return false;
    endHeap = dvmGetRelativeTimeUsec();
    startQuit = dvmGetRelativeTimeUsec();
    /* start signal catcher thread that dumps stacks on SIGQUIT */
    if (!gDvm.reduceSignals && !gDvm.noQuitHandler) {
        if (!dvmSignalCatcherStartup())
            return false;
    }
    /* start stdout/stderr copier, if requested */
    if (gDvm.logStdio) {
        if (!dvmStdioConverterStartup())
            return false;
    }
    endQuit = dvmGetRelativeTimeUsec();
    startJdwp = dvmGetRelativeTimeUsec();
    /*
     * Start JDWP thread.  If the command-line debugger flags specified
     * "suspend=y", this will pause the VM.  We probably want this to
     * come last.
     */
    if (!dvmInitJDWP()) {
        LOGD("JDWP init failed; continuing anyway\n");
    }
    endJdwp = dvmGetRelativeTimeUsec();
    LOGV("thread-start heap=%d quit=%d jdwp=%d total=%d usec\n",
        (int)(endHeap-startHeap), (int)(endQuit-startQuit),
```

```
        (int)(endJdwp-startJdwp), (int)(endJdwp-startHeap));
#ifdef WITH_JIT
    if (gDvm.executionMode == kExecutionModeJit) {
        if (!dvmCompilerStartup())
            return false;
    }
#endif
    return true;
}
```

函数 dvmSignalCatcherStartup()的实现代码如下所示：

```
bool dvmSignalCatcherStartup(void)
{
gDvm.haltSignalCatcher= false;
if(!dvmCreateInternalThread(&gDvm.signalCatcherHandle,
"SignalCatcher", signalCatcherThreadStart,NULL))
returnfalse;
returntrue;
}
```

通过上面的这段代码，就可以看到线程运行函数是 signalCatcherThreadStart()，在这个函数里就会调用函数 dvmCollectGarbage()来进行垃圾回收。函数 dvmCollectGarbage 的具体实现代码如下所示：

```
void dvmCollectGarbage(bool collectSoftReferences)
{
dvmLockHeap();
LOGVV("ExplicitGC\n");
dvmCollectGarbageInternal(collectSoftReferences);
dvmUnlockHeap();
}
```

此函数主要通过锁来锁住多线程访问的堆空间相关对象，然后直接就调用函数 dvmCollect GarbageInternal 来进行垃圾回收过程了，也就调用上面标记删除算法的函数。

另一种方式通过调用运行库的 GC 来回收，例如下面的代码。

```
staticvoidDalvik_java_lang_Runtime_gc(constu4* args,JValue*pResult)
{
UNUSED_PARAMETER(args);
dvmCollectGarbage(false);
RETURN_VOID();
}
```

此处也是调用了函数 dvmCollectGarbage()来进行垃圾回收。手动的方式适合当需要内存但线程又没有调用时进行。

13.5 调试信息

一般来说，Java 虚拟机要求支持 verbosegc 选项，输出详细的垃圾收集调试信息。Dalvik 虚拟机可以接收 verbosegc 选项，然后什么都不做。Dalvik 虚拟机使用自己的一套 LOG 机制来输出调试信息。

如果在 Linux 下运行 adb logcat 命令，会看到输出如下信息：

```
D/dalvikvm( 745): GC_CONCURRENT freed 199K, 53% free 3023K/6343K,external 0K/0K, paused 2ms+2ms
```

（1）D/dalvikvm：表示由 Dalvik VM 输出的调试信息，括号后的数字代表 Dalvik VM 所在进程的 pid。GC_CONCURRENT 有以下几种表示触发垃圾收集的原因。

- GC_MALLOC：内存分配失败时触发。
- GC_CONCURRENT：当分配的对象大小超过 384 KB 时触发。

- GC_EXPLICIT：对垃圾收集的显式调用（System.gc）。
- GC_EXTERNAL_ALLOC：外部内存分配失败时触发。

（2）freed 199 K：表示本次垃圾收集释放了 199 KB 的内存。

（3）53% free 3023 K/6343 K：其中 6343 K 表示当前内存总量，3023 K 表示可用内存，53% 表示可用内存占总内存的比例。

（4）external 0 K/0 K：表示可用外部内存/外部内存总量。

（5）paused 2 ms+2 ms：第一个时间值表示 markrootset 的时间，第二个时间值表示第二次 mark 的时间。如果触发原因不是 GC_CONCURRENT，这一行为单个时间值，表示垃圾收集的耗时时间。

由此可以得出如下 3 个结论。

（1）虽然 Dalvik 虚拟机提供了一些调试信息，但是还缺乏一些关键信息，比如说 mark 和 sweep 的时间，分配内存失败时是因为分配多大的内存失败，还有对于 SoftReference、WeakReference 和 PhantomReference 的处理，每次垃圾收集处理了多少个这些引用等。

（2）目前 Dalvik 所有线程共享一个内存堆，这样在分配内存时必须在线程之间互斥，可以考虑为每个内存分配一个线程局部存储堆，一些小的内存分配可以直接从该堆中分配而无须互斥锁。

（3）Dalvik 虚拟机中引入了 concurrentmark，但是对于多核 CPU，可以实现 parrelmark，即可以使用多个线程同时运行 mark 阶段。

这些都是目前 Dalvik 虚拟机内存管理可以做出的改进。

13.6 Dalvik VM 和 JVM 垃圾收集机制的区别

Java VM 是一个规范，或者符合该规范的实现，或者这样的实现运行的实例。JVM 规范中并没有规定要使用各种 GC 机制；或者说，JVM 规范写明了符合规范的 JVM 实现要提供自动内存管理的功能，但并不一定要有某种特定的“GC”。

而 Dalvik VM 是一个具体的实现，即便是在同一个 JVM 中也会有多个 GC 实现。例如 Oracle 的 HotSpot VM 中根据不同的使用场景而实现了不同算法/不同调教的 GC。在 Dalvik VM 1.0 版本中时，使用的 GC 算法是没有分代的“标记—清除”（Mark Sweep），对堆上数据进行准确式（exact/precise）标记，对栈/寄存器上数据进行保守式（conservative）标记。标记的内容可以参考文件“dalvik/vm/alloc/MarkSweep.c”中的如下注释：

```
/* Mark the set of root objects.
 *
 * Things we need to scan:
 * - System classes defined by root classloader
 * - For each thread:
 *   - Interpreted stack, from top to "curFrame"
 *     - Dalvik registers (args + local vars)
 *   - JNI local references
 *   - Automatic VM local references (TrackedAlloc)
 *   - Associated Thread/VMThread object
 *   - ThreadGroups (could track & start with these instead of working
 *     upward from Threads)
 *   - Exception currently being thrown, if present
 * - JNI global references
 * - Interned string table
 * - Primitive classes
 * - Special objects
 *   - gDvm.outOfMemoryObj
 * - Objects allocated with ALLOC_NO_GC
 * - Objects pending finalization (but not yet finalized)
 * - Objects in debugger object registry
 *
```

```
 * Don't need:
 * - Native stack (for in-progress stuff in the VM)
 *   - The TrackedAlloc stuff watches all native VM references.
 */
```

许多 GC 实现都是在对象开头的地方留一小块空间给 GC 标记用，而 Dalvik VM 则不同，在进行 GC 的时候会单独申请一块空间，以位图的形式来保存整个堆上的对象的标记，在 GC 结束后就释放该空间。

在标记阶段，从根集合开始，沿着对象用的引用进行标记直到没有更多可标记的对象为止。标记结束后，被标记的就是活着的对象，没被标记到的就是“垃圾”。在清除阶段，Dalvik VM 并不直接对堆做什么操作，而是在一个记录分配状况的位图上把被认为是垃圾的对象所在位置的分配标记清零。为了不让这个位图太大，位图中并不是每一位对应到堆上的一个字节，而是对应到一块固定大小的空间。为此，堆空间的分配也是有一定对齐的。

只进行“标记—清除”工作，在经过多次 GC 后可能会使堆被碎片化。Android 所实现的 libc（称为 Bionic）对这种情况有特别的实现，可以避免碎片化这一问题的发生。其实 Dalvik VM 的根源还是在 JVM 上，只要能符合规范正确执行 Java 的.class 文件的就是 JVM。那么 Android 开发包中的 dx 与 Dalvik VM 结合起来，就可以看成是一个 JVM 了（要把一个东西称为“JVM”必须要通过 JCK（Java Compliance Kit）的测试并获得授权后才能行，所以严格来说“dx + Dalvik VM”不能叫做 JVM，因为没授权）。如果去阅读 Dalvik VM 的文档，会发现其中有很多引用到 JVM 规范的地方，而且整体设计都考虑到了与 JVM 的兼容性的。它与 JVM 规范的规定最大的不同在于它采用了基于寄存器的指令集，而 JVM 采用了基于栈的指令集。这可以看作是专门为 ARM 而优化的设计。Dalvik VM 要省内存和省电，有很多设计都是围绕这两个目标来进行的。

第 14 章　Dalvik VM 内存优化机制详解

在 Android 系统中，使用垃圾回收机制的方式达到节约内存的目的，并最终实现提高手机的处理效率的目的。在 Android 系统中，sp 和 wp 被称为智能指针（android refbase 类（sp 和 wp））。其实 sp 和 wp 就是 Android 为其 C++实现的自动垃圾回收机制。在本章的内容中，将详细讲解 Dalvik VM 内存优化机制的基本知识，为读者步入本书后面知识的学习打下基础。

14.1　sp 和 wp 简介

在传统的 C++编程语言中，指针一直是程序员的最大学习障碍。使用指针时需要十分谨慎，一旦使用不当就会造成内存泄漏的问题。例如用“new”新建一个对象并使用完之后，经常忘记 delete（删除）这个对象，这样长期下去会造成系统崩溃。在 Android 系统中，因为其运行时库这一层代码是用 C++来编写的，所以也会因为使用指针的原因而造成内存泄漏问题。为此 Android 特意提供了智能指针机制，通过使用 sp 命令和 wp 命令来解决指针问题。其实 sp 和 wp 就是 Android 为其 C++实现的自动垃圾回收机制。如果具体到内部实现，sp 和 wp 实际上只是一个实现垃圾回收功能的接口而已，而真正实现垃圾回收的是 refbase 这个基类。这部分代码位于如下文件中。

```
/frameworks/base/include/utils/RefBase.h
```

在此所有的类都会虚继承于 refbase 类，因为它实现了达到 Android 垃圾回收所需要的所有 function，因此实际上所有的对象声明出来以后都具备了自动释放自己的能力，也就是说实际上智能指针就是对象本身，它会维持一个对本身强引用和弱引用的计数，一旦强引用计数为 0 它就会释放掉自己。

14.1.1　sp 基础

在 Android 系统中，sp 实际上不是 smart pointer 的缩写，而是 strong pointer，它内部实际上就包含了一个指向对象的指针而已。可以简单看看 sp 的一个构造函数：

```
template< typename T>
sp< T>::sp(T* other)
: m_ptr(other)
{
if (other) other->incStrong(this);
}
```

比如说声明一个对象：

```
sp< CameraHardwareInterface> hardware(new CameraHal());
```

实际上 sp 指针对本身没有进行什么操作，就是一个指针的赋值，包含了一个指向对象的指针，但是对象会对对象本身增加一个强引用计数，这个 incStrong 的实现就在 refbase 类里面。新“new”出来一个 CameraHal 对象，将它的值给 sp< CameraHardwareInterface>的时候，它的强引用计数就会从 0 变为 1。因此每次将对象赋值给一个 sp 指针的时候，对象的强引用计数都会加 1，下面再看看 sp 的析构函数：

```
template< typename T>
sp< T>::~sp()
{
if (m_ptr) m_ptr->decStrong(this);
}
```

实际上每次删除一个 sp 对象的时候，sp 指针指向的对象的强引用计数就会减 1，当对象的强引用计数为 0 时，这个对象就会被自动释放掉。

14.1.2 wp 基础

在 Android 系统中，wp 就是 Weak Pointer 的缩写，弱引用指针的原理，就是为了应用 Android 垃圾回收来减少对那些“胖子”对象对内存的占用，首先来看 wp 的一个构造函数：

```
wp< T>::wp(T* other)
: m_ptr(other)
{
if (other) m_refs = other->createWeak(this);
}
```

在 Android 系统中，wp 和 sp 一样，实际上也就是仅仅对指针进行了赋值而已，对象本身会增加一个对自身的弱引用计数，同时 wp 还包含一个 m_ref 指针，这个指针主要是用来将 wp 升级为 sp 时候使用的，如下所示：

```
template< typename T>
sp< T> wp< T>::promote() const
{
return sp< T>(m_ptr, m_refs);
}
template< typename T>
sp< T>::sp(T* p, weakref_type* refs)
: m_ptr((p && refs->attemptIncStrong(this)) ? p : 0)
{
}
```

实际上对 wp 指针唯一能做的就是将 wp 指针升级为一个 sp 指针，然后判断是否升级成功，如果成功说明对象依旧存在，如果失败说明对象已经被释放掉了。wp 指针现在在单例中使用很多，确保 mhardware 对象只有一个，如以下代码：

```
wp< CameraHardwareInterface> CameraHardwareStub::singleton;
sp< CameraHardwareInterface> CameraHal::createInstance()
{
LOG_FUNCTION_NAME
if (singleton != 0) {
sp< CameraHardwareInterface> hardware = singleton.promote();
if (hardware != 0) {
return hardware;
}
}
sp< CameraHardwareInterface> hardware(new CameraHal()); //强引用加 1
singleton = hardware;//弱引用加 1
return hardware;//赋值构造函数，强引用加 1
}
//hardware 被删除，强引用减 1
```

14.2 智能指针详解

在 Android 的源代码中，经常会看到形如：sp<xxx>、wp<xxx>形式的类型定义，这其实是

Android 中的智能指针。Android 的智能指针相关的源代码在如下两个文件中：

```
frameworks/base/include/utils/RefBase.h
frameworks/base/libs/utils/RefBase.cpp
```

涉及的类以及类之间的关系如图 14-1 所示。

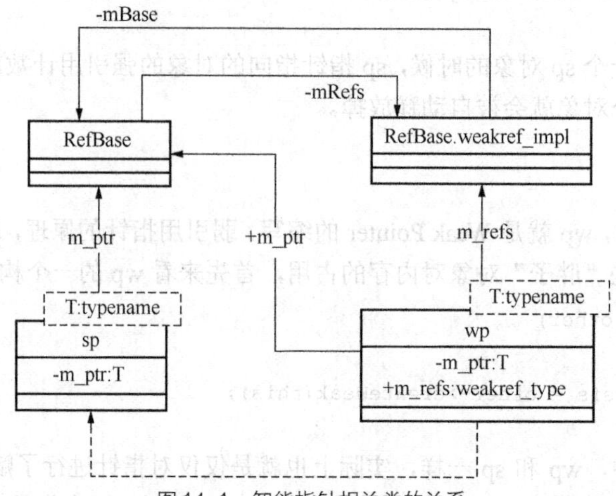

图 14-1　智能指针相关类的关系

14.2.1　智能指针基础

Android 中定义了 3 种智能指针类型，分别是强指针（Strong Pointer）、弱指针（Weak Pointer）和轻量级指针（Light Pointer）。强指针与一般意义的智能指针概念相同，通过引用计数来记录有多少使用者在使用一个对象，如果所有使用者都放弃了对该对象的引用，则该对象将被自动销毁。

弱指针也指向一个对象，但是弱指针仅仅记录该对象的地址，不能通过弱指针来访问该对象，也就是说不能通过弱指针来调用对象的成员函数或访问对象的成员变量。要想访问弱指针所指向的对象，需首先将弱指针升级为强指针（通过 wp 类所提供的 promote()方法）。弱指针所指向的对象是有可能在其他地方被销毁的，如果对象已经被销毁，wp 的 promote()方法将返回空指针，这样就能避免出现地址访问错的情况。

究竟指针是怎么做到这一点的呢？其实一点也不复杂，原因就在于每一个可以被智能指针引用的对象，都同时被附加了另外一个 weakref_impl 类型的对象，这个对象中负责记录对象的强指针引用计数和弱指针引用计数。这个对象是智能指针的实现内部使用的，智能指针的使用者看不到这个对象。弱指针操作的就是这个对象，只有当强引用计数和弱引用计数都为 0 时，这个对象才会被销毁。

接下来开始分析到底该怎么使用智能指针。假设现在有一个类 MyClass，如果要使用智能指针来引用这个类的对象，那么这个类需满足下列两个前提条件。

（1）这个类是基类 RefBase 的子类或间接子类。

（2）这个类必须定义虚构造函数，即它的构造函数需要这样定义：

```
virtual ~MyClass();
```

满足了上述条件的类后就可以定义智能指针，定义方法和普通指针类似。比如普通指针是这样定义：

```
MyClass* p_obj;
```

智能指针是这样定义：

```
sp<MyClass> p_obj;
```

注意不要定义成 sp<MyClass>* p_obj。初学者容易犯这种错误，这样实际上相当于定义了一个指针的指针。尽管在语法上没有问题，但是最好永远不要使用这样的定义。

定义了一个智能指针的变量，就可以像普通指针那样使用它，包括赋值、访问对象成员、作为函数的返回值、作为函数的参数等。例如：

```
p_obj = new MyClass(); // 注意不要写成 p_obj = new sp<MyClass>
sp<MyClass> p_obj2 = p_obj;
p_obj->func();
p_obj = create_obj();
some_func(p_obj);
```

注意不要试图 delete（删除）一个智能指针，即执行 delete p_obj 操作。无需担心对象的销毁问题，智能指针的最大作用就是自动销毁不再使用的对象。当不需要再使用一个对象后，只需直接将指针赋值为 NULL 即可：

```
p_obj = NULL;
```

上面说的都是强指针，弱指针的定义方法和强指针类似，但是不能通过弱指针来访问对象的成员。下面是弱指针的示例。

```
wp<MyClass> wp_obj = new MyClass();
p_obj = wp_obj.promote(); // 升级为强指针。不过这里要用.而不是->，真是有负其指针之名啊
wp_obj = NULL;
```

由此可见，智能指针用起来很方便，在一般情况下最好使用智能指针来代替普通指针。但是需要知道一个智能指针其实是一个对象，而不是一个真正的指针，因此其运行效率是远远比不上普通指针的。所以在对运行效率敏感的地方，最好还是不要使用智能指针为好。

14.2.2 轻量级指针

在 Android 系统中，轻量级指针通过引用计数技术来维护对象的声明周期。支持轻量级指针的对象必须继承自基类 LightRefBase，类 LightRefBase 在文件“frameworks\native\include\utils\RefBase.h”中定义，具体实现代码如下所示：

```
template <class T>
class LightRefBase
{
public:
    inline LightRefBase() : mCount(0) { }
    inline void incStrong(const void* id) const {
        android_atomic_inc(&mCount);
    }
    inline void decStrong(const void* id) const {
        if (android_atomic_dec(&mCount) == 1) {
            delete static_cast<const T*>(this);
        }
    }
    //! DEBUGGING ONLY: Get current strong ref count.
    inline int32_t getStrongCount() const {
        return mCount;
    }
    typedef LightRefBase<T> basetype;
protected:
    inline ~LightRefBase() { }
private:
    friend class ReferenceMover;
    inline static void moveReferences(void* d, void const* s, size_t n,
            const ReferenceConverterBase& caster) { }
private:
    mutable volatile int32_t mCount;
};
```

由上述代码可以看出，类 LightRefBase 它只一个引用计数器成员变量 mCount，其初始化值为 0。另外，类 LightRefBase 还通过成员函数 incStrong 和 decStrong 维护引用计数器的值，这两个函数被智能指针调用。在函数 decStrong 中，如果当前引用计数值为 1，那么当减 1 后就会变为 0，这表示 delete（删除）这个对象。

在 Android 系统中，和 LightRefBase 引用计数配套使用的智能指针类是 sp，sp 是轻量级指针的实现类。在文件“frameworks\native\include\utils\RefBase.h”中，sp 的具体实现代码如下所示：

```
template <typename T>
class sp
{
public:
    typedef typename RefBase::weakref_type weakref_type;
    inline sp() : m_ptr(0) { }

    sp(T* other);//T 表示对象的实际类型，
    sp(const sp<T>& other);
    template<typename U> sp(U* other);
    template<typename U> sp(const sp<U>& other);

    ~sp();

    // Assignment

    sp& operator = (T* other);
    sp& operator = (const sp<T>& other);

    template<typename U> sp& operator = (const sp<U>& other);
    template<typename U> sp& operator = (U* other);

    //! Special optimization for use by ProcessState (and nobody else).
    void force_set(T* other);

    // Reset

    void clear();

    // Accessors

    inline  T&      operator* () const  { return *m_ptr; }
    inline  T*      operator-> () const { return m_ptr;  }
    inline  T*      get() const         { return m_ptr; }

    // Operators

    COMPARE(==)
        COMPARE(!=)
        COMPARE(>)
        COMPARE(<)
        COMPARE(<=)
        COMPARE(>=)

private:
    template<typename Y> friend class sp;
    template<typename Y> friend class wp;

    // Optimization for wp::promote().
    sp(T* p, weakref_type* refs);

    T*              m_ptr;
};
```

类 sp 有如下两个构造函数：

❑ 普通构造函数。

- 拷贝构造函数。

上述两个构造函数在文件“frameworks\native\include\utils\RefBase.h”中实现，具体实现代码如下所示：

```
template<typename T>
sp<T>::sp(T* other)
    : m_ptr(other)
{
    if (other) other->incStrong(this);
}

template<typename T>
sp<T>::sp(const sp<T>& other)
    : m_ptr(other.m_ptr)
{
    if (m_ptr) m_ptr->incStrong(this);
}
```

类 sp 中包含了析构函数，功能是调用 m_ptr 的成员函数 decStrong 减少对象的引用计数值。函数 decStrong 在类 LightRefBase 中定义，当引用计数减 1 后变成 0 时会自动 delete（删除）这个对象。定义析构函数的实现代码如下所示：

```
template<typename T>
sp<T>::~sp()
{
    if (m_ptr) m_ptr->decStrong(this);
}
```

14.2.3 强指针

在 Android 系统中，强指针使用的引用计数类是 RefBase，类 RefBase 比类 LightRefBase 要复杂。但是其功能和类 LightRefBase 一样，也提供了 incStrong 和 decStrong 成员函数来操作它的引用计数器。类 RefBase 与类 LightRefBase 的最大区别是，它不像类 LightRefBase 一样直接提供一个整型值（mutable volatile int32_t mCount）来维护对象的引用计数。原因是复杂的引用计数技术同时支持强引用计数和弱引用计数。所以在类 RefBase 的具体实现中，强引用计数和弱引用计数功能是通过其成员变量 mRefs 提供的。

类 RefBase 在文件“frameworks\native\include\utils\RefBase.h”中定义，具体实现代码如下所示：

```
class RefBase
{
public:
            void            incStrong(const void* id) const;
            void            decStrong(const void* id) const;
            void            forceIncStrong(const void* id) const;
            //! DEBUGGING ONLY: Get current strong ref count.
            int32_t         getStrongCount() const;
    class weakref_type
    {
    public:
        RefBase*            refBase() const;

        void                incWeak(const void* id);
        void                decWeak(const void* id);

        // acquires a strong reference if there is already one.
        bool                attemptIncStrong(const void* id);

        // acquires a weak reference if there is already one.
        // This is not always safe. see ProcessState.cpp and BpBinder.cpp
        // for proper use.
        bool                attemptIncWeak(const void* id);
        //! DEBUGGING ONLY: Get current weak ref count.
```

```
        int32_t          getWeakCount() const;
        //! DEBUGGING ONLY: Print references held on object.
        void             printRefs() const;
        //! DEBUGGING ONLY: Enable tracking for this object.
        // enable -- enable/disable tracking
        // retain -- when tracking is enable, if true, then we save a stack trace
        //         for each reference and dereference; when retain == false, we
        //         match up references and dereferences and keep only the
        //         outstanding ones.

        void             trackMe(bool enable, bool retain);
    };
            weakref_type*   createWeak(const void* id) const;

            weakref_type*   getWeakRefs() const;
            //! DEBUGGING ONLY: Print references held on object.
    inline  void            printRefs() const { getWeakRefs()->printRefs(); }
            //! DEBUGGING ONLY: Enable tracking of object.
    inline  void            trackMe(bool enable, bool retain)
    {
        getWeakRefs()->trackMe(enable, retain);
    }
    typedef RefBase basetype;
protected:
                            RefBase();
    virtual                 ~RefBase();
    //! Flags for extendObjectLifetime()
    enum {
        OBJECT_LIFETIME_STRONG  = 0x0000,
        OBJECT_LIFETIME_WEAK    = 0x0001,
        OBJECT_LIFETIME_MASK    = 0x0001
    };
            void            extendObjectLifetime(int32_t mode);

    //! Flags for onIncStrongAttempted()
    enum {
        FIRST_INC_STRONG = 0x0001
    };
    virtual void            onFirstRef();
    virtual void            onLastStrongRef(const void* id);
    virtual bool            onIncStrongAttempted(uint32_t flags, const void* id);
    virtual void            onLastWeakRef(const void* id);
private:
    friend class ReferenceMover;
    static void moveReferences(void* d, void const* s, size_t n,
            const ReferenceConverterBase& caster);
private:
    friend class weakref_type;
    class weakref_impl;
                            RefBase(const RefBase& o);
            RefBase&        operator=(const RefBase& o);
        weakref_impl* const mRefs;
};
```

在类 RefBase 中，其成员变量 mRefs 的类型为 weakref_impl 指针。类 RefBase 的具体实现在文件“frameworks\native\libs\utils\RefBase.cpp”中定义，具体实现代码如下所示：

```
class RefBase::weakref_impl : public RefBase::weakref_type
{
public:
    volatile int32_t    mStrong;
    volatile int32_t    mWeak;
    RefBase* const      mBase;
    volatile int32_t    mFlags;

#if !DEBUG_REFS

    weakref_impl(RefBase* base)
```

```
        : mStrong(INITIAL_STRONG_VALUE)
        , mWeak(0)
        , mBase(base)
        , mFlags(0)
    {
    }

    void addStrongRef(const void* /*id*/) { }
    void removeStrongRef(const void* /*id*/) { }
    void renameStrongRefId(const void* /*old_id*/, const void* /*new_id*/) { }
    void addWeakRef(const void* /*id*/) { }
    void removeWeakRef(const void* /*id*/) { }
    void renameWeakRefId(const void* /*old_id*/, const void* /*new_id*/) { }
    void printRefs() const { }
    void trackMe(bool, bool) { }
#else

    weakref_impl(RefBase* base)
        : mStrong(INITIAL_STRONG_VALUE)
        , mWeak(0)
        , mBase(base)
        , mFlags(0)
        , mStrongRefs(NULL)
        , mWeakRefs(NULL)
        , mTrackEnabled(!!DEBUG_REFS_ENABLED_BY_DEFAULT)
        , mRetain(false)
    {
    }

    ~weakref_impl()
    {
        bool dumpStack = false;
        if (!mRetain && mStrongRefs != NULL) {
            dumpStack = true;
#if DEBUG_REFS_FATAL_SANITY_CHECKS
            LOG_ALWAYS_FATAL("Strong references remain!");
#else
            ALOGE("Strong references remain:");
#endif
            ref_entry* refs = mStrongRefs;
            while (refs) {
                char inc = refs->ref >= 0 ? '+' : '-';
                ALOGD("\t%c ID %p (ref %d):", inc, refs->id, refs->ref);
#if DEBUG_REFS_CALLSTACK_ENABLED
                refs->stack.dump();
#endif
                refs = refs->next;
            }
        }

        if (!mRetain && mWeakRefs != NULL) {
            dumpStack = true;
#if DEBUG_REFS_FATAL_SANITY_CHECKS
            LOG_ALWAYS_FATAL("Weak references remain:");
#else
            ALOGE("Weak references remain!");
#endif
            ref_entry* refs = mWeakRefs;
            while (refs) {
                char inc = refs->ref >= 0 ? '+' : '-';
                ALOGD("\t%c ID %p (ref %d):", inc, refs->id, refs->ref);
#if DEBUG_REFS_CALLSTACK_ENABLED
                refs->stack.dump();
#endif
                refs = refs->next;
            }
        }
        if (dumpStack) {
```

```
        ALOGE("above errors at:");
        CallStack stack;
        stack.update();
        stack.dump();
    }
}

void addStrongRef(const void* id) {
    //ALOGD_IF(mTrackEnabled,
    //        "addStrongRef: RefBase=%p, id=%p", mBase, id);
    addRef(&mStrongRefs, id, mStrong);
}

void removeStrongRef(const void* id) {
    //ALOGD_IF(mTrackEnabled,
    //        "removeStrongRef: RefBase=%p, id=%p", mBase, id);
    if (!mRetain) {
        removeRef(&mStrongRefs, id);
    } else {
        addRef(&mStrongRefs, id, -mStrong);
    }
}

void renameStrongRefId(const void* old_id, const void* new_id) {
    //ALOGD_IF(mTrackEnabled,
    //        "renameStrongRefId: RefBase=%p, oid=%p, nid=%p",
    //        mBase, old_id, new_id);
    renameRefsId(mStrongRefs, old_id, new_id);
}

void addWeakRef(const void* id) {
    addRef(&mWeakRefs, id, mWeak);
}

void removeWeakRef(const void* id) {
    if (!mRetain) {
        removeRef(&mWeakRefs, id);
    } else {
        addRef(&mWeakRefs, id, -mWeak);
    }
}

void renameWeakRefId(const void* old_id, const void* new_id) {
    renameRefsId(mWeakRefs, old_id, new_id);
}

void trackMe(bool track, bool retain)
{
    mTrackEnabled = track;
    mRetain = retain;
}

void printRefs() const
{
    String8 text;

    {
        Mutex::Autolock _l(mMutex);
        char buf[128];
        sprintf(buf, "Strong references on RefBase %p (weakref_type %p):\n", mBase, this);
        text.append(buf);
        printRefsLocked(&text, mStrongRefs);
        sprintf(buf, "Weak references on RefBase %p (weakref_type %p):\n", mBase, this);
        text.append(buf);
        printRefsLocked(&text, mWeakRefs);
    }

    {
        char name[100];
```

```
            snprintf(name, 100, "/data/%p.stack", this);
            int rc = open(name, O_RDWR | O_CREAT | O_APPEND);
            if (rc >= 0) {
                write(rc, text.string(), text.length());
                close(rc);
                ALOGD("STACK TRACE for %p saved in %s", this, name);
            }
            else ALOGE("FAILED TO PRINT STACK TRACE for %p in %s: %s", this,
                    name, strerror(errno));
        }
    }

private:
    struct ref_entry
    {
        ref_entry* next;
        const void* id;
#if DEBUG_REFS_CALLSTACK_ENABLED
        CallStack stack;
#endif
        int32_t ref;
    };

    void addRef(ref_entry** refs, const void* id, int32_t mRef)
    {
        if (mTrackEnabled) {
            AutoMutex _l(mMutex);

            ref_entry* ref = new ref_entry;
            // Reference count at the time of the snapshot, but before the
            // update.  Positive value means we increment, negative--we
            // decrement the reference count.
            ref->ref = mRef;
            ref->id = id;
#if DEBUG_REFS_CALLSTACK_ENABLED
            ref->stack.update(2);
#endif
            ref->next = *refs;
            *refs = ref;
        }
    }

    void removeRef(ref_entry** refs, const void* id)
    {
        if (mTrackEnabled) {
            AutoMutex _l(mMutex);

            ref_entry* const head = *refs;
            ref_entry* ref = head;
            while (ref != NULL) {
                if (ref->id == id) {
                    *refs = ref->next;
                    delete ref;
                    return;
                }
                refs = &ref->next;
                ref = *refs;
            }

#if DEBUG_REFS_FATAL_SANITY_CHECKS
            LOG_ALWAYS_FATAL("RefBase: removing id %p on RefBase %p"
                    "(weakref_type %p) that doesn't exist!",
                    id, mBase, this);
#endif

            ALOGE("RefBase: removing id %p on RefBase %p"
                    "(weakref_type %p) that doesn't exist!",
                    id, mBase, this);
```

```
            ref = head;
            while (ref) {
                char inc = ref->ref >= 0 ? '+' : '-';
                ALOGD("\t%c ID %p (ref %d):", inc, ref->id, ref->ref);
                ref = ref->next;
            }

            CallStack stack;
            stack.update();
            stack.dump();
        }
    }

    void renameRefsId(ref_entry* r, const void* old_id, const void* new_id)
    {
        if (mTrackEnabled) {
            AutoMutex _l(mMutex);
            ref_entry* ref = r;
            while (ref != NULL) {
                if (ref->id == old_id) {
                    ref->id = new_id;
                }
                ref = ref->next;
            }
        }
    }

    void printRefsLocked(String8* out, const ref_entry* refs) const
    {
        char buf[128];
        while (refs) {
            char inc = refs->ref >= 0 ? '+' : '-';
            sprintf(buf, "\t%c ID %p (ref %d):\n",
                    inc, refs->id, refs->ref);
            out->append(buf);
#if DEBUG_REFS_CALLSTACK_ENABLED
            out->append(refs->stack.toString("\t\t"));
#else
            out->append("\t\t(call stacks disabled)");
#endif
            refs = refs->next;
        }
    }

    mutable Mutex mMutex;
    ref_entry* mStrongRefs;
    ref_entry* mWeakRefs;

    bool mTrackEnabled;
    // Collect stack traces on addref and removeref, instead of deleting the stack references
    // on removeref that match the address ones.
    bool mRetain;

#endif
};

// ---------------------------------------------------------------------------

void RefBase::incStrong(const void* id) const
{
    weakref_impl* const refs = mRefs;
    refs->incWeak(id);

    refs->addStrongRef(id);
    const int32_t c = android_atomic_inc(&refs->mStrong);
    ALOG_ASSERT(c > 0, "incStrong() called on %p after last strong ref", refs);
#if PRINT_REFS
    ALOGD("incStrong of %p from %p: cnt=%d\n", this, id, c);
#endif
```

```
    if (c != INITIAL_STRONG_VALUE) {
        return;
    }

    android_atomic_add(-INITIAL_STRONG_VALUE, &refs->mStrong);
    refs->mBase->onFirstRef();
}

void RefBase::decStrong(const void* id) const
{
    weakref_impl* const refs = mRefs;
    refs->removeStrongRef(id);
    const int32_t c = android_atomic_dec(&refs->mStrong);
#if PRINT_REFS
    ALOGD("decStrong of %p from %p: cnt=%d\n", this, id, c);
#endif
    ALOG_ASSERT(c >= 1, "decStrong() called on %p too many times", refs);
    if (c == 1) {
        refs->mBase->onLastStrongRef(id);
        if ((refs->mFlags&OBJECT_LIFETIME_MASK) == OBJECT_LIFETIME_STRONG) {
            delete this;
        }
    }
    refs->decWeak(id);
}

void RefBase::forceIncStrong(const void* id) const
{
    weakref_impl* const refs = mRefs;
    refs->incWeak(id);

    refs->addStrongRef(id);
    const int32_t c = android_atomic_inc(&refs->mStrong);
    ALOG_ASSERT(c >= 0, "forceIncStrong called on %p after ref count underflow",
           refs);
#if PRINT_REFS
    ALOGD("forceIncStrong of %p from %p: cnt=%d\n", this, id, c);
#endif

    switch (c) {
    case INITIAL_STRONG_VALUE:
        android_atomic_add(-INITIAL_STRONG_VALUE, &refs->mStrong);
        // fall through...
    case 0:
        refs->mBase->onFirstRef();
    }
}

int32_t RefBase::getStrongCount() const
{
    return mRefs->mStrong;
}

RefBase* RefBase::weakref_type::refBase() const
{
    return static_cast<const weakref_impl*>(this)->mBase;
}

void RefBase::weakref_type::incWeak(const void* id)
{
    weakref_impl* const impl = static_cast<weakref_impl*>(this);
    impl->addWeakRef(id);
    const int32_t c = android_atomic_inc(&impl->mWeak);
    ALOG_ASSERT(c >= 0, "incWeak called on %p after last weak ref", this);
}

void RefBase::weakref_type::decWeak(const void* id)
{
```

```
    weakref_impl* const impl = static_cast<weakref_impl*>(this);
    impl->removeWeakRef(id);
    const int32_t c = android_atomic_dec(&impl->mWeak);
    ALOG_ASSERT(c >= 1, "decWeak called on %p too many times", this);
    if (c != 1) return;

    if ((impl->mFlags&OBJECT_LIFETIME_WEAK) == OBJECT_LIFETIME_STRONG) {
        // This is the regular lifetime case. The object is destroyed
        // when the last strong reference goes away. Since weakref_impl
        // outlive the object, it is not destroyed in the dtor, and
        // we'll have to do it here.
        if (impl->mStrong == INITIAL_STRONG_VALUE) {
            // Special case: we never had a strong reference, so we need to
            // destroy the object now.
            delete impl->mBase;
        } else {
            // ALOGV("Freeing refs %p of old RefBase %p\n", this, impl->mBase);
            delete impl;
        }
    } else {
        // less common case: lifetime is OBJECT_LIFETIME_{WEAK|FOREVER}
        impl->mBase->onLastWeakRef(id);
        if ((impl->mFlags&OBJECT_LIFETIME_MASK) == OBJECT_LIFETIME_WEAK) {
            // this is the OBJECT_LIFETIME_WEAK case. The last weak-reference
            // is gone, we can destroy the object.
            delete impl->mBase;
        }
    }
}

bool RefBase::weakref_type::attemptIncStrong(const void* id)
{
    incWeak(id);

    weakref_impl* const impl = static_cast<weakref_impl*>(this);

    int32_t curCount = impl->mStrong;
    ALOG_ASSERT(curCount >= 0, "attemptIncStrong called on %p after underflow",
            this);
    while (curCount > 0 && curCount != INITIAL_STRONG_VALUE) {
        if (android_atomic_cmpxchg(curCount, curCount+1, &impl->mStrong) == 0) {
            break;
        }
        curCount = impl->mStrong;
    }

    if (curCount <= 0 || curCount == INITIAL_STRONG_VALUE) {
        bool allow;
        if (curCount == INITIAL_STRONG_VALUE) {
            // Attempting to acquire first strong reference...  this is allowed
            // if the object does NOT have a longer lifetime (meaning the
            // implementation doesn't need to see this), or if the implementation
            // allows it to happen.
            allow = (impl->mFlags&OBJECT_LIFETIME_WEAK) != OBJECT_LIFETIME_WEAK
                  || impl->mBase->onIncStrongAttempted(FIRST_INC_STRONG, id);
        } else {
            // Attempting to revive the object...  this is allowed
            // if the object DOES have a longer lifetime (so we can safely
            // call the object with only a weak ref) and the implementation
            // allows it to happen.
            allow = (impl->mFlags&OBJECT_LIFETIME_WEAK) == OBJECT_LIFETIME_WEAK
                  && impl->mBase->onIncStrongAttempted(FIRST_INC_STRONG, id);
        }
        if (!allow) {
            decWeak(id);
            return false;
        }
        curCount = android_atomic_inc(&impl->mStrong);
```

```
        // If the strong reference count has already been incremented by
        // someone else, the implementor of onIncStrongAttempted() is holding
        // an unneeded reference.  So call onLastStrongRef() here to remove it.
        // (No, this is not pretty.)  Note that we MUST NOT do this if we
        // are in fact acquiring the first reference.
        if (curCount > 0 && curCount < INITIAL_STRONG_VALUE) {
            impl->mBase->onLastStrongRef(id);
        }
    }

    impl->addStrongRef(id);

#if PRINT_REFS
    ALOGD("attemptIncStrong of %p from %p: cnt=%d\n", this, id, curCount);
#endif

    if (curCount == INITIAL_STRONG_VALUE) {
        android_atomic_add(-INITIAL_STRONG_VALUE, &impl->mStrong);
        impl->mBase->onFirstRef();
    }

    return true;
}

bool RefBase::weakref_type::attemptIncWeak(const void* id)
{
    weakref_impl* const impl = static_cast<weakref_impl*>(this);

    int32_t curCount = impl->mWeak;
    ALOG_ASSERT(curCount >= 0, "attemptIncWeak called on %p after underflow",
            this);
    while (curCount > 0) {
        if (android_atomic_cmpxchg(curCount, curCount+1, &impl->mWeak) == 0) {
            break;
        }
        curCount = impl->mWeak;
    }

    if (curCount > 0) {
        impl->addWeakRef(id);
    }

    return curCount > 0;
}

int32_t RefBase::weakref_type::getWeakCount() const
{
    return static_cast<const weakref_impl*>(this)->mWeak;
}

void RefBase::weakref_type::printRefs() const
{
    static_cast<const weakref_impl*>(this)->printRefs();
}

void RefBase::weakref_type::trackMe(bool enable, bool retain)
{
    static_cast<weakref_impl*>(this)->trackMe(enable, retain);
}

RefBase::weakref_type* RefBase::createWeak(const void* id) const
{
    mRefs->incWeak(id);
    return mRefs;
}

RefBase::weakref_type* RefBase::getWeakRefs() const
{
    return mRefs;
```

```
}

RefBase::RefBase()
    : mRefs(new weakref_impl(this))
{
}

RefBase::~RefBase()
{
    if (mRefs->mStrong == INITIAL_STRONG_VALUE) {
        // we never acquired a strong (and/or weak) reference on this object.
        delete mRefs;
    } else {
        // life-time of this object is extended to WEAK or FOREVER, in
        // which case weakref_impl doesn't out-live the object and we
        // can free it now.
        if ((mRefs->mFlags & OBJECT_LIFETIME_MASK) != OBJECT_LIFETIME_STRONG) {
            // It's possible that the weak count is not 0 if the object
            // re-acquired a weak reference in its destructor
            if (mRefs->mWeak == 0) {
                delete mRefs;
            }
        }
    }
    // for debugging purposes, clear this.
    const_cast<weakref_impl*&>(mRefs) = NULL;
}

void RefBase::extendObjectLifetime(int32_t mode)
{
    android_atomic_or(mode, &mRefs->mFlags);
}

void RefBase::onFirstRef()
{
}

void RefBase::onLastStrongRef(const void* /*id*/)
{
}

bool RefBase::onIncStrongAttempted(uint32_t flags, const void* id)
{
    return (flags&FIRST_INC_STRONG) ? true : false;
}

void RefBase::onLastWeakRef(const void* /*id*/)
{
}

// ---------------------------------------------------------------------------

void RefBase::moveReferences(void* dst, void const* src, size_t n,
        const ReferenceConverterBase& caster)
{
#if DEBUG_REFS
    const size_t itemSize = caster.getReferenceTypeSize();
    for (size_t i=0 ; i<n ; i++) {
        void*       d = reinterpret_cast<void       *>(intptr_t(dst) + i*itemSize);
        void const* s = reinterpret_cast<void const*>(intptr_t(src) + i*itemSize);
        RefBase* ref(reinterpret_cast<RefBase*>(caster.getReferenceBase(d)));
        ref->mRefs->renameStrongRefId(s, d);
        ref->mRefs->renameWeakRefId(s, d);
    }
#endif
}
// ---------------------------------------------------------------------------
TextOutput& printStrongPointer(TextOutput& to, const void* val)
{
```

```
    to << "sp<>(" << val << ")";
    return to;
}
TextOutput& printWeakPointer(TextOutput& to, const void* val)
{
    to << "wp<>(" << val << ")";
    return to;
}
}; // namespace android
```

整个上述代码被分为了如下所示的两大部分：

（1）用如下 DEBUG_REFS 标记标识的部分，表示类 weakref_impl 被编译成调试版本。Debug 版本的源代码的成员函数都是有实现的，实现这些函数的目的都是便于开发人员调试引用计数用。

```
#if !DEBUG_REFS
......
#else
```

（2）用如下标记标识的部分，表示类 weakref_ impl 被编译成非调试版本。

```
#else
......
#endif
```

14.2.4 弱指针

在 Android 系统中，弱指针和强指针使用一样的引用计数类：RefBase 类。和强指针类一样，弱指针也有一个指向目标对象的成员变量 m_ptr。另外，弱指针还有一个类型是 weakref_type 指针的额外的成员变量 m_refs。类 wp 在文件“frameworks\native\include\utils\RefBase.h”中定义，具体实现代码如下所示：

```
template <typename T>
class wp
{
public:
    typedef typename RefBase::weakref_type weakref_type;

    inline wp() : m_ptr(0) { }

    wp(T* other);
    wp(const wp<T>& other);
    wp(const sp<T>& other);
    template<typename U> wp(U* other);
    template<typename U> wp(const sp<U>& other);
    template<typename U> wp(const wp<U>& other);

    ~wp();

    // Assignment

    wp& operator = (T* other);
    wp& operator = (const wp<T>& other);
    wp& operator = (const sp<T>& other);

    template<typename U> wp& operator = (U* other);
    template<typename U> wp& operator = (const wp<U>& other);
    template<typename U> wp& operator = (const sp<U>& other);

    void set_object_and_refs(T* other, weakref_type* refs);

    // promotion to sp

    sp<T> promote() const;

    // Reset
```

```
    void clear();

    // Accessors

    inline  weakref_type* get_refs() const { return m_refs; }

    inline  T* unsafe_get() const { return m_ptr; }

    // Operators

    COMPARE_WEAK(==)
    COMPARE_WEAK(!=)
    COMPARE_WEAK(>)
    COMPARE_WEAK(<)
    COMPARE_WEAK(<=)
    COMPARE_WEAK(>=)

    inline bool operator == (const wp<T>& o) const {
        return (m_ptr == o.m_ptr) && (m_refs == o.m_refs);
    }
    template<typename U>
    inline bool operator == (const wp<U>& o) const {
        return m_ptr == o.m_ptr;
    }

    inline bool operator > (const wp<T>& o) const {
        return (m_ptr == o.m_ptr) ? (m_refs > o.m_refs) : (m_ptr > o.m_ptr);
    }
    template<typename U>
    inline bool operator > (const wp<U>& o) const {
        return (m_ptr == o.m_ptr) ? (m_refs > o.m_refs) : (m_ptr > o.m_ptr);
    }

    inline bool operator < (const wp<T>& o) const {
        return (m_ptr == o.m_ptr) ? (m_refs < o.m_refs) : (m_ptr < o.m_ptr);
    }
    template<typename U>
    inline bool operator < (const wp<U>& o) const {
        return (m_ptr == o.m_ptr) ? (m_refs < o.m_refs) : (m_ptr < o.m_ptr);
    }
                         inline bool operator != (const wp<T>& o) const { return m_refs != o.m_refs; }
    template<typename U> inline bool operator != (const wp<U>& o) const { return !operator == (o); }
                         inline bool operator <= (const wp<T>& o) const { return !operator > (o); }
    template<typename U> inline bool operator <= (const wp<U>& o) const { return !operator > (o); }
                         inline bool operator >= (const wp<T>& o) const { return !operator < (o); }
    template<typename U> inline bool operator >= (const wp<U>& o) const { return !operator < (o); }

private:
    template<typename Y> friend class sp;
    template<typename Y> friend class wp;

    T*              m_ptr;
    weakref_type*   m_refs;
};
```

类 wp 的构造函数的实现代码如下所示：

```
template<typename T>
wp<T>::wp(T* other)
    : m_ptr(other)
{
    if (other) m_refs = other->createWeak(this);
}
```

在上述代码中，参数 other 类继承于类 RefBase，并调用了类 RefBase 的成员函数 createWeak。函数 createWeak 在文件"frameworks\native\libs\utils\RefBase.cpp"中定义，具体实现代码如下所示：

```
RefBase::weakref_type* RefBase::createWeak(const void* id) const
{
   mRefs->incWeak(id);
   return mRefs;// mRefs 的类型为 weakref_impl 指针
}
```

再看类 wp 的析构函数，此函数直接调用目标对象的 weakref_impl 对象的函数 decWeak，目的是减少弱引用计数。当弱引用计数为 0 时，根据在目标对象的标志位（0、OBJECT_LIFETIME_WEAK 或者 OBJECT_LIFETIME_FOREVER）来决定是否要 delete（删除）目标对象。下面是析构函数的实现代码。

```
template<typename T>
wp<T>::~wp()
{
   if (m_ptr) m_refs->decWeak(this);
}
```

弱指针的最大特点是不能直接操作目标对象，原因是弱指针类没有重载“*”和“->”操作符号，而强指针重载了这两个操作符号。如果坚持要操作目标对象，则需要把弱指针升级为强指针。升级方法是使用成员变量 m_ptr 和 m_refs 构造一个强指针 sp，m_ptr 是指目标对象的一个指针，m_refs 是指指向目标对象里面的 weakref_impl 对象。升级代码如下：

```
template<typename T>
sp<T> wp<T>::promote() const
{
   return sp<T>(m_ptr, m_refs);
}
```

与之对应的强指针构造代码如下所示：

```
template<typename T>
sp<T>::sp(T* p, weakref_type* refs)
   : m_ptr((p && refs->attemptIncStrong(this)) ? p : 0)
{
}
```

在上述构造代码中，初始化指向了目标对象的成员变量 m_ptr。如果还存在目标对象，则 m_ptr 指向目标对象。如果这个目标对象已经不存在，则 m_ptr 为 NULL。是否升级成功需要参考函数 attemptIncStrong 的返回结果。函数 attemptIncStrong 的具体实现代码如下所示：

```
bool RefBase::weakref_type::attemptIncStrong(const void* id)
{
   incWeak(id);

   weakref_impl* const impl = static_cast<weakref_impl*>(this);

   int32_t curCount = impl->mStrong;
   LOG_ASSERT(curCount >= 0, "attemptIncStrong called on %p after underflow",
      this);
   while (curCount > 0 && curCount != INITIAL_STRONG_VALUE) {
      if (android_atomic_cmpxchg(curCount, curCount+1, &impl->mStrong) == 0) {
         break;
      }
      curCount = impl->mStrong;
   }

   if (curCount <= 0 || curCount == INITIAL_STRONG_VALUE) {
      bool allow;
      if (curCount == INITIAL_STRONG_VALUE) {
         // Attempting to acquire first strong reference... this is allowed
         // if the object does NOT have a longer lifetime (meaning the
         // implementation doesn't need to see this), or if the implementation
         // allows it to happen.
         allow = (impl->mFlags&OBJECT_LIFETIME_WEAK) != OBJECT_LIFETIME_WEAK
```

```
            || impl->mBase->onIncStrongAttempted(FIRST_INC_STRONG, id);
        } else {
            // Attempting to revive the object... this is allowed
            // if the object DOES have a longer lifetime (so we can safely
            // call the object with only a weak ref) and the implementation
            // allows it to happen.
            allow = (impl->mFlags&OBJECT_LIFETIME_WEAK) == OBJECT_LIFETIME_WEAK
                && impl->mBase->onIncStrongAttempted(FIRST_INC_STRONG, id);
        }
        if (!allow) {
            decWeak(id);
            return false;
        }
        curCount = android_atomic_inc(&impl->mStrong);

        // If the strong reference count has already been incremented by
        // someone else, the implementor of onIncStrongAttempted() is holding
        // an unneeded reference.  So call onLastStrongRef() here to remove it.
        // (No, this is not pretty.)  Note that we MUST NOT do this if we
        // are in fact acquiring the first reference.
        if (curCount > 0 && curCount < INITIAL_STRONG_VALUE) {
            impl->mBase->onLastStrongRef(id);
        }
    }

    impl->addWeakRef(id);
    impl->addStrongRef(id);

#if PRINT_REFS
    LOGD("attemptIncStrong of %p from %p: cnt=%d\n", this, id, curCount);
#endif

    if (curCount == INITIAL_STRONG_VALUE) {
        android_atomic_add(-INITIAL_STRONG_VALUE, &impl->mStrong);
        impl->mBase->onFirstRef();
    }

    return true;
}
```

第 15 章　分析 Dalvik VM 的启动过程

经过本书前面内容的学习可知，Android 系统中的应用程序进程是由 Zygote 进程孕育出来的，而 Zygote 进程又是由 Init 进程启动的。在启动 Zygote 进程时会创建一个 Dalvik VM 实例，每当 Zygote 进程孕育出一个新的应用程序进程时，会将这个 Dalvik VM 实例复制到新的应用程序中，这样可以使每个应用程序进程都拥有一个独立的 Dalvik VM 实例。在本章的内容中，将详细分析启动 Dalvik VM 的具体过程，为读者步入本书后面知识的学习打下基础。

15.1 Dalvik VM 启动流程概览

在启动 Zygote 进程的过程中，以调用类 AndroidRuntime 的成员函数 start 作为启动 Dalvik VM 的第一步。在启动 Zygote 进程的过程中，除了会创建一个 Dalvik 虚拟机实例之外，还会将 Java 运行时库加载到进程中来，以及注册一些 Android 核心类的 JNI 方法来前面创建的 Dalvik 虚拟机实例中去。当一个应用程序进程被 Zygote 进程孵化出来，不仅会获得 Zygote 进程中的 Dalvik 虚拟机实例拷贝，还会与 Zygote 一起共享 Java 运行时库，这完全得益于 Linux 内核的进程创建机制（fork）。这种 Zygote 孵化机制的优点是不仅可以快速地启动一个应用程序进程，而且还可以节省整体的内存消耗。但是缺点也十分明显，那便是影响开机速度，因为 Zygote 是在开机过程中启动的。但是毕竟整个系统只有一个 Zygote 进程，并且可能有无数个应用程序进程，而且人们不会经常去关闭手机，在大多数情况下只是让它进入休眠状态。由此可见，总体来说 Zygote 孵化机制是利大于弊的。

在 Android 系统中启动 Zygote 进程的过程中，会调用到类 AndroidRuntime 的成员函数 start 开启 Dalvik 虚拟机启动相关的过程。具体过程如图 15-1 所示。

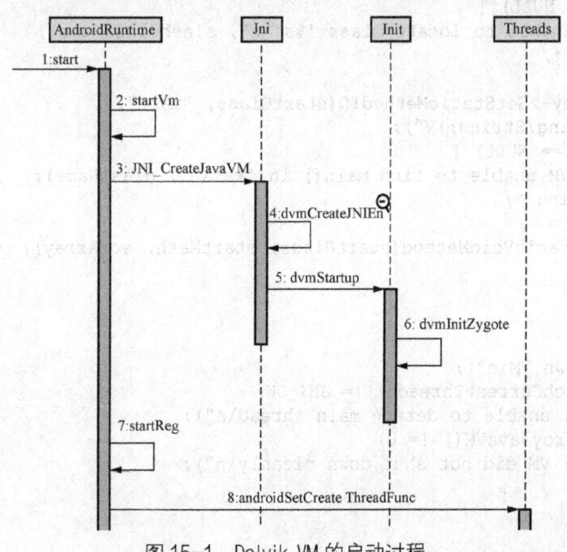

图 15-1　Dalvik VM 的启动过程

15.2 Dalvik VM 启动过程详解

通过图 15-1 可以大体了解 Dalvik VM 的启动过程，在本节的内容中，将详细讲解启动 Dalvik VM 的具体过程，详细剖析每一个步骤的具体实现。

15.2.1 创建 Dalvik VM 实例

在文件“frameworks/base/core/jni/AndroidRuntime.cpp”中，调用函数 start 作为启动 Dalvik VM 的第一步。函数 start 调用函数 startVm 创建一个 Dalvik VM 实例，并且保存在成员变量 mJavaVM 中。然后调用成员函数 startReg 注册 Android 核心类的 JNI 方法，并调用参数 className 所对应的一个 Java 函数 main 作为 Zygote 进程的 Java 层入口。函数 start 的具体实现代码如下所示：

```
void AndroidRuntime::start(const char* className, const bool startSystemServer)
{
    /* 函数 startVm 用于创建一个 Dalvik 虚拟机实例，并且保存在成员变量 mJavaVM 中 */
    if (startVm(&mJavaVM, &env) != 0)
        goto bail;
    /*
     *函数 startReg 用于注册 Android 核心类的 JNI 方法
     */
    if (startReg(env) < 0) {
        LOGE("Unable to register all android natives\n");
        goto bail;
    }
    ......
    /*
     * 开始启动虚拟机
     */
    jclass startClass;
    jmethodID startMeth;

    slashClassName = strdup(className);
    for (cp = slashClassName; *cp != '\0'; cp++)
        if (*cp == '.')
            *cp = '/';

    startClass = env->FindClass(slashClassName);
    if (startClass == NULL) {
        LOGE("JavaVM unable to locate class '%s'\n", slashClassName);
        /* keep going */
    } else {
        startMeth = env->GetStaticMethodID(startClass, "main",
            "([Ljava/lang/String;)V");
        if (startMeth == NULL) {
            LOGE("JavaVM unable to find main() in '%s'\n", className);
            /* keep going */
        } else {
            env->CallStaticVoidMethod(startClass, startMeth, strArray);
            ......
        }
    }

    LOGD("Shutting down VM\n");
    if (mJavaVM->DetachCurrentThread() != JNI_OK)
        LOGW("Warning: unable to detach main thread\n");
    if (mJavaVM->DestroyJavaVM() != 0)
        LOGW("Warning: VM did not shut down cleanly\n");

    ......
}
```

在退出上述函数代码之前，需要调用刚刚创建的 Dalvik VM 实例的如下两个成员函数。

- DetachCurrentThread：功能是将 Zygote 进程的主线程脱离刚刚创建的 Dalvik 虚拟机实例。
- DestroyJavaVM：用于销毁刚刚创建的 Dalvik VM 实例。

15.2.2 指定一系列控制选项

函数 startVm 在文件“frameworks/base/core/jni/AndroidRuntime.cpp”中定义，功能是在启动 Dalvik VM 时指定一系列有用的控制选项。函数 startVm 的具体实现代码如下所示：

```
int AndroidRuntime::startVm(JavaVM** pJavaVM, JNIEnv** pEnv)
{
   int result = -1;
   JavaVMInitArgs initArgs;
   JavaVMOption opt;
   char propBuf[PROPERTY_VALUE_MAX];
   char stackTraceFileBuf[PROPERTY_VALUE_MAX];
   char dexoptFlagsBuf[PROPERTY_VALUE_MAX];
   char enableAssertBuf[sizeof("-ea:")-1 + PROPERTY_VALUE_MAX];
   char jniOptsBuf[sizeof("-Xjniopts:")-1 + PROPERTY_VALUE_MAX];
   char heapstartsizeOptsBuf[sizeof("-Xms")-1 + PROPERTY_VALUE_MAX];
   char heapsizeOptsBuf[sizeof("-Xmx")-1 + PROPERTY_VALUE_MAX];
   char heapgrowthlimitOptsBuf[sizeof("-XX:HeapGrowthLimit=")-1 + PROPERTY_VALUE_MAX];
   char heapminfreeOptsBuf[sizeof("-XX:HeapMinFree=")-1 + PROPERTY_VALUE_MAX];
   char heapmaxfreeOptsBuf[sizeof("-XX:HeapMaxFree=")-1 + PROPERTY_VALUE_MAX];
   char heaptargetutilizationOptsBuf[sizeof("-XX:HeapTargetUtilization=")-1 + PROPERTY_VALUE_MAX];
   char extraOptsBuf[PROPERTY_VALUE_MAX];
   char* stackTraceFile = NULL;
   bool checkJni = false;
   bool checkDexSum = false;
   bool logStdio = false;
   enum {
     kEMDefault,
     kEMIntPortable,
     kEMIntFast,
     kEMJitCompiler,
   } executionMode = kEMDefault;


   property_get("dalvik.vm.checkjni", propBuf, "");
   if (strcmp(propBuf, "true") == 0) {
      checkJni = true;
   } else if (strcmp(propBuf, "false") != 0) {
      /* property is neither true nor false; fall back on kernel parameter */
      property_get("ro.kernel.android.checkjni", propBuf, "");
      if (propBuf[0] == '1') {
         checkJni = true;
      }
   }

   property_get("dalvik.vm.execution-mode", propBuf, "");
   if (strcmp(propBuf, "int:portable") == 0) {
      executionMode = kEMIntPortable;
   } else if (strcmp(propBuf, "int:fast") == 0) {
      executionMode = kEMIntFast;
   } else if (strcmp(propBuf, "int:jit") == 0) {
      executionMode = kEMJitCompiler;
   }

   property_get("dalvik.vm.stack-trace-file", stackTraceFileBuf, "");

   property_get("dalvik.vm.check-dex-sum", propBuf, "");
   if (strcmp(propBuf, "true") == 0) {
      checkDexSum = true;
   }

   property_get("log.redirect-stdio", propBuf, "");
```

```
    if (strcmp(propBuf, "true") == 0) {
        logStdio = true;
    }

    strcpy(enableAssertBuf, "-ea:");
    property_get("dalvik.vm.enableassertions", enableAssertBuf+4, "");

    strcpy(jniOptsBuf, "-Xjniopts:");
    property_get("dalvik.vm.jniopts", jniOptsBuf+10, "");

    /* route exit() to our handler */
    opt.extraInfo = (void*) runtime_exit;
    opt.optionString = "exit";
    mOptions.add(opt);

    /* route fprintf() to our handler */
    opt.extraInfo = (void*) runtime_vfprintf;
    opt.optionString = "vfprintf";
    mOptions.add(opt);

    /* register the framework-specific "is sensitive thread" hook */
    opt.extraInfo = (void*) runtime_isSensitiveThread;
    opt.optionString = "sensitiveThread";
    mOptions.add(opt);

    opt.extraInfo = NULL;

    /* enable verbose; standard options are { jni, gc, class } */
    //options[curOpt++].optionString = "-verbose:jni";
    opt.optionString = "-verbose:gc";
    mOptions.add(opt);
    //options[curOpt++].optionString = "-verbose:class";

    /*
     * The default starting and maximum size of the heap.  Larger
     * values should be specified in a product property override.
     */
    strcpy(heapstartsizeOptsBuf, "-Xms");
    property_get("dalvik.vm.heapstartsize", heapstartsizeOptsBuf+4, "4m");
    opt.optionString = heapstartsizeOptsBuf;
    mOptions.add(opt);
    strcpy(heapsizeOptsBuf, "-Xmx");
    property_get("dalvik.vm.heapsize", heapsizeOptsBuf+4, "16m");
    opt.optionString = heapsizeOptsBuf;
    mOptions.add(opt);

    // Increase the main thread's interpreter stack size for bug 6315322.
    opt.optionString = "-XX:mainThreadStackSize=24K";
    mOptions.add(opt);

    strcpy(heapgrowthlimitOptsBuf, "-XX:HeapGrowthLimit=");
    property_get("dalvik.vm.heapgrowthlimit", heapgrowthlimitOptsBuf+20, "");
    if (heapgrowthlimitOptsBuf[20] != '\0') {
        opt.optionString = heapgrowthlimitOptsBuf;
        mOptions.add(opt);
    }

    strcpy(heapminfreeOptsBuf, "-XX:HeapMinFree=");
    property_get("dalvik.vm.heapminfree", heapminfreeOptsBuf+16, "");
    if (heapminfreeOptsBuf[16] != '\0') {
        opt.optionString = heapminfreeOptsBuf;
        mOptions.add(opt);
    }

    strcpy(heapmaxfreeOptsBuf, "-XX:HeapMaxFree=");
    property_get("dalvik.vm.heapmaxfree", heapmaxfreeOptsBuf+16, "");
    if (heapmaxfreeOptsBuf[16] != '\0') {
        opt.optionString = heapmaxfreeOptsBuf;
        mOptions.add(opt);
```

```
}

strcpy(heaptargetutilizationOptsBuf, "-XX:HeapTargetUtilization=");
property_get("dalvik.vm.heaptargetutilization", heaptargetutilizationOptsBuf+26, "");
if (heaptargetutilizationOptsBuf[26] != '\0') {
    opt.optionString = heaptargetutilizationOptsBuf;
    mOptions.add(opt);
}

/*
 * Enable or disable dexopt features, such as bytecode verification and
 * calculation of register maps for precise GC.
 */
property_get("dalvik.vm.dexopt-flags", dexoptFlagsBuf, "");
if (dexoptFlagsBuf[0] != '\0') {
    const char* opc;
    const char* val;

    opc = strstr(dexoptFlagsBuf, "v=");    /* verification */
    if (opc != NULL) {
        switch (*(opc+2)) {
        case 'n':   val = "-Xverify:none";    break;
        case 'r':   val = "-Xverify:remote";  break;
        case 'a':   val = "-Xverify:all";     break;
        default:    val = NULL;               break;
        }

        if (val != NULL) {
            opt.optionString = val;
            mOptions.add(opt);
        }
    }

    opc = strstr(dexoptFlagsBuf, "o=");    /* optimization */
    if (opc != NULL) {
        switch (*(opc+2)) {
        case 'n':   val = "-Xdexopt:none";     break;
        case 'v':   val = "-Xdexopt:verified"; break;
        case 'a':   val = "-Xdexopt:all";      break;
        case 'f':   val = "-Xdexopt:full";     break;
        default:    val = NULL;                break;
        }

        if (val != NULL) {
            opt.optionString = val;
            mOptions.add(opt);
        }
    }

    opc = strstr(dexoptFlagsBuf, "m=y");   /* register map */
    if (opc != NULL) {
        opt.optionString = "-Xgenregmap";
        mOptions.add(opt);

        /* turn on precise GC while we're at it */
        opt.optionString = "-Xgc:precise";
        mOptions.add(opt);
    }
}

/* enable debugging; set suspend=y to pause during VM init */
/* use android ADB transport */
opt.optionString =
    "-agentlib:jdwp=transport=dt_android_adb,suspend=n,server=y";
mOptions.add(opt);

ALOGD("CheckJNI is %s\n", checkJni ? "ON" : "OFF");
if (checkJni) {
    /* extended JNI checking */
```

```
    opt.optionString = "-Xcheck:jni";
    mOptions.add(opt);

    /* set a cap on JNI global references */
    opt.optionString = "-Xjnigreflimit:2000";
    mOptions.add(opt);

    /* with -Xcheck:jni, this provides a JNI function call trace */
    //opt.optionString = "-verbose:jni";
    //mOptions.add(opt);
}

char lockProfThresholdBuf[sizeof("-Xlockprofthreshold:") + sizeof(propBuf)];
property_get("dalvik.vm.lockprof.threshold", propBuf, "");
if (strlen(propBuf) > 0) {
  strcpy(lockProfThresholdBuf, "-Xlockprofthreshold:");
  strcat(lockProfThresholdBuf, propBuf);
  opt.optionString = lockProfThresholdBuf;
  mOptions.add(opt);
}

/* Force interpreter-only mode for selected opcodes. Eg "1-0a,3c,f1-ff" */
char jitOpBuf[sizeof("-Xjitop:") + PROPERTY_VALUE_MAX];
property_get("dalvik.vm.jit.op", propBuf, "");
if (strlen(propBuf) > 0) {
    strcpy(jitOpBuf, "-Xjitop:");
    strcat(jitOpBuf, propBuf);
    opt.optionString = jitOpBuf;
    mOptions.add(opt);
}

/* Force interpreter-only mode for selected methods */
char jitMethodBuf[sizeof("-Xjitmethod:") + PROPERTY_VALUE_MAX];
property_get("dalvik.vm.jit.method", propBuf, "");
if (strlen(propBuf) > 0) {
    strcpy(jitMethodBuf, "-Xjitmethod:");
    strcat(jitMethodBuf, propBuf);
    opt.optionString = jitMethodBuf;
    mOptions.add(opt);
}

if (executionMode == kEMIntPortable) {
    opt.optionString = "-Xint:portable";
    mOptions.add(opt);
} else if (executionMode == kEMIntFast) {
    opt.optionString = "-Xint:fast";
    mOptions.add(opt);
} else if (executionMode == kEMJitCompiler) {
    opt.optionString = "-Xint:jit";
    mOptions.add(opt);
}

if (checkDexSum) {
    /* perform additional DEX checksum tests */
    opt.optionString = "-Xcheckdexsum";
    mOptions.add(opt);
}

if (logStdio) {
    /* convert stdout/stderr to log messages */
    opt.optionString = "-Xlog-stdio";
    mOptions.add(opt);
}

if (enableAssertBuf[4] != '\0') {
    /* accept "all" to mean "all classes and packages" */
    if (strcmp(enableAssertBuf+4, "all") == 0)
        enableAssertBuf[3] = '\0';
    ALOGI("Assertions enabled: '%s'\n", enableAssertBuf);
```

```
        opt.optionString = enableAssertBuf;
        mOptions.add(opt);
    } else {
        ALOGV("Assertions disabled\n");
    }

    if (jniOptsBuf[10] != '\0') {
        ALOGI("JNI options: '%s'\n", jniOptsBuf);
        opt.optionString = jniOptsBuf;
        mOptions.add(opt);
    }

    if (stackTraceFileBuf[0] != '\0') {
        static const char* stfOptName = "-Xstacktracefile:";

        stackTraceFile = (char*) malloc(strlen(stfOptName) +
            strlen(stackTraceFileBuf) +1);
        strcpy(stackTraceFile, stfOptName);
        strcat(stackTraceFile, stackTraceFileBuf);
        opt.optionString = stackTraceFile;
        mOptions.add(opt);
    }

    /* extra options; parse this late so it overrides others */
    property_get("dalvik.vm.extra-opts", extraOptsBuf, "");
    parseExtraOpts(extraOptsBuf);

    /* Set the properties for locale */
    {
        char langOption[sizeof("-Duser.language=") + 3];
        char regionOption[sizeof("-Duser.region=") + 3];
        strcpy(langOption, "-Duser.language=");
        strcpy(regionOption, "-Duser.region=");
        readLocale(langOption, regionOption);
        opt.extraInfo = NULL;
        opt.optionString = langOption;
        mOptions.add(opt);
        opt.optionString = regionOption;
        mOptions.add(opt);
    }

    /*
     * We don't have /tmp on the device, but we often have an SD card. Apps
     * shouldn't use this, but some test suites might want to exercise it.
     */
    opt.optionString = "-Djava.io.tmpdir=/sdcard";
    mOptions.add(opt);

    initArgs.version = JNI_VERSION_1_4;
    initArgs.options = mOptions.editArray();
    initArgs.nOptions = mOptions.size();
    initArgs.ignoreUnrecognized = JNI_FALSE;

    /*
     * 初始化VM.
     *
     * The JavaVM* is essentially per-process, and the JNIEnv* is per-thread.
     * If this call succeeds, the VM is ready, and we can start issuing
     * JNI calls.
     */
    if (JNI_CreateJavaVM(pJavaVM, pEnv, &initArgs) < 0) {
        ALOGE("JNI_CreateJavaVM failed\n");
        goto bail;
    }
    result = 0;

bail:
    free(stackTraceFile);
```

```
    return result;
}
```

在上述代码中，设置了一系列的控制 Dalvik VM 的选项，开发者可以通过特定的系统属性来指定这些选项。其中最为常用的选项有如下 4 个。

- -Xcheck:jni：功能是启动 JNI 检查方法。通过使用这个选项可以检查要访问的 Java 对象的成员变量或者成员函数的合法性，例如检查类型是否匹配。在日常应用中，可以使用系统属性 dalvik.vm.checkjni 或者 ro.kernel.android.checkjni 来指定是否要启用-Xcheck:jni 选项。
- -Xint:portable/-Xint:fast/-Xint:jit：功能是指定 Dalvik 虚拟机的执行模式，可以通过系统属性 dalvik.vm.execution-mode 来指定 Dalvik VM 的解释模式。在 Dalvik VM 中支持如下 3 种运行模式。
  - Portable：表示 Dalvik VM 以可移植的方式来进行编译，编译出来的虚拟机可以在任意平台上运行。
  - Fast：可以针对当前平台的类型来编译 Dalvik VM。
  - Jit：将代码动态编译成本地语言后再执行。
- Xstacktracefile：功能是指定调用堆栈输出文件。当指定了-Xstacktracefile 选项之后，可以将线程的调用堆栈输出到指定的文件中去。
- -Xmx：功能是指定 Java 对象堆的最大值，Dalvik VM 的 Java 对象堆的默认最大值是 16 MB。

15.2.3　创建并初始化 Dalvik VM 实例

调用函数 JNI_CreateJavaVM 创建并初始化一个 Dalvik VM 实例，此函数在文件“dalvik\vm\Jni.cpp”中定义，具体实现代码如下所示：

```
jint JNI_CreateJavaVM(JavaVM** p_vm, JNIEnv** p_env, void* vm_args) {
    const JavaVMInitArgs* args = (JavaVMInitArgs*) vm_args;
    if (args->version < JNI_VERSION_1_2) {
        return JNI_EVERSION;
    }

    // TODO: don't allow creation of multiple VMs -- one per customer for now

    /* zero globals; not strictly necessary the first time a VM is started */
    memset(&gDvm, 0, sizeof(gDvm));

    /*
     *为当前进程创建 Dalvik 虚拟机实例，即一个 JavaVMExt 对象。
     */
    JavaVMExt* pVM = (JavaVMExt*) calloc(1, sizeof(JavaVMExt));
    pVM->funcTable = &gInvokeInterface;
    pVM->envList = NULL;
    dvmInitMutex(&pVM->envListLock);

    UniquePtr<const char*[]> argv(new const char*[args->nOptions]);
    memset(argv.get(), 0, sizeof(char*) * (args->nOptions));

    /*
     *为当前线程创建和初始化一个 JNI 环境
     *即一个 JNIEnvExt 对象
     *此功能是通过调用函数 dvmCreateJNIEnv 来完成的
     */
    int argc = 0;
    for (int i = 0; i < args->nOptions; i++) {
        const char* optStr = args->options[i].optionString;
        if (optStr == NULL) {
            dvmFprintf(stderr, "ERROR: CreateJavaVM failed: argument %d was NULL\n", i);
            return JNI_ERR;
        } else if (strcmp(optStr, "vfprintf") == 0) {
            gDvm.vfprintfHook = (int (*)(FILE *, const char*, va_list))args->options[i].extraInfo;
        } else if (strcmp(optStr, "exit") == 0) {
```

```
            gDvm.exitHook = (void (*)(int)) args->options[i].extraInfo;
        } else if (strcmp(optStr, "abort") == 0) {
            gDvm.abortHook = (void (*)(void))args->options[i].extraInfo;
        } else if (strcmp(optStr, "sensitiveThread") == 0) {
            gDvm.isSensitiveThreadHook = (bool (*)(void))args->options[i].extraInfo;
        } else if (strcmp(optStr, "-Xcheck:jni") == 0) {
            gDvmJni.useCheckJni = true;
        } else if (strncmp(optStr, "-Xjniopts:", 10) == 0) {
            char* jniOpts = strdup(optStr + 10);
            size_t jniOptCount = 1;
            for (char* p = jniOpts; *p != 0; ++p) {
                if (*p == ',') {
                    ++jniOptCount;
                    *p = 0;
                }
            }
            char* jniOpt = jniOpts;
            for (size_t i = 0; i < jniOptCount; ++i) {
                if (strcmp(jniOpt, "warnonly") == 0) {
                    gDvmJni.warnOnly = true;
                } else if (strcmp(jniOpt, "forcecopy") == 0) {
                    gDvmJni.forceCopy = true;
                } else if (strcmp(jniOpt, "logThirdPartyJni") == 0) {
                    gDvmJni.logThirdPartyJni = true;
                } else {
                    dvmFprintf(stderr, "ERROR: CreateJavaVM failed: unknown -Xjniopts option '%s'\n",
                            jniOpt);
                    return JNI_ERR;
                }
                jniOpt += strlen(jniOpt) + 1;
            }
            free(jniOpts);
        } else {
            /* regular option */
            argv[argc++] = optStr;
        }
    }

    if (gDvmJni.useCheckJni) {
        dvmUseCheckedJniVm(pVM);
    }

    if (gDvmJni.jniVm != NULL) {
        dvmFprintf(stderr, "ERROR: Dalvik only supports one VM per process\n");
        return JNI_ERR;
    }
    gDvmJni.jniVm = (JavaVM*) pVM;

    /*
     * Create a JNIEnv for the main thread.  We need to have something set up
     * here because some of the class initialization we do when starting
     * up the VM will call into native code.
     */
    JNIEnvExt* pEnv = (JNIEnvExt*) dvmCreateJNIEnv(NULL);

    /*将参数 vm_args 所描述的 Dalvik VM 启动选项拷贝到变量 argv 所描述的一个字符串数组中去
    *调用函数 dvmStartup 来初始化前面所创建的 Dalvik 虚拟机实例
    */
    gDvm.initializing = true;
    std::string status =
            dvmStartup(argc, argv.get(), args->ignoreUnrecognized, (JNIEnv*)pEnv);
    gDvm.initializing = false;

    if (!status.empty()) {
        free(pEnv);
        free(pVM);
        ALOGW("CreateJavaVM failed: %s", status.c_str());
        return JNI_ERR;
    }
```

```
    /*
     * 调用函数dvmChangeStatus将当前线程的状态设置为正在执行NATIVE代码
     * 通过输出参数p_vm和p_env
     * 将刚刚创建和初始化好的JavaVMExt对象和JNIEnvExt对象返回给调用者.
     */
    dvmChangeStatus(NULL, THREAD_NATIVE);
    *p_env = (JNIEnv*) pEnv;
    *p_vm = (JavaVM*) pVM;
    ALOGV("CreateJavaVM succeeded");
    return JNI_OK;
}
```

15.2.4　创建 JNIEnvExt 对象

再分析函数 dvmCreateJNIEnv，其功能是创建 JNIEnvExt 对象以描述当前的 JNI 环境，并设置此 JNIEnvExt 对象的宿主 Dalvik VM 和所使用的本地接口表。此处的宿主 Dalvik VM 就是当前进程的 Dalvik VM，被保存在全局变量 gDvm 的成员变量 vmList 中。函数 dvmCreateJNIEnv 在文件“dalvik\vm\Jni.cpp”中定义，具体实现代码如下所示：

```
JNIEnv* dvmCreateJNIEnv(Thread* self) {
    JavaVMExt* vm = (JavaVMExt*) gDvmJni.jniVm;

    //if (self != NULL)
    //    ALOGI("Ent CreateJNIEnv: threadid=%d %p", self->threadId, self);

    assert(vm != NULL);

    JNIEnvExt* newEnv = (JNIEnvExt*) calloc(1, sizeof(JNIEnvExt));
    newEnv->funcTable = &gNativeInterface;
    if (self != NULL) {
        dvmSetJniEnvThreadId((JNIEnv*) newEnv, self);
        assert(newEnv->envThreadId != 0);
    } else {
        /* make it obvious if we fail to initialize these later */
        newEnv->envThreadId = 0x77777775;
        newEnv->self = (Thread*) 0x77777779;
    }
    if (gDvmJni.useCheckJni) {
        dvmUseCheckedJniEnv(newEnv);
    }

    ScopedPthreadMutexLock lock(&vm->envListLock);

    /* insert at head of list */
    newEnv->next = vm->envList;
    assert(newEnv->prev == NULL);
    if (vm->envList == NULL) {
        // rare, but possible
        vm->envList = newEnv;
    } else {
        vm->envList->prev = newEnv;
    }
    vm->envList = newEnv;

    //if (self != NULL)
    //    ALOGI("Xit CreateJNIEnv: threadid=%d %p", self->threadId, self);
    return (JNIEnv*) newEnv;
}
```

在上述代码中，参数 self 表示前面创建的 JNIEnvExt 对象要关联的线程，通过源码分析可知可以通过调用函数 dvmSetJniEnvThreadId 来将它们关联起来。

再看函数 dvmStartup，功能是初始化 Dalvik VM，并设置处理 Dalvik VM 的启动选项。这些启动选项都被保存在参数 argv 中，并且被保存选项的个数为 argc。在处理启动选项之前会执行如下两个操作。

- 调用函数 setCommandLineDefaults 给 Dalvik VM 设置默认参数。
- 调用函数 processOptions 分配足够的内存空间，以容纳由参数 argv 和 argc 所描述的启动选项。

函数 dvmStartup 在文件“dalvik\vm\Init.cpp”中定义，具体实现代码如下所示：

```
std::string dvmStartup(int argc, const char* const argv[],
        bool ignoreUnrecognized, JNIEnv* pEnv)
{
    ScopedShutdown scopedShutdown;

    assert(gDvm.initializing);

    ALOGV("VM init args (%d):", argc);
    for (int i = 0; i < argc; i++) {
        ALOGV("  %d: '%s'", i, argv[i]);
    }
//调用函数 setCommandLineDefaults 来给 Dalvik 虚拟机设置默认参数
//因为启动选项不一定会指定 Dalvik 虚拟机的所有属性
    setCommandLineDefaults();

    /*
     * 调用函数 processOptions 来分配足够的内存空间来容纳由参数 argv 和 argc 所描述的启动选项。
     */
    int cc = processOptions(argc, argv, ignoreUnrecognized);
    if (cc != 0) {
        if (cc < 0) {
            dvmFprintf(stderr, "\n");
            usage("dalvikvm");
        }
        return "syntax error";
    }

#if WITH_EXTRA_GC_CHECKS > 1
    /* only "portable" interp has the extra goodies */
    if (gDvm.executionMode != kExecutionModeInterpPortable) {
        ALOGI("Switching to 'portable' interpreter for GC checks");
        gDvm.executionMode = kExecutionModeInterpPortable;
    }
#endif

    /* 配置调度选项 */
    if (!access("/dev/cpuctl/tasks", F_OK)) {
        ALOGV("Using kernel group scheduling");
        gDvm.kernelGroupScheduling = 1;
    } else {
        ALOGV("Using kernel scheduler policies");
    }

    /* 配置处理标志 */
    if (!gDvm.reduceSignals)
        blockSignals();

    /* 验证系统页面大小 */
    if (sysconf(_SC_PAGESIZE) != SYSTEM_PAGE_SIZE) {
        return StringPrintf("expected page size %d, got %d",
                SYSTEM_PAGE_SIZE, (int) sysconf(_SC_PAGESIZE));
    }

    /* mterp 设置 */
    ALOGV("Using executionMode %d", gDvm.executionMode);
    dvmCheckAsmConstants();

    /*
     *从下面的代码开始初始化 Dalvik 虚拟机的各个子模块
     */
    dvmQuasiAtomicsStartup();
```

```
    if (!dvmAllocTrackerStartup()) {//用于初始化 Davlik 虚拟机的对象分配记录子模块
        return "dvmAllocTrackerStartup failed";
    }
    if (!dvmGcStartup()) {//用来初始化 Davlik 虚拟机的垃圾收集（GC）子模块
        return "dvmGcStartup failed";
    }
// dvmThreadStartup 用于初始化 Davlik 虚拟机的线程列表
// 为主线程创建一个 Thread 对象以及为主线程初始化执行环境
    if (!dvmThreadStartup()) {
        return "dvmThreadStartup failed";
    }
//用于初始化 Davlik 虚拟机的内建 Native 函数表
    if (!dvmInlineNativeStartup()) {
        return "dvmInlineNativeStartup";
    }
//用于初始化寄存器映射集（Register Map）子模块
    if (!dvmRegisterMapStartup()) {
        return "dvmRegisterMapStartup failed";
    }
//用于初始化 instanceof 操作符子模块
    if (!dvmInstanceofStartup()) {
        return "dvmInstanceofStartup failed";
    }
//用于初始化启动类加载器（Bootstrap Class Loader），同时初始化 java.lang.Class 类
    if (!dvmClassStartup()) {
        return "dvmClassStartup failed";
    }

    /*
     * At this point, the system is guaranteed to be sufficiently
     * initialized that we can look up classes and class members. This
     * call populates the gDvm instance with all the class and member
     * references that the VM wants to use directly.
     */
    if (!dvmFindRequiredClassesAndMembers()) {
        return "dvmFindRequiredClassesAndMembers failed";
    }
//用于初始化 java.lang.String 类内部私有一个字符串池，这样当 Dalvik 虚拟机运行起来之后
// 就可以调用 java.lang.String 类的成员函数 intern 来访问这个字符串池里面的字符串
    if (!dvmStringInternStartup()) {
        return "dvmStringInternStartup failed";
    }
//用于初始化 Native Shared Object 库加载表，即 SO 库加载表
    if (!dvmNativeStartup()) {
        return "dvmNativeStartup failed";
    }
// 用于初始化内部 Native 函数表
// 所有需要直接访问 Dalvik 虚拟机内部函数或者数据结构的 Native 函数都定义在这张表中
//因为它们如果定义在外部的其他 SO 文件中，就无法直接访问 Dalvik 虚拟机的内部函数或者数据结构
    if (!dvmInternalNativeStartup()) {
        return "dvmInternalNativeStartup failed";
    }
//用于初始化全局引用表，以及加载一些与 Direct Buffer 相关的类
//例如 DirectBuffer、PhantomReference 和 ReferenceQueue 等
    if (!dvmJniStartup()) {
        return "dvmJniStartup failed";
    }
//用于初始化 Dalvik 虚拟机的性能分析子模块，以及加载 dalvik.system.VMDebug 类等
    if (!dvmProfilingStartup()) {
        return "dvmProfilingStartup failed";
    }

    /*
     * Create a table of methods for which we will substitute an "inline"
     * version for performance.
     */
    if (!dvmCreateInlineSubsTable()) {
        return "dvmCreateInlineSubsTable failed";
    }
```

```
    /*
     *用于验证 Dalvik 虚拟机中存在相应的装箱类，并且这些装箱类有且仅有一个成员变量
     *这个成员变量是用来描述对应的数字值的。这些装箱类包括 java.lang.Boolean、java.lang.Character、
     *java.lang.Float、java.lang.Double、 java.lang.Byte、java.lang.Short、java.lang.Integer
和 java.lang.Long
     */
    if (!dvmValidateBoxClasses()) {
        return "dvmValidateBoxClasses failed";
    }

    /*
     * 用于准备主线程的 JNI 环境虽然我们已经为当前线程创建好一个 JNI 环境了,
     *但是还没有将该 JNI 环境与主线程关联，也就是还没有将主线程的 ID 设置到该 JNI 环境中去
     */
    if (!dvmPrepMainForJni(pEnv)) {
        return "dvmPrepMainForJni failed";
    }

    /*
     * Explicitly initialize java.lang.Class.  This doesn't happen
     * automatically because it's allocated specially (it's an instance
     * of itself).  Must happen before registration of system natives,
     * which make some calls that throw assertions if the classes they
     * operate on aren't initialized.
     */
    if (!dvmInitClass(gDvm.classJavaLangClass)) {
        return "couldn't initialized java.lang.Class";
    }

    /*
     * 调用另外一个函数 jniRegisterSystemMethods，后者接着又调用了函数 registerCoreLibrariesJni 来
     *为 Java 核心类注册 JNI 方法
     */
    if (!registerSystemNatives(pEnv)) {
        return "couldn't register system natives";
    }

    /*
     * 用于预创建一些与内存分配有关的异常对象，并且将它们缓存起来，以便以后可以快速使用
     *这些异常对象包括 java.lang.OutOfMemoryError、java.lang.InternalError
     *和 java.lang.NoClassDefFoundError
     */
    if (!dvmCreateStockExceptions()) {
        return "dvmCreateStockExceptions failed";
    }

    /*
     * 用于为主线程创建一个 java.lang.ThreadGroup 对象、java.lang.Thread 对角和
java.lang.VMThread 对象。
     *这些 Java 对象和在前面 Step 5 中创建的 C++层 Thread 对象关联一起,
     *共同用来描述 Dalvik 虚拟机的主线程
     */
    if (!dvmPrepMainThread()) {
        return "dvmPrepMainThread failed";
    }

    /*
     *用于确保主线程当前不引用有任何 Java 对象，是为了保证主线程接下来以干净的方式来执行程序入口
     */
    if (dvmReferenceTableEntries(&dvmThreadSelf()->internalLocalRefTable) != 0)
    {
        ALOGW("Warning: tracked references remain post-initialization");
        dvmDumpReferenceTable(&dvmThreadSelf()->internalLocalRefTable, "MAIN");
    }

    /*用于初始化 Dalvik 虚拟机的调试环境,
    *Dalvik VM 与 Java VM 一样，都是通过 JDWP 协议来支持远程调试的
```

```
    */
    if (!dvmDebuggerStartup()) {
        return "dvmDebuggerStartup failed";
    }
    if (!dvmGcStartupClasses()) {
        return "dvmGcStartupClasses failed";
    }

    /*
     * Init for either zygote mode or non-zygote mode.  The key difference
     * is that we don't start any additional threads in Zygote mode.
     */
    if (gDvm.zygote) {
        if (!initZygote()) {
            return "initZygote failed";
        }
    } else {
        if (!dvmInitAfterZygote()) {
            return "dvmInitAfterZygote failed";
        }
    }


#ifndef NDEBUG
    if (!dvmTestHash())
        ALOGE("dvmTestHash FAILED");
    if (false /*noisy!*/ && !dvmTestIndirectRefTable())
        ALOGE("dvmTestIndirectRefTable FAILED");
#endif

    if (dvmCheckException(dvmThreadSelf())) {
        dvmLogExceptionStackTrace();
        return "Exception pending at end of VM initialization";
    }

    scopedShutdown.disarm();
    return "";
}
```

15.2.5　设置当前进程和进程组 ID

再看函数 dvmInitZygote，其功能是调用系统中的 setpgid 函数来设置当前进程，设置 Zyogte 进程的进程组 ID。当系统在调用 setpgid 时会传递两个都为 0 的参数，这表示 Zyogte 进程的进程组 ID 与进程 ID 相同，即 Zyogte 进程运行在一个单独的进程组里面。函数 dvmInitZygote 在文件“dalvik/vm/Init.c”中定义，具体实现代码如下所示：

```
static bool dvmInitZygote(void)
{
    /* zygote goes into its own process group */
    setpgid(0,0);

    return true;
}
```

15.2.6　注册 Android 核心类的 JNI 方法

接下来看函数 startReg，在文件 AndroidRuntime.java 中，函数 start 调用了类 AndroidRuntime 中的成员函数 startReg 来注册 Android 核心类的 JNI 方法。函数 startReg 在文件“frameworks/base/core/jni/AndroidRuntime.cpp”中定义，具体实现代码如下所示：

```
/*
 * Register android native functions with the VM.
 */
/*static*/
```

```
int AndroidRuntime::startReg(JNIEnv* env)
{
    /*
     *调用函数 androidSetCreateThreadFunc 设置一个线程创建钩子 javaCreateThreadEtc。
     *此线程创建钩子是用来初始化一个 Native 线程的 JNI 环境的，
     *即当我们在 C++代码中创建一个 Native 线程的时候，
     *函数 javaCreateThreadEtc 会被调用来初始化该 Native 线程的 JNI 环境
     */
    androidSetCreateThreadFunc((android_create_thread_fn) javaCreateThreadEtc);

    LOGV("--- registering native functions ---\n");

    env->PushLocalFrame(200);
    /*
     *调用函数 register_jni_procs 来注册 Android 核心类的 JNI 方法。在注册 JNI 方法的过程中，
     *需要在 Native 代码中引用到一些 Java 对象，
     *这些 Java 对象引用需要记录在当前线程的一个 Native 堆栈中。
     *但是此时 Dalvik 虚拟机还没有真正运行起来，也就是当前线程的 Native 堆栈还没有准备就绪。
     *此时需要在注册 JNI 方法之前，手动地将在当前线程的 Native 堆栈中压入一个帧（Frame），
     *并且在注册 JNI 方法之后，手动地将该帧弹出来
     */
    if (register_jni_procs(gRegJNI, NELEM(gRegJNI), env) < 0) {
        env->PopLocalFrame(NULL);
        return -1;
    }
    env->PopLocalFrame(NULL);
    //createJavaThread("fubar", quickTest, (void*) "hello");
    return 0;
}
```

在上述代码中，参数 env 指向了一个 JNIEnv 对象，此对象表示当前线程的 JNI 环境。通过调用 JNIEnv 对象的成员函数 PushLocalFrame 和 PopLocalFrame，可以用手动的方式向当前线程的 Native 堆栈中分别压入和弹出一个帧。这里的帧是一个本地帧，只能保存 Java 对象在 Native（本地）代码中的本地引用。函数 register_jni_procs 的具体实现代码如下所示：

```
static int register_jni_procs(const RegJNIRec array[], size_t count, JNIEnv* env)
{
    for (size_t i = 0; i < count; i++) {
        if (array[i].mProc(env) < 0) {
#ifndef NDEBUG
            ALOGD("----------!!! %s failed to load\n", array[i].mName);
#endif
            return -1;
        }
    }
    return 0;
}
```

在上述代码中，参数 array 指向了全局变量 gRegJNI 所描述的 JNI 方法注册函数表，每一个表的选项都用一个 RegJNIRec 对象来描述，而每一个 RegJNIRec 对象都有一个成员变量 mProc 指向一个 JNI 方法注册函数，通过依次调用这些注册函数的方式即可将 JNI 方法注册到创建的 Dalvik VM 中去。

在文件“frameworks/base/core/jni/AndroidRuntime.cpp”中一一列出了全局变量 gRegJNI 所描述的 JNI 方法注册函数表，在此可以了解注册了哪些 Android 核心类的 JNI 方法。对应代码如下所示：

```
static const RegJNIRec gRegJNI[] = {
    REG_JNI(register_android_debug_JNITest),
    REG_JNI(register_com_android_internal_os_RuntimeInit),
    REG_JNI(register_android_os_SystemClock),
    REG_JNI(register_android_util_EventLog),
    REG_JNI(register_android_util_Log),
    REG_JNI(register_android_util_FloatMath),
    REG_JNI(register_android_text_format_Time),
```

```
    REG_JNI(register_android_content_AssetManager),
    REG_JNI(register_android_content_StringBlock),
    REG_JNI(register_android_content_XmlBlock),
    REG_JNI(register_android_emoji_EmojiFactory),
    REG_JNI(register_android_text_AndroidCharacter),
    REG_JNI(register_android_text_AndroidBidi),
    REG_JNI(register_android_view_InputDevice),
    REG_JNI(register_android_view_KeyCharacterMap),
    REG_JNI(register_android_os_Process),
    REG_JNI(register_android_os_SystemProperties),
    REG_JNI(register_android_os_Binder),
    REG_JNI(register_android_os_Parcel),
    REG_JNI(register_android_view_DisplayEventReceiver),
    REG_JNI(register_android_nio_utils),
    REG_JNI(register_android_graphics_PixelFormat),
    REG_JNI(register_android_graphics_Graphics),
    REG_JNI(register_android_view_GLES20DisplayList),
    REG_JNI(register_android_view_GLES20Canvas),
    REG_JNI(register_android_view_HardwareRenderer),
    REG_JNI(register_android_view_Surface),
    REG_JNI(register_android_view_SurfaceControl),
    REG_JNI(register_android_view_SurfaceSession),
    REG_JNI(register_android_view_TextureView),
    REG_JNI(register_com_google_android_gles_jni_EGLImpl),
    REG_JNI(register_com_google_android_gles_jni_GLImpl),
    REG_JNI(register_android_opengl_jni_EGL14),
    REG_JNI(register_android_opengl_jni_EGLExt),
    REG_JNI(register_android_opengl_jni_GLES10),
    REG_JNI(register_android_opengl_jni_GLES10Ext),
    REG_JNI(register_android_opengl_jni_GLES11),
    REG_JNI(register_android_opengl_jni_GLES11Ext),
    REG_JNI(register_android_opengl_jni_GLES20),
    REG_JNI(register_android_opengl_jni_GLES30),

    REG_JNI(register_android_graphics_Bitmap),
    REG_JNI(register_android_graphics_BitmapFactory),
    REG_JNI(register_android_graphics_BitmapRegionDecoder),
    REG_JNI(register_android_graphics_Camera),
    REG_JNI(register_android_graphics_Canvas),
    REG_JNI(register_android_graphics_ColorFilter),
    REG_JNI(register_android_graphics_DrawFilter),
    REG_JNI(register_android_graphics_Interpolator),
    REG_JNI(register_android_graphics_LayerRasterizer),
    REG_JNI(register_android_graphics_MaskFilter),
    REG_JNI(register_android_graphics_Matrix),
    REG_JNI(register_android_graphics_Movie),
    REG_JNI(register_android_graphics_NinePatch),
    REG_JNI(register_android_graphics_Paint),
    REG_JNI(register_android_graphics_Path),
    REG_JNI(register_android_graphics_PathMeasure),
    REG_JNI(register_android_graphics_PathEffect),
    REG_JNI(register_android_graphics_Picture),
    REG_JNI(register_android_graphics_PorterDuff),
    REG_JNI(register_android_graphics_Rasterizer),
    REG_JNI(register_android_graphics_Region),
    REG_JNI(register_android_graphics_Shader),
    REG_JNI(register_android_graphics_SurfaceTexture),
    REG_JNI(register_android_graphics_Typeface),
    REG_JNI(register_android_graphics_Xfermode),
    REG_JNI(register_android_graphics_YuvImage),

    REG_JNI(register_android_database_CursorWindow),
    REG_JNI(register_android_database_SQLiteConnection),
    REG_JNI(register_android_database_SQLiteGlobal),
    REG_JNI(register_android_database_SQLiteDebug),
    REG_JNI(register_android_os_Debug),
    REG_JNI(register_android_os_FileObserver),
    REG_JNI(register_android_os_FileUtils),
    REG_JNI(register_android_os_MessageQueue),
```

```
    REG_JNI(register_android_os_ParcelFileDescriptor),
    REG_JNI(register_android_os_SELinux),
    REG_JNI(register_android_os_Trace),
    REG_JNI(register_android_os_UEventObserver),
    REG_JNI(register_android_net_LocalSocketImpl),
    REG_JNI(register_android_net_NetworkUtils),
    REG_JNI(register_android_net_TrafficStats),
    REG_JNI(register_android_net_wifi_WifiManager),
    REG_JNI(register_android_os_MemoryFile),
    REG_JNI(register_com_android_internal_os_ZygoteInit),
    REG_JNI(register_android_hardware_Camera),
    REG_JNI(register_android_hardware_SensorManager),
    REG_JNI(register_android_hardware_SerialPort),
    REG_JNI(register_android_hardware_UsbDevice),
    REG_JNI(register_android_hardware_UsbDeviceConnection),
    REG_JNI(register_android_hardware_UsbRequest),
    REG_JNI(register_android_media_AudioRecord),
    REG_JNI(register_android_media_AudioSystem),
    REG_JNI(register_android_media_AudioTrack),
    REG_JNI(register_android_media_JetPlayer),
    REG_JNI(register_android_media_RemoteDisplay),
    REG_JNI(register_android_media_ToneGenerator),

    REG_JNI(register_android_opengl_classes),
    REG_JNI(register_android_server_NetworkManagementSocketTagger),
    REG_JNI(register_android_server_Watchdog),
    REG_JNI(register_android_ddm_DdmHandleNativeHeap),
    REG_JNI(register_android_backup_BackupDataInput),
    REG_JNI(register_android_backup_BackupDataOutput),
    REG_JNI(register_android_backup_FileBackupHelperBase),
    REG_JNI(register_android_backup_BackupHelperDispatcher),
    REG_JNI(register_android_app_backup_FullBackup),
    REG_JNI(register_android_app_ActivityThread),
    REG_JNI(register_android_app_NativeActivity),
    REG_JNI(register_android_view_InputChannel),
    REG_JNI(register_android_view_InputEventReceiver),
    REG_JNI(register_android_view_InputEventSender),
    REG_JNI(register_android_view_InputQueue),
    REG_JNI(register_android_view_KeyEvent),
    REG_JNI(register_android_view_MotionEvent),
    REG_JNI(register_android_view_PointerIcon),
    REG_JNI(register_android_view_VelocityTracker),

    REG_JNI(register_android_content_res_ObbScanner),
    REG_JNI(register_android_content_res_Configuration),

    REG_JNI(register_android_animation_PropertyValuesHolder),
    REG_JNI(register_com_android_internal_content_NativeLibraryHelper),
    REG_JNI(register_com_android_internal_net_NetworkStatsFactory),
};
```

15.2.7 创建 javaCreateThreadEtc 钩子

在文件 AndroidRuntime.java 中，函数 start 调用了类 AndroidRuntime 类中的成员函数 androidSetCreateThreadFunc，这样可以使用线程创建 javaCreateThreadEtc 钩子。函数 androidSetCreateThreadFunc 在文件“frameworks/base/core/jni/AndroidRuntime.cpp”中定义，具体实现代码如下所示：

```
void androidSetCreateThreadFunc(android_create_thread_fn func)
{
    gCreateThreadFn = func;
}
```

由上述实现代码可知，在函数指针 gCreateThreadFn 中保存了线程创建的 javaCreateThreadEtc 钩子。函数指针 gCreateThreadFn 默认是指向函数 androidCreateRawThreadEtc，如果不设置线程创建钩子，则函数 androidCreateRawThreadEtc 就是默认使用的线程创建函数。函数 androidCreateRawThreadEtc 在文件“frameworks\native\libs\utils\Threads.cpp”中定义，具体实现代码如下所示：

```
int androidCreateRawThreadEtc(android_thread_func_t entryFunction,
                               void *userData,
                               const char* threadName,
                               int32_t threadPriority,
                               size_t threadStackSize,
                               android_thread_id_t *threadId)
{
    pthread_attr_t attr;
    pthread_attr_init(&attr);
    pthread_attr_setdetachstate(&attr, PTHREAD_CREATE_DETACHED);

#ifdef HAVE_ANDROID_OS  /* valgrind is rejecting RT-priority create reqs */
    if (threadPriority != PRIORITY_DEFAULT || threadName != NULL) {
        // Now that the pthread_t has a method to find the associated
        // android_thread_id_t (pid) from pthread_t, it would be possible to avoid
        // this trampoline in some cases as the parent could set the properties
        // for the child.  However, there would be a race condition because the
        // child becomes ready immediately, and it doesn't work for the name.
        // prctl(PR_SET_NAME) only works for self; prctl(PR_SET_THREAD_NAME) was
        // proposed but not yet accepted.
        thread_data_t* t = new thread_data_t;
        t->priority = threadPriority;
        t->threadName = threadName ? strdup(threadName) : NULL;
        t->entryFunction = entryFunction;
        t->userData = userData;
        entryFunction = (android_thread_func_t)&thread_data_t::trampoline;
        userData = t;
    }
#endif

    if (threadStackSize) {
        pthread_attr_setstacksize(&attr, threadStackSize);
    }

    errno = 0;
    pthread_t thread;
    int result = pthread_create(&thread, &attr,
                    (android_pthread_entry)entryFunction, userData);
    pthread_attr_destroy(&attr);
    if (result != 0) {
        ALOGE("androidCreateRawThreadEtc failed (entry=%p, res=%d, errno=%d)\n"
             "(android threadPriority=%d)",
            entryFunction, result, errno, threadPriority);
        return 0;
    }

    // Note that *threadID is directly available to the parent only, as it is
    // assigned after the child starts.  Use memory barrier / lock if the child
    // or other threads also need access.
    if (threadId != NULL) {
        *threadId = (android_thread_id_t)thread; // XXX: this is not portable
    }
    return 1;
}
```

第 16 章　注册 Dalvik VM 并创建线程

在 Android 系统中，如果 Dalvik VM 调用的成员函数是一个 JNI 方法，则会直接跳到这个函数的地址去执行。由此可见，Android 会直接在本地操作系统上执行 JNI 方法，而不是由 Dalvik VM 解释器执行。另外，在 Android 系统中的 Dalvik VM 不但可以执行 Java 代码，而且还可以执行 Native（本地）代码，也就是 C/C++函数。在执行这些 C/C++函数的过程中，可以通过本地操作系统提供的系统调用创建 Linux 进程和线程。如果在 Native 代码中创建出来的线程能够执行 Java 代码，那么它实际上又可以看作是一个 Dalvik VM 线程。在本章的内容中，将详细讲解注册 Dalvik VM 并创建线程的过程，为读者步入本书后面知识的学习打下基础。

16.1 注册 Dalvik VM 的 JNI 方法

在 Android 系统中，当 Dalvik VM 调用一个成员函数时，如果这个成员函数是一个 JNI 方法，则会直接跳到这个函数的地址去执行。由此可见，直接在本地操作系统上执行 JNI 方法，而不是由 Dalvik VM 解释器执行。在 Android 系统中，以 C/C++语言实现 JNI 方法，然后将其编译在一个 SO 文件中，在调用 JNI 方法前需要先加载到当前应用程序进程的地址空间。

16.1.1 设置加载程序

在文件“libcore/luni/src/main/java/java/lang/System.java”中，通过函数 loadLibrary 来设置加载程序，具体实现代码如下所示：

```
public static void loadLibrary(String libName) {
    //调用函数 loadLibrary 来加载名称为 libName 的 so 文件
    Runtime.getRuntime().loadLibrary(libName, VMStack.getCallingClassLoader());
}
```

在 Dalvik VM 系统中，类加载器不但知道它要加载的类所在的文件路径，而且还知道该类所属的 APK 能够保存的 so 文件的路径。正因如此，所以当给定一个 so 文件名称后，类加载器可以判断它是否存在自己的 so 文件目录中。

16.1.2 加载 so 文件并验证

再看在类 Runtime 中的成员函数 loadLibrary，功能是加载名称为 libName 的 so 文件。其参数 libraryName 表示要加载的 so 文件，参数 loader 表示与要加载的 so 文件所关联的类的一个类加载器。函数 loadLibrary 在文件“libcore/luni/src/main/java/java/lang/Runtime.java”中定义，具体实现代码如下所示：

```
void loadLibrary(String libraryName, ClassLoader loader) {
    if (loader != null) {
        String filename = loader.findLibrary(libraryName);
        if (filename == null) {
```

```
            throw new UnsatisfiedLinkError("Couldn't load " + libraryName +
                                           " from loader " + loader +
                                           ": findLibrary returned null");
        }
        String error = doLoad(filename, loader);
        if (error != null) {
            throw new UnsatisfiedLinkError(error);
        }
        return;
    }

    String filename = System.mapLibraryName(libraryName);
    List<String> candidates = new ArrayList<String>();
    String lastError = null;
    for (String directory : mLibPaths) {
        String candidate = directory + filename;
        candidates.add(candidate);

        if (IoUtils.canOpenReadOnly(candidate)) {
            String error = doLoad(candidate, loader);
            if (error == null) {
                return; // We successfully loaded the library. Job done.
            }
            lastError = error;
        }
    }

    if (lastError != null) {
        throw new UnsatisfiedLinkError(lastError);
    }
    throw new UnsatisfiedLinkError("Library " + libraryName + " not found; tried " + candidates);
}
```

在上述代码中，如果参数 loader 的值等于 null，那么就表示当前要加载的 so 文件要在系统范围的 so 文件目录查找。在这些指定的系统范围的 so 文件目录中，被保存在类 Runtime 中定义的变量 mLibPaths 所描述的 String 数组中。在此需要依次检查这些目录是否存在与参数 libraryName 对应的 so 文件中的方法，就可以确定参数 libraryName 所指定加载的 so 文件是否是一个合法的 so 文件。如果合法，则调用类 Runtime 中的函数 nativeLoad 将其加载到当前进程的 Dalvik VM 中。在检查参数 libraryName 所表示的 so 文件是否存在于系统范围的 so 文件目录前，需要将其转换为 lib<name>.so 的形式，这一功能是通过调用类 System 的静态成员函数 mapLibraryName 来实现的。

在类 Runtime 中，函数 nativeLoad 的功能是加载参数 libraryName 描述的 so 文件。函数 nativeLoad 在文件“libcore/luni/src/main/java/java/lang/Runtime.java”中定义，具体实现代码如下所示：

```
private static native String nativeLoad(String filename, ClassLoader loader, String ldLibraryPath);
```

函数 nativeLoad 是一个 JNI 方法，在启动 Dalvik VM 时就已经在内部注册了此 JNI 方法。可以直接调用它注册其他的 JNI 方法，也就是 so 文件 filename 里面所指定的 JNI 方法。函数 nativeLoad 在 C++层对应的实现函数是 Dalvik_java_lang_Runtime_nativeLoad，功能是先调用函数 dvmCreateCstrFromString 将它转换成一个 C++层的字符串 fileName，然后再调用函数 dvmLoadNativeCode 来加载 so 文件。函数 Dalvik_java_lang_Runtime_nativeLoad 在文件“dalvik/vm/ native/java_lang_Runtime.c”中定义，具体实现代码如下所示：

```
static void Dalvik_java_lang_Runtime_nativeLoad(const u4* args,
    JValue* pResult)
{
    StringObject* fileNameObj = (StringObject*) args[0];
    Object* classLoader = (Object*) args[1];
    StringObject* ldLibraryPathObj = (StringObject*) args[2];

    assert(fileNameObj != NULL);
```

```
    char* fileName = dvmCreateCstrFromString(fileNameObj);

    if (ldLibraryPathObj != NULL) {
        char* ldLibraryPath = dvmCreateCstrFromString(ldLibraryPathObj);
        void* sym = dlsym(RTLD_DEFAULT, "android_update_LD_LIBRARY_PATH");
        if (sym != NULL) {
            typedef void (*Fn)(const char*);
            Fn android_update_LD_LIBRARY_PATH = reinterpret_cast<Fn>(sym);
            (*android_update_LD_LIBRARY_PATH)(ldLibraryPath);
        } else {
            ALOGE("android_update_LD_LIBRARY_PATH not found; .so dependencies will not work!");
        }
        free(ldLibraryPath);
    }

    StringObject* result = NULL;
    char* reason = NULL;
    bool success = dvmLoadNativeCode(fileName, classLoader, &reason);
    if (!success) {
        const char* msg = (reason != NULL) ? reason : "unknown failure";
        result = dvmCreateStringFromCstr(msg);
        dvmReleaseTrackedAlloc((Object*) result, NULL);
    }

    free(reason);
    free(fileName);
    RETURN_PTR(result);
}
```

在上述代码中，参数 args[0]保存了 Java 层的一个 String 对象，此 String 对象描述了要加载的 so 文件。

接下来看函数 dvmLoadNativeCode，其功能是调用函数 findSharedLibEntry 检查是否已经加载参数 pathName 指定的 so 文件。如果已经加载过，则可以获得一个 SharedLib 对象 pEntry，在此 SharedLib 对象 pEntry 中，描述了参数 pathName 所指定的和 so 文件相关的加载信息。如果上次加载它的类加载器不等于当前所使用的类加载器，或者上次没有加载成功，那么函数 dvmLoadNativeCode 直接给调用者返回 false，表示不能在当前进程中加载参数 pathName 所描述的 so 文件。函数 dvmLoadNativeCode 在文件“dalvik/vm/Native.cpp”中定义，具体实现代码如下所示：

```
bool dvmLoadNativeCode(const char* pathName, Object* classLoader,
        char** detail)
{
    SharedLib* pEntry;
    void* handle;
    bool verbose;

    /* reduce noise by not chattering about system libraries */
    verbose = !!strncmp(pathName, "/system", sizeof("/system")-1);
    verbose = verbose && !!strncmp(pathName, "/vendor", sizeof("/vendor")-1);

    if (verbose)
        ALOGD("Trying to load lib %s %p", pathName, classLoader);

    *detail = NULL;

    /*
     * See if we've already loaded it.  If we have, and the class loader
     * matches, return successfully without doing anything.
     */
    pEntry = findSharedLibEntry(pathName);
    if (pEntry != NULL) {
        if (pEntry->classLoader != classLoader) {
            ALOGW("Shared lib '%s' already opened by CL %p; can't open in %p",
                pathName, pEntry->classLoader, classLoader);
```

```
        return false;
    }
    if (verbose) {
        ALOGD("Shared lib '%s' already loaded in same CL %p",
            pathName, classLoader);
    }
    if (!checkOnLoadResult(pEntry))
        return false;
    return true;
}
/*
 * Open the shared library.  Because we're using a full path, the system
 * doesn't have to search through LD_LIBRARY_PATH.  (It may do so to
 * resolve this library's dependencies though.)
 *
 * Failures here are expected when java.library.path has several entries
 * and we have to hunt for the lib.
 *
 * The current version of the dynamic linker prints detailed information
 * about dlopen() failures.  Some things to check if the message is
 * cryptic:
 *   - make sure the library exists on the device
 *   - verify that the right path is being opened (the debug log message
 *     above can help with that)
 *   - check to see if the library is valid (e.g. not zero bytes long)
 *   - check config/prelink-linux-arm.map to ensure that the library
 *     is listed and is not being overrun by the previous entry (if
 *     loading suddenly stops working on a prelinked library, this is
 *     a good one to check)
 *   - write a trivial app that calls sleep() then dlopen(), attach
 *     to it with "strace -p <pid>" while it sleeps, and watch for
 *     attempts to open nonexistent dependent shared libs
 *
 * This can execute slowly for a large library on a busy system, so we
 * want to switch from RUNNING to VMWAIT while it executes.  This allows
 * the GC to ignore us.
 */
Thread* self = dvmThreadSelf();
ThreadStatus oldStatus = dvmChangeStatus(self, THREAD_VMWAIT);
handle = dlopen(pathName, RTLD_LAZY);
dvmChangeStatus(self, oldStatus);

if (handle == NULL) {
    *detail = strdup(dlerror());
    ALOGE("dlopen(\"%s\") failed: %s", pathName, *detail);
    return false;
}

/* create a new entry */
SharedLib* pNewEntry;
pNewEntry = (SharedLib*) calloc(1, sizeof(SharedLib));
pNewEntry->pathName = strdup(pathName);
pNewEntry->handle = handle;
pNewEntry->classLoader = classLoader;
dvmInitMutex(&pNewEntry->onLoadLock);
pthread_cond_init(&pNewEntry->onLoadCond, NULL);
pNewEntry->onLoadThreadId = self->threadId;

/* try to add it to the list */
SharedLib* pActualEntry = addSharedLibEntry(pNewEntry);

if (pNewEntry != pActualEntry) {
    ALOGI("WOW: we lost a race to add a shared lib (%s CL=%p)",
        pathName, classLoader);
    freeSharedLibEntry(pNewEntry);
    return checkOnLoadResult(pActualEntry);
} else {
    if (verbose)
```

```
        ALOGD("Added shared lib %s %p", pathName, classLoader);

    bool result = true;
    void* vonLoad;
    int version;

    vonLoad = dlsym(handle, "JNI_OnLoad");
    if (vonLoad == NULL) {
        ALOGD("No JNI_OnLoad found in %s %p, skipping init",
            pathName, classLoader);
    } else {
        /*
         * Call JNI_OnLoad.  We have to override the current class
         * loader, which will always be "null" since the stuff at the
         * top of the stack is around Runtime.loadLibrary().  (See
         * the comments in the JNI FindClass function.)
         */
        OnLoadFunc func = (OnLoadFunc)vonLoad;
        Object* prevOverride = self->classLoaderOverride;

        self->classLoaderOverride = classLoader;
        oldStatus = dvmChangeStatus(self, THREAD_NATIVE);
        if (gDvm.verboseJni) {
            ALOGI("[Calling JNI_OnLoad for \"%s\"]", pathName);
        }
        version = (*func)(gDvmJni.jniVm, NULL);
        dvmChangeStatus(self, oldStatus);
        self->classLoaderOverride = prevOverride;

        if (version != JNI_VERSION_1_2 && version != JNI_VERSION_1_4 &&
            version != JNI_VERSION_1_6)
        {
            ALOGW("JNI_OnLoad returned bad version (%d) in %s %p",
                version, pathName, classLoader);
            /*
             * It's unwise to call dlclose() here, but we can mark it
             * as bad and ensure that future load attempts will fail.
             *
             * We don't know how far JNI_OnLoad got, so there could
             * be some partially-initialized stuff accessible through
             * newly-registered native method calls.  We could try to
             * unregister them, but that doesn't seem worthwhile.
             */
            result = false;
        } else {
            if (gDvm.verboseJni) {
                ALOGI("[Returned from JNI_OnLoad for \"%s\"]", pathName);
            }
        }
    }

    if (result)
        pNewEntry->onLoadResult = kOnLoadOkay;
    else
        pNewEntry->onLoadResult = kOnLoadFailed;

    pNewEntry->onLoadThreadId = 0;

    /*
     * Broadcast a wakeup to anybody sleeping on the condition variable.
     */
    dvmLockMutex(&pNewEntry->onLoadLock);
    pthread_cond_broadcast(&pNewEntry->onLoadCond);
    dvmUnlockMutex(&pNewEntry->onLoadLock);
    return result;
  }
}
```

16.1.3 获取描述类

再看函数 jniRegisterNativeMethods，通过调用参数 env 所指向的结构体 JNIEnv 来获得参数 className 所描述的类。此类就是将要注册的 JNI 的类，它所要注册的 JNI 就是由参数 gMethods 来描述的。函数 jniRegisterNativeMethods 在文件"libnativehelper/dalvik/vm/JNIHelp.cpp"中定义，具体实现代码如下所示：

```
extern "C" int jniRegisterNativeMethods(C_JNIEnv* env, const char* className,
    const JNINativeMethod* gMethods, int numMethods)
{
    JNIEnv* e = reinterpret_cast<JNIEnv*>(env);

    ALOGV("Registering %s natives", className);

    scoped_local_ref<jclass> c(env, findClass(env, className));
    if (c.get() == NULL) {
        char* msg;
        asprintf(&msg, "Native registration unable to find class '%s', aborting", className);
        e->FatalError(msg);
    }

    if ((*env)->RegisterNatives(e, c.get(), gMethods, numMethods) < 0) {
        char* msg;
        asprintf(&msg, "RegisterNatives failed for '%s', aborting", className);
        e->FatalError(msg);
    }

    return 0;
}
```

16.1.4 注册 JNI 方法

再看函数 RegisterNatives，其功能是调用 env 所指向的结构体 JNIEnv 的成员来注册参数 gMethods 所描述的 JNI 方法。函数 RegisterNatives 在文件"dalvik/libnativehelper/include/nativehelper/jni.h"中定义，具体实现代码如下所示：

```
jint RegisterNatives(jclass clazz, const JNINativeMethod* methods,
        jint nMethods)
    {
 return functions->RegisterNatives(this, clazz, methods, nMethods);
    }
```

在上述代码中，函数 RegisterNatives 的指针指向了在 Dalvik 虚拟机内部定义的函数 RegisterNatives。函数 RegisterNatives 的功能是先调用函数 dvmDecodeIndirectRef 来获得要注册 JNI 方法的类对象，然后通过 for 循环依次调用函数 dvmRegisterJNIMethod 注册参数 methods 描述所描述的每一个 JNI 方法。函数 RegisterNatives 在文件"dalvik/vm/Jni.cpp"中定义，具体实现代码如下所示：

```
static jint RegisterNatives(JNIEnv* env, jclass jclazz,
    const JNINativeMethod* methods, jint nMethods)
{
    ScopedJniThreadState ts(env);

    ClassObject* clazz = (ClassObject*) dvmDecodeIndirectRef(ts.self(), jclazz);

    if (gDvm.verboseJni) {
        ALOGI("[Registering JNI native methods for class %s]",
            clazz->descriptor);
    }

    for (int i = 0; i < nMethods; i++) {
        if (!dvmRegisterJNIMethod(clazz, methods[i].name,
```

```
            methods[i].signature, methods[i].fnPtr))
        {
            return JNI_ERR;
        }
    }
    return JNI_OK;
}
```

在上述代码中，每一个 JNI 方法都由名称、签名和地址来描述。

再看函数 dvmRegisterJNIMethod，功能是注册参数 methods 描述所描述的每一个 JNI 方法。函数 dvmRegisterJNIMethod 在文件"dalvik/vm/Jni.cpp"中定义，具体实现代码如下所示：

```
static bool dvmRegisterJNIMethod(ClassObject* clazz, const char* methodName,
    const char* signature, void* fnPtr)
{
    if (fnPtr == NULL) {
        return false;
    }

    // If a signature starts with a '!', we take that as a sign that the native code doesn't
    // need the extra JNI arguments (the JNIEnv* and the jclass).
    bool fastJni = false;
    if (*signature == '!') {
        fastJni = true;
        ++signature;
        ALOGV("fast JNI method %s.%s:%s detected", clazz->descriptor, methodName, signature);
    }
//调用函数 dvmFindDirectMethodByDescriptor 来检查 methodName 是否是 clazz 的一个非虚成员函数
    Method* method = dvmFindDirectMethodByDescriptor(clazz, methodName, signature);
    if (method == NULL) {
//调用函数 dvmFindVirtualMethodByDescriptor 来检查 methodName 是否是 clazz 的一个虚成员函数
        method = dvmFindVirtualMethodByDescriptor(clazz, methodName, signature);
    }
    if (method == NULL) {
        dumpCandidateMethods(clazz, methodName, signature);
        return false;
    }
//过调用函数 dvmIsNativeMethod 判断类 clazz 的成员函数 methodName 确实是声明为 JNI 方法,
//即带有 native 修饰符
    if (!dvmIsNativeMethod(method)) {
        ALOGW("Unable to register: not native: %s.%s:%s", clazz->descriptor, methodName, signature);
        return false;
    }

    if (fastJni) {
        // In this case, we have extra constraints to check...
        if (dvmIsSynchronizedMethod(method)) {
            // Synchronization is usually provided by the JNI bridge,
            // but we won't have one.
            ALOGE("fast JNI method %s.%s:%s cannot be synchronized",
                    clazz->descriptor, methodName, signature);
            return false;
        }
        if (!dvmIsStaticMethod(method)) {
            // There's no real reason for this constraint, but since we won't
            // be supplying a JNIEnv* or a jobject 'this', you're effectively
            // static anyway, so it seems clearer to say so.
            ALOGE("fast JNI method %s.%s:%s cannot be non-static",
                    clazz->descriptor, methodName, signature);
            return false;
        }
    }

    if (method->nativeFunc != dvmResolveNativeMethod) {
        /* this is allowed, but unusual */
        ALOGV("Note: %s.%s:%s was already registered", clazz->descriptor, methodName, signature);
```

```
    method->fastJni = fastJni;
    dvmUseJNIBridge(method, fnPtr);

    ALOGV("JNI-registered %s.%s:%s", clazz->descriptor, methodName, signature);
    return true;
}
```

在上述实现代码中，在注册参数 methodName 描述的 JNI 方法之前会进行如下所示的检查工作。

- ❑ 确保在参数 clazz 所描述类中存在成员函数 methodName。
- ❑ 确保类 clazz 中的成员函数 methodName 被声明为 JNI 方法，即带有 native 修饰符。

16.1.5　实现 JNI 操作

接下来看函数 dvmUseJNIBridge，其功能是在调用 JNI 方法前实现相关的初始化操作。函数 dvmUseJNIBridge 在文件“dalvik/vm/Jni.cpp”中定义，具体实现代码如下所示：

```
void dvmUseJNIBridge(Method* method, void* func) {
    method->shouldTrace = shouldTrace(method);
    // Does the method take any reference arguments?
    method->noRef = true;
    const char* cp = method->shorty;
    while (*++cp != '\0') { // Pre-increment to skip return type.
        if (*cp == 'L') {
            method->noRef = false;
            break;
        }
    }
    DalvikBridgeFunc bridge = gDvmJni.useCheckJni ? dvmCheckCallJNIMethod : dvmCallJNIMethod;
    dvmSetNativeFunc(method, bridge, (const u2*) func);
}
```

再看函数函数 dvmSetNativeFunc，其功能是执行真正的 JNI 方法注册操作。函数 dvmSetNativeFunc 在文件“dalvik/vm/oo/Class.cpp”中定义，具体实现代码如下所示：

```
void dvmSetNativeFunc(Method* method, DalvikBridgeFunc func,
    const u2* insns)
{
    ClassObject* clazz = method->clazz;

    assert(func != NULL);

    /* just open up both; easier that way */
    dvmLinearReadWrite(clazz->classLoader, clazz->virtualMethods);
    dvmLinearReadWrite(clazz->classLoader, clazz->directMethods);

    if (insns != NULL) {
        /* update both, ensuring that "insns" is observed first */
        method->insns = insns;
        android_atomic_release_store((int32_t) func,
            (volatile int32_t*)(void*) &method->nativeFunc);
    } else {
        /* only update nativeFunc */
        method->nativeFunc = func;
    }

    dvmLinearReadOnly(clazz->classLoader, clazz->virtualMethods);
    dvmLinearReadOnly(clazz->classLoader, clazz->directMethods);
}
```

在上述代码中，各个参数的具体说明如下所示。

- ❑ method：表示要注册 JNI 方法的 Java 类成员函数。
- ❑ 参数 func：表示 JNI 方法的 Bridge 函数。
- ❑ 参数 insns：表示要注册的 JNI 方法的函数地址。当参数 insns 的值不等于 NULL 时，函数

dvmSetNativeFunc 会分别将参数 insns 和 func 的值保存在参数 method 所指向的一个 Method 对象的成员变量 insns 和 nativeFunc 中。当 insns 的值等于 NULL 时，函数 dvmSetNativeFunc 只将参数 func 的值保存在参数 method 所指向的一个 Method 对象成员变量 nativeFunc 中。

16.2 创建 Dalvik VM 进程

在 Android 系统中，通过类 android.os.Process 的静态成员函数 start 来创建 Dalvik VM 进程。即由 ActivityManagerService 服务通过类 android.os.Process 的静态成员函数 start 来请求 Zygote 进程的方式创建，而 Zyogte 进程又是通过类 dalvik.system.Zygote 的静态成员函数 forkAndSpecialize 创建该 Android 应用程序进程的。

16.2.1 分析底层启动过程

首先看函数 forkAndSpecialize，这是一个 JNI 函数，此函数在文件“libcore/dalvik/src/main/java/dalvik/system/Zygote.java”中定义，具体实现代码如下所示：

```
    native public static int forkAndSpecialize(int uid, int gid, int[] gids,
          int debugFlags, int[][] rlimits);
    ......
}
```

函数 forkAndSpecialize 是由 C++层的函数 Dalvik_dalvik_system_Zygote_forkAndSpecialize 来实现的，此函数在文件“dalvik/vm/native/dalvik_system_Zygote.cpp”中定义，具体实现代码如下所示：

```
static void Dalvik_dalvik_system_Zygote_forkAndSpecialize(const u4* args,
   JValue* pResult)
{
    pid_t pid;
    pid = forkAndSpecializeCommon(args, false);
    RETURN_INT(pid);
}
```

各个参数的具体说明如下所示。

- 参数 args：指向了一个 u4 数组，其中包含了由 Dalvik 虚拟机封装的所有从 Java 层传递进来的参数。
- 参数 pResult：用于保存 JNI 方法调用结果，这是通过宏 RETURN_INT 来实现的。

16.2.2 创建 Dalvik VM 进程

在函数 Dalvik_dalvik_system_Zygote_forkAndSpec 中调用了函数 forkAndSpecializeCommon，功能是创建一个 Dalvik VM 进程。函数 forkAndSpecializeCommon 在文件“dalvik/vm/native/dalvik_system_Zygote.cpp”中定义，具体实现代码如下所示：

```
static pid_t forkAndSpecializeCommon(const u4* args, bool isSystemServer)
{
    pid_t pid;

    uid_t uid = (uid_t) args[0];
    gid_t gid = (gid_t) args[1];
    ArrayObject* gids = (ArrayObject *)args[2];
    u4 debugFlags = args[3];
    ArrayObject *rlimits = (ArrayObject *)args[4];
    u4 mountMode = MOUNT_EXTERNAL_NONE;
    int64_t permittedCapabilities, effectiveCapabilities;
    char *seInfo = NULL;
    char *niceName = NULL;
```

```
    if (isSystemServer) {
        /*
         * Don't use GET_ARG_LONG here for now. gcc is generating code
         * that uses register d8 as a temporary, and that's coming out
         * scrambled in the child process. b/3138621
         */
        //permittedCapabilities = GET_ARG_LONG(args, 5);
        //effectiveCapabilities = GET_ARG_LONG(args, 7);
        permittedCapabilities = args[5] | (int64_t) args[6] << 32;
        effectiveCapabilities = args[7] | (int64_t) args[8] << 32;
    } else {
        mountMode = args[5];
        permittedCapabilities = effectiveCapabilities = 0;
        StringObject* seInfoObj = (StringObject*)args[6];
        if (seInfoObj) {
            seInfo = dvmCreateCstrFromString(seInfoObj);
            if (!seInfo) {
                ALOGE("seInfo dvmCreateCstrFromString failed");
                dvmAbort();
            }
        }
        StringObject* niceNameObj = (StringObject*)args[7];
        if (niceNameObj) {
            niceName = dvmCreateCstrFromString(niceNameObj);
            if (!niceName) {
                ALOGE("niceName dvmCreateCstrFromString failed");
                dvmAbort();
            }
        }
    }

    if (!gDvm.zygote) {
        dvmThrowIllegalStateException(
            "VM instance not started with -Xzygote");

        return -1;
    }

    if (!dvmGcPreZygoteFork()) {
        ALOGE("pre-fork heap failed");
        dvmAbort();
    }

    setSignalHandler();

    dvmDumpLoaderStats("zygote");
    pid = fork();

    if (pid == 0) {
        int err;
        /* The child process */

#ifdef HAVE_ANDROID_OS
        extern int gMallocLeakZygoteChild;
        gMallocLeakZygoteChild = 1;

        /* keep caps across UID change, unless we're staying root */
        if (uid != 0) {
            err = prctl(PR_SET_KEEPCAPS, 1, 0, 0, 0);

            if (err < 0) {
                ALOGE("cannot PR_SET_KEEPCAPS: %s", strerror(errno));
                dvmAbort();
            }
        }

        for (int i = 0; prctl(PR_CAPBSET_READ, i, 0, 0, 0) >= 0; i++) {
            err = prctl(PR_CAPBSET_DROP, i, 0, 0, 0);
            if (err < 0) {
                if (errno == EINVAL) {
                    ALOGW("PR_CAPBSET_DROP %d failed: %s. "
```

```
                "Please make sure your kernel is compiled with "
                "file capabilities support enabled.",
                i, strerror(errno));
        } else {
            ALOGE("PR_CAPBSET_DROP %d failed: %s.", i, strerror(errno));
            dvmAbort();
        }
    }
}

#endif /* HAVE_ANDROID_OS */

    if (mountMode != MOUNT_EXTERNAL_NONE) {
        err = mountEmulatedStorage(uid, mountMode);
        if (err < 0) {
            ALOGE("cannot mountExternalStorage(): %s", strerror(errno));

            if (errno == ENOTCONN || errno == EROFS) {
                // When device is actively encrypting, we get ENOTCONN here
                // since FUSE was mounted before the framework restarted.
                // When encrypted device is booting, we get EROFS since
                // FUSE hasn't been created yet by init.
                // In either case, continue without external storage.
            } else {
                dvmAbort();
            }
        }
    }

    err = setgroupsIntarray(gids);
    if (err < 0) {
        ALOGE("cannot setgroups(): %s", strerror(errno));
        dvmAbort();
    }

    err = setrlimitsFromArray(rlimits);
    if (err < 0) {
        ALOGE("cannot setrlimit(): %s", strerror(errno));
        dvmAbort();
    }

    err = setresgid(gid, gid, gid);
    if (err < 0) {
        ALOGE("cannot setresgid(%d): %s", gid, strerror(errno));
        dvmAbort();
    }

    err = setresuid(uid, uid, uid);
    if (err < 0) {
        ALOGE("cannot setresuid(%d): %s", uid, strerror(errno));
        dvmAbort();
    }

    if (needsNoRandomizeWorkaround()) {
        int current = personality(0xffffFFFF);
        int success = personality((ADDR_NO_RANDOMIZE | current));
        if (success == -1) {
            ALOGW("Personality switch failed. current=%d error=%d\n", current, errno);
        }
    }

    err = setCapabilities(permittedCapabilities, effectiveCapabilities);
    if (err != 0) {
        ALOGE("cannot set capabilities (%llx,%llx): %s",
            permittedCapabilities, effectiveCapabilities, strerror(err));
        dvmAbort();
    }

    err = set_sched_policy(0, SP_DEFAULT);
    if (err < 0) {
        ALOGE("cannot set_sched_policy(0, SP_DEFAULT): %s", strerror(-err));
        dvmAbort();
```

```
        }

        err = setSELinuxContext(uid, isSystemServer, seInfo, niceName);
        if (err < 0) {
            ALOGE("cannot set SELinux context: %s\n", strerror(errno));
            dvmAbort();
        }
        // These free(3) calls are safe because we know we're only ever forking
        // a single-threaded process, so we know no other thread held the heap
        // lock when we forked.
        free(seInfo);
        free(niceName);

        /*
         * Our system thread ID has changed.  Get the new one.
         */
        Thread* thread = dvmThreadSelf();
        thread->systemTid = dvmGetSysThreadId();

        /* configure additional debug options */
        enableDebugFeatures(debugFlags);

        unsetSignalHandler();
        gDvm.zygote = false;
        if (!dvmInitAfterZygote()) {
            ALOGE("error in post-zygote initialization");
            dvmAbort();
        }
    } else if (pid > 0) {
        /* the parent process */
        free(seInfo);
        free(niceName);
    }

    return pid;
}
```

当函数 forkAndSpecializeCommon 创建 System 进程时，参数 isSystemServer 的值等于 true，此时在参数列表 args 会包含两个额外的参数 permittedCapabilities 和 effectiveCapabilities。其中前者表示 System 进程允许的特权，而后者表示 System 进程当前的有效特权。

16.2.3 初始化运行的 Dalvik VM

另外，在函数 forkAndSpecializeCommon 中还调用了函数 dvmInitAfterZygote，功能是进一步初始化在新创建的进程中运行的 Dalvik VM。函数 dvmInitAfterZygote 在文件“dalvik/vm/Init.cpp”中定义，具体实现代码如下所示：

```
bool dvmInitAfterZygote()
{
    u8 startHeap, startQuit, startJdwp;
    u8 endHeap, endQuit, endJdwp;

    startHeap = dvmGetRelativeTimeUsec();
    /*
     * Post-zygote heap initialization, including starting
     * the HeapWorker thread.
     */
    if (!dvmGcStartupAfterZygote())
        return false;

    endHeap = dvmGetRelativeTimeUsec();
    startQuit = dvmGetRelativeTimeUsec();

    /* start signal catcher thread that dumps stacks on SIGQUIT */
    if (!gDvm.reduceSignals && !gDvm.noQuitHandler) {
        if (!dvmSignalCatcherStartup())
            return false;
```

```
    }

    /* start stdout/stderr copier, if requested */
    if (gDvm.logStdio) {
        if (!dvmStdioConverterStartup())
            return false;
    }

    endQuit = dvmGetRelativeTimeUsec();
    startJdwp = dvmGetRelativeTimeUsec();

    /*
     * Start JDWP thread.  If the command-line debugger flags specified
     * "suspend=y", this will pause the VM.  We probably want this to
     * come last.
     */
    if (!initJdwp()) {
        ALOGD("JDWP init failed; continuing anyway");
    }

    endJdwp = dvmGetRelativeTimeUsec();

    ALOGV("thread-start heap=%d quit=%d jdwp=%d total=%d usec",
        (int)(endHeap-startHeap), (int)(endQuit-startQuit),
        (int)(endJdwp-startJdwp), (int)(endJdwp-startHeap));

#ifdef WITH_JIT
    if (gDvm.executionMode == kExecutionModeJit) {
        if (!dvmCompilerStartup())
            return false;
    }
#endif

    return true;
}
```

到此为止，一个 Dalvik VM 进程的创建工作就完成了，由此可以看出：Dalvik VM 进程的实质就是 Linux 进程。

16.3 创建 Dalvik VM 线程

在 Java 层的应用代码中，通过类 java.lang.Thread 的成员函数 start 创建 Dalvik VM 线程。在本节的内容中，将从函数 start 开始分析创建 Dalvik VM 线程的源码。

16.3.1 检查状态值

在文件“libcore/luni/src/main/java/java/lang/Thread.java”中，函数 start 会先检查变量 hasBeenStarted 的值是否为 true。如果为 true 则说明现已经启动了正在处理的 Thread 对象所描述的 Java 线程；如果不为 true 则表示不能重复启动 Java 线程，否则类 Thread 中的函数 start 会抛出 IllegalThreadStateException 异常。函数 start 的对应实现代码如下所示：

```
public class Thread implements Runnable {
    public synchronized void start() {
        if (hasBeenStarted) {
            throw new IllegalThreadStateException("Thread already started.");
        }
        hasBeenStarted = true;
        VMThread.create(this, stackSize);
    }
}
```

16.3.2 创建线程

在类 Thread 的函数 start 中，调用了类 VMThread 中的静态成员函数 create，其功能是创建一个新的运行线程。函数 create 在文件“libcore\luni\src\main\java\java\lang\VMThread.java”中定义，具体实现代码如下所示：

```
native static void create(Thread t, long stacksize);
```

上述函数 create 是一个 JNI 函数，是通过 C++层的函数 Dalvik_java_lang_VMThread_create 来实现的。函数 create 在文件“dalvik\vm\native\java_lang_VMThread.cpp”中定义，具体实现代码如下所示：

```
static void Dalvik_java_lang_VMThread_create(const u4* args, JValue* pResult)
{
    Object* threadObj = (Object*) args[0];
    s8 stackSize = GET_ARG_LONG(args, 1);

    /* copying collector will pin threadObj for us since it was an argument */
    dvmCreateInterpThread(threadObj, (int) stackSize);
    RETURN_VOID();
}
```

在上述代码中，获取 Java 层传递过来的参数后，调用函数 dvmCreateInterpThread 执行具体的创建线程工作。函数 dvmCreateInterpThread 在文件“dalvik\vm\Thread.cpp”中定义，具体实现代码如下所示：

```
bool dvmCreateInterpThread(Object* threadObj, int reqStackSize)
{
    assert(threadObj != NULL);

    Thread* self = dvmThreadSelf();
    int stackSize;
    if (reqStackSize == 0)
        stackSize = gDvm.stackSize;
    else if (reqStackSize < kMinStackSize)
        stackSize = kMinStackSize;
    else if (reqStackSize > kMaxStackSize)
        stackSize = kMaxStackSize;
    else
        stackSize = reqStackSize;

    pthread_attr_t threadAttr;
    pthread_attr_init(&threadAttr);
    pthread_attr_setdetachstate(&threadAttr, PTHREAD_CREATE_DETACHED);

    /*
     * To minimize the time spent in the critical section, we allocate the
     * vmThread object here.
     */
    Object* vmThreadObj = dvmAllocObject(gDvm.classJavaLangVMThread, ALLOC_DEFAULT);
    if (vmThreadObj == NULL)
        return false;

    Thread* newThread = allocThread(stackSize);
    if (newThread == NULL) {
        dvmReleaseTrackedAlloc(vmThreadObj, NULL);
        return false;
    }

    newThread->threadObj = threadObj;

    assert(newThread->status == THREAD_INITIALIZING);

    /*
     * We need to lock out other threads while we test and set the
     * "vmThread" field in java.lang.Thread, because we use that to determine
     * if this thread has been started before.  We use the thread list lock
     * because it's handy and we're going to need to grab it again soon
     * anyway.
     */
    dvmLockThreadList(self);
```

```
    if (dvmGetFieldObject(threadObj, gDvm.offJavaLangThread_vmThread) != NULL) {
        dvmUnlockThreadList();
        dvmThrowIllegalThreadStateException(
            "thread has already been started");
        freeThread(newThread);
        dvmReleaseTrackedAlloc(vmThreadObj, NULL);
    }

    /*
     * There are actually three data structures: Thread (object), VMThread
     * (object), and Thread (C struct). All of them point to at least one
     * other.
     *
     * As soon as "VMThread.vmData" is assigned, other threads can start
     * making calls into us (e.g. setPriority).
     */
    dvmSetFieldInt(vmThreadObj, gDvm.offJavaLangVMThread_vmData, (u4)newThread);
    dvmSetFieldObject(threadObj, gDvm.offJavaLangThread_vmThread, vmThreadObj);

    /*
     * Thread creation might take a while, so release the lock.
     */
    dvmUnlockThreadList();

    ThreadStatus oldStatus = dvmChangeStatus(self, THREAD_VMWAIT);
    pthread_t threadHandle;
    int cc = pthread_create(&threadHandle, &threadAttr, interpThreadStart, newThread);
    pthread_attr_destroy(&threadAttr);
    dvmChangeStatus(self, oldStatus);

    if (cc != 0) {
        /*
         * Failure generally indicates that we have exceeded system
         * resource limits.  VirtualMachineError is probably too severe,
         * so use OutOfMemoryError.
         */

        dvmSetFieldObject(threadObj, gDvm.offJavaLangThread_vmThread, NULL);

        ALOGE("pthread_create (stack size %d bytes) failed: %s", stackSize, strerror(cc));
        dvmThrowExceptionFmt(gDvm.exOutOfMemoryError,
                             "pthread_create (stack size %d bytes) failed: %s",
                             stackSize, strerror(cc));
        goto fail;
    }
    dvmLockThreadList(self);
    assert(self->status == THREAD_RUNNING);
    self->status = THREAD_VMWAIT;
    while (newThread->status != THREAD_STARTING)
        pthread_cond_wait(&gDvm.threadStartCond, &gDvm.threadListLock);

    LOG_THREAD("threadid=%d: adding to list", newThread->threadId);
    newThread->next = gDvm.threadList->next;
    if (newThread->next != NULL)
        newThread->next->prev = newThread;
    newThread->prev = gDvm.threadList;
    gDvm.threadList->next = newThread;

    /* Add any existing global modes to the interpBreak control */
    dvmInitializeInterpBreak(newThread);

    if (!dvmGetFieldBoolean(threadObj, gDvm.offJavaLangThread_daemon))
        gDvm.nonDaemonThreadCount++;        // guarded by thread list lock

    dvmUnlockThreadList();
    /* change status back to RUNNING, self-suspending if necessary */
    dvmChangeStatus(self, THREAD_RUNNING);
    dvmLockThreadList(self);
    assert(newThread->status == THREAD_STARTING);
    newThread->status = THREAD_VMWAIT;
    pthread_cond_broadcast(&gDvm.threadStartCond);
```

```
    dvmUnlockThreadList();
    dvmReleaseTrackedAlloc(vmThreadObj, NULL);
    return true;
fail:
    freeThread(newThread);
    dvmReleaseTrackedAlloc(vmThreadObj, NULL);
    return false;
}
```

在上述代码中，通过函数 pthread_create 来创建一个线程。当函数 pthread_create 创建一个线程的时候，需要知道该线程的属性。这些属性通过结构体 pthread_attr_t 来描述。结构体 pthread_attr_t 可以调用函数 pthread_attr_init 实现初始化处理，并调用函数 pthread_attr_setdetachstate 来设置它的分离状态。函数 dvmCreateInterpThread 的参数的具体说明如下所示。

- 参数 reqStackSize：表示要创建的 Dalvik VM 线程的 Java 栈大小，当值等于 0 时会使用保存在 gDvm.stackSize 中的值作为后面要创建的线程的 Java 栈大小。如果参数 reqStackSize 值不为 0，那么它必须大于等于 kMinStackSize（512+768），并且小于等于 kMaxStackSize（256×1024+768），否则就会被修改。
- 参数 threadObj：是 Java 层的一个 Thread 对象，在 Dalvik VM 中对应有一个 Native 层的 Thread 对象。

16.3.3　分析启动过程

接下来分析函数 interpThreadStart，功能是启动新创建的 Dalvik VM 线程。函数 interpThreadStart 在文件“dalvik\vm\Thread.cpp”中定义，具体实现代码如下所示：

```
static void* interpThreadStart(void* arg)
{
    Thread* self = (Thread*) arg;

    std::string threadName(dvmGetThreadName(self));
    setThreadName(threadName.c_str());

    /*
     * Finish initializing the Thread struct.
     */
    dvmLockThreadList(self);
    prepareThread(self);

    LOG_THREAD("threadid=%d: created from interp", self->threadId);

    /*
     * Change our status and wake our parent, who will add us to the
     * thread list and advance our state to VMWAIT.
     */
    self->status = THREAD_STARTING;
    pthread_cond_broadcast(&gDvm.threadStartCond);
    while (self->status != THREAD_VMWAIT)
        pthread_cond_wait(&gDvm.threadStartCond, &gDvm.threadListLock);

    dvmUnlockThreadList();

    /*
     * Add a JNI context.
     */
    self->jniEnv = dvmCreateJNIEnv(self);

    /*
     * Change our state so the GC will wait for us from now on.  If a GC is
     * in progress this call will suspend us.
     */
    dvmChangeStatus(self, THREAD_RUNNING);

    /*
```

```
     * Notify the debugger & DDM.  The debugger notification may cause
     * us to suspend ourselves (and others).  The thread state may change
     * to VMWAIT briefly if network packets are sent.
     */
    if (gDvm.debuggerConnected)
        dvmDbgPostThreadStart(self);

    /*
     * Set the system thread priority according to the Thread object's
     * priority level.  We don't usually need to do this, because both the
     * Thread object and system thread priorities inherit from parents.  The
     * tricky case is when somebody creates a Thread object, calls
     * setPriority(), and then starts the thread.  We could manage this with
     * a "needs priority update" flag to avoid the redundant call.
     */
    int priority = dvmGetFieldInt(self->threadObj,
                        gDvm.offJavaLangThread_priority);
    dvmChangeThreadPriority(self, priority);

    /*
     * Execute the "run" method.
     *
     * At this point our stack is empty, so somebody who comes looking for
     * stack traces right now won't have much to look at.  This is normal.
     */
    Method* run = self->threadObj->clazz->vtable[gDvm.voffJavaLangThread_run];
    JValue unused;

    ALOGV("threadid=%d: calling run()", self->threadId);
    assert(strcmp(run->name, "run") == 0);
    dvmCallMethod(self, run, self->threadObj, &unused);
    ALOGV("threadid=%d: exiting", self->threadId);

    /*
     * Remove the thread from various lists, report its death, and free
     * its resources.
     */
    dvmDetachCurrentThread();

    return NULL;
}
```

在上述代码中，参数 arg 指向了 Native 层的 Thread 对象，此 Thread 对象最后被保存在变量 self 中，用来描述新创建的 Dalvik VM 线程。

再看函数 prepareThread，其功能是初始化新创建的 Dalvik VM 线程。函数 interpThreadStart 在文件“dalvik\vm\Thread.cpp”中定义，具体实现代码如下所示：

```
static bool prepareThread(Thread* thread)
{
    assignThreadId(thread);
    thread->handle = pthread_self();
    thread->systemTid = dvmGetSysThreadId();

    //ALOGI("SYSTEM TID IS %d (pid is %d)", (int) thread->systemTid,
    //    (int) getpid());
    /*
     * If we were called by dvmAttachCurrentThread, the self value is
     * already correctly established as "thread".
     */
    setThreadSelf(thread);

    ALOGV("threadid=%d: interp stack at %p",
        thread->threadId, thread->interpStackStart - thread->interpStackSize);

    /*
     * Initialize invokeReq.
     */
    dvmInitMutex(&thread->invokeReq.lock);
    pthread_cond_init(&thread->invokeReq.cv, NULL);
```

```
    /*
     * Initialize our reference tracking tables.
     *
     * Most threads won't use jniMonitorRefTable, so we clear out the
     * structure but don't call the init function (which allocs storage).
     */
    if (!thread->jniLocalRefTable.init(kJniLocalRefMin,
            kJniLocalRefMax, kIndirectKindLocal)) {
        return false;
    }
    if (!dvmInitReferenceTable(&thread->internalLocalRefTable,
            kInternalRefDefault, kInternalRefMax))
        return false;

    memset(&thread->jniMonitorRefTable, 0, sizeof(thread->jniMonitorRefTable));

    pthread_cond_init(&thread->waitCond, NULL);
    dvmInitMutex(&thread->waitMutex);

    /* Initialize safepoint callback mechanism */
    dvmInitMutex(&thread->callbackMutex);

    return true;
}
```

在函数 interpThreadStart 中，接下来会调用函数 dvmCreateJNIEnv 为新创建的 Dalvik VM 线程创建 JNI 环境。然后调用函数 dvmCallMethod 通知 Dalvik VM 解释器去执行类 java.lang.Thread 的成员函数 run。函数 dvmCallMethod 在文件“dalvik/vm/interp/Stack.c”中定义，具体实现代码如下所示：

```
void dvmCallMethod(Thread* self, const Method* method, Object* obj,
    JValue* pResult, ...)
{
    va_list args;
    va_start(args, pResult);
    dvmCallMethodV(self, method, obj, false, pResult, args);
    va_end(args);
}
```

而函数 run 的实现则比较简单，具体实现代码如下所示：

```
    public void run() {
        if (target != null) {
            target.run();
        }
    }
```

在上述代码中，首先检查成员变量 target 的值是否为 null，如果不是则会调用它所指向的 Runnable 对象的成员函数 run 来作为新创建的 Dalvik 虚拟机线程的执行主体，否则什么也不做。

16.3.4　清理线程

函数 dvmDetachCurrentThread 的功能是执行清理工作，能够调用函数 dvmThreadSelf 获得用来描述当前即将要退出的 Dalvik VM 线程的 Native 层的 Thread 对象 self。函数 dvmDetachCurrentThread 在文件“dalvik\vm\Thread.cpp”中定义，具体实现代码如下所示：

```
void dvmDetachCurrentThread()
{
    Thread* self = dvmThreadSelf();
    Object* vmThread;
    Object* group;

    /*
     * Make sure we're not detaching a thread that's still running.  (This
     * could happen with an explicit JNI detach call.)
```

```
 *
 * A thread created by interpreted code will finish with a depth of
 * zero, while a JNI-attached thread will have the synthetic "stack
 * starter" native method at the top.
 */
int curDepth = dvmComputeExactFrameDepth(self->interpSave.curFrame);
if (curDepth != 0) {
    bool topIsNative = false;

    if (curDepth == 1) {
        /* not expecting a lingering break frame; just look at curFrame */
        assert(!dvmIsBreakFrame((u4*)self->interpSave.curFrame));
        StackSaveArea* ssa = SAVEAREA_FROM_FP(self->interpSave.curFrame);
        if (dvmIsNativeMethod(ssa->method))
            topIsNative = true;
    }

    if (!topIsNative) {
        ALOGE("ERROR: detaching thread with interp frames (count=%d)",
            curDepth);
        dvmDumpThread(self, false);
        dvmAbort();
    }
}

group = dvmGetFieldObject(self->threadObj, gDvm.offJavaLangThread_group);
LOG_THREAD("threadid=%d: detach (group=%p)", self->threadId, group);

/*
 * 释放在 JNI 方法中持有的 Monitor，也就是 MonitorExit 那些被 MonitorEnter 了的对象
 */
dvmReleaseJniMonitors(self);

/*
 * 检查线程是否有未处理异常。如果有则调用函数 threadExitUncaughtException
 * 将异常交给 thread-exit-uncaught-exception handler 处理.
 */
if (dvmCheckException(self))
    threadExitUncaughtException(self, group);

/*
 *从线程组移除线程.
 */
if (group != NULL) {
    Method* removeThread =
        group->clazz->vtable[gDvm.voffJavaLangThreadGroup_removeThread];
    JValue unused;
    dvmCallMethod(self, removeThread, group, &unused, self->threadObj);
}

/*
 * Clear the vmThread reference in the Thread object.  Interpreted code
 * will now see that this Thread is not running.  As this may be the
 * only reference to the VMThread object that the VM knows about, we
 * have to create an internal reference to it first.
 */
vmThread = dvmGetFieldObject(self->threadObj,
            gDvm.offJavaLangThread_vmThread);
dvmAddTrackedAlloc(vmThread, self);
dvmSetFieldObject(self->threadObj, gDvm.offJavaLangThread_vmThread, NULL);

/* clear out our struct Thread pointer, since it's going away */
dvmSetFieldObject(vmThread, gDvm.offJavaLangVMThread_vmData, NULL);
/*
 * 检查 gDvm.debuggerConnected 的值是否等于 true,
 * 如果被调试器连接，则通知调试器当前线程要退出
 */
if (gDvm.debuggerConnected)
    dvmDbgPostThreadDeath(self);
/*
 * 向调用了 Thread.join 等待当前线程结束的线程发送通知
```

```
     */
    dvmLockObject(self, vmThread);
    dvmObjectNotifyAll(self, vmThread);
    dvmUnlockObject(self, vmThread);
    dvmReleaseTrackedAlloc(vmThread, self);
    vmThread = NULL;

    /*
     * We're done manipulating objects, so it's okay if the GC runs in
     * parallel with us from here out.  It's important to do this if
     * profiling is enabled, since we can wait indefinitely.
     */
    volatile void* raw = reinterpret_cast<volatile void*>(&self->status);
    volatile int32_t* addr = reinterpret_cast<volatile int32_t*>(raw);
    android_atomic_release_store(THREAD_VMWAIT, addr);

    MethodTraceState* traceState = &gDvm.methodTrace;

    dvmLockMutex(&traceState->startStopLock);
    if (traceState->traceEnabled) {
        ALOGI("threadid=%d: waiting for method trace to finish",
            self->threadId);
        while (traceState->traceEnabled) {
            dvmWaitCond(&traceState->threadExitCond,
                        &traceState->startStopLock);
        }
    }
    dvmUnlockMutex(&traceState->startStopLock);

    dvmLockThreadList(self);

    /*
     *销毁用来描述当前 JNI 上下文环境的一个 JNIEnvExt 对象
     */
    dvmDestroyJNIEnv(self->jniEnv);
    self->jniEnv = NULL;

    self->status = THREAD_ZOMBIE;

    /*
     *从 Dalvik VM 的线程列表中移除当前线程
     */
    unlinkThread(self);

    /*
     * If we're the last one standing, signal anybody waiting in
     * DestroyJavaVM that it's okay to exit.
     */
    if (!dvmGetFieldBoolean(self->threadObj, gDvm.offJavaLangThread_daemon)) {
        gDvm.nonDaemonThreadCount--;        // guarded by thread list lock

        if (gDvm.nonDaemonThreadCount == 0) {
            ALOGV("threadid=%d: last non-daemon thread", self->threadId);
            //dvmDumpAllThreads(false);
            // cond var guarded by threadListLock, which we already hold
            int cc = pthread_cond_signal(&gDvm.vmExitCond);
            if (cc != 0) {
                ALOGE("pthread_cond_signal(&gDvm.vmExitCond) failed: %s", strerror(cc));
                dvmAbort();
            }
        }
    }

    ALOGV("threadid=%d: bye!", self->threadId);
    releaseThreadId(self);
    dvmUnlockThreadList();

    setThreadSelf(NULL);
//释放当前线程的 Java 栈和各个引用表（即前面 Step 5 所描述的三个引用表）所占用的内存
//释放 Thread 对象 self 所占用的内存。
    freeThread(self);
}
```

第 17 章　Dalvik VM 异常处理详解

异常是指程序在运行时发生的错误或者不正常的情况。对程序员来说，异常是一件很麻烦的事情，需要程序员来进行检测和处理。但是 Java VM 和 Dalvik VM 都非常人性化，可以自动检测异常，并对异常进行捕获，并且通过程序可以对异常进行处理。在本章的内容中，将详细讲解 Dalvik VM 处理异常的基本知识。

17.1 Java 异常处理机制

在编写 Java 程序时，要尽量做到这个程序的健壮性。所谓程序的健壮性，就是指程序在多数情况下能够正常运行，返回预期的正确结果；如果偶尔遇到异常情况，程序也能采取周到的解决措施。而不健壮的程序则没有事先充分预计到可能出现的异常，或者没有提供强有力的异常解决措施，导致程序在运行时，经常莫名其妙地终止，或者返回错误的运行结果，而且难以检测出现异常的原因。

17.1.1　方法调用栈

Java VM 用方法调用栈（Method Invocation Stack）来跟踪一系列的方法调用过程。该堆栈保存了每个调用方法的本地信息（比如方法的局部变量）。当一个新方法被调用时，Java 虚拟机把描述该方法的栈结构置入栈顶，位于栈顶的方法为正在执行的方法。图 17-1 描述了方法调用栈的结构，图中方法的调用顺序为：main()方法调用 methodB()方法，methodB()方法调用 methodA()方法。

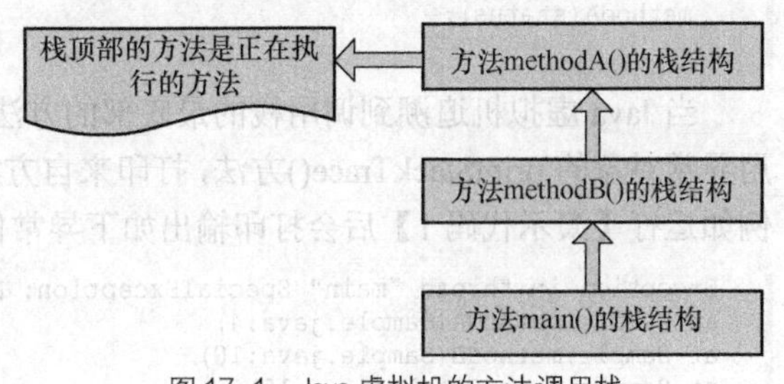

图 17-1　Java 虚拟机的方法调用栈

当方法 methodB()调用方法 methodA()时，如果方法中的代码块抛出异常，有如下两种处理办法。

（1）如果当前方法有能力自己解决异常，就在当前方法中通过 try-catch 语句捕获并处理异常，例如下面的代码：

```
public void methodA(int status){
  try{
//以下代码可能会抛出 SpecialException
if(status==-1)
  throw new SpecialException("Monster");
  }catch(SpecialException e){
处理异常
  }
}
```

（2）如果当前方法没能力自己解决异常，就在方法的声明处通过 throws 语句声明抛出异常，例如下面的代码。

```
public void methodA(int status) throws SpecialException{
  //以下代码可能会抛出 SpecialException
  if(status==-1)
throw new SpecialException("Monster");
}
```

当一个方法正常执行完毕，Java 虚拟机会从调用栈中弹出该方法的栈结构，然后继续处理前一个方法。如果在执行方法的过程中抛出异常，Java 虚拟机必须找到能捕获该异常的 catch 代码块。它首先察看当前方法是否存在这样的 catch 代码块，如果存在，就执行该 catch 代码块；否则，Java 虚拟机会从调用栈中弹出该方法的栈结构，继续到前一个方法中查找合适的 catch 代码块。

例如，当方法 methodA()抛出 SpecialException 异常时，如果在该方法中提供了捕获 SpecialException 的 catch 代码块，就执行这个异常处理代码块。如果方法 methodA()未捕获该异常，而是采用第二种方式声明抛出 SpecialException，那么 Java 虚拟机的处理流程将退回到上层调用方法 methodB()，再察看方法 methodB()中有没有捕获 SpecialException。如果在方法 methodB()中存在捕获该异常的 catch 代码块，就执行这个 catch 代码块，此时定义方法 methodB()的代码如下所示：

```
public void methodB(int status){
  try{
methodA(status);
  }catch(SpecialException e){
处理异常
  }
}
```

由此可见，在回溯过程中，如果 Java 虚拟机在某个方法中找到了处理该异常的代码块，则该方法的栈结构将成为栈顶元素，程序流程将转到该方法的异常处理代码部分继续执行。

如果方法 methodB()也没有捕获 SpecialException，而是声明抛出该异常，那么 Java 虚拟机的处理流程将退回到 main()方法，此时定义方法 methodB()的代码如下所示：

```
public void methodB(int status) throws SpecialException{
  methodA(status);
}
```

当 Java 虚拟机追溯到调用栈的最底部的方法，如果仍然没有找到处理该异常的代码块，将调用异常对象的 printStackTrace()方法，打印来自方法调用栈的异常信息，随后整个应用程序被终止。例如运行【演示代码 1】后会打印输出如下异常信息。

```
Exception in thread "main" SpecialException: Monster
at Sample.methodA(Sample.java:4)
at Sample.methodB(Sample.java:10)
at Sample.main(Sample.java:15)
```

【演示代码 1】的具体代码如下所示：

```
public class Sample{
  public void methodA(int status)throws SpecialException{
if(status==-1)
  throw new SpecialException("Monster");
System.out.println("methodA");
  }

  public void methodB(int status)throws SpecialException{
methodA(status);
System.out.println("methodB");
  }
  public static void main(String args[])throws SpecialException{
new Sample().methodB(-1);
  }
}
```

在上述代码中，类 SpecialException 表示某一种异常，【演示代码 2】是它的源程序，【演示代码 2】的具体代码如下所示：

```
public class SpecialException extends Exception{
  public SpecialException(){}

  public SpecialException(String msg){
  super(msg);
    }
}
```

17.1.2 Java 提供的异常处理类

在 Java 中有一个 lang 包，在此包里面有一个专门处理异常的类——Throwable，此类是所有异常的父类，每一个异常的类都是它的子类。其中 Error 和 Exception 这两个类十分重要，用得也较多，前者是用来定义那些通常情况下不希望被捕获的异常，而后者是程序能够捕获的异常情况。Java 中常用异常类的信息如表 17-1 所示。

表 17-1　　Java 中的异常类

异常类名称	异常类含义
ArithmeticExeption	算术异常类
ArratIndexOutOfBoundsExeption	数组小标越界异常类
ArrayStroeException	将与数组类型不兼容的值赋值给数组元素时抛出的异常
ClassCastException	类型强制转换异常类
ClassNotFoundException	为找到相应大类异常
EOFEException	文件已结束异常类
FileNotFoundException	文件未找到异常类
IllegalAccessException	访问某类被拒绝时抛出的异常类
InstantiationException	试图通过 newInstance()方法创建一个抽象类或抽象接口的实例时抛出异常类
IOEException	输入输出抛出异常类
NegativeArraySizeException	建立元素个数为负数的异常类
NullPointerException	空指针异常
NumberFormatException	字符串转换为数字异常类
NoSuchFieldException	字段未找到异常类
NoSuchMethodException	方法未找到异常类
SecurityException	小应用程序执行浏览器的安全设置禁止动作时抛出的异常类
SQLException	操作数据库异常类
StringIndexOutOfBoundsException	字符串索引超出范围异常类

17.2 Java VM 异常处理机制详解

异常（exception）是可被硬件或软件检测到的要求进行特殊处理的异常事件。异常处理作为程序设计语言的组成部分，为开发可靠性软件系统提供了强有力的支持。Java 作为一种优秀的程序设计语言，不仅具有面向对象、并发、平台中立等特点，同时也定义了灵活的异常处理机制，Java 虚拟机是这一机制的具体实施者。异常处理作为现代程序设计语言的特点已被广泛采纳，它提高了程序运行的可靠性。但由于 Java 语言的特点，异常处理也带来了不小的麻烦。这主要体现在及时编译情况下，异常处理的实现较为复杂，更重要的一点是由于字节码编译与及时编译是两个独立的过程，因此使及时编译的优化设计受到限制，当机器指令生成后，方法内代码的优化必然涉及字节码与机器码之间的对应关系的调整、异常处理表的调整。在异常处理语句内部，机器代码的优化更要慎重，代码的删除和外提都可能会造成异常处理范围的改变。其次，异常处理也降低了程序的运行效率，为了提高运行效率有些及时编译器中删除了数组越界检查。

在本节将详细讲解 Java 语言的异常处理机制，并结合国产开放系统平台 COSIX 虚拟机异常处理的设计，深入探讨了在解释执行和及时编译执行两种不同的情况下，异常处理设计与实现的关键技术。

17.2.1 Java 语言及虚拟机的异常处理机制

Java 语言在设计上与 C++有许多相同之处，它的异常处理机制基本上沿袭了 C++的规则。Java 的异常处理语句及抛出异常语句与相应的 C++的语句完全一样，它的异常处理是静态绑定的，没被处理的异常是沿着方法调用栈向上传播，异常处理完毕后程序的执行点转移到异常处理句柄的下一条语句。

与 C++不同的是，Java 具有完善的异常类定义，Java 的异常类可分为三大类：Error、一般异常及 RuntimeException。当类动态连接失败或产生其他硬件错误时，虚拟机产生 Error 异常，一般的 Java 程序不会产生该异常，也不必对该类异常进行处理；RuntimeException 类的异常是虚拟机在运行时产生的，如算术运算异常、数组索引异常、引用异常等。由于该类异常在程序中普遍存在，因此用户没必要对它们进行检测、处理，编译程序在编译时也不会去检查该类异常，这些异常由虚拟机在运行时检测并对其进行处理；通常用户程序需要产生并提供异常处理句柄的异常为一般异常，一般异常与 Errors 不同，它不是严重的系统异常，也不是在程序中普遍存在的，因此在编译时编译器就会提示用户提供异常处理句柄，否则编译不会通过，例如 I/O 异常。

编译程序为含有异常处理语句的方法生成一个异常处理表，该表指明了异常处理语句产生异常的字节码范围、异常处理句柄的地址以及产生的异常类型。

虚拟机是 Java 语言异常处理机制的微观实现，当虚拟机产生异常或程序执行时由字节码指令 athrow 产生异常时，异常处理程序都会根据所产生的异常类型及产生异常的当前程序点，在方法异常表中查找对应的异常处理句柄，方法异常表给出了某一程序段代码所产生的异常类型及其对应的异常处理句柄。若查到对应的异常处理句柄则执行该句柄，执行完毕后返回异常处理句柄的下一条语句；若没找到，则异常沿方法调用栈向上传播，将产生的异常传播给该方法的调用者。

17.2.2 COSIX 虚拟机异常处理的设计与实现

一般可以采用如下 3 种方法实现异常处理。

（1）动态建立异常处理语句的数据结构链表。

（2）使用静态的异常处理表，运行时查找该表，以搜寻异常处理句柄。

（3）使用有两个返回值的函数。

其中 C++和 Ada 的异常处理使用了第 1 种方法，第 2 种方法较第 1 种可显著提高异常处理的效率，它在 C++及 Java 语言中得到了应用。在该方式下，编译器为每一方法提供异常处理表，由于有了异常处理表，就没有必要再为每一条异常处理语句建立数据链表。当异常产生时，异常处理程序可直接查询异常表，快速定位异常处理句柄；若没找到，则将异常传播给上层方法，在它的异常查询表中继续查询。

第 3 种方法在某些 Java 虚拟机的及时编译设计时被采用。该处理形式虽然简单，但是编译过程较为复杂。由于方法间没有共享方法异常表，而使编译后的代码冗长，该方法适用于异常发生频繁的程序。

1. 解释执行时的异常处理

解释器的异常处理实现较为简单，由于对异常处理的编译是直接针对字节码的，因此异常的查找和传播都较及时编译方便得多。

Java 编译程序为有异常处理句柄的方法生成异常处理表，因此在实现时可直接利用这一信息生成方法的异常表，方法异常表的数据结构如下所示：

```
MexceptionTable {
  int start-pc;//产生异常的指令起始范围
  int end-pc;//产生异常的指令终止范围
  int handler-pc;//异常处理句柄指针
  Hjava-lang-Class* catch-type;//异常类型
}
```

当产生异常时，异常处理程序根据产生该异常的指令位置及其产生的异常类型，在方法异常表中查找符合这两个条件的异常处理句柄。

Java 异常处理机制规定没有处理的异常要沿方法调用栈传播给上层调用方法，因此在程序的执行时应建立方法调用关系链表，以实现异常在方法调用栈中的逆向传播。方法调用链表的数据结构应能恢复方法的运行环境并记录方法的运行状态。方法调用链表的数据结构如下所示：

```
…MethodCallList{
…u4 pc; //该方法的当前字节码指令指针
…ObjLock* lock; //该方法所在对象的对象锁
…methods* meth; //指向该方法在类中的位置
…jmp-buf jbuf; //设置解释器解释执行该方法时的运行环境
…MethodCallList* prev; //指向该方法的上层调用方法
…}
```

当异常产生时，异常处理程序根据 pc 值及产生的异常类型，在该方法的异常处理表（由 meth 获得）中查询对应的异常处理句柄，若没找到则异常传播给该方法的调用者（prev 指向的方法），异常处理程序进行同样的查找，当找到对应的异常处理句柄时，异常处理程序恢复该句柄所在方法解释执行时的运行环境，并将该方法的字节码指令指针指向异常处理句柄，运行环境被恢复后解释器执行该异常处理句柄，之后，解释器继续执行其下面的其他语句。

一般用操作系统提供的库函数 setjmp 及 longjmp 来实现运行环境的保存与恢复。在方法调用链表数据结构中，lock 为该方法所属对象的对象锁，若该方法为静态方法，则 lock 为其所属类的类锁。该域是为同步方法设计的，对于同步方法异常处理完成后，应释放该对象（类）锁。

异常处理程序在进行异常处理时，将异常分为两大类，即内部异常与外部异常。内部异常主要指虚拟机在运行时检测到的异常，它们主要为 Error 及 RuntimeException。例如虚拟机在类加载时需要进行类文件格式的检查，若类文件格式错误，虚拟机将产生 ClassFormatError 异常；虚拟机执行字节码指令时自动对数组下标进行检查，当数组范围越界时，虚拟机会产生 ArrayIndexOutOfBoundsException。当虚拟机检测到异常时，其触发异常程序的执行。外部异常一般指用户自定义的异常，字节码指令 athrow 产生外部异常并触发异常处理程序的执行，该异常存放在操作数栈的栈顶。如果在当前方法中找到了异常处理句柄，athrow 指令抛弃操作数栈上的所有数据，然后将抛出的异常对象压入栈；如果在当前方法中没找到异常处理句柄，则异常沿方法调用栈传播，直到找到处理该异常的方法。此时，处理该异常的方法的操作数栈被清除，并将异常对象压入到这个空的操作数栈，该传播过程中的其他方法栈都将被抛弃.不管内部异常还是外部异常，异常处理句柄都得由用户提供。

对于内部异常，有两个异常是比较特殊的，它们是 NullPointerException(空指针引用)、ArithmeticException（浮点溢出），这两个异常可由硬件检测到，操作系统提供这两个异常的中断信号，因此在异常初始化时，可由库函数 signal 设置这两个异常的服务程序。

在对内部异常进行处理时，异常处理程序除查找异常处理句柄并执行该句柄外，异常处理程序还应提供方法调用过程的全部信息，该信息包括方法调用栈中所有方法的当前字节码指令指针所在的类名、方法名及所处的源文件行号。该过程同样用逆向遍历方法调用链表的全部数据单元，由方法调用表数据结构中的 pc 值在方法的行号表中查找该 pc 值所对应的源文件行号，行号表内记录着字节码生成时字节码指令与源文件行号的对应关系，该表保存在类的方法表中，类名、方法名也可在方法表中查到。

以上介绍了解释执行时异常处理程序的设计，由于编译器在编译程序时提供了充足的异常处理信息，如方法异常处理表、方法行号表等，使得异常处理过程较为简单。

2. 及时编译执行时的异常处理

在及时编译异常处理设计上，一般采用了与解释器异常处理相同的设计方法。在解释执行时，由于解释器控制方法的全部活动，包括栈空间的分配、指令的执行，因此可显式地创建一个方法调用表（解释器创建），供异常处理程序使用。但由于及时编译将字节码编译成本地码，方法的栈空间直接分配在线程的本地栈中，无法创建一显式的数据结构去反映方法调用关系及记录方法的运行状态。因此，要处理异常在方法调用栈中的传播及异常句柄的查询，必须掌握本地方法栈的结构及字节码与本地代码之间的对应关系。

在及时编译时，首先应建立字节码与机器码之间的对应关系，在编译每一条字节码时，将字节码所对应的机器码在本地方法代码中的偏移记录下来，可利用这一信息进行方法异常处理表的翻译，将异常处理表中字节码的指令偏移替换为对应的机器码的内存位置。同理，可翻译方法的行号表将源文件行号与字节码的对应关系转换为源文件行号同机器码之间的对应关系。下面的程序给出了异常处理表的翻译过程。

```
if(meth -> exception-table!=null){ //方法异常表非空
  for(I=0;I<meth->exception-table-len;I++){//exception-table-len 为异常表长度
    e=&meth->exception-table[I]; //获取第 I 个异常表
    e->start-pc=nativeOffset[e->start-pc]+(int)nativeCodeBase;
    e->end-pc=nativeOffset[e->end-pc]+(int)nativeCodeBase;
    e->handler-pc=nativeOffset[e->handler-pc]+(int)nativeCodeBase;
  }
}
```

在上面的程序中，nativeCodeBase 为一指针，其指向编译后本地代码的起始位置，nativeOffset 为一整型数组存放每一字节码在本地方法代码中的偏移。经转换后，e->start－pc，e->end－pc，e->handler－pc 被转换为对应字节码的本地码在内存中的地址。同理，可进行方法行号表转换。方法异常表及行号表的转换为异常处理提供了方便，异常处理程序可直接由这两个表查询异常处理句柄及异常信息，其查询过程与解释异常处理相同。

及时编译执行时，异常的传播处理较为复杂，这需要对本地方法栈的结构有一个清楚的认识。线程在创建时 JVM 为其分配一个固定的运行空间（可由用户指定），线程的一切活动所需的空间都被分配在该空间中。对于及时编译，该空间存放本地方法栈，本地方法栈实际上就是传统的 C 栈。要实现异常在方法调用栈中的传播，就得解决如何确定产生异常方法栈的位置及调用该方法的上层方法的栈位置。例如在图 17-2 中，给出了线程栈空间中方法栈存放的示意图，图 17-2 的方法调用过程为方法 1 调用方法 2，方法 2 调用方法 3。方法栈在线程空间中是连续存放的。线程空间也是一个连续的空间，它的起始与结束地址分别存放在线程背景数据结构中。

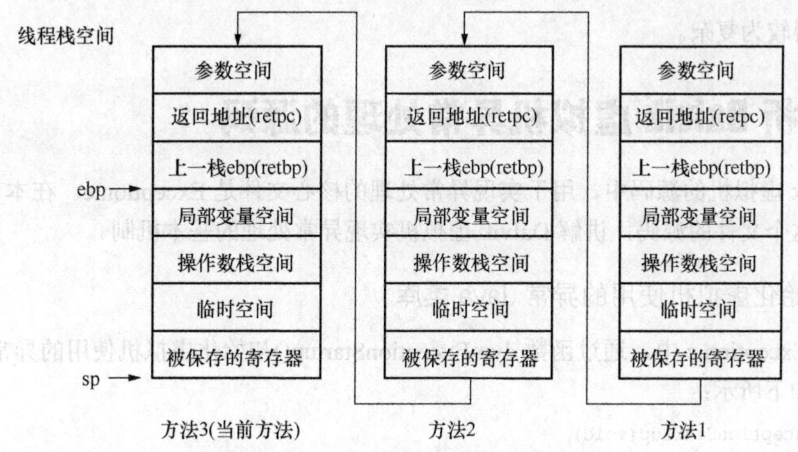

图 17-2 线程栈空间中方法栈存放的示意图

从图 17-2 可知，方法栈空间的连接及方法的当前执行点分别是由方法栈中的数据 retbp、retpc 建立的。只要能获取这两个数据就能实现异常在方法中的传播。对于内部异常，虚拟机直接调用异常处理程序执行异常处理；用户产生的异常则在及时编译器编译异常产生指令 athrow 时产生调用异常处理程序的机器指令。异常处理程序执行时，异常处理程序的方法栈为当前栈，异常处理程序可根据其第 1 个方法参数的存储位置，减去 8 字节偏移得到其调用者方法栈的基址指针 retbp，第 1 个方法参数的存储位置减去 4 字节偏移得到调用方法的返回地址 retpc，retpc-1 即为调用点的指令地址；同理，在调用方法的 retbp 处又可以得到该方法的调用者的栈基址及返回地址。采用这种方式可实现异常在方法调用栈中的传播。因此，及时编译异常处理时方法调用关系链表的数据结构如下所示：

```
MethodCallList{
  int retbp; //上一方法栈的基址指针
  int retpc; //调用方法的返回地址
};
```

及时编译时异常处理的另一个繁琐的过程是，如何确定调用点指令所属的类及方法。从以上的叙述中可知，retpc-1 为某方法调用其子方法的方法调用指令地址，应根据该地址信息去查询方法的异常处理表以获取异常处理句柄，但这一前提是如何先找到该方法，为了找到地址 retpc-1 所处的方法，不得不遍历所有的类及其方法，以判断该地址是否在某一方法代码内。显然这一方法较为费时，当然也可设计复杂的数据结构记录指令与方法的查询关系以简化这一过程，但这确实没有必要，毕竟异常产生的次数很少，即便产生异常，大多数情况下也要终止程序的执行，因此花费在这方面的时间可忽略不记。

在确定了异常如何传播及异常处理句柄如何查找后，下一步的任务就是如何调用异常处理句柄。在异常处理表中提供的只是异常句柄的地址，必须用汇编语言实现异常处理句柄的调用，调用程序如下所示，下段代码遵循的是汇编调用 C 子程序的规则。异常处理句柄 catch 的调用基本上与 C 函数的调用相同，异常对象可看成是它的方法参数，但不同的是该参数并不被压入参数空间，它是在执行 catch 的第一条指令时被压入指定的局部变量空间。在及时编译时一般都是以基址寄存器的地址为基准进行数据的访问，因此应恢复异常句柄所属方法的基址寄存器，最后是执行异常处理句柄。

```
movl eobj,%%eax                              //异常对象引用存入 eax
movl retbp,%%ebp                             //恢复调用者方法的基址
jmp*handler pc                               //调用异常处理句柄
```

本节讨论了及时编译执行时异常处理的关键技术，可以看出由于本地代码的存在，使得异常

处理过程变得较为复杂。

17.3 分析 Dalvik 虚拟机异常处理的源码

在 Dalvik 虚拟机的源码中，用于实现异常处理的核心文件是 Exception.c。在本节的内容中，将简要分析这个文件的源码，讲解 Dalvik 虚拟机实现异常处理的基本机制。

17.3.1　初始化虚拟机使用的异常 Java 类库

在文件 Exception.c 中，通过函数 dvmExceptionStartup()初始化虚拟机使用的异常 Java 类库。其实现源码如下所示：

```
bool dvmExceptionStartup(void)
{
    gDvm.classJavaLangThrowable =
        dvmFindSystemClassNoInit("Ljava/lang/Throwable;");
    gDvm.classJavaLangRuntimeException =
        dvmFindSystemClassNoInit("Ljava/lang/RuntimeException;");
    gDvm.classJavaLangError =
        dvmFindSystemClassNoInit("Ljava/lang/Error;");
    gDvm.classJavaLangStackTraceElement =
        dvmFindSystemClassNoInit("Ljava/lang/StackTraceElement;");
    gDvm.classJavaLangStackTraceElementArray =
        dvmFindArrayClass("[Ljava/lang/StackTraceElement;", NULL);
    if (gDvm.classJavaLangThrowable == NULL ||
        gDvm.classJavaLangStackTraceElement == NULL ||
        gDvm.classJavaLangStackTraceElementArray == NULL)
    {
        LOGE("Could not find one or more essential exception classes\n");
        return false;
    }
    /*
     * Find the constructor.  Note that, unlike other saved method lookups,
     * we're using a Method* instead of a vtable offset.  This is because
     * constructors don't have vtable offsets.  (Also, since we're creating
     * the object in question, it's impossible for anyone to sub-class it.)
     */
    Method* meth;
    meth = dvmFindDirectMethodByDescriptor(gDvm.classJavaLangStackTraceElement,
        "<init>", "(Ljava/lang/String;Ljava/lang/String;Ljava/lang/String;I)V");
    if (meth == NULL) {
        LOGE("Unable to find constructor for StackTraceElement\n");
        return false;
    }
    gDvm.methJavaLangStackTraceElement_init = meth;

    /* grab an offset for the stackData field */
    gDvm.offJavaLangThrowable_stackState =
        dvmFindFieldOffset(gDvm.classJavaLangThrowable,
            "stackState", "Ljava/lang/Object;");
    if (gDvm.offJavaLangThrowable_stackState < 0) {
        LOGE("Unable to find Throwable.stackState\n");
        return false;
    }
    /* and one for the message field, in case we want to show it */
    gDvm.offJavaLangThrowable_message =
        dvmFindFieldOffset(gDvm.classJavaLangThrowable,
            "detailMessage", "Ljava/lang/String;");
    if (gDvm.offJavaLangThrowable_message < 0) {
        LOGE("Unable to find Throwable.detailMessage\n");
        return false;
    }

    /* and one for the cause field, just 'cause */
```

```
    gDvm.offJavaLangThrowable_cause =
        dvmFindFieldOffset(gDvm.classJavaLangThrowable,
            "cause", "Ljava/lang/Throwable;");
    if (gDvm.offJavaLangThrowable_cause < 0) {
        LOGE("Unable to find Throwable.cause\n");
        return false;
    }
    return true;
}
```

17.3.2 抛出一个线程异常

在文件 Exception.c 中，通过函数 dvmThrowChainedException()创建一个抛出并抛出一个异常，在当前线程，抛出的正是设置线程的例外指针。如果有一个坏的异常层次正在抛出，如果初始化“丢失”，然后试图抛出一个异常将导致另一个例外情况。严重的是通常会允许“一连串”例外的情况，所以很难自动检测这一问题。因为这只发生在破碎的系统类中，所以不值得花周期检测。

函数 dvmThrowChainedException()的实现源码如下所示：

```
void dvmThrowChainedException(const char* exceptionDescriptor, const char* msg,
    Object* cause)
{
    ClassObject* excepClass;

    LOGV("THROW '%s' msg='%s' cause=%s\n",
        exceptionDescriptor, msg,
        (cause != NULL) ? cause->clazz->descriptor : "(none)");

    if (gDvm.initializing) {
        if (++gDvm.initExceptionCount >= 2) {
            LOGE("Too many exceptions during init (failed on '%s' '%s')\n",
                exceptionDescriptor, msg);
            dvmAbort();
        }
    }

    excepClass = dvmFindSystemClass(exceptionDescriptor);
    if (excepClass == NULL) {
        /*
         * We couldn't find the exception class.  The attempt to find a
         * nonexistent class should have raised an exception.  If no
         * exception is currently raised, then we're pretty clearly unable
         * to throw ANY sort of exception, and we need to pack it in.
         *
         * If we were able to throw the "class load failed" exception,
         * stick with that.  Ideally we'd stuff the original exception
         * into the "cause" field, but since we can't find it we can't
         * do that.  The exception class name should be in the "message"
         * field.
         */
        if (!dvmCheckException(dvmThreadSelf())) {
            LOGE("FATAL: unable to throw exception (failed on '%s' '%s')\n",
                exceptionDescriptor, msg);
            dvmAbort();
        }
        return;
    }
    dvmThrowChainedExceptionByClass(excepClass, msg, cause);
}
```

17.3.3 持续抛出进程

在文件 Exception.c 中，函数 dvmThrowChainedExceptionByClass()的功能是：如果当前有一个类的引用，则“开始/继续”抛出进程。其实现源码如下所示：

```
void dvmThrowChainedExceptionByClass(ClassObject* excepClass, const char* msg,
    Object* cause)
```

```
{
    Thread* self = dvmThreadSelf();
    Object* exception;
    /* make sure the exception is initialized */
    if (!dvmIsClassInitialized(excepClass) && !dvmInitClass(excepClass)) {
        LOGE("ERROR: unable to initialize exception class '%s'\n",
            excepClass->descriptor);
        if (strcmp(excepClass->descriptor, "Ljava/lang/InternalError;") == 0)
            dvmAbort();
        dvmThrowChainedException("Ljava/lang/InternalError;",
            "failed to init original exception class", cause);
        return;
    }
    exception = dvmAllocObject(excepClass, ALLOC_DEFAULT);
    if (exception == NULL) {
        /*
         * We're in a lot of trouble. We might be in the process of
         * throwing an out-of-memory exception, in which case the
         * pre-allocated object will have been thrown when our object alloc
         * failed. So long as there's an exception raised, return and
         * allow the system to try to recover. If not, something is broken
         * and we need to bail out.
         */
        if (dvmCheckException(self))
            goto bail;
        LOGE("FATAL: unable to allocate exception '%s' '%s'\n",
            excepClass->descriptor, msg != NULL ? msg : "(no msg)");
        dvmAbort();
    }
    /*
     * 初始化异常.
     */
    if (gDvm.optimizing) {
        /* need the exception object, but can't invoke interpreted code */
        LOGV("Skipping init of exception %s '%s'\n",
            excepClass->descriptor, msg);
    } else {
        assert(excepClass == exception->clazz);
        if (!initException(exception, msg, cause, self)) {
            /*
             * Whoops. If we can't initialize the exception, we can't use
             * it. If there's an exception already set, the constructor
             * probably threw an OutOfMemoryError.
             */
            if (!dvmCheckException(self)) {
                /*
                 * We're required to throw something, so we just
                 * throw the pre-constructed internal error.
                 */
                self->exception = gDvm.internalErrorObj;
            }
            goto bail;
        }
    }
    self->exception = exception;

bail:
    dvmReleaseTrackedAlloc(exception, self);
}
```

17.3.4 找出异常原因

在文件 Exception.c 中，函数 initException()的功能是使用虚线形式类描述符抛出异常名，并描述这一异常的发生原因。而函数 dvmThrowExceptionByClassWithClassMessage()和函数 dvmthrowexceptionwithmessagefromdescriptor()类似，但是它采取的是一个类的对象，而不是一个名字。这两个函数的实现源码如下所示：

```
void dvmThrowChainedExceptionWithClassMessage(const char* exceptionDescriptor,
    const char* messageDescriptor, Object* cause)
{
    char* message = dvmDescriptorToDot(messageDescriptor);
    dvmThrowChainedException(exceptionDescriptor, message, cause);
    free(message);
}
void dvmThrowExceptionByClassWithClassMessage(ClassObject* exceptionClass,
    const char* messageDescriptor)
{
    char* message = dvmDescriptorToName(messageDescriptor);
    dvmThrowExceptionByClass(exceptionClass, message);
    free(message);
}
```

17.3.5 找出异常原因

在文件 Exception.c 中，函数 initException()的功能是使用构造函数的方式初始化一个异常。如果初始化异常会导致另一个例外（例如，outofmemoryerror）被抛出，则返回一个错误。其实现源码如下所示：

```
static bool initException(Object* exception, const char* msg, Object* cause,
    Thread* self)
{
    enum {
        kInitUnknown,
        kInitNoarg,
        kInitMsg,
        kInitMsgThrow,
        kInitThrow
    } initKind = kInitUnknown;
    Method* initMethod = NULL;
    ClassObject* excepClass = exception->clazz;
    StringObject* msgStr = NULL;
    bool result = false;
    bool needInitCause = false;

    assert(self != NULL);
    assert(self->exception == NULL);

    /* if we have a message, create a String */
    if (msg == NULL)
        msgStr = NULL;
    else {
        msgStr = dvmCreateStringFromCstr(msg, ALLOC_DEFAULT);
        if (msgStr == NULL) {
            LOGW("Could not allocate message string \"%s\" while "
                    "throwing internal exception (%s)\n",
                    msg, excepClass->descriptor);
            goto bail;
        }
    }

    if (cause != NULL) {
        if (!dvmInstanceof(cause->clazz, gDvm.classJavaLangThrowable)) {
            LOGE("Tried to init exception with cause '%s'\n",
                cause->clazz->descriptor);
            dvmAbort();
        }
    }

    /*
     * The Throwable class has four public constructors:
     *  (1) Throwable()
     *  (2) Throwable(String message)
     *  (3) Throwable(String message, Throwable cause)  (added in 1.4)
     *  (4) Throwable(Throwable cause)                   (added in 1.4)
     *
```

```
     * The first two are part of the original design, and most exception
     * classes should support them. The third prototype was used by
     * individual exceptions. e.g. ClassNotFoundException added it in 1.2.
     * The general "cause" mechanism was added in 1.4. Some classes,
     * such as IllegalArgumentException, initially supported the first
     * two, but added the second two in a later release.
     *
     * Exceptions may be picky about how their "cause" field is initialized.
     * If you call ClassNotFoundException(String), it may choose to
     * initialize its "cause" field to null. Doing so prevents future
     * calls to Throwable.initCause().
     *
     * So, if "cause" is not NULL, we need to look for a constructor that
     * takes a throwable. If we can't find one, we fall back on calling
     * #1/#2 and making a separate call to initCause(). Passing a null ref
     * for "message" into Throwable(String, Throwable) is allowed, but we
     * prefer to use the Throwable-only version because it has different
     * behavior.
     *
     * java.lang.TypeNotPresentException is a strange case -- it has #3 but
     * not #2. (Some might argue that the constructor is actually not #3,
     * because it doesn't take the message string as an argument, but it
     * has the same effect and we can work with it here.)
     */
    if (cause == NULL) {
        if (msgStr == NULL) {
            initMethod = dvmFindDirectMethodByDescriptor(excepClass, "<init>", "()V");
            initKind = kInitNoarg;
        } else {
            initMethod = dvmFindDirectMethodByDescriptor(excepClass, "<init>",
                            "(Ljava/lang/String;)V");
            if (initMethod != NULL) {
                initKind = kInitMsg;
            } else {
                /* no #2, try #3 */
                initMethod = dvmFindDirectMethodByDescriptor(excepClass, "<init>",
                                "(Ljava/lang/String;Ljava/lang/Throwable;)V");
                if (initMethod != NULL)
                    initKind = kInitMsgThrow;
            }
        }
    } else {
        if (msgStr == NULL) {
            initMethod = dvmFindDirectMethodByDescriptor(excepClass, "<init>",
                            "(Ljava/lang/Throwable;)V");
            if (initMethod != NULL) {
                initKind = kInitThrow;
            } else {
                initMethod = dvmFindDirectMethodByDescriptor(excepClass, "<init>", "()V");
                initKind = kInitNoarg;
                needInitCause = true;
            }
        } else {
            initMethod = dvmFindDirectMethodByDescriptor(excepClass, "<init>",
                            "(Ljava/lang/String;Ljava/lang/Throwable;)V");
            if (initMethod != NULL) {
                initKind = kInitMsgThrow;
            } else {
                initMethod = dvmFindDirectMethodByDescriptor(excepClass, "<init>",
                                "(Ljava/lang/String;)V");
                initKind = kInitMsg;
                needInitCause = true;
            }
        }
    }

    if (initMethod == NULL) {
        /*
         * We can't find the desired constructor. This can happen if a
```

```
         * subclass of java/lang/Throwable doesn't define an expected
         * constructor, e.g. it doesn't provide one that takes a string
         * when a message has been provided.
         */
        LOGW("WARNING: exception class '%s' missing constructor "
            "(msg='%s' kind=%d)\n",
            excepClass->descriptor, msg, initKind);
        assert(strcmp(excepClass->descriptor,
                    "Ljava/lang/RuntimeException;") != 0);
        dvmThrowChainedException("Ljava/lang/RuntimeException;",
            "re-throw on exception class missing constructor", NULL);
        goto bail;
    }

    /*
     * Call the constructor with the appropriate arguments.
     */
    JValue unused;
    switch (initKind) {
    case kInitNoarg:
        LOGVV("+++ exc noarg (ic=%d)\n", needInitCause);
        dvmCallMethod(self, initMethod, exception, &unused);
        break;
    case kInitMsg:
        LOGVV("+++ exc msg (ic=%d)\n", needInitCause);
        dvmCallMethod(self, initMethod, exception, &unused, msgStr);
        break;
    case kInitThrow:
        LOGVV("+++ exc throw");
        assert(!needInitCause);
        dvmCallMethod(self, initMethod, exception, &unused, cause);
        break;
    case kInitMsgThrow:
        LOGVV("+++ exc msg+throw");
        assert(!needInitCause);
        dvmCallMethod(self, initMethod, exception, &unused, msgStr, cause);
        break;
    default:
        assert(false);
        goto bail;
    }

    /*
     * It's possible the constructor has thrown an exception.  If so, we
     * return an error and let our caller deal with it.
     */
    if (self->exception != NULL) {
        LOGW("Exception thrown (%s) while throwing internal exception (%s)\n",
            self->exception->clazz->descriptor, exception->clazz->descriptor);
        goto bail;
    }

    /*
     * If this exception was caused by another exception, and we weren't
     * able to find a cause-setting constructor, set the "cause" field
     * with an explicit call.
     */
    if (needInitCause) {
        Method* initCause;
        initCause = dvmFindVirtualMethodHierByDescriptor(excepClass, "initCause",
            "(Ljava/lang/Throwable;)Ljava/lang/Throwable;");
        if (initCause != NULL) {
            dvmCallMethod(self, initCause, exception, &unused, cause);
            if (self->exception != NULL) {
                /* initCause() threw an exception; return an error and
                 * let the caller deal with it.
                 */
                LOGW("Exception thrown (%s) during initCause() "
                    "of internal exception (%s)\n",
```

```
                self->exception->clazz->descriptor,
                exception->clazz->descriptor);
            goto bail;
        }
    } else {
        LOGW("WARNING: couldn't find initCause in '%s'\n",
            excepClass->descriptor);
    }
  }
  result = true;
bail:
  dvmReleaseTrackedAlloc((Object*) msgStr, self);    // NULL is ok
  return result;
}
```

17.3.6　清除挂起的异常和等待初始化的异常

在文件 Exception.c 中，函数 dvmClearOptException()的功能是清除挂起的异常和等待初始化的异常。在此使用了优化和验证码机制，如果运行中没有发生异常，则将“初始化”工作设置为避免进入“death-spin”模式。其实现源码如下所示：

```
void dvmClearOptException(Thread* self)
{
  self->exception = NULL;
  gDvm.initExceptionCount = 0;
}
```

17.3.7　包装“现在等待”异常的不同例外

在文件 Exception.c 中，函数 dvmWrapException()的功能是包装“现在等待”异常的不同例外。在此使用一个未经声明的方法来检查异常，一个异常（未检查的）和代替挂起的失败相关。其实现源码如下所示：

```
void dvmWrapException(const char* newExcepStr)
{
  Thread* self = dvmThreadSelf();
  Object* origExcep;
  ClassObject* iteClass;
  origExcep = dvmGetException(self);
  dvmAddTrackedAlloc(origExcep, self);    // don't let the GC free it
  dvmClearException(self);                // clear before class lookup
  iteClass = dvmFindSystemClass(newExcepStr);
  if (iteClass != NULL) {
    Object* iteExcep;
    Method* initMethod;
    iteExcep = dvmAllocObject(iteClass, ALLOC_DEFAULT);
    if (iteExcep != NULL) {
      initMethod = dvmFindDirectMethodByDescriptor(iteClass, "<init>",
          "(Ljava/lang/Throwable;)V");
      if (initMethod != NULL) {
        JValue unused;
        dvmCallMethod(self, initMethod, iteExcep, &unused,
          origExcep);
        /* if <init> succeeded, replace the old exception */
        if (!dvmCheckException(self))
          dvmSetException(self, iteExcep);
      }
      dvmReleaseTrackedAlloc(iteExcep, NULL);

      /* if initMethod doesn't exist, or failed... */
      if (!dvmCheckException(self))
        dvmSetException(self, origExcep);
    } else {
      /* leave OutOfMemoryError pending */
    }
  } else {
```

```
        /* leave ClassNotFoundException pending */
    }
    assert(dvmCheckException(self));
    dvmReleaseTrackedAlloc(origExcep, self);
}
```

17.3.8 输出跟踪当前异常的错误信息

在文件 Exception.c 中，通过函数 dvmPrintExceptionStackTrace()输出跟踪当前异常的错误信息，这是通过呼叫 JNI 异常描述实现的。其实现源码如下所示：

```
void dvmPrintExceptionStackTrace(void)
{
    Thread* self = dvmThreadSelf();
    Object* exception;
    Method* printMethod;

    exception = self->exception;
    if (exception == NULL)
        return;

    self->exception = NULL;
    printMethod = dvmFindVirtualMethodHierByDescriptor(exception->clazz,
            "printStackTrace", "()V");
    if (printMethod != NULL) {
        JValue unused;
        dvmCallMethod(self, printMethod, exception, &unused);
    } else {
        LOGW("WARNING: could not find printStackTrace in %s\n",
            exception->clazz->descriptor);
    }
    if (self->exception != NULL) {
        LOGI("NOTE: exception thrown while printing stack trace: %s\n",
            self->exception->clazz->descriptor);
    }
    self->exception = exception;
}
```

17.3.9 搜索和当前异常相匹配的方法

在文件 Exception.c 中，通过函数 findCatchInMethod()在方法列表中搜索和当前异常相匹配的方法。其实现源码如下所示：

```
static int findCatchInMethod(Thread* self, const Method* method, int relPc,
    ClassObject* excepClass)
{
    /*
     * Need to clear the exception before entry.  Otherwise, dvmResolveClass
     * might think somebody threw an exception while it was loading a class.
     */
    assert(!dvmCheckException(self));
    assert(!dvmIsNativeMethod(method));

    LOGVV("findCatchInMethod %s.%s excep=%s depth=%d\n",
        method->clazz->descriptor, method->name, excepClass->descriptor,
        dvmComputeExactFrameDepth(self->curFrame));

    DvmDex* pDvmDex = method->clazz->pDvmDex;
    const DexCode* pCode = dvmGetMethodCode(method);
    DexCatchIterator iterator;

    if (dexFindCatchHandler(&iterator, pCode, relPc)) {
        for (;;) {
            DexCatchHandler* handler = dexCatchIteratorNext(&iterator);

            if (handler == NULL) {
                break;
            }
```

```
        if (handler->typeIdx == kDexNoIndex) {
            /* catch-all */
            LOGV("Match on catch-all block at 0x%02x in %s.%s for %s\n",
                    relPc, method->clazz->descriptor,
                    method->name, excepClass->descriptor);
            return handler->address;
        }
        ClassObject* throwable =
            dvmDexGetResolvedClass(pDvmDex, handler->typeIdx);
        if (throwable == NULL) {
            /*
             * TODO: this behaves badly if we run off the stack
             * while trying to throw an exception.  The problem is
             * that, if we're in a class loaded by a class loader,
             * the call to dvmResolveClass has to ask the class
             * loader for help resolving any previously-unresolved
             * classes.  If this particular class loader hasn't
             * resolved StackOverflowError, it will call into
             * interpreted code, and blow up.
             *
             * We currently replace the previous exception with
             * the StackOverflowError, which means they won't be
             * catching it *unless* they explicitly catch
             * StackOverflowError, in which case we'll be unable
             * to resolve the class referred to by the "catch"
             * block.
             *
             * We end up getting a huge pile of warnings if we do
             * a simple synthetic test, because this method gets
             * called on every stack frame up the tree, and it
             * fails every time.
             *
             * This eventually bails out, effectively becoming an
             * uncatchable exception, so other than the flurry of
             * warnings it's not really a problem.  Still, we could
             * probably handle this better.
             */
            throwable = dvmResolveClass(method->clazz, handler->typeIdx,
                true);
            if (throwable == NULL) {
                /*
                 * We couldn't find the exception they wanted in
                 * our class files (or, perhaps, the stack blew up
                 * while we were querying a class loader). Cough
                 * up a warning, then move on to the next entry.
                 * Keep the exception status clear.
                 */
                LOGW("Could not resolve class ref'ed in exception "
                        "catch list (class index %d, exception %s)\n",
                        handler->typeIdx,
                        (self->exception != NULL) ?
                        self->exception->clazz->descriptor : "(none)");
                dvmClearException(self);
                continue;
            }
        }
        //LOGD("ADDR MATCH, check %s instanceof %s\n",
        //    excepClass->descriptor, pEntry->excepClass->descriptor);
        if (dvmInstanceof(excepClass, throwable)) {
            LOGV("Match on catch block at 0x%02x in %s.%s for %s\n",
                    relPc, method->clazz->descriptor,
                    method->name, excepClass->descriptor);
            return handler->address;
        }
    }
}
LOGV("No matching catch block at 0x%02x in %s for %s\n",
    relPc, method->name, excepClass->descriptor);
return -1;
}
```

17.3.10 获取匹配的捕获块

在文件 Exception.c 中，通过函数 dvmFindCatchBlock()获取匹配的捕获块。其实现源码如下所示：

```
int dvmFindCatchBlock(Thread* self, int relPc, Object* exception,
    bool scanOnly, void** newFrame)
{
    void* fp = self->curFrame;
    int catchAddr = -1;

    assert(!dvmCheckException(self));

    while (true) {
        StackSaveArea* saveArea = SAVEAREA_FROM_FP(fp);
        catchAddr = findCatchInMethod(self, saveArea->method, relPc,
                exception->clazz);
        if (catchAddr >= 0)
            break;

        /*
         * Normally we'd check for ACC_SYNCHRONIZED methods and unlock
         * them as we unroll.  Dalvik uses what amount to generated
         * "finally" blocks to take care of this for us.
         */

        /* output method profiling info */
        if (!scanOnly) {
            TRACE_METHOD_UNROLL(self, saveArea->method);
        }

        /*
         * Move up one frame.  If the next thing up is a break frame,
         * break out now so we're left unrolled to the last method frame.
         * We need to point there so we can roll up the JNI local refs
         * if this was a native method.
         */
        assert(saveArea->prevFrame != NULL);
        if (dvmIsBreakFrame(saveArea->prevFrame)) {
            if (!scanOnly)
                break;      // bail with catchAddr == -1

            /*
             * We're scanning for the debugger.  It needs to know if this
             * exception is going to be caught or not, and we need to figure
             * out if it will be caught *ever* not just between the current
             * position and the next break frame.  We can't tell what native
             * code is going to do, so we assume it never catches exceptions.
             *
             * Start by finding an interpreted code frame.
             */
            fp = saveArea->prevFrame;           // this is the break frame
            saveArea = SAVEAREA_FROM_FP(fp);
            fp = saveArea->prevFrame;           // this may be a good one
            while (fp != NULL) {
                if (!dvmIsBreakFrame(fp)) {
                    saveArea = SAVEAREA_FROM_FP(fp);
                    if (!dvmIsNativeMethod(saveArea->method))
                        break;
                }

                fp = SAVEAREA_FROM_FP(fp)->prevFrame;
            }
            if (fp == NULL)
                break;      // bail with catchAddr == -1

            /*
             * Now fp points to the "good" frame.  When the interp code
             * invoked the native code, it saved a copy of its current PC
             * into xtra.currentPc.  Pull it out of there.
             */
```

```
            relPc =
                saveArea->xtra.currentPc - SAVEAREA_FROM_FP(fp)->method->insns;
        } else {
            fp = saveArea->prevFrame;

            /* savedPc in was-current frame goes with method in now-current */
            relPc = saveArea->savedPc - SAVEAREA_FROM_FP(fp)->method->insns;
        }
    }

    if (!scanOnly)
        self->curFrame = fp;

    /*
     * The class resolution in findCatchInMethod() could cause an exception.
     * Clear it to be safe.
     */
    self->exception = NULL;

    *newFrame = fp;
    return catchAddr;
}
```

17.3.11　进行堆栈跟踪

在文件 Exception.c 中，通过函数 dvmFillInStackTraceInternal()对已经进行的异常进行堆栈跟踪。在许多情况下这一过程不会被检查，这使它保持在一个紧凑状态。当每次执行的时候，会清空原来栈内的 trace 信息。然后在当前的调用位置处重新建立 trace 信息。函数 dvmFillInStackTraceInternal()的实现源码如下所示：

```
void* dvmFillInStackTraceInternal(Thread* thread, bool wantObject, int* pCount)
{
    ArrayObject* stackData = NULL;
    int* simpleData = NULL;
    void* fp;
    void* startFp;
    int stackDepth;
    int* intPtr;
    if (pCount != NULL)
        *pCount = 0;
    fp = thread->curFrame;
    assert(thread == dvmThreadSelf() || dvmIsSuspended(thread));
    /*
     * We're looking at a stack frame for code running below a Throwable
     * constructor.  We want to remove the Throwable methods and the
     * superclass initializations so the user doesn't see them when they
     * read the stack dump.
     *
     * TODO: this just scrapes off the top layers of Throwable.  Might not do
     * the right thing if we create an exception object or cause a VM
     * exception while in a Throwable method.
     */
    while (fp != NULL) {
        const StackSaveArea* saveArea = SAVEAREA_FROM_FP(fp);
        const Method* method = saveArea->method;
        if (dvmIsBreakFrame(fp))
            break;
        if (!dvmInstanceof(method->clazz, gDvm.classJavaLangThrowable))
            break;
        //LOGD("EXCEP: ignoring %s.%s\n",
        //         method->clazz->descriptor, method->name);
        fp = saveArea->prevFrame;
    }
    startFp = fp;
    /*
     * Compute the stack depth.
     */
    stackDepth = 0;
```

```
    while (fp != NULL) {
        const StackSaveArea* saveArea = SAVEAREA_FROM_FP(fp);
        if (!dvmIsBreakFrame(fp))
            stackDepth++;
        assert(fp != saveArea->prevFrame);
        fp = saveArea->prevFrame;
    }
    //LOGD("EXCEP: stack depth is %d\n", stackDepth);
    if (!stackDepth)
        goto bail;
    /*
     * We need to store a pointer to the Method and the program counter.
     * We have 4-byte pointers, so we use '[I'.
     */
    if (wantObject) {
        assert(sizeof(Method*) == 4);
        stackData = dvmAllocPrimitiveArray('I', stackDepth*2, ALLOC_DEFAULT);
        if (stackData == NULL) {
            assert(dvmCheckException(dvmThreadSelf()));
            goto bail;
        }
        intPtr = (int*) stackData->contents;
    } else {
        /* array of ints; first entry is stack depth */
        assert(sizeof(Method*) == sizeof(int));
        simpleData = (int*) malloc(sizeof(int) * stackDepth*2);
        if (simpleData == NULL)
            goto bail;
        assert(pCount != NULL);
        intPtr = simpleData;
    }
    if (pCount != NULL)
        *pCount = stackDepth;
    fp = startFp;
    while (fp != NULL) {
        const StackSaveArea* saveArea = SAVEAREA_FROM_FP(fp);
        const Method* method = saveArea->method;
        if (!dvmIsBreakFrame(fp)) {
            //LOGD("EXCEP keeping %s.%s\n", method->clazz->descriptor,
            //         method->name);
            *intPtr++ = (int) method;
            if (dvmIsNativeMethod(method)) {
                *intPtr++ = 0;      /* no saved PC for native methods */
            } else {
                assert(saveArea->xtra.currentPc >= method->insns &&
                        saveArea->xtra.currentPc <
                        method->insns + dvmGetMethodInsnsSize(method));
                *intPtr++ = (int) (saveArea->xtra.currentPc - method->insns);
            }
            stackDepth--;       // for verification
        }
        assert(fp != saveArea->prevFrame);
        fp = saveArea->prevFrame;
    }
    assert(stackDepth == 0);
bail:
    if (wantObject) {
        dvmReleaseTrackedAlloc((Object*) stackData, dvmThreadSelf());
        return stackData;
    } else {
        return simpleData;
    }
}
```

17.3.12 生成堆栈跟踪元素

函数 dvmGetStackTraceRaw()的功能是通过调用原整数数据编码函数 dvmfillinstacktrace()生成堆栈跟踪元素。函数 dvmGetStackTraceRaw()的实现源码如下所示：

```
ArrayObject* dvmGetStackTraceRaw(const int* intVals, int stackDepth)
{
    ArrayObject* steArray = NULL;
    Object** stePtr;
    int i;
    /* init this if we haven't yet */
    if (!dvmIsClassInitialized(gDvm.classJavaLangStackTraceElement))
        dvmInitClass(gDvm.classJavaLangStackTraceElement);
    /* allocate a StackTraceElement array */
    steArray = dvmAllocArray(gDvm.classJavaLangStackTraceElementArray,
                    stackDepth, kObjectArrayRefWidth, ALLOC_DEFAULT);
    if (steArray == NULL)
        goto bail;
    stePtr = (Object**) steArray->contents;
    /*
     * Allocate and initialize a StackTraceElement for each stack frame.
     * We use the standard constructor to configure the object.
     */
    for (i = 0; i < stackDepth; i++) {
        Object* ste;
        Method* meth;
        StringObject* className;
        StringObject* methodName;
        StringObject* fileName;
        int lineNumber, pc;
        const char* sourceFile;
        char* dotName;
        ste = dvmAllocObject(gDvm.classJavaLangStackTraceElement,ALLOC_DEFAULT);
        if (ste == NULL)
            goto bail;
        meth = (Method*) *intVals++;
        pc = *intVals++;
        if (pc == -1)      // broken top frame?
            lineNumber = 0;
        else
            lineNumber = dvmLineNumFromPC(meth, pc);
        dotName = dvmDescriptorToDot(meth->clazz->descriptor);
        className = dvmCreateStringFromCstr(dotName, ALLOC_DEFAULT);
        free(dotName);
        methodName = dvmCreateStringFromCstr(meth->name, ALLOC_DEFAULT);
        sourceFile = dvmGetMethodSourceFile(meth);
        if (sourceFile != NULL)
            fileName = dvmCreateStringFromCstr(sourceFile, ALLOC_DEFAULT);
        else
            fileName = NULL;
        /*
         * Invoke:
         *  public StackTraceElement(String declaringClass, String methodName,
         *      String fileName, int lineNumber)
         * (where lineNumber==-2 means "native")
         */
        JValue unused;
        dvmCallMethod(dvmThreadSelf(), gDvm.methJavaLangStackTraceElement_init,
            ste, &unused, className, methodName, fileName, lineNumber);
        dvmReleaseTrackedAlloc(ste, NULL);
        dvmReleaseTrackedAlloc((Object*) className, NULL);
        dvmReleaseTrackedAlloc((Object*) methodName, NULL);
        dvmReleaseTrackedAlloc((Object*) fileName, NULL);
        if (dvmCheckException(dvmThreadSelf()))
            goto bail;
        *stePtr++ = ste;
    }
bail:
    dvmReleaseTrackedAlloc((Object*) steArray, NULL);
    return steArray;
}
```

17.3.13　将内容添加到堆栈跟踪日志中

在文件 Exception.c 中，函数 dvmLogRawStackTrace()的功能是将获取的异常信息内容添加到堆栈跟踪日志中。函数 dvmLogRawStackTrace()的实现源码如下所示：

```
void dvmLogRawStackTrace(const int* intVals, int stackDepth)
{
    int i;
    /*
     * Run through the array of stack frame data.
     */
    for (i = 0; i < stackDepth; i++) {
        Method* meth;
        int lineNumber, pc;
        const char* sourceFile;
        char* dotName;
        meth = (Method*) *intVals++;
        pc = *intVals++;
        if (pc == -1)      // broken top frame?
            lineNumber = 0;
        else
            lineNumber = dvmLineNumFromPC(meth, pc);
        // probably don't need to do this, but it looks nicer
        dotName = dvmDescriptorToDot(meth->clazz->descriptor);
        if (dvmIsNativeMethod(meth)) {
            LOGI("\tat %s.%s(Native Method)\n", dotName, meth->name);
        } else {
            LOGI("\tat %s.%s(%s:%d)\n",
                dotName, meth->name, dvmGetMethodSourceFile(meth),
                dvmLineNumFromPC(meth, pc));
        }
        free(dotName);
        sourceFile = dvmGetMethodSourceFile(meth);
    }
}
```

17.3.14　将内容添加到堆栈跟踪日志中

在文件 Exception.c 中，函数 logStackTraceOf()的功能是将异常日志信息直接打印输出为堆栈跟踪信息。函数 logStackTraceOf()的实现源码如下所示：

```
static void logStackTraceOf(Object* exception)
{
    const ArrayObject* stackData;
    StringObject* messageStr;
    int stackSize;
    const int* intVals;
    messageStr = (StringObject*) dvmGetFieldObject(exception,
                    gDvm.offJavaLangThrowable_message);
    if (messageStr != NULL) {
        char* cp = dvmCreateCstrFromString(messageStr);
        LOGI("%s: %s\n", exception->clazz->descriptor, cp);
        free(cp);
    } else {
        LOGI("%s:\n", exception->clazz->descriptor);
    }

    stackData = (const ArrayObject*) dvmGetFieldObject(exception,
                    gDvm.offJavaLangThrowable_stackState);
    if (stackData == NULL) {
        LOGI("  (no stack trace data found)\n");
        return;
    }

    stackSize = stackData->length / 2;
    intVals = (const int*) stackData->contents;

    dvmLogRawStackTrace(intVals, stackSize);
}
```

17.4 常见异常的类型与原因

因为 Android 应用程序是使用 Java 语言编写的，所以本节将详细讲解 Java 应用程序中的常见异常。在具体讲解之前，一定要知道有哪些常见的 Java 应用程序异常，并知道哪些原因可能会造成这个异常。这不仅需要程序管理人员在日常工作中注意积累，在必要的情况下还需要从其他渠道收集资料。笔者对此就进行一个分析，希望能够对各位程序开发人员有一定的帮助。

17.4.1 SQLException：操作数据库异常类

当前的 Java 应用程序大部分都是依赖于数据库运行的，如果 Java 应用程序与数据库进行沟通时产生了错误，这时就会触发这个类。同时，会将数据库的错误信息通过这个类显示给用户。也就是说，这个操作数据库异常类是数据库与用户之间异常信息传递的桥梁。如现在用户往系统中插入数据，而在数据库中规定某个字段必须唯一。当用户插入数据的时候，如果这个字段的值与现有的纪录重复了，违反了数据库的唯一性约束，此时数据库就会跑出一个异常信息。这个信息一般用户可能看不到，因为其发生在数据库层面的。此时这个操作数据库异常类就会捕捉到数据库的这个异常信息，并将这个异常信息传递到前台。这样前台用户就可以根据这个异常信息来分析发生错误的原因。这就是这个操作数据库异常类的主要用途。在 Java 应用程序中，所有数据库操作发生异常时，都会触发这一个类。此时 Java 应用程序本身的所有提示信息往往过于笼统，只是说与数据库交互出现错误，没有多大的参考价值。此时反而是数据库的提示信息更加有使用价值。

17.4.2 ClassCastException：数据类型转换异常

在 Java 应用程序中，有时候需要对数据类型进行转换，这个转换包括显式的转换与隐式的转换。但是无论怎么转换，都必须要符合“数据类型的兼容性”这个前提条件。如果在数据转换的过程中违反了这个原则，那么就会触发数据类型转换异常。例如在应用程序中，开发人员需要将一个字符型的日期数据转换为数据库所能够接受的日期型数据，此时只需要在前台应用程序中进行控制，这时一般不会有什么问题。但是，如果前台应用程序缺乏相关的控制，如用户在输入日期的时候只输入月、日信息，而没有年份的信息。此时应用程序在进行数据类型转换的时候，就会出现异常。根据笔者的经验，数据类型转换异常在应用程序开发中是一个出现的比较多的异常，也是一个比较低级的异常。因为大部分情况下，都可以在应用程序窗口中对数据类型进行一些强制的控制。即在数据类型进行转换之前，就保证数据类型的兼容性。这样就不容易造成数据类型的转换异常。如在只允许数值类型的字段中，可以设置不允许用户输入数值以外的字符。虽然说有了异常处理机制，可以保证应用程序不会被错误地运行。但是在实际开发中需要尽可能多地预见错误发生的原因，应该尽量避免发生异常。

17.4.3 NumberFormatException：字符串转换为数字类型时抛出的异常

在数据类型转换过程中，如果是字符型转换为数字型过程中出现的问题，对于这个异常，在 Java 程序中采用了一个独立的异常，即 NumberFormatException。如现在将字符型的数据“123456”转换为数值型数据时，是允许的。但是如果字符型数据中包含了非数字型的字符，如 123#56，此时转换为数值型时就会出现异常。系统就会捕捉到这个异常，并进行处理。

Java 应用程序中常见的异常类还有很多，如未找到相应类异常、不允许访问某些类异常、文件已经结束异常、文件未找到异常、字段未找到异常等。一般系统开发人员都可以根据这个异常名来判断当前异常的类型。虽然不错，但是好记性不如烂笔头。程序开发人员在必要的时候（特别是存在自定义异常的时候），最后手头有一份异常明细表。这样，无论是应用程序在调试过程中发现问题，还是运行过程中接到用户的投诉，都可以及时根据异常名字找到异常发生的原因，从而可以在最短时间内解决异常，恢复应用程序的正常运行。

17.5 调用堆栈跟踪分析异常

和其他桌面虚拟机一样，当 Dalvik 虚拟机收到 SIGQUIT（Ctrl-\ 或者 kill -3）指令时，会为所有的 dump 生成堆栈追踪信息，这些信息会被默认写入 Android 系统中的 log（日志）中，或被写入一个文件中。在处理 Dalvik VM 异常信息时，可以通过调用堆栈跟踪信息的方法来分析并解决异常。

17.5.1 解决段错误

在 Android 开发过程中，经常会遇到段错误，也就是 SIGSEGV(11)错误。此时 libc 的 backtrace 会打印出对应的堆栈信息，这些信息是由一连串数字构成的。例如下面的堆栈信息。

```
ActivityManager( 1105): Displayed activity com.android.browser/.BrowserActivity: 2460 ms (total
2460 ms)
I/DEBUG   (13002): *** *** *** *** *** *** *** *** *** *** *** *** *** *** *** ***
I/DEBUG   (13002): Build fingerprint: 'unknown'
I/DEBUG   (13002): pid: 20363, tid: 20375  >>> com.android.browser <<<
I/DEBUG   (13002): signal 11 (SIGSEGV), fault addr ffc00000
I/DEBUG   (13002):  r0 059fc2a0  r1 4a3bcef8  r2 e59fc2a0  r3 4a3bcc58
I/DEBUG   (13002):  r4 4a3bc101  r5 4ebe0a3c  r6 4a3bc120  r7 012fff10
I/DEBUG   (13002):  r8 500de101  r9 500ee12d  10 a87dfb20  fp 4ebe58e0
I/DEBUG   (13002):  ip ffc00000  sp 4ebe0a30  lr 4a3bcc58  pc a862f3a0  cpsr 00000030
I/DEBUG   (13002):  d0  0000001100000011  d1  0000001100000011
I/DEBUG   (13002):  d2  0000001100000011  d3  0000001100000011
I/DEBUG   (13002):  d4  0000001100000011  d5  0000001100000011
I/DEBUG   (13002):  d6  0000001100000011  d7  4060000000000080
I/DEBUG   (13002):  d8  41d3d1762e40d70a  d9  41d3d1762e440a3d
I/DEBUG   (13002):  d10 0000000000000000  d11 0000000000000000
I/DEBUG   (13002):  d12 0000000000000000  d13 0000000000000000
I/DEBUG   (13002):  d14 0000000000000000  d15 0000000000000000
I/DEBUG   (13002):  d16 3ff0000000000000  d17 3ff0000000000000
I/DEBUG   (13002):  d18 40cd268000000000  d19 3f3b9cc1b0bac000
I/DEBUG   (13002):  d20 3ff0000000000000  d21 8000000000000000
I/DEBUG   (13002):  d22 0000000000000000  d23 0000000000000000
I/DEBUG   (13002):  d24 3ff0000000000000  d25 0000000000000000
I/DEBUG   (13002):  d26 0000000000000000  d27 0000000000000000
I/DEBUG   (13002):  d28 0000000000000000  d29 3ff0000000000000
I/DEBUG   (13002):  d30 0000000000000000  d31 3ff0000000000000
I/DEBUG   (13002):  scr 60000013
I/DEBUG   (13002):
I/DEBUG   (13002):          #00  pc 0032f3a0  /system/lib/libwebcore.so
I/DEBUG   (13002):          #01  pc 003243b0  /system/lib/libwebcore.so
I/DEBUG   (13002):          #02  pc 003167b2  /system/lib/libwebcore.so
I/DEBUG   (13002):          #03  pc 0038f2de  /system/lib/libwebcore.so
I/DEBUG   (13002):          #04  pc 0038f416  /system/lib/libwebcore.so
I/DEBUG   (13002):          #05  pc 0030d392  /system/lib/libwebcore.so
I/DEBUG   (13002):          #06  pc 003796e2  /system/lib/libwebcore.so
I/DEBUG   (13002):          #07  pc 0038e36a  /system/lib/libwebcore.so
I/DEBUG   (13002):          #08  pc 003189f0  /system/lib/libwebcore.so
I/DEBUG   (13002):          #09  pc 00377f82  /system/lib/libwebcore.so
I/DEBUG   (13002):          #10  pc 0037ae0c  /system/lib/libwebcore.so
I/DEBUG   (13002):          #11  pc 0038e254  /system/lib/libwebcore.so
I/DEBUG   (13002):          #12  pc 003189f0  /system/lib/libwebcore.so
I/DEBUG   (13002):          #13  pc 0031cf2c  /system/lib/libwebcore.so
I/DEBUG   (13002):          #14  pc 0038e52a  /system/lib/libwebcore.so
I/DEBUG   (13002):          #15  pc 0038c2d0  /system/lib/libwebcore.so
I/DEBUG   (13002):          #16  pc 0031cf76  /system/lib/libwebcore.so
I/DEBUG   (13002):          #17  pc 0038e546  /system/lib/libwebcore.so
I/DEBUG   (13002):          #18  pc 003189f0  /system/lib/libwebcore.so
I/DEBUG   (13002):          #19  pc 0031ca40  /system/lib/libwebcore.so
I/DEBUG   (13002):          #20  pc 0038e3be  /system/lib/libwebcore.so
I/DEBUG   (13002):          #21  pc 0038c2d0  /system/lib/libwebcore.so
```

```
I/DEBUG   (13002):           #22  pc 0031cf76  /system/lib/libwebcore.so
I/DEBUG   (13002):           #23  pc 0038e546  /system/lib/libwebcore.so
I/DEBUG   (13002):           #24  pc 0038c2d0  /system/lib/libwebcore.so
I/DEBUG   (13002):           #25  pc 00379054  /system/lib/libwebcore.so
I/DEBUG   (13002):           #26  pc 0031d254  /system/lib/libwebcore.so
I/DEBUG   (13002):           #27  pc 0030d5d6  /system/lib/libwebcore.so
I/DEBUG   (13002):           #28  pc 0030d7d2  /system/lib/libwebcore.so
I/DEBUG   (13002):           #29  pc 0031e354  /system/lib/libwebcore.so
I/DEBUG   (13002):           #30  pc 0034ab3c  /system/lib/libwebcore.so
I/DEBUG   (13002):
I/DEBUG   (13002): code around pc:
I/DEBUG   (13002): a862f380 469e4694 cc04f853 0e04f1a3 510cea4f
I/DEBUG   (13002): a862f390 f41c0d09 bf080f00 44714249 c008f8d1
I/DEBUG   (13002): a862f3a0 e000f8dc 0c1ff10e bf0842b8 2d04f853
I/DEBUG   (13002): a862f3b0 0d010510 0f00f412 4249bf08 f8c2185a
I/DEBUG   (13002): a862f3c0 e006c008 d1042b0c 99019b05 18426818
I/DEBUG   (13002):
I/DEBUG   (13002): code around lr:
I/DEBUG   (13002): 4a3bcc38 e58d0000 e49d0004 e598200b e582002f
I/DEBUG   (13002): 4a3bcc48 e52d0004 e3100001 0a000018 e3a03030
I/DEBUG   (13002): 4a3bcc58 e59fc2a0 e002100c e59fc29c e151000c
I/DEBUG   (13002): 4a3bcc68 0a000012 e59fc294 e002100c e0813003
I/DEBUG   (13002): 4a3bcc78 e1a03123 e1c2200c e3530b02 ba000004
I/DEBUG   (13002):
I/DEBUG   (13002): stack:
I/DEBUG   (13002):     4ebe09f0  50bfd848
I/DEBUG   (13002):     4ebe09f4  50bfd858
I/DEBUG   (13002):     4ebe09f8  50bfd834
I/DEBUG   (13002):     4ebe09fc  afd19a05  /system/lib/libc.so
I/DEBUG   (13002):     4ebe0a00  50bd3264
I/DEBUG   (13002):     4ebe0a04  a86510ef  /system/lib/libwebcore.so
I/DEBUG   (13002):     4ebe0a08  00000004
I/DEBUG   (13002):     4ebe0a0c  50bfd854
I/DEBUG   (13002):     4ebe0a10  002ece20  [heap]
I/DEBUG   (13002):     4ebe0a14  4a3ba000
I/DEBUG   (13002):     4ebe0a18  4ebe0a3c
I/DEBUG   (13002):     4ebe0a1c  4ebe0a3c
I/DEBUG   (13002):     4ebe0a20  4a3bc101
I/DEBUG   (13002):     4ebe0a24  4ebe0a3c
I/DEBUG   (13002):     4ebe0a28  df002777
I/DEBUG   (13002):     4ebe0a2c  e3a070ad
I/DEBUG   (13002): #00 4ebe0a30  002ece20  [heap]
I/DEBUG   (13002):     4ebe0a34  49f627d0
I/DEBUG   (13002):     4ebe0a38  a87d63c0  /system/lib/libwebcore.so
I/DEBUG   (13002):     4ebe0a3c  4a3bd0e7
I/DEBUG   (13002):     4ebe0a40  4a3bd0b8
I/DEBUG   (13002):     4ebe0a44  4a3bcc58
I/DEBUG   (13002):     4ebe0a48  00000003
I/DEBUG   (13002):     4ebe0a4c  00000000
I/DEBUG   (13002):     4ebe0a50  00001100
I/DEBUG   (13002):     4ebe0a54  0000001f
I/DEBUG   (13002):     4ebe0a58  00001074
I/DEBUG   (13002):     4ebe0a5c  4ebe0b04
I/DEBUG   (13002):     4ebe0a60  a87d63c0  /system/lib/libwebcore.so
I/DEBUG   (13002):     4ebe0a64  4ebe0acc
I/DEBUG   (13002):     4ebe0a68  4a3bc101
I/DEBUG   (13002):     4ebe0a6c  a86243b5  /system/lib/libwebcore.so
I/DEBUG   (13002): #01 4ebe0a70  4ebe0b38
I/DEBUG   (13002):     4ebe0a74  00000064
I/DEBUG   (13002):     4ebe0a78  003f0914  [heap]
I/DEBUG   (13002):     4ebe0a7c  fffffc00
I/DEBUG   (13002):     4ebe0a80  50bfd834
I/DEBUG   (13002):     4ebe0a84  a87d63c0  /system/lib/libwebcore.so
I/DEBUG   (13002):     4ebe0a88  4ebe0b38
I/DEBUG   (13002):     4ebe0a8c  4ebe0b04
I/DEBUG   (13002):     4ebe0a90  4ebe0acc
I/DEBUG   (13002):     4ebe0a94  a86167b7  /system/lib/libwebcore.so
```

但是对于很多底层开发用户来说，因为开发板经常剥离各种 lib，所以不会显示上述符号信息，这样更是无法跟踪分析具体原因了。在这个时候，可以通过编译时生成的库来获取对应的符号信息。编译器提供了相应的 addr2line 工具，此工具的全名为 arm-eabi-addr2line，可以在对应板子源码目录找到这个工具。

通过对上面堆栈信息的分析，可以看出库 system/lib/libwebcore.so 出现了断错误，可以将其 pull（拖）下来正逐行进行分析。例如拖到桌面中的命令为：

```
arm-eabi-addr2line -f -e ~/桌面/libwebcore.so 0038f2de
```

此时就可以简单地查找具体原因了，这种分析法同样适用于分析使用 JNI 开发的库。

17.5.2 跟踪 Android Callback 调用堆栈

在调试 Android 系统时，可以通过打印调用堆栈 Callback Stack（回调栈）来分析和解决 Android 问题。

1. Java 应用输出堆栈信息

要想在 Java 应用输出 Callback Stack 信息，可以借助 catch exception 语句，并使用 Log.w（LOGTAG, Log.getStackTraceString（throwable））来输出调用堆栈信息，例如下面的代码。

```
Throwable throwable = new Throwable();
    Log.w(LOGTAG, Log.getStackTraceString(throwable));
```

或者：

```
try {
   wait();
} catch (InterruptedException e) {
   Log.e(LOGTAG, "Caught exception while waiting for overrideUrl");
   Log.e(LOGTAG, Log.getStackTraceString(e));
}
```

2. Android 底层开发过程获取堆栈信息

如果是在 C/C++应用，通常可以通过 segment fault 等错误即信号 SIGSEGV(11) 做出相应处理。也就是设置 SIGSEGV 的 handler 调用 libc 的 backtrace，这样就可以输出对应的 callback stack，定位得到问题所在。但是在 Android 系统中，bionic 不能提供上述类似的功能，而且只能通过 logcat 才能看到 log 信息。其实完全可以根据 Android 的出错信息获得调用堆栈信息，例如下面的出错信息。

```
    D/CallStack( 2029): #00  pc 00008156  /system/lib/hw/audio.primary.tf4.so
    D/CallStack( 2029): #01  pc 000089e8  /system/lib/hw/audio.primary.tf4.so (android_audio_legacy::
AudioHardware::AudioStreamOutALSA::setParameters(android::String const&)+139)
    D/CallStack( 2029): #02  pc 0000b2ca  /system/lib/hw/audio.primary.tf4.so
    D/CallStack( 2029): #03  pc 0003ac6a  /system/lib/libaudioflinger.so (android::AudioFlinger::
MixerThread::checkForNewParameters_l()+377)
    D/CallStack( 2029): #04  pc 0003960a  /system/lib/libaudioflinger.so (android::AudioFlinger::
PlaybackThread::threadLoop()+145)
    D/CallStack( 2029): #05 pc 00011264 /system/lib/libutils.so (android::Thread::_threadLoop(void*)+111)
    D/CallStack( 2029): #06  pc 00010dca  /system/lib/libutils.so
    D/CallStack( 2029): #07  pc 0000e3f8  /system/lib/libc.so (__thread_entry+72)
    D/CallStack( 2029): #08  pc 0000dae4  /system/lib/libc.so (pthread_create+160)
```

在 Android 底层开发应用中，可以通过如下所示的方式获取堆栈信息。

（1）arm-linux-addr2line。

使用 arm-linux-addr2line 获得调用堆栈信息的输出。

```
arm-eabi-addr2line -C -f -e symbols/system/lib/*.so addr
```

（2）ndk-stack 工具。

也可以使用 ndk-stack 工具保存出错 log 为 logcat.log 的方式也可以输出调用堆栈信息。

```
cat logcat..log | ndk-stack -sym ~/[SOURCE-DIR]/out/target/product/[PROJECT]/symbols/system/lib/
```

（3）panic.py 脚本。

也可以使用 panic.py 脚本分析来打印调用堆栈：

```
/panic.py logcat.log
```

此时 logcat 必须被转换成如下所示的格式。

```
D/CallStack( 2029): #00  pc 00008156  /system/lib/hw/audio.primary.tf4.so
D/CallStack( 2029): #01  pc 000089e8  /system/lib/hw/audio.primary.tf4.so
D/CallStack( 2029): #02  pc 0000b2ca  /system/lib/hw/audio.primary.tf4.so
D/CallStack( 2029): #03  pc 0003ac6a  /system/lib/libaudioflinger.so
D/CallStack( 2029): #04  pc 0003960a  /system/lib/libaudioflinger.so
D/CallStack( 2029): #05  pc 00011264  /system/lib/libutils.so
D/CallStack( 2029): #06  pc 00010dca  /system/lib/libutils.so
D/CallStack( 2029): #07  pc 0000e3f8  /system/lib/libc.so
D/CallStack( 2029): #08  pc 0000dae4  /system/lib/libc.so
```

此时便可以执行脚本打印出定位信息：

```
./panic.py setincallpath_l.txt,
```

打印出的信息如下所示：

```
w@w:/mmm/JellyBean-4.3.1/trunk/out/target/product/tf4$ ./panic.py ./backtrack/setincall_path.txt
    read file ok
    AudioHardware.cpp:829 android_audio_legacy::AudioHardware::setIncallPath_l(unsigned int)
    AudioHardware.cpp:1537
android_audio_legacy::AudioHardware::AudioStreamOutALSA::setParameters(android::String8
const&)
    audio_hw_hal.cpp:197  out_set_parametersAudioFlinger.cpp:3535  android::AudioFlinger::MixerThread::
checkForNewParameters_l()
    AudioFlinger.cpp:2586          android::AudioFlinger::PlaybackThread::threadLoop()
    Threads.cpp:793                android::Thread::_threadLoop(void*)
    Threads.cpp:132                thread_data_t::trampoline(thread_data_t const*)
    pthread.c:204                  __thread_entry
    pthread.c:348                 pthread_create
```

（4）python 脚本。

Google 公司为开发人员提供了一个 python 脚本，读者可以从如下地址得到这个 python 脚本。

```
http://code.google.com/p/android-ndk-stacktrace-analyzer/
```

然后可以使用 adb logcat -d > logfile 导 crash 的 log，并使用“build/prebuilt/linux-x86/arm-eabi-4.3.1/”目录中的 binarm-eabi-objdump 把 so 或 exe 转换成汇编代码，例如：

```
arm-eabi-objdump -S mylib.so > mylib.asm,
```

然后使用下面的脚本命令：

```
python parse_stack.py <asm-file> <logcat-file>
```

接下来设置 panic.py 的环境，具体代码如下所示，这样也可以获取堆栈信息。

```
#!/usr/bin/python
# stack symbol parser
import os
import string
import sys
#define android product name
#ANDROID_PRODUCT_NAME = 'generic'
ANDROID_PRODUCT_NAME = 'ok'
ANDROID_WORKSPACE = os.getcwd()+"/"
# addr2line tool path and symbol path
addr2line_tool = 'arm-linux-addr2line'
symbol_dir = ANDROID_WORKSPACE + '/symbols'
symbol_bin = symbol_dir + '/system/bin/'
symbol_lib = symbol_dir + '/system/lib/'
    class ReadLog:
        def __init__(self,filename):
            self.logname = filename
        def parse(self):
```

```
        f = file(self.logname,'r')
        lines = f.readlines()
        if lines != []:
            print 'read file ok'
        else:
            print 'read file failed'
        result =[]
        for line in lines:
            if line.find('stack') != -1:
                print 'stop search'
                break
            elif line.find('system') != -1:
                #print 'find one item' + line
                result.append(line)
        return result

class ParseContent:
    def __init__(self,addr,lib):
        self.address = addr # pc address
        self.exename = lib  # executable or shared library
    def addr2line(self):
        cmd = addr2line_tool + " -C -f -s -e " + symbol_dir + self.exename + " " + self.address
        #print cmd
        stream = os.popen(cmd)
        lines = stream.readlines();
        list = map(string.strip,lines)
        return list

inputarg = sys.argv
if len(inputarg) < 2:
    print 'Please input panic log'
    exit()

filename = inputarg[1]
readlog = ReadLog(filename)
inputlist = readlog.parse()

for item in inputlist:
    itemsplit = item.split()
    test = ParseContent(itemsplit[-2],itemsplit[-1])
    list = test.addr2line()
    print "%-30s%s" % (list[1],list[0])
```

在前面介绍的都是程序出错的情形，其实除了上述系统主动输出出错信息之外，还可以通过代码在系统不出错的情况下输出调用信息，然后通过 panic.py 打印调用堆栈。具体方法是在 cpp 文件中添加如下代码。

```
#include <utils/CallStack.h>
...
status_t AudioHardware::setIncallPath_l(uint32_t device) {
  ...
  #ifdef _ARM_
      android::CallStack stack;
      stack.update(1, 100);
      stack.dump("");
  #endif
  ...
}
```

在 Android.mk 中加入如下所示的代码。

```
LOCAL_CFLAGS += -D_ARM_
LOCAL_SHARED_LIBRARIES += libutils
```

此时就可以输出上面所描述的调用信息，这样能够便于开发人员分析代码和 debug，以便实现问题定位的功能。

第 18 章　JIT 编译

从 Android 2.2 版本开始，在 Dalvik VM 中新增了 JIT 编译器，此编译器能提高 Android 上的 Java 程序的性能。在本章的内容中，将详细讲解 JIT 编译器的基本知识。

18.1 JIT 简介

谷歌声称：自从 Android 虚拟机 Dalvik 使用了 JIT 技术后，使其运行速度快了 5 倍。在本节的内容中，将简要介绍 JIT 技术的基本知识，为读者步入本书后面知识的学习打下基础。

18.1.1 JIT 概述

Dalvik 解释并执行程序，JIT 技术主要是对多次运行的代码进行编译，当再次调用时使用编译之后的机器码，而不是每次都解释，以节约时间。5 倍是测试程序测出的值，并不是说程序运行速度也能达到 5 倍，这是因为测试程序有很多的重复调用和循环，而一般程序主要是顺序执行的，而且它是一边运行，一边编译，一开始的时候提速不多，所以真正运行程序速度提高不是特别明显。

JIT 的全称是 just-in-time compilation，是对代码的动态编译/翻译。JIT 编译器是一个 tracing JIT（也叫 trace-based JIT），主要以“trace”为单位来决定要编译的内容；目前也同时支持以整个方法为单位的编译。在程序运行的时候才将某种形式的源翻译为目标代码，“源”可能是高级语言的文本形式的源码，不过更常见的是字节码或者说虚拟机指令形式的；而目标代码一般是实际机器的指令。

JIT 与传统的静态编译器最大的不同在于，前者是在用户程序运行过程中进行编译的，而传统静态编译器则是在用户程序运行之前先完成编译。为此在许多取舍上两者都有所不同。JIT 能够承受开销较小的编译工作，而传统静态编译器可以看作能够承受无限的编译开销。早期的 JIT 编译器受到编译开销的限制，就只能生成质量一般的代码了。

现在许多动态编译器已经算不上原本意义上的“JIT”了——“just-in-time”是指“刚好赶上”，一般就是说在某个函数/方法（也可能是别的编译单元）初次执行的时候就对它进行编译，生成的目标代码刚好赶上该函数/方法的执行。而现在的一些混合执行模式的虚拟机/动态编译系统中，函数/方法或许一开始是在解释器里执行的，等一阵子才被动态编译，按原本“JIT”的意思就已经太迟了，不是“刚好赶上”，不过 JIT 已经更多地成为一种惯用称谓，也就不必那么在意这个小细节了。

JIT 编译器是一个连续体，一端是编译速度非常快，但只能生成质量一般的代码的编译器；另一端是编译速度较慢，而生成高度优化代码的编译器。

其中可以分为许多小类别，最简单的一类快速 JIT 编译器可以是基于模板的，也就是每个字节码对应一个固定的目标代码模式。所谓“编译”是指在这种 JIT 编译器中，简单地根据源程序把预置的模板串在一起的过程。整个过程基本上不对代码做分析工作，也不做优化工作。而其实解释器也是一个连续体，从最简单最慢的行解释器到相对高速的 context-threading、inline-threading 等。高速的解释器与基于模板的 JIT 编译器正好接上，都会在运行时生成新的目标代码用于执行用户程序。因此，如果把视野再放开一点，从解释器到动态编译器其实也构成一个连续体。解释器与动态编译器的分水岭，笔者的观点是，在于用户程序中的指令分派（instruction dispatch）是用软件来做还是直接由硬件实现。一个解释器如果优化到完全消除了用户程序的指令分派在软件一侧的开销，就可以称之为编译系统了。

在一个从 Java 源码编译到 JVM 字节码的编译器（如 javac、ECJ）的过程中，一个“编译单元”（Compilation Unit）指的是一个 Java 源文件。而在 Dalvik VM 的 JIT 中也有一个名为“CompilationUnit”的结构体，这个千万不能跟 Java 源码级的编译单元弄混了——它在这里指的就是一个“trace”。

许多早期的 JIT 编译器以“函数”或者“方法”为单位进行编译，并通过函数/方法内联来降低调用成本、扩大优化的作用域。但一个函数/方法中也可能存在热路径与冷路径的区别，如果以函数/方法为粒度来编译，很可能会在冷路径上浪费了编译的时间和空间，却没有得到执行速度的提升。为此，许多 JIT 编译器会记录方法内分支的执行频率，在 JIT 编译时只对热路径编译，将冷路径生成为“uncommon trap”，等真的执行到冷路径时跳回到解释器或其他备用实行方式继续。

Tracing JIT 能够更简单有效地获取到涉及循环的热代码中的执行路径，该编译器的中间表示分为两种，分别是 MIR（middle-level intermediate representation）与 LIR（low-level intermediate representation）。MIR 与 LIR 节点各自形成链表，分别被组织在 BasicBlock 与 ConpilationUnit 中。具体编译流程如下所示。

（1）创建 CompilationUnit 对象来存放一次编译中需要的信息。

（2）将 dex 文件中的 Dalvik 字节码解码为 DecodedInstruction，并创建对应的 MIR 节点。

（3）定位基本块的边界，并创建相应的 BasicBlock 对象，将 MIR 塞进去。

（4）确定控制流关系，将基本块连接起来构成控制流图（CFG），并添加恢复解释器状态和异常处理用的基本块。

（5）将基本块都加到 CompilationUnit 里去。

（6）将 MIR 转换为 LIR（带有局部优化和全局优化）。

（7）从 LIR 生成机器码。

从上述编译流程来看，真正与 CPU 架构相关的是第（5）～（7）个步骤，因为 LIR 是与 CPU 密切相关的汇编级指令集。而（1）～（4）则是 dex 字节码到 MIR 的转换，可以视为 JIT 中与 CPU 架构无关的部分。因此，将 JIT 移植到一个新的架构上去，应该模拟实现一套 LIR 指令集，以及重新现实对 LIR 指令进行处理的函数。这里，工作量集中体现在“compiler/codegen/TARGET_ARCH/”，可以从 MIPS 代码中看到移植该部分的工作量（C 代码）。另外，从上述分析的相关的汇编也是工作量的一部分。

Dalvik 虚拟机运行流程（含 JIT）如图 18-1 所示。

要想让 Java 程序能够在 Android 上运行，在由源码编译为 JVM 字节码之后，还需要经过 dx 的处理，从 JVM 字节码转换为 Dalvik VM 字节码。dx 在转换过程中已经做过一些优化了，所以可以理解为什么 Dalvik VM 中的 JIT 在 MIR 层面基本上没做优化。

不过现在 Dalvik VM 的 JIT 的完成度还很低，局部优化只有冗余 load/store 消除，全局优化只有冗余分支消除。dx 本身并没有做像是循环不变量外提之类的优化，因此就算有了 JIT，Dalvik 生成出来的代码质量也不会很好，目前能看到的最明显的效果只是把解释器中指令分派的开销给消除掉了而已。

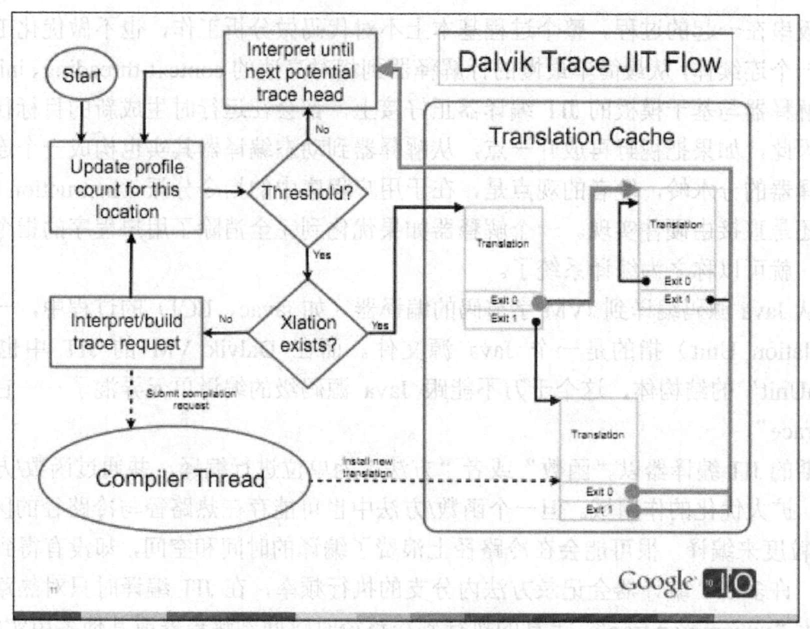

图 18-1 Dalvik 虚拟机的运行流程

18.1.2 Java 虚拟机主要的优化技术

（1）JIT。

最开始指在执行前编译，但是到现在已经发展成为，一开始解释执行，只有被多次调用的程序段才被编译，编译后存放在内存中，下次直接执行编译后的机器码。

- method 方式：以函数或方法为单位进行编译。
- trace 方式：以 trace 为单位进行编译（可以把循环中的内容作为单位编译），此方法也包含 method。

（2）AOT（Ahead Of Time）。

在程序下载到本地时就编译成机器码，并存储在本地硬盘上，以加快运行程度，用此种方式，可执行的程序会变大 4～5 倍。

18.1.3 Dalvik 中 JIT 的实现

每启动一个应用程序，都会相应地启动一个 Dalvik 虚拟机，启动时会建立 JIT 线程，一直在后台运行。当某段代码被调用时，虚拟机会判断它是否需要编译成机器码，如果需要，就做一个标记，JIT 线程不断判断此标记，如果发现被设定就把它编译成机器码，并将其机器码地址及相关信息放入 entry table 中，下次执行到此就跳到机器码段执行，而不再解释执行，从而提高速度。

18.2 Dalvik VM 对 JIT 的支持

为了对 JIT 编译器提供良好支持，在 Dalvik VM 原本的解释器里进行了相应地修改，添加了新的方法入口和 profile 处理。例如在文件“vm/mterp/armv5te/header.S”中添加了如下粗体代码，这表示新添加了一些宏：

```
#define GOTO_OPCODE(_reg)       add    pc, rIBASE, _reg, lsl #${handler_size_bits}
#define GOTO_OPCODE_IFEQ(_reg)  addeq  pc, rIBASE, _reg, lsl #${handler_size_bits}
#define GOTO_OPCODE_IFNE(_reg)  addne  pc, rIBASE, _reg, lsl #${handler_size_bits}
```

```
/*
 * Get/set the 32-bit value from a Dalvik register.
 */
#define GET_VREG(_reg, _vreg)   ldr    _reg, [rFP, _vreg, lsl #2]
#define SET_VREG(_reg, _vreg)   str    _reg, [rFP, _vreg, lsl #2]

#if defined(WITH_JIT)
#define GET_JIT_ENABLED(_reg)        ldr     _reg,[rGLUE,#offGlue_jitEnabled]
#define GET_JIT_PROF_TABLE(_reg)     ldr     _reg,[rGLUE,#offGlue_pJitProfTable]
#endif
```

并且在一些指令的处理代码中也有对它的应用，例如下面的比较指令。

```
%verify "branch taken"
%verify "branch not taken"
   /*
    * Generic two-operand compare-and-branch operation.  Provide a "revcmp"
    * fragment that specifies the *reverse* comparison to perform, e.g.
    * for "if-le" you would use "gt" .
    *
    * For: if-eq, if-ne, if-lt, if-ge, if-gt, if-le
    */
   /* if-cmp vA, vB, +CCCC */
   mov    r0, rINST, lsr #8              @ r0<- A+
   mov    r1, rINST, lsr #12             @ r1<- B
   and    r0, r0, #15
   GET_VREG(r3, r1)                      @ r3<- vB
   GET_VREG(r2, r0)                      @ r2<- vA
   mov    r9, #4                         @ r0<- BYTE branch dist for not-taken
   cmp    r2, r3                         @ compare (vA, vB)
   b${revcmp} 1f                         @ branch to 1 if comparison failed
   FETCH_S(r9, 1 )                       @ r9<- branch offset, in code units
   movs   r9, r9, asl #1                 @ convert to bytes, check sign
   bmi    common_backwardBranch          @ yes, do periodic checks
1 :
#if defined(WITH_JIT)
   GET_JIT_PROF_TABLE(r0)
   FETCH_ADVANCE_INST_RB(r9)             @ update rPC, load rINST
   b      common_testUpdateProfile
#else
   FETCH_ADVANCE_INST_RB(r9)             @ update rPC, load rINST
   GET_INST_OPCODE(ip)                   @ extract opcode from rINST
   GOTO_OPCODE(ip)                       @ jump to next instruction
#endif
```

并且在文件“vm/interp/InterpDefs.h”中，新添加了如下对方法入口的注释。

```
/*
 * There are six entry points from the compiled code to the interpreter:
 * 1) dvmJitToInterpNormal: find if there is a corresponding compilation for
 *    the new dalvik PC. If so, chain the originating compilation with the
 *    target then jump to it.
 * 2) dvmJitToInterpInvokeNoChain: similar to 1) but don't chain. This is
 *    for handling 1-to-many mappings like virtual method call and
 *    packed switch.
 * 3) dvmJitToInterpPunt: use the fast interpreter to execute the next
 *    instruction(s) and stay there as long as it is appropriate to return
 *    to the compiled land. This is used when the jit'ed code is about to
 *    throw an exception.
 * 4) dvmJitToInterpSingleStep: use the portable interpreter to execute the
 *    next instruction only and return to pre-specified location in the
 *    compiled code to resume execution. This is mainly used as debugging
 *    feature to bypass problematic opcode implementations without
 *    disturbing the trace formation.
 * 5) dvmJitToTraceSelect: if there is a single exit from a translation that
 *    has already gone hot enough to be translated, we should assume that
 *    the exit point should also be translated (this is a common case for
 *    invokes). This trace exit will first check for a chaining
```

```
 *    opportunity, and if none is available will switch to the debug
 *    interpreter immediately for trace selection (as if threshold had
 *    just been reached).
 * 6) dvmJitToPredictedChain: patch the chaining cell for a virtual call site
 *    to a predicted callee.
 */
```

mterp 解释器的统一入口是 dvmMterpStdRun()。这个在 mterp 的平台相关代码的 entry.S 里定义。

18.3 汇编代码和改动

在本节的内容中，将简要讲解 JIT 汇编代码的内容和对 C 文件的改动过程，为读者深入了解 JIT 的知识打下基础。

18.3.1 汇编部分代码

与架构相关的编译模板文件如下。

```
/template/out/Compiler/TemplateAsm-unicore32S.S
```

该文件内容比较多，工作量比较大。文件 TemplateAsm-unicore32S.S 会产生脚本，主要工作量体现在生产该文件的源文件，源码目录是：

```
/dalvik/vm/compiler/template
```

该部分内容的处理与 mterp 解释器中汇编的处理方式相似，用一个脚本把所有汇编内容汇集到一个文件中。

再看如下目录：

```
dalvik/vm/mterp/##ARCH##/
```

该目录是解释器的实现，因为 JIT 与解释器并不是独立工作的，因此解释器中有不少针对 JIT 状态处理的函数，代码块使用#if defined(WITH_JIT)来控制。代码量主要体现在文件 footer.S 中。

18.3.2 对 C 文件的改动

JIT 对 C 代码的改动有很多，其中主要集中有如下几个文件。

- ArchUtility.c。
- Assemble.c。
- CodegenDriver.c。
- MipsLIR.h（重写）。
- MipsFP.c（重写）。
- RallocUtil.c。

这些改动是 MIPS 对 JIT 的优化，与 MIPS 架构是紧密相连的。从架构上看，UNICORE 与 ARM 架构比较相近，但 JIT 中有大量 16 位 Thumb 指令。MIPS 指令长度为 32 位，但架构相似度较低。

MIPS 可借鉴的是 compiler 模拟 32 位指令集的工作量评估及后期移植指导。在 MIPS 代码中使用了大量的 assert()，初步判断为调试所用，在移植过程中可以借鉴这些调试点。

从注释和数据命名特征分析，MIPS 是以 ARM 为原型对 JIT 进行的移植，移植过程中加入了 MIPS 架构，鉴于 UNICORE 架构与 ARM 架构的相近性，初步调研结果为以参考 ARM 为主，借鉴 MIPS 为辅对 JIT 进行移植，认真阅读每一行注释，仔细推敲每个函数命名。

18.4 Dalvik VM 中的 JIT 源码

在 Dalvik VM 的“vm\mterp”目录下，保存了和 JIT 编译相关的核心源码。在本节的内容中，将简要分析 Dalvik VM 中的 JIT 源码，为读者步入本书后面知识的学习打下基础。

18.4.1 入口文件

首先看文件“vm/interp/Jit.c”，此文件提供了外界调用的入口。接下来开始分析这个文件的源码。

（1）函数 dvmJitStartup(void)，用于创建 JIT 编译，为编译工作做好准备。主要代码如下所示：

```
int dvmJitStartup(void)
{
    unsigned int i;
    bool res = true;  /* Assume success */

    // 开始创建编译器*/
    res &= dvmCompilerStartup();

    dvmInitMutex(&gDvmJit.tableLock);
    if (res && gDvm.executionMode == kExecutionModeJit) {
        JitEntry *pJitTable = NULL;
        unsigned char *pJitProfTable = NULL;
        assert(gDvm.jitTableSize &&
            !(gDvm.jitTableSize & (gDvmJit.jitTableSize - 1))); // Power of 2?
        dvmLockMutex(&gDvmJit.tableLock);
        pJitTable = (JitEntry*)
                  calloc(gDvmJit.jitTableSize, sizeof(*pJitTable));
        if (!pJitTable) {
            LOGE("jit table allocation failed\n");
            res = false;
            goto done;
        }
        /*
         * NOTE: 该文件表必须只分配一次，在全局范围内
         * Profiling is turned on and off by nulling out gDvm.pJitProfTable
         * and then restoring its original value.  However, this action
         * is not syncronized for speed so threads may continue to hold
         * and update the profile table after profiling has been turned
         * off by null'ng the global pointer.  Be aware.
         */
        pJitProfTable = (unsigned char *)malloc(JIT_PROF_SIZE);
        if (!pJitProfTable) {
            LOGE("jit prof table allocation failed\n");
            res = false;
            goto done;
        }
        memset(pJitProfTable,0,JIT_PROF_SIZE);
        for (i=0; i < gDvmJit.jitTableSize; i++) {
           pJitTable[i].u.info.chain = gDvmJit.jitTableSize;
        }
        /* Is chain field wide enough for termination pattern? */
        assert(pJitTable[0].u.info.chain == gDvm.maxJitTableEntries);

done:
        gDvmJit.pJitEntryTable = pJitTable;
        gDvmJit.jitTableMask = gDvmJit.jitTableSize - 1;
        gDvmJit.jitTableEntriesUsed = 0;
        gDvmJit.pProfTableCopy = gDvmJit.pProfTable = pJitProfTable;
        dvmUnlockMutex(&gDvmJit.tableLock);
    }
    return res;
}
```

（2）停止编译函数 dvmJitStopTranslationRequests()，如果一个固定的表或翻译缓冲区填满，

则调用这个函数停止编译，专业可以避免周期浪费未来的翻译要求。主要代码如下所示：

```
void dvmJitStopTranslationRequests()
{
    /*
     * Note 1: This won't necessarily stop all translation requests, and
     * operates on a delayed mechanism.  Running threads look to the copy
     * of this value in their private InterpState structures and won't see
     * this change until it is refreshed (which happens on interpreter
     * entry).
     * Note 2: This is a one-shot memory leak on this table. Because this is a
     * permanent off switch for Jit profiling, it is a one-time leak of 1K
     * bytes, and no further attempt will be made to re-allocate it. Can't
     * free it because some thread may be holding a reference.
     */
    gDvmJit.pProfTable = gDvmJit.pProfTableCopy = NULL;
}
```

（3）通过函数 dvmJitStats()转储调试和优化数据日志，主要代码如下所示：

```
void dvmJitStats()
{
    int i;
    int hit;
    int not_hit;
    int chains;
    if (gDvmJit.pJitEntryTable) {
        for (i=0, chains=hit=not_hit=0;
             i < (int) gDvmJit.jitTableSize;
             i++) {
            if (gDvmJit.pJitEntryTable[i].dPC != 0)
                hit++;
            else
                not_hit++;
            if (gDvmJit.pJitEntryTable[i].u.info.chain != gDvmJit.jitTableSize)
                chains++;
        }
        LOGD(
         "JIT: %d traces, %d slots, %d chains, %d maxQ, %d thresh, %s",
         hit, not_hit + hit, chains, gDvmJit.compilerMaxQueued,
         gDvmJit.threshold, gDvmJit.blockingMode ? "Blocking" : "Non-blocking");
#if defined(EXIT_STATS)
        LOGD(
         "JIT: Lookups: %d hits, %d misses; %d NoChain, %d normal, %d punt",
         gDvmJit.addrLookupsFound, gDvmJit.addrLookupsNotFound,
         gDvmJit.noChainExit, gDvmJit.normalExit, gDvmJit.puntExit);
#endif
        LOGD("JIT: %d Translation chains", gDvmJit.translationChains);
#if defined(INVOKE_STATS)
        LOGD("JIT: Invoke: %d chainable, %d pred. chain, %d native, "
             "%d return",
             gDvmJit.invokeChain, gDvmJit.invokePredictedChain,
             gDvmJit.invokeNative, gDvmJit.returnOp);
#endif
        if (gDvmJit.profile) {
            dvmCompilerSortAndPrintTraceProfiles();
        }
    }
}
```

（4）通过函数 dvmJitShutdown(void)实现准时关机功能，注意只做一次关机操作，不要试图重新启动。主要代码如下所示：

```
void dvmJitShutdown(void)
{
    /*编译线程关闭*/
    dvmCompilerShutdown();
    dvmCompilerDumpStats();
    dvmDestroyMutex(&gDvmJit.tableLock);
```

```
    if (gDvmJit.pJitEntryTable) {
        free(gDvmJit.pJitEntryTable);
        gDvmJit.pJitEntryTable = NULL;
    }
    if (gDvmJit.pProfTable) {
        free(gDvmJit.pProfTable);
        gDvmJit.pProfTable = NULL;
    }
}
```

（5）核心函数 dvmJitShutdown(void)，其被 define 成 CHECK_JIT()，此函数的功能是判断在什么条件时认为需要编译。主要代码如下所示：

```
int dvmCheckJit(const u2* pc, Thread* self, InterpState* interpState)
{
    int flags,i,len;
    int switchInterp = false;
    int debugOrProfile = (gDvm.debuggerActive || self->suspendCount
#if defined(WITH_PROFILER)
                  || gDvm.activeProfilers
#endif
          );

    switch (interpState->jitState) {
        char* nopStr;
        int target;

        int offset;
        DecodedInstruction decInsn;
        case kJitTSelect:
            dexDecodeInstruction(gDvm.instrFormat, pc, &decInsn);
#if defined(SHOW_TRACE)
            LOGD("TraceGen: adding %s",getOpcodeName(decInsn.opCode));
#endif
            flags = dexGetInstrFlags(gDvm.instrFlags, decInsn.opCode);
            len = dexGetInstrOrTableWidthAbs(gDvm.instrWidth, pc);
            offset = pc - interpState->method->insns;
            if (pc != interpState->currRunHead + interpState->currRunLen) {
                int currTraceRun;
                /* We need to start a new trace run */
                currTraceRun = ++interpState->currTraceRun;
                interpState->currRunLen = 0;
                interpState->currRunHead = (u2*)pc;
                interpState->trace[currTraceRun].frag.startOffset = offset;
                interpState->trace[currTraceRun].frag.numInsts = 0;
                interpState->trace[currTraceRun].frag.runEnd = false;
                interpState->trace[currTraceRun].frag.hint = kJitHintNone;
            }
            interpState->trace[interpState->currTraceRun].frag.numInsts++;
            interpState->totalTraceLen++;
            interpState->currRunLen += len;
            if (  ((flags & kInstrUnconditional) == 0) &&
                  /* don't end trace on INVOKE_DIRECT_EMPTY  */
                  (decInsn.opCode != OP_INVOKE_DIRECT_EMPTY) &&
                  ((flags & (kInstrCanBranch |
                             kInstrCanSwitch |
                             kInstrCanReturn |
                             kInstrInvoke)) != 0)) {
                    interpState->jitState = kJitTSelectEnd;
#if defined(SHOW_TRACE)
            LOGD("TraceGen: ending on %s, basic block end",
                 getOpcodeName(decInsn.opCode));
#endif
        }
            if (decInsn.opCode == OP_THROW) {
                interpState->jitState = kJitTSelectEnd;
            }
            if (interpState->totalTraceLen >= JIT_MAX_TRACE_LEN) {
                interpState->jitState = kJitTSelectEnd;
```

```
        }
        if (debugOrProfile) {
           interpState->jitState = kJitTSelectAbort;
           switchInterp = !debugOrProfile;
           break;
        }
        if ((flags & kInstrCanReturn) != kInstrCanReturn) {
           break;
        }
        /* NOTE: intentional fallthrough for returns */
     case kJitTSelectEnd:
        {
           if (interpState->totalTraceLen == 0) {
              switchInterp = !debugOrProfile;
              break;
           }
           JitTraceDescription* desc =
             (JitTraceDescription*)malloc(sizeof(JitTraceDescription) +
              sizeof(JitTraceRun) * (interpState->currTraceRun+1));
           if (desc == NULL) {
              LOGE("Out of memory in trace selection");
              dvmJitStopTranslationRequests();
              interpState->jitState = kJitTSelectAbort;
              switchInterp = !debugOrProfile;
              break;
           }
           interpState->trace[interpState->currTraceRun].frag.runEnd =
              true;
           interpState->jitState = kJitNormal;
           desc->method = interpState->method;
           memcpy((char*)&(desc->trace[0]),
              (char*)&(interpState->trace[0]),
              sizeof(JitTraceRun) * (interpState->currTraceRun+1));
#if defined(SHOW_TRACE)
           LOGD("TraceGen:  trace done, adding to queue");
#endif
           dvmCompilerWorkEnqueue(
                 interpState->currTraceHead,kWorkOrderTrace,desc);
           if (gDvmJit.blockingMode) {
              dvmCompilerDrainQueue();
           }
           switchInterp = !debugOrProfile;
        }
        break;
     case kJitSingleStep:
        interpState->jitState = kJitSingleStepEnd;
        break;
     case kJitSingleStepEnd:
        interpState->entryPoint = kInterpEntryResume;
        switchInterp = !debugOrProfile;
        break;
     case kJitTSelectAbort:
#if defined(SHOW_TRACE)
        LOGD("TraceGen:  trace abort");
#endif
        interpState->jitState = kJitNormal;
        switchInterp = !debugOrProfile;
        break;
     case kJitNormal:
        switchInterp = !debugOrProfile;
        break;
     default:
        dvmAbort();
  }
  return switchInterp;
}
```

在上述函数中，增加了目前跟踪请求的一个指令的时间，这只是在指令的解释工作。一般来说，指示“建议”被添加到对当前跟踪之前解释。如果解释器成功地完成指令动作，则将被视为请求的一部分，这使得能够审查这之前的状态。如果中止查询指令则会发生意想不到的事情，然而返回指令将会导致立即结束翻译工作，这一切都是在编译器回归前完成。这样做的目的是针对特殊处理返回一定的解释，并对问题根源进行了描述。

（6）通过 dvmJitGetCodeAddr(const u2* dPC)，如果在虚拟机指针中存在翻译的代码地址，则加速翻译这个进程。主要代码如下所示：

```
void* dvmJitGetCodeAddr(const u2* dPC)
{
    int idx = dvmJitHash(dPC);

    /* If anything is suspended, don't re-enter the code cache */
    if (gDvm.sumThreadSuspendCount > 0) {
        return NULL;
    }

    /* Expect a high hit rate on 1st shot */
    if (gDvmJit.pJitEntryTable[idx].dPC == dPC) {
#if defined(EXIT_STATS)
        gDvmJit.addrLookupsFound++;
#endif
        return gDvmJit.pJitEntryTable[idx].codeAddress;
    } else {
        int chainEndMarker = gDvmJit.jitTableSize;
        while (gDvmJit.pJitEntryTable[idx].u.info.chain != chainEndMarker) {
            idx = gDvmJit.pJitEntryTable[idx].u.info.chain;
            if (gDvmJit.pJitEntryTable[idx].dPC == dPC) {
#if defined(EXIT_STATS)
                gDvmJit.addrLookupsFound++;
#endif
                return gDvmJit.pJitEntryTable[idx].codeAddress;
            }
        }
    }
#if defined(EXIT_STATS)
    gDvmJit.addrLookupsNotFound++;
#endif
    return NULL;
}
```

（7）通过 dvmJitLookupAndAdd(const u2* dPC) 判断是否有相对某段程序的字节码可用。如果有则返回其地址；如果没有，则做标记，以通知编译线程对其进行编译。主要代码如下所示：

```
JitEntry *dvmJitLookupAndAdd(const u2* dPC)
{
    u4 chainEndMarker = gDvmJit.jitTableSize;
    u4 idx = dvmJitHash(dPC);

    /* Walk the bucket chain to find an exact match for our PC */
    while ((gDvmJit.pJitEntryTable[idx].u.info.chain != chainEndMarker) &&
        (gDvmJit.pJitEntryTable[idx].dPC != dPC)) {
        idx = gDvmJit.pJitEntryTable[idx].u.info.chain;
    }

    if (gDvmJit.pJitEntryTable[idx].dPC != dPC) {
        /*
         * No match.  Aquire jitTableLock and find the last
         * slot in the chain. Possibly continue the chain walk in case
         * some other thread allocated the slot we were looking
         * at previuosly (perhaps even the dPC we're trying to enter).
         */
        dvmLockMutex(&gDvmJit.tableLock);
        /*
         * At this point, if .dPC is NULL, then the slot we're
         * looking at is the target slot from the primary hash
         * (the simple, and common case).  Otherwise we're going
         * to have to find a free slot and chain it.
```

```
     */
    MEM_BARRIER(); /* Make sure we reload [].dPC after lock */
    if (gDvmJit.pJitEntryTable[idx].dPC != NULL) {
        u4 prev;
        while (gDvmJit.pJitEntryTable[idx].u.info.chain != chainEndMarker) {
            if (gDvmJit.pJitEntryTable[idx].dPC == dPC) {
                /* Another thread got there first for this dPC */
                dvmUnlockMutex(&gDvmJit.tableLock);
                return &gDvmJit.pJitEntryTable[idx];
            }
            idx = gDvmJit.pJitEntryTable[idx].u.info.chain;
        }
        /* Here, idx should be pointing to the last cell of an
         * active chain whose last member contains a valid dPC */
        assert(gDvmJit.pJitEntryTable[idx].dPC != NULL);
        /* Linear walk to find a free cell and add it to the end */
        prev = idx;
        while (true) {
            idx++;
            if (idx == chainEndMarker)
                idx = 0;  /* Wraparound */
            if ((gDvmJit.pJitEntryTable[idx].dPC == NULL) ||
                (idx == prev))
                break;
        }
        if (idx != prev) {
            JitEntryInfoUnion oldValue;
            JitEntryInfoUnion newValue;
            /*
             * Although we hold the lock so that noone else will
             * be trying to update a chain field, the other fields
             * packed into the word may be in use by other threads.
             */
            do {
                oldValue = gDvmJit.pJitEntryTable[prev].u;
                newValue = oldValue;
                newValue.info.chain = idx;
            } while (!ATOMIC_CMP_SWAP(
                    &gDvmJit.pJitEntryTable[prev].u.infoWord,
                    oldValue.infoWord, newValue.infoWord));
        }
    }
    if (gDvmJit.pJitEntryTable[idx].dPC == NULL) {
       /* Allocate the slot */
        gDvmJit.pJitEntryTable[idx].dPC = dPC;
        gDvmJit.jitTableEntriesUsed++;
    } else {
        /* Table is full */
        idx = chainEndMarker;
    }
    dvmUnlockMutex(&gDvmJit.tableLock);
  }
  return (idx == chainEndMarker) ? NULL : &gDvmJit.pJitEntryTable[idx];
}
```

（8）通过 dvmJitSetCodeAddr（const u2* dPC, void *nPC, JitInstructionSetType set）将代码指针注册为 JIT 码。如果被编译的代码地址为空，则不能停止所有的线程工作，这样的程序被称为编译线程。主要代码如下所示：

```
void dvmJitSetCodeAddr(const u2* dPC, void *nPC, JitInstructionSetType set) {
    JitEntryInfoUnion oldValue;
    JitEntryInfoUnion newValue;
    JitEntry *jitEntry = dvmJitLookupAndAdd(dPC);
    assert(jitEntry);
    /* Note: order of update is important */
    do {
        oldValue = jitEntry->u;
        newValue = oldValue;
```

```
        newValue.info.instructionSet = set;
    } while (!ATOMIC_CMP_SWAP(
            &jitEntry->u.infoWord,
            oldValue.infoWord, newValue.infoWord));
    jitEntry->codeAddress = nPC;
}
```

（9）通过 dvmJitCheckTraceRequest（Thread* self, InterpState* interpState）确定是否存在有效的“trace-bulding”活跃，如果需要中止和切换到快速翻译则返回真，否则返回假。主要代码如下所示：

```
bool dvmJitCheckTraceRequest(Thread* self, InterpState* interpState)
{
    bool res = false;         /* Assume success */
    int i;
    if (gDvmJit.pJitEntryTable != NULL) {
        /* Two-level filtering scheme */
        for (i=0; i< JIT_TRACE_THRESH_FILTER_SIZE; i++) {
            if (interpState->pc == interpState->threshFilter[i]) {
                break;
            }
        }
        if (i == JIT_TRACE_THRESH_FILTER_SIZE) {
            /*
             * Use random replacement policy - otherwise we could miss a large
             * loop that contains more traces than the size of our filter array.
             */
            i = rand() % JIT_TRACE_THRESH_FILTER_SIZE;
            interpState->threshFilter[i] = interpState->pc;
            res = true;
        }
        /*
         * If the compiler is backlogged, or if a debugger or profiler is
         * active, cancel any JIT actions
         */
        if ( res || (gDvmJit.compilerQueueLength >= gDvmJit.compilerHighWater) ||
            gDvm.debuggerActive || self->suspendCount
#if defined(WITH_PROFILER)
                || gDvm.activeProfilers
#endif
                                             ) {
            if (interpState->jitState != kJitOff) {
                interpState->jitState = kJitNormal;
            }
        } else if (interpState->jitState == kJitTSelectRequest) {
            JitEntry *slot = dvmJitLookupAndAdd(interpState->pc);
            if (slot == NULL) {
                /*
                 * Table is full.  This should have been
                 * detected by the compiler thread and the table
                 * resized before we run into it here.  Assume bad things
                 * are afoot and disable profiling.
                 */
                interpState->jitState = kJitTSelectAbort;
                LOGD("JIT: JitTable full, disabling profiling");
                dvmJitStopTranslationRequests();
            } else if (slot->u.info.traceRequested) {
                /* Trace already requested - revert to interpreter */
                interpState->jitState = kJitTSelectAbort;
            } else {
                /* Mark request */
                JitEntryInfoUnion oldValue;
                JitEntryInfoUnion newValue;
                do {
                    oldValue = slot->u;
                    newValue = oldValue;
                    newValue.info.traceRequested = true;
                } while (!ATOMIC_CMP_SWAP( &slot->u.infoWord,
                        oldValue.infoWord, newValue.infoWord));
            }
```

```
    }
    switch (interpState->jitState) {
        case kJitTSelectRequest:
            interpState->jitState = kJitTSelect;
            interpState->currTraceHead = interpState->pc;
            interpState->currTraceRun = 0;
            interpState->totalTraceLen = 0;
            interpState->currRunHead = interpState->pc;
            interpState->currRunLen = 0;
            interpState->trace[0].frag.startOffset =
                interpState->pc - interpState->method->insns;
            interpState->trace[0].frag.numInsts = 0;
            interpState->trace[0].frag.runEnd = false;
            interpState->trace[0].frag.hint = kJitHintNone;
            break;
        case kJitTSelect:
        case kJitTSelectAbort:
            res = true;
        case kJitSingleStep:
        case kJitSingleStepEnd:
        case kJitOff:
        case kJitNormal:
            break;
        default:
            dvmAbort();
    }
  }
  return res;
}
```

（10）通过 dvmJitResizeJitTable(unsigned int size)调整 JIT 码，此处必须是 2 的幂。如果失败则返回真，并停止所有的线程。主要代码如下所示：

```
bool dvmJitResizeJitTable( unsigned int size )
{
    JitEntry *pNewTable;
    JitEntry *pOldTable;
    u4 newMask;
    unsigned int oldSize;
    unsigned int i;

    assert(gDvm.pJitEntryTable != NULL);
    assert(size && !(size & (size - 1)));   /* Is power of 2? */

    LOGD("Jit: resizing JitTable from %d to %d", gDvmJit.jitTableSize, size);

    newMask = size - 1;

    if (size <= gDvmJit.jitTableSize) {
        return true;
    }

    pNewTable = (JitEntry*)calloc(size, sizeof(*pNewTable));
    if (pNewTable == NULL) {
        return true;
    }
    for (i=0; i< size; i++) {
        pNewTable[i].u.info.chain = size;  /* Initialize chain termination */
    }

    /* Stop all other interpreting/jit'ng threads */
    dvmSuspendAllThreads(SUSPEND_FOR_JIT);

    pOldTable = gDvmJit.pJitEntryTable;
    oldSize = gDvmJit.jitTableSize;

    dvmLockMutex(&gDvmJit.tableLock);
    gDvmJit.pJitEntryTable = pNewTable;
    gDvmJit.jitTableSize = size;
```

```
    gDvmJit.jitTableMask = size - 1;
    gDvmJit.jitTableEntriesUsed = 0;
    dvmUnlockMutex(&gDvmJit.tableLock);

    for (i=0; i < oldSize; i++) {
        if (pOldTable[i].dPC) {
            JitEntry *p;
            u2 chain;
            p = dvmJitLookupAndAdd(pOldTable[i].dPC);
            p->dPC = pOldTable[i].dPC;
            /*
             * Compiler thread may have just updated the new entry's
             * code address field, so don't blindly copy null.
             */
            if (pOldTable[i].codeAddress != NULL) {
                p->codeAddress = pOldTable[i].codeAddress;
            }
            /* We need to preserve the new chain field, but copy the rest */
            dvmLockMutex(&gDvmJit.tableLock);
            chain = p->u.info.chain;
            p->u = pOldTable[i].u;
            p->u.info.chain = chain;
            dvmUnlockMutex(&gDvmJit.tableLock);
        }
    }

    free(pOldTable);

    /* Restart the world */
    dvmResumeAllThreads(SUSPEND_FOR_JIT);

    return false;
}
```

（11）通过 dvmJitd2l(double d)和 dvmJitf2l(float f)进行格式转换，分别将 double 格式和 float 格式设置为最大和最小的整数格式。主要代码如下所示：

```
s8 dvmJitd2l(double d)
{
    static const double kMaxLong = (double)(s8)0x7fffffffffffffffULL;
    static const double kMinLong = (double)(s8)0x8000000000000000ULL;
    if (d >= kMaxLong)
        return (s8)0x7fffffffffffffffULL;
    else if (d <= kMinLong)
        return (s8)0x8000000000000000ULL;
    else if (d != d) // NaN case
        return 0;
    else
        return (s8)d;
}

s8 dvmJitf2l(float f)
{
    static const float kMaxLong = (float)(s8)0x7fffffffffffffffULL;
    static const float kMinLong = (float)(s8)0x8000000000000000ULL;
    if (f >= kMaxLong)
        return (s8)0x7fffffffffffffffULL;
    else if (f <= kMinLong)
        return (s8)0x8000000000000000ULL;
    else if (f != f) // NaN case
        return 0;
    else
        return (s8)f;
}
```

18.4.2 核心函数

在文件“vm/compiler/compiler.c”中定义了核心函数的具体实现，接下来开始分析这个文件的源码。

（1）在虚拟机启动时调用函数 dvmCompilerStartup(void)，主要代码如下所示：

```
bool dvmCompilerStartup(void)
{
    /* Make sure the BBType enum is in sane state */
    assert(CHAINING_CELL_NORMAL == 0);

    /* Architecture-specific chores to initialize */
    if (!dvmCompilerArchInit())
        goto fail;

    /*
     * Setup the code cache if it is not done so already. For apps it should be
     * done by the Zygote already, but for command-line dalvikvm invocation we
     * need to do it here.
     */
    if (gDvmJit.codeCache == NULL) {
        if (!dvmCompilerSetupCodeCache())
            goto fail;
    }

    /* Allocate the initial arena block */
    if (dvmCompilerHeapInit() == false) {
        goto fail;
    }

    dvmInitMutex(&gDvmJit.compilerLock);
    pthread_cond_init(&gDvmJit.compilerQueueActivity, NULL);
    pthread_cond_init(&gDvmJit.compilerQueueEmpty, NULL);

    dvmLockMutex(&gDvmJit.compilerLock);

    gDvmJit.haltCompilerThread = false;

    /* Reset the work queue */
    memset(gDvmJit.compilerWorkQueue, 0,
           sizeof(CompilerWorkOrder) * COMPILER_WORK_QUEUE_SIZE);
    gDvmJit.compilerWorkEnqueueIndex = gDvmJit.compilerWorkDequeueIndex = 0;
    gDvmJit.compilerQueueLength = 0;
    gDvmJit.compilerHighWater =
        COMPILER_WORK_QUEUE_SIZE - (COMPILER_WORK_QUEUE_SIZE/4);

    assert(gDvmJit.compilerHighWater < COMPILER_WORK_QUEUE_SIZE);
    if (!dvmCreateInternalThread(&gDvmJit.compilerHandle, "Compiler",
                                 compilerThreadStart, NULL)) {
        dvmUnlockMutex(&gDvmJit.compilerLock);
        goto fail;
    }
    /* Track method-level compilation statistics */
    gDvmJit.methodStatsTable =  dvmHashTableCreate(32, NULL);
    dvmUnlockMutex(&gDvmJit.compilerLock);
    return true;
fail:
    return false;
}
```

（2）在虚拟机关闭时调用函数 dvmCompilerShutdown(void)，主要代码如下所示：

```
void dvmCompilerShutdown(void)
{
    void *threadReturn;
    if (gDvmJit.compilerHandle) {
        gDvmJit.haltCompilerThread = true;
        dvmLockMutex(&gDvmJit.compilerLock);
        pthread_cond_signal(&gDvmJit.compilerQueueActivity);
        dvmUnlockMutex(&gDvmJit.compilerLock);
        if (pthread_join(gDvmJit.compilerHandle, &threadReturn) != 0)
            LOGW("Compiler thread join failed\n");
```

```
    else
        LOGD("Compiler thread has shut down\n");
  }
}
```

（3）再看 compilerThreadStart(void *arg)，这是一个线程函数，被 dvmCompilerStartup()调用，在虚拟机运行过程中一直生存的线程，while 循环判断是否有代码需要编译，如果有，则调用 dvmCompilerDoWork()对其进行编译。主要代码如下所示：

```
static void *compilerThreadStart(void *arg)
{
    dvmLockMutex(&gDvmJit.compilerLock);
    /*
     * Since the compiler thread will not touch any objects on the heap once
     * being created, we just fake its state as VMWAIT so that it can be a
     * bit late when there is suspend request pending.
     */
    dvmChangeStatus(NULL, THREAD_VMWAIT);
    while (!gDvmJit.haltCompilerThread) {
        if (workQueueLength() == 0) {
            int cc;
            cc = pthread_cond_signal(&gDvmJit.compilerQueueEmpty);
            assert(cc == 0);
            pthread_cond_wait(&gDvmJit.compilerQueueActivity,
                              &gDvmJit.compilerLock);
            continue;
        } else {
            do {
                CompilerWorkOrder work = workDequeue();
                dvmUnlockMutex(&gDvmJit.compilerLock);
                /* Check whether there is a suspend request on me */
                dvmCheckSuspendPending(NULL);
                /* Is JitTable filling up? */
                if (gDvmJit.jitTableEntriesUsed >
                    (gDvmJit.jitTableSize - gDvmJit.jitTableSize/4)) {
                    dvmJitResizeJitTable(gDvmJit.jitTableSize * 2);
                }
                if (gDvmJit.haltCompilerThread) {
                    LOGD("Compiler shutdown in progress - discarding request");
                } else {
                    /* Compilation is successful */
                    if (dvmCompilerDoWork(&work)) {
                        dvmJitSetCodeAddr(work.pc, work.result.codeAddress,
                                          work.result.instructionSet);
                    }
                }
                free(work.info);
                dvmLockMutex(&gDvmJit.compilerLock);
            } while (workQueueLength() != 0);
        }
    }
    pthread_cond_signal(&gDvmJit.compilerQueueEmpty);
    dvmUnlockMutex(&gDvmJit.compilerLock);

    LOGD("Compiler thread shutting down\n");
    return NULL;
}
```

（4）再看函数 dvmCompilerSetupCodeCache(void)，功能是存储编译后的字节码。具体代码如下所示：

```
bool dvmCompilerSetupCodeCache(void)
{
    extern void dvmCompilerTemplateStart(void);
    extern void dmvCompilerTemplateEnd(void);

    /* Allocate the code cache */
    gDvmJit.codeCache = mmap(0, CODE_CACHE_SIZE,
```

```
                        PROT_READ | PROT_WRITE | PROT_EXEC,
                        MAP_PRIVATE | MAP_ANONYMOUS, -1, 0);
    if (gDvmJit.codeCache == MAP_FAILED) {
        LOGE("Failed to create the code cache: %s\n", strerror(errno));
        return false;
    }

    /* Copy the template code into the beginning of the code cache */
    int templateSize = (intptr_t) dmvCompilerTemplateEnd -
                       (intptr_t) dvmCompilerTemplateStart;
    memcpy((void *) gDvmJit.codeCache,
           (void *) dvmCompilerTemplateStart,
           templateSize);

    gDvmJit.templateSize = templateSize;
    gDvmJit.codeCacheByteUsed = templateSize;

    /* Flush dcache and invalidate the icache to maintain coherence */
    cacheflush((intptr_t) gDvmJit.codeCache,
               (intptr_t) gDvmJit.codeCache + CODE_CACHE_SIZE, 0);
    return true;
}
```

对上述函数有如下 4 点说明。

- gDvmJit.codeCache：使用 mmap 分配（1024x1024），用于存在编译后的代码。
- gDvmJit.codeCacheByteUsed：codeCache 的使用情况。
- gDvmJit.codeCacheFull：codeCache 是否已写满。
- gDvmJit.pJitEntryTable：entry 表，每个 trace 对应一个 entry。

18.4.3 编译文件

在文件“vm/compiler/Frongend.c”中定义了两个方法，分别实现了两种编译方法。

（1）通过函数 dvmCompileMethod()实现 method 方式，即以函数或方法为单位进行编译。实现代码如下所示：

```
bool dvmCompileMethod(const Method *method, JitTranslationInfo *info)
{
    const DexCode *dexCode = dvmGetMethodCode(method);
    const u2 *codePtr = dexCode->insns;
    const u2 *codeEnd = dexCode->insns + dexCode->insnsSize;
    int blockID = 0;
    unsigned int curOffset = 0;

    BasicBlock *firstBlock = dvmCompilerNewBB(DALVIK_BYTECODE);
    firstBlock->id = blockID++;

    /* Allocate the bit-vector to track the beginning of basic blocks */
    BitVector *bbStartAddr = dvmAllocBitVector(dexCode->insnsSize+1, false);
    dvmSetBit(bbStartAddr, 0);

    /*
     * Sequentially go through every instruction first and put them in a single
     * basic block. Identify block boundaries at the mean time.
     */
    while (codePtr < codeEnd) {
        MIR *insn = dvmCompilerNew(sizeof(MIR), false);
        insn->offset = curOffset;
        int width = parseInsn(codePtr, &insn->dalvikInsn, false);
        bool isInvoke = false;
        const Method *callee;
        insn->width = width;

        /* Terminate when the data section is seen */
        if (width == 0)
            break;
```

```
        dvmCompilerAppendMIR(firstBlock, insn);
        /*
         * Check whether this is a block ending instruction and whether it
         * suggests the start of a new block
         */
        unsigned int target = curOffset;

        /*
         * If findBlockBoundary returns true, it means the current instruction
         * is terminating the current block. If it is a branch, the target
         * address will be recorded in target.
         */
        if (findBlockBoundary(method, insn, curOffset, &target, &isInvoke,
                              &callee)) {
            dvmSetBit(bbStartAddr, curOffset + width);
            if (target != curOffset) {
                dvmSetBit(bbStartAddr, target);
            }
        }

        codePtr += width;
        /* each bit represents 16-bit quantity */
        curOffset += width;
    }

    /*
     * The number of blocks will be equal to the number of bits set to 1 in the
     * bit vector minus 1, because the bit representing the location after the
     * last instruction is set to one.
     */
    int numBlocks = dvmCountSetBits(bbStartAddr);
    if (dvmIsBitSet(bbStartAddr, dexCode->insnsSize)) {
        numBlocks--;
    }

    CompilationUnit cUnit;
    BasicBlock **blockList;

    memset(&cUnit, 0, sizeof(CompilationUnit));
    cUnit.method = method;
    blockList = cUnit.blockList =
        dvmCompilerNew(sizeof(BasicBlock *) * numBlocks, true);

    /*
     * Register the first block onto the list and start split it into block
     * boundaries from there.
     */
    blockList[0] = firstBlock;
    cUnit.numBlocks = 1;

    int i;
    for (i = 0; i < numBlocks; i++) {
        MIR *insn;
        BasicBlock *curBB = blockList[i];
        curOffset = curBB->lastMIRInsn->offset;

        for (insn = curBB->firstMIRInsn->next; insn; insn = insn->next) {
            /* Found the beginning of a new block, see if it is created yet */
            if (dvmIsBitSet(bbStartAddr, insn->offset)) {
                int j;
                for (j = 0; j < cUnit.numBlocks; j++) {
                    if (blockList[j]->firstMIRInsn->offset == insn->offset)
                        break;
                }

                /* Block not split yet - do it now */
                if (j == cUnit.numBlocks) {
                    BasicBlock *newBB = dvmCompilerNewBB(DALVIK_BYTECODE);
                    newBB->id = blockID++;
```

```
                newBB->firstMIRInsn = insn;
                newBB->startOffset = insn->offset;
                newBB->lastMIRInsn = curBB->lastMIRInsn;
                curBB->lastMIRInsn = insn->prev;
                insn->prev->next = NULL;
                insn->prev = NULL;

                /*
                 * If the insn is not an unconditional branch, set up the
                 * fallthrough link.
                 */
                if (!isUnconditionalBranch(curBB->lastMIRInsn)) {
                    curBB->fallThrough = newBB;
                }

                /* enqueue the new block */
                blockList[cUnit.numBlocks++] = newBB;
                break;
            }
        }
    }
}

if (numBlocks != cUnit.numBlocks) {
    LOGE("Expect %d vs %d basic blocks\n", numBlocks, cUnit.numBlocks);
    dvmAbort();
}

dvmFreeBitVector(bbStartAddr);

/* Connect the basic blocks through the taken links */
for (i = 0; i < numBlocks; i++) {
    BasicBlock *curBB = blockList[i];
    MIR *insn = curBB->lastMIRInsn;
    unsigned int target = insn->offset;
    bool isInvoke;
    const Method *callee;

    findBlockBoundary(method, insn, target, &target, &isInvoke, &callee);

    /* Found a block ended on a branch */
    if (target != insn->offset) {
        int j;
        /* Forward branch */
        if (target > insn->offset) {
            j = i + 1;
        } else {
            /* Backward branch */
            j = 0;
        }
        for (; j < numBlocks; j++) {
            if (blockList[j]->firstMIRInsn->offset == target) {
                curBB->taken = blockList[j];
                break;
            }
        }

        /* Don't create dummy block for the callee yet */
        if (j == numBlocks && !isInvoke) {
            LOGE("Target not found for insn %x: expect target %x\n",
                 curBB->lastMIRInsn->offset, target);
            dvmAbort();
        }
    }
}
/* Set the instruction set to use (NOTE: later components may change it) */
cUnit.instructionSet = dvmCompilerInstructionSet(&cUnit);
dvmCompilerMIR2LIR(&cUnit);
dvmCompilerAssembleLIR(&cUnit, info);
```

```
    dvmCompilerDumpCompilationUnit(&cUnit);
    dvmCompilerArenaReset();
    return info->codeAddress != NULL;
}
```

（2）通过函数 dvmCompileMethod()实现 trace 方式，即以 trace 为单位进行编译（可以把循环中的内容作为单位编译），此方法也包含 method。实现代码如下所示：

```
bool dvmCompileTrace(JitTraceDescription *desc, int numMaxInsts,
                     JitTranslationInfo *info)
{
    const DexCode *dexCode = dvmGetMethodCode(desc->method);
    const JitTraceRun* currRun = &desc->trace[0];
    unsigned int curOffset = currRun->frag.startOffset;
    unsigned int numInsts = currRun->frag.numInsts;
    const u2 *codePtr = dexCode->insns + curOffset;
    int traceSize = 0;  // # of half-words
    const u2 *startCodePtr = codePtr;
    BasicBlock *startBB, *curBB, *lastBB;
    int numBlocks = 0;
    static int compilationId;
    CompilationUnit cUnit;
    CompilerMethodStats *methodStats;

    compilationId++;
    memset(&cUnit, 0, sizeof(CompilationUnit));

    /* Locate the entry to store compilation statistics for this method */
    methodStats = analyzeMethodBody(desc->method);

    cUnit.registerScoreboard.nullCheckedRegs =
        dvmAllocBitVector(desc->method->registersSize, false);

    /* Initialize the printMe flag */
    cUnit.printMe = gDvmJit.printMe;

    /* Initialize the profile flag */
    cUnit.executionCount = gDvmJit.profile;

    /* Identify traces that we don't want to compile */
    if (gDvmJit.methodTable) {
        int len = strlen(desc->method->clazz->descriptor) +
                  strlen(desc->method->name) + 1;
        char *fullSignature = dvmCompilerNew(len, true);
        strcpy(fullSignature, desc->method->clazz->descriptor);
        strcat(fullSignature, desc->method->name);
        int hashValue = dvmComputeUtf8Hash(fullSignature);
        /*
         * Doing three levels of screening to see whether we want to skip
         * compiling this method
         */
        /* First, check the full "class;method" signature */
        bool methodFound =
            dvmHashTableLookup(gDvmJit.methodTable, hashValue,
                               fullSignature, (HashCompareFunc) strcmp,
                               false) !=
            NULL;

        /* Full signature not found - check the enclosing class */
        if (methodFound == false) {
            int hashValue = dvmComputeUtf8Hash(desc->method->clazz->descriptor);
            methodFound =
                dvmHashTableLookup(gDvmJit.methodTable, hashValue,
                               (char *) desc->method->clazz->descriptor,
                               (HashCompareFunc) strcmp, false) !=
                NULL;
            /* Enclosing class not found - check the method name */
            if (methodFound == false) {
                int hashValue = dvmComputeUtf8Hash(desc->method->name);
```

```
                methodFound =
                    dvmHashTableLookup(gDvmJit.methodTable, hashValue,
                                    (char *) desc->method->name,
                                    (HashCompareFunc) strcmp, false) !=
                    NULL;
            }
        }

        /*
         * Under the following conditions, the trace will be *conservatively*
         * compiled by only containing single-step instructions to and from the
         * interpreter.
         * 1) If includeSelectedMethod == false, the method matches the full or
         *    partial signature stored in the hash table.
         *
         * 2) If includeSelectedMethod == true, the method does not match the
         *    full and partial signature stored in the hash table.
         */
        if (gDvmJit.includeSelectedMethod != methodFound) {
            cUnit.allSingleStep = true;
        } else {
            /* Compile the trace as normal */
            /* Print the method we cherry picked */
            if (gDvmJit.includeSelectedMethod == true) {
                cUnit.printMe = true;
            }
        }
    }

    /* Allocate the first basic block */
    lastBB = startBB = curBB = dvmCompilerNewBB(DALVIK_BYTECODE);
    curBB->startOffset = curOffset;
    curBB->id = numBlocks++;
    if (cUnit.printMe) {
        LOGD("--------\nCompiler: Building trace for %s, offset 0x%x\n",
            desc->method->name, curOffset);
    }
    /*
     * Analyze the trace descriptor and include up to the maximal number
     * of Dalvik instructions into the IR.
     */
    while (1) {
        MIR *insn;
        int width;
        insn = dvmCompilerNew(sizeof(MIR),false);
        insn->offset = curOffset;
        width = parseInsn(codePtr, &insn->dalvikInsn, cUnit.printMe);
        /* The trace should never incude instruction data */
        assert(width);
        insn->width = width;
        traceSize += width;
        dvmCompilerAppendMIR(curBB, insn);
        cUnit.numInsts++;
        /* Instruction limit reached - terminate the trace here */
        if (cUnit.numInsts >= numMaxInsts) {
            break;
        }
        if (--numInsts == 0) {
            if (currRun->frag.runEnd) {
                break;
            } else {
                curBB = dvmCompilerNewBB(DALVIK_BYTECODE);
                lastBB->next = curBB;
                lastBB = curBB;
                curBB->id = numBlocks++;
                currRun++;
                curOffset = currRun->frag.startOffset;
                numInsts = currRun->frag.numInsts;
                curBB->startOffset = curOffset;
```

```
            codePtr = dexCode->insns + curOffset;
        }
    } else {
        curOffset += width;
        codePtr += width;
    }
}
/* Convert # of half-word to bytes */
methodStats->compiledDalvikSize += traceSize * 2;
/*
 * Now scan basic blocks containing real code to connect the
 * taken/fallthrough links. Also create chaining cells for code not included
 * in the trace.
 */
for (curBB = startBB; curBB; curBB = curBB->next) {
    MIR *lastInsn = curBB->lastMIRInsn;
    /* Hit a pseudo block - exit the search now */
    if (lastInsn == NULL) {
        break;
    }
    curOffset = lastInsn->offset;
    unsigned int targetOffset = curOffset;
    unsigned int fallThroughOffset = curOffset + lastInsn->width;
    bool isInvoke = false;
    const Method *callee = NULL;
    findBlockBoundary(desc->method, curBB->lastMIRInsn, curOffset,
                      &targetOffset, &isInvoke, &callee);
    /* Link the taken and fallthrough blocks */
    BasicBlock *searchBB;
    /* No backward branch in the trace - start searching the next BB */
    for (searchBB = curBB->next; searchBB; searchBB = searchBB->next) {
        if (targetOffset == searchBB->startOffset) {
            curBB->taken = searchBB;
        }
        if (fallThroughOffset == searchBB->startOffset) {
            curBB->fallThrough = searchBB;
        }
    }
    int flags = dexGetInstrFlags(gDvm.instrFlags,
                                 lastInsn->dalvikInsn.opCode);
    /*
     * Some blocks are ended by non-control-flow-change instructions,
     * currently only due to trace length constraint. In this case we need
     * to generate an explicit branch at the end of the block to jump to
     * the chaining cell.
     *
     * NOTE: INVOKE_DIRECT_EMPTY is actually not an invoke but a nop
     */
    curBB->needFallThroughBranch =
        ((flags & (kInstrCanBranch | kInstrCanSwitch | kInstrCanReturn |
                   kInstrInvoke)) == 0) ||
        (lastInsn->dalvikInsn.opCode == OP_INVOKE_DIRECT_EMPTY);
    /* Target block not included in the trace */
    if (curBB->taken == NULL &&
        (isInvoke || (targetOffset != curOffset))) {
        BasicBlock *newBB;
        if (isInvoke) {
            /* Monomorphic callee */
            if (callee) {
                newBB = dvmCompilerNewBB(CHAINING_CELL_INVOKE_SINGLETON);
                newBB->startOffset = 0;
                newBB->containingMethod = callee;
            /* Will resolve at runtime */
            } else {
                newBB = dvmCompilerNewBB(CHAINING_CELL_INVOKE_PREDICTED);
                newBB->startOffset = 0;
            }
        /* For unconditional branches, request a hot chaining cell */
        } else {
```

```
            newBB = dvmCompilerNewBB(flags & kInstrUnconditional ?
                                     CHAINING_CELL_HOT :
                                     CHAINING_CELL_NORMAL);
            newBB->startOffset = targetOffset;
        }
        newBB->id = numBlocks++;
        curBB->taken = newBB;
        lastBB->next = newBB;
        lastBB = newBB;
    }
    /* Fallthrough block not included in the trace */
    if (!isUnconditionalBranch(lastInsn) && curBB->fallThrough == NULL) {
        /*
         * If the chaining cell is after an invoke or
         * instruction that cannot change the control flow, request a hot
         * chaining cell.
         */
        if (isInvoke || curBB->needFallThroughBranch) {
            lastBB->next = dvmCompilerNewBB(CHAINING_CELL_HOT);
        } else {
            lastBB->next = dvmCompilerNewBB(CHAINING_CELL_NORMAL);
        }
        lastBB = lastBB->next;
        lastBB->id = numBlocks++;
        lastBB->startOffset = fallThroughOffset;
        curBB->fallThrough = lastBB;
    }
}
/* Now create a special block to host PC reconstruction code */
lastBB->next = dvmCompilerNewBB(PC_RECONSTRUCTION);
lastBB = lastBB->next;
lastBB->id = numBlocks++;
/* And one final block that publishes the PC and raise the exception */
lastBB->next = dvmCompilerNewBB(EXCEPTION_HANDLING);
lastBB = lastBB->next;
lastBB->id = numBlocks++;

if (cUnit.printMe) {
    LOGD("TRACEINFO (%d): 0x%08x %s%s 0x%x %d of %d, %d blocks",
        compilationId,
        (intptr_t) desc->method->insns,
        desc->method->clazz->descriptor,
        desc->method->name,
        desc->trace[0].frag.startOffset,
        traceSize,
        dexCode->insnsSize,
        numBlocks);
}
BasicBlock **blockList;
cUnit.method = desc->method;
cUnit.traceDesc = desc;
cUnit.numBlocks = numBlocks;
dvmInitGrowableList(&cUnit.pcReconstructionList, 8);
blockList = cUnit.blockList =
    dvmCompilerNew(sizeof(BasicBlock *) * numBlocks, true);
int i;
for (i = 0, curBB = startBB; i < numBlocks; i++) {
    blockList[i] = curBB;
    curBB = curBB->next;
}
/* Make sure all blocks are added to the cUnit */
assert(curBB == NULL);
if (cUnit.printMe) {
    dvmCompilerDumpCompilationUnit(&cUnit);
}
/* Set the instruction set to use (NOTE: later components may change it) */
cUnit.instructionSet = dvmCompilerInstructionSet(&cUnit);
/* Convert MIR to LIR, etc. */
dvmCompilerMIR2LIR(&cUnit);
```

```
    /* Convert LIR into machine code. */
    dvmCompilerAssembleLIR(&cUnit, info);
    if (cUnit.printMe) {
        if (cUnit.halveInstCount) {
            LOGD("Assembler aborted");
        } else {
            dvmCompilerCodegenDump(&cUnit);
        }
        LOGD("End %s%s, %d Dalvik instructions",
             desc->method->clazz->descriptor, desc->method->name,
             cUnit.numInsts);
    }
    /* Reset the compiler resource pool */
    dvmCompilerArenaReset();
    /* Free the bit vector tracking null-checked registers */
    dvmFreeBitVector(cUnit.registerScoreboard.nullCheckedRegs);
    if (!cUnit.halveInstCount) {
    /* Success */
        methodStats->nativeSize += cUnit.totalSize;
        return info->codeAddress != NULL;
    /* Halve the instruction count and retry again */
    } else {
        return dvmCompileTrace(desc, cUnit.numInsts / 2, info);
    }
}
```

（3）通过 parseInsn()解析 dex 字节码，实现代码如下所示：

```
static inline int parseInsn(const u2 *codePtr, DecodedInstruction *decInsn,
                            bool printMe)
{
    u2 instr = *codePtr;
    OpCode opcode = instr & 0xff;
    int insnWidth;
    // Don't parse instruction data
    if (opcode == OP_NOP && instr != 0) {
        return 0;
    } else {
        insnWidth = gDvm.instrWidth[opcode];
        if (insnWidth < 0) {
            insnWidth = -insnWidth;
        }
    }

    dexDecodeInstruction(gDvm.instrFormat, codePtr, decInsn);
    if (printMe) {
        LOGD("%p: %#06x %s\n", codePtr, opcode, getOpcodeName(opcode));
    }
    return insnWidth;
}
```

（4）使用 CompilerMethodStats *analyzeMethodBody()解析被追踪过的方法判断其是否被调用，分析出热路径(hot method)。实现代码如下所示：

```
static CompilerMethodStats *analyzeMethodBody(const Method *method)
{
    const DexCode *dexCode = dvmGetMethodCode(method);
    const u2 *codePtr = dexCode->insns;
    const u2 *codeEnd = dexCode->insns + dexCode->insnsSize;
    int insnSize = 0;
    int hashValue = dvmComputeUtf8Hash(method->name);
    CompilerMethodStats dummyMethodEntry; // For hash table lookup
    CompilerMethodStats *realMethodEntry; // For hash table storage
    /* For lookup only */
    dummyMethodEntry.method = method;
    realMethodEntry = dvmHashTableLookup(gDvmJit.methodStatsTable, hashValue,
                                         &dummyMethodEntry,
                                         (HashCompareFunc) compareMethod,
```

```
                false);
    /* Part of this method has been compiled before - just return the entry */
    if (realMethodEntry != NULL) {
        return realMethodEntry;
    }
    /*
     * First time to compile this method - set up a new entry in the hash table
     */
    realMethodEntry =
        (CompilerMethodStats *) calloc(1, sizeof(CompilerMethodStats));
    realMethodEntry->method = method;
    dvmHashTableLookup(gDvmJit.methodStatsTable, hashValue,
                       realMethodEntry,
                       (HashCompareFunc) compareMethod,
                       true);
    /* Count the number of instructions */
    while (codePtr < codeEnd) {
        DecodedInstruction dalvikInsn;
        int width = parseInsn(codePtr, &dalvikInsn, false);
        /* Terminate when the data section is seen */
        if (width == 0)
            break;
        insnSize += width;
        codePtr += width;
    }
    realMethodEntry->dalvikSize = insnSize * 2;
    return realMethodEntry;
}
```

18.4.4 BasicBlock 处理

文件“vm\compiler\IntermediateRep.c”的功能是申请 BasicBlock 空间，组织 MIR instruction 到 BasicBlock 中尾或头位置以及 LIR instruction 到 LIR 链表中的位置。主要代码如下所示：

```
#include "Dalvik.h"
#include "CompilerInternals.h"

/* Allocate a new basic block */
BasicBlock *dvmCompilerNewBB(BBType blockType)
{
    BasicBlock *bb = dvmCompilerNew(sizeof(BasicBlock), true);
    bb->blockType = blockType;
    return bb;
}

/* Insert an MIR instruction to the end of a basic block */
void dvmCompilerAppendMIR(BasicBlock *bb, MIR *mir)
{
    if (bb->firstMIRInsn == NULL) {
        assert(bb->firstMIRInsn == NULL);
        bb->lastMIRInsn = bb->firstMIRInsn = mir;
        mir->prev = mir->next = NULL;
    } else {
        bb->lastMIRInsn->next = mir;
        mir->prev = bb->lastMIRInsn;
        mir->next = NULL;
        bb->lastMIRInsn = mir;
    }
}

/*
 * Append an LIR instruction to the LIR list maintained by a compilation
 * unit
 */
void dvmCompilerAppendLIR(CompilationUnit *cUnit, LIR *lir)
{
    if (cUnit->firstLIRInsn == NULL) {
        assert(cUnit->lastLIRInsn == NULL);
```

```
        cUnit->lastLIRInsn = cUnit->firstLIRInsn = lir;
        lir->prev = lir->next = NULL;
    } else {
        cUnit->lastLIRInsn->next = lir;
        lir->prev = cUnit->lastLIRInsn;
        lir->next = NULL;
        cUnit->lastLIRInsn = lir;
    }
}

/*
 * Insert an LIR instruction before the current instruction, which cannot be the
 * first instruction.
 *
 * prevLIR <-> newLIR <-> currentLIR
 */
void dvmCompilerInsertLIRBefore(LIR *currentLIR, LIR *newLIR)
{
    if (currentLIR->prev == NULL)
        dvmAbort();
    LIR *prevLIR = currentLIR->prev;

    prevLIR->next = newLIR;
    newLIR->prev = prevLIR;
    newLIR->next = currentLIR;
    currentLIR->prev = newLIR;
}
```

18.4.5 内存初始化

文件“vm\compiler\Utility.c”的功能是实现内存初始化工作，具体来说有如下 3 个功能。

- 实现 compilation tasks 内存的分配。
- 实现对 GrowableList 的管理。
- 提供了 compilation unit 调试等一系列工具函数。

文件 Utility.c 的主要代码如下所示：

```
#include "Dalvik.h"
#include "CompilerInternals.h"
static ArenaMemBlock *arenaHead, *currentArena;
static int numArenaBlocks;

/* Allocate the initial memory block for arena-based allocation */
bool dvmCompilerHeapInit(void)
{
    assert(arenaHead == NULL);
    arenaHead =
        (ArenaMemBlock *) malloc(sizeof(ArenaMemBlock) + ARENA_DEFAULT_SIZE);
    if (arenaHead == NULL) {
        LOGE("No memory left to create compiler heap memory\n");
        return false;
    }
    currentArena = arenaHead;
    currentArena->bytesAllocated = 0;
    currentArena->next = NULL;
    numArenaBlocks = 1;
    return true;
}
/* Arena-based malloc for compilation tasks */
void * dvmCompilerNew(size_t size, bool zero)
{
    size = (size + 3) & ~3;
retry:
    /* Normal case - space is available in the current page */
    if (size + currentArena->bytesAllocated <= ARENA_DEFAULT_SIZE) {
        void *ptr;
        ptr = &currentArena->ptr[currentArena->bytesAllocated];
```

```
        currentArena->bytesAllocated += size;
        if (zero) {
            memset(ptr, 0, size);
        }
        return ptr;
    } else {
        /*
         * See if there are previously allocated arena blocks before the last
         * reset
         */
        if (currentArena->next) {
            currentArena = currentArena->next;
            goto retry;
        }
        /*
         * If we allocate really large variable-sized data structures that
         * could go above the limit we need to enhance the allocation
         * mechanism.
         */
        if (size > ARENA_DEFAULT_SIZE) {
            LOGE("Requesting %d bytes which exceed the maximal size allowed\n",
                 size);
            return NULL;
        }
        /* Time to allocate a new arena */
        ArenaMemBlock *newArena = (ArenaMemBlock *)
            malloc(sizeof(ArenaMemBlock) + ARENA_DEFAULT_SIZE);
        newArena->bytesAllocated = 0;
        newArena->next = NULL;
        currentArena->next = newArena;
        currentArena = newArena;
        numArenaBlocks++;
        LOGD("Total arena pages for JIT: %d", numArenaBlocks);
        goto retry;
    }
    return NULL;
}
/* Reclaim all the arena blocks allocated so far */
void dvmCompilerArenaReset(void)
{
    ArenaMemBlock *block;
    for (block = arenaHead; block; block = block->next) {
        block->bytesAllocated = 0;
    }
    currentArena = arenaHead;
}
/* Growable List initialization */
void dvmInitGrowableList(GrowableList *gList, size_t initLength)
{
    gList->numAllocated = initLength;
    gList->numUsed = 0;
    gList->elemList = (void **) dvmCompilerNew(sizeof(void *) * initLength,
                                               true);
}
/* Expand the capacity of a growable list */
static void expandGrowableList(GrowableList *gList)
{
    int newLength = gList->numAllocated;
    if (newLength < 128) {
        newLength <<= 1;
    } else {
        newLength += 128;
    }
    void *newArray = dvmCompilerNew(sizeof(void *) * newLength, true);
    memcpy(newArray, gList->elemList, sizeof(void *) * gList->numAllocated);
    gList->numAllocated = newLength;
    gList->elemList = newArray;
}
/* Insert a new element into the growable list */
```

```
void dvmInsertGrowableList(GrowableList *gList, void *elem)
{
    if (gList->numUsed == gList->numAllocated) {
        expandGrowableList(gList);
    }
    gList->elemList[gList->numUsed++] = elem;
}
/* Debug Utility - dump a compilation unit */
void dvmCompilerDumpCompilationUnit(CompilationUnit *cUnit)
{
    int i;
    BasicBlock *bb;
    LOGD("%d blocks in total\n", cUnit->numBlocks);
    for (i = 0; i < cUnit->numBlocks; i++) {
        bb = cUnit->blockList[i];
        LOGD("Block %d (insn %04x - %04x%s)\n",
             bb->id, bb->startOffset,
             bb->lastMIRInsn ? bb->lastMIRInsn->offset : bb->startOffset,
             bb->lastMIRInsn ? "" : " empty");
        if (bb->taken) {
            LOGD("  Taken branch: block %d (%04x)\n",
                 bb->taken->id, bb->taken->startOffset);
        }
        if (bb->fallThrough) {
            LOGD("  Fallthrough : block %d (%04x)\n",
                 bb->fallThrough->id, bb->fallThrough->startOffset);
        }
    }
}
/*
 * dvmHashForeach callback.
 */
static int dumpMethodStats(void *compilerMethodStats, void *totalMethodStats)
{
    CompilerMethodStats *methodStats =
        (CompilerMethodStats *) compilerMethodStats;
    CompilerMethodStats *totalStats =
        (CompilerMethodStats *) totalMethodStats;
    const Method *method = methodStats->method;
    totalStats->dalvikSize += methodStats->dalvikSize;
    totalStats->compiledDalvikSize += methodStats->compiledDalvikSize;
    totalStats->nativeSize += methodStats->nativeSize;
    /* Enable the following when fine-tuning the JIT performance */
#if 0
    int limit = (methodStats->dalvikSize >> 2) * 3;
    /* If over 3/4 of the Dalvik code is compiled, print something */
    if (methodStats->compiledDalvikSize >= limit) {
        LOGD("Method stats: %s%s, %d/%d (compiled/total Dalvik), %d (native)",
             method->clazz->descriptor, method->name,
             methodStats->compiledDalvikSize,
             methodStats->dalvikSize,
             methodStats->nativeSize);
    }
#endif
    return 0;
}
/*
 * Dump the current stats of the compiler, including number of bytes used in
 * the code cache, arena size, and work queue length, and various JIT stats.
 */
void dvmCompilerDumpStats(void)
{
    CompilerMethodStats totalMethodStats;

    memset(&totalMethodStats, 0, sizeof(CompilerMethodStats));
    LOGD("%d compilations using %d + %d bytes",
         gDvmJit.numCompilations,
         gDvmJit.templateSize,
         gDvmJit.codeCacheByteUsed - gDvmJit.templateSize);
```

```
    LOGD("Compiler arena uses %d blocks (%d bytes each)",
        numArenaBlocks, ARENA_DEFAULT_SIZE);
    LOGD("Compiler work queue length is %d/%d", gDvmJit.compilerQueueLength,
        gDvmJit.compilerMaxQueued);
    dvmJitStats();
    dvmCompilerArchDump();
    dvmHashForeach(gDvmJit.methodStatsTable, dumpMethodStats,
                &totalMethodStats);
    LOGD("Code size stats: %d/%d (compiled/total Dalvik), %d (native)",
        totalMethodStats.compiledDalvikSize,
        totalMethodStats.dalvikSize,
        totalMethodStats.nativeSize);
}
```

18.4.6　对 JIT 源码的总结

经过前面对 JIT 编译源码的讲解，已经对此编译方式有了一个大体的了解。由此可以总结出如下从主函数到 JIT 的调用流程。

（1）AndroidRuntime::Start()。

（2）startVm()。

（3）_JNIEnv::CallStaticVoidMethod()。

（4）Check_CallStaticVoidMethodV()。

（5）CallStaticVoidMethodV()。

（6）dvmCallMethodV()。

（7）dvmInterpret()。

（8）dvmMterpStd()。

（9）dalvik_mterp()。

（10）Dalvik_java_lang_reflect_Method_invokeNative()。

（11）dvmInvokeMethod()。

（12）dvmInterpret()。

（13）dvmJitCheckTraceRequest()。

（14）dvmjitLookupAndAdd()。

从目前 JIT 的工作调研结果来看，可以划分为 3 个大的方面，分别是汇编部分的移植、C 部分的移植以及 Dalvik 调试。之所以将 dalvik 调试作为一个大的方面，原因有 3 点。

- JIT 移植的工作量较大，涉及代码较多，找到一个好的调试方法对工作的进展与完成有着举足轻重的作用（必须性）。
- 在调研过程中发现，Google 官方一些文档中涉及 Dalvik 的调试，并有简单的实例。
- 在 Dalvik 源码中有调试的部分，包括："dalvik/tools/"下的 gdbjithelper 与 dmtracedump，"dalvik/vm"下的 Debug.c，"vm/mterp/armv5te"中的 debug.c，"/dalvik/vm/compiler/Loop.c"中用于调试的函数以及 MIPS 中的大量使用的 assert()。

第19章 Dalvik VM内存优化

在本书前面的内容中，已经讲解了 Android 系统内存的运作原理和机制，了解了 Dalvik VM 内存运行机制的基本知识。通过这些内容，为本章将要讲解的内存优化知识做好了铺垫。希望读者认真学习本章中关于内存优化的知识。

19.1 Android 内存优化的作用

Android 系统以其开源免费、界面优美，而得到了大众的青睐。各种为其量身定做的软件层出不穷。其中不乏系统优化类软件，但是所谓的内存优化真的有用么？

Android 应用程序使用 Java 作为开发语言，aapt（Android Asset Packaging Tool。保存在 SDK 的 build-tools 目录下）工具可以把编译后的 Java 代码连同其他应用程序需要的数据和资源文件一起打包到一个 Android 包文件中。这个 Android 包文件使用“.apk”格式作为扩展名，是分发应用程序并安装到移动设备的媒介，用户只需下载并安装此文件到他们的设备。单一.apk 文件中的所有代码被认为是一个应用程序。

从多方面来看，每个 Android 应用程序都存在于它自己的世界中，具体说明如下所示。

- 在默认情况下，每个应用程序均运行于它自己的 Linux 进程中。当应用程序中的任意代码开始执行时，Android 启动一个进程，而当不再需要此进程而其他应用程序又需要系统资源时，则关闭这个进程。
- 每个进程都运行于自己的 Java 虚拟机（VM）中。因此应用程序代码实际上与其他应用程序的代码是隔绝的。
- 默认情况下，每个应用程序均被赋予一个唯一的 Linux 用户 ID，并加以权限设置，使得应用程序的文件仅对这个用户、这个应用程序可见。当然，也有其他的方法使得这些文件同样能为别的应用程序所访问。

使两个应用程序共用同一个用户 ID 是可行的，这种情况下它们可以看到彼此的文件。从系统资源维护的角度来看，拥有同一个 ID 的应用程序也将在运行时使用同一个 Linux 进程以及同一个虚拟机。

Android 系统使用了 Java 语言，与传统的 C/C++等编程语言相比，它的一个明显优点是解决了内存泄漏的问题。Java 程序的内存分配与回收都是由 JRE 在后台自动进行的。JRE 会负责回收那些不再使用的内存，也就是大家听说的内存回收机制（Garbage Collection，GC）。

Java 的堆内存是一个运行时（Runtime）数据区，用以保存对象，Java 虚拟机堆内存中存储着正在运行的应用程序所建立的所有对象，这些对象不需要程序通过代码来显式的释放。垃圾回收是一种动态存储管理技术，它自动释放不再被程序引用的对象，按照特定的垃圾回收算法来实现内存资源的自动回收功能。

总之一句话，Java 的内存都是由 JVM（Java 虚拟机）控制的。有些人可能会说，既然 Java

会自动回收内存，那为什么手机总显示占用内存达到 70%，甚至 80%呢？先不说这个原因，笔者就问一句：内存占用 80%的时候，和你用所谓的内存优化软件优化之后，手机运行速度有改变么？至于内存占用较高的问题，就是 JVM 的一个缺点了（毕竟人无完人）。JVM 必须跟踪程序中的有用对象，才可以确定哪些对象是无用的，并最终释放这些无用对象。所以需要额外占用一部分内存。

19.2 查看 Android 内存和 CPU 使用情况

在本节的内容中，将首先简单介绍查看 Android 系统单个进程内存和 CPU 的使用情况。具体来说有 4 种方法，在本节的内容中将一一为大家介绍。

19.2.1 利用 Android API 函数查看

利用 ActivityManager 可以查看可用内存，例如下面的代码：

```
ActivityManager.MemoryInfo outInfo = new ActivityManager.MemoryInfo();
am.getMemoryInfo(outInfo);
```

在上述代码中，outInfo.availMem 表示可用的空闲内存。

另外，可以使用 Android.os.Debug 来查询 PSS、VSS 和 USS 等单个进程使用内存信息。例如下面的代码：

```
MemoryInfo[] memoryInfoArray = am.getProcessMemoryInfo(pids);
MemoryInfo pidMemoryInfo=memoryInfoArray[0];
pidMemoryInfo.getTotalPrivateDirty();
getTotalPrivateDirty()
Return total private dirty memory usage in kB. USS
getTotalPss()
Return total PSS memory usage in kB.
PSS
getTotalSharedDirty()
Return total shared dirty memory usage in kB. RSS
```

19.2.2 直接对 Android 文件进行解析查询

Android 实际上是一个 Linux 的衍生系统，虽然 Google 曾经在 Linux kernel 之上增加了一些专用的内存分配驱动，但是关于一般应用程序的内存使用还是采用 Linux 的传统方法，因此通过 Linux 的“/proc”文件系统的 meminfo 来分析这个系统的内存使用情况更客观。之所以这么说，是因为通过这种方法可以绕开繁琐的 Dalvik 实现机制，以系统的层面来分析。具体来说，在“/proc/cpuinfo 系统”中保存了 CPU 等的多种信息，而在“/proc/meminfo”中保存了系统内存的使用信息。

例如在“/proc/meminfo”中存在如下信息：

```
MemTotal: 16344972 kB
MemFree: 13634064 kB
Buffers: 3656 kB
Cached: 1195708 kB
```

查看机器内存时会发现 MemFree 的值很小。这主要是因为，在 Linux 中有这么一种思想，内存不用白不用，因此它尽可能地 cache 和 buffer 一些数据，以方便下次使用。但实际上这些内存也是可以立刻拿来使用的。所以：

```
空闲内存=free+buffers+cached=total-used
```

通过读取文件“/proc/meminfo”的信息获取 Memory 的总量。通过 ActivityManager.getMemoryInfo(ActivityManager.MemoryInfo)可以获取当前的可用 Memory 量。

19.2.3 通过 Runtime 类实现

通过 Android 系统提供的 Runtime 类，然后执行 adb 命令（top,procrank,ps...等命令）即可实现查询功能。通过对执行结果的标准控制台输出进行解析，这样就大大的扩展了 Android 查询功能。例如下面的演示。

```
final Process m_process = Runtime.getRuntime().exec("/system/bin/top -n 1");
final StringBuilder sbread = new StringBuilder();
BufferedReader bufferedReader = new BufferedReader(new InputStreamReader(m_process.getInputStream()),
8192);

# procrank
Runtime.getRuntime().exec("/system/xbin/procrank");
```

内存耗用分别用 VSS、RSS、PSS 或 USS 来表示，具体说明如下所示。

- VSS：是 Virtual Set Size 的缩写，表示虚拟耗用内存（包含共享库占用的内存）。
- RSS：是 Resident Set Size 的缩写，实际使用物理内存（包含共享库占用的内存）。
- PSS：是 Proportional Set Size 的缩写，实际使用的物理内存（比例分配共享库占用的内存）。
- USS：是 Unique Set Size 的缩写，进程独自占用的物理内存（不包含共享库占用的内存）。

一般来说，内存占用大小有如下规律。

```
VSS >= RSS >= PSS >= USS
```

例如下面是读取 Android 设备的内存数据（USS、PSS 和 RSS）的演示代码。

```
final ActivityManager am = (ActivityManager) getSystemService(ACTIVITY_SERVICE);
Android.os.Debug.MemoryInfo[]          memoryInfoArray          =          am.getProcessMemoryInfo(new
int[]{android.os.Process.myPid()});
```

其中类 MemoryInfo 提供了 API 接口帮助我们获取内存数据，获取各种数据的对应函数如下所示：

- 函数 getTotalPrivateDirty()：获取 USS 数据。
- 函数 getTotalSharedDirty()：获取 RSS 数据。
- 函数 getTotalPss()：获取 PSS 数据。

19.2.4 使用 DDMS 工具获取

Google 为开发人员提供了一个专用工具 DDMS，通过此工具可以获取内存的使用状况。DDMS 的全称是 Dalvik Debug Monitor Service，它提供了如下所示的功能。

- 为测试设备截屏。
- 针对特定的进程查看正在运行的线程以及堆信息。
- Logcat。
- 广播状态信息。
- 模拟电话呼叫。
- 接收 SMS。
- 虚拟地理坐标。

在接下来的内容中，将详细讲解使用 DDMS 查看 Android 内存和 CPU 使用情况的知识。

1. 启动 DDMS

DDMS 工具存放在“SDK - tools/”路径下，启动 DDMS 的方法如下所示。

（1）直接双击 ddms.bat 运行。

（2）在 Eclipes 调试程序的过程中启动 DDMS，在 Eclipes 中的界面如图 19-1 所示。然后选择“Other”，如图 19-2 所示。

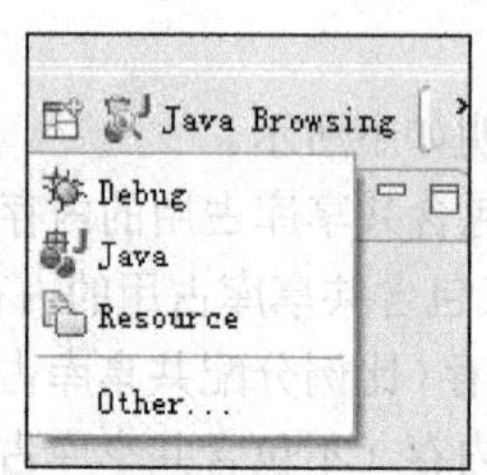

图 19-1 Eclipes 中的界面

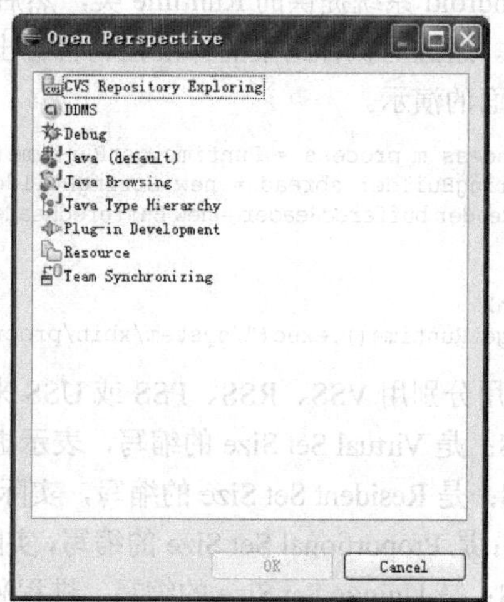

图 19-2 选择“Other”后的界面

此时双击 DDMS 就可以启动了。DDMS 对 Emulator 和外接测试机有同等效用。如果系统检测到它们（VM)）同时运行，那么 DDMS 将会默认指向 Emulator。以上两种启动后的操作有些不一样，建议分别尝试一下。

2. DDMS 的工作原理

DDMS 将搭建起 IDE 与测试终端（Emulator 或者 connected device）的链接，它们应用各自独立的端口监听调试器的信息，DDMS 可以实时监测到测试终端的连接情况。当有新的测试终端连接后，DDMS 将捕捉到终端的 ID（即运行进程），如图 19-3 所示，并通过 adb 建立调试器，从而实现发送指令到测试终端的目的。

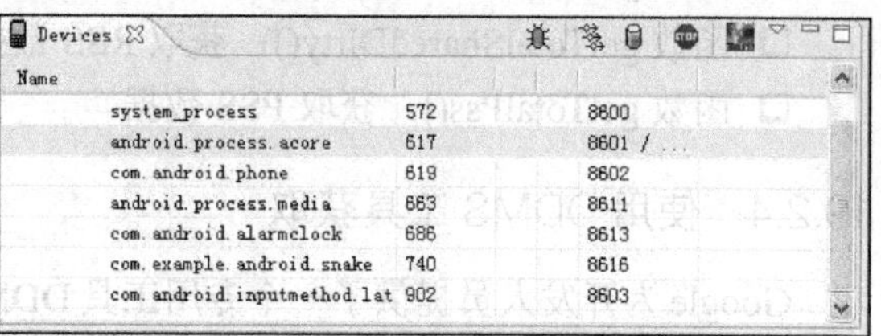

图 19-3 捕捉到终端的 ID

DDMS 监听第一个终端 App 进程的端口为 8600，APP 进程将分配 8601，如果有更多终端或者更多 APP 进程，则将按照这个顺序依次类推。DDMS 通过 8700 端口（“base port”）接收所有终端的指令。

在 GUI 的左上角可以看到标签为“Devices”的面板，这里可以查看到所有与 DDMS 连接的终端的详细信息，以及每个终端正在运行的 APP 进程，每个进程最右边相对应的是与调试器链接的端口。因为 Android 是基于 Linux 内核开发的操作平台，同时也保留了 Linux 中特有的进程 ID，它介于进程名和端口号之间，如图 19-4 所示。

在面板的右上角有一排很重要的按键，它们分别是 Debug the selected process、Update Threads、Update Heap、Stop Process 和 ScreenShot。

3. Emulator Control

通过使用 Emulator Control 面板中的一些功能，可以非常容易地使测试终端模拟真实手机所具备的一些交互功能，比如接听电话，根据选项模拟各种不同网络情况，模拟接收 SMS 消息和发送虚拟地址坐标用于测试 GPS 功能等，如图 19-5 所示。

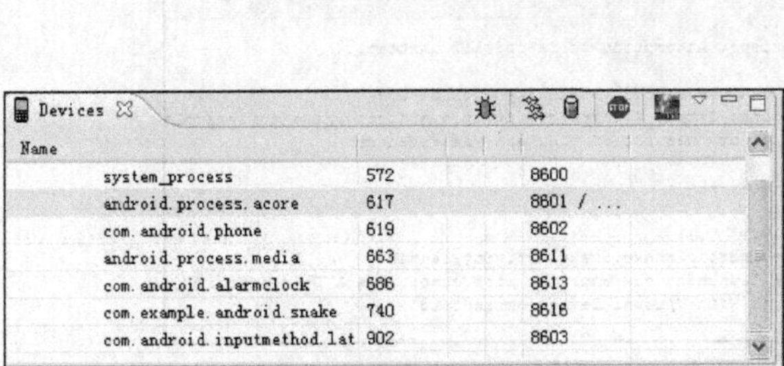

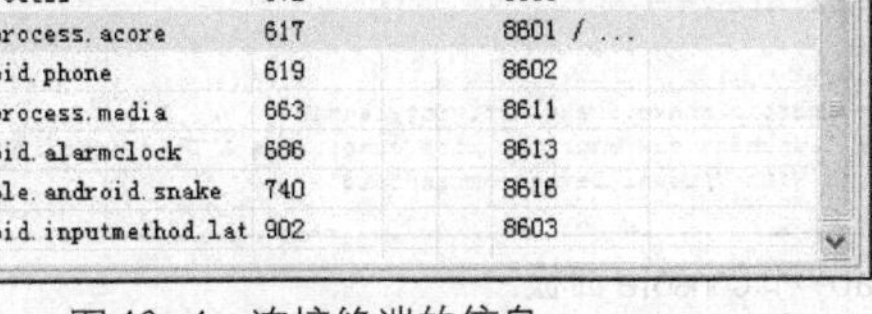

图 19-4　连接终端的信息

图 19-5　Emulator Control 面板

图 19-5 中所示的 Emulator Control 面板中，各个选项的具体说明如下所示。

- ❑ Telephony Status：通过选项模拟语音质量以及信号连接模式。
- ❑ Telephony Actions：模拟电话接听和发送 SMS 到测试终端。
- ❑ Location Control：模拟地理坐标或者模拟动态的路线坐标变化并显示预设的地理标识，可以通过以下 3 种方式。
  - Manual：手动为终端发送二维经纬坐标。
  - GPX：通过 GPX 文件导入序列动态变化地理坐标，从而模拟行进中 GPS 变化的数值。
  - KML：　通过 KML 文件导入独特的地理标识，并以动态形式根据变化的地理坐标显示在测试终端。

4. Threads、Heap、File Exporler

Threads、Heap、File Exporler 属于同一面板，例如 Heap 的界面如图 19-6 所示。

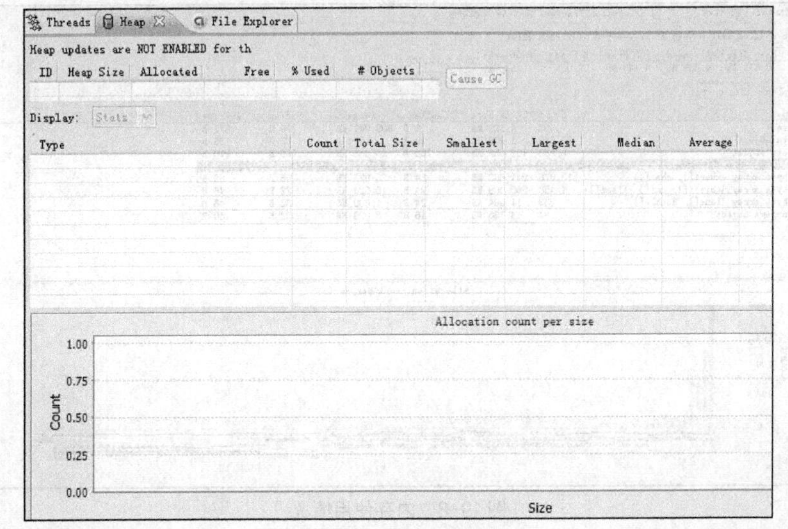

图 19-6　Heap 的界面

通过 File Exporler 可以查看 Android 模拟器中的文件，可以很方便地导入/出文件。

5. Locate、Console

Locate 用于显示输出的调试信息，Console 是 Android 模拟器输出的信息，加载程序等信息。界面如图 19-7 所示。

```
LogCat | Console
Android
[2009-09-11 13:42:57 - Snake] Installing Snake.apk...
[2009-09-11 13:43:24 - Snake] Application already exists. Attempting to re-install instead...
[2009-09-11 13:43:37 - Snake] Success!
[2009-09-11 13:43:37 - Snake] Starting activity com.example.android.snake.Snake on device
[2009-09-11 13:43:41 - Snake] ActivityManager: Starting: Intent { comp={com.example.android.snake/com.example
[2009-09-11 13:57:04 - Emulator] emulator: emulator window was out of view and was recentred
[2009-09-11 13:57:04 - Emulator]
[2009-09-11 13:57:11 - Snake] ------------------------------
[2009-09-11 13:57:11 - Snake] Android Launch!
[2009-09-11 13:57:11 - Snake] adb is running normally.
[2009-09-11 13:57:11 - Snake] Performing com.example.android.snake.Snake activity launch
[2009-09-11 13:57:11 - Snake] Automatic Target Mode: launching new emulator with compatible AVD 'moandroid'
[2009-09-11 13:57:11 - Snake] Launching a new emulator with Virtual Device 'moandroid'
```

图 19-7　Locate 和 Console 面板

6. 使用 DDMS 获取内存数据

在 DDMS 中有一个很不错的内存监测工具 Heap，使用 Heap 可以监测应用进程使用内存情况，具体操作步骤如下。

（1）启动 Eclipse 后，切换到 DDMS 透视图，并确认 Devices 视图、Heap 视图都是打开的。

（2）将手机通过 USB 链接至电脑，链接时需要确认手机是处于“USB 调试”模式，而不是作为“Mass Storage”。

（3）链接成功后，在 DDMS 的 Devices 视图中将会显示手机设备的序列号，以及设备中正在运行的部分进程信息。

（4）单击选中想要监测的进程，比如 system_process 进程。

（5）单击选中 Devices 视图界面中最上方一排图标中的“Update Heap”图标。

（6）单击 Heap 视图中的“Cause GC”按钮。

（7）此时在 Heap 视图中就会看到当前选中进程的内存使用量的详细情况，如图 19-8 所示。

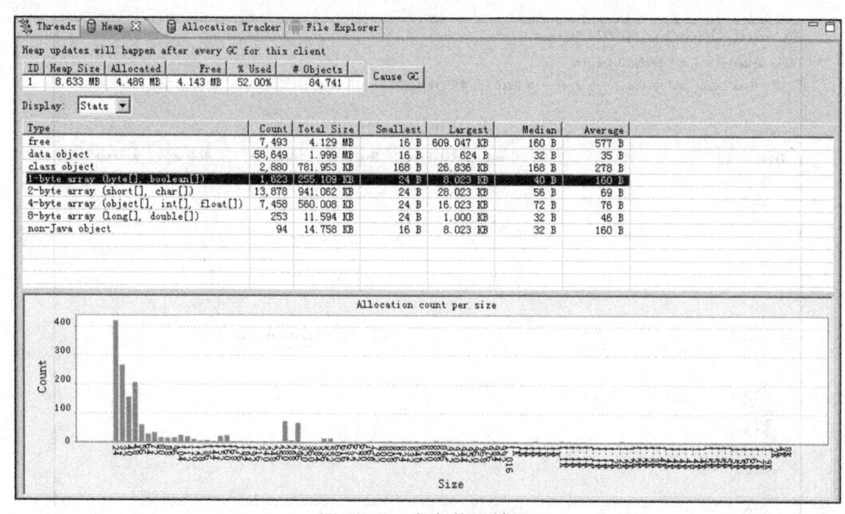

ID	Heap Size	Allocated	Free	% Used	# Objects
1	8.633 MB	4.489 MB	4.143 MB	52.00%	84,741

Type	Count	Total Size	Smallest	Largest	Median	Average
free	7,493	4.129 MB	16 B	609.047 KB	160 B	577 B
data object	58,649	1.999 MB	16 B	624 B	32 B	35 B
class object	2,880	781.953 KB	168 B	26.836 KB	168 B	278 B
1-byte array (byte[], boolean[])	1,623	255.109 KB	24 B	8.023 KB	40 B	160 B
2-byte array (short[], char[])	13,878	941.062 KB	24 B	28.023 KB	56 B	69 B
4-byte array (object[], int[], float[])	7,458	560.008 KB	24 B	16.023 KB	72 B	76 B
8-byte array (long[], double[])	253	11.594 KB	24 B	1.000 KB	32 B	46 B
non-Java object	94	14.758 KB	16 B	8.023 KB	32 B	160 B

图 19-8　内存使用情况

在图 19-8 中列出了现在系统的一些进程和使用情况。其中系统随时可以用的两项内存是 Free 和 Buffers，因为笔者设置的系统只有 128 MB 的内存，所以看上去这部分可用内存已经很少了。

笔者在此系统试着跑很占内存的游戏等应用程序的时候，并没有发现内存不足的问题。鉴于这个原因，认为这张图并不能反应笔者要得到的系统内存资源信息，因此只能从另一个角度去分析。

接下来笔者用本章前面 19.2.2 的方法进行获取，请看“/proc/meminfo”数据的截图，如图 19-9 所示。

```
ricky@ricky-laptop:~$ adb shell
* daemon not running. starting it now on port 5037 *
* daemon started successfully *
# cat /proc/meminfo
MemTotal:         107756 kB
MemFree:            1984 kB
Buffers:            2464 kB
Cached:            51988 kB
SwapCached:            0 kB
Active:            43476 kB
Inactive:          49976 kB
Active(anon):      34096 kB
Inactive(anon):     5080 kB
Active(file):       9380 kB
Inactive(file):    44896 kB
SwapTotal:             0 kB
SwapFree:              0 kB
Dirty:                 0 kB
Writeback:             0 kB
AnonPages:         39012 kB
Mapped:            26772 kB
Slab:               5164 kB
SReclaimable:       2424 kB
SUnreclaim:         2740 kB
PageTables:         3536 kB
NFS_Unstable:          0 kB
Bounce:                0 kB
WritebackTmp:          0 kB
CommitLimit:       53876 kB
Committed_AS:    1069292 kB
VmallocTotal:     385024 kB
VmallocUsed:       33996 kB
VmallocChunk:     344068 kB
#
```

图 19-9　“/proc/meminfo”数据的截图

在图 19-9 所示的截图中，对于 Linux 系统来说，可以立即使用的内存是：

```
MemFree+Buffers+Cache=556436kB
```

系统总共可用的内存为：

```
MemTotal = 107756kB
```

通过运算可以发现实际上系统目前还有 52%的内存处于空闲状态，和从 DDMS 中拿到的图差很多。或者说 Google 隐藏了 cache，没有给出笔者想要的东西。

由此可见，Android 系统为了加快系统的运行速度会在系统允许的情况下，大量地使用内存作为应用程序的 cache。而当系统内存紧张的时候，会首先释放 cache 的内存，这也就是笔者依然能跑占内存比较大的游戏的原因。由此可知，如果想得到每个 Android APP 的内存比例可以用 DDMS 来得到，如果想判断系统内存更详细的信息可以用 Linux 的“proc/meminfo”。

19.2.5　其他方法

除了本节前面介绍的 4 种方法外，接下来还介绍几种其他查看内存的方法。

1. Running Services 方式

可以通过手机上 Running Services 的 Activity 查看内存，依次单击 Setting->Applications->Running services 来实现。

2. 使用 ActivityManager 的 getMemoryInfo(ActivityManager.MemoryInfo outInfo)

函数 ActivityManager.getMemoryInfo()的主要功能是，得到当前系统剩余内存及判断是否处于低内存运行。例如下面的演示代码：

```
private void displayBriefMemory() {
    final ActivityManager activityManager = (ActivityManager) getSystemService(ACTIVITY_SERVICE);
    ActivityManager.MemoryInfo info = new ActivityManager.MemoryInfo();
    activityManager.getMemoryInfo(info);
    Log.i(tag,"系统剩余内存:"+(info.availMem >> 10)+"k");
    Log.i(tag,"系统是否处于低内存运行: "+info.lowMemory);
    Log.i(tag,"当系统剩余内存低于"+info.threshold+"时就看成低内存运行");
}
```

函数 ActivityManager.getMemoryInfo()是用 ActivityManager.MemoryInfo 返回结果，而不是 Debug.MemoryInfo。

ActivityManager.MemoryInfo 只有如下 3 个 Field。

- ❑ availMem：表示系统剩余内存；
- ❑ lowMemory：是 boolean 值，表示系统是否处于低内存运行；
- ❑ hreshold：表示当系统剩余内存低于多少时就看成低内存运行。

3. 在代码中使用 Debug 的 getMemoryInfo(Debug.MemoryInfo memoryInfo)或 ActivityManager 的 MemoryInfo[] getProcessMemoryInfo(int[] pids)

该方式能够得到比较详细的 MemoryInfo 所描述的内存使用情况，数据单位是 KB。Android 和 Linux 一样有大量内存在进程之间共享。某个进程具体使用多少内存实际上是很难统计的。

因为有 Paging out to Disk（换页）的存在，所以如果把所有映射到进程的内存相加，它可能大于内存的实际物理大小。此方法 MemoryInfo 的 Field 说明如下。

- dalvik：是指 dalvik 所使用的内存。
- native：是被 Native（本地）堆使用的内存，应该指使用 C\C++在堆上分配的内存。
- other：是指除 dalvik 和 native 使用的内存。但是具体是指什么呢？至少包括在 C\C++分配的非堆内存，比如分配在栈上的内存。
- private：是指私有的、非共享的。
- share：是指共享的内存。
- PSS：实际使用的物理内存（比例分配共享库占用的内存）。
- Pss：是把共享内存根据一定比例分摊到共享它的各个进程来计算进程使用内存。网上又说是比例分配共享库占用的内存，那么至于这里的共享是否只是库的共享，还是不清楚。
- PrivateDirty：是指非共享的，又不能换页出去（Can not be paged to Disk ）的内存的大小。比如 Linux 为了提高分配内存速度而缓冲的小对象，即使进程结束，该内存也不会释放掉，它只是又重新回到缓冲中而已。
- SharedDirty：是指共享的，又不能换页出去（Can not be paged to Disk）的内存的大小。比如 Linux 为了提高分配内存速度而缓冲的小对象，即使所有共享它的进程结束，该内存也不会释放掉，它只是又重新回到缓冲中而已。

MemoryInfo 所描述的内存使用情况都可以通过命令 adb shell "dumpsys meminfo %curProcess Name%" 得到。如果想在代码中同时得到多个进程的内存使用或非本进程的内存使用情况请使用 ActivityManager 的 MemoryInfo[] getProcessMemoryInfo(int[] pids)，否则 Debug 的 getMemoryInfo (Debug.MemoryInfo memoryInfo)就可以了。

可以通过 ActivityManager 的 List<ActivityManager.RunningAppProcessInfo> getRunningApp Processes()得到当前所有运行的进程信息。ActivityManager.RunningAppProcessInfo 中就有进程的 id，名字以及该进程包括的所有 apk 包名列表等。

4. 使用 Debug 的方法

这里的 Debug 的方法是指 getNativeHeapSize()、getNativeHeapAllocatedSize()、getNative HeapFreeSize()3 个方法。该方式只能得到 Native 堆的内存大概情况，数据单位为字节。

- static long getNativeHeapAllocatedSize()：返回的是当前进程 navtive 堆中已使用的内存大小。
- static long getNativeHeapFreeSize()：返回的是当前进程 navtive 堆中已经剩余的内存大小。
- static long getNativeHeapSize()：返回的是当前进程 navtive 堆本身总的内存大小。

例如下面的演示代码：

```
Log.i(tag,"NativeHeapSizeTotal:"+(Debug.getNativeHeapSize()>>10));
Log.i(tag,"NativeAllocatedHeapSize:"+(Debug.getNativeHeapAllocatedSize()>>10));
Log.i(tag,"NativeAllocatedFree:"+(Debug.getNativeHeapFreeSize()>>10));
```

5. 使用 dumpsys meminfo 命令

可以在 adb shell 中运行 dumpsys meminfo 命令来得到进程的内存信息，在该命令的后面需要加上进程的名字，以确定是哪个进程。比如命令"adb shell dumpsys meminfo com.teleca.robin.test"会得到com.teleca.robin.test进程使用的内存的信息：

```
Applications Memory Usage (kB):
Uptime: 12101826 Realtime: 270857936
** MEMINFO in pid 3407 [com.teleca.robin.test] **
                    native   dalvik    other    total
            size:    3456     3139      N/A      6595
       allocated:     3432     2823      N/A     6255
            free:      23      316     N/A       339
           (Pss):     724     1101     1070     2895
  (shared dirty):      1584     4540     1668     7792
    (priv dirty):      644      608      688     1940
Objects
           Views:      0         ViewRoots:       0
     AppContexts:       0      Activities:        0
          Assets:        3    AssetManagers:       3
   Local Binders:        5    Proxy Binders:       11
Death Recipients:        0
 OpenSSL Sockets:        0
SQL
            heap:        0        memoryUsed:       0
pageCacheOverflo:         0 largestMemAlloc:        0
Asset Allocations
    zip:/data/app/com.teleca.robin.test-1.apk:/resources.arsc: 1K
```

在上述信息中，“size”表示的是总内存大小（KB），"allocated"表示的是已使用了的内存大小（KB），“free”表示的是剩余的内存大小（KB）。

6. 使用"adb shell procrank"命令

如果想查看所有进程的内存使用情况，可以使用"adb shell procrank"命令。此命令会返回如下信息：

```
  PID      Vss      Rss      Pss      Uss  cmdline
  188   75832K   51628K   24824K   19028K  system_server
  308   50676K   26476K    9839K    6844K  system_server
 2834   35896K   31892K    9201K    6740K  com.sec.android.app.twlauncher
  265   28536K   28532K    7985K    5824K  com.android.phone
  100   29052K   29048K    7299K    4984K  zygote
  258   27128K   27124K    7067K    5248K  com.swype.android.inputmethod
  270   25820K   25816K    6752K    5420K  com.android.kineto
 1253   27004K   27000K    6489K    4880K  com.google.android.voicesearch
 2898   26620K   26616K    6204K    3408K  com.google.android.apps.maps:FriendService
  297   26180K   26176K    5886K    4548K  com.google.process.gapps
 3157   24140K   24136K    5191K    4272K  android.process.acore
 2854   23304K   23300K    4067K    2788K  com.android.vending
 3604   22844K   22840K    4036K    3060K  com.wssyncmldm
  592   23372K   23368K    3987K    2812K  com.google.android.googlequicksearchbox
 3000   22768K   22764K    3844K    2724K  com.tmobile.selfhelp
  101    8128K    8124K    3649K    2996K  /system/bin/mediaserver
 3473   21792K   21784K    3103K    2164K  com.android.providers.calendar
 3407   22092K   22088K    2982K    1980K  com.teleca.robin.test
 2840   21380K   21376K    2953K    1996K  com.sec.android.app.controlpanel
```

其实这里的PSS和方式4PSS的total并不一致，这是因为procrank 命令和meminfo命令使用的内核机制不太一样，所以结果会有细微差别。

另外这里的USS和方式4的Priv Dirtyd的total几乎相等，它们似乎表示的是同一个意义。但是现在得到的关于它们的意义的解释却不太相同。

7. 使用"adb shell cat/proc/meminfo" 命令

该方式只能得出系统整个内存的大概使用情况，例如下面的格式：

```
MemTotal:        395144 kB
MemFree:         184936 kB
Buffers:            880 kB
Cached:           84104 kB
SwapCached:           0 kB
```

上述格式中的具体说明如下。

- MemTotal：可供系统和用户使用的总内存大小，它比实际的物理内存要小，因为还有些内存要用于 radio、DMA buffers 等。
- MemFree：剩余的可用内存大小。该值比较大，实际上一般 Android system 的该值通常都很小，因为尽量让进程都保持运行，这样会耗掉大量内存。
- Cached：是系统用于文件缓冲等的内存，通常 Systems 需要 20 MB 大小以避免寻呼状态不好。当内存紧张时，Android 的 Memory Killer 会“杀死”一些后台进程，以避免它们消耗过多的 Cached RAM。

19.3 Android 的内存泄露

虽然 Dalvik VM 支持垃圾收集，但是这不并意味着可以不用关心内存管理。其实更应该格外注意移动设备的内存使用，毕竟其内存空间是受到限制的。在实际应用中，一些内存使用问题是很明显的，例如在每次用户触摸屏幕的时候如果应用程序有内存泄露，将会有可能触发 OutOfMemoryError，最终会导致程序崩溃。另外一些问题却很微妙，也许只是降低应用程序和整个系统的性能（当高频率和长时间地运行垃圾收集器的时候）。在本节的内容中，将详细讲解 Android 系统的内存泄露问题。

19.3.1 什么是内存泄漏

内存泄漏指由于疏忽或错误造成程序未能释放已经不再使用的内存的情况。内存泄漏并非指内存在物理上的消失，而是应用程序分配某段内存后，由于设计错误，导致在释放该段内存之前就失去了对该段内存的控制，从而造成了内存的浪费。内存泄漏与许多其他问题有着相似的症状，并且通常情况下只能由那些可以获得程序源代码的程序员才可以分析出来。然而，有不少人习惯于把任何不需要的内存使用的增加描述为内存泄漏，即使严格意义上来说这是不准确的。内存泄漏会因为减少可用内存的数量从而降低计算机的性能。最终，在最糟糕的情况下，过多的可用内存被分配掉，导致全部或部分设备停止正常工作，或者应用程序崩溃。

内存泄漏可能不严重，甚至能够被常规的手段检测出来。在现代操作系统中，一个应用程序使用的常规内存在程序终止时被释放。这表示一个短暂运行的应用程序中的内存泄漏不会导致严重后果。

在以下情况，内存泄漏导致较严重的后果。

- 程序运行后置之不理，并且随着时间的流失消耗越来越多的内存。比如服务器上的后台任务，尤其是嵌入式系统中的后台任务，这些任务可能被运行后很多年内都置之不理。
- 新的内存被频繁地分配，比如当显示电脑游戏或动画视频画面时。
- 程序能够请求未被释放的内存，例如共享内存，甚至是在程序终止的时候。
- 泄漏在操作系统内部发生。
- 泄漏在系统关键驱动中发生。
- 内存非常有限，比如在嵌入式系统或便携设备中。
- 当运行于一个终止时内存并不自动释放的操作系统（比如 AmigaOS）之上，而且一旦丢失只能通过重启来恢复。

19.3.2 为什么会发生内存泄露

JVM 会根据 generation（代）来进行 GC，如图 19-10 所示，一共被分为 Young Generation（年轻代）、Tenured Generation（老年代）、Permanent Generation（永久代，Perm Gen），Perm Gen（或称 Non-Heap 非堆）是个异类。注意，Heap 空间不包括 Perm Gen。

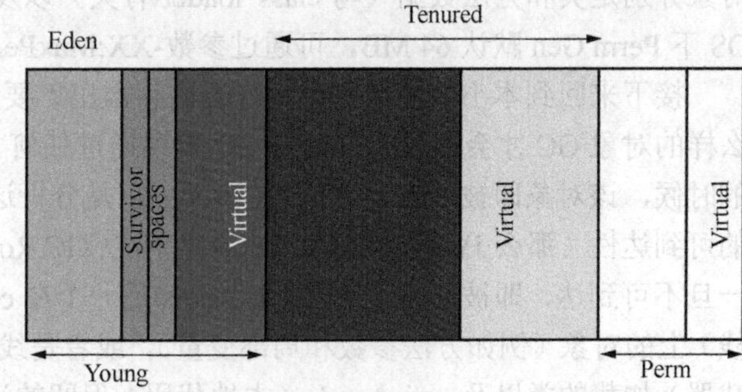

图 19-10 JVM 根据 Generation（代）来进行 GC

绝大多数的对象都在 Young Generation 被分配，也在 Young Generation 被收回。当 Young Generation 的空间被填满时，GC 会进行 Minor Collection（次回收），这样子的次回收不涉及 Heap 中的其他 Generation。Minor Collection 会根据 Weak Generational Hypothesis（弱年代假设）来假设 Young Generation 中大量的对象都是垃圾需要回收，Minor Collection 的过程会非常快。在 Young Generation 中，没有被回收的对象被转移到 Tenured Generation，然而 Tenured Generation 也会被填满，最终触发 Major Collection（主回收），这次回收针对整个 Heap，由于涉及到大量对象，所以比 Minor Collection 慢得多。

JVM 有如下 3 种垃圾回收器。

- throughput collector：用来做并行 Young Generation 回收，由参数-XX:+UseParallelGC 启动。
- concurrent low pause collector：用来做 Tenured Generation 并发回收，由参数-XX:+UseConcMarkSweepGC 启动。
- incremental low pause collector：是默认的垃圾回收器。

不建议直接使用某种垃圾回收器，最好让 JVM 自己决断，除非自己有足够的把握。

Heap 中各 Generation 空间是如何划分的呢？通过 JVM 的-Xmx=n 参数可指定最大 Heap 空间，而-Xms=n 则是指定最小 Heap 空间。在 JVM 初始化的时候，如果最小 Heap 空间小于最大 Heap 空间，如图 19-10 所示，JVM 会把未用到的空间标注为 Virtual。除了这两个参数还有-XX:MinHeapFreeRatio=n 和-XX:MaxHeapFreeRatio=n 来分别控制最大、最小的剩余空间与活动对象之比例。在 32 位 Solaris SPARC 操作系统下，默认值如表 19-1 所示，在 32 位 Windows XP 下，默认值也差不多。

表 19-1 各个参数的默认值

参 数	默 认 值
MinHeapFreeRatio	40
MaxHeapFreeRatio	70
-Xms	3670 k
-Xmx	64 m

由于 Tenured Generation 的 Major Collection 过程较慢，所以如果 Tenured Generation 空间小于 Young Generation，会造成频繁的 Major Collection，会影响效率。Server JVM 默认的 Young Generation 和 Tenured Generation 空间比例为 1:2，也就是说 Young Generation 的 Eden 和 Survivor 空间之和是整个 Heap（当然不包括 perm gen）的 1/3，该比例可以通过-XX:NewRatio=n 参数来控制，而 Client JVM 默认的-XX:NewRatio 是 8。

Young Generation 中幸存的对象被转移到 Tenured Generation，但是 Concurrent Collector 线程在这里进行 Major Collection，而在回收任务结束前空间被耗尽了，这时将会发生 Full Collections（Full GC），整个应用程序都会停止下来直到回收完成。由此可见，Full GC 是高负载生产环境的噩梦。

在此还需要说一说异类 Perm Gen，它是 JVM 用来存储无法在 Java 语言级描述的对象，这些对象分别是类和方法数据（与 class loader 有关）以及 interned strings（字符串驻留）。一般 32 位 OS 下 Perm Gen 默认 64 MB，可通过参数-XX:MaxPermSize=n 指定。

接下来回到本小节的问题：为何会内存溢出？要回答这个问题又要引出另外一个话题，即什么样的对象 GC 才会回收？当然是 GC 发现通过任何 reference chain（引用链）无法访问某个对象的时候，该对象即被回收。名词 GC Roots 正是分析这一过程的起点，例如 JVM 自己确保了对象的可到达性（那么 JVM 就是 GC Roots），所以 GC Roots 就是这样在内存中保持对象可到达性的，一旦不可到达，即被回收。通常 GC Roots 是一个在 current thread（当前线程）的 call stack（调用栈）上的对象（例如方法参数和局部变量），或者是线程自身或者是 system class loader（系统类加载器）加载的类以及 native code（本地代码）保留的活动对象。所以 GC Roots 是分析对象为何还存活于内存中的利器。

从最强到最弱，不同的引用（可到达性）级别反映了对象的生命周期。

- ❑ Strong Ref（强引用）：通常编写的代码都是 Strong Ref，与此对应的是强可达性，只有去掉强可达，对象才被回收。
- ❑ Soft Ref（软引用）：对应软可达性，只要有足够的内存，就一直保持对象，直到发现内存吃紧且没有 Strong Ref 时才回收对象。一般可用来实现缓存，通过 java.lang.ref.SoftReference 类实现。
- ❑ Weak Ref（弱引用）：比 Soft Ref 更弱，当发现不存在 Strong Ref 时，立刻回收对象而不必等到内存吃紧的时候。通过 java.lang.ref.WeakReference 和 java.util.WeakHashMap 类实现。
- ❑ Phantom Ref（虚引用）：根本不会在内存中保持任何对象，只能使用 Phantom Ref 本身。一般用于在进入 finalize()方法后进行特殊的清理过程，通过 java.lang.ref.PhantomReference 实现。

19.3.3 shallow size、retained size

shallow size 是指对象本身占用内存的大小，不包含对其他对象的引用，也就是对象头加成员变量（不是成员变量的值）的总和。在 32 位系统上，对象头占用 8 字节，int 占用 4 字节，不管成员变量（对象或数组）是否引用了其他对象（实例）或者赋值为 null 它始终占用 4 字节。因此，对于 String 对象实例来说，它有 3 个 int 成员（3×4=12 字节）、一个 char[]成员（1×4=4 字节）以及一个对象头（8 字节），总共 3×4 +1×4+8=24 字节。根据这一原则，对 String a="rosen jiang"来说，实例 a 的 shallow size 也是 24 字节。

Retained size 是指该对象自己的 shallow size，加上从该对象能直接或间接访问到对象的 shallow size 之和。换句话说，retained size 是该对象被 GC 之后所能回收到内存的总和。为了更好地理解 retained size，不妨看个例子。

把内存中的对象看成图 19-11 中的节点，并且对象和对象之间互相引用。这里有一个特殊的节点 GC Roots，这就是 reference chain 的起点。

利用 Strong Ref 存储大量数据，直到 heap 撑破，利用 interned strings（或者 class loader 加载大量的类）把 perm gen 撑破。在图 19-11 中，从 obj1 入手，灰色节点代表仅仅只有通过 obj1 才能直接或间接访问的对象。因为可以通过 GC Roots 访问，所以左图的 obj3 不是灰色节点；而在右图却是灰色，因为它已经被包含在 retained 集合内。所以对于图 19-11 中的左图来说，obj1 的 retained size 是 obj1、obj2、obj4 的 shallow size 总和；而右图的 retained size 是 obj1、obj2、obj3、 obj4 的 shallow size 总和。obj2 的 retained size 可以通过相同的方式计算。

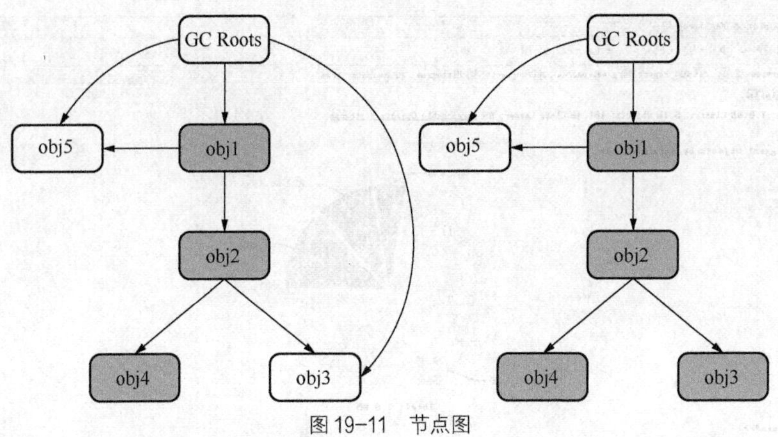

图 19-11　节点图

19.3.4　查看 Android 内存泄露的工具——MAT

在开发应用过程中，可以使用现成的工具来查看内存泄露情况。例如 DDMS 和 MAT。有关 DDMS 的知识在本章前面的内容中已经介绍过了，在接下来将讲解 MAT 工具的基本知识。

MAT 是 Memory Analyzer Tool 的缩写，是一个Eclipse插件，同时也有单独的RCP客户端。笔者使用的是 MAT 的 eclipse 插件，使用插件要比 RCP 稍微方便一些。下载后的目录结构如图 19-12 所示。

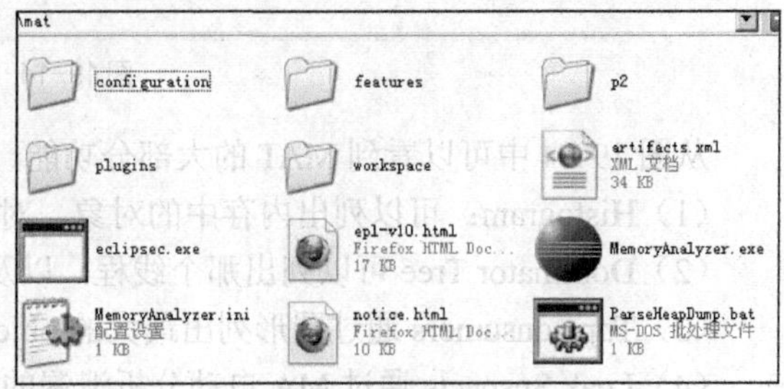

图 19-12　MAT 的文件目录

双击图 19-12 中的 MemoryAnalyzer.exe 可以打开 MAT，打开后的界面如图 19-13 所示。

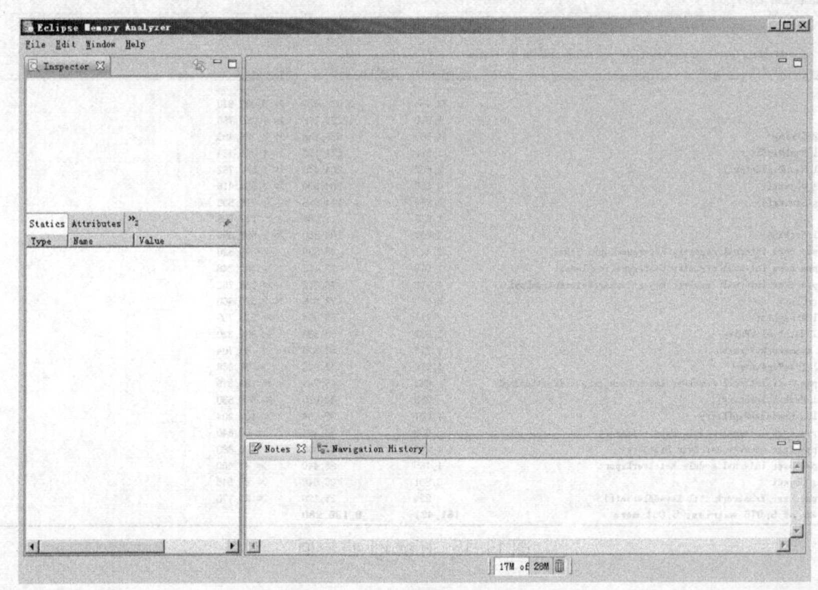

图 19-13　打开 MAT 后的界面

这样通过图 19-13 中的“File”菜单可以打开用 DDMS 生成的.hprof 文件，具体生成.hprof 文件的方法请读者参阅本章后面 19.3.2 中的内容。例如打开一个.hprof 文件后的界面如图 19-14 所示。

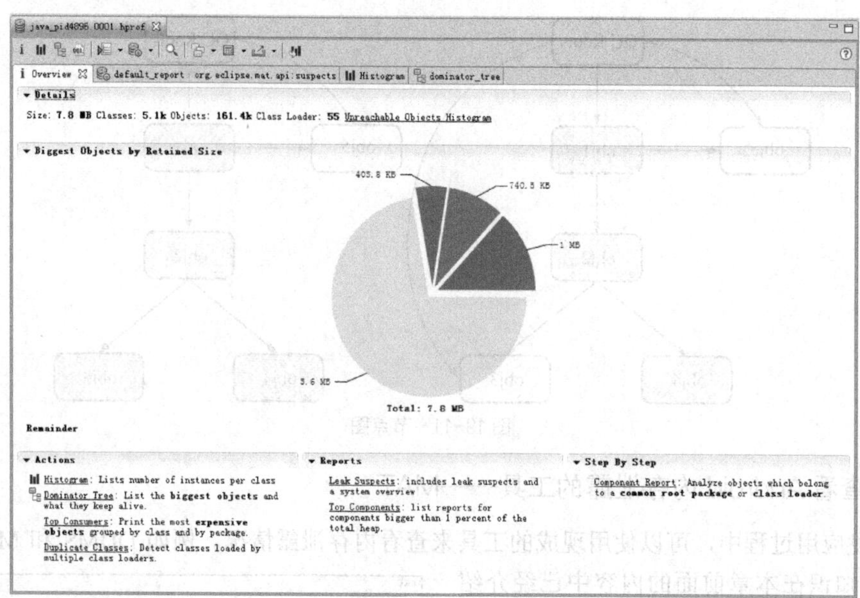

图 19-14　分析界面

从图 19-14 中可以看到 MAT 的大部分功能，具体说明如下。

（1）Histogram：可以列出内存中的对象、对象的个数以及大小。

（2）Dominator Tree 可以列出那个线程，以及线程下面的那些对象占用的空间。

（3）Top consumers 通过图形列出最大的 object。

（4）Leak Suspects 通过 MA 自动分析泄漏的原因。

单击 Histogram 选项后的界面如图 19-15 所示。

java_pid4896.0001.hprof

Overview | default_report org.eclipse.mat.api:suspects | Histogram | dominator_tree

Class Name	Objects	Shallow Heap	Retained Heap
<Regex>	<Numeric>	<Numeric>	<Numeric>
char[]	33,495	3,035,920	>= 3,035,920
byte[]	1,346	1,293,768	>= 1,293,768
java.lang.String	34,589	830,136	>= 3,395,488
java.util.HashMap$Entry	11,299	271,176	>= 1,348,424
java.util.HashMap$Entry[]	2,657	204,976	>= 1,522,752
java.lang.Object[]	4,655	204,800	>= 1,557,416
java.lang.String[]	3,569	124,656	>= 716,536
int[]	1,938	117,096	>= 117,096
java.util.HashMap	2,607	104,280	>= 1,552,096
org.eclipse.core.internal.registry.ReferenceMap$SoftRef	2,413	96,520	>= 96,520
org.eclipse.core.internal.registry.ConfigurationElement	1,919	92,112	>= 482,248
org.eclipse.osgi.internal.resolver.ExportPackageDescriptionImpl	1,193	76,352	>= 168,752
java.lang.Class	5,085	72,816	>= 2,390,400
java.util.ArrayList	2,711	65,064	>= 223,736
java.util.Hashtable$Entry	2,537	60,888	>= 309,320
org.osgi.framework.Version	1,775	56,800	>= 83,704
java.util.TreeMap$Entry	1,526	48,832	>= 56,888
org.eclipse.osgi.internal.resolver.ImportPackageSpecificationImpl	953	45,744	>= 183,376
java.util.Hashtable$Entry[]	255	32,680	>= 550,680
java.util.LinkedHashMap$Entry	1,007	32,224	>= 107,904
org.eclipse.osgi.internal.module.ResolverImport	953	30,496	>= 60,840
org.eclipse.osgi.service.resolver.StateWire	1,245	29,880	>= 29,880
org.eclipse.osgi.internal.module.ResolverExport	1,185	28,440	>= 47,400
java.lang.Object	3,231	25,848	>= 25,848
org.eclipse.osgi.framework.util.KeyedElement[]	234	25,120	>= 27,176
Σ Total: 25 of 5,076 entries; 5,051 more	161,421	8,136,960	

图 19-15　Histogram 界面

图 19-15 中主要选项的说明如下所示。

- Objects：类的对象的数量。

❑ Shallow size：就是对象本身占用内存的大小，不包含对其他对象的引用，也就是对象头加成员变量（不是成员变量的值）的总和。

❑ Retained size：是该对象自己的 shallow size，加上从该对象能直接或间接访问到对象的 shallow size 之和。换句话说，retained size 是该对象被 GC 之后所能回收到内存的总和。

单击 Dominator Tree 选项后的界面如图 19-16 所示。

i Overview | default_report org.eclipse.mat.api:suspects | Histogram | dominator_tree

Class Name	Shallow Heap	Retained ...	Percentage
<Regex>	<Numeric>	<Numeric>	<Numeric>
class java.lang.ref.Finalizer @ 0x18032900 System Class	16	1,086,984	13.36%
org.eclipse.core.internal.registry.RegistryObjectManager @ 0x183062c8	64	758,232	9.32%
sun.net.www.protocol.jar.URLJarFile @ 0x18094890	72	415,544	5.11%
org.eclipse.osgi.internal.loader.BundleLoaderProxy @ 0x18443068	32	407,856	5.01%
java.util.jar.JarFile @ 0x18034080	56	407,080	5.00%
org.eclipse.osgi.internal.resolver.SystemState @ 0x1812dc00	96	274,376	3.37%
org.eclipse.osgi.internal.baseadaptor.DefaultClassLoader @ 0x18433260 org.e	72	156,408	1.92%
sun.util.resources.TimeZoneNames @ 0x185fa760	40	102,880	1.26%
org.eclipse.osgi.internal.baseadaptor.DefaultClassLoader @ 0x3350888 org.ec	72	82,040	1.01%
sun.nio.cs.ext.ExtendedCharsets @ 0x18077830	32	68,224	0.84%
class com.sun.org.apache.xerces.internal.util.XMLChar @ 0x18309c50 System C	40	65,592	0.81%
char[28672] @ 0x18034f68 \ufffd\ufffd\ufffd\ufffd\ufffd\ufffd\ufffd\ufffd\u	57,360	57,360	0.70%
class sun.nio.cs.ext.GBK @ 0x1802faa8 System Class	32	55,048	0.68%
org.eclipse.osgi.internal.baseadaptor.DefaultClassLoader @ 0x1856f970 org.e	72	54,664	0.67%
char[256][] @ 0x180343a8	1,040	51,440	0.63%
org.eclipse.equinox.launcher.Main$StartupClassLoader @ 0x1800e518 Equinox S	80	45,400	0.56%
org.eclipse.osgi.framework.internal.core.BundleHost @ 0x181cb9c0	48	44,896	0.55%
org.eclipse.osgi.framework.internal.core.BundleHost @ 0x182798a8	48	40,488	0.50%
org.eclipse.osgi.internal.baseadaptor.DefaultClassLoader @ 0x1856f890 org.e	72	37,936	0.47%
org.eclipse.osgi.framework.internal.core.BundleHost @ 0x182a4c10	48	34,416	0.42%
class java.lang.System @ 0x1801ce90 System Class	32	33,336	0.41%
class java.lang.CharacterData00 @ 0x1801d210 System Class	32	30,992	0.38%
java.lang.Thread @ 0x1800ec58 main Thread	104	30,720	0.38%
org.eclipse.osgi.internal.baseadaptor.DefaultClassLoader @ 0x18433548 org.e	72	28,088	0.35%
org.eclipse.mat.hprof.acquire.LocalJavaProcessesUtils$StreamCollector @ 0x2f	112	26,008	0.32%
Σ Total: 25 of 17,255 entries; 17,230 more			

图 19-16 Dominator Tree 界面

单击 Overview 选项后的界面如图 19-17 所示。

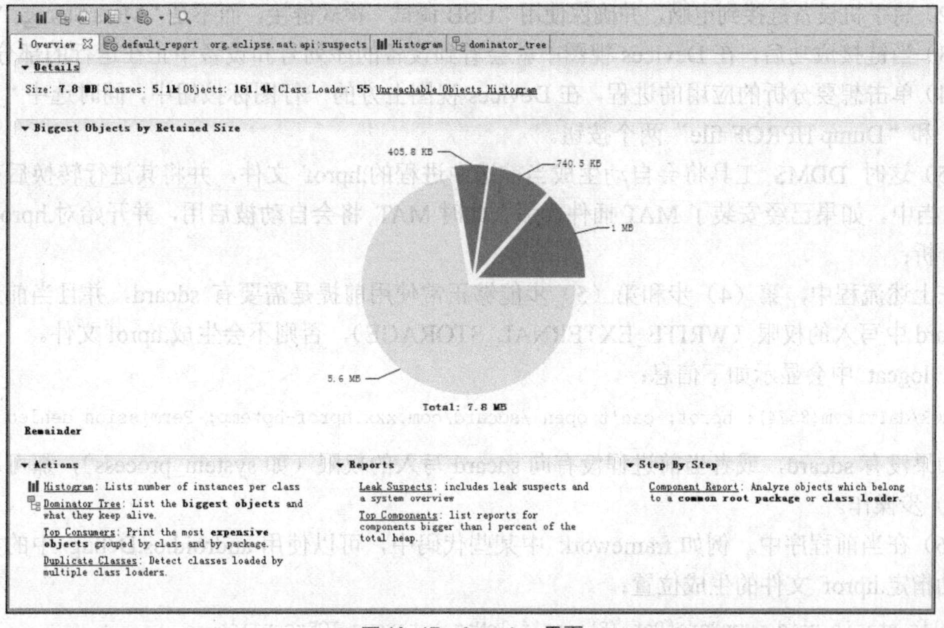

图 19-17 Overview 界面

单击图 19-17 下方的“Leak Suspects”链接后，可以查看详细的内存报表，如图 19-18 所示。

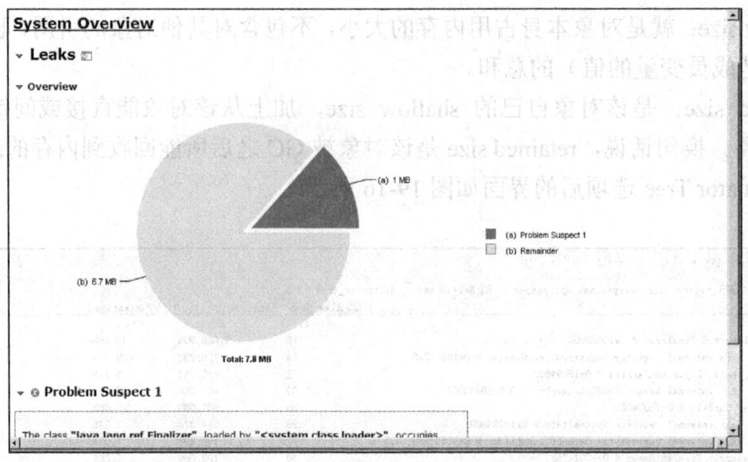

图 19-18　Leak Suspects 查看详细的内存报表

19.3.5　查看 Android 内存泄露的方法

在日常应用中，通常有如下 3 种查看 Android 内存泄露的方法。

1. 生成.hprof 文件

生成.hprof 文件的方法有很多，而且 Android 的不同版本中生成.hprof 的方式也稍有差别，各个版本中生成.prof 文件的方法请参考如下官方网址：

```
http://android.git.kernel.org/?p=platform/dalvik.git;a=blob_plain;f=docs/heapprofiling.html;hb=HEAD
```

下面以 2.1 版本为例，具体生成流程如下所示。

（1）打开 Eclipse，切换到 DDMS 透视图，同时确认已经打开了 Devices、Heap 和 logcat 视图。

（2）将手机设备链接到电脑，并确保使用“USB 调试”模式链接，而不是“Mass Storage“模式。

（3）当链接成功后，在 Devices 视图中就会看到设备的序列号和设备中正在运行的部分进程。

（4）单击想要分析的应用的进程，在 Devices 视图上方的一行图标按钮中，同时选中“Update Heap”和“Dump HPROF file”两个按钮。

（5）这时 DDMS 工具将会自动生成当前选中进程的.hprof 文件，并将其进行转换后存放在 sdcard 当中。如果已经安装了 MAT 插件，那么此时 MAT 将会自动被启用，并开始对.hprof 文件进行分析；

在上述流程中，第（4）步和第（5）步能够正常使用前提是需要有 sdcard，并且当前进程有向 sdcard 中写入的权限（WRITE_EXTERNAL_STORAGE），否则不会生成.hprof 文件。

在 logcat 中会显示如下信息：

```
ERROR/dalvikvm(8574): hprof: can't open /sdcard/com.xxx.hprof-hptemp: Permission denied
```

如果没有 sdcard，或者当前进程没有向 sdcard 写入的权限（如 system_process），那可以进行第（6）步操作。

（6）在当前程序中，例如 framework 中某些代码中，可以使用 android.os.Debug 中的如下方法手动指定.hprof 文件的生成位置：

```
public static void dumpHprofData(String fileName) throws IOException
```

例如：

```
xxxButton.setOnClickListener(new View.OnClickListener() {
 public void onClick(View view)
 {
```

```
android.os.Debug.dumpHprofData("/data/temp/myapp.hprof");
... ... }
}
```

上述代码的功能是，希望在某个按钮被点击的时候开始抓取内存使用信息，并保存在指定的位置"/data/temp/myapp.hprof"，这样就没有权限的限制了，而且也无须用 sdcard。但是这样做的前提是要保证"/data/temp"目录是存在的。这个路径可以自己定义，当然也可以写成 sdcard 当中的某个路径。

2. 使用 MAT 导入.hprof 文件

如果是 Eclipse 自动生成的.hprof 文件，则可以使用 MAT 插件（在后面将讲解这个插件）直接打开（可能是比较新的 ADT 才支持）。如果 Eclipse 自动生成的.hprof 文件不能被 MAT 直接打开，或者是使用 android.os.Debug.dumpHprofData()方法手动生成的.hprof 文件，则需要将.hprof 文件进行转换。为了讲解具体的转换方法，笔者举一个例子，例如将.hprof 文件拷贝到 PC 上的"/ANDROID_SDK/tools"目录下，并输入命令 hprofconv xxx.hprof yyy.hprof，其中 xxx.hprof 表示原始文件，yyy.hprof 为转换后的文件。转换后的文件自动放在"/ANDROID_SDK/tools"目录下。到此为止，.hprof 文件处理完毕，此时就可以用来分析内存泄露情况了。

在 Eclipse 中依次单击 Windows->Open Perspective->Other->Memory Analyzer，或者打 Memory Analyzer Tool 的 RCP。在 MAT 中单击 File->Open File，浏览并导入刚刚转换而得到的.hprof 文件。

3. 使用 MAT 的视图工具分析内存

导入.hprof 文件以后，MAT 会自动解析并生成报告，单击 Dominator Tree，并按照 Package 分组，选择自己所定义的 Package 类然后单击右键，在弹出菜单中依次选择 List objects->With incoming references。这时会列出所有可疑类，右键单击某一项，并依次选择 Path to GC Roots -> exclude weak/soft references，会进一步筛选出跟程序相关的所有有内存泄露的类。据此，可以追踪到代码中的某一个产生泄露的类。

具体的分析方法在此不做说明了，因为在 MAT 的官方网站和客户端的帮助文档中有十分详尽的介绍。了解 MAT 中各个视图的作用很重要，例如 www.eclipse.org/mat/about/screenshots.php 中介绍的。

总之使用 MAT 分析内存查找内存泄漏的根本思路，就是找到哪个类的对象的引用没有被释放，找到没有被释放的原因，也就可以很容易定位代码中哪些片段的逻辑有问题了。

另外在测试过程首先需要分析如何操作一个应用会产生内存泄露，然后在不断的操作中抓取该进程产生的 hhprof 文件使用 MAT 工具分析。目前查看内存分析内存泄露还有以下几种方法。

（1）使用 top 命令查看某个进程的内存。例如创建一个脚本文件 music.sh，该文件的内容是指定程序每隔一秒输出某个进程的内存使用情况，在此具体实现如下：

```
#!/bin/bash
 while true; do
adb shell procrank | grep "com.android.music"
 sleep 1
 done
```

此外，配合使用 procank 工具可以查看 music 进程每一秒的内存使用情况。

（2）另外使用 top 命令也可是查看内存具体为：

```
adb shell top -m 10//查看使用资源最多的10个进程
adb shell top|grep com.android.music//查看music进程的内存
```

（3）ree 命令

free 命令用来显示内存的使用情况，使用权限是所有用户。格式如下：

```
free [-b|-k|-m] [-o] [-s delay] [-t] [-V]
```

- 参数-b/-k/-m：表示分别以字节（KB、MB）为单位显示内存使用情况。
- 参数-s delay：显示每隔多少秒数来显示一次内存使用情况。
- 参数-t：显示内存总和列。
- 参数-o：不显示缓冲区调节列。

19.3.6 Android（Java）中常见的容易引起内存泄漏的不良代码

Android 主要应用在嵌入式设备当中，而嵌入式设备由于一些众所周知的条件限制，通常都不会有很高的配置，特别是内存是比较有限的。如果编写的代码中有太多的对内存使用不当的地方，难免会使得设备运行缓慢，甚至是死机。为了能够使得 Android 应用程序安全且快速地运行，Android 的每个应用程序都会使用一个专有的 Dalvik 虚拟机实例来运行，它是由 Zygote 服务进程孵化出来的，也就是说每个应用程序都是在属于自己的进程中运行的。一方面，如果程序在运行过程中出现了内存泄漏的问题，仅仅会使得自己的进程被 kill 掉，而不会影响其他进程（如果是 system_process 等系统进程出问题，则会引起系统重启）。另一方面，Android 为不同类型的进程分配了不同的内存使用上限，如果应用进程使用的内存超过了这个上限，则会被系统视为内存泄漏，从而被 kill 掉。Android 为应用进程分配的内存上限保存在"ANDROID_SOURCE/system/core/rootdir/init.rc"脚本中，例如下面的部分脚本代码：

```
# Define the oom_adj values for the classes of processes that can be
# killed by the kernel.  These are used in ActivityManagerService.
    setprop ro.FOREGROUND_APP_ADJ 0
    setprop ro.VISIBLE_APP_ADJ 1
    setprop ro.SECONDARY_SERVER_ADJ 2
    setprop ro.BACKUP_APP_ADJ 2
    setprop ro.HOME_APP_ADJ 4
    setprop ro.HIDDEN_APP_MIN_ADJ 7
    setprop ro.CONTENT_PROVIDER_ADJ 14
    setprop ro.EMPTY_APP_ADJ 15
# Define the memory thresholds at which the above process classes will
# be killed.  These numbers are in pages (4k).
    setprop ro.FOREGROUND_APP_MEM 1536
    setprop ro.VISIBLE_APP_MEM 2048
    setprop ro.SECONDARY_SERVER_MEM 4096
    setprop ro.BACKUP_APP_MEM 4096
    setprop ro.HOME_APP_MEM 4096
    setprop ro.HIDDEN_APP_MEM 5120
    setprop ro.CONTENT_PROVIDER_MEM 5632
    setprop ro.EMPTY_APP_MEM 6144
# Write value must be consistent with the above properties.
# Note that the driver only supports 6 slots, so we have HOME_APP at the
# same memory level as services.
    write /sys/module/lowmemorykiller/parameters/adj 0,1,2,7,14,15
    write /proc/sys/vm/overcommit_memory 1
    write /proc/sys/vm/min_free_order_shift 4
    write /sys/module/lowmemorykiller/parameters/minfree 1536,2048,4096,5120,5632,6144
    # Set init its forked children's oom_adj.
    write /proc/1/oom_adj -16
```

正因为应用程序能够使用的内存有限，所以在编写代码的时候需要特别注意内存使用问题。以下是一些常见的内存使用不当的情况。

19.4 常见的引起内存泄露的坏习惯

在本节的内容中，将简单讲解在开发 Android 项目时，因为程序员自身坏习惯而造成内存泄露的几种情形。希望读者认真学习，为步入本书后面知识的学习打下基础。

19.4.1　查询数据库时忘记关闭游标

程序中经常会进行查询数据库的操作，但是经常会有使用完毕 Cursor 后没有关闭的情况。如果查询结果集比较小，对内存的消耗不容易被发现，只有在长时间大量操作的情况下才会复现内存问题，这样就会给以后的测试和问题排查带来困难和风险。例如下面的代码：

```
Cursor cursor = getContentResolver().query(uri ...);
if (cursor.moveToNext()) {
    ... ...
}
```

上述代码就是在查询数据库时没有关闭游标，优化修改后的代码如下：

```
Cursor cursor = null;
try {
    cursor = getContentResolver().query(uri ...);
    if (cursor != null && cursor.moveToNext()) {
        ... ...
    }
} finally {
    if (cursor != null) {
        try {
            cursor.close();
        } catch (Exception e) {
            //ignore this
        }
    }
}
```

19.4.2　构造 Adapter 时不习惯使用缓存的 convertView

这也是大多数初学者容易忽视的问题，以构造 ListView 的 BaseAdapter 为例，在 BaseAdapter 中用如下方法向 ListView 提供每一个 item 所需要的 view 对象。

```
public View getView(int position, View convertView, ViewGroup parent)
```

初始时，ListView 会从 BaseAdapter 中根据当前的屏幕布局实例化一定数量的 view 对象，同时 ListView 会将这些 view 对象缓存起来。当向上滚动 ListView 时，原先位于最上面的 list item 的 view 对象会被回收，然后被用来构造新出现的最下面的 list item。这个构造过程就是由 getView() 方法完成的，getView()的第二个形参 View convertView 就是被缓存起来的 list item 的 view 对象（初始化时缓存中没有 view 对象则 convertView 是 null）。

由此可以看出，如果不使用 convertView，而是每次都在 getView()中重新实例化一个 View 对象，不但浪费资源，而且也浪费时间，也会使得内存占用越来越大。ListView 回收 list item 的 view 对象的过程可以查看 android.widget.AbsListView.java -> void addScrapView(View scrap)方法。

例如下面的演示代码就是不科学的：

```
public View getView(int position, View convertView, ViewGroup parent) {
    View view = new Xxx(...);
    ... ...
    return view;
}
```

优化后的代码如下所示：

```
public View getView(int position, View convertView, ViewGroup parent) {
    View view = null;
    if (convertView != null) {
        view = convertView;
        populate(view, getItem(position));
        ...
    } else {
        view = new Xxx(...);
```

```
        ...
    }
    return view;
}
```

19.4.3 没有及时释放对象的引用

这种情况描述起来比较麻烦，为了说明问题，接下来举两个演示进行说明。

（1）演示 A。

假设有如下操作：

```
public class DemoActivity extends Activity {
    ...
    private Handler mHandler = ...
    private Object obj;
    public void operation() {
     obj = initObj();
     ...
     [Mark]
     mHandler.post(new Runnable() {
            public void run() {
             useObj(obj);
            }
     });
    }
}
```

在上述代码中有一个成员变量 obj，在 operation()中希望能够将处理 obj 实例的操作 post 到某个线程的 MessageQueue 中。在上述代码中，即便 mHandler 所在的线程使用完了 obj 所引用的对象，但这个对象仍然不会被垃圾回收掉，因为 DemoActivity.obj 还保有这个对象的引用。所以如果在 DemoActivity 中不再使用这个对象了，可以在[Mark]的位置释放对象的引用，代码可以修改为如下：

```
public void operation() {
    obj = initObj();
    ...
    final Object o = obj;
    obj = null;
    mHandler.post(new Runnable() {
        public void run() {
            useObj(o);
        }
    }
}
```

（2）演示 B。

假设希望在锁屏界面（LockScreen）中监听系统中的电话服务以获取一些信息（如信号强度等），则可以在 LockScreen 中定义一个 PhoneStateListener 的对象，同时将它注册到 TelephonyManager 服务中。对于 LockScreen 对象，当需要显示锁屏界面的时候就会创建一个 LockScreen 对象，而当锁屏界面消失的时候，LockScreen 对象就会被释放掉。

但是如果在释放 LockScreen 对象时，忘记取消之前注册的 PhoneStateListener 对象，则会导致 LockScreen 无法被垃圾回收。如果不断地使锁屏界面显示和消失，则最终会由于大量的 LockScreen 对象没有办法被回收而引起 OutOfMemory，使得 system_process 进程挂掉。

由此可见，当一个生命周期较短的对象 A 被一个生命周期较长的对象 B 保有其引用的情况下，在 A 的生命周期结束时，要及时在 B 中清除掉对 A 的引用。

19.4.4 不在使用 Bitmap 对象时调用 recycle()释放内存

有时会手工操作 Bitmap 对象，如果一个 Bitmap 对象比较占用内存，当它不在被使用的时候，可以调用函数 Bitmap.recycle()回收此对象的像素所占用的内存。但这种做法并不是必须的，需要视具体情况而定。

> **注意** 除了上述 4 种常见的情形外，Android 应用程序中最典型的需要注意释放资源的情况是在 Activity 的生命周期中，在 onPause()、onStop()、 onDestroy()方法中需要适当地释放资源的情况。由于此情况很基础，在此不详细说明，具体可以查看官方文档对 Activity 生命周期的介绍，以明确何时应该释放哪些资源。

19.5 解决内存泄露实践

Java 编程中经常容易被忽视、但本身又十分重要的一个问题就是内存使用的问题。Android 应用主要使用 Java 语言编写，因此这个问题也同样会在 Android 开发中出现。本节不对 Java 编程问题做探讨，而是对于在 Android 中，特别是应用开发中的此类问题进行整理。

19.5.1 使用 MAT 根据 heap dump 分析 Java 代码内存泄漏的根源

在接下来的内容中，将介绍 MAT 如何根据 heap dump 分析泄漏根源。因为绝大多数 Android 应用程序是用 Java 语言编写的，所以本小节先用一段 Java 代码来测试内存泄露。这段测试代码非常简单，很容易找出问题，希望读者能够借此举一反三。

一开始不得不说说 ClassLoader，本质上，它的工作就是把磁盘上的类文件读入内存，然后调用 java.lang.ClassLoader.defineClass 方法告诉系统把内存镜像处理成合法的字节码。Java 提供了抽象类 ClassLoader，所有用户自定义类装载器都实例化自 ClassLoader 的子类。system class loader 在没有指定装载器的情况下默认装载用户类，在 Sun Java 1.5 中即 sun.misc.Launcher$AppClassLoader。

（1）准备 heap dump。

请看下面的 Pilot 类的演示代码：

```
package org.rosenjiang.bo;
public class Pilot{
   String name;
   int age;

   public Pilot(String a, int b){
      name = a;
      age = b;
   }
}
```

然后再看类 OOMHeapTest 是如何撑破 heap dump 的：

```
package org.rosenjiang.test;

import java.util.Date;
import java.util.HashMap;
import java.util.Map;
import org.rosenjiang.bo.Pilot;

public class OOMHeapTest {
   public static void main(String[] args){
      oom();
   }
   private static void oom(){
      Map<String, Pilot> map = new HashMap<String, Pilot>();
      Object[] array = new Object[1000000];
      for(int i=0; i<1000000; i++){
```

```
            String d = new Date().toString();
            Pilot p = new Pilot(d, i);
            map.put(i+"rosen jiang", p);
            array[i]=p;
        }
    }
}
```

在上面构造了很多的 Pilot 类实例，然后向数组和 map 中存放。由于是 Strong Ref，GC 自然不会回收这些对象，一直放在 heap 中直到溢出。当然在运行前，先要在 Eclipse 中配置 VM 参数 -XX:+HeapDumpOnOutOfMemoryError。好了，一会儿功夫内存溢出，控制台打出如下信息：

```
java.lang.OutOfMemoryError: Java heap space
Dumping heap to java_pid3600.hprof
Heap dump file created [78233961 bytes in 1.995 secs]
Exception in thread "main" java.lang.OutOfMemoryError: Java heap space
```

文件 java_pid3600.hprof 就是所需的 heap dump，读者可以在 OOMHeapTest 类所在的工程根目录下找到。

（2）使用 MAT。

使用 MAT 解析 hprof 文件，弹出向导后直接单击 Finish 按钮后会看到如图 19-19 所示的界面。

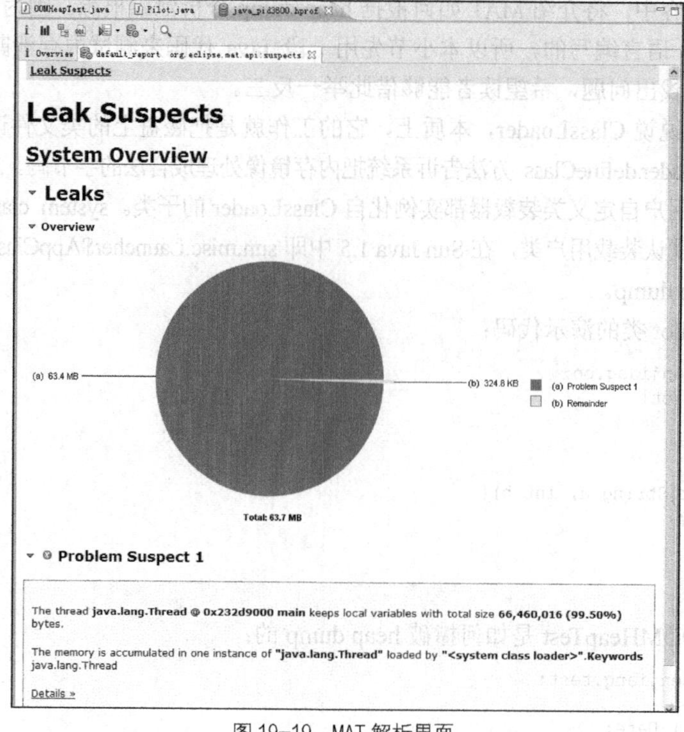

图 19-19 MAT 解析界面

由此可见，通过使用 MAT 工具分析了 heap dump 后，会在界面上非常直观地展示了一个饼图，该图深色区域被怀疑有内存泄漏。可以发现整个 heap 才 64 MB 内存，深色区域就占了 99.5%。接下来是一个简短的描述，说明 main()线程占用了大量内存，并且明确指出 system class loader 加载的"java.lang.Thread"实例有内存聚集，并建议用关键字"java.lang.Thread"进行检查。所以，MAT 通过简单的两句话就说明了问题所在，就算使用者没什么处理内存问题的经验。在下面还有一个"Details"链接，在点开之前不妨考虑一个问题：为何对象实例会聚集在内存中，为何存活（而未被 GC）？是因为 Strong Ref，如图 19-20 所示。

i Overview default_report org.eclipse.mat.api:suspects

The memory is accumulated in one instance of "java.lang.Thread" loaded by "<system class loader>".Keywords java.lang.Thread

▼ Shortest Paths To the Accumulation Point

Class Name	Shallow Heap	Retained Heap
java.lang.Thread @ 0x232d9000 main Thread	104	66,460,016

▼ Accumulated Objects

Class Name	Shallow Heap	Retained Heap	Percentage
java.lang.Thread @ 0x232d9000 main	104	66,460,016	99.50%
├ java.util.HashMap @ 0x232d8f20	40	29,951,768	44.84%
├ java.lang.Object[1000000] @ 0x22ec0000	4,000,016	4,000,016	5.99%
├ java.lang.ThreadLocal$ThreadLocalMap @ 0x232deda0	24	232	0.00%
├ java.lang.ThreadLocal$ThreadLocalMap @ 0x232dedb8	24	136	0.00%
├ org.rosenjiang.bo.Pilot @ 0x229e00c0	16	112	0.00%
├ org.rosenjiang.bo.Pilot @ 0x229e0190	16	112	0.00%
├ org.rosenjiang.bo.Pilot @ 0x229e0260	16	112	0.00%

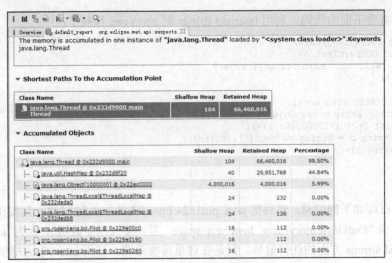

图 19-20 "Details"界面

由此可见，单击"Details"链接之后，除了在上一页看到的描述外，还有“Shortest Paths To the Accumulation Point”和“Accumulated Objects”部分，这里说明了从 GC root 到聚集点的最短路径，以及完整的 reference chain。观察 Accumulated Objects 部分，java.util.HashMap 和 java.lang.Object[1000000]实例的 retained heap(size)最大，知道 retained heap 代表从该类实例沿着 reference chain 往下所能收集到的其他类实例的 shallow heap(size)总和，所以明显类实例都聚集在 HashMap 和 Object 数组中了。这里发现一个有趣的现象，即 Object 数组的 shallow heap 和 retained heap 一样，数组的 shallow heap 和一般对象（非数组）不同，依赖于数组的长度和里面元素的类型，对数组求 shallow heap，也就是求数组集合内所有对象的 shallow heap 之和。接下来再来看 org.rosenjiang.bo.Pilot 对象实例的 shallow heap 为何是 16，因为对象头是 8 字节，成员变量 int 是 4 字节，String 引用是 4 字节，所以总共 16 字节。

接下来再来看 Accumulated Objects by Class 区域，如图 19-21 所示。

▼ Accumulated Objects by Class

Label	Number Of Objects	Used Heap Size	Retained Heap Size
org.rosenjiang.bo.Pilot	290,235	4,643,760	32,506,320
java.util.HashMap	1	40	29,951,768
java.lang.Object[]	1	4,000,016	4,000,016
java.lang.String[]	38	1,200	1,200
java.lang.ThreadLocal$ThreadLocalMap	2	48	368
sun.util.calendar.Gregorian$Date	1	96	96
java.lang.StringBuilder	1	16	88
java.security.AccessControlContext	1	24	24
char[]	1	24	24
java.lang.Object	1	8	8
java.lang.Class	1	0	0
∑ Total: 11 entries	290,283	8,645,232	66,459,912

图 19-21 Accumulated Objects by Class 区域

顾名思义，在“Accumulated Objects by Class”区域能找到被聚集的对象实例的类名。此处的类 org.rosenjiang.bo.Pilot 是头条，被实例化了 290325 次，再返回去看程序，其实是笔者故意为之的。还有很多有用的报告可用来协助分析问题，只是本文中的例子太简单，所以也用不上。

（3）perm gen。

perm gen 是一个异类，在里面存储了类和方法数据（与 class loader 有关）以及 interned strings（字符串驻留）。在 heap dump 中没有包含太多的 perm gen 信息。那么就用这些少量的信息来解决问题吧。请读者看下面的代码，利用 interned strings 把 perm gen 撑破了：

```
package org.rosenjiang.test;
public class OOMPermTest {
    public static void main(String[] args){
        oom();
    }
    private static void oom(){
        Object[] array = new Object[10000000];
        for(int i=0; i<10000000; i++){
            String d = String.valueOf(i).intern();
            array[i]=d;
        }
    }
}
```

控制台会打印如下的信息，然后把 java_pid1824.hprof 文件导入到 MAT。其实在 MAT 里，看到的状况应该和“OutOfMemoryError: Java heap space”差不多（用了数组），因为 heap dump 并没有包含 interned strings 方面的任何信息。只是在这里需要强调，使用 intern()方法的时候应该多加注意：

```
java.lang.OutOfMemoryError: PermGen space
Dumping heap to java_pid1824.hprof
Heap dump file created [121273334 bytes in 2.845 secs]
Exception in thread "main" java.lang.OutOfMemoryError: PermGen space
```

开始思考如何把 class loader 撑破，经过尝试会发现使用 ASM 来动态生成类才能达到目的。ASM（http://asm.objectweb.org）的主要作用是处理已编译类（compiled class），能对已编译类进行生成、转换、分析（功能之一是实现动态代理），而且它运行起来足够快和小巧，文档也全面。ASM 提供了 core API 和 tree API，前者是基于事件的方式，后者是基于对象的方式，类似于 XML 的 SAX、DOM 解析，但是使用 tree API 性能会有损失。到此为止，已编译类的结构如下。

- 修饰符（例如 public、private）、类名、父类名、接口和 annotation 部分。
- 类成员变量声明，包括每个成员的修饰符、名字、类型和 annotation。
- 方法和构造函数描述，包括修饰符、名字、返回和传入参数类型以及 annotation，当然还包括这些方法或构造函数的具体 Java 字节码。
- 常量池（constant pool）部分，constant pool 是一个包含类中出现的数字、字符串、类型常量的数组。

已编译类和原来的类源码区别在于，已编译类只包含类本身，内部类不会在已编译类中出现，而是生成另外一个已编译类文件；其二，已编译类中没有注释；其三，已编译类没有 package 和 import 部分。这里还得说说已编译类对 Java 类型的描述，对于原始类型由单个大写字母表示，Z 代表 boolean、C 代表 char、B 代表 byte、S 代表 short、I 代表 int、F 代表 float、J 代表 long、D 代表 double；而对类类型的描述使用内部名（internal name）外加前缀 L 和后面的分号共同来表示，所谓内部名就是带全包路径的表示法，例如 String 的内部名是“java/lang/String”；对于数组类型，使用单方括号加上数据元素类型的方式描述。最后对于方法的描述，用圆括号来表示，如果返回是 void 用 V 表示，具体参考图 19-22。

Java type	Type descriptor
boolean	Z
char	C
byte	B
short	S
int	I
float	F
long	J
double	D
Object	Ljava/lang/Object;
int[]	[I
Object[][]	[[Ljava/lang/Object;

图 19-22 Java 类型的描述

而在下面的代码中会使用 ASM core API，在此需要注意接口 ClassVisitor 是核心，FieldVisitor、MethodVisitor 都是辅助接口。ClassVisitor 应该按照这样的方式来调用：visit visitSource? visitOuterClass? (visitAnnotation | visitAttribute)*(visitInnerClass |

visitField | visitMethod)* visitEnd。就是说方法 visit 必须首先调用，再调用最多一次的 visitSource，再调用最多一次的 visitOuterClass 方法，接下来再多次调用 visitAnnotation 和 visitAttribute 方法，最后是多次调用 visitInnerClass、visitField 和 visitMethod 方法。调用完后再调用 visitEnd 方法作为结尾。

另外还需要注意 ClassWriter 类，该类实现了 ClassVisitor 接口，通过 toByteArray 方法可以把已编译类直接构建成二进制形式。由于要动态生成子类，所以这里只对 ClassWriter 感兴趣。首先是抽象类原型：

```
package org.rosenjiang.test;
public abstract class MyAbsClass {
    int LESS = -1;
    int EQUAL = 0;
    int GREATER = 1;
    abstract int absTo(Object o);
}
```

其次是自定义类加载器，因为 ClassLoader 的 defineClass 方法都是 protected 的，所以要想加载字节数组形式（因为 toByteArray 了）的类，只有通过继承自己后再实现：

```
package org.rosenjiang.test;

public class MyClassLoader extends ClassLoader {
    public Class defineClass(String name, byte[] b) {
        return defineClass(name, b, 0, b.length);
    }
}
```

最后看测试类的演示代码如下：

```
package org.rosenjiang.test;
import java.util.ArrayList;
import java.util.List;
import org.objectweb.asm.ClassWriter;
import org.objectweb.asm.Opcodes;

public class OOMPermTest {
    public static void main(String[] args) {
        OOMPermTest o = new OOMPermTest();
        o.oom();
    }

    private void oom() {
        try {
            ClassWriter cw = new ClassWriter(0);
            cw.visit(Opcodes.V1_5, Opcodes.ACC_PUBLIC + Opcodes.ACC_ABSTRACT,
            "org/rosenjiang/test/MyAbsClass", null, "java/lang/Object",
            new String[] {});
            cw.visitField(Opcodes.ACC_PUBLIC + Opcodes.ACC_FINAL + Opcodes.ACC_STATIC, "LESS", "I",
            null, new Integer(-1)).visitEnd();
            cw.visitField(Opcodes.ACC_PUBLIC + Opcodes.ACC_FINAL + Opcodes.ACC_STATIC, "EQUAL", "I",
            null, new Integer(0)).visitEnd();
            cw.visitField(Opcodes.ACC_PUBLIC + Opcodes.ACC_FINAL + Opcodes.ACC_STATIC, "GREATER", "I",
            null, new Integer(1)).visitEnd();
            cw.visitMethod(Opcodes.ACC_PUBLIC + Opcodes.ACC_ABSTRACT, "absTo",
            "(Ljava/lang/Object;)I", null, null).visitEnd();
            cw.visitEnd();
            byte[] b = cw.toByteArray();

            List<ClassLoader> classLoaders = new ArrayList<ClassLoader>();
            while (true) {
                MyClassLoader classLoader = new MyClassLoader();
                classLoader.defineClass("org.rosenjiang.test.MyAbsClass", b);
                classLoaders.add(classLoader);
            }
        } catch (Exception e) {
```

```
                e.printStackTrace();
            }
        }
    }
```

运行后控制台会报错，输出如下信息：

```
java.lang.OutOfMemoryError: PermGen space
Dumping heap to java_pid3023.hprof
Heap dump file created [92593641 bytes in 2.405 secs]
Exception in thread "main" java.lang.OutOfMemoryError: PermGen space
```

打开文件 java_pid3023.hprof，如图 19-23 所示。着重看图中的 Classes: 88.1 k 和 Class Loader: 87.7 k 部分，从这点可看出 class loader 加载了大量的类。

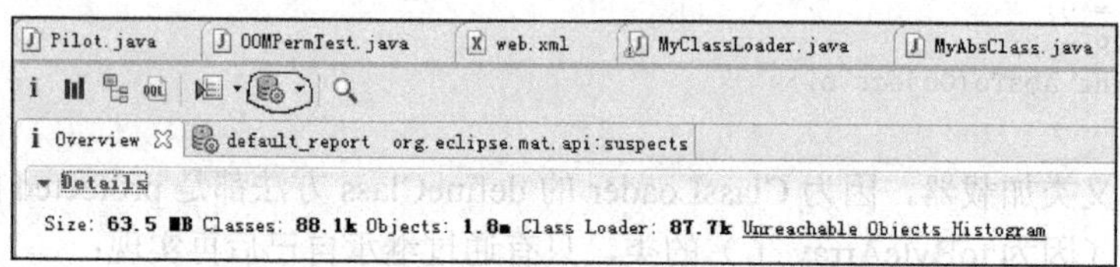

图 19-23 打开文件 java_pid3023.hprof

更进一步分析，需要点击图 19-23 中的按钮，然后选择 Java Basics——Class Loader Explorer 功能。打开后能看到图 19-24 所示的界面，第一列是 Class Loader 名字；第二列是 class loader 已定义类（defined classes）的个数，这里要说一下已定义类和已加载类（loaded classes）了，当需要加载类的时候，相应的 class loader 会首先把请求委派给父 Class Loader，只有当父 Class Loader 加载失败后，该 Class Lloader 才会自己定义并加载类，这就是 Java 自己的"双亲委派加载链"结构；第三列是 Class Loader 所加载的类的实例数目。

i Overview | default_report org.eclipse.mat.api:suspects | classloaderexplorerquery

Class Name	Defined Cla...	No. of Instances
<Regex>	<Numeric>	<Numeric>
<system class loader>	444	1,670,445
sun.misc.Launcher$AppClassLoader @ 0x22efe608	11	87,657
org.rosenjiang.test.MyClassLoader @ 0x229e0148	1	0
parent sun.misc.Launcher$AppClassLoader @ 0x22efe608	11	87,657
org.rosenjiang.test.MyAbsClass		0
Total: 2 entries		
org.rosenjiang.test.MyClassLoader @ 0x229e0428	1	0
org.rosenjiang.test.MyClassLoader @ 0x229e0708	1	0
org.rosenjiang.test.MyClassLoader @ 0x229e09e8	1	0
org.rosenjiang.test.MyClassLoader @ 0x229e0cc8	1	0
org.rosenjiang.test.MyClassLoader @ 0x229e0fa8	1	0
org.rosenjiang.test.MyClassLoader @ 0x229e1288	1	0
org.rosenjiang.test.MyClassLoader @ 0x229e1568	1	0
org.rosenjiang.test.MyClassLoader @ 0x229e1848	1	0
org.rosenjiang.test.MyClassLoader @ 0x229e1b28	1	0
org.rosenjiang.test.MyClassLoader @ 0x229e1e08	1	0
org.rosenjiang.test.MyClassLoader @ 0x229e20e8	1	0
org.rosenjiang.test.MyClassLoader @ 0x229e23c8	1	0
org.rosenjiang.test.MyClassLoader @ 0x229e26a8	1	0
org.rosenjiang.test.MyClassLoader @ 0x229e2988	1	0
org.rosenjiang.test.MyClassLoader @ 0x229e2c68	1	0
org.rosenjiang.test.MyClassLoader @ 0x229e2f48	1	0
org.rosenjiang.test.MyClassLoader @ 0x229e3228	1	0
org.rosenjiang.test.MyClassLoader @ 0x229e3508	1	0
org.rosenjiang.test.MyClassLoader @ 0x229e37e8	1	0
org.rosenjiang.test.MyClassLoader @ 0x229e3ac8	1	0
org.rosenjiang.test.MyClassLoader @ 0x229e3da8	1	0
Total: 24 of 87,659 entries	88,110	1,758,102

图 19-24 Class Loader Explorer 功能

在 Class Loader Explorer 面板会发现 Class Loader 是否加载了过多的类。另外，还有 Duplicate Classes 功能，也能协助分析重复加载的类。在此可以肯定的是，MyAbsClass 被重复加载了很多次。

> **注意** 其实MAT工具已经非常强大了，上述演示根本用不到MAT的其他分析功能。在上述演示中，对于OOM不只列举了两种溢出错误，其实还有多种其他错误，但对于perm gen来说，如果实在找不出问题所在，建议使用JVM的-verbose参数，该参数会在后台打印出日志，可以用来查看哪个class loader加载了什么类，例如："[Loaded org.rosenjiang.test.MyAbsClass from org.rosenjiang.test.MyClassLoader]"。

19.5.2 演练Android中内存泄露代码优化及检测

在接下来的内容中，将演示测试一个Android应用项目内存泄露的过程。

源码路径	\daima\19\MAT_Test
功能	演练Android中内存泄露代码优化及检测

（1）创建工程。

新建Android工程"com.devdiv.test.mat_test"，新建如下所示的类代码，然后运行该工程：

```
package com.devdiv.test.mat_test;

import java.util.ArrayList;
import java.util.List;
import android.app.Activity;
import android.os.Bundle;

public class MainActivity extends Activity {
    List<String> list = new ArrayList<String>();
//    private PersonInfo person = new PersonInfo();
    @Override
    public void onCreate(Bundle savedInstanceState) {
        super.onCreate(savedInstanceState);
        setContentView(R.layout.activity_main);

        new Thread () {
            @Override
            public void run() {
                while (true){
                    MainActivity.this.list.add("OutOfMemoryError soon");
                }
            }
        }.start();
    }
}
```

（2）分析内存。

在此需要获得.hprof内存镜像文件，可以在进程运行过程中切换到DDMS的透视图页面，然后选中要查看内存镜像的进程，并单击"Dump HPROF file"即可，如图19-25所示。

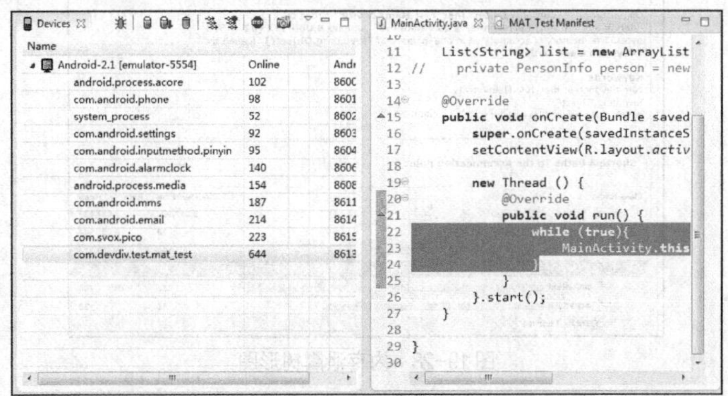

图19-25 内存镜像文件分析

生成的 hprof 文件会默认使用 MAT 打开，选择“Leak Suspects Report”后，单击 Finish 按钮。如图 19-26 所示。

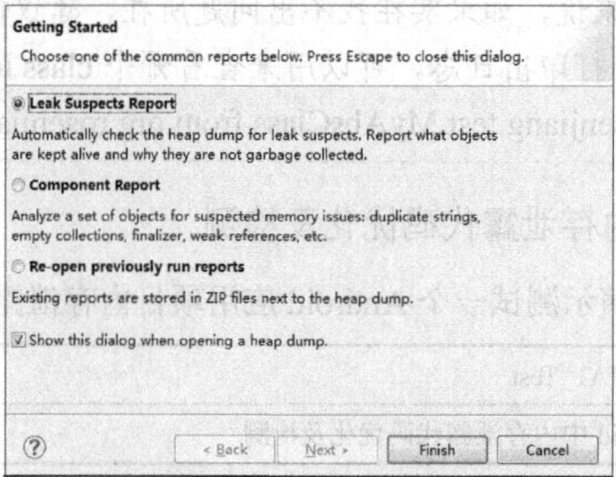

图 19-26　用 MAT 打开内存镜像文件

经过一段时间的初始化后，就能够直观地看到关于内存泄露的饼图，如图 19-27 所示。

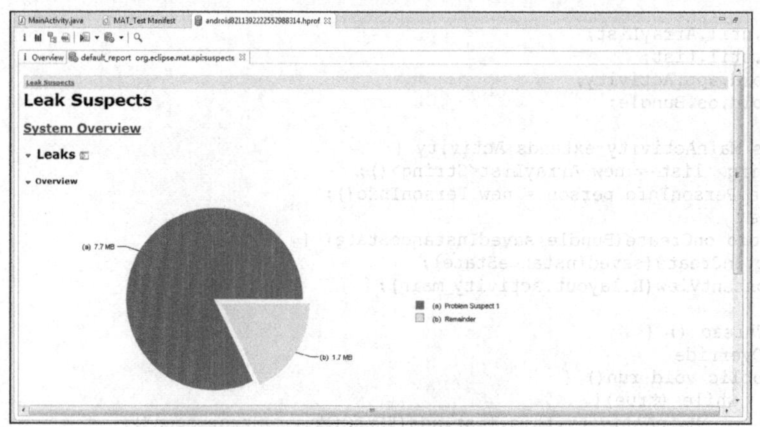

图 19-27　内存泄露饼图

然后就可以查看相关的内存泄露，如图 19-28 所示。

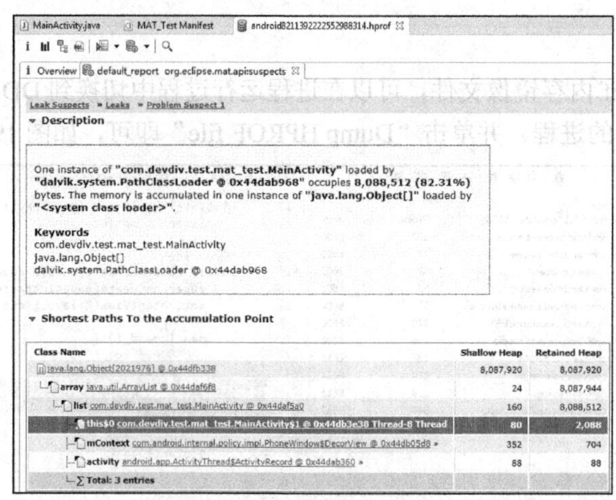

图 19-28　内存泄露树形图

这样通过提示就可以找到内存泄露所在，然后就可以根据本章前面介绍的知识实现优化。

第 20 章　Dalvik VM 性能优化

在科学术语中，性能是指测量仪器仪表实现预期功能的能力的特性。在 Android 应用中，各个软件和内在系统的能力特性就是性能，例如资源存储、加载能力、虚拟机和渲染机制等。在本章的内容中，将详细讲解 Dalvik VM 性能优化的基本知识。

20.1 加载 APK/DEX 文件优化

在 Android 系统中，对编译出来的 DEX 字节码和 APK 文件的加载过程实现了尽可能的优化。具体来说有如下两点优化工作。

- 对于预置应用：Android 会在系统编译后，生成优化文件，以 ODEX 后缀结尾，这样在发布时除 APK 文件（不包含 DEX）外，还有一个相应的 ODEX 文件。
- 对于非预置应用：在运行前，Android 会优化 DEX 文件，在第一次启动应用时，执行文件的 DEX 被优化成 DEY 文件并放在“/data/dalvik-cache”目录。如果应用的 APK 文件不发生变化，DEX 文件不会被重新生成，加快了以后的启动速度。加载过程如图 20-1 所示。

DEX 文件由 header、string_ids、type_ids、proto_ids、field_ids、method_ids、class_defs、data 等几部分构成。图 20-2 显示了这几部分内容在 DEX 文件中的布局。

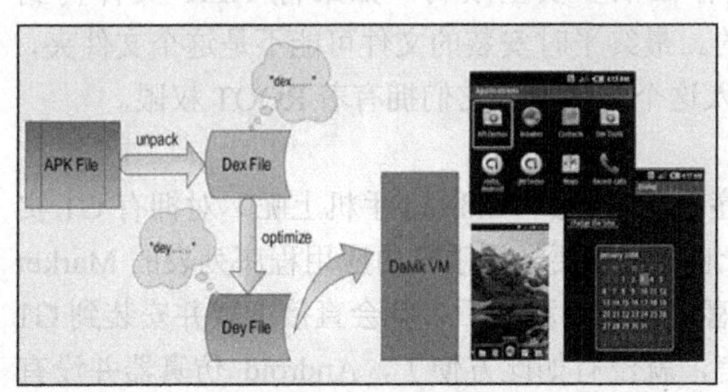

图 20-1　加载过程

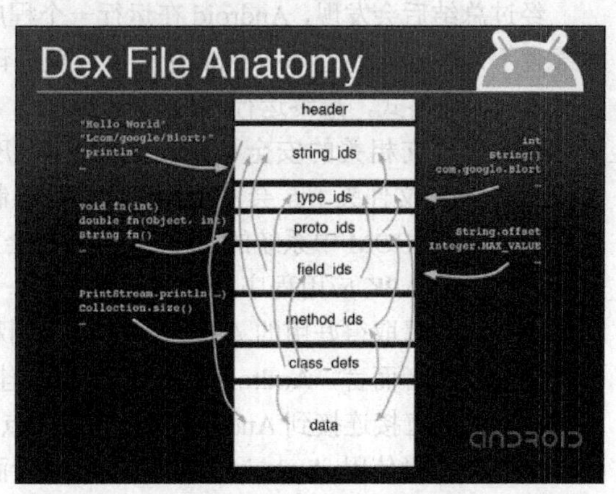

图 20-2　在 DEX 文件中的布局

在 Java 中，每一个类会被编译成相应的 CLASS 文件，一个应用会定义若干个类，这就导致同一个应用的多个 CLASS 文件中会存在冗余信息，而在 Android 中，“dx”工具会将同一个应用的所有 CLASS 文件内容整合到一个 DEX 文件中，这样就减小了整体的文件尺寸，I/O 操作也提高了类的查找速度。原来每个 CLASS 文件中的常量池，在 DEX 文件中由一个常量池来统一管理。“dx”工具整合 CLASS 文件的过程如图 20-3 所示。

具体到 DEX 文件，经过“dx”工具优化后的内部逻辑如图 20-4 所示。

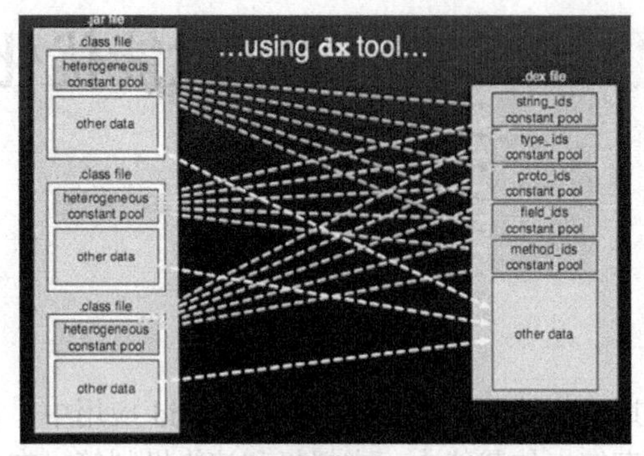

图 20-3　“dx”工具整合 CLASS 文件的过程

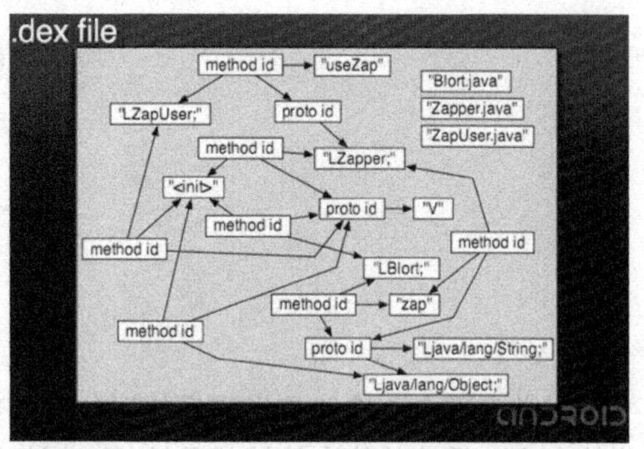

图 20-4　经过“dx”工具优化后的内部逻辑

20.1.1　APK 文件介绍

APK 是 Android Package 的缩写，即 Android 安装包。APK 是类似 Symbian Sis 或 Sisx 的文件格式。通过将 APK 文件直接传到 Android 模拟器或 Android 手机中执行即可安装。

（1）APK 文件的结构。

APK 文件和 SIS 一样最终把 Android SDK 编译的工程打包成一个格式为 APK 的安装程序文件。APK 文件其实是 zip 格式，但后缀名被修改为 APK，通过 UnZip 解压后，可以看到 Dex 文件，Dex 是 Dalvik VM executes 的缩写，即 Android Dalvik 执行程序，并非 Java ME 的字节码而是 Dalvik 字节码。一个 APK 文件结构如下。

- “res\”：存放资源文件的目录。
- AndroidManifest.xml：程序全局配置文件。
- classes.dex：Dalvik 字节码。
- resources.arsc：编译后的二进制资源文件。

经过总结后会发现，Android 在运行一个程序时首先需要 UnZip，这样做对于程序的保密性和可靠性不是很高，通过 dexdump 命令可以反编译。在 Android 平台中，Dalvik VM 的执行文件被打包为 APK 格式，最终运行时加载器会解压，然后获取编译后的 androidmanifest.xml 文件中的 permission 分支相关的安全访问。但是这样仍然存在很多安全限制，如果将 APK 文件传到“/system/app”文件夹下，会发现执行是不受限制的。最终平时安装的文件可能不是这个文件夹，而在 Android ROOM 中系统的 APK 文件默认会放入这个文件夹，它们拥有着 ROOT 权限。

（2）下载 APK 应用程序。

可以从哪里取得好用的 Android APK 应用程序，并安装到 Android 手机上呢？对拥有 G1 实体手机的使用者而言，Android Market 就是最佳的地方，只要使用手机内应用程序列表的 Market 程序，就可以直接连接到 Android Market，而点选喜爱的应用程序后，就会直接下载并安装到 G1 手机上。不过对使用 Android 仿真器的使用者而言，就没有如此方便了，Android 仿真器并没有 Android Market 这个应用程序，只能使用内附的浏览器浏览 Android Market，为何说是浏览呢？因为 Android Market 不是采用通用网页浏览方式来下载文件，虽然可以使用常见的浏览器看到 Android Market 上的应用程序，但是没有办法下载到 Android 仿真器或一般的计算机上，原因是 Android Market 采用特有的网页 API，使用 native UI 的方式来访问，唯有通过内建在 G1 手机内的 Market 应用程序，才能下载 Android Market 网页中的应用程序，并自动安装到 G1 手机上。

所以 Android 仿真器的使用者只好浏览该网页上的应用程序，然后通过搜索引擎去看是否有开发人员将应用程序放到 Android Market 之后，还另外将 APK 文件放置在一般网页上了。到此为止，使用 Android 仿真器的您也不需要这么灰心，因为有太多的人遇到同样的问题，也就生成非常多的 Android 应用程序网页，您可以浏览这些网页并把上面的 APK 文件下载到一般计算机上，再安装到 Android 仿真器上。

20.1.2 DEX 文件优化

DEX 即 Android Dalvik 执行程序，Google 在新发布的 Android 平台上使用了自己的 Dalvik 虚拟机来定义。这种虚拟机执行的并非 Java 字节码，而是另一种字节码：dex 格式的字节码。在编译 Java 代码之后，通过 Android 平台上的工具可以将 Java 字节码转换成 Dex 字节码。这个 DalvikVM 针对手机程序程序的 CPU 进行了优化处理，可以同时执行许多 VM，而不会占用太多 Resource（资源）。

对于 Android DEX 文件进行优化，需要注意的一点是 DEX 文件的结构是紧凑的，但是还是要想方设法提高程序的运行速度，就仍然需要对 DEX 文件进行进一步优化。

调整所有字段的字节序（LITTLE_ENDIAN）和对齐结构中的每一个域验证 DEX 文件中的所有类 对一些特定的类进行优化，对方法里的操作码进行优化。优化后的文件大小会有所增加，应该是原 Android DEX 文件的 1～4 倍。优化发生的时机有两个：对于预置应用，可以在系统编译后，生成优化文件，以 ODEX 结尾。

这样在发布时除 APK 文件（不包含 DEX）以外，还有一个相应的 Android DEX 文件；对于非预置应用，包含在 APK 文件里的 DEX 文件会在运行时被优化，优化后的文件将被保存在缓存中。每一个 Android 应用都运行在一个 Dalvik 虚拟机实例里，而每一个虚拟机实例都是一个独立的进程空间。虚拟机的线程机制、内存分配和管理、Mutex 等等都是依赖底层操作系统而实现的。

所有 Android 应用的线程都对应一个 Linux 线程，虚拟机因而可以更多地依赖操作系统的线程调度和管理机制。不同的应用在不同的进程空间里运行，加之对不同来源的应用都使用不同的 Linux 用户来运行，可以最大程度地保护应用的安全和独立运行。

Zygote 是一个虚拟机进程，同时也是一个虚拟机实例的孵化器，每当系统要求执行一个 Android 应用程序，Zygote 就会 FORK 出一个子进程来执行该应用程序。这样做的好处显而易见：Zygote 进程是在系统启动时产生的，它会完成虚拟机的初始化、库的加载、预置类库的加载和初始化等操作，而在系统需要一个新的虚拟机实例时，Zygote 通过复制自身，最快速地提供给系统。另外，对于一些只读的系统库，所有虚拟机实例都和 Zygote 共享一块内存区域，大大节省给内存开销。Android 应用开发和 Dalvik 虚拟机 Android 应用所使用的编程语言是 Java 语言，和 Java SE 一样，编译时使用 Sun JDK 将 Java 源程序编程为标准的 Java 字节码文件（.class 文件）。

而后通过工具软件 DX 把所有的字节码文件转成 Android DEX 文件（classes.dex）。最后使用 Android 打包工具（aapt）将 DEX 文件、资源文件以及 AndroidManifest.xml 文件（二进制格式）组合成一个应用程序包（APK）。应用程序包可以被发布到手机上运行。

20.1.3 使用类动态加载技术实现加密优化

在加载 APK 文件和 DEX 文件时，使用了类动态加载技术。在 Android 应用开发过程中，通常常规的开发方式和代码架构就能满足普通的需求。但是有些特殊问题，常常引发人们进一步的沉思。考虑下面的问题。

- 如何开发一个可以自定义控件的 Android 应用，就像 Eclipse 一样，可以动态加载插件？
- 如何让 Android 应用执行服务器上的不可预知的代码？
- 如何对 Android 应用加密，而只在执行时自解密，从而防止被破解？

熟悉 Java 技术的读者会想到，需要使用类加载器灵活地加载执行的类。这在 Java 中已经算是一项比较成熟的技术了，但是在 Android 应用中，人们都还比较陌生。

（1）类加载机制。

Dalvik 虚拟机如同其他 Java 虚拟机一样，在运行程序时首先需要将对应的类加载到内存中。而在 Java 标准的虚拟机中，类加载可以从 class 文件中读取，也可以是其他形式的二进制流。因此，常常利用这一点，在程序运行时手动加载 Class，从而达到代码动态加载执行的目的。

但是Dalvik虚拟机毕竟不算是标准的Java虚拟机，因此在类加载机制方面它们有相同的地方，也有不同之处，必须区别对待。例如，在使用标准 Java 虚拟机时，经常自定义继承自 ClassLoader 的类加载器。然后通过 defineClass()方法来从一个二进制流中加载 Class，但是这在 Android 里是行不通的，这一点可以从 Android 源码知道。Android 中 ClassLoader 的 defineClass()方法，具体是调用 VMClassLoader 的 defineClass()本地静态方法。而这个本地方法除了抛出一个“UnsupportedOperationException”异常之外，什么都没做，甚至连返回值都为空。演示代码如下：

```
static void Dalvik_java_lang_VMClassLoader_defineClass(const u4* args,JValue* pResult)
{
    Object* loader = (Object*) args[0];
    StringObject* nameObj = (StringObject*) args[1];
    const u1* data = (const u1*) args[2];
    int offset = args[3];
    int len = args[4];
    Object* pd = (Object*) args[5];
    char* name = NULL;
    name = dvmCreateCstrFromString(nameObj);
    LOGE("ERROR: defineClass(%p, %s, %p, %d, %d, %p)\n",
        loader, name, data, offset, len, pd);
    dvmThrowException("Ljava/lang/UnsupportedOperationException;",
        "can't load this type of class file");
    free(name);
    RETURN_VOID();
}
```

（2）Dalvik VM 类的加载机制。

如果在 Dalvik VM 中的 ClassLoader 不好用，此时应该如何实现动态加载类呢？Android 从 ClassLoader 派生出了两个类：DexClassLoader 和 PathClassLoader。其中需要特别说明的是，PathClassLoader 中如下被注释掉的代码：

```
/* --this doesn't work in current version of Dalvik--
    if (data != null) {
        System.out.println("--- Found class " + name
            + " in zip[" + i + "] '" + mZips[i].getName() + "'");
        int dotIndex = name.lastIndexOf('.');
        if (dotIndex != -1) {
            String packageName = name.substring(0, dotIndex);
            synchronized (this) {
                Package packageObj = getPackage(packageName);
                if (packageObj == null) {
                    definePackage(packageName, null, null,
                        null, null, null, null, null);
                }
            }
        }
        return defineClass(name, data, 0, data.length);
    }
*/
```

这可以从另一方面证明了 defineClass()函数在 Dalvik 虚拟机中被“阉割”了。而在这两个继承自 ClassLoader 的类加载器，本质上是重载了 ClassLoader 的 findClass 方法。在执行 loadClass 时可以参照 ClassLoader 的部分源码：

```
protected Class<?> loadClass(String className, boolean resolve)
throws ClassNotFoundException {
Class<?> clazz = findLoadedClass(className);
    if (clazz == null) {
        try {
            clazz = parent.loadClass(className, false);
        } catch (ClassNotFoundException e) {
            // Don't want to see this.
        }
        if (clazz == null) {
            clazz = findClass(className);
        }
    }
    return clazz;
}
```

由此可见，DexClassLoader 和 PathClassLoader 都属于符合双亲委派模型的类加载器（因为它们没有重载 loadClass 方法）。也就是说，它们在加载一个类之前，会检查自己以及自己以上的类加载器是否已经加载了这个类。如果已经加载过了，则会直接将之返回，而不会重复加载。

DexClassLoader 和 PathClassLoader 其实都是通过类 DexFile 实现类加载功能的。这里需要顺便提一下的是，Dalvik 虚拟机识别的是 dex 文件，而不是 class 文件。因此，供类加载的文件也只能是 dex 文件，或者包含有 dex 文件的.apk 或.jar 文件。

PathClassLoader 是通过构造函数 new DexFile(path)来生成 DexFile 对象的；而 DexClassLoader 则是通过其静态方法 loadDex（path, outpath, 0）得到 DexFile 对象的。这两者的区别在于 DexClassLoader 需要提供一个可写的 outpath 路径，用来释放.apk 包或者.jar 包中的 dex 文件。也就是说，PathClassLoader 不能主动从 zip 包中释放出 dex，因此只支持直接操作 dex 格式文件，或者已经安装的 apk（因为已经安装的 apk 在 cache 中存在缓存的 dex 文件）。而 DexClassLoader 可以支持.apk、.jar 和.dex 文件，并且会在指定的 outpath 路径释放出 dex 文件。

当 PathClassLoader 在加载类时，调用的是 DexFile 的 loadClassBinaryName，而 DexClassLoader 调用的是 loadClass。所以在使用 PathClassLoader 时，类的全名需要用“/”替换“.”。

（3）具体操作。

在具体操作时，可能需要使用到的工具有：javac、dx、eclipse 等。其中在使用 dx 工具时，最好指明：--no-strict，因为 class 文件的路径可能不匹配。当加载好类后，通常可以通过 Java 反射机制来使用这个类。但是这样做的效率相对不高，而且老用反射代码也比较复杂凌乱。更好的做法是定义一个 interface，并将这个 interface 写进容器端。待加载的类，继承自这个 interface，并且有一个参数为空的构造函数，使得能够通过 Class 的 newInstance 方法产生对象。然后将对象强制转换为 interface 对象，于是就可以直接调用成员方法了。

（4）代码加密。

在加密代码时，最初设想将 dex 文件加密，然后通过 JNI 将解密代码写在 Native 层。解密之后直接传入二进制流，再通过 defineClass 将类加载到内存中。但是由于不能直接使用 defineClass，而必须传文件路径给 Dalvik 虚拟机内核，因此解密后的文件需要写到磁盘上，增加了被破解的风险。

Dalvik 虚拟机内核仅支持从 dex 文件加载类的方式是不灵活的，由于没有非常深入地研究内核，笔者不能确定是 Dalvik 虚拟机本身不支持还是 Android 在移植时将其阉割了。不过坚信的是，Dalvik 或 Android 开源项目都正在向能够支持 raw 数据定义类的方向努力。

在 RawDexFile 出来之前，只能使用这种存在一定风险的加密方式。需要注意释放的 dex 文件路径及权限管理。另外在类加载完毕之后，除非出于其他目的，否则应该马上删除临时的解密文件。

20.2 SD 卡优化

SD 卡作为手机的扩展存储设备，在手机中充当硬盘角色，可以让手机存放更多的数据以及多媒体等大体积文件。因此查看 SD 卡的内存就与查看硬盘的剩余空间一样，是人们经常操作的一件事，那么在 Android 开发中，如何能获取 SD 卡的内存容量呢？

（1）要获取 SD 卡上面的信息，必须先对 SD 卡有访问的权限，因此第一件事就是需要添加访问扩展设备的权限。

```
<uses-permission android:name="android.permission.WRITE_EXTERNAL_STORAGE">
</uses-permission>
```

（2）需要判断手机上面 SD 卡是否插好，如果有 SD 卡的情况下才可以访问得到并获取到它的相关信息，当然以下这个语句需要用 if 做判断：

```
Environment.getExternalStorageState().equals(Environment.MEDIA_MOUNTED)
```

取得 SD 卡文件路径的代码如下：

```
File path = Environment.getExternalStorageDirectory();
StatFs statfs = new StatFs(path.getPath());
```

获取 block 大小的代码如下：

```
long blocSize = statfs.getBlockSize();
```

获取 BLOCK 数量的代码如下：

```
long totalBlocks = statfs.getBlockCount();
```

获取空闲的 Block 数量的代码如下：

```
long availaBlock = statfs.getAvailableBlocks();
```

计算总空间大小和空闲的空间大小的代码如下：

```
/**
* 取得空闲 sd 卡空间大小
* @return
*/
public long getAvailaleSize(){
File path = Environment.getExternalStorageDirectory(); //取得 sdcard 文件路径
StatFs stat = new StatFs(path.getPath());
/*获取 block 的 SIZE*/
long blockSize = stat.getBlockSize();
/*空闲的 Block 的数量*/
long availableBlocks = stat.getAvailableBlocks();
/* 返回 bit 大小值*/
return availableBlocks * blockSize/1024/1024;
//(availableBlocks * blockSize)/1024 KIB 单位
//(availableBlocks * blockSize)/1024 /1024 MIB 单位
}
/**
* SD 卡大小
* @return
*/
public long getAllSize(){
File path = Environment.getExternalStorageDirectory();
StatFs stat = new StatFs(path.getPath());
/*获取 block 的 SIZE*/
long blockSize = stat.getBlockSize();
/*块数量*/
long availableBlocks = stat.getBlockCount();
/* 返回 bit 大小值*/
return availableBlocks * blockSize/1024/1024;
}
```

（1）加载优化。

Android 的图片浏览器等多媒体应用可以加载整个 SD 卡内的所有图像，在加载前会把数据做成数据库，不用每次扫描，这大大加快了启动速度。事实上扫描操作是通过 MediaScanner 来实现的，目前支持的文件类型在 MediaFile.java 中定义。主要包括音频、MIDI、视频、图片、播放列表等。MediaScannerService 服务的启动仅在收到如下权限后才会启动。

- android.intent.action.BOOT_COMPLETED。
- android.intent.action.MEDIA_MOUNTED。
- android.intent.action.MEDIA_SCANNER_SCAN_FILE。

当然，在 SD 卡容量较大且文件较多时，MediaScannerService 服务将会运行一段不短的时间，这对电池的持续能力会造成一定的影响，尤其是在电池技术始终不能有显著突破的前提下。

（2）分区优化。

在将 SD 卡分区时，通常把第一分区的簇的大小设置为 16 KB 或者更大之后，都会得到更高的 PC 测试得分，而且手机也会增加流畅度。这是因为 fat/fat32/vfat 系统采用扇区+簇的方式来存储文件，一个扇区一般是 512 KB，一个簇就是一组扇区的集合。在默认状态下，4 GB 以上的 FAT32 分区应该是每簇 16 个扇区，也就是 16×512=8 KB，这个字节对于 Android 的使用来说偏小了，当然也会提高些空间利用率。通过调大簇的大小，例如调到 64 个扇区（32 KB），可以提高大文件的存取效率。假设一个文件的大小是 1 024 KB，如果是 8 KB 的簇则最坏情况需要 1024/8=128 次 IO；如果是 32 KB 的簇，则最坏情况只需要 1 024/32=32 次 IO。当然实际的 IO 次数可能比这些都少，因为操作系统有自己的优化方法，会尽量多读一些进来。最坏情况指的是 1 024 KB 的数据真的被分别存在 128 个互不相邻的簇上，这样就是真的 128 次 IO 了。因此更大的簇对于大文件是有非常好的优化效果的，现在我们日常用的文件其实大部分都大于 1 024 KB 了，比如一个 MP3 至少也要 3 MB 才算可听，而导航数据就更大了。因此尽量使用更大的簇是很有必要的。

所以在优化分区时，建议在格式化 SD 卡第一分区（FAT32）的时候设置簇大小为 32 KB，其实最高可以到 64 BK，但是 64 KB 是 FAT32 设计的极限，从软件角度来说，在极限状态运行是不可靠的。因此使用较低一档的大小，格式化之后把数据复制回去。

为什么提高了 FAT32 分区的效率就会提高手机的整体效率呢？这是因为这两个分区是在一个硬件上，如果 FAT32 占用的 IO 负载大，则 Ext 分区分到的 IO 带宽自然就小了，而 Android 手机在第一次运行的时候其实是非常频繁地访问 FAT32 分区的，因为“Media Scaner”在做数据搜集扫描工作时，为 Android 特有的手机全局搜索准备数据。因此，对 FAT 32 的优化可以对整个手机的运行效率有所提高的。当然可以等“Media Scaner”扫描完后再用手机，这样会很流畅，那时候手机流畅度就跟 FAT 32 是否优化无关了。

20.3 虚拟机优化详解

在虚拟机中的指令的解释事件主要分为 3 个方面，分别是分发指令、访问运算数、执行运算。其中“分发指令”这个环节对性能的影响最大，为了加快运行速度，必须提高分发指令的速度。在 Android 虚拟机和原生库层面，Android 同样进行了很多的优化。在本节的内容中，将详细讲解 Android 虚拟机优化的基本知识。

20.3.1 平台优化——ARM 的流水线技术

Android 虚拟机充分挖掘了 CPU 的性能，针对 armv5te 进行了优化，充分利用 armv5te 的执行流水线来提高执行的效率。在 Android 刚诞生的时候，虽然支持 ARM CPU，其实实际上只支持 armv5te 的指令集，因为 Android 系统专门为 armv5te 进行了优化，充分利用 armv5te 的执行流水线来提高执行的效率，这也是在 500 MHz 的三星 2440 运行效果不是很好，而在 200 MHz 的 OMAP CPU 上运行比较流畅的原因了，所以在最新的代码中有专门针对 x86 和 armv4 的优化部分。

1. 什么是流水线技术

流水线技术通过多个功能部件并行工作来缩短程序执行时间，提高了处理器核的效率和吞吐率，从而成为微处理器设计中最为重要的技术之一。ARM 7 处理器核使用了典型三级流水线的冯•诺伊曼结构，ARM9 系列则采用了基于五级流水线的哈佛结构。通过增加流水线级数简化了流水线各级的逻辑，进一步提高了处理器的性能。

ARM7 的三级流水线在执行单元完成了大量的工作，包括与操作数相关的寄存器和存储器读写操作、ALU 操作以及相关器件之间的数据传输。执行单元的工作往往占用多个时钟周期，从而成为系统性能的瓶颈。ARM9 采用了更为高效的五级流水线设计，增加了 2 个功能部件分别访问存储器并写回结果，且将读寄存器的操作转移到译码部件上，使流水线各部件在功能上更平衡；同时其哈佛架构避免了数据访问和取指的总线冲突。

然而不论是三级流水线还是五级流水线，当出现多周期指令、跳转分支指令和中断发生的时候，流水线都会发生阻塞，而且相邻指令之间也可能因为寄存器冲突导致流水线阻塞，降低流水线的效率。本章节在对流水线原理及运行情况详细分析的基础上，研究通过调整多个功能部件并行工作来缩短程序执行时间，提高处理器核的效率和吞吐率，从而成为微处理器设计中最为重要的技术之一。

2. ARM7 流水线技术

ARM7 系列处理器中每条指令分取指、译码、执行 3 个阶段，分别在不同的功能部件上依次独立完成。取指部件完成从存储器装载一条指令，通过译码部件产生下一周期数据路径需要的控制信号，完成寄存器的解码，再送到执行单元完成寄存器的读取、ALU 运算及运算结果的写回，需要访问存储器的指令完成存储器的访问。流水线上虽然一条指令仍需 3 个时钟周期来完成，但通过多个部件并行，使得处理器的吞吐率约为每个周期一条指令，提高了流式指令的处理速度，从而可达到 0.9 MIPS/MHz 的指令执行速度。

在三级流水线下，通过 R15 访问程序计数器（PC）时会出现取指位置和执行位置不同的现象。这须结合流水线的执行情况考虑，取指部件根据 PC 取指，取指完成后 PC+4 送到 PC，并把取到的指令传递给译码部件，然后取指部件根据新的 PC 取指。因为每条指令 4 字节，故 PC 值等于当前程序执行位置+8。

3. ARM 9 流水线技术

ARM 9 系列处理器的流水线分为取指、译码、执行、访存、回写。取指部件完成从指令存储器取指；译码部件读取寄存器操作数，与三级流水线中不占有数据路径区别很大；执行部件产生 ALU 运算结果或产生存储器地址（对于存储器访问指令而言）；访存部件访问数据存储器；回写部件完成执行结果写回寄存器。把三级流水线中的执行单元进一步细化，减少了在每个时钟周期内必须完成的工作量，进而允许使用较高的时钟频率，且具有分开的指令和数据存储器，减少了冲突的发生，每条指令的平均周期数明显减少。

4. 三级流水线运行情况分析

三级流水线在处理简单的寄存器操作指令时，吞吐率为平均每个时钟周期一条指令。但是在存储器访问指令、跳转指令的情况下会出现流水线阻断情况，导致流水线的性能下降。图 20-5 给出了流水线的最佳运行情况，图中的 MOV、ADD、SUB 指令为单周期指令。从 T_1 开始，用 3 个时钟周期执行了 3 条指令，指令平均周期数（CPI）等于 1 个时钟周期。

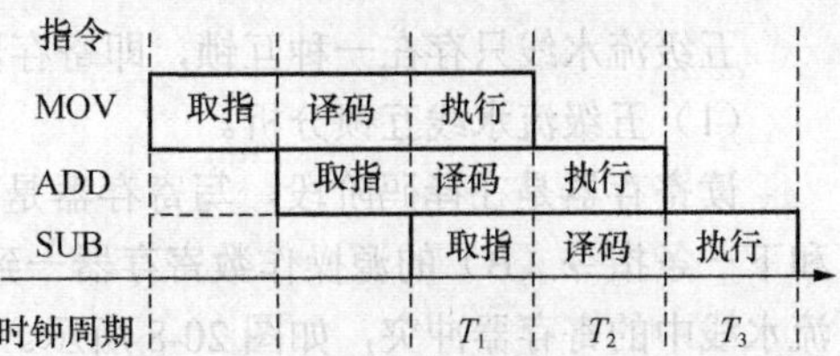

图 20-5 ARM7 单周期指令的最佳流水线

流水线中阻断现象也十分普遍，下面就各种阻断情况下的流水线性能进行详细分析。

（1）带有存储器访问指令的流水线。

对存储器的访问指令 LDR 就是非单周期指令，如图 20-6 所示。这类指令在执行阶段，首先要进行存储器的地址计算，占用控制信号线，而译码的过程同样需要占用控制信号线，所以下一条指令（第一个 SUB）的译码被阻断，并且由于 LDR 访问存储器和回写寄存器的过程中需要继续占用执行单元，所以下一条（第一个 SUB）的执行也被阻断。由于采用冯·诺伊曼体系结构，不能够同时访问数据存储器和指令存储器，当 LDR 处于访存周期的过程中时，MOV 指令的取指被阻断。因此处理器用 8 个时钟周期执行了 6 条指令，指令平均周期数（CPI）=1.3 个时钟周期。

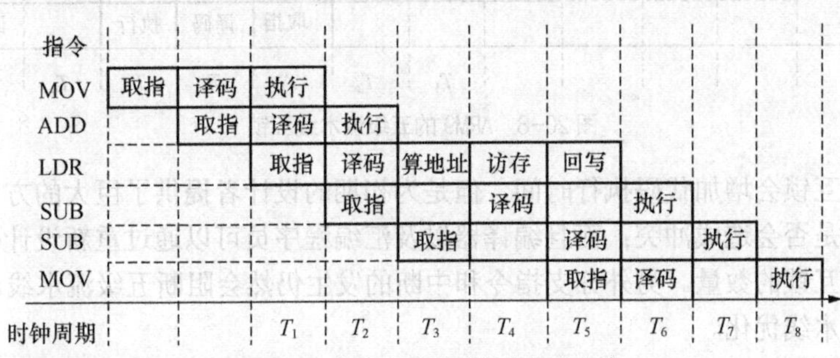

图 20-6 带有存储器访指令的流水线

（2）带有分支指令的流水线。

当指令序列中含有具有分支功能的指令（如 BL 等）时，流水线也会被阻断，如图 20-7 所示。分支指令在执行时，其后第 1 条指令被译码，其后第 2 条指令进行取指，但是这两步操作的指令并不被执行。因为分支指令执行完毕后，程序应该转到跳转的目标地址处执行，因此在流水线上需要丢弃这两条指令，同时程序计数器就会转移到新的位置接着进行取指、译码和执行。此外还有一些特殊的转移指令需要在跳转完成的同时进行写链接寄存器、程序计数寄存器，如 BL 执行过程中包括两个附加操作——写链接寄存器和调整程序指针。这两个操作仍然占用执行单元，这时处于译码和取指的流水线被阻断了。

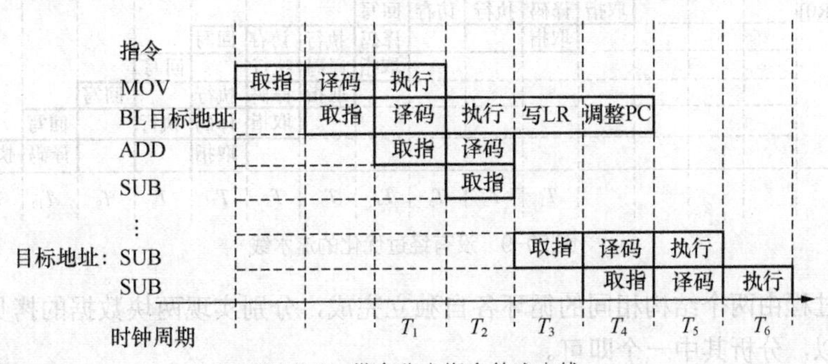

图 20-7 带有分支指令的流水线

5. 五级流水线技术

五级流水线只存在一种互锁，即寄存器冲突。

（1）五级流水线互锁分析。

读寄存器是在译码阶段，写寄存器是在回写阶段。如果当前指令（A）的目的操作数寄存器和下一条指令（B）的源操作数寄存器一致，B 指令就需要等 A 回写之后才能译码。这就是五级流水线中的寄存器冲突，如图 20-8 所示。LDR 指令写 R9 是在回写阶段，而 MOV 中需要用到的 R9 正是 LDR 在回写阶段将会重新写入的寄存器值，MOV 译码需要等待，直到 LDR 指令的寄存器回写操作完成。在当前处理器设计中，可以通过寄存器旁路技术对流水线进行优化，解决流水线的寄存器冲突问题。

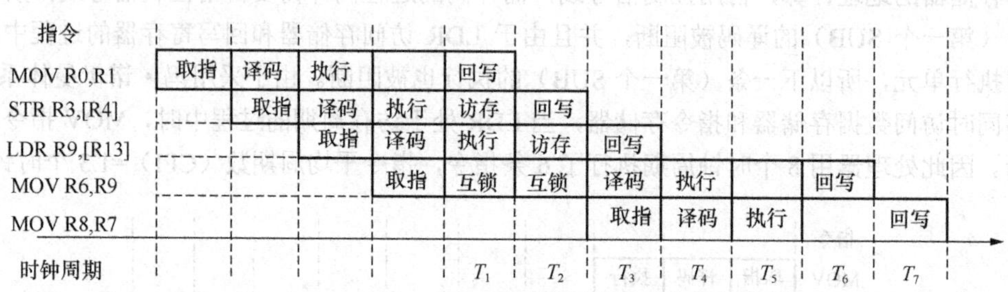

图 20-8　ARM9 的五级流水线互锁

虽然流水线互锁会增加代码执行时间，但是为初期的设计者提供了巨大的方便，可以不必考虑使用的寄存器是否会造成冲突；而且编译器以及汇编程序员可以通过重新设计代码的顺序或者其他方法来减少互锁的数量。另外分支指令和中断的发生仍然会阻断五级流水线。

（2）五级流水线优化。

采用重新设计代码顺序在很多情况下可以很好地减少流水线的阻塞，使流水线的运行流畅。下面详细分析代码优化对流水线的优化和效率的提高。

假设要实现把内存地址 0x1000 和 0x2000 处的数据分别拷贝到 0x8000 和 0x9000 处。

0xl000 处的内容为：1，2，3，4，5，6，7，8，9，10；

0x2000 处的内容为：H，e，l，l，o，W，o，r，l，d；

实现第一个拷贝过程的程序代码及指令的执行时空图如图 20-9 所示。

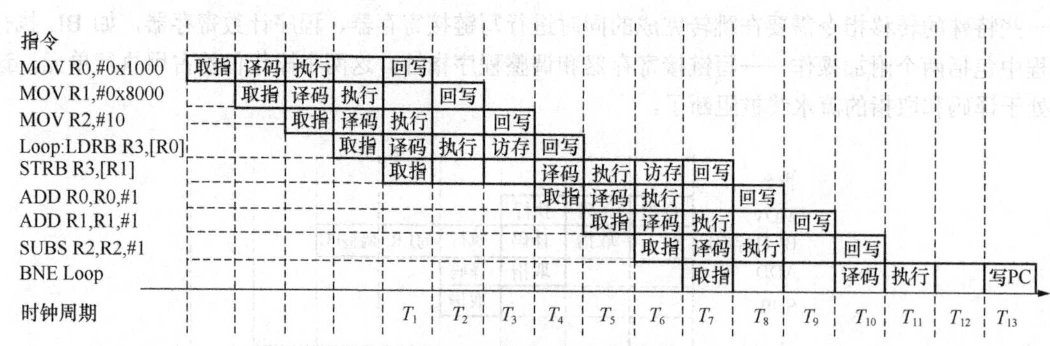

图 20-9　没有经过优化的流水线

全部拷贝过程由两个结构相同的循环各自独立完成，分别实现两块数据的拷贝，并且两个拷贝过程极为类似，分析其中一个即可。

在图 20-13 中，T_1～T_3 是 3 个单独的时钟周期；T_4～T_{11} 是一个循环，在时空图中描述了第一次循环的执行情况。在 T_{12} 写 LR 的同时，开始对循环的第一条语句进行取指，所以总的流水线周期数为 3+10×10+2×9=121。整个拷贝过程需要 121×2+2=244 个时钟周期完成。

考虑到通过减少流水线的冲突可以提高流水线的执行效率，而流水线的冲突主要来自寄存器冲突和分支指令，因此对代码作如下两方面调整。

- 将两个循环合并成一个循环能够充分减少循环跳转的次数，减少跳转带来的流水线停滞；
- 调整代码的顺序，将带有与临近指令不相关的寄存器插到带有相关寄存器的指令之间，能够充分地避免寄存器冲突导致的流水线阻塞。

对代码调整和流水线的时空图如图 20-10 所示。

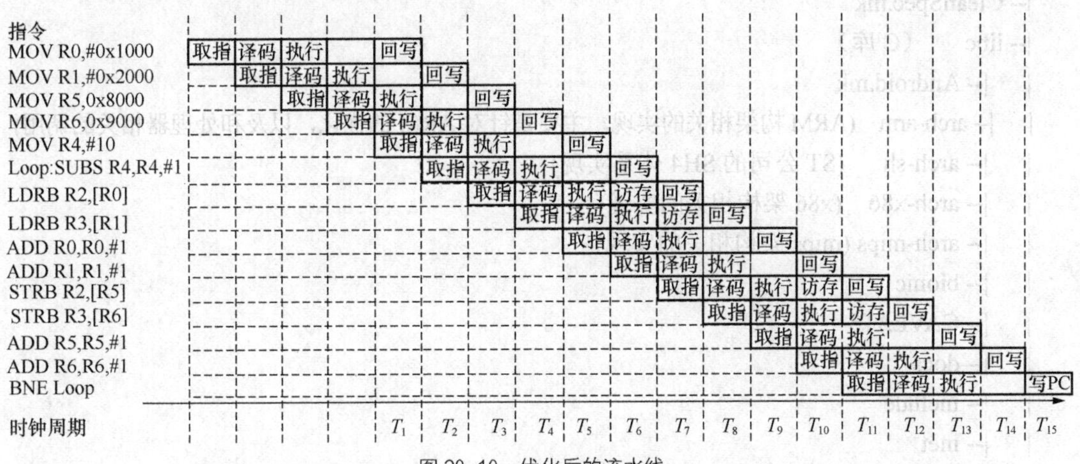

图 20-10 优化后的流水线

由此可见，在调整之后，T_1～T_5 是 5 个单独的时钟周期，T_6～T_{13} 是一个循环，同样在 T_{14} 的时候 BNE 指令在写 LR 的同时，循环的第一条指令开始取指，所以总的 指令周期数为 5+10×10+2×9+2=125。

通过两段代码的比较可看出：调整之前整个拷贝过程总共使用了 244 个时钟周期，调整了循环内指令的顺序后，总共使用了 125 个时钟周期就完成了同样的工作，时钟周期减少了 119 个，缩短了 119/244=420.8%，效率提升十分明显。

代码优化前后执行周期数对比的情况如表 20-1 所示。

表 20-1 代码优化前后执行周期数对比

	优化前周期数	优化后周期数	提高比例/%
顺序语句	6	5	16.7
循环 1	118	60	49.2
循环 2	120	60	50
总周期数	244	125	48.8

所以流水线的优化问题主要应该从如下两个方面考虑。

- 通过合并循环等方式减少分支指令的个数，从而减少流水线的浪费；
- 通过交换指令的顺序，避免寄存器冲突造成的流水线停滞。

20.3.2 Android 对 C 库优化

在 Android 系统中，通过优化和裁剪的 libc 库 Bionic，拥有更高的效率、低内存占用、非常快速和小的线程实现、内置了对 Android 特有服务的支持等特点。

Bionic 是 Android 的 C/C++ library，libc 是 GNU/Linux 以及其他类 UNIX 系统的基础函数库，最常用的就是 GNU 的 libc，也叫 glibc。Android 之所以采用 Bionic 而不是 glibc，有如下 3 个原因。

（1）版权问题，因为 glibc 是 LGPL。

（2）库的体积和速度，Bionic 要比 glibc 小很多。

（3）提供了一些 Android 特定的函数，getprop LOGI 等。

Bionic 的主要目录结构及主要功能的说明如下所示：

```
|-- Android.mk
|-- CleanSpec.mk
|-- libc     （C 库）
|   |-- Android.mk
|   |-- arch-arm  (ARM 构架相关的实现，主要是针对 ARM 的优化，以及和处理器相关的调用)
|   |-- arch-sh    (ST 公司的 SH4 体系实现)
|   |-- arch-x86   (x86 架构相关的实现)
|   |-- arch-mips (mips 架构相关的实现)
|   |-- bionic
|   |-- CAVEATS
|   |-- docs
|   |-- include
|   |-- inet
|   |-- Jamfile
|   |-- kernel
|   |-- MODULE_LICENSE_BSD
|   |-- netbsd
|   |-- NOTICE
|   |-- private
|   |-- README
|   |-- regex
|   |-- stdio
|   |-- stdlib
|   |-- string
|   |-- SYSCALLS.TXT
|   |-- tools
|   |-- tzcode
|   |-- unistd
|   |-- wchar
|   `-- zoneinfo
|-- libdl          （动态链接库访问接口 dlopen dlsym dlerror dlclose dladdr 的实现）
|   |-- Android.mk
|   |-- arch-sh
|   |-- dltest.c
|   |-- libdl.c
|   |-- MODULE_LICENSE_BSD
```

```
|     `-- NOTICE
|-- libm     （C 数学函数库，提供了常见的数序函数和浮点运算）
|     |-- alpha
|     |-- amd64
|     |-- Android.mk
|     |-- arm
|     |-- bsdsrc
|     |-- fpclassify.c
|     |-- i386
|     |-- i387
|     |-- ia64
|     |-- include
|     |-- isinf.c
|     |-- Makefile-orig
|     |-- man
|     |-- MODULE_LICENSE_BSD_LIKE
|     |-- NOTICE
|     |-- powerpc
|     |-- sh
|     |-- sincos.c
|     |-- sparc64
|     `-- src
|-- libstdc++    （standard c++ lib）
|     |-- Android.mk
|     |-- include
|     |-- MODULE_LICENSE_BSD
|     |-- NOTICE
|     `-- src
|-- libthread_db (线程调试库，可以利用此库对多线程程序进行调试)
|     |-- Android.mk
|     |-- include
|     |-- libthread_db.c
|     |-- MODULE_LICENSE_BSD
|     `-- NOTICE
|-- linker (Android dynamic linker)
|     |-- Android.mk
|     |-- arch
|     |-- ba.c
|     |-- ba.h
|     |-- debugger.c
|     |-- dlfcn.c
|     |-- linker.c
```

```
|    |-- linker_debug.h
|    |-- linker_format.c
|    |-- linker_format.h
|    |-- linker.h
|    |-- MODULE_LICENSE_APACHE2
|    |-- NOTICE
|    |-- README.TXT
|    `-- rt.c
|-- MAINTAINERS
```

另外，在系统移植时也经常用到 Bionic。因为本书讲解的 Android 应用开发，所以不介绍相关的内容。

20.3.3　优化创建的进程

Linux 在创建一个新进程时利用了写时拷贝（Copy-on-Write）机制，使得创建一个新的进程非常高效。Android 中每个进程都是基于虚拟机的，并且也要加载基本的库，实际上这些都是可以共享的。基于这方面的考虑，Andoid 引入了写时拷贝机制，使得 Android 启动一个新的进程，实际上并不消耗很多的内存和 CPU 资源。

另外，Android 在后台一直有个 Zygote 虚拟机在运行，实际上是一个虚拟机实例的孵化器。如果要启动一个新的应用，Zygote 就会创建出一个新的子进程来执行该应用程序，十分高效。图 20-11 显示了 Zygote 创建子进程的过程。

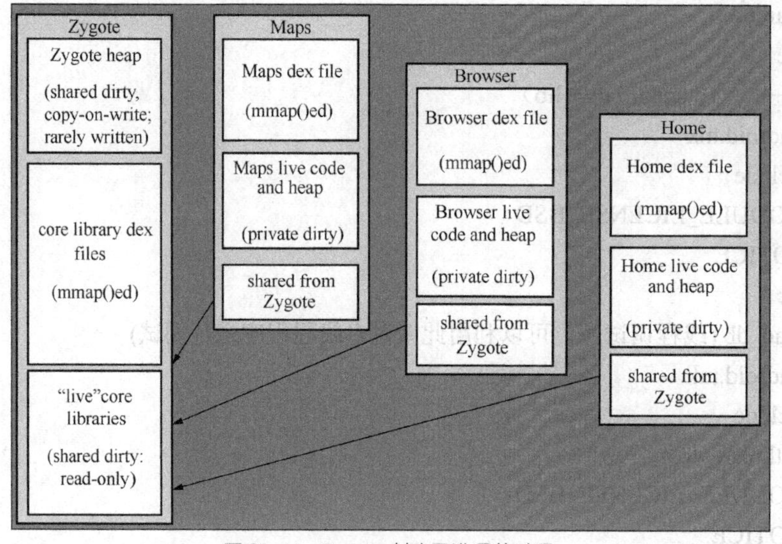

图 20-11　Zygote 创建子进程的过程

当 Android 开机后，会首先启动 Zygote 虚拟机，而不是先启动系统服务器（System Server），这是出于利用写时拷贝机制创建进程比较高效的考虑。Zygote 虚拟机在启动后会完成虚拟机的初始化、库的加载、预置类库的加载和初始化等操作。在系统需要一个新的虚拟机对象时，Zygote 还可以通过写时拷贝机制高效地创建出新的虚拟机对象。

20.3.4　渲染优化

在进行渲染时，Android 会根据变化的部分进行局部更新，并不是每次都需要重绘整个屏幕。

首先计算需要重绘的区域（mInvalidRegion），如果 DisplayHardware::UPDATE_ ON_DEMAND，则通过设定需要重绘的区域的边界来进行局部重绘。

1. Android Browser 快速渲染优化

对于内容繁多复杂的页面来说，当 Browser 进行 scroll 和 zoom 操作时，会显得不流畅，通过阅读 webkit 相关代码发现，webkit 在进行上述操作过程的每一次绘制时，都是整个重绘，没有进行任何缓存操作，所以考虑将绘制的页面缓存起来，这样下一次绘制时大部分绘制内容只用将上一次的缓存页面进行一个移位（translate）或者缩放（scale）操作，然后再绘制上有内容更新区域上的内容即可。

Android Browser 快速渲染优化所涉及的代码文件如下所示。

- \frameworks\base\core\java\android\webkit\WebViewCore.java。
- \external\webkit\WebKit\android\jni\WebViewCore.h。
- \external\webkit\WebKit\android\jni\WebViewCore.cpp。

（1）修改 Java 层。

Browser 中的绘制都是调用 WebViewCore.java 中的 drawContentPicture 方法，而 drawContentPicture 又会通过 jni 调用到 webkit 的 C++层。共有 3 种绘制：normal、zoom、scroll，官方代码是通过对 zoom 绘制和 scroll 绘制 setDrawFilter 的方法进行了小量的优化，本章节修改后定义一个 flags 参数区分 3 种绘制并通过 JNI 调用传到 C++层。

（2）修改 C++层。

在文件 WebViewCore.java 中的，drawContentPicture 最终会调用 webkit 中文件 WebViewCore.cpp 的 drawContent()方法，在该方法中我们调用优化后的绘制实现。在类 WebViewCore 中增加私有成员 int m_contentModSeq，其中前者用来标识不同的绘制请求，每次新的绘制请求都会使该变量加 1。

（3）实现快速渲染。

采用双缓存机制，每个缓存对应一个绘制序号，该序号越大表明该缓存越新，具体实现如下。

- normal 绘制：通过 PictureSet 绘制内容到缓存，并从缓存绘制到 canvas。
- zoom 绘制：使用最新的缓存进行 scale 操作，并从缓存绘制到 canvas，不更新缓存。
- scroll 绘制：使用最新的缓存进行 translate 操作，并将其绘制到旧的缓存，通过 PictureSet 绘制更新内容到旧的缓存，最后从该旧的缓存绘制到 canvas，更新旧缓存的绘制序号为最新。

2. 三维场景的渲染优化

对于任何一个 3D 应用程序来说，追求场景画面真实感是一个无止尽的目标，其结果就是让我们的场景越来越复杂，模型更加精细，这必然给图形硬件带来极大的负荷以致于无法达到实时绘制帧率。因此，渲染优化是必不可少的。在渲染优化工作之前，要对应用程序性能进行系统的评测，找出瓶颈，然后对症下药。对于 3D 应用程序来说，影响性能的十分多，同时不同的硬件配置条件下，瓶径也会有所不同。因此，对应用程序进行有效的性能评测，不仅需要对整个渲染管线原理有深入地了解，此外借助一些评测工具能让工作事半功倍。

渲染流水线的速度是由最慢的阶段决定的，因此对一个 3D 应用程序进行评测，首先要分析影响渲染性能的瓶颈是在 CPU 端还是 GPU 端，由此来决定优化的对象。由于目前的图形加速硬件都很强大，这个瓶径往往出现在 CPU 端，可以通过一些工具获得这个信息，如 Nvidia 的 NVPerfHUD。在评测选项中，可以查看 CPU 和 GPU 繁忙度这项，当 CPU 繁忙度是 100%时，GPU 还不是时，可以知道性能的瓶颈在 CPU 端，必须删减 CPU 端的操作，同时尽量“喂饱”GPU，把一些费事的计算移值到 GPU 上，例如硬件骨骼蒙皮。当 GPU 端是瓶颈时，说明 GPU 超荷负载，有可能是因为有过多的渲染填充，也就是多边形数量太多（当前强大的 GPU 使得这种情况并不多见）。

CPU 上的瓶颈产生有两个方面，一是因为复杂 AI 计算或低效的代码，二是由于不好的渲染批处理或资源管理。对于第一种情况，可以利用 VTurn 这类的工具，把应用程序中所有函数调用时间从大到小的排列出来，就很容易知道问题所在。对第二种情况来说，同样利用 NVPerfHUD，可以查看每帧的 DP 数目，看看批的数量是否过多（有一个具体的换算公式），查看纹理内存的数目，是否消耗了过多的显存。利用这些工具，基本上能够定位应用程序的瓶颈。在应用程序内部，编写一个内嵌的 profiler 功能，能更加便利地进行评测，此外利用 Lua 这样的脚本程序，可以运行时调试，也能提高评测的效率。

静态场景包括了地形、植被、建筑物等一般不改变位置的实体集合，对它的优化是场景优化中最重主要的内容。在接下来的内容中，将详细讲解静态场景优化的常见问题。

（1）批的优化。

批是场景优化中最重要的概念之一，它指的是一次渲染调用（DP），批的尺寸是这次渲染调用所能渲染的多边形数量。每个批的调用都会消耗一定的 CPU 时间，对于显卡来说，一个批里的多边形数量远达不到最大绘制数量。因此尽可能将更多的多边形放在一个批里渲染，以此来减少批的数目，最终降低 CPU 时间，是批的优化基本原则。然而事情往往不尽如人意，有些情况下原有的批会被打破，造成额外的开销，如纹理的改变或不同的矩阵状态。针对这些问题，可以采用一些方法来尽量避免它，已达到批尺寸的最大化。

- 合并多个小纹理为一张大纹理。

在某个场景中，地面上有十多种不同的植被，它们除了纹理不同外，渲染状态都是一样。可以把它们的纹理打包成一个大纹理，再为每个植被模型指定 UV，这样就可以用一个渲染调用来渲染所有的物体，批的数量就从十多个降为一个。这种方法比较适合对纹理精度要求不高、面数不会太多的物体。

- 利用顶点 shader 来统一不同矩阵的情况。即使场景中的所有物体材质都一样，如果它们的矩阵状态不同（特别是场景图管理的引擎），也会打碎原有的批。利用顶点 shader 技术可以避免这种情况，因为可以把要乘的变换矩阵通过常量寄存器传到 shader 程序中，这样统一了物体的矩阵状态，可以放在一个批里渲染。

（2）渲染状态管理。

渲染状态是用来控制渲染器的渲染行为，在 D3D 中是 setRenderState，通过改变渲染状态，可以设置纹理状态、深度写入等等。改变渲染状态对显卡来说，是一个比较耗时的工作，因为显卡执行 API 必须严格按照渲染路径。当渲染状态变化时，显卡就必须执行浮点运算来改变渲染路径，因此给 CPU 和 GPU 带来时间消耗（CPU 必须等待），渲染状态变化越大，所要进行的浮点运算越多。因此将渲染状态进行有效的管理，尽可能减少其变化，对渲染性能影响巨大（新六代的显卡 Geforce8 系列中将一些常见的状态参数集存储在显卡核心中，当渲染状态状态发生变化可以直接读取保存的参数集，以消除不必要的开销）。绝大部分的 3D 引擎都会按照渲染状态对 PASS 进行分组渲染。

（3）LOD。

LOD 这个已经被人讨论烂掉的技术就不多说了，简单说一些实际应用。地形的 LOD 就不多说了，方法太多，不过目前情况下最实用的还是连锁分片的方法。对于模型 LOD，自动减面的算法，如渐近网格子（VDPM）并不少见，但是效果都很一般。常规的做法还是让美工做低模进行替换，对于复杂场景来说，模型 LOD 的效果还是比较明显的。材质 LOD 就需要一些技巧，例如可以将雾后的物体，包括地形等统一成一种材质，采用雾的颜色，这样就统一了渲染状态，至于是否要打包成一个 DP 就要看具体情况了（这个统一的材质最好把光照影响关掉，这也是比较费时的)。至于角色模型的 LOD 和普通模型 LOD 相类似，低模减少了顶点数，自然减少了蒙皮计算量。个人认为骨骼 LOD 不是特别的必要，看具体的情况。

(4) 场景管理的优化。

场景管理的优化包括场景分割和可见性剔除等。现在的室外场景一般采用 quadtree 或 octree，当在性能评测时发现遍历树的过程比较慢时，可能有两个原因。一是树的深度设置的不合理，可以很容易寻找到一个最佳的深度。另一个原因可能是为太多数量众多但体积很小的物体分配了节点，造成节点数量的冗余。解决方法是把这些小物体划分到它们所在的大的节点中。

可见性剔除是最常见优化方法，常用的是视锥裁减，这也是非常有效的。视锥裁减也是许多优化方法，这里就不详说了。遮挡裁减也是经常被用到的方法，常见的有地平线裁减。但是在有些情况下，遮挡裁减的效果并不明显，如当 CPU 使用率已经是 100%时，CPU 端是瓶颈，这时进行遮挡裁减计算消耗 CPU 时间，效果就不明显。但是有些情况下利用一些预生成信息的方法，降低遮挡裁减计算的复杂度，提高遮挡裁减计算的效率，对场景性能会一定的改善。

3. 优化浏览器

为页面中所有图片指定宽度和高度，可以消除不必要的 reflows 和重新绘制页面，使页面渲染速度更快。当浏览器勾画页面时，它需要能够流动的，如图片这样的可替换的元素提供了图片尺寸，浏览器知道去环绕附近的不可替换元素，甚至可以在图片下载之前开始渲染页面。如果没有指定的图片尺寸，或者如果指定的尺寸不符合图片的实际尺寸，一旦图片下载，浏览器将需要回流和重新绘制页面。为了防止 reflows，在 HTML 的<img>标签中或在 CSS 中为所有图片指定宽度和高度。

因此笔者在此提出如下建议。

- ❑ 务必指定与图片本身相一致的尺寸。
- ❑ 不要使用非图片原始尺寸来缩放图片。如果一个图片文件实际上的大小是 60×60 像素，不要在 HTML 或 CSS 里设置尺寸为 30 像素×30 像素。如果图片需要较小的尺寸，在图像编辑软件中，设置成相一致的尺寸。
- ❑ 一定要指定图片或它的块级父元素的尺寸。
- ❑ 一定要设置<img>元素本身，或它的块级父元素的尺寸。如果父元素不是块级元素，尺寸将被忽略。不要在一个非最近父元素的祖先元素上设置尺寸。

第 21 章　分析 ART 的启动过程

经过本书前面内容的学习可知，Android 系统中的应用程序进程是由 Zygote 进程孕育出来的，而 Zygote 进程又是由 Init 进程启动的。在启动 Zygote 进程时会创建一个 Dalvik VM 或 ART 实例。和 Dalvik VM 模式相比，ART 模式更加高效，更加合理。在本章的内容中，将详细分析启动 ART 的具体过程，为读者步入本书后面知识的学习打下基础。

21.1 运行环境的转换

传统的 Dalvik 虚拟机其实是一个 Java 虚拟机，只不过它执行的不是 class 文件，而是 dex 文件。因此，ART 运行时最理想的方式也是实现为一个 Java 虚拟机的形式，这样就可以很容易地将 Dalvik 虚拟机替换掉。ART 运行时就是真的与 Dalvik 虚拟机一样，实现了一套完全兼容 Java 虚拟机的接口。为了方便描述，接下来就将 ART 运行时称为 ART 虚拟机，它和 Dalvik 虚拟机、Java 虚拟机的关系如图 21-1 所示。

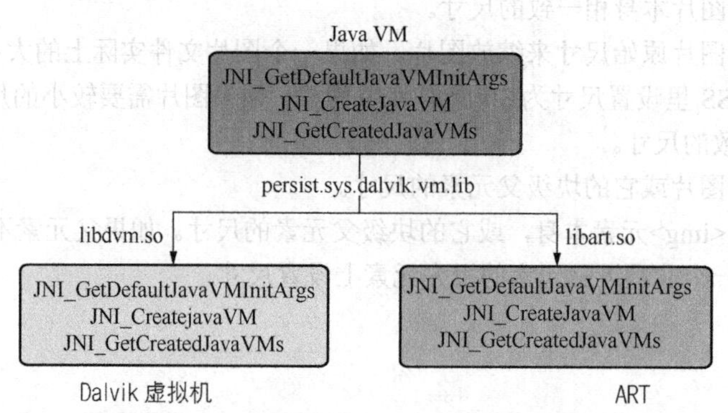

图 21-1　ART、Dalvik 和 Java 虚拟机的关系

从图 21-1 可知，Dalvik 虚拟机和 ART 虚拟机都实现了如下 3 个用来抽象 Java 虚拟机的接口。

（1）JNI_GetDefaultJavaVMInitArgs：获取虚拟机的默认初始化参数。

（2）JNI_CreateJavaVM：在进程中创建虚拟机实例。

（3）JNI_GetCreatedJavaVMs：获取进程中创建的虚拟机实例。

在 Android 系统中，Davik 虚拟机实现在 libdvm.so 文件中，ART 虚拟机实现在 libart.so 文件中。也就是说，文件 libdvm.so 和文件 libart.so 导出了如下 3 个接口供外界调用。

- JNI_GetDefaultJavaVMInitArgs。
- JNI_CreateJavaVM。
- JNI_GetCreatedJavaVMs。

此外，Android 系统还提供了一个系统属性 persist.sys.dalvik.vm.lib，其值等于 libdvm.s 或

libart.so，具体说明如下所示。

- 当等于 libdvm.so 时，表示当前用的是 Dalvik 虚拟机。
- 当等于 libart.so 时，表示当前用的是 ART 虚拟机。

上面介绍了 Dalvik 虚拟机和 ART 虚拟机的共同之处，当然它们之间最显著还是不同之处。Dalvik 虚拟机执行的是 dex 字节码，ART 虚拟机执行的是本地机器码。这意味着 Dalvik 虚拟机包含一个解释器，用来执行 dex 字节码。当然，Android 从 2.2 开始，也包含 JIT（Just-In-Time），用来在运行时动态地将执行频率很高的 dex 字节码翻译成本地机器码，然后再执行。通过 JIT，就可以有效地提高 Dalvik 虚拟机的执行效率。但是，将 dex 字节码翻译成本地机器码是发生在应用程序的运行过程中的，并且应用程序每一次重新运行的时候，都要做重这个翻译工作的。因此，即使采用了 JIT，Dalvik 虚拟机的总体性能还是不能与直接执行本地机器码的 ART 虚拟机相比。

21.2 运行 app_process 进程

启动过程从 init.rc 文件开始，在文件 init.rc 中由这一行表示启动 zygote：

```
service zygote /system/bin/app_process -Xzygote /system/bin --zygote--start-system-server
```

init 进程根据它执行 app_process（frameworks/base/cmds/app_process/app_main.cpp），也就是 Zygote 了。当 Android 系统启动时会创建一个 Zygote 进程，作为应用程序的进程孵化器，并且在启动 Zygote 进程的过程中会创建一个 Dalvik 虚拟机。Zygote 进程是通过复制自己来创建新的应用程序进程的，这意味着 Zygote 进程会将自己的 Dalvik 虚拟机复制给应用程序进程。上述方式可以大大地提高应用程序的启动速度，因为这种方式避免了每一个应用程序进程在启动的时候都要去创建一个 Dalvik。事实上，Zygote 进程通过自我复制的方式来创建应用程序进程，省去的不仅仅是应用程序进程创建 Dalvik 虚拟机的时间，还能省去应用程序进程加载各种系统库和系统资源的时间，因为它们在 Zygote 进程中已经加载过了，并且也会连同 Dalvik 虚拟机一起复制到应用程序进程中去。这也就是 ART 优于 Dalvik 的原因所在。

当 Android 系统启动 init 进程时会运行 app_process 进程，在文件“/frameworks/base/cmds/app_process/app_main.cpp”中定义了 app_process 进程的具体实现，在主函数 main 中会启动 Zygote，对应代码如下所示的加粗部分：

```
    if (niceName && *niceName) {
        setArgv0(argv0, niceName);
        set_process_name(niceName);
    }

    runtime.mParentDir = parentDir;

    if (zygote) {
        runtime.start("com.android.internal.os.ZygoteInit",
                startSystemServer ? "start-system-server" : "");
    } else if (className) {
        // Remainder of args get passed to startup class main()
        runtime.mClassName = className;
        runtime.mArgC = argc - i;
        runtime.mArgV = argv + i;
        runtime.start("com.android.internal.os.RuntimeInit",
                application ? "application" : "tool");
    } else {
        fprintf(stderr, "Error: no class name or --zygote supplied.\n");
        app_usage();
        LOG_ALWAYS_FATAL("app_process: no class name or --zygote supplied.");
        return 10;
    }
}
```

在上述代码中，runtime 是 AppRuntime 的实例，AppRuntime 继承自 AndroidRuntime。类 AndroidRuntime 中的函数 start 在文件“frameworks/base/core/jni/AndroidRuntime.cpp”中定义，具体实现代码如下所示：

```
void AndroidRuntime::start(const char* className, const char* options)
{
    ALOGD("\n>>>>>> AndroidRuntime START %s <<<<<<\n",
            className != NULL ? className : "(unknown)");

    /*
     * 'startSystemServer == true' means runtime is obsolete and not run from
     * init.rc anymore, so we print out the boot start event here.
     */
    if (strcmp(options, "start-system-server") == 0) {
        /* track our progress through the boot sequence */
        const int LOG_BOOT_PROGRESS_START = 3000;
        LOG_EVENT_LONG(LOG_BOOT_PROGRESS_START,
                       ns2ms(systemTime(SYSTEM_TIME_MONOTONIC)));
    }

    const char* rootDir = getenv("ANDROID_ROOT");
    if (rootDir == NULL) {
        rootDir = "/system";
        if (!hasDir("/system")) {
            LOG_FATAL("No root directory specified, and /android does not exist.");
            return;
        }
        setenv("ANDROID_ROOT", rootDir, 1);
    }

    //const char* kernelHack = getenv("LD_ASSUME_KERNEL");
    //ALOGD("Found LD_ASSUME_KERNEL='%s'\n", kernelHack);

    /* start the virtual machine */
    JniInvocation jni_invocation;
    jni_invocation.Init(NULL);
    JNIEnv* env;
    if (startVm(&mJavaVM, &env) != 0) {
        return;
    }
    onVmCreated(env);

    /*
     * Register android functions.
     */
    if (startReg(env) < 0) {
        ALOGE("Unable to register all android natives\n");
        return;
    }

    /*
     * We want to call main() with a String array with arguments in it.
     * At present we have two arguments, the class name and an option string.
     * Create an array to hold them.
     */
    jclass stringClass;
    jobjectArray strArray;
    jstring classNameStr;
    jstring optionsStr;

    stringClass = env->FindClass("java/lang/String");
    assert(stringClass != NULL);
    strArray = env->NewObjectArray(2, stringClass, NULL);
    assert(strArray != NULL);
    classNameStr = env->NewStringUTF(className);
    assert(classNameStr != NULL);
    env->SetObjectArrayElement(strArray, 0, classNameStr);
```

```
    optionsStr = env->NewStringUTF(options);
    env->SetObjectArrayElement(strArray, 1, optionsStr);

    /*
     * Start VM.  This thread becomes the main thread of the VM, and will
     * not return until the VM exits.
     */
    char* slashClassName = toSlashClassName(className);
    jclass startClass = env->FindClass(slashClassName);
    if (startClass == NULL) {
        ALOGE("JavaVM unable to locate class '%s'\n", slashClassName);
        /* keep going */
    } else {
        jmethodID startMeth = env->GetStaticMethodID(startClass, "main",
            "([Ljava/lang/String;)V");
        if (startMeth == NULL) {
            ALOGE("JavaVM unable to find main() in '%s'\n", className);
            /* keep going */
        } else {
            env->CallStaticVoidMethod(startClass, startMeth, strArray);

#if 0
            if (env->ExceptionCheck())
                threadExitUncaughtException(env);
#endif
        }
    }
    free(slashClassName);

    ALOGD("Shutting down VM\n");
    if (mJavaVM->DetachCurrentThread() != JNI_OK)
        ALOGW("Warning: unable to detach main thread\n");
    if (mJavaVM->DestroyJavaVM() != 0)
        ALOGW("Warning: VM did not shut down cleanly\n");
}
```

在上述代码中，“JniInvocation jni_invocation;”用于声明类 JniInvocation 的变量，“jni_invocation.Init(NULL);”用于调用类 JniInvocation 中的函数 Init。由此可见，类 AndroidRutime 的成员函数 start 最主要实现了如下所示的 3 个功能。

- 创建一个 JniInvocation 实例，并且调用它的成员函数 init 来初始化 JNI 环境。
- 调用 AndroidRutime 类的成员函数 startVm 来创建一个虚拟机及其对应的 JNI 接口，即创建一个 JavaVM 接口和一个 JNIEnv 接口。
- 通过上述 JavaVM 接口和 JNIEnv 接口在 Zygote 进程中加载指定的 class。

其中，上述第 1 个功能和第 2 个功能又是最关键的。因此，接下来继续分析它们所对应的函数的实现。

类 JniInvocation 在文件“/libnativehelper/JniInvocation.cpp”中定义，函数 Init 的具体实现代码如下所示：

```
bool JniInvocation::Init(const char* library) {
#ifdef HAVE_ANDROID_OS
  char default_library[PROPERTY_VALUE_MAX];
  property_get("persist.sys.dalvik.vm.lib", default_library, "libdvm.so");
#else
  const char* default_library = "libdvm.so";
#endif
  if (library == NULL) {
    library = default_library;
  }

  handle_ = dlopen(library, RTLD_NOW);
  if (handle_ == NULL) {
    ALOGE("Failed to dlopen %s: %s", library, dlerror());
    return false;
```

```
    }
    if (!FindSymbol(reinterpret_cast<void**>(&JNI_GetDefaultJavaVMInitArgs_),
                    "JNI_GetDefaultJavaVMInitArgs")) {
      return false;
    }
    if (!FindSymbol(reinterpret_cast<void**>(&JNI_CreateJavaVM_),
                    "JNI_CreateJavaVM")) {
      return false;
    }
    if (!FindSymbol(reinterpret_cast<void**>(&JNI_GetCreatedJavaVMs_),
                    "JNI_GetCreatedJavaVMs")) {
      return false;
    }
    return true;
  }
```

在上述代码中，函数 init 首先读取系统属性 persist.sys.dalvik.vm.lib 的值。因为系统属性 persist.sys.dalvik.vm.lib 的值要么等于 libdvm.so，要么等于 libart.so。所以接下来通过函数 dlopen 加载到进程来的要么是 libdvm.so，要么是 libart.so。无论加载的是哪一个，都要求它导出 JNI_GetDefaultJavaVMInitArgs、JNI_CreateJavaVM 和 JNI_GetCreatedJavaVMs 这 3 个接口，并且分别保存在 JniInvocation 类的 3 个成员变量 JNI_GetDefaultJavaVMInitArgs_、JNI_CreateJavaVM_和 JNI_GetCreatedJavaVMs_中，这 3 个接口也就是前面提到的用来抽象 Java 虚拟机的 3 个接口。

21.3 准备启动

回到函数 AndroidRuntime::start，"if (startVm(&mJavaVM, &env) != 0) {"用于调用函数 startVm 启动虚拟机。也就是说，类 JniInvocation 的成员函数 init 实际上就是根据系统属性 persist.sys.dalvik.vm.lib 来初始化 Dalvik 虚拟机或者 ART 虚拟机环境。类 AndroidRutime 的成员函数 AndroidRuntime::startVm 的具体实现代码如下所示：

```
int AndroidRuntime::startVm(JavaVM** pJavaVM, JNIEnv** pEnv)
{
    int result = -1;
    JavaVMInitArgs initArgs;
    JavaVMOption opt;
    char propBuf[PROPERTY_VALUE_MAX];
    char stackTraceFileBuf[PROPERTY_VALUE_MAX];
    char dexoptFlagsBuf[PROPERTY_VALUE_MAX];
    char enableAssertBuf[sizeof("-ea:")-1 + PROPERTY_VALUE_MAX];
    char jniOptsBuf[sizeof("-Xjniopts:")-1 + PROPERTY_VALUE_MAX];
    char heapstartsizeOptsBuf[sizeof("-Xms")-1 + PROPERTY_VALUE_MAX];
    char heapsizeOptsBuf[sizeof("-Xmx")-1 + PROPERTY_VALUE_MAX];
    char heapgrowthlimitOptsBuf[sizeof("-XX:HeapGrowthLimit=")-1 + PROPERTY_VALUE_MAX];
    char heapminfreeOptsBuf[sizeof("-XX:HeapMinFree=")-1 + PROPERTY_VALUE_MAX];
    char heapmaxfreeOptsBuf[sizeof("-XX:HeapMaxFree=")-1 + PROPERTY_VALUE_MAX];
    char heaptargetutilizationOptsBuf[sizeof("-XX:HeapTargetUtilization=")-1 + PROPERTY_VALUE_MAX];
    char jitcodecachesizeOptsBuf[sizeof("-Xjitcodecachesize:")-1 + PROPERTY_VALUE_MAX];
    char extraOptsBuf[PROPERTY_VALUE_MAX];
    char* stackTraceFile = NULL;
    bool checkJni = false;
    bool checkDexSum = false;
    bool logStdio = false;
    enum {
      kEMDefault,
      kEMIntPortable,
      kEMIntFast,
      kEMJitCompiler,
    } executionMode = kEMDefault;


    property_get("dalvik.vm.checkjni", propBuf, "");
```

```
if (strcmp(propBuf, "true") == 0) {
    checkJni = true;
} else if (strcmp(propBuf, "false") != 0) {
    /* property is neither true nor false; fall back on kernel parameter */
    property_get("ro.kernel.android.checkjni", propBuf, "");
    if (propBuf[0] == '1') {
        checkJni = true;
    }
}

property_get("dalvik.vm.execution-mode", propBuf, "");
if (strcmp(propBuf, "int:portable") == 0) {
    executionMode = kEMIntPortable;
} else if (strcmp(propBuf, "int:fast") == 0) {
    executionMode = kEMIntFast;
} else if (strcmp(propBuf, "int:jit") == 0) {
    executionMode = kEMJitCompiler;
}

property_get("dalvik.vm.stack-trace-file", stackTraceFileBuf, "");

property_get("dalvik.vm.check-dex-sum", propBuf, "");
if (strcmp(propBuf, "true") == 0) {
    checkDexSum = true;
}

property_get("log.redirect-stdio", propBuf, "");
if (strcmp(propBuf, "true") == 0) {
    logStdio = true;
}

strcpy(enableAssertBuf, "-ea:");
property_get("dalvik.vm.enableassertions", enableAssertBuf+4, "");

strcpy(jniOptsBuf, "-Xjniopts:");
property_get("dalvik.vm.jniopts", jniOptsBuf+10, "");

/* exit()线程处理 */
opt.extraInfo = (void*) runtime_exit;
opt.optionString = "exit";
mOptions.add(opt);

/* fprintf()线程处理*/
opt.extraInfo = (void*) runtime_vfprintf;
opt.optionString = "vfprintf";
mOptions.add(opt);

/* 注册敏感线程框架 */
opt.extraInfo = (void*) runtime_isSensitiveThread;
opt.optionString = "sensitiveThread";
mOptions.add(opt);

opt.extraInfo = NULL;

/* enable verbose; standard options are { jni, gc, class } */
//options[curOpt++].optionString = "-verbose:jni";
opt.optionString = "-verbose:gc";
mOptions.add(opt);
//options[curOpt++].optionString = "-verbose:class";

/*
 *默认的启动和堆的最大尺寸
 */
strcpy(heapstartsizeOptsBuf, "-Xms");
property_get("dalvik.vm.heapstartsize", heapstartsizeOptsBuf+4, "4m");
opt.optionString = heapstartsizeOptsBuf;
mOptions.add(opt);
strcpy(heapsizeOptsBuf, "-Xmx");
property_get("dalvik.vm.heapsize", heapsizeOptsBuf+4, "16m");
```

```
opt.optionString = heapsizeOptsBuf;
mOptions.add(opt);

//增加错误主线程的解释器的堆栈大小：6315322
opt.optionString = "-XX:mainThreadStackSize=24K";
mOptions.add(opt);

//设置最大 JIT 代码缓存大小。注： 0 表示将禁用 JIT
strcpy(jitcodecachesizeOptsBuf, "-Xjitcodecachesize:");
property_get("dalvik.vm.jit.codecachesize", jitcodecachesizeOptsBuf+19,  NULL);
if (jitcodecachesizeOptsBuf[19] != '\0') {
 opt.optionString = jitcodecachesizeOptsBuf;
 mOptions.add(opt);
}

strcpy(heapgrowthlimitOptsBuf, "-XX:HeapGrowthLimit=");
property_get("dalvik.vm.heapgrowthlimit", heapgrowthlimitOptsBuf+20, "");
if (heapgrowthlimitOptsBuf[20] != '\0') {
  opt.optionString = heapgrowthlimitOptsBuf;
  mOptions.add(opt);
}

strcpy(heapminfreeOptsBuf, "-XX:HeapMinFree=");
property_get("dalvik.vm.heapminfree", heapminfreeOptsBuf+16, "");
if (heapminfreeOptsBuf[16] != '\0') {
  opt.optionString = heapminfreeOptsBuf;
  mOptions.add(opt);
}

strcpy(heapmaxfreeOptsBuf, "-XX:HeapMaxFree=");
property_get("dalvik.vm.heapmaxfree", heapmaxfreeOptsBuf+16, "");
if (heapmaxfreeOptsBuf[16] != '\0') {
  opt.optionString = heapmaxfreeOptsBuf;
  mOptions.add(opt);
}

strcpy(heaptargetutilizationOptsBuf, "-XX:HeapTargetUtilization=");
property_get("dalvik.vm.heaptargetutilization", heaptargetutilizationOptsBuf+26, "");
if (heaptargetutilizationOptsBuf[26] != '\0') {
  opt.optionString = heaptargetutilizationOptsBuf;
  mOptions.add(opt);
}

property_get("ro.config.low_ram", propBuf, "");
if (strcmp(propBuf, "true") == 0) {
 opt.optionString = "-XX:LowMemoryMode";
 mOptions.add(opt);
}

/*
 *启用或禁用 dexopt 特征，如字节码校验和为精确计算 GC 寄存器映射
 */
property_get("dalvik.vm.dexopt-flags", dexoptFlagsBuf, "");
if (dexoptFlagsBuf[0] != '\0') {
  const char* opc;
  const char* val;

  opc = strstr(dexoptFlagsBuf, "v=");    /* verification */
  if (opc != NULL) {
    switch (*(opc+2)) {
    case 'n':  val = "-Xverify:none";     break;
    case 'r':  val = "-Xverify:remote";   break;
    case 'a':  val = "-Xverify:all";      break;
    default:   val = NULL;                break;
    }

    if (val != NULL) {
      opt.optionString = val;
      mOptions.add(opt);
```

```
        }
    }

    opc = strstr(dexoptFlagsBuf, "o=");    /* optimization */
    if (opc != NULL) {
        switch (*(opc+2)) {
        case 'n':   val = "-Xdexopt:none";      break;
        case 'v':   val = "-Xdexopt:verified";  break;
        case 'a':   val = "-Xdexopt:all";       break;
        case 'f':   val = "-Xdexopt:full";      break;
        default:    val = NULL;                 break;
        }

        if (val != NULL) {
            opt.optionString = val;
            mOptions.add(opt);
        }
    }

    opc = strstr(dexoptFlagsBuf, "m=y");    /* register map */
    if (opc != NULL) {
        opt.optionString = "-Xgenregmap";
        mOptions.add(opt);

        /* turn on precise GC while we're at it */
        opt.optionString = "-Xgc:precise";
        mOptions.add(opt);
    }
}

/*启用调试; 设置暂停= Y, 暂停 VM 初始化*/
/* use android ADB transport */
opt.optionString =
    "-agentlib:jdwp=transport=dt_android_adb,suspend=n,server=y";
mOptions.add(opt);

ALOGD("CheckJNI is %s\n", checkJni ? "ON" : "OFF");
if (checkJni) {
    /*扩展的 JNI 检查*/
    opt.optionString = "-Xcheck:jni";
    mOptions.add(opt);

    /* 设置 JNI 全局引用*/
    opt.optionString = "-Xjnigreflimit:2000";
    mOptions.add(opt);

    /* with -Xcheck:jni, this provides a JNI function call trace */
    //opt.optionString = "-verbose:jni";
    //mOptions.add(opt);
}

char lockProfThresholdBuf[sizeof("-Xlockprofthreshold:") + sizeof(propBuf)];
property_get("dalvik.vm.lockprof.threshold", propBuf, "");
if (strlen(propBuf) > 0) {
  strcpy(lockProfThresholdBuf, "-Xlockprofthreshold:");
  strcat(lockProfThresholdBuf, propBuf);
  opt.optionString = lockProfThresholdBuf;
  mOptions.add(opt);
}

/* Force interpreter-only mode for selected opcodes. Eg "1-0a,3c,f1-ff" */
char jitOpBuf[sizeof("-Xjitop:") + PROPERTY_VALUE_MAX];
property_get("dalvik.vm.jit.op", propBuf, "");
if (strlen(propBuf) > 0) {
    strcpy(jitOpBuf, "-Xjitop:");
    strcat(jitOpBuf, propBuf);
    opt.optionString = jitOpBuf;
    mOptions.add(opt);
}
```

```
    /* Force interpreter-only mode for selected methods */
    char jitMethodBuf[sizeof("-Xjitmethod:") + PROPERTY_VALUE_MAX];
    property_get("dalvik.vm.jit.method", propBuf, "");
    if (strlen(propBuf) > 0) {
        strcpy(jitMethodBuf, "-Xjitmethod:");
        strcat(jitMethodBuf, propBuf);
        opt.optionString = jitMethodBuf;
        mOptions.add(opt);
    }

    if (executionMode == kEMIntPortable) {
        opt.optionString = "-Xint:portable";
        mOptions.add(opt);
    } else if (executionMode == kEMIntFast) {
        opt.optionString = "-Xint:fast";
        mOptions.add(opt);
    } else if (executionMode == kEMJitCompiler) {
        opt.optionString = "-Xint:jit";
        mOptions.add(opt);
    }

    if (checkDexSum) {
        /* perform additional DEX checksum tests */
        opt.optionString = "-Xcheckdexsum";
        mOptions.add(opt);
    }

    if (logStdio) {
        /* convert stdout/stderr to log messages */
        opt.optionString = "-Xlog-stdio";
        mOptions.add(opt);
    }

    if (enableAssertBuf[4] != '\0') {
        /* accept "all" to mean "all classes and packages" */
        if (strcmp(enableAssertBuf+4, "all") == 0)
            enableAssertBuf[3] = '\0';
        ALOGI("Assertions enabled: '%s'\n", enableAssertBuf);
        opt.optionString = enableAssertBuf;
        mOptions.add(opt);
    } else {
        ALOGV("Assertions disabled\n");
    }

    if (jniOptsBuf[10] != '\0') {
        ALOGI("JNI options: '%s'\n", jniOptsBuf);
        opt.optionString = jniOptsBuf;
        mOptions.add(opt);
    }

    if (stackTraceFileBuf[0] != '\0') {
        static const char* stfOptName = "-Xstacktracefile:";

        stackTraceFile = (char*) malloc(strlen(stfOptName) +
            strlen(stackTraceFileBuf) +1);
        strcpy(stackTraceFile, stfOptName);
        strcat(stackTraceFile, stackTraceFileBuf);
        opt.optionString = stackTraceFile;
        mOptions.add(opt);
    }

    /* extra options; parse this late so it overrides others */
    property_get("dalvik.vm.extra-opts", extraOptsBuf, "");
    parseExtraOpts(extraOptsBuf);

    /* 设置本地属性 */
    {
        char langOption[sizeof("-Duser.language=") + 3];
```

```
        char regionOption[sizeof("-Duser.region=") + 3];
        strcpy(langOption, "-Duser.language=");
        strcpy(regionOption, "-Duser.region=");
        readLocale(langOption, regionOption);
        opt.extraInfo = NULL;
        opt.optionString = langOption;
        mOptions.add(opt);
        opt.optionString = regionOption;
        mOptions.add(opt);
    }
    opt.optionString = "-Djava.io.tmpdir=/sdcard";
    mOptions.add(opt);

    initArgs.version = JNI_VERSION_1_4;
    initArgs.options = mOptions.editArray();
    initArgs.nOptions = mOptions.size();
    initArgs.ignoreUnrecognized = JNI_FALSE;

    /*
     * 初始化VM.

     */
    if (JNI_CreateJavaVM(pJavaVM, pEnv, &initArgs) < 0) {
        ALOGE("JNI_CreateJavaVM failed\n");
        goto bail;
    }

    result = 0;

bail:
    free(stackTraceFile);
    return result;
}
```

由上述实现代码可知，函数 AndroidRuntime::startVm 最终会调用 JNI_CreateJavaVM 函数。此处的函数 JNI_ CreateJavaVM 在文件“art/runtime/jni_internal.cc”中定义，具体实现代码如下所示：

```
extern "C" jint JNI_CreateJavaVM(JavaVM** p_vm, JNIEnv** p_env, void* vm_args) {
  const JavaVMInitArgs* args = static_cast<JavaVMInitArgs*>(vm_args);
  if (IsBadJniVersion(args->version)) {
    LOG(ERROR) << "Bad JNI version passed to CreateJavaVM: " << args->version;
    return JNI_EVERSION;
  }
  Runtime::Options options;
  for (int i = 0; i < args->nOptions; ++i) {
    JavaVMOption* option = &args->options[i];
    options.push_back(std::make_pair(std::string(option->optionString), option->extraInfo));
  }
  bool ignore_unrecognized = args->ignoreUnrecognized;
  if (!Runtime::Create(options, ignore_unrecognized)) {
    return JNI_ERR;
  }
  Runtime* runtime = Runtime::Current();
  bool started = runtime->Start();
  if (!started) {
    delete Thread::Current()->GetJniEnv();
    delete runtime->GetJavaVM();
    LOG(WARNING) << "CreateJavaVM failed";
    return JNI_ERR;
  }
  *p_env = Thread::Current()->GetJniEnv();
  *p_vm = runtime->GetJavaVM();
  return JNI_OK;
}
```

类 JniInvocation 的静态成员函数 GetJniInvocation 返回的便是前面所创建的 JniInvocation 实

例。有了这个 JniInvocation 实例之后，就继续调用它的成员函数 JNI_CreateJavaVM 来创建一个 JavaVM 接口及其对应的 JNIEnv 接口。

函数 GetJniInvocation 定义在文件“libnativehelper/JniInvocation.cpp”中，具体实现代码如下所示：

```
jint JniInvocation::JNI_CreateJavaVM(JavaVM** p_vm, JNIEnv** p_env, void* vm_args) {
  return JNI_CreateJavaVM_(p_vm, p_env, vm_args);
}
```

类 JniInvocation 的成员变量 JNI_CreateJavaVM_指向的就是前面所加载的 libdvm.so 或者 libart.so 所导出的函数 JNI_CreateJavaVM。类 JniInvocation 的成员函数 JNI_CreateJavaVM 返回的 JavaVM 接口指向的要么是 Dalvik 虚拟机，要么是 ART 虚拟机。

21.4 创建运行实例

在文件“art/runtime/jni_internal.cc”中，函数 JNI_CreateJavaVM 会调用函数 Create 创建 Runtime 的实例。函数 Create 在文件“art/runtime/runtime.cc”中定义，具体实现代码如下所示：

```
bool Runtime::Create(const Options& options, bool ignore_unrecognized) {
  // TODO: acquire a static mutex on Runtime to avoid racing.
  if (Runtime::instance_ != NULL) {
    return false;
  }
  InitLogging(NULL);  //初始化 Log 系统.
  instance_ = new Runtime; //创建 Runtime 实例
  if (!instance_->Init(options, ignore_unrecognized)) {
    delete instance_;
    instance_ = NULL;
    return false;//初始化 Runtime
  }
  return true;
}
```

再次回到函数 JNI_ Create JavaVM 中，“Runtime* runtime = Runtime::Current();”用于获得 Runtime 当前实例，Runtime 使用单例模式实现。“bool started = runtime->Start();”用于调用 Start 函数，该函数在文件“art/runtime/runtime.cc”中定义，具体实现代码如下所示：

```
bool Runtime::Start() {
  VLOG(startup) << "Runtime::Start entering";
  CHECK(host_prefix_.empty()) << host_prefix_;
  Thread* self = Thread::Current();//获得当前运行线程
  self->TransitionFromRunnableToSuspended(kNative);// 将该线程状态从 Runnable 切换到 Suspend

  started_ = true;
  // 完成 Native 函数的初始化工作
  InitNativeMethods();

  // Initialize well known thread group values that may be accessed threads while attaching.
  InitThreadGroups(self);

  Thread::FinishStartup();

  if (is_zygote_) {
    if (!InitZygote()) {
      return false;
    }
  } else {
    DidForkFromZygote();
  }
```

```
  StartDaemonThreads();

  system_class_loader_ = CreateSystemClassLoader();

  self->GetJniEnv()->locals.AssertEmpty();

  VLOG(startup) << "Runtime::Start exiting";

  finished_starting_ = true;

  return true;
}
```

函数 Runtime::InitNativeMethods 在文件“art/runtime/runtime.cc”中定义，具体实现代码如下所示：

```
void Runtime::InitNativeMethods() {
  VLOG(startup) << "Runtime::InitNativeMethods entering";
  Thread* self = Thread::Current();
  JNIEnv* env = self->GetJniEnv();//获取 JNI 环境

  // Must be in the kNative state for calling native methods (JNI_OnLoad code).
  CHECK_EQ(self->GetState(), kNative);

  // First set up JniConstants, which is used by both the runtime's built-in native
  // methods and libcore.
  JniConstants::init(env);
  WellKnownClasses::Init(env);

  //调用 RegisterRuntimeNativeMethods 函数完成 Native 函数的注册
  RegisterRuntimeNativeMethods(env);

  // Then set up libcore, which is just a regular JNI library with a regular JNI_OnLoad.
  // Most JNI libraries can just use System.loadLibrary, but libcore can't because it's
  // the library that implements System.loadLibrary!
  {
    std::string mapped_name(StringPrintf(OS_SHARED_LIB_FORMAT_STR, "javacore"));
    std::string reason;
    self->TransitionFromSuspendedToRunnable();
    if (!instance_->java_vm_->LoadNativeLibrary(mapped_name, NULL, reason)) {
      LOG(FATAL) << "LoadNativeLibrary failed for \"" << mapped_name << "\": " << reason;
    }
    self->TransitionFromRunnableToSuspended(kNative);
  }

  // Initialize well known classes that may invoke runtime native methods.
  WellKnownClasses::LateInit(env);

  VLOG(startup) << "Runtime::InitNativeMethods exiting";
}
```

21.5 注册本地 JNI 函数

在文件“art/runtime/runtime.cc”中的函数 Runtime::InitNativeMethods 中，通过代码行“RegisterRuntimeNativeMethods(env);”调用函数 RegisterRuntimeNativeMethods 来注册 Native 函数，函数 RegisterRuntimeNativeMethods 具体实现代码如下所示：

```
void Runtime::RegisterRuntimeNativeMethods(JNIEnv* env) {
#define REGISTER(FN) extern void FN(JNIEnv*); FN(env)
  // Register Throwable first so that registration of other native methods can throw exceptions
  REGISTER(register_java_lang_Throwable);
  REGISTER(register_dalvik_system_DexFile);
  REGISTER(register_dalvik_system_VMDebug);
  REGISTER(register_dalvik_system_VMRuntime);
```

```
  REGISTER(register_dalvik_system_VMStack);
  REGISTER(register_dalvik_system_Zygote);
  REGISTER(register_java_lang_Class);
  REGISTER(register_java_lang_DexCache);
  REGISTER(register_java_lang_Object);
  REGISTER(register_java_lang_Runtime);
  REGISTER(register_java_lang_String);
  REGISTER(register_java_lang_System);
  REGISTER(register_java_lang_Thread);
  REGISTER(register_java_lang_VMClassLoader);
  REGISTER(register_java_lang_reflect_Array);
  REGISTER(register_java_lang_reflect_Constructor);
  REGISTER(register_java_lang_reflect_Field);
  REGISTER(register_java_lang_reflect_Method);
  REGISTER(register_java_lang_reflect_Proxy);
  REGISTER(register_java_util_concurrent_atomic_AtomicLong);
  REGISTER(register_org_apache_harmony_dalvik_ddmc_DdmServer);
  REGISTER(register_org_apache_harmony_dalvik_ddmc_DdmVmInternal);
  REGISTER(register_sun_misc_Unsafe);
#undef REGISTER
}
```

在上述代码中列出了需要注册的函数列表，有关上述函数的具体实现请读者自行分析，例如 register_java_lang_Throwable 在文件“runtime/native/java_lang_Throwable.cc”中定义，具体实现代码如下所示：

```
void register_java_lang_Throwable(JNIEnv* env) {
  REGISTER_NATIVE_METHODS("java/lang/Throwable");
}
```

21.6 启动守护进程

再次返回到 Runtime:: Start 函数，“if (!InitZygote()) {”代码行用于调用 InitZygote 完成一些文件文件系统的 mount 工作。然后通过“StartDaemonThreads();”代码行调用 java.lang.Daemons.start()函数启动守护进程。函数 StartDaemonThreads()的具体实现代码如下所示：

```
void Runtime::StartDaemonThreads() {
  VLOG(startup) << "Runtime::StartDaemonThreads entering";

  Thread* self = Thread::Current();

  // Must be in the kNative state for calling native methods.
  CHECK_EQ(self->GetState(), kNative);

  JNIEnv* env = self->GetJniEnv();
  env->CallStaticVoidMethod(WellKnownClasses::java_lang_Daemons,
                            WellKnownClasses::java_lang_Daemons_start);
  if (env->ExceptionCheck()) {
    env->ExceptionDescribe();
    LOG(FATAL) << "Error starting java.lang.Daemons";
  }

  VLOG(startup) << "Runtime::StartDaemonThreads exiting";
}
```

综上所述，Android 系统通过将 ART 运行时抽象成一个 Java 虚拟机，以及通过系统属性 persist.sys.dalvik.vm.lib 和一个适配层 JniInvocation，就可以无缝地将 Dalvik 虚拟机替换为 ART 运行时。这个替换过程设计非常巧妙，因为涉及的代码修改是非常少的。涉及类的具体关系如图 21-2 所示。

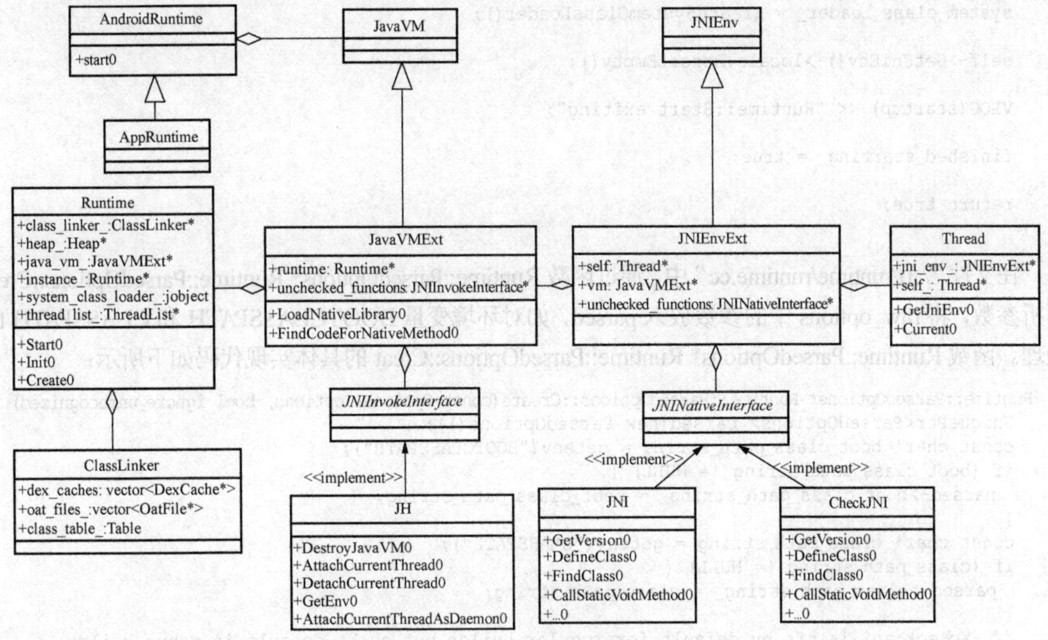

图 21-2　启动 ART 涉及的类

21.7 解析参数

在函数 JNI_CreateJavaVM()中，先调用 Create()函数创建 Runtime。Runtime 是一个单例，创建后会马上调用文件“/art/runtime/runtime.cc”中的 Init()函数。函数 Init()的功能是解析参数，初始化 Heap 和 JavaVMExt 结构，实现线程和信号处理，并创建 ClassLinker 等。函数 Init()的具体实现代码如下所示：

```
bool Runtime::Start() {
  VLOG(startup) << "Runtime::Start entering";

  CHECK(host_prefix_.empty()) << host_prefix_;

  // Restore main thread state to kNative as expected by native code.
  Thread* self = Thread::Current();
  self->TransitionFromRunnableToSuspended(kNative);

  started_ = true;

  // InitNativeMethods needs to be after started_ so that the classes
  // it touches will have methods linked to the oat file if necessary.
  InitNativeMethods();

  // Initialize well known thread group values that may be accessed threads while attaching.
  InitThreadGroups(self);

  Thread::FinishStartup();

  if (is_zygote_) {
    if (!InitZygote()) {
      return false;
    }
  } else {
    DidForkFromZygote();
  }

  StartDaemonThreads();
```

```
    system_class_loader_ = CreateSystemClassLoader();

    self->GetJniEnv()->locals.AssertEmpty();

    VLOG(startup) << "Runtime::Start exiting";

    finished_starting_ = true;

    return true;
}
```

在文件“/art/runtime/runtime.cc”中，通过函数 Runtime::ParsedOptions* Runtime::ParsedOptions::Creat 解析参数，将 raw_options 中的参数放入 parsed，如对环境变量 BOOTCLASSPATH 和 CLASSPATH 的处理。函数 Runtime::ParsedOptions* Runtime::ParsedOptions::Creat 的具体实现代码如下所示：

```
Runtime::ParsedOptions* Runtime::ParsedOptions::Create(const Options& options, bool ignore_unrecognized) {
  UniquePtr<ParsedOptions> parsed(new ParsedOptions());
  const char* boot_class_path_string = getenv("BOOTCLASSPATH");
  if (boot_class_path_string != NULL) {
    parsed->boot_class_path_string_ = boot_class_path_string;
  }
  const char* class_path_string = getenv("CLASSPATH");
  if (class_path_string != NULL) {
    parsed->class_path_string_ = class_path_string;
  }
  // -Xcheck:jni is off by default for regular builds but on by default in debug builds.
  parsed->check_jni_ = kIsDebugBuild;

  parsed->heap_initial_size_ = gc::Heap::kDefaultInitialSize;
  parsed->heap_maximum_size_ = gc::Heap::kDefaultMaximumSize;
  parsed->heap_min_free_ = gc::Heap::kDefaultMinFree;
  parsed->heap_max_free_ = gc::Heap::kDefaultMaxFree;
  parsed->heap_target_utilization_ = gc::Heap::kDefaultTargetUtilization;
  parsed->heap_growth_limit_ = 0;  // 0 means no growth limit.
  // Default to number of processors minus one since the main GC thread also does work.
  parsed->parallel_gc_threads_ = sysconf(_SC_NPROCESSORS_CONF) - 1;
  // Only the main GC thread, no workers.
  parsed->conc_gc_threads_ = 0;
  parsed->stack_size_ = 0;  // 0 means default.
  parsed->low_memory_mode_ = false;
……
```

然后初始化 Monitor（相当于 mutex+conditional variable，可用于多个线程同步）和线程链表等，再实现比较重要的 Heap 及 GC 的初始化工作。其中，gc::Heap 功能通过文件“art/runtime/gc/heap.cc”中的函数 Heap::Heap 实现，具体实现代码如下所示：

```
Heap::Heap(size_t initial_size, size_t growth_limit, size_t min_free, size_t max_free,
           double target_utilization, size_t capacity, const std::string& original_image_file_name,
           bool concurrent_gc, size_t parallel_gc_threads, size_t conc_gc_threads,
           bool low_memory_mode, size_t long_pause_log_threshold, size_t long_gc_log_threshold,
           bool ignore_max_footprint)
    : alloc_space_(NULL),
      card_table_(NULL),
      concurrent_gc_(concurrent_gc),
      parallel_gc_threads_(parallel_gc_threads),
      conc_gc_threads_(conc_gc_threads),
      low_memory_mode_(low_memory_mode),
      long_pause_log_threshold_(long_pause_log_threshold),
      long_gc_log_threshold_(long_gc_log_threshold),
      ignore_max_footprint_(ignore_max_footprint),
      have_zygote_space_(false),
      soft_ref_queue_lock_(NULL),
      weak_ref_queue_lock_(NULL),
      finalizer_ref_queue_lock_(NULL),
      phantom_ref_queue_lock_(NULL),
      is_gc_running_(false),
      last_gc_type_(collector::kGcTypeNone),
```

```
      next_gc_type_(collector::kGcTypePartial),
      capacity_(capacity),
      growth_limit_(growth_limit),
      max_allowed_footprint_(initial_size),
      native_footprint_gc_watermark_(initial_size),
      native_footprint_limit_(2 * initial_size),
      activity_thread_class_(NULL),
      application_thread_class_(NULL),
      activity_thread_(NULL),
      application_thread_(NULL),
      last_process_state_id_(NULL),
      /* Initially care about pauses in case we never get notified of process states, or if the JNI*/
      // code becomes broken
      care_about_pause_times_(true),
      concurrent_start_bytes_(concurrent_gc_ ? initial_size - kMinConcurrentRemainingBytes
          :  std::numeric_limits<size_t>::max()),
      total_bytes_freed_ever_(0),
      total_objects_freed_ever_(0),
      large_object_threshold_(3 * kPageSize),
      num_bytes_allocated_(0),
      native_bytes_allocated_(0),
      gc_memory_overhead_(0),
      verify_missing_card_marks_(false),
      verify_system_weaks_(false),
      verify_pre_gc_heap_(false),
      verify_post_gc_heap_(false),
      verify_mod_union_table_(false),
      min_alloc_space_size_for_sticky_gc_(2 * MB),
      min_remaining_space_for_sticky_gc_(1 * MB),
      last_trim_time_ms_(0),
      allocation_rate_(0),
      /* For GC a lot mode, we limit the allocations stacks to be kGcAlotInterval allocations. This
       * causes a lot of GC since we do a GC for alloc whenever the stack is full. When heap
       * verification is enabled, we limit the size of allocation stacks to speed up their
       * searching.
       */
      max_allocation_stack_size_(kGCALotMode ? kGcAlotInterval
          : (kDesiredHeapVerification > kNoHeapVerification) ? KB : MB),
      reference_referent_offset_(0),
      reference_queue_offset_(0),
      reference_queueNext_offset_(0),
      reference_pendingNext_offset_(0),
      finalizer_reference_zombie_offset_(0),
      min_free_(min_free),
      max_free_(max_free),
      target_utilization_(target_utilization),
      total_wait_time_(0),
      total_allocation_time_(0),
      verify_object_mode_(kHeapVerificationNotPermitted),
      running_on_valgrind_(RUNNING_ON_VALGRIND) {
  if (VLOG_IS_ON(heap) || VLOG_IS_ON(startup)) {
    LOG(INFO) << "Heap() entering";
  }

  live_bitmap_.reset(new accounting::HeapBitmap(this));
  mark_bitmap_.reset(new accounting::HeapBitmap(this));

  // Requested begin for the alloc space, to follow the mapped image and oat files
  byte* requested_alloc_space_begin = NULL;
  std::string image_file_name(original_image_file_name);
  if (!image_file_name.empty()) {
    space::ImageSpace* image_space = space::ImageSpace::Create(image_file_name);
    CHECK(image_space != NULL) << "Failed to create space for " << image_file_name;
    AddContinuousSpace(image_space);
    // Oat files referenced by image files immediately follow them in memory, ensure alloc space
    // isn't going to get in the middle
    byte* oat_file_end_addr = image_space->GetImageHeader().GetOatFileEnd();
    CHECK_GT(oat_file_end_addr, image_space->End());
    if (oat_file_end_addr > requested_alloc_space_begin) {
```

```
    requested_alloc_space_begin =
        reinterpret_cast<byte*>(RoundUp(reinterpret_cast<uintptr_t>(oat_file_end_addr),
                                        kPageSize));
    }
  }

  alloc_space_ = space::DlMallocSpace::Create(Runtime::Current()->IsZygote() ? "zygote space" :
"alloc space",
                                              initial_size,
                                              growth_limit, capacity,
                                              requested_alloc_space_begin);
  CHECK(alloc_space_ != NULL) << "Failed to create alloc space";
  alloc_space_->SetFootprintLimit(alloc_space_->Capacity());
  AddContinuousSpace(alloc_space_);

  // Allocate the large object space.
  const bool kUseFreeListSpaceForLOS = false;
  if (kUseFreeListSpaceForLOS) {
    large_object_space_ = space::FreeListSpace::Create("large object space", NULL, capacity);
  } else {
    large_object_space_ = space::LargeObjectMapSpace::Create("large object space");
  }
  CHECK(large_object_space_ != NULL) << "Failed to create large object space";
  AddDiscontinuousSpace(large_object_space_);

  // Compute heap capacity. Continuous spaces are sorted in order of Begin().
  byte* heap_begin = continuous_spaces_.front()->Begin();
  size_t heap_capacity = continuous_spaces_.back()->End() - continuous_spaces_.front()->Begin();
  if (continuous_spaces_.back()->IsDlMallocSpace()) {
    heap_capacity += continuous_spaces_.back()->AsDlMallocSpace()->NonGrowthLimitCapacity();
  }

  // Allocate the card table.
  card_table_.reset(accounting::CardTable::Create(heap_begin, heap_capacity));
  CHECK(card_table_.get() != NULL) << "Failed to create card table";

  image_mod_union_table_.reset(new accounting::ModUnionTableToZygoteAllocspace(this));
  CHECK(image_mod_union_table_.get() != NULL) << "Failed to create image mod-union table";

  zygote_mod_union_table_.reset(new accounting::ModUnionTableCardCache(this));
  CHECK(zygote_mod_union_table_.get() != NULL) << "Failed to create Zygote mod-union table";

  // TODO: Count objects in the image space here.
  num_bytes_allocated_ = 0;

  // Default mark stack size in bytes.
  static const size_t default_mark_stack_size = 64 * KB;
  mark_stack_.reset(accounting::ObjectStack::Create("mark stack", default_mark_stack_size));
  allocation_stack_.reset(accounting::ObjectStack::Create("allocation stack",
                                                         max_allocation_stack_size_));
  live_stack_.reset(accounting::ObjectStack::Create("live stack",
                                                   max_allocation_stack_size_));

  // It's still too early to take a lock because there are no threads yet, but we can create locks
  // now. We don't create it earlier to make it clear that you can't use locks during heap
  // initialization.
  gc_complete_lock_ = new Mutex("GC complete lock");
  gc_complete_cond_.reset(new ConditionVariable("GC complete condition variable",
                                                *gc_complete_lock_));

  // Create the reference queue locks, this is required so for parallel object scanning in the GC.
  soft_ref_queue_lock_ = new Mutex("Soft reference queue lock");
  weak_ref_queue_lock_ = new Mutex("Weak reference queue lock");
  finalizer_ref_queue_lock_ = new Mutex("Finalizer reference queue lock");
  phantom_ref_queue_lock_ = new Mutex("Phantom reference queue lock");

  last_gc_time_ns_ = NanoTime();
  last_gc_size_ = GetBytesAllocated();
```

```
if (ignore_max_footprint_) {
  SetIdealFootprint(std::numeric_limits<size_t>::max());
  concurrent_start_bytes_ = max_allowed_footprint_;
}

// Create our garbage collectors.
for (size_t i = 0; i < 2; ++i) {
  const bool concurrent = i != 0;
  mark_sweep_collectors_.push_back(new collector::MarkSweep(this, concurrent));
  mark_sweep_collectors_.push_back(new collector::PartialMarkSweep(this, concurrent));
  mark_sweep_collectors_.push_back(new collector::StickyMarkSweep(this, concurrent));
}

CHECK_NE(max_allowed_footprint_, 0U);
if (VLOG_IS_ON(heap) || VLOG_IS_ON(startup)) {
  LOG(INFO) << "Heap() exiting";
}
}
```

在上述代码中，函数 ImageSpace::Create()会检测 image 文件，如果没有就调用 GenerateImage()来创建。正因为上述操作过程，所以 log 中会有如下所示的信息：

```
I/art    ( 161): GenerateImage: /system/bin/dex2oat--image=/data/dalvik-cache/system@framework@boot.art
--runtime-arg    -Xms64m--runtime-arg    -Xmx64m    --dex-file=/system/framework/core-libart.jar   …
--oat-file=/data/dalvik-cache/system@framework@boot.oat
--base=0x60000000--image-classes-zip=/system/ framework/framework...
```

上述过程调用了 dex2oat，把 BOOTCLASSPATH 里的包打成 image 文件，它最后会生成 boot.art 和 boot.oat 两个文件。其中前者是 image 文件，后者是 elf 文件。这个 image 会被放到创建的 Heap 中。在函数 Heap::Heap 中，接下来会为一些数据结构分配空间，创建各种互斥量及初始化 GC。其中 MarkSweep、PartialMarkSweep 和 StickyMarkSweep 都是 art::gc::collector::GarbageCollector 的继承类，和几个子类应用了 TemplateMethod 模式。在函数 GarbageCollector::Run()中实现了主要算法，此函数在文件“art/runtime/gc/collector/garbage_collector.cc”中定义，具体实现代码如下所示：

```
void GarbageCollector::Run() {
  ThreadList* thread_list = Runtime::Current()->GetThreadList();
  uint64_t start_time = NanoTime();
  pause_times_.clear();
  duration_ns_ = 0;

  InitializePhase();

  if (!IsConcurrent()) {
    // Pause is the entire length of the GC.
    uint64_t pause_start = NanoTime();
    ATRACE_BEGIN("Application threads suspended");
    thread_list->SuspendAll();
    MarkingPhase();
    ReclaimPhase();
    thread_list->ResumeAll();
    ATRACE_END();
    uint64_t pause_end = NanoTime();
    pause_times_.push_back(pause_end - pause_start);
  } else {
    Thread* self = Thread::Current();
    {
      ReaderMutexLock mu(self, *Locks::mutator_lock_);
      MarkingPhase();
    }
    bool done = false;
    while (!done) {
      uint64_t pause_start = NanoTime();
      ATRACE_BEGIN("Suspending mutator threads");
      thread_list->SuspendAll();
```

```
      ATRACE_END();
      ATRACE_BEGIN("All mutator threads suspended");
      done = HandleDirtyObjectsPhase();
      ATRACE_END();
      uint64_t pause_end = NanoTime();
      ATRACE_BEGIN("Resuming mutator threads");
      thread_list->ResumeAll();
      ATRACE_END();
      pause_times_.push_back(pause_end - pause_start);
    }
    {
      ReaderMutexLock mu(self, *Locks::mutator_lock_);
      ReclaimPhase();
    }
  }

  uint64_t end_time = NanoTime();
  duration_ns_ = end_time - start_time;

  FinishPhase();
}
```

在上述代码中调用了 InitializePhase()、MarkingPhase()、ReclaimPhase()和 FinishPhase()等虚函数，这几个虚函数在 MarkSweep 等几个子类中有具体实现。

再次回到函数 Runtime::Init()，通过如下代码实现 ClassLinker 的初始化操作，其主要功能是调用 CreateFromImage()函数实现的：

```
ClassLinker* ClassLinker::CreateFromImage(InternTable* intern_table) {
  UniquePtr<ClassLinker> class_linker(new ClassLinker(intern_table));
  class_linker->InitFromImage();
  return class_linker.release();
}
```

在文件"art\art\runtime\class_linker.cc"中，通过函数 InitFromImage()从 Heap 中得到 image 的空间，然后得到 dex caches 数组，接着把这些 dex caches 对应的 dex file 信息注册到 BootClassPath 中去。函数 InitFromImage()的具体实现代码如下所示：

```
void ClassLinker::InitFromImage() {
  VLOG(startup) << "ClassLinker::InitFromImage entering";
  CHECK(!init_done_);

  gc::Heap* heap = Runtime::Current()->GetHeap();
  gc::space::ImageSpace* space = heap->GetImageSpace();
  dex_cache_image_class_lookup_required_ = true;
  CHECK(space != NULL);
  OatFile& oat_file = GetImageOatFile(space);
  CHECK_EQ(oat_file.GetOatHeader().GetImageFileLocationOatChecksum(), 0U);
  CHECK_EQ(oat_file.GetOatHeader().GetImageFileLocationOatDataBegin(), 0U);
  CHECK(oat_file.GetOatHeader().GetImageFileLocation().empty());
  portable_resolution_trampoline_ = oat_file.GetOatHeader().GetPortableResolutionTrampoline();
  quick_resolution_trampoline_ = oat_file.GetOatHeader().GetQuickResolutionTrampoline();
  mirror::Object*  dex_caches_object  =  space->GetImageHeader().GetImageRoot  (ImageHeader::
kDexCaches);
  mirror::ObjectArray<mirror::DexCache>* dex_caches =
      dex_caches_object->AsObjectArray<mirror::DexCache>();

  mirror::ObjectArray<mirror::Class>* class_roots =

space->GetImageHeader().GetImageRoot(ImageHeader::kClassRoots)->AsObjectArray<mirror::Class>();
  class_roots_ = class_roots;

  // Special case of setting up the String class early so that we can test arbitrary objects
  // as being Strings or not
  mirror::String::SetClass(GetClassRoot(kJavaLangString));

  CHECK_EQ(oat_file.GetOatHeader().GetDexFileCount(),
```

```
        static_cast<uint32_t>(dex_caches->GetLength()));
    Thread* self = Thread::Current();
    for (int32_t i = 0; i < dex_caches->GetLength(); i++) {
      SirtRef<mirror::DexCache> dex_cache(self, dex_caches->Get(i));
      const std::string& dex_file_location(dex_cache->GetLocation()->ToModifiedUtf8());
      const OatFile::OatDexFile* oat_dex_file = oat_file.GetOatDexFile(dex_file_location, NULL);
      CHECK(oat_dex_file != NULL) << oat_file.GetLocation() << " " << dex_file_location;
      const DexFile* dex_file = oat_dex_file->OpenDexFile();
      if (dex_file == NULL) {
        LOG(FATAL) << "Failed to open dex file " << dex_file_location
                   << " from within oat file " << oat_file.GetLocation();

      }

      CHECK_EQ(dex_file->GetLocationChecksum(), oat_dex_file->GetDexFileLocationChecksum());

      AppendToBootClassPath(*dex_file, dex_cache);
    }

    // Set classes on AbstractMethod early so that IsMethod tests can be performed during the live
    // bitmap walk.
    mirror::ArtMethod::SetClass(GetClassRoot(kJavaLangReflectArtMethod));

    // Set entry point to interpreter if in InterpretOnly mode.
    if (Runtime::Current()->GetInstrumentation()->InterpretOnly()) {
      ReaderMutexLock mu(self, *Locks::heap_bitmap_lock_);
      heap->FlushAllocStack();
      heap->GetLiveBitmap()->Walk(InitFromImageInterpretOnlyCallback, this);
    }

    // reinit class_roots_
    mirror::Class::SetClassClass(class_roots->Get(kJavaLangClass));
    class_roots_ = class_roots;

    // reinit array_iftable_ from any array class instance, they should be ==
    array_iftable_ = GetClassRoot(kObjectArrayClass)->GetIfTable();
    DCHECK(array_iftable_ == GetClassRoot(kBooleanArrayClass)->GetIfTable());
    // String class root was set above
    mirror::ArtField::SetClass(GetClassRoot(kJavaLangReflectArtField));
    mirror::BooleanArray::SetArrayClass(GetClassRoot(kBooleanArrayClass));
    mirror::ByteArray::SetArrayClass(GetClassRoot(kByteArrayClass));
    mirror::CharArray::SetArrayClass(GetClassRoot(kCharArrayClass));
    mirror::DoubleArray::SetArrayClass(GetClassRoot(kDoubleArrayClass));
    mirror::FloatArray::SetArrayClass(GetClassRoot(kFloatArrayClass));
    mirror::IntArray::SetArrayClass(GetClassRoot(kIntArrayClass));
    mirror::LongArray::SetArrayClass(GetClassRoot(kLongArrayClass));
    mirror::ShortArray::SetArrayClass(GetClassRoot(kShortArrayClass));
    mirror::Throwable::SetClass(GetClassRoot(kJavaLangThrowable));
    mirror::StackTraceElement::SetClass(GetClassRoot(kJavaLangStackTraceElement));

    FinishInit();

    VLOG(startup) << "ClassLinker::InitFromImage exiting";
}
```

在文件“art\art\runtime\class_linker.cc”中，通过函数 AppendToBootClassPath()和 RegisterDexFileLocked()将 dex cache 和 dex file 关联起来，同时把 dex file 注册到 boot_class_path_，将 dex cache 注册到 dex_caches_。函数 AppendToBootClassPath()和 RegisterDexFileLocked()的具体实现代码如下所示：

```
void ClassLinker::AppendToBootClassPath(const DexFile& dex_file, SirtRef<mirror::DexCache>&
dex_cache) {
  CHECK(dex_cache.get() != NULL) << dex_file.GetLocation();
  boot_class_path_.push_back(&dex_file);
  RegisterDexFile(dex_file, dex_cache);
}
void ClassLinker::RegisterDexFileLocked(const DexFile& dex_file, SirtRef<mirror::DexCache>&
dex_cache) {
  dex_lock_.AssertExclusiveHeld(Thread::Current());
```

```
  CHECK(dex_cache.get() != NULL) << dex_file.GetLocation();
  CHECK(dex_cache->GetLocation()->Equals(dex_file.GetLocation()))
      << dex_cache->GetLocation()->ToModifiedUtf8() << " " << dex_file.GetLocation();
  dex_caches_.push_back(dex_cache.get());
  dex_cache->SetDexFile(&dex_file);
  dex_caches_dirty_ = true;
}
```

当注册上述信息后，在 ClassLinker 调用 FindClass()函数时会用到。执行完 Runtime 中的函数 Create()和 Init()后，在 JNI_CreateJavaVM 函数中 Runtime 的 Start()函数被调用。

21.8 初始化类、方法和域

在文件 runtime.cc 的函数 InitNativeMethods()中分别调用函数 JniConstants::init()和函数 WellKnownClasses::Init()。函数 InitNativeMethods()的具体实现代码如下所示：

```
void Runtime::InitNativeMethods() {
  VLOG(startup) << "Runtime::InitNativeMethods entering";
  Thread* self = Thread::Current();
  JNIEnv* env = self->GetJniEnv();

  // Must be in the kNative state for calling native methods (JNI_OnLoad code).
  CHECK_EQ(self->GetState(), kNative);

  // First set up JniConstants, which is used by both the runtime's built-in native
  // methods and libcore.
  JniConstants::init(env);
  WellKnownClasses::Init(env);

  // Then set up the native methods provided by the runtime itself.
  RegisterRuntimeNativeMethods(env);

  // Then set up libcore, which is just a regular JNI library with a regular JNI_OnLoad.
  // Most JNI libraries can just use System.loadLibrary, but libcore can't because it's
  // the library that implements System.loadLibrary!
  {
    std::string mapped_name(StringPrintf(OS_SHARED_LIB_FORMAT_STR, "javacore"));
    std::string reason;
    self->TransitionFromSuspendedToRunnable();
    if (!instance_->java_vm_->LoadNativeLibrary(mapped_name, NULL, reason)) {
      LOG(FATAL) << "LoadNativeLibrary failed for \"" << mapped_name << "\": " << reason;
    }
    self->TransitionFromRunnableToSuspended(kNative);
  }

  // Initialize well known classes that may invoke runtime native methods.
  WellKnownClasses::LateInit(env);

  VLOG(startup) << "Runtime::InitNativeMethods exiting";
}
```

函数 JniConstants::init()在文件“libnativehelper/JniConstants.cpp”中定义，WellKnownClasses::Init()在文件“art/runtime/well_known_classes.cc”中定义，这两个函数的具体实现代码如下所示：

```
void JniConstants::init(JNIEnv* env) {
    bidiRunClass = findClass(env, "java/text/Bidi$Run");
    bigDecimalClass = findClass(env, "java/math/BigDecimal");
    booleanClass = findClass(env, "java/lang/Boolean");
    byteClass = findClass(env, "java/lang/Byte");
    byteArrayClass = findClass(env, "[B");
    calendarClass = findClass(env, "java/util/Calendar");
    characterClass = findClass(env, "java/lang/Character");
    charsetICUClass = findClass(env, "java/nio/charset/CharsetICU");
```

```
    constructorClass = findClass(env, "java/lang/reflect/Constructor");
    floatClass = findClass(env, "java/lang/Float");
    deflaterClass = findClass(env, "java/util/zip/Deflater");
    doubleClass = findClass(env, "java/lang/Double");
    errnoExceptionClass = findClass(env, "libcore/io/ErrnoException");
    fieldClass = findClass(env, "java/lang/reflect/Field");
    fieldPositionIteratorClass = findClass(env, "libcore/icu/NativeDecimalFormat$FieldPositionIterator");
    fileDescriptorClass = findClass(env, "java/io/FileDescriptor");
    gaiExceptionClass = findClass(env, "libcore/io/GaiException");
    inet6AddressClass = findClass(env, "java/net/Inet6Address");
    inetAddressClass = findClass(env, "java/net/InetAddress");
    inetSocketAddressClass = findClass(env, "java/net/InetSocketAddress");
    inetUnixAddressClass = findClass(env, "java/net/InetUnixAddress");
    inflaterClass = findClass(env, "java/util/zip/Inflater");
    inputStreamClass = findClass(env, "java/io/InputStream");
    integerClass = findClass(env, "java/lang/Integer");
    localeDataClass = findClass(env, "libcore/icu/LocaleData");
    longClass = findClass(env, "java/lang/Long");
    methodClass = findClass(env, "java/lang/reflect/Method");
    mutableIntClass = findClass(env, "libcore/util/MutableInt");
    mutableLongClass = findClass(env, "libcore/util/MutableLong");
    objectClass = findClass(env, "java/lang/Object");
    objectArrayClass = findClass(env, "[Ljava/lang/Object;");
    outputStreamClass = findClass(env, "java/io/OutputStream");
    parsePositionClass = findClass(env, "java/text/ParsePosition");
    patternSyntaxExceptionClass = findClass(env, "java/util/regex/PatternSyntaxException");
    realToStringClass = findClass(env, "java/lang/RealToString");
    referenceClass = findClass(env, "java/lang/ref/Reference");
    shortClass = findClass(env, "java/lang/Short");
    socketClass = findClass(env, "java/net/Socket");
    socketImplClass = findClass(env, "java/net/SocketImpl");
    stringClass = findClass(env, "java/lang/String");
    structAddrinfoClass = findClass(env, "libcore/io/StructAddrinfo");
    structFlockClass = findClass(env, "libcore/io/StructFlock");
    structGroupReqClass = findClass(env, "libcore/io/StructGroupReq");
    structLingerClass = findClass(env, "libcore/io/StructLinger");
    structPasswdClass = findClass(env, "libcore/io/StructPasswd");
    structPollfdClass = findClass(env, "libcore/io/StructPollfd");
    structStatClass = findClass(env, "libcore/io/StructStat");
    structStatVfsClass = findClass(env, "libcore/io/StructStatVfs");
    structTimevalClass = findClass(env, "libcore/io/StructTimeval");
    structUcredClass = findClass(env, "libcore/io/StructUcred");
    structUtsnameClass = findClass(env, "libcore/io/StructUtsname");
  }
  void WellKnownClasses::Init(JNIEnv* env) {
    com_android_dex_Dex = CacheClass(env, "com/android/dex/Dex");
    dalvik_system_PathClassLoader = CacheClass(env, "dalvik/system/PathClassLoader");
    java_lang_ClassLoader = CacheClass(env, "java/lang/ClassLoader");
    java_lang_ClassNotFoundException = CacheClass(env, "java/lang/ClassNotFoundException");
    java_lang_Daemons = CacheClass(env, "java/lang/Daemons");
    java_lang_Object = CacheClass(env, "java/lang/Object");
    java_lang_Error = CacheClass(env, "java/lang/Error");
    java_lang_reflect_AbstractMethod = CacheClass(env, "java/lang/reflect/AbstractMethod");
    java_lang_reflect_ArtMethod = CacheClass(env, "java/lang/reflect/ArtMethod");
    java_lang_reflect_Constructor = CacheClass(env, "java/lang/reflect/Constructor");
    java_lang_reflect_Field = CacheClass(env, "java/lang/reflect/Field");
    java_lang_reflect_Method = CacheClass(env, "java/lang/reflect/Method");
    java_lang_reflect_Proxy = CacheClass(env, "java/lang/reflect/Proxy");
    java_lang_RuntimeException = CacheClass(env, "java/lang/RuntimeException");
    java_lang_StackOverflowError = CacheClass(env, "java/lang/StackOverflowError");
    java_lang_System = CacheClass(env, "java/lang/System");
    java_lang_Thread = CacheClass(env, "java/lang/Thread");
    java_lang_Thread$UncaughtExceptionHandler = CacheClass(env, "java/lang/Thread$UncaughtExceptionHandler");
    java_lang_ThreadGroup = CacheClass(env, "java/lang/ThreadGroup");
    java_lang_Throwable = CacheClass(env, "java/lang/Throwable");
    java_nio_DirectByteBuffer = CacheClass(env, "java/nio/DirectByteBuffer");
    org_apache_harmony_dalvik_ddmc_Chunk = CacheClass(env, "org/apache/harmony/dalvik/ddmc/Chunk");
```

```
org_apache_harmony_dalvik_ddmc_DdmServer = CacheClass(env, "org/apache/harmony/dalvik/ddmc/DdmServer");

com_android_dex_Dex_create = CacheMethod(env, com_android_dex_Dex, true, "create", "(Ljava/nio/ByteBuffer;)Lcom/android/dex/Dex;");
java_lang_ClassNotFoundException_init = CacheMethod(env, java_lang_ClassNotFoundException, false, "<init>", "(Ljava/lang/String;Ljava/lang/Throwable;)V");
java_lang_ClassLoader_loadClass = CacheMethod(env, java_lang_ClassLoader, false, "loadClass", "(Ljava/lang/String;)Ljava/lang/Class;");

java_lang_Daemons_requestGC = CacheMethod(env, java_lang_Daemons, true, "requestGC", "()V");
java_lang_Daemons_requestHeapTrim = CacheMethod(env, java_lang_Daemons, true, "requestHeapTrim", "()V");
java_lang_Daemons_start = CacheMethod(env, java_lang_Daemons, true, "start", "()V");

ScopedLocalRef<jclass> java_lang_ref_FinalizerReference(env, env->FindClass("java/lang/ref/FinalizerReference"));
java_lang_ref_FinalizerReference_add = CacheMethod(env, java_lang_ref_FinalizerReference.get(), true, "add", "(Ljava/lang/Object;)V");
ScopedLocalRef<jclass> java_lang_ref_ReferenceQueue(env, env->FindClass("java/lang/ref/ReferenceQueue"));
java_lang_ref_ReferenceQueue_add = CacheMethod(env, java_lang_ref_ReferenceQueue.get(), true, "add", "(Ljava/lang/ref/Reference;)V");

java_lang_reflect_Proxy_invoke = CacheMethod(env, java_lang_reflect_Proxy, true, "invoke", "(Ljava/lang/reflect/Proxy;Ljava/lang/reflect/ArtMethod;[Ljava/lang/Object;)Ljava/lang/Object;"];
java_lang_Thread_init = CacheMethod(env, java_lang_Thread, false, "<init>", "(Ljava/lang/ThreadGroup;Ljava/lang/String;IZ)V");
java_lang_Thread_run = CacheMethod(env, java_lang_Thread, false, "run", "()V");
java_lang_Thread$UncaughtExceptionHandler_uncaughtException = CacheMethod(env, java_lang_Thread$UncaughtExceptionHandler, false, "uncaughtException", "(Ljava/lang/Thread; Ljava/lang/Throwable;)V");
java_lang_ThreadGroup_removeThread = CacheMethod(env, java_lang_ThreadGroup, false, "removeThread", "(Ljava/lang/Thread;)V");
java_nio_DirectByteBuffer_init = CacheMethod(env, java_nio_DirectByteBuffer, false, "<init>", "(JI)V");
org_apache_harmony_dalvik_ddmc_DdmServer_broadcast = CacheMethod(env, org_apache_harmony_dalvik_ddmc_DdmServer, true, "broadcast", "(I)V");
org_apache_harmony_dalvik_ddmc_DdmServer_dispatch = CacheMethod(env, org_apache_harmony_dalvik_ddmc_DdmServer, true, "dispatch", "(I[BII)Lorg/apache/harmony/dalvik/ddmc/Chunk;");

java_lang_Thread_daemon = CacheField(env, java_lang_Thread, false, "daemon", "Z");
java_lang_Thread_group = CacheField(env, java_lang_Thread, false, "group", "Ljava/lang/ThreadGroup;");
java_lang_Thread_lock = CacheField(env, java_lang_Thread, false, "lock", "Ljava/lang/Object;");
java_lang_Thread_name = CacheField(env, java_lang_Thread, false, "name", "Ljava/lang/String;");
java_lang_Thread_priority = CacheField(env, java_lang_Thread, false, "priority", "I");
java_lang_Thread_uncaughtHandler = CacheField(env, java_lang_Thread, false, "uncaughtHandler", "Ljava/lang/Thread$UncaughtExceptionHandler;");
java_lang_Thread_nativePeer = CacheField(env, java_lang_Thread, false, "nativePeer", "I");
java_lang_ThreadGroup_mainThreadGroup = CacheField(env, java_lang_ThreadGroup, true, "mainThreadGroup", "Ljava/lang/ThreadGroup;");
java_lang_ThreadGroup_name = CacheField(env, java_lang_ThreadGroup, false, "name", "Ljava/lang/String;");
java_lang_ThreadGroup_systemThreadGroup = CacheField(env, java_lang_ThreadGroup, true, "systemThreadGroup", "Ljava/lang/ThreadGroup;");
java_lang_reflect_AbstractMethod_artMethod = CacheField(env, java_lang_reflect_AbstractMethod, false, "artMethod", "Ljava/lang/reflect/ArtMethod;");
java_lang_reflect_Field_artField = CacheField(env, java_lang_reflect_Field, false, "artField", "Ljava/lang/reflect/ArtField;");
java_lang_reflect_Proxy_h = CacheField(env, java_lang_reflect_Proxy, false, "h", "Ljava/lang/reflect/InvocationHandler;");
java_nio_DirectByteBuffer_capacity = CacheField(env, java_nio_DirectByteBuffer, false, "capacity", "I");
java_nio_DirectByteBuffer_effectiveDirectAddress = CacheField(env, java_nio_DirectByteBuffer, false, "effectiveDirectAddress", "J");
org_apache_harmony_dalvik_ddmc_Chunk_data = CacheField(env, org_apache_harmony_dalvik_ddmc_Chunk, false, "data", "[B");
```

```
  org_apache_harmony_dalvik_ddmc_Chunk_length = CacheField(env, org_apache_harmony_dalvik_ddmc_Chunk,
false, "length", "I");
  org_apache_harmony_dalvik_ddmc_Chunk_offset = CacheField(env, org_apache_harmony_dalvik_ddmc_Chunk,
false, "offset", "I");
  org_apache_harmony_dalvik_ddmc_Chunk_type = CacheField(env, org_apache_harmony_dalvik_ddmc_Chunk,
false, "type", "I");

  java_lang_Boolean_valueOf = CachePrimitiveBoxingMethod(env, 'Z', "java/lang/Boolean");
  java_lang_Byte_valueOf = CachePrimitiveBoxingMethod(env, 'B', "java/lang/Byte");
  java_lang_Character_valueOf = CachePrimitiveBoxingMethod(env, 'C', "java/lang/Character");
  java_lang_Double_valueOf = CachePrimitiveBoxingMethod(env, 'D', "java/lang/Double");
  java_lang_Float_valueOf = CachePrimitiveBoxingMethod(env, 'F', "java/lang/Float");
  java_lang_Integer_valueOf = CachePrimitiveBoxingMethod(env, 'I', "java/lang/Integer");
  java_lang_Long_valueOf = CachePrimitiveBoxingMethod(env, 'J', "java/lang/Long");
  java_lang_Short_valueOf = CachePrimitiveBoxingMethod(env, 'S', "java/lang/Short");
}
```

通过上述代码可知，通过 FindClass()、GetStaticFieldID()和 GetStaticMethodID()等函数分别初始化了系统基本类、方法和域，这一些都是最基本的类。

然后 RegisterRuntimeNativeMethods()函数注册了系列系统类中的 Native 函数：

```
void Runtime::RegisterRuntimeNativeMethods(JNIEnv* env) {
#define REGISTER(FN) extern void FN(JNIEnv*); FN(env)
  // Register Throwable first so that registration of other native methods can throw exceptions
  REGISTER(register_java_lang_Throwable);
  …
```

接着函数 InitNativeMethods()会载入 libjavacore.so 这个库，单独载入是因为它本身包含了System.loadLibrary()实现，不先载入会导致顺序紊乱问题。

```
if (!instance_->java_vm_->LoadNativeLibrary(mapped_name, NULL, reason)) {
```

再次回到 Runtime::Start()函数进行线程初始化，再判断是否为 Zygote 进程。如果是则调用InitZygote()进行初始化（其中主要是 mount 一些文件系统）操作，否则调用 DidForkFromZygote()函数。函数 DidForkFromZygote()会创建线程池，创建 signalcatcher 线程和启动 JDWP 调试线程。这个函数主要工作是调用 Heap 对象的 CreateThreadPool()函数来创建线程池。函数 DidForkFromZygote 在文件 Runtime:.cc 中定义，具体实现代码如下所示：

```
void Runtime::DidForkFromZygote() {
  is_zygote_ = false;

  // Create the thread pool.
  heap_->CreateThreadPool();

  StartSignalCatcher();

  // Start the JDWP thread. If the command-line debugger flags specified "suspend=y",
  // this will pause the runtime, so we probably want this to come last.
  Dbg::StartJdwp();
}
```

最后 Start()函数中调用了 StartDaemonThreads()函数，这个函数的工作是调用 Java 类 Daemons 的 start()方法来启动一些 Deamon 线程，这个过程实际上和 Dalvik 启动时完成的最后一项工作相同。函数 Runtime::在文件 Runtime.cc 中定义，具体实现代码如下所示：

```
void Runtime::StartDaemonThreads() {
  VLOG(startup) << "Runtime::StartDaemonThreads entering";

  Thread* self = Thread::Current();

  // Must be in the kNative state for calling native methods.
  CHECK_EQ(self->GetState(), kNative);
```

```
  JNIEnv* env = self->GetJniEnv();
  env->CallStaticVoidMethod(WellKnownClasses::java_lang_Daemons,
                            WellKnownClasses::java_lang_Daemons_start);
  if (env->ExceptionCheck()) {
    env->ExceptionDescribe();
    LOG(FATAL) << "Error starting java.lang.Daemons";
  }

  VLOG(startup) << "Runtime::StartDaemonThreads exiting";
}
```

然后启动后台线程，再用 JNI 调用 java.lang.ClassLoader.getSystemClassLoader()得到系统的 ClassLoader（由 createSystemClassLoader()创建），一会调用 com.android.internal.os.ZygoteInit.main()时用的就是它：

```
StartDaemonThreads();
system_class_loader_ = CreateSystemClassLoader();
```

从 StartVM()返回后，AndroidRuntime 执行 startReg()在创建线程时加一个 hook 函数，这样每个 Thread 起来时会先去执行 AndroidRuntime::javaThreadShell()，而该函数会初始化 Java 虚拟机环境，这样新建的线程就可以调用 Java 层了。函数 startReg 在文件 AndroidRuntime.cpp 中定义，具体实现代码如下所示：

```
int AndroidRuntime::startReg(JNIEnv* env)
{
    /*
     * This hook causes all future threads created in this process to be
     * attached to the JavaVM.  (This needs to go away in favor of JNI
     * Attach calls.)
     */
    androidSetCreateThreadFunc((android_create_thread_fn) javaCreateThreadEtc);

    ALOGV("--- registering native functions ---\n");

    /*
     * Every "register" function calls one or more things that return
     * a local reference (e.g. FindClass).  Because we haven't really
     * started the VM yet, they're all getting stored in the base frame
     * and never released.  Use Push/Pop to manage the storage.
     */
    env->PushLocalFrame(200);

    if (register_jni_procs(gRegJNI, NELEM(gRegJNI), env) < 0) {
       env->PopLocalFrame(NULL);
       return -1;
    }
    env->PopLocalFrame(NULL);

    //createJavaThread("fubar", quickTest, (void*) "hello");

    return 0;
}

AndroidRuntime* AndroidRuntime::getRuntime()
{
    return gCurRuntime;
}
```

到此为止，AndroidRuntime 中的 start()函数的执行过程全部讲解完毕。总结的执行流程如图 21-3 所示。

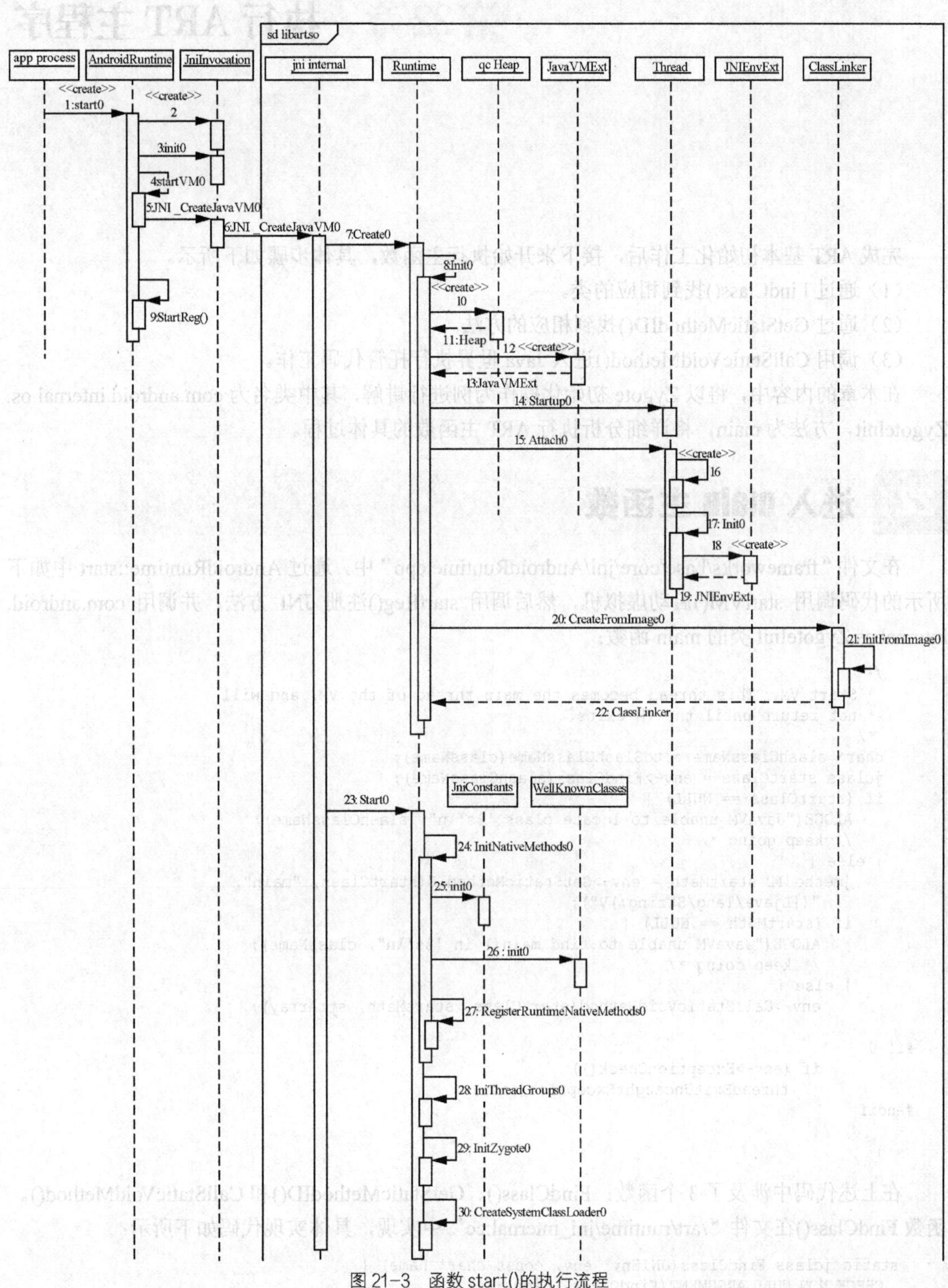

图 21-3　函数 start()的执行流程

第 22 章　执行 ART 主程序

完成 ART 基本初始化工作后，接下来开始执行主函数，具体步骤如下所示。

（1）通过 FindClass()找到相应的类。

（2）通过 GetStaticMethodID()找到相应的方法。

（3）调用 CallStaticVoidMethod()进入 Java 世界执行托管代码工作。

在本章的内容中，将以 Zygote 初始化操作为例进行讲解，其中类名为 com.android.internal.os.ZygoteInit，方法为 main，将详细分析执行 ART 主函数的具体过程。

22.1 进入 main 主函数

在文件“frameworks/base/core/jni/AndroidRuntime.cpp”中，通过 AndroidRuntime::start 中如下所示的代码调用 startVM()启动虚拟机，然后调用 startReg()注册 JNI 方法，并调用 com.android.internal.os.ZygoteInit 类的 main 函数：

```
    /*
     * Start VM.  This thread becomes the main thread of the VM, and will
     * not return until the VM exits.
     */
    char* slashClassName = toSlashClassName(className);
    jclass startClass = env->FindClass(slashClassName);
    if (startClass == NULL) {
        ALOGE("JavaVM unable to locate class '%s'\n", slashClassName);
        /* keep going */
    } else {
        jmethodID startMeth = env->GetStaticMethodID(startClass, "main",
            "([Ljava/lang/String;)V");
        if (startMeth == NULL) {
            ALOGE("JavaVM unable to find main() in '%s'\n", className);
            /* keep going */
        } else {
            env->CallStaticVoidMethod(startClass, startMeth, strArray);

#if 0
            if (env->ExceptionCheck())
                threadExitUncaughtException(env);
#endif
        }
    }
```

在上述代码中涉及了 3 个函数：FindClass()、GetStaticMethodID()和 CallStaticVoidMethod()。函数 FindClass()在文件“/art/runtime/jni_internal.cc”中实现，具体实现代码如下所示：

```
static jclass FindClass(JNIEnv* env, const char* name) {
  CHECK_NON_NULL_ARGUMENT(FindClass, name);
  Runtime* runtime = Runtime::Current();
  ClassLinker* class_linker = runtime->GetClassLinker();
  std::string descriptor(NormalizeJniClassDescriptor(name));
  ScopedObjectAccess soa(env);
```

```
  Class* c = NULL;
  if (runtime->IsStarted()) {
    ClassLoader* cl = GetClassLoader(soa);
    c = class_linker->FindClass(descriptor.c_str(), cl);
  } else {
    c = class_linker->FindSystemClass(descriptor.c_str());
  }
  return soa.AddLocalReference<jclass>(c);
}
```

函数 GetClassLoader()也是在文件“art/runtime/jni_internal.cc”中定义，功能是调用 GetSystemClassLoader()得到前面初始化好的系统 ClassLoader，具体实现代码如下所示：

```
static ClassLoader* GetClassLoader(const ScopedObjectAccess& soa)
    SHARED_LOCKS_REQUIRED(Locks::mutator_lock_) {
  ArtMethod* method = soa.Self()->GetCurrentMethod(NULL);
  // If we are running Runtime.nativeLoad, use the overriding ClassLoader it set.
  if (method == soa.DecodeMethod(WellKnownClasses::java_lang_Runtime_nativeLoad)) {
    return soa.Self()->GetClassLoaderOverride();
  }
  // If we have a method, use its ClassLoader for context.
  if (method != NULL) {
    return method->GetDeclaringClass()->GetClassLoader();
  }
  // We don't have a method, so try to use the system ClassLoader.
  ClassLoader*                              class_loader                          =
soa.Decode<ClassLoader*>(Runtime::Current()->GetSystemClassLoader());
  if (class_loader != NULL) {
    return class_loader;
  }
  // See if the override ClassLoader is set for gtests.
  class_loader = soa.Self()->GetClassLoaderOverride();
  if (class_loader != NULL) {
    // If so, CommonTest should have set UseCompileTimeClassPath.
    CHECK(Runtime::Current()->UseCompileTimeClassPath());
    return class_loader;
  }
  // Use the BOOTCLASSPATH.
  return NULL;
}
```

22.2 查找目标类

开始调用 ClassLinker 的函数 FindClass()查找目标类，在这一过程中涉及的关键函数有：LookupClass()、DefineClass()、InsertClass()、LoadClass()和 LinkClass()，上述关键函数的具体说明如下所示。

22.2.1 函数 LookupClass()

函数 LookupClass()在文件“art/runtime/class_linker.cc”中定义，先在 ClassLinker 的成员变量 class_table_中找指定类，找到就返回，找不到则看是否要在 image 中找（class_loader 为 NULL 且 dex_cache_image_class_lookup_required 为 true）。如果要则调用 LookupClassFromImage()在 Image 中进行查找，找到了就调用 InsertClass()将找到的类放入 class_table_中方便下次查找。

函数 LookupClass()的具体实现代码如下所示：

```
mirror::Class* ClassLinker::FindClass(const char* descriptor, mirror::ClassLoader* class_loader)
{
  DCHECK_NE(*descriptor, '\0') << "descriptor is empty string";
  Thread* self = Thread::Current();
  DCHECK(self != NULL);
  self->AssertNoPendingException();
  if (descriptor[1] == '\0') {
```

```
    // only the descriptors of primitive types should be 1 character long, also avoid class lookup
    // for primitive classes that aren't backed by dex files.
    return FindPrimitiveClass(descriptor[0]);
  }
  // Find the class in the loaded classes table.
  mirror::Class* klass = LookupClass(descriptor, class_loader);
  if (klass != NULL) {
    return EnsureResolved(self, klass);
  }
  // Class is not yet loaded.
  if (descriptor[0] == '[') {
    return CreateArrayClass(descriptor, class_loader);

  } else if (class_loader == NULL) {
    DexFile::ClassPathEntry pair = DexFile::FindInClassPath(descriptor, boot_class_path_);
    if (pair.second != NULL) {
      return DefineClass(descriptor, NULL, *pair.first, *pair.second);
    }

  } else if (Runtime::Current()->UseCompileTimeClassPath()) {
    // First try the boot class path, we check the descriptor first to avoid an unnecessary
    // throw of a NoClassDefFoundError.
    if (IsInBootClassPath(descriptor)) {
      mirror::Class* system_class = FindSystemClass(descriptor);
      CHECK(system_class != NULL);
      return system_class;
    }
    // Next try the compile time class path.
    const std::vector<const DexFile*>* class_path;
    {
      ScopedObjectAccessUnchecked soa(self);
      ScopedLocalRef<jobject> jclass_loader(soa.Env(), soa.AddLocalReference<jobject>(class_loader));
      class_path = &Runtime::Current()->GetCompileTimeClassPath(jclass_loader.get());
    }

    DexFile::ClassPathEntry pair = DexFile::FindInClassPath(descriptor, *class_path);
    if (pair.second != NULL) {
      return DefineClass(descriptor, class_loader, *pair.first, *pair.second);
    }

  } else {
    ScopedObjectAccessUnchecked soa(self->GetJniEnv());
    ScopedLocalRef<jobject> class_loader_object(soa.Env(),
                                                soa.AddLocalReference<jobject>(class_loader));
    std::string class_name_string(DescriptorToDot(descriptor));
    ScopedLocalRef<jobject> result(soa.Env(), NULL);
    {
      ScopedThreadStateChange tsc(self, kNative);
      ScopedLocalRef<jobject> class_name_object(soa.Env(),

 soa.Env()->NewStringUTF(class_name_string.c_str()));
      if (class_name_object.get() == NULL) {
        return NULL;
      }
      CHECK(class_loader_object.get() != NULL);
      result.reset(soa.Env()->CallObjectMethod(class_loader_object.get(),
                                               WellKnownClasses::java_lang_ClassLoader_loadClass,
                                               class_name_object.get()));
    }
    if (soa.Self()->IsExceptionPending()) {
      // If the ClassLoader threw, pass that exception up.
      return NULL;
    } else if (result.get() == NULL) {
      // broken loader - throw NPE to be compatible with Dalvik
      ThrowNullPointerException(NULL, StringPrintf("ClassLoader.loadClass returned null for %s",
                                                   class_name_string.c_str()).c_str());
      return NULL;
    } else {
      // success, return mirror::Class*
```

```
      return soa.Decode<mirror::Class*>(result.get());
    }
  }

  ThrowNoClassDefFoundError("Class %s not found", PrintableString(descriptor).c_str());
  return NULL;
}
```

函数 LookupClassFromImage()也在文件“art/runtime/class_linker.cc”中定义，具体实现代码如下所示：

```
mirror::Class* ClassLinker::LookupClassFromImage(const char* descriptor) {
  Thread* self = Thread::Current();
  const char* old_no_suspend_cause =
      self->StartAssertNoThreadSuspension("Image class lookup");
  mirror::ObjectArray<mirror::DexCache>* dex_caches = GetImageDexCaches();
  for (int32_t i = 0; i < dex_caches->GetLength(); ++i) {
    mirror::DexCache* dex_cache = dex_caches->Get(i);
    const DexFile* dex_file = dex_cache->GetDexFile();
    // First search using the class def map, but don't bother for non-class types.
    if (descriptor[0] == 'L') {
      const DexFile::StringId* descriptor_string_id = dex_file->FindStringId(descriptor);
      if (descriptor_string_id != NULL) {
        const DexFile::TypeId* type_id =
            dex_file->FindTypeId(dex_file->GetIndexForStringId(*descriptor_string_id));
        if (type_id != NULL) {
          mirror::Class* klass = dex_cache->GetResolvedType(dex_file->GetIndexForTypeId(*type_id));
          if (klass != NULL) {
            self->EndAssertNoThreadSuspension(old_no_suspend_cause);
            return klass;
          }
        }
      }
    }
    // Now try binary searching the string/type index.
    const DexFile::StringId* string_id = dex_file->FindStringId(descriptor);
    if (string_id != NULL) {
      const DexFile::TypeId* type_id =
          dex_file->FindTypeId(dex_file->GetIndexForStringId(*string_id));
      if (type_id != NULL) {
        uint16_t type_idx = dex_file->GetIndexForTypeId(*type_id);
        mirror::Class* klass = dex_cache->GetResolvedType(type_idx);
        if (klass != NULL) {
          self->EndAssertNoThreadSuspension(old_no_suspend_cause);
          return klass;
        }
      }
    }
  }
  self->EndAssertNoThreadSuspension(old_no_suspend_cause);
  return NULL;
}
```

22.2.2 函数 DefineClass()

函数 DefineClass()在文件“art/runtime/class_linker.cc”中定义，实现 LoadClass()、InsertClass()和 LinkClass()等动作。其中，LoadClass()调用 LoadField()和 LoadMethod()等函数把类中的域和方法数据从 dex 文件中读出来，填入 Class 结构。

函数 DefineClass()的具体实现代码如下所示：

```
mirror::Class* ClassLinker::DefineClass(const char* descriptor,
                                        mirror::ClassLoader* class_loader,
                                        const DexFile& dex_file,
                                        const DexFile::ClassDef& dex_class_def) {
  Thread* self = Thread::Current();
  SirtRef<mirror::Class> klass(self, NULL);
```

```
  // Load the class from the dex file.
  if (UNLIKELY(!init_done_)) {
    // finish up init of hand crafted class_roots_
    if (strcmp(descriptor, "Ljava/lang/Object;") == 0) {
      klass.reset(GetClassRoot(kJavaLangObject));
    } else if (strcmp(descriptor, "Ljava/lang/Class;") == 0) {
      klass.reset(GetClassRoot(kJavaLangClass));
    } else if (strcmp(descriptor, "Ljava/lang/String;") == 0) {
      klass.reset(GetClassRoot(kJavaLangString));
    } else if (strcmp(descriptor, "Ljava/lang/DexCache;") == 0) {
      klass.reset(GetClassRoot(kJavaLangDexCache));
    } else if (strcmp(descriptor, "Ljava/lang/reflect/ArtField;") == 0) {
      klass.reset(GetClassRoot(kJavaLangReflectArtField));
    } else if (strcmp(descriptor, "Ljava/lang/reflect/ArtMethod;") == 0) {
      klass.reset(GetClassRoot(kJavaLangReflectArtMethod));
    } else {
      klass.reset(AllocClass(self, SizeOfClass(dex_file, dex_class_def)));
    }
  } else {
    klass.reset(AllocClass(self, SizeOfClass(dex_file, dex_class_def)));
  }
  if (UNLIKELY(klass.get() == NULL)) {
    CHECK(self->IsExceptionPending());  // Expect an OOME.
    return NULL;
  }
  klass->SetDexCache(FindDexCache(dex_file));
  LoadClass(dex_file, dex_class_def, klass, class_loader);
  // Check for a pending exception during load
  if (self->IsExceptionPending()) {
    klass->SetStatus(mirror::Class::kStatusError, self);
    return NULL;
  }
  ObjectLock lock(self, klass.get());
  klass->SetClinitThreadId(self->GetTid());
  {
    // Add the newly loaded class to the loaded classes table.
    mirror::Class* existing = InsertClass(descriptor, klass.get(), Hash(descriptor));
    if (existing != NULL) {
      // We failed to insert because we raced with another thread. Calling EnsureResolved may cause
      // this thread to block.
      return EnsureResolved(self, existing);
    }
  }
  // Finish loading (if necessary) by finding parents
  CHECK(!klass->IsLoaded());
  if (!LoadSuperAndInterfaces(klass, dex_file)) {
    // Loading failed.
    klass->SetStatus(mirror::Class::kStatusError, self);
    return NULL;
  }
  CHECK(klass->IsLoaded());
  // Link the class (if necessary)
  CHECK(!klass->IsResolved());
  if (!LinkClass(klass, NULL, self)) {
    // Linking failed.
    klass->SetStatus(mirror::Class::kStatusError, self);
    return NULL;
  }
  CHECK(klass->IsResolved());
  Dbg::PostClassPrepare(klass.get());

  return klass.get();
}
```

函数 LoadClass()也是在文件“art/runtime/class_linker.cc”中定义的，具体实现代码如下所示：

```
void ClassLinker::LoadClass(const DexFile& dex_file,
                            const DexFile::ClassDef& dex_class_def,
                            SirtRef<mirror::Class>& klass,
                            mirror::ClassLoader* class_loader) {
```

```
CHECK(klass.get() != NULL);
CHECK(klass->GetDexCache() != NULL);
CHECK_EQ(mirror::Class::kStatusNotReady, klass->GetStatus());
const char* descriptor = dex_file.GetClassDescriptor(dex_class_def);
CHECK(descriptor != NULL);

klass->SetClass(GetClassRoot(kJavaLangClass));
uint32_t access_flags = dex_class_def.access_flags_;
// Make sure that none of our runtime-only flags are set.
CHECK_EQ(access_flags & ~kAccJavaFlagsMask, 0U);
klass->SetAccessFlags(access_flags);
klass->SetClassLoader(class_loader);
DCHECK_EQ(klass->GetPrimitiveType(), Primitive::kPrimNot);
klass->SetStatus(mirror::Class::kStatusIdx, NULL);

klass->SetDexClassDefIndex(dex_file.GetIndexForClassDef(dex_class_def));
klass->SetDexTypeIndex(dex_class_def.class_idx_);

// Load fields fields.
const byte* class_data = dex_file.GetClassData(dex_class_def);
if (class_data == NULL) {
  return;  // no fields or methods - for example a marker interface
}
ClassDataItemIterator it(dex_file, class_data);
Thread* self = Thread::Current();
if (it.NumStaticFields() != 0) {
  mirror::ObjectArray<mirror::ArtField>* statics = AllocArtFieldArray(self, it.NumStaticFields());
  if (UNLIKELY(statics == NULL)) {
    CHECK(self->IsExceptionPending());  // OOME.
    return;
  }
  klass->SetSFields(statics);
}
if (it.NumInstanceFields() != 0) {
  mirror::ObjectArray<mirror::ArtField>* fields =
      AllocArtFieldArray(self, it.NumInstanceFields());
  if (UNLIKELY(fields == NULL)) {
    CHECK(self->IsExceptionPending());  // OOME.
    return;
  }
  klass->SetIFields(fields);
}
for (size_t i = 0; it.HasNextStaticField(); i++, it.Next()) {
  SirtRef<mirror::ArtField> sfield(self, AllocArtField(self));
  if (UNLIKELY(sfield.get() == NULL)) {
    CHECK(self->IsExceptionPending());  // OOME.
    return;
  }
  klass->SetStaticField(i, sfield.get());
  LoadField(dex_file, it, klass, sfield);
}
for (size_t i = 0; it.HasNextInstanceField(); i++, it.Next()) {
  SirtRef<mirror::ArtField> ifield(self, AllocArtField(self));
  if (UNLIKELY(ifield.get() == NULL)) {
    CHECK(self->IsExceptionPending());  // OOME.
    return;
  }
  klass->SetInstanceField(i, ifield.get());
  LoadField(dex_file, it, klass, ifield);
}

UniquePtr<const OatFile::OatClass> oat_class;
if (Runtime::Current()->IsStarted() && !Runtime::Current()->UseCompileTimeClassPath()) {
  oat_class.reset(GetOatClass(dex_file, klass->GetDexClassDefIndex()));
}

// Load methods.
if (it.NumDirectMethods() != 0) {
```

```
    // TODO: append direct methods to class object
    mirror::ObjectArray<mirror::ArtMethod>* directs =
        AllocArtMethodArray(self, it.NumDirectMethods());
    if (UNLIKELY(directs == NULL)) {
      CHECK(self->IsExceptionPending());  // OOME.
      return;
    }
    klass->SetDirectMethods(directs);
  }
  if (it.NumVirtualMethods() != 0) {
    // TODO: append direct methods to class object
    mirror::ObjectArray<mirror::ArtMethod>* virtuals =
       AllocArtMethodArray(self, it.NumVirtualMethods());
    if (UNLIKELY(virtuals == NULL)) {
      CHECK(self->IsExceptionPending());  // OOME.
      return;
    }
    klass->SetVirtualMethods(virtuals);
  }
  size_t class_def_method_index = 0;
  for (size_t i = 0; it.HasNextDirectMethod(); i++, it.Next()) {
    SirtRef<mirror::ArtMethod> method(self, LoadMethod(self, dex_file, it, klass));
    if (UNLIKELY(method.get() == NULL)) {
      CHECK(self->IsExceptionPending());  // OOME.
      return;
    }
    klass->SetDirectMethod(i, method.get());
    if (oat_class.get() != NULL) {
      LinkCode(method, oat_class.get(), class_def_method_index);
    }
    method->SetMethodIndex(class_def_method_index);
    class_def_method_index++;
  }
  for (size_t i = 0; it.HasNextVirtualMethod(); i++, it.Next()) {
    SirtRef<mirror::ArtMethod> method(self, LoadMethod(self, dex_file, it, klass));
    if (UNLIKELY(method.get() == NULL)) {
      CHECK(self->IsExceptionPending());  // OOME.
      return;
    }
    klass->SetVirtualMethod(i, method.get());
    DCHECK_EQ(class_def_method_index, it.NumDirectMethods() + i);
    if (oat_class.get() != NULL) {
      LinkCode(method, oat_class.get(), class_def_method_index);
    }
    class_def_method_index++;
  }
  DCHECK(!it.HasNext());
}
```

22.2.3　函数 InsertClass()

函数 InsertClass()在文件“art/runtime/class_linker.cc”中定义，主要功能是把该类写入 class_table_中方便下次查找。函数 InsertClass()的具体实现代码如下所示：

```
mirror::Class* ClassLinker::InsertClass(const char* descriptor, mirror::Class* klass,
                                        size_t hash) {
  if (VLOG_IS_ON(class_linker)) {
    mirror::DexCache* dex_cache = klass->GetDexCache();
    std::string source;
    if (dex_cache != NULL) {
      source += " from ";
      source += dex_cache->GetLocation()->ToModifiedUtf8();
    }
    LOG(INFO) << "Loaded class " << descriptor << source;
  }
  WriterMutexLock mu(Thread::Current(), *Locks::classlinker_classes_lock_);
  mirror::Class* existing =
      LookupClassFromTableLocked(descriptor, klass->GetClassLoader(), hash);
```

```
  if (existing != NULL) {
    return existing;
  }
  if (kIsDebugBuild && klass->GetClassLoader() == NULL && dex_cache_image_class_lookup_required_)
{
    // Check a class loaded with the system class loader matches one in the image if the class
    // is in the image.
    existing = LookupClassFromImage(descriptor);
    if (existing != NULL) {
      CHECK(klass == existing);
    }
  }
  Runtime::Current()->GetHeap()->VerifyObject(klass);
  class_table_.insert(std::make_pair(hash, klass));
  class_table_dirty_ = true;
  return NULL;
}
```

22.2.4 函数 LinkClass()

函数 LinkClass()在文件“art/runtime/class_linke.cc”中定义，功能是动态绑定虚函数和接口函数，其调用结构如下所示：

```
LinkSuperClass() // 检查父类
LinkMethods()
LinkVirtualMethods() // 结合父类进行虚函数绑定，填写 Class 中的虚函数表 vtable_
LinkInterfaceMethods() /*处理接口类函数信息 iftable_。注意接口类中的虚函数也会影响虚函数表，因此会更新
vtable_*/
LinkInstanceFields() & LinkStaticFields() // 更新域信息，如域中的 Offset 和类的对象大小等
```

函数 LinkClass()的具体实现代码如下所示：

```
bool ClassLinker::LinkClass(SirtRef<mirror::Class>& klass,
                           mirror::ObjectArray<mirror::Class>* interfaces, Thread* self) {
  CHECK_EQ(mirror::Class::kStatusLoaded, klass->GetStatus());
  if (!LinkSuperClass(klass)) {
    return false;
  }
  if (!LinkMethods(klass, interfaces)) {
    return false;
  }
  if (!LinkInstanceFields(klass)) {
    return false;
  }
  if (!LinkStaticFields(klass)) {
    return false;
  }
  CreateReferenceInstanceOffsets(klass);
  CreateReferenceStaticOffsets(klass);
  CHECK_EQ(mirror::Class::kStatusLoaded, klass->GetStatus());
  klass->SetStatus(mirror::Class::kStatusResolved, self);
  return true;
}
```

对于函数 FindClass()来说，总共包含了内置类、启动类、系统类和其他类。其中内置类是很基本的类，一般是初始化时预加载好的（如 WellKnownClasses 和 JniConstants 里那些），它们可以通过 LookupClassFromImage()函数找到。启动类是在 BOOTCLASSPATH 里的类，由于它们是启动类，所以这里还没有 ClassLoader。除掉前面的内置类，其余的通过 DexFile::FindInClassPath()查找得到。而系统类与其他类的加载过程是类似的，都是通过 ClassLoader 的 loadClass 方法，区别在于前者通过特殊的 SystemClassLoader 进行加载。例如对于一个还没被加载过的启动类来说，一般流程如图 22-1 所示。

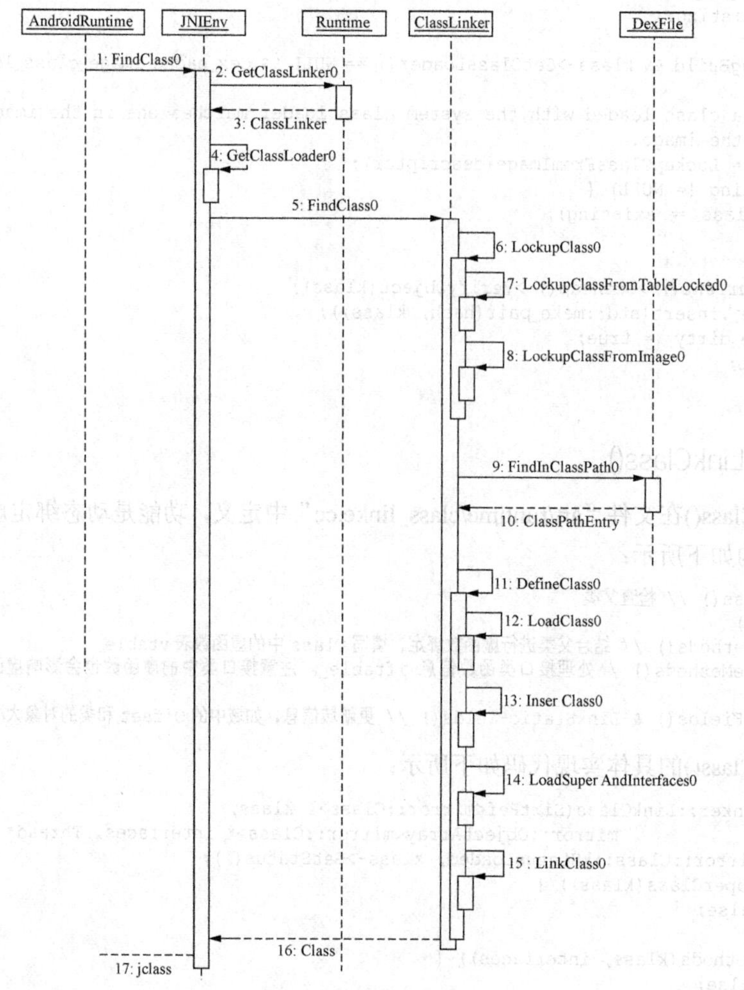

图 22-1　加载启动类的流程

整个过程涉及到很多类，其中最主要的是 Class 类，具体结构如图 22-2 所示。

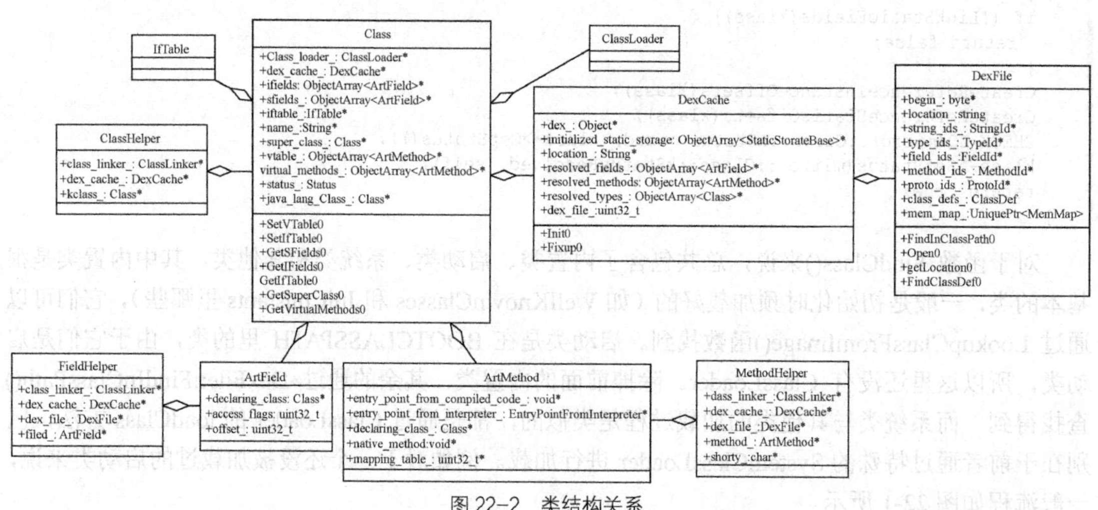

图 22-2　类结构关系

22.3 类操作

再次回到 FindClass()函数，因为在调用 ZygoteInit.main()时，所需的类在初始化时都已经装载链接好了，所以此处不按照上面的流程进行，而是直接通过 JNI 调用 ClassLoader.loadClass()进行装载。完成后将找到的类转为 jclass 返回给 AndroidRuntime。类的查找工作结束后可以找相应的方法。GetStaticMethodID 会调用 FindMethodID()函数，它首先对该类进行验证，保证这个类是初始化好的，再调用其他函数进行目标函数的查找。

其中函数 GetStaticMethodID()在文件 jni_internal.cc 中定义，能够将未初始化的类初始化，获取静态函数 main 的 ID。具体实现代码如下所示：

```
static jmethodID GetStaticMethodID(JNIEnv* env, jclass java_class, const char* name,
                                    const char* sig) {
  CHECK_NON_NULL_ARGUMENT(GetStaticMethodID, java_class);
  CHECK_NON_NULL_ARGUMENT(GetStaticMethodID, name);
  CHECK_NON_NULL_ARGUMENT(GetStaticMethodID, sig);
  ScopedObjectAccess soa(env);
  return FindMethodID(soa, java_class, name, sig, true);
}
```

FindMethodID()函数也在文件 jni_internal.cc 中定义，具体实现代码如下所示：

```
static jmethodID FindMethodID(ScopedObjectAccess& soa, jclass jni_class,
                              const char* name, const char* sig, bool is_static)
    SHARED_LOCKS_REQUIRED(Locks::mutator_lock_) {
  Class* c = soa.Decode<Class*>(jni_class);
  if (!Runtime::Current()->GetClassLinker()->EnsureInitialized(c, true, true)) {
    return NULL;
  }

  ArtMethod* method = NULL;
  if (is_static) {
    method = c->FindDirectMethod(name, sig);
  } else {
    method = c->FindVirtualMethod(name, sig);
    if (method == NULL) {
      // No virtual method matching the signature.  Search declared
      // private methods and constructors.
      method = c->FindDeclaredDirectMethod(name, sig);
    }
  }

  if (method == NULL || method->IsStatic() != is_static) {
    ThrowNoSuchMethodError(soa, c, name, sig, is_static ? "static" : "non-static");
    return NULL;
  }

  return soa.EncodeMethod(method);
}
```

通过上述实现代码可知，会根据不同的需求执行不同的函数，具体说明如下所示。

- 如果要找的是静态函数（通过 GetStaticMethodID()过来的），则调用 FindDirectMethod()查找该类及其父类的非虚函数（通过 Class 的成员变量 direct_methods_）。
- 否则调用 FindVirtualMethod()查找该类及其父类的虚函数（通过 Class 的成员变量 virtual_methods_），如果没找到，再调用 FindDeclaredDirectMethod()查找该类的非虚函数。找到的条件是函数名和函数签名相同，如这里是“main”和“([Ljava/lang/String;)V”。找到目标函数后，就可以执行了。

函数 FindDirectMethod()的具体定义代码如下所示：

```
-Method* Class::FindDirectMethod(const StringPiece& name,
-                                 const StringPiece& signature) {
-  for (Class* klass = this; klass != NULL; klass = klass->GetSuperClass()) {
+Method* Class::FindDeclaredDirectMethod(const DexCache* dex_cache, uint32_t dex_method_idx) const
{
+  if (GetDexCache() == dex_cache) {
+    for (size_t i = 0; i < NumDirectMethods(); ++i) {
+    Method* method = GetDirectMethod(i);
+      if (method->GetDexMethodIndex() == dex_method_idx) {
+      return method;
+    }
+    }
+  }
+  return NULL;
+}
```

函数 FindDeclaredDirectMethod()的具体定义代码如下所示：

```
-Method* Class::FindDeclaredDirectMethod(const StringPiece& name,
-                                         const StringPiece& signature) {
+Method* Class::FindInterfaceMethod(const DexCache* dex_cache, uint32_t dex_method_idx) const {
+  // Check the current class before checking the interfaces.
+  Method* method = FindDeclaredVirtualMethod(dex_cache, dex_method_idx);
+  if (method != NULL) {
+    return method;
+  }
+
+  int32_t iftable_count = GetIfTableCount();
+  ObjectArray<InterfaceEntry>* iftable = GetIfTable();
+  for (int32_t i = 0; i < iftable_count; i++) {
+    method = iftable->Get(i)->GetInterface()->FindVirtualMethod(dex_cache, dex_method_idx);
+    if (method != NULL) {
+      return method;
+    }
+  }
+  return NULL;
+}
```

22.4 实现托管操作

开始执行托管操作，函数 InvokeMain()会验证 Java 的 main 方法，并最终调用 CallStaticVoidMethod 来运行 main 方法。CallStaticVoidMethod(jni.h)在结构_JNIEnv 中实现，函数 CallStaticVoidMethod()在文件 jni_internal.cc 中定义，具体实现代码如下所示：

```
static void CallStaticVoidMethod(JNIEnv* env, jclass, jmethodID mid, ...) {
  va_list ap;
  va_start(ap, mid);
  CHECK_NON_NULL_ARGUMENT(CallStaticVoidMethod, mid);
  ScopedObjectAccess soa(env);
  InvokeWithVarArgs(soa, NULL, mid, ap);
  va_end(ap);
}
```

在上述代码中，va_list 用于处理不定传参数，function 表示 JNINativeInterface 结构指针表，用于保存 JNI 接口函数，例如调用 method 等。

函数 CallStaticVoidMethodV 在文件 JNI_interl.cc 中定义，具体实现代码如下所示：

```
static void CallStaticVoidMethodV(JNIEnv* env, jclass, jmethodID mid, va_list args) {
  CHECK_NON_NULL_ARGUMENT(CallStaticVoidMethodV, mid);
  ScopedObjectAccess soa(env);
  InvokeWithVarArgs(soa, NULL, mid, args);
}
```

函数 InvokeWithArgArray()也在文件 JNI_interl.cc 中定义，具体实现代码如下所示：

```
void InvokeWithArgArray(const ScopedObjectAccess& soa, ArtMethod* method,
                   ArgArray* arg_array, JValue* result, char result_type)
    SHARED_LOCKS_REQUIRED(Locks::mutator_lock_) {
  uint32_t* args = arg_array->GetArray();
  if (UNLIKELY(soa.Env()->check_jni)) {
    CheckMethodArguments(method, args);
  }
  method->Invoke(soa.Self(), args, arg_array->GetNumBytes(), result, result_type);
}
```

接下来执行文件“art/runtime/mirrorart_method.cc”中的函数 invoke，具体实现代码如下所示：

```
void ArtMethod::Invoke(Thread* self, uint32_t* args, uint32_t args_size, JValue* result,
                  char result_type) {
  if (kIsDebugBuild) {
    self->AssertThreadSuspensionIsAllowable(); //设定 debug 时线程可以被 hold
    CHECK_EQ(kRunnable, self->GetState());
  }

  // Push a transition back into managed code onto the linked list in thread.
  ManagedStack fragment;
  self->PushManagedStackFragment(&fragment); //管理栈帧: this 放入 fragment,this 清空, 保存现场

  Runtime* runtime = Runtime::Current();
  // Call the invoke stub, passing everything as arguments.
  if (UNLIKELY(!runtime->IsStarted())) {
    LOG(INFO) << "Not invoking " << PrettyMethod(this) << " for a runtime that isn't started";
    if (result != NULL) {
      result->SetJ(0);
    }
  } else {
    const bool kLogInvocationStartAndReturn = false;
    if (GetEntryPointFromCompiledCode() != NULL) {    //存在被编译的 code
      if (kLogInvocationStartAndReturn) {
        LOG(INFO) << StringPrintf("Invoking '%s' code=%p", PrettyMethod(this).c_str(), GetEntry
PointFromCompiledCode());
      }
#ifdef ART_USE_PORTABLE_COMPILER
      (*art_portable_invoke_stub)(this, args, args_size, self, result, result_type);
#else
      (*art_quick_invoke_stub)(this, args, args_size, self, result, result_type);
#endif
      if (UNLIKELY(reinterpret_cast<int32_t>(self->GetException(NULL)) == -1)) {
        // Unusual case where we were running LLVM generated code and an
        // exception was thrown to force the activations to be removed from the
        // stack. Continue execution in the interpreter.
       //llvm 生成代码过程中异常会进入解释器?
        self->ClearException();
        ShadowFrame* shadow_frame = self->GetAndClearDeoptimizationShadowFrame(result);
        self->SetTopOfStack(NULL, 0);
        self->SetTopOfShadowStack(shadow_frame);
       //stack & shadow frame 设置
        interpreter::EnterInterpreterFromDeoptimize(self, shadow_frame, result);
       //在解释器继续执行
      }
      if (kLogInvocationStartAndReturn) {
        LOG(INFO) << StringPrintf("Returned '%s' code=%p", PrettyMethod(this).c_str(), GetEntry
PointFromCompiledCode());
      }
    } else {
      LOG(INFO) << "Not invoking '" << PrettyMethod(this)
          << "' code=" << reinterpret_cast<const void*>(GetEntryPointFromCompiledCode());
      if (result != NULL) {
        result->SetJ(0);
      }
    }
  }
```

```
    // Pop transition.
    self->PopManagedStackFragment(fragment); //恢复现场
}
```

在上述代码中，前后分别实现了对托管代码栈的保存和恢复工作。

接下来进入解释器的函数 EnterInterpreterFromDeoptimize()，具体实现代码如下所示：

```
void EnterInterpreterFromDeoptimize(Thread* self, ShadowFrame* shadow_frame, JValue* ret_val)
    SHARED_LOCKS_REQUIRED(Locks::mutator_lock_) {
  JValue value;
  value.SetJ(ret_val->GetJ());  /* Set value to last known result in case the shadow frame chain
is empty*/
  MethodHelper mh;                                                    //method 操作的 class
  while (shadow_frame != NULL) {
    self->SetTopOfShadowStack(shadow_frame);
    mh.ChangeMethod(shadow_frame->GetMethod());                       /获取 method
    const DexFile::CodeItem* code_item = mh.GetCodeItem();            //code 传递
    value = Execute(self, mh, code_item, *shadow_frame, value);       //执行解释器
    ShadowFrame* old_frame = shadow_frame;
    shadow_frame = shadow_frame->GetLink();          //下一个 frame 赋值给 shadow_frame
    delete old_frame;                                //删掉前一个 frame
  }
  ret_val->SetJ(value.GetJ());
}
```

函数 Execute()在文件 interpretor.cc 中定义，具体实现代码如下所示：

```
static inline JValue Execute(Thread* self, MethodHelper& mh, const DexFile::CodeItem* code_item,
                             ShadowFrame& shadow_frame, JValue result_register) {
  DCHECK(shadow_frame.GetMethod() == mh.GetMethod() ||
       shadow_frame.GetMethod()->GetDeclaringClass()->IsProxyClass());
  DCHECK(!shadow_frame.GetMethod()->IsAbstract());
  DCHECK(!shadow_frame.GetMethod()->IsNative());
  if (shadow_frame.GetMethod()->IsPreverified()) { //是否提前做过 method access verify
    // Enter the "without access check" interpreter.
    return ExecuteImpl<false>(self, mh, code_item, shadow_frame, result_register);
    /*进入具体实现函数，可以发现是个 C 语言解释器*/
  } else {
    // Enter the "with access check" interpreter.
    return ExecuteImpl<true>(self, mh, code_item, shadow_frame, result_register);
    /*进入具体实现函数，是一个 C 语言解释器*/
  }
}
```

再回到前面文件“art/runtime/mirrorart_method.cc”中的函数 invoke，

```
    const bool kLogInvocationStartAndReturn = false;
    if (GetEntryPointFromCompiledCode() != NULL) {        //存在被编译的 code
      if (kLogInvocationStartAndReturn) {
        LOG(INFO) << StringPrintf("Invoking '%s' code=%p", PrettyMethod(this).c_str(), GetEntryPoint
FromCompiledCode());
      }
#ifdef ART_USE_PORTABLE_COMPILER           //portable 编译器
      (*art_portable_invoke_stub)(this, args, args_size, self, result, result_type);
#else
      (*art_quick_invoke_stub)(this, args, args_size, self, result, result_type);
#endif
      if (UNLIKELY(reinterpret_cast<int32_t>(self->GetException(NULL)) == -1)) {
        // Unusual case where we were running LLVM generated code and an
        // exception was thrown to force the activations to be removed from the
        // stack. Continue execution in the interpreter.     //llvm 生成代码过程中异常会进入解释器
```

上述两个分支分别代表函数 art_portable_invoke_stub()和 art_quick_invoke_stub()，ART_USE_PORTABLE_COMPILER 是一个重要的宏。

在执行托管代码前，要先为其创建栈。这些栈通过 ManagedStack 的成员 link_形成一个先入后出

的链表。当执行完托管代码后，只要将最近放入的托管代码栈恢复回来即可。中间是目标函数的执行，但在跳入目标函数体前还需要先执行一些 ABI 层的上下文处理代码，这段代码称为 stub。首先按 ART_USE_PORTABLE_COMPILER 来决定是用 art_quick_invok_stub 还是 art_portable_invok_stub。由于它们是由汇编写成，平台相关，所以每个体系结构（X86、ARM、MPS）都有其实现。以 X86 体系为例，art_portable_invoke_stub 定义在文件 portable_entrypoints_x86.S 中，具体实现代码如下所示：

```
DEFINE_FUNCTION art_portable_invoke_stub
    PUSH ebp                          // save ebp
    PUSH ebx                          // save ebx
    mov %esp, %ebp                    // copy value of stack pointer into base pointer
    .cfi_def_cfa_register ebp
    mov 20(%ebp), %ebx                // get arg array size
    addl LITERAL(28), %ebx
// reserve space for return addr, method*, ebx, and ebp in frame
    andl LITERAL(0xFFFFFFF0), %ebx  // align frame size to 16 bytes
    subl LITERAL(12), %ebx          // remove space for return address, ebx, and ebp
    subl %ebx, %esp                 // reserve stack space for argument array
    lea  4(%esp), %eax              // use stack pointer + method ptr as dest for memcpy
    pushl 20(%ebp)                  // push size of region to memcpy
    pushl 16(%ebp)                  // push arg array as source of memcpy
    pushl %eax                      // push stack pointer as destination of memcpy
    call SYMBOL(memcpy)             // (void*, const void*, size_t)
    addl LITERAL(12), %esp          // pop arguments to memcpy
    mov 12(%ebp), %eax              // move method pointer into eax
    mov %eax, (%esp)                // push method pointer onto stack
    call *METHOD_CODE_OFFSET(%eax)  // call the method
    mov %ebp, %esp                  // restore stack pointer
    POP ebx                         // pop ebx
    POP ebp                         // pop ebp
    mov 20(%esp), %ecx              // get result pointer
    cmpl LITERAL(68), 24(%esp)      // test if result type char == 'D'
    je return_double_portable
    cmpl LITERAL(70), 24(%esp)      // test if result type char == 'F'
    je return_float_portable
    mov %eax, (%ecx)                // store the result
    mov %edx, 4(%ecx)               // store the other half of the result
    ret
return_double_portable:
    fstpl (%ecx)                    // store the floating point result as double
    ret
return_float_portable:
    fstps (%ecx)                    // store the floating point result as float
    ret
END_FUNCTION art_portable_invoke_stub
```

由此可见这是 X86 体系中的函数调用过程。首先保存栈帧等信息，然后把参数数组拷贝到栈中，再执行 call 指令跳转到要执行的目标函数。METHOD_CODE_OFFSET 指向 ArtMethod 中的成员变量 entry_point_from_compiled_code_，也就是编译好的目标函数的地址。接下来，就是等目标函数愉快地执行完，然后恢复上下文，保存返回值，最后执行 ret 指令返回。查找目标函数和执行的过程比较直观，如图 22-3 所示。

到此为止，执行完托管代码后返回到 AndroidRuntimie::start()函数，调用函数 DetachCurrentThread()和函数 DestroyJavaVM()来做清理工作，并关闭虚拟机，完成整个工作过程。

```
ALOGD("Shutting down VM\n");
if (mJavaVM->DetachCurrentThread() != JNI_OK)
    ALOGW("Warning: unable to detach main thread\n");
if (mJavaVM->DestroyJavaVM() != 0)
    ALOGW("Warning: VM did not shut down cleanly\n");
```

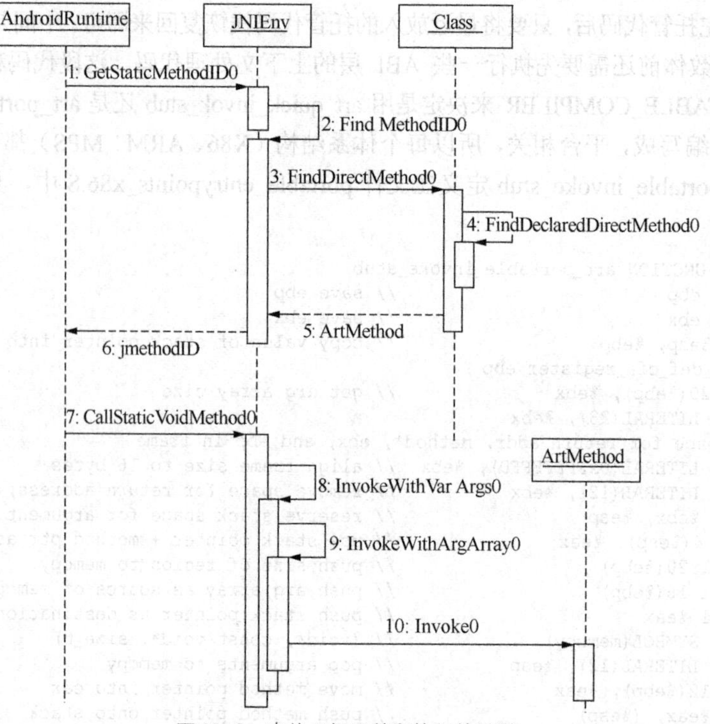

图 22-3　查找目标函数和执行的过程

函数 DetachCurrentThread()和函数 DestroyJavaVM()在文件 jni_internal.cc 中定义，具体实现代码如下所示：

```
  static jint DetachCurrentThread(JavaVM* vm) {
    if (vm == NULL || Thread::Current() == NULL) {
      return JNI_ERR;
    }
    JavaVMExt* raw_vm = reinterpret_cast<JavaVMExt*>(vm);
    Runtime* runtime = raw_vm->runtime;
    runtime->DetachCurrentThread();
    return JNI_OK;
  }
public:
  static jint DestroyJavaVM(JavaVM* vm) {
    if (vm == NULL) {
      return JNI_ERR;
    }
    JavaVMExt* raw_vm = reinterpret_cast<JavaVMExt*>(vm);
    delete raw_vm->runtime;
    return JNI_OK;
  }
```

在 Android 系统中，当在一个线程里面调用 AttachCurrentThread 后，在不需要用的时候一定要 DetachCurrentThread 处理，否则线程无法正常退出。

第 23 章　安装 APK 应用程序

安装系统的应用程序文件的最终格式是 APK 文件，PackageManagerService 在 Android 系统中负责 APK 包的管理以及应用程序的安装和卸载工作，同时提供 APK 包的相关信息查询功能。在传统的 Dalvik VM 环境下，开发者开发出的应用程序经过编译和打包之后，是一个包含 dex 字节码的 APK 文件。而在 ART 环境下需要运行的是本地机器码，所以必须在应用程序安装时，通过 dex2oat 工具将应用的 dex 字节码翻译成本地机器码。在本章的内容中，将首先详细分析在 Dalvik VM 环境下安装 APK 应用程序的过程，为读者步入本书后面知识的学习打下基础。

23.1 PackageManagerService 概述

在启动 Android 系统时会启动一个应用程序管理服务 PackageManagerService，这个服务负责扫描系统中特定的目录，寻找里面的 APK 格式的应用程序文件，然后对这些文件进行解析，得到应用程序的相关信息，并最终完成应用程序的安装过程。

应用程序管理服务 PackageManagerService 在安装应用程序的过程中，全程解析了应用程序配置文件 AndroidManifest.xml，并从里面得到应用程序的相关信息，例如得到了当前安装应用程序的 Activity、Service、Broadcast Receiver 和 Content Provider 等信息。根据这些信息，通过 ActivityManagerService 服务就可以在系统中正常地使用这些应用程序了。

在 Android 系统中，当系统启动时由 SystemServer 组件启动 PackageManagerService 应用程序管理服务，启动后会执行应用程序安装的过程。在本章的内容中，将从 SystemServer 启动 PackageManagerService 服务的过程开始，详细讲解在 Android 系统中安装应用程序的过程。

23.2 主函数 main

在 Android 系统中，当通过 Zygote 进程启动 SystemServer 组件时会调用系统服务主函数 main，此函数在文件“frameworks/base/services/java/com/android/server/SystemServer.java”中定义，功能是调用 JNI 方法来实现一些系统初始化方面的工作。函数 main 的具体实现代码如下所示：

```
public static void main(String[] args) {
  if (System.currentTimeMillis() < EARLIEST_SUPPORTED_TIME) {
      // If a device's clock is before 1970 (before 0), a lot of
      // APIs crash dealing with negative numbers, notably
      // java.io.File#setLastModified, so instead we fake it and
      // hope that time from cell towers or NTP fixes it
      // shortly.
      Slog.w(TAG, "System clock is before 1970; setting to 1970.");
      SystemClock.setCurrentTimeMillis(EARLIEST_SUPPORTED_TIME);
  }

  if (SamplingProfilerIntegration.isEnabled()) {
```

```
        SamplingProfilerIntegration.start();
        timer = new Timer();
        timer.schedule(new TimerTask() {
            @Override
            public void run() {
                SamplingProfilerIntegration.writeSnapshot("system_server", null);
            }
        }, SNAPSHOT_INTERVAL, SNAPSHOT_INTERVAL);
    }

    // Mmmmmm... more memory!
    dalvik.system.VMRuntime.getRuntime().clearGrowthLimit();

    // The system server has to run all of the time, so it needs to be
    // as efficient as possible with its memory usage.
    VMRuntime.getRuntime().setTargetHeapUtilization(0.8f);

    Environment.setUserRequired(true);

    System.loadLibrary("android_servers");

    Slog.i(TAG, "Entered the Android system server!");
    // Initialize native services.
    nativeInit();

    // This used to be its own separate thread, but now it is
    // just the loop we run on the main thread.
    ServerThread thr = new ServerThread();
    thr.initAndLoop();
}
```

通过上述代码可知，主函数会首先调用 nativeInit()实现本地初始化。

23.3 调用初始化函数

函数 nativeInit()是一个本地 JNI 函数，在文件“frameworks/base/services/jni/com_android_server_SystemServer.cpp”中定义，具体实现代码如下所示：

```
static void android_server_SystemServer_nativeInit(JNIEnv* env, jobject clazz) {
    char propBuf[PROPERTY_VALUE_MAX];
    property_get("system_init.startsensorservice", propBuf, "1");
    if (strcmp(propBuf, "1") == 0) {
        // Start the sensor service
        SensorService::instantiate();
    }
}
/*
 * JNI registration.
 */
static JNINativeMethod gMethods[] = {
    /* name, signature, funcPtr */
    { "nativeInit", "()V", (void*) android_server_SystemServer_nativeInit },
};

int register_android_server_SystemServer(JNIEnv* env)
{
    return jniRegisterNativeMethods(env, "com/android/server/SystemServer",
            gMethods, NELEM(gMethods));
}

}; // namespace android
```

上述函数代码非常简单，只是调用了 system_init.startsensorservice 函数来进一步执行操作。函数 system_init 在 libsystem_server 库中实现，其源代码在文件“frameworks/base/cmds/system_ server/library/system_init.cpp”定义，具体实现代码如下所示：

```
extern "C" status_t system_init()
{
    LOGI("Entered system_init()");

    sp<ProcessState> proc(ProcessState::self());

    sp<IServiceManager> sm = defaultServiceManager();
    LOGI("ServiceManager: %p\n", sm.get());

    sp<GrimReaper> grim = new GrimReaper();
    sm->asBinder()->linkToDeath(grim, grim.get(), 0);

    char propBuf[PROPERTY_VALUE_MAX];
    property_get("system_init.startsurfaceflinger", propBuf, "1");
    if (strcmp(propBuf, "1") == 0) {
        // Start the SurfaceFlinger
        SurfaceFlinger::instantiate();
    }

    // Start the sensor service
    SensorService::instantiate();

    // On the simulator, audioflinger et al don't get started the
    // same way as on the device, and we need to start them here
    if (!proc->supportsProcesses()) {

        // Start the AudioFlinger
        AudioFlinger::instantiate();

        // Start the media playback service
        MediaPlayerService::instantiate();

        // Start the camera service
        CameraService::instantiate();

        // Start the audio policy service
        AudioPolicyService::instantiate();
    }

    // And now start the Android runtime. We have to do this bit
    // of nastiness because the Android runtime initialization requires
    // some of the core system services to already be started.
    // All other servers should just start the Android runtime at
    // the beginning of their processes's main(), before calling
    // the init function.
    LOGI("System server: starting Android runtime.\n");

    AndroidRuntime* runtime = AndroidRuntime::getRuntime();

    LOGI("System server: starting Android services.\n");
    runtime->callStatic("com/android/server/SystemServer", "init2");

    // If running in our own process, just go into the thread
    // pool. Otherwise, call the initialization finished
    // func to let this process continue its initilization.
    if (proc->supportsProcesses()) {
        LOGI("System server: entering thread pool.\n");
        ProcessState::self()->startThreadPool();
        IPCThreadState::self()->joinThreadPool();
        LOGI("System server: exiting thread pool.\n");
    }

    return NO_ERROR;
}
```

在上述代码中，首先会初始化 SurfaceFlinger、SensorService、AudioFlinger、MediaPlayerService、

CameraService 和 AudioPolicyService 等服务，然后通过系统全局唯一的 AndroidRuntime 实例变量 runtime 的 callStatic 来调用 SystemServer 的 init2 函数。

函数 AndroidRuntime::callMain()在文件“frameworks/base/core/jni/AndroidRuntime.cpp”中定义，具体实现代码如下所示：

```
status_t AndroidRuntime::callMain(const char* className,
    jclass clazz, int argc, const char* const argv[])
{
    JNIEnv* env;
    jmethodID methodId;

    ALOGD("Calling main entry %s", className);

    env = getJNIEnv();
    if (clazz == NULL || env == NULL) {
        return UNKNOWN_ERROR;
    }

    methodId = env->GetStaticMethodID(clazz, "main", "([Ljava/lang/String;)V");
    if (methodId == NULL) {
        ALOGE("ERROR: could not find method %s.main(String[])\n", className);
        return UNKNOWN_ERROR;
    }

    /*
     * We want to call main() with a String array with our arguments in it.
     * Create an array and populate it.
     */
    jclass stringClass;
    jobjectArray strArray;

    stringClass = env->FindClass("java/lang/String");
    strArray = env->NewObjectArray(argc, stringClass, NULL);

    for (int i = 0; i < argc; i++) {
        jstring argStr = env->NewStringUTF(argv[i]);
        env->SetObjectArrayElement(strArray, i, argStr);
    }

    env->CallStaticVoidMethod(clazz, methodId, strArray);
    return NO_ERROR;
}
```

在上述代码中，通过参数 className 指定了 java 类的静态成员函数。上面传进来的参数 className 的值为“com/android/server/SystemServer”，接下来就会调用 SystemServer 类的 run 函数。

函数 ServerThread.run 在文件“frameworks/base/services/java/com/android/server/SystemServer.java”文件中定义，具体实现代码如下所示：

```
            public void run() {
                //Looper.myLooper().setMessageLogging(new LogPrinter(
                //        android.util.Log.DEBUG, TAG, android.util.Log.LOG_ID_SYSTEM));
                android.os.Process.setThreadPriority(
                        android.os.Process.THREAD_PRIORITY_DISPLAY);
                android.os.Process.setCanSelfBackground(false);

                // For debug builds, log event loop stalls to dropbox for analysis.
                if (StrictMode.conditionallyEnableDebugLogging()) {
                    Slog.i(TAG, "Enabled StrictMode logging for WM Looper");
                }
            }
        });
......
```

上述代码除了启动 PackageManagerService 服务之外，还启动了其他很多的服务，例如 ActivityManagerService 服务。

23.4 创建 PackageManagerService 服务

在文件“frameworks\base\services\java\com\android\server\pm\PackageManagerService.java”中，通过函数 main 创建一个 PackageManagerService 服务实例，然后把这个服务添加到 ServiceManager 中去。ServiceManager 是 Android 系统 Binder 进程间通信机制的守护进程，负责管理系统中的 Binder 对象。函数 PackageManagerService.main 的具体实现代码如下所示：

```
public static final IPackageManager main(Context context, Installer installer,
        boolean factoryTest, boolean onlyCore) {
    PackageManagerService m = new PackageManagerService(context, installer,
            factoryTest, onlyCore);
    ServiceManager.addService("package", m);
    return m;
}
```

上述代码在创建这个 PackageManagerService 服务实例时，会在 PackageManagerService 类的构造函数中开始执行安装应用程序的过程，在文件 PackageManagerService.java 中的对应实现代码如下所示：

```
class PackageManagerService extends IPackageManager.Stub {
  ......

  public PackageManagerService(Context context, boolean factoryTest) {
    ......

    synchronized (mInstallLock) {
      synchronized (mPackages) {
        ......

        File dataDir = Environment.getDataDirectory();
        mAppDataDir = new File(dataDir, "data");
        mSecureAppDataDir = new File(dataDir, "secure/data");
        mDrmAppPrivateInstallDir = new File(dataDir, "app-private");

        ......

        mFrameworkDir = new File(Environment.getRootDirectory(), "framework");
        mDalvikCacheDir = new File(dataDir, "dalvik-cache");

        ......

        // Find base frameworks (resource packages without code).
        mFrameworkInstallObserver = new AppDirObserver(
        mFrameworkDir.getPath(), OBSERVER_EVENTS, true);
        mFrameworkInstallObserver.startWatching();
        scanDirLI(mFrameworkDir, PackageParser.PARSE_IS_SYSTEM
            | PackageParser.PARSE_IS_SYSTEM_DIR,
            scanMode | SCAN_NO_DEX, 0);

        // Collect all system packages.
        mSystemAppDir = new File(Environment.getRootDirectory(), "app");
        mSystemInstallObserver = new AppDirObserver(
            mSystemAppDir.getPath(), OBSERVER_EVENTS, true);
        mSystemInstallObserver.startWatching();
        scanDirLI(mSystemAppDir, PackageParser.PARSE_IS_SYSTEM
            | PackageParser.PARSE_IS_SYSTEM_DIR, scanMode, 0);

        // Collect all vendor packages.
        mVendorAppDir = new File("/vendor/app");
        mVendorInstallObserver = new AppDirObserver(
            mVendorAppDir.getPath(), OBSERVER_EVENTS, true);
        mVendorInstallObserver.startWatching();
        scanDirLI(mVendorAppDir, PackageParser.PARSE_IS_SYSTEM
```

```
            | PackageParser.PARSE_IS_SYSTEM_DIR, scanMode, 0);
        mAppInstallObserver = new AppDirObserver(
            mAppInstallDir.getPath(), OBSERVER_EVENTS, false);
        mAppInstallObserver.startWatching();
        scanDirLI(mAppInstallDir, 0, scanMode, 0);
        mDrmAppInstallObserver = new AppDirObserver(
            mDrmAppPrivateInstallDir.getPath(), OBSERVER_EVENTS, false);
        mDrmAppInstallObserver.startWatching();
        scanDirLI(mDrmAppPrivateInstallDir, PackageParser.PARSE_FORWARD_LOCK,
            scanMode, 0);
        ......
      }
    }
  }
  ......
}
```

在上述代码中，会调用函数scanDirLI来扫描移动设备中如下5个目录中的APK文件。

- /system/framework。
- /system/app。
- /vendor/app。
- /data/app。
- /data/app-private。

23.5 扫描并解析

函数PackageManagerService.scanDirLI在文件“frameworks/base/services/java/com/android/server/PackageManagerService.java”中定义，具体实现代码如下所示：

```
private void scanDirLI(File dir, int flags, int scanMode, long currentTime) {
    String[] files = dir.list();
    if (files == null) {
        Log.d(TAG, "No files in app dir " + dir);
        return;
    }

    if (DEBUG_PACKAGE_SCANNING) {
        Log.d(TAG, "Scanning app dir " + dir + " scanMode=" + scanMode
                + " flags=0x" + Integer.toHexString(flags));
    }

    int i;
    for (i=0; i<files.length; i++) {
        File file = new File(dir, files[i]);
        if (!isPackageFilename(files[i])) {
            // Ignore entries which are not apk's
            continue;
        }
        PackageParser.Package pkg = scanPackageLI(file,
                flags|PackageParser.PARSE_MUST_BE_APK, scanMode, currentTime, null);
        // Don't mess around with apps in system partition.
        if (pkg == null && (flags & PackageParser.PARSE_IS_SYSTEM) == 0 &&
                mLastScanError == PackageManager.INSTALL_FAILED_INVALID_APK) {
            // Delete the apk
            Slog.w(TAG, "Cleaning up failed install of " + file);
            file.delete();
        }
    }
}
```

通过上述实现代码，对于目录中的每一个文件用“.apk”作为后缀名，那么就调用函数scanPackageLI来对它进行解析和安装。函数PackageParser.parsePackage在文件“frameworks/base/core/java/

android/content/pm/PackageParser.java”中定义，首先会为这个安装的 APK 文件创建一个 PackageParser 实例，然后调用这个实例的 parsePackage 函数对这个 APK 文件进行解析。最后还会调用另外一个版本的 scanPackageLI 函数把来解析后得到的应用程序信息保存在 PackageManagerService 中。函数 PackageParser.parsePackage 的具体实现代码如下所示：

```
private PackageParser.Package scanPackageLI(File scanFile,
        int parseFlags, int scanMode, long currentTime, UserHandle user) {
    mLastScanError = PackageManager.INSTALL_SUCCEEDED;
    String scanPath = scanFile.getPath();
    if (DEBUG_INSTALL) Slog.d(TAG, "Parsing: " + scanPath);
    parseFlags |= mDefParseFlags;
    PackageParser pp = new PackageParser(scanPath);
    pp.setSeparateProcesses(mSeparateProcesses);
    pp.setOnlyCoreApps(mOnlyCore);
    final PackageParser.Package pkg = pp.parsePackage(scanFile,
            scanPath, mMetrics, parseFlags);

    if (pkg == null) {
        mLastScanError = pp.getParseError();
        return null;
    }

    PackageSetting ps = null;
    PackageSetting updatedPkg;
    // reader
    synchronized (mPackages) {
        // Look to see if we already know about this package.
        String oldName = mSettings.mRenamedPackages.get(pkg.packageName);
        if (pkg.mOriginalPackages != null && pkg.mOriginalPackages.contains(oldName)) {
            // This package has been renamed to its original name.  Let's
            // use that.
            ps = mSettings.peekPackageLPr(oldName);
        }
        // If there was no original package, see one for the real package name.
        if (ps == null) {
            ps = mSettings.peekPackageLPr(pkg.packageName);
        }
        // Check to see if this package could be hiding/updating a system
        // package.  Must look for it either under the original or real
        // package name depending on our state.
        updatedPkg = mSettings.getDisabledSystemPkgLPr(ps != null ? ps.name : pkg.packageName);
        if (DEBUG_INSTALL && updatedPkg != null) Slog.d(TAG, "updatedPkg = " + updatedPkg);
    }
    // First check if this is a system package that may involve an update
    if (updatedPkg != null && (parseFlags&PackageParser.PARSE_IS_SYSTEM) != 0) {
        if (ps != null && !ps.codePath.equals(scanFile)) {
            // The path has changed from what was last scanned...  check the
            // version of the new path against what we have stored to determine
            // what to do.
            if (DEBUG_INSTALL) Slog.d(TAG, "Path changing from " + ps.codePath);
            if (pkg.mVersionCode < ps.versionCode) {
                // The system package has been updated and the code path does not match
                // Ignore entry. Skip it.
                Log.i(TAG, "Package " + ps.name + " at " + scanFile
                        + " ignored: updated version " + ps.versionCode
                        + " better than this " + pkg.mVersionCode);
                if (!updatedPkg.codePath.equals(scanFile)) {
                    Slog.w(PackageManagerService.TAG, "Code path for hidden system pkg : "
                            + ps.name + " changing from " + updatedPkg.codePathString
                            + " to " + scanFile);
                    updatedPkg.codePath = scanFile;
                    updatedPkg.codePathString = scanFile.toString();
                    // This is the point at which we know that the system-disk APK
                    // for this package has moved during a reboot (e.g. due to an OTA),
                    // so we need to reevaluate it for privilege policy.
                    if (locationIsPrivileged(scanFile)) {
                        updatedPkg.pkgFlags |= ApplicationInfo.FLAG_PRIVILEGED;
```

```
                }
            }
            updatedPkg.pkg = pkg;
            mLastScanError = PackageManager.INSTALL_FAILED_DUPLICATE_PACKAGE;
            return null;
        } else {
            synchronized (mPackages) {
                // Just remove the loaded entries from package lists.
                mPackages.remove(ps.name);
            }
            Slog.w(TAG, "Package " + ps.name + " at " + scanFile
                    + "reverting from " + ps.codePathString
                    + ": new version " + pkg.mVersionCode
                    + " better than installed " + ps.versionCode);

            InstallArgs args = createInstallArgs(packageFlagsToInstallFlags(ps),
                    ps.codePathString, ps.resourcePathString, ps.nativeLibraryPathString);
            synchronized (mInstallLock) {
                args.cleanUpResourcesLI();
            }
            synchronized (mPackages) {
                mSettings.enableSystemPackageLPw(ps.name);
            }
        }
    }
}

if (updatedPkg != null) {
    // An updated system app will not have the PARSE_IS_SYSTEM flag set
    // initially
    parseFlags |= PackageParser.PARSE_IS_SYSTEM;
}
// Verify certificates against what was last scanned
if (!collectCertificatesLI(pp, ps, pkg, scanFile, parseFlags)) {
    Slog.w(TAG, "Failed verifying certificates for package:" + pkg.packageName);
    return null;
}
boolean shouldHideSystemApp = false;
if (updatedPkg == null && ps != null
        && (parseFlags & PackageParser.PARSE_IS_SYSTEM_DIR) != 0 && !isSystemApp(ps)) {
    if (compareSignatures(ps.signatures.mSignatures, pkg.mSignatures)
            != PackageManager.SIGNATURE_MATCH) {
        if (DEBUG_INSTALL) Slog.d(TAG, "Signature mismatch!");
        deletePackageLI(pkg.packageName, null, true, null, null, 0, null, false);
        ps = null;
    } else {
        if (pkg.mVersionCode < ps.versionCode) {
            shouldHideSystemApp = true;
        } else {
            Slog.w(TAG, "Package " + ps.name + " at " + scanFile + "reverting from "
                    + ps.codePathString + ": new version " + pkg.mVersionCode
                    + " better than installed " + ps.versionCode);
            InstallArgs args = createInstallArgs(packageFlagsToInstallFlags(ps),
                    ps.codePathString, ps.resourcePathString, ps.nativeLibraryPathString);
            synchronized (mInstallLock) {
                args.cleanUpResourcesLI();
            }
        }
    }
}
if ((parseFlags & PackageParser.PARSE_IS_SYSTEM_DIR) == 0) {
    if (ps != null && !ps.codePath.equals(ps.resourcePath)) {
        parseFlags |= PackageParser.PARSE_FORWARD_LOCK;
    }
}

String codePath = null;
String resPath = null;
```

```
        if ((parseFlags & PackageParser.PARSE_FORWARD_LOCK) != 0) {
            if (ps != null && ps.resourcePathString != null) {
                resPath = ps.resourcePathString;
            } else {
                // Should not happen at all. Just log an error.
                Slog.e(TAG, "Resource path not set for pkg : " + pkg.packageName);
            }
        } else {
            resPath = pkg.mScanPath;
        }

        codePath = pkg.mScanPath;
        // Set application objects path explicitly.
        setApplicationInfoPaths(pkg, codePath, resPath);
        // Note that we invoke the following method only if we are about to unpack an application
        PackageParser.Package scannedPkg = scanPackageLI(pkg, parseFlags, scanMode
                | SCAN_UPDATE_SIGNATURE, currentTime, user);

        if (shouldHideSystemApp) {
            synchronized (mPackages) {
                /*
                 * We have to grant systems permissions before we hide, because
                 * grantPermissions will assume the package update is trying to
                 * expand its permissions.
                 */
                grantPermissionsLPw(pkg, true);
                mSettings.disableSystemPackageLPw(pkg.packageName);
            }
        }

        return scannedPkg;
    }
```

在Android系统中，每一个APK文件都是一个归档文件，在里面包含了Android应用程序的配置文件AndroidManifest.xml，在此主要就是对这个配置文件进行解析。当从APK归档文件中得到这个配置文件后，接下来调用文件“frameworks/base/core/java/android/content/pm/PackageParser.java”中的parsePackage函数对这个应用程序进行解析，此parsePackage函数的具体实现代码如下所示：

```
    private Package parsePackage(
        Resources res, XmlResourceParser parser, int flags, String[] outError)
        throws XmlPullParserException, IOException {
        AttributeSet attrs = parser;

        mParseInstrumentationArgs = null;
        mParseActivityArgs = null;
        mParseServiceArgs = null;
        mParseProviderArgs = null;

        String pkgName = parsePackageName(parser, attrs, flags, outError);
        if (pkgName == null) {
            mParseError = PackageManager.INSTALL_PARSE_FAILED_BAD_PACKAGE_NAME;
            return null;
        }
        int type;

        if (mOnlyCoreApps) {
            boolean core = attrs.getAttributeBooleanValue(null, "coreApp", false);
            if (!core) {
                mParseError = PackageManager.INSTALL_SUCCEEDED;
                return null;
            }
        }

        final Package pkg = new Package(pkgName);
        boolean foundApp = false;

        TypedArray sa = res.obtainAttributes(attrs,
                com.android.internal.R.styleable.AndroidManifest);
```

```
pkg.mVersionCode = sa.getInteger(
        com.android.internal.R.styleable.AndroidManifest_versionCode, 0);
pkg.mVersionName = sa.getNonConfigurationString(
        com.android.internal.R.styleable.AndroidManifest_versionName, 0);
if (pkg.mVersionName != null) {
    pkg.mVersionName = pkg.mVersionName.intern();
}
String str = sa.getNonConfigurationString(
        com.android.internal.R.styleable.AndroidManifest_sharedUserId, 0);
if (str != null && str.length() > 0) {
    String nameError = validateName(str, true);
    if (nameError != null && !"android".equals(pkgName)) {
        outError[0] = "<manifest> specifies bad sharedUserId name \""
            + str + "\": " + nameError;
        mParseError = PackageManager.INSTALL_PARSE_FAILED_BAD_SHARED_USER_ID;
        return null;
    }
    pkg.mSharedUserId = str.intern();
    pkg.mSharedUserLabel = sa.getResourceId(
            com.android.internal.R.styleable.AndroidManifest_sharedUserLabel, 0);
}
sa.recycle();

pkg.installLocation = sa.getInteger(
        com.android.internal.R.styleable.AndroidManifest_installLocation,
        PARSE_DEFAULT_INSTALL_LOCATION);
pkg.applicationInfo.installLocation = pkg.installLocation;

/* Set the global "forward lock" flag */
if ((flags & PARSE_FORWARD_LOCK) != 0) {
    pkg.applicationInfo.flags |= ApplicationInfo.FLAG_FORWARD_LOCK;
}

/* Set the global "on SD card" flag */
if ((flags & PARSE_ON_SDCARD) != 0) {
    pkg.applicationInfo.flags |= ApplicationInfo.FLAG_EXTERNAL_STORAGE;
}

// Resource boolean are -1, so 1 means we don't know the value.
int supportsSmallScreens = 1;
int supportsNormalScreens = 1;
int supportsLargeScreens = 1;
int supportsXLargeScreens = 1;
int resizeable = 1;
int anyDensity = 1;

int outerDepth = parser.getDepth();
while ((type = parser.next()) != XmlPullParser.END_DOCUMENT
        && (type != XmlPullParser.END_TAG || parser.getDepth() > outerDepth)) {
    if (type == XmlPullParser.END_TAG || type == XmlPullParser.TEXT) {
        continue;
    }

    String tagName = parser.getName();
    if (tagName.equals("application")) {
        if (foundApp) {
            if (RIGID_PARSER) {
                outError[0] = "<manifest> has more than one <application>";
                mParseError = PackageManager.INSTALL_PARSE_FAILED_MANIFEST_MALFORMED;
                return null;
            } else {
                Slog.w(TAG, "<manifest> has more than one <application>");
                XmlUtils.skipCurrentTag(parser);
                continue;
            }
        }

        foundApp = true;
        if (!parseApplication(pkg, res, parser, attrs, flags, outError)) {
```

```
                return null;
            }
        } else if (tagName.equals("keys")) {
            if (!parseKeys(pkg, res, parser, attrs, outError)) {
                return null;
            }
        } else if (tagName.equals("permission-group")) {
            if (parsePermissionGroup(pkg, flags, res, parser, attrs, outError) == null) {
                return null;
            }
        } else if (tagName.equals("permission")) {
            if (parsePermission(pkg, res, parser, attrs, outError) == null) {
                return null;
            }
        } else if (tagName.equals("permission-tree")) {
            if (parsePermissionTree(pkg, res, parser, attrs, outError) == null) {
                return null;
            }
        } else if (tagName.equals("uses-permission")) {
            if (!parseUsesPermission(pkg, res, parser, attrs, outError)) {
                return null;
            }

        } else if (tagName.equals("uses-configuration")) {
            ConfigurationInfo cPref = new ConfigurationInfo();
            sa = res.obtainAttributes(attrs,
                    com.android.internal.R.styleable.AndroidManifestUsesConfiguration);
            cPref.reqTouchScreen = sa.getInt(
                    com.android.internal.R.styleable.AndroidManifestUsesConfiguration_
                    reqTouchScreen, Configuration.TOUCHSCREEN_UNDEFINED);
            cPref.reqKeyboardType = sa.getInt(
                    com.android.internal.R.styleable.AndroidManifestUsesConfiguration_
                    reqKeyboardType, Configuration.KEYBOARD_UNDEFINED);
            if (sa.getBoolean(
                    com.android.internal.R.styleable.AndroidManifestUsesConfiguration_
                    reqHardKeyboard, false)) {
                cPref.reqInputFeatures |= ConfigurationInfo.INPUT_FEATURE_HARD_KEYBOARD;
            }
            cPref.reqNavigation = sa.getInt(
                    com.android.internal.R.styleable.AndroidManifestUsesConfiguration_
                    reqNavigation, Configuration.NAVIGATION_UNDEFINED);
            if (sa.getBoolean(
                    com.android.internal.R.styleable.AndroidManifestUsesConfiguration_
                    reqFiveWayNav, false)) {
                cPref.reqInputFeatures |= ConfigurationInfo.INPUT_FEATURE_FIVE_WAY_NAV;
            }
            sa.recycle();
            pkg.configPreferences.add(cPref);

            XmlUtils.skipCurrentTag(parser);

        } else if (tagName.equals("uses-feature")) {
            FeatureInfo fi = new FeatureInfo();
            sa = res.obtainAttributes(attrs,
                    com.android.internal.R.styleable.AndroidManifestUsesFeature);
            // Note: don't allow this value to be a reference to a resource
            // that may change.
            fi.name = sa.getNonResourceString(
                    com.android.internal.R.styleable.AndroidManifestUsesFeature_name);
            if (fi.name == null) {
                fi.reqGlEsVersion = sa.getInt(
                        com.android.internal.R.styleable.AndroidManifestUsesFeature_
                        glEsVersion, FeatureInfo.GL_ES_VERSION_UNDEFINED);
```

```
        }
        if (sa.getBoolean(
                com.android.internal.R.styleable.AndroidManifestUsesFeature_required,
                true)) {
            fi.flags |= FeatureInfo.FLAG_REQUIRED;
        }
        sa.recycle();
        if (pkg.reqFeatures == null) {
            pkg.reqFeatures = new ArrayList<FeatureInfo>();
        }
        pkg.reqFeatures.add(fi);

        if (fi.name == null) {
            ConfigurationInfo cPref = new ConfigurationInfo();
            cPref.reqGlEsVersion = fi.reqGlEsVersion;
            pkg.configPreferences.add(cPref);
        }

        XmlUtils.skipCurrentTag(parser);

    } else if (tagName.equals("uses-sdk")) {
        if (SDK_VERSION > 0) {
            sa = res.obtainAttributes(attrs,
                    com.android.internal.R.styleable.AndroidManifestUsesSdk);

            int minVers = 0;
            String minCode = null;
            int targetVers = 0;
            String targetCode = null;

            TypedValue val = sa.peekValue(
                    com.android.internal.R.styleable.AndroidManifestUsesSdk_
                    minSdkVersion);
            if (val != null) {
                if (val.type == TypedValue.TYPE_STRING && val.string != null) {
                    targetCode = minCode = val.string.toString();
                } else {
                    // If it's not a string, it's an integer.
                    targetVers = minVers = val.data;
                }
            }

            val = sa.peekValue(
                    com.android.internal.R.styleable.AndroidManifestUsesSdk_
                    targetSdkVersion);
            if (val != null) {
                if (val.type == TypedValue.TYPE_STRING && val.string != null) {
                    targetCode = minCode = val.string.toString();
                } else {
                    // If it's not a string, it's an integer.
                    targetVers = val.data;
                }
            }

            sa.recycle();

            if (minCode != null) {
                if (!minCode.equals(SDK_CODENAME)) {
                    if (SDK_CODENAME != null) {
                        outError[0] = "Requires development platform " + minCode
                                + " (current platform is " + SDK_CODENAME + ")";
                    } else {
                        outError[0] = "Requires development platform " + minCode
                                + " but this is a release platform.";
                    }
                    mParseError = PackageManager.INSTALL_FAILED_OLDER_SDK;
                    return null;
                }
            } else if (minVers > SDK_VERSION) {
```

```
                    outError[0] = "Requires newer sdk version #" + minVers
                            + " (current version is #" + SDK_VERSION + ")";
                    mParseError = PackageManager.INSTALL_FAILED_OLDER_SDK;
                    return null;
                }

                if (targetCode != null) {
                    if (!targetCode.equals(SDK_CODENAME)) {
                        if (SDK_CODENAME != null) {
                            outError[0] = "Requires development platform " + targetCode
                                    + " (current platform is " + SDK_CODENAME + ")";
                        } else {
                            outError[0] = "Requires development platform " + targetCode
                                    + " but this is a release platform.";
                        }
                        mParseError = PackageManager.INSTALL_FAILED_OLDER_SDK;
                        return null;
                    }
                    // If the code matches, it definitely targets this SDK.
                    pkg.applicationInfo.targetSdkVersion
                            = android.os.Build.VERSION_CODES.CUR_DEVELOPMENT;
                } else {
                    pkg.applicationInfo.targetSdkVersion = targetVers;
                }
            }

            XmlUtils.skipCurrentTag(parser);

        } else if (tagName.equals("supports-screens")) {
            sa = res.obtainAttributes(attrs,
                    com.android.internal.R.styleable.AndroidManifestSupportsScreens);

            pkg.applicationInfo.requiresSmallestWidthDp = sa.getInteger(
                    com.android.internal.R.styleable.AndroidManifestSupportsScreens_
                    requiresSmallestWidthDp, 0);
            pkg.applicationInfo.compatibleWidthLimitDp = sa.getInteger(
                    com.android.internal.R.styleable.AndroidManifestSupportsScreens_
                    compatibleWidthLimitDp, 0);
            pkg.applicationInfo.largestWidthLimitDp = sa.getInteger(
                    com.android.internal.R.styleable.AndroidManifestSupportsScreens_
                    largestWidthLimitDp, 0);

            // This is a trick to get a boolean and still able to detect
            // if a value was actually set.
            supportsSmallScreens = sa.getInteger(
                    com.android.internal.R.styleable.AndroidManifestSupportsScreens_
                    smallScreens,
                    supportsSmallScreens);
            supportsNormalScreens = sa.getInteger(
                    com.android.internal.R.styleable.AndroidManifestSupportsScreens_
                    normalScreens,
                    supportsNormalScreens);
            supportsLargeScreens = sa.getInteger(
                    com.android.internal.R.styleable.AndroidManifestSupportsScreens_
                    largeScreens,
                    supportsLargeScreens);
            supportsXLargeScreens = sa.getInteger(
                    com.android.internal.R.styleable.AndroidManifestSupportsScreens_
                    xlargeScreens,
                    supportsXLargeScreens);
            resizeable = sa.getInteger(
                    com.android.internal.R.styleable.AndroidManifestSupportsScreens_
                    resizeable,
                    resizeable);
            anyDensity = sa.getInteger(
```

```
                    com.android.internal.R.styleable.AndroidManifestSupportsScreens_
anyDensity,
                    anyDensity);

            sa.recycle();

            XmlUtils.skipCurrentTag(parser);

        } else if (tagName.equals("protected-broadcast")) {
            sa = res.obtainAttributes(attrs,
                    com.android.internal.R.styleable.AndroidManifestProtectedBroadcast);

            // Note: don't allow this value to be a reference to a resource
            // that may change.
            String name = sa.getNonResourceString(
                    com.android.internal.R.styleable.AndroidManifestProtectedBroadcast_name);

            sa.recycle();

            if (name != null && (flags&PARSE_IS_SYSTEM) != 0) {
                if (pkg.protectedBroadcasts == null) {
                    pkg.protectedBroadcasts = new ArrayList<String>();
                }
                if (!pkg.protectedBroadcasts.contains(name)) {
                    pkg.protectedBroadcasts.add(name.intern());
                }
            }

            XmlUtils.skipCurrentTag(parser);

        } else if (tagName.equals("instrumentation")) {
            if (parseInstrumentation(pkg, res, parser, attrs, outError) == null) {
                return null;
            }

        } else if (tagName.equals("original-package")) {
            sa = res.obtainAttributes(attrs,
                    com.android.internal.R.styleable.AndroidManifestOriginalPackage);

            String orig =sa.getNonConfigurationString(
                    com.android.internal.R.styleable.AndroidManifestOriginalPackage_name, 0);
            if (!pkg.packageName.equals(orig)) {
                if (pkg.mOriginalPackages == null) {
                    pkg.mOriginalPackages = new ArrayList<String>();
                    pkg.mRealPackage = pkg.packageName;
                }
                pkg.mOriginalPackages.add(orig);
            }

            sa.recycle();

            XmlUtils.skipCurrentTag(parser);

        } else if (tagName.equals("adopt-permissions")) {
            sa = res.obtainAttributes(attrs,
                    com.android.internal.R.styleable.AndroidManifestOriginalPackage);

            String name = sa.getNonConfigurationString(
                    com.android.internal.R.styleable.AndroidManifestOriginalPackage_name, 0);

            sa.recycle();

            if (name != null) {
                if (pkg.mAdoptPermissions == null) {
                    pkg.mAdoptPermissions = new ArrayList<String>();
                }
                pkg.mAdoptPermissions.add(name);
            }

            XmlUtils.skipCurrentTag(parser);
```

```
    } else if (tagName.equals("uses-gl-texture")) {
        // Just skip this tag
        XmlUtils.skipCurrentTag(parser);
        continue;
    } else if (tagName.equals("compatible-screens")) {
        // Just skip this tag
        XmlUtils.skipCurrentTag(parser);
        continue;
    } else if (tagName.equals("supports-input")) {
        XmlUtils.skipCurrentTag(parser);
        continue;

    } else if (tagName.equals("eat-comment")) {
        // Just skip this tag
        XmlUtils.skipCurrentTag(parser);
        continue;

    } else if (RIGID_PARSER) {
        outError[0] = "Bad element under <manifest>: "
            + parser.getName();
        mParseError = PackageManager.INSTALL_PARSE_FAILED_MANIFEST_MALFORMED;
        return null;

    } else {
        Slog.w(TAG, "Unknown element under <manifest>: " + parser.getName()
              + " at " + mArchiveSourcePath + " "
              + parser.getPositionDescription());
        XmlUtils.skipCurrentTag(parser);
        continue;
    }
}
if (!foundApp && pkg.instrumentation.size() == 0) {
    outError[0] = "<manifest> does not contain an <application> or <instrumentation>";
    mParseError = PackageManager.INSTALL_PARSE_FAILED_MANIFEST_EMPTY;
}

final int NP = PackageParser.NEW_PERMISSIONS.length;
StringBuilder implicitPerms = null;
for (int ip=0; ip<NP; ip++) {
    final PackageParser.NewPermissionInfo npi
        = PackageParser.NEW_PERMISSIONS[ip];
    if (pkg.applicationInfo.targetSdkVersion >= npi.sdkVersion) {
        break;
    }
    if (!pkg.requestedPermissions.contains(npi.name)) {
        if (implicitPerms == null) {
            implicitPerms = new StringBuilder(128);
            implicitPerms.append(pkg.packageName);
            implicitPerms.append(": compat added ");
        } else {
            implicitPerms.append(' ');
        }
        implicitPerms.append(npi.name);
        pkg.requestedPermissions.add(npi.name);
        pkg.requestedPermissionsRequired.add(Boolean.TRUE);
    }
}
if (implicitPerms != null) {
    Slog.i(TAG, implicitPerms.toString());
}

final int NS = PackageParser.SPLIT_PERMISSIONS.length;
for (int is=0; is<NS; is++) {
    final PackageParser.SplitPermissionInfo spi
        = PackageParser.SPLIT_PERMISSIONS[is];
    if (pkg.applicationInfo.targetSdkVersion >= spi.targetSdk
```

```
                || !pkg.requestedPermissions.contains(spi.rootPerm)) {
                continue;
            }
            for (int in=0; in<spi.newPerms.length; in++) {
                final String perm = spi.newPerms[in];
                if (!pkg.requestedPermissions.contains(perm)) {
                    pkg.requestedPermissions.add(perm);
                    pkg.requestedPermissionsRequired.add(Boolean.TRUE);
                }
            }
        }

        if (supportsSmallScreens < 0 || (supportsSmallScreens > 0
                && pkg.applicationInfo.targetSdkVersion
                        >= android.os.Build.VERSION_CODES.DONUT)) {
            pkg.applicationInfo.flags |= ApplicationInfo.FLAG_SUPPORTS_SMALL_SCREENS;
        }
        if (supportsNormalScreens != 0) {
            pkg.applicationInfo.flags |= ApplicationInfo.FLAG_SUPPORTS_NORMAL_SCREENS;
        }
        if (supportsLargeScreens < 0 || (supportsLargeScreens > 0
                && pkg.applicationInfo.targetSdkVersion
                        >= android.os.Build.VERSION_CODES.DONUT)) {
            pkg.applicationInfo.flags |= ApplicationInfo.FLAG_SUPPORTS_LARGE_SCREENS;
        }
        if (supportsXLargeScreens < 0 || (supportsXLargeScreens > 0
                && pkg.applicationInfo.targetSdkVersion
                        >= android.os.Build.VERSION_CODES.GINGERBREAD)) {
            pkg.applicationInfo.flags |= ApplicationInfo.FLAG_SUPPORTS_XLARGE_SCREENS;
        }
        if (resizeable < 0 || (resizeable > 0
                && pkg.applicationInfo.targetSdkVersion
                        >= android.os.Build.VERSION_CODES.DONUT)) {
            pkg.applicationInfo.flags |= ApplicationInfo.FLAG_RESIZEABLE_FOR_SCREENS;
        }
        if (anyDensity < 0 || (anyDensity > 0
                && pkg.applicationInfo.targetSdkVersion
                        >= android.os.Build.VERSION_CODES.DONUT)) {
            pkg.applicationInfo.flags |= ApplicationInfo.FLAG_SUPPORTS_SCREEN_DENSITIES;
        }

        /*
         * b/8528162: Ignore the <uses-permission android:required> attribute if
         * targetSdkVersion < JELLY_BEAN_MR2. There are lots of apps in the wild
         * which are improperly using this attribute, even though it never worked.
         */
        if (pkg.applicationInfo.targetSdkVersion < Build.VERSION_CODES.JELLY_BEAN_MR2) {
            for (int i = 0; i < pkg.requestedPermissionsRequired.size(); i++) {
                pkg.requestedPermissionsRequired.set(i, Boolean.TRUE);
            }
        }

        return pkg;
    }
```

上述 parsePackage 函数的实现代码比较复杂，其核心功能是对文件 AndroidManifest.xml 中的各个标签进行解析，各个标签的具体含义在官方文档中进行了讲解：http://developer.android.com/guide/topics/manifest/manifest-intro.html。例如 application 标签的解析功能是通过调用文件“frameworks/base/core/java/android/content/pm/PackageParser.java”中的函数 parseApplication 实现的，具体实现代码如下所示：

```
    private boolean parseApplication(Package owner, Resources res,
            XmlPullParser parser, AttributeSet attrs, int flags, String[] outError)
        throws XmlPullParserException, IOException {
        final ApplicationInfo ai = owner.applicationInfo;
        final String pkgName = owner.applicationInfo.packageName;
```

```
TypedArray sa = res.obtainAttributes(attrs,
        com.android.internal.R.styleable.AndroidManifestApplication);

String name = sa.getNonConfigurationString(
        com.android.internal.R.styleable.AndroidManifestApplication_name, 0);
if (name != null) {
    ai.className = buildClassName(pkgName, name, outError);
    if (ai.className == null) {
        sa.recycle();
        mParseError = PackageManager.INSTALL_PARSE_FAILED_MANIFEST_MALFORMED;
        return false;
    }
}

String manageSpaceActivity = sa.getNonConfigurationString(
        com.android.internal.R.styleable.AndroidManifestApplication_manageSpaceActivity,
        Configuration.NATIVE_CONFIG_VERSION);
if (manageSpaceActivity != null) {
    ai.manageSpaceActivityName = buildClassName(pkgName, manageSpaceActivity,
            outError);
}

boolean allowBackup = sa.getBoolean(
        com.android.internal.R.styleable.AndroidManifestApplication_allowBackup, true);
if (allowBackup) {
    ai.flags |= ApplicationInfo.FLAG_ALLOW_BACKUP;

    // backupAgent, killAfterRestore, and restoreAnyVersion are only relevant
    // if backup is possible for the given application.
    String backupAgent = sa.getNonConfigurationString(
            com.android.internal.R.styleable.AndroidManifestApplication_backupAgent,
            Configuration.NATIVE_CONFIG_VERSION);
    if (backupAgent != null) {
        ai.backupAgentName = buildClassName(pkgName, backupAgent, outError);
        if (DEBUG_BACKUP) {
            Slog.v(TAG, "android:backupAgent = " + ai.backupAgentName
                    + " from " + pkgName + "+" + backupAgent);
        }

        if (sa.getBoolean(
                com.android.internal.R.styleable.AndroidManifestApplication_
                killAfterRestore, true)) {
            ai.flags |= ApplicationInfo.FLAG_KILL_AFTER_RESTORE;
        }
        if (sa.getBoolean(
                com.android.internal.R.styleable.AndroidManifestApplication_
                restoreAnyVersion, false)) {
            ai.flags |= ApplicationInfo.FLAG_RESTORE_ANY_VERSION;
        }
    }
}

TypedValue v = sa.peekValue(
        com.android.internal.R.styleable.AndroidManifestApplication_label);
if (v != null && (ai.labelRes=v.resourceId) == 0) {
    ai.nonLocalizedLabel = v.coerceToString();
}

ai.icon = sa.getResourceId(
        com.android.internal.R.styleable.AndroidManifestApplication_icon, 0);
ai.logo = sa.getResourceId(
        com.android.internal.R.styleable.AndroidManifestApplication_logo, 0);
ai.theme = sa.getResourceId(
        com.android.internal.R.styleable.AndroidManifestApplication_theme, 0);
ai.descriptionRes = sa.getResourceId(
        com.android.internal.R.styleable.AndroidManifestApplication_description, 0);
```

```
        if ((flags&PARSE_IS_SYSTEM) != 0) {
            if (sa.getBoolean(
                    com.android.internal.R.styleable.AndroidManifestApplication_persistent,
                    false)) {
                ai.flags |= ApplicationInfo.FLAG_PERSISTENT;
            }
        }

        if (sa.getBoolean(
                com.android.internal.R.styleable.AndroidManifestApplication_requiredForAllUsers,
                false)) {
            owner.mRequiredForAllUsers = true;
        }

        String restrictedAccountType = sa.getString(com.android.internal.R.styleable
                .AndroidManifestApplication_restrictedAccountType);
        if (restrictedAccountType != null && restrictedAccountType.length() > 0) {
            owner.mRestrictedAccountType = restrictedAccountType;
        }

        String requiredAccountType = sa.getString(com.android.internal.R.styleable
                .AndroidManifestApplication_requiredAccountType);
        if (requiredAccountType != null && requiredAccountType.length() > 0) {
            owner.mRequiredAccountType = requiredAccountType;
        }

        if (sa.getBoolean(
                com.android.internal.R.styleable.AndroidManifestApplication_debuggable,
                false)) {
            ai.flags |= ApplicationInfo.FLAG_DEBUGGABLE;
        }

        if (sa.getBoolean(
                com.android.internal.R.styleable.AndroidManifestApplication_vmSafeMode,
                false)) {
            ai.flags |= ApplicationInfo.FLAG_VM_SAFE_MODE;
        }

        boolean hardwareAccelerated = sa.getBoolean(
                com.android.internal.R.styleable.AndroidManifestApplication_hardwareAccelerated,
                owner.applicationInfo.targetSdkVersion >= Build.VERSION_CODES.ICE_CREAM_SANDWICH);

        if (sa.getBoolean(
                com.android.internal.R.styleable.AndroidManifestApplication_hasCode,
                true)) {
            ai.flags |= ApplicationInfo.FLAG_HAS_CODE;
        }

        if (sa.getBoolean(
com.android.internal.R.styleable.AndroidManifestApplication_allowTaskReparenting,
                false)) {
            ai.flags |= ApplicationInfo.FLAG_ALLOW_TASK_REPARENTING;
        }

        if (sa.getBoolean(
                com.android.internal.R.styleable.AndroidManifestApplication_allowClearUserData,
                true)) {
            ai.flags |= ApplicationInfo.FLAG_ALLOW_CLEAR_USER_DATA;
        }

        if (sa.getBoolean(
                com.android.internal.R.styleable.AndroidManifestApplication_testOnly,
                false)) {
            ai.flags |= ApplicationInfo.FLAG_TEST_ONLY;
        }

        if (sa.getBoolean(
                com.android.internal.R.styleable.AndroidManifestApplication_largeHeap,
```

```
            false)) {
        ai.flags |= ApplicationInfo.FLAG_LARGE_HEAP;
    }

    if (sa.getBoolean(
            com.android.internal.R.styleable.AndroidManifestApplication_supportsRtl,
            false /* default is no RTL support*/)) {
        ai.flags |= ApplicationInfo.FLAG_SUPPORTS_RTL;
    }

    String str;
    str = sa.getNonConfigurationString(
            com.android.internal.R.styleable.AndroidManifestApplication_permission, 0);
    ai.permission = (str != null && str.length() > 0) ? str.intern() : null;

    if (owner.applicationInfo.targetSdkVersion >= Build.VERSION_CODES.FROYO) {
        str = sa.getNonConfigurationString(
                com.android.internal.R.styleable.AndroidManifestApplication_taskAffinity,
                Configuration.NATIVE_CONFIG_VERSION);
    } else {
        // Some older apps have been seen to use a resource reference
        // here that on older builds was ignored (with a warning). We
        // need to continue to do this for them so they don't break.
        str = sa.getNonResourceString(
                com.android.internal.R.styleable.AndroidManifestApplication_taskAffinity);
    }
    ai.taskAffinity = buildTaskAffinityName(ai.packageName, ai.packageName,
            str, outError);

    if (outError[0] == null) {
        CharSequence pname;
        if (owner.applicationInfo.targetSdkVersion >= Build.VERSION_CODES.FROYO) {
            pname = sa.getNonConfigurationString(
                    com.android.internal.R.styleable.AndroidManifestApplication_process,
                    Configuration.NATIVE_CONFIG_VERSION);
        } else {
            // Some older apps have been seen to use a resource reference
            // here that on older builds was ignored (with a warning). We
            // need to continue to do this for them so they don't break.
            pname = sa.getNonResourceString(
                    com.android.internal.R.styleable.AndroidManifestApplication_process);
        }
        ai.processName = buildProcessName(ai.packageName, null, pname,
                flags, mSeparateProcesses, outError);

        ai.enabled = sa.getBoolean(
                com.android.internal.R.styleable.AndroidManifestApplication_enabled, true);

        if (false) {
            if (sa.getBoolean(
                    com.android.internal.R.styleable.AndroidManifestApplication_cantSaveState,
                    false)) {
                ai.flags |= ApplicationInfo.FLAG_CANT_SAVE_STATE;

                // A heavy-weight application can not be in a custom process.
                // We can do direct compare because we intern all strings.
                if (ai.processName != null && ai.processName != ai.packageName) {
                    outError[0] = "cantSaveState applications can not use custom processes";
                }
            }
        }
    }

    ai.uiOptions = sa.getInt(
            com.android.internal.R.styleable.AndroidManifestApplication_uiOptions, 0);

    sa.recycle();

    if (outError[0] != null) {
```

```
            mParseError = PackageManager.INSTALL_PARSE_FAILED_MANIFEST_MALFORMED;
            return false;
        }

        final int innerDepth = parser.getDepth();
        int type;
        while ((type = parser.next()) != XmlPullParser.END_DOCUMENT
                && (type != XmlPullParser.END_TAG || parser.getDepth() > innerDepth)) {
            if (type == XmlPullParser.END_TAG || type == XmlPullParser.TEXT) {
                continue;
            }

            String tagName = parser.getName();
            if (tagName.equals("activity")) {
                Activity a = parseActivity(owner, res, parser, attrs, flags, outError, false,
                        hardwareAccelerated);
                if (a == null) {
                    mParseError = PackageManager.INSTALL_PARSE_FAILED_MANIFEST_MALFORMED;
                    return false;
                }

                owner.activities.add(a);

            } else if (tagName.equals("receiver")) {
                Activity a = parseActivity(owner, res, parser, attrs, flags, outError, true, false);
                if (a == null) {
                    mParseError = PackageManager.INSTALL_PARSE_FAILED_MANIFEST_MALFORMED;
                    return false;
                }

                owner.receivers.add(a);

            } else if (tagName.equals("service")) {
                Service s = parseService(owner, res, parser, attrs, flags, outError);
                if (s == null) {
                    mParseError = PackageManager.INSTALL_PARSE_FAILED_MANIFEST_MALFORMED;
                    return false;
                }

                owner.services.add(s);

            } else if (tagName.equals("provider")) {
                Provider p = parseProvider(owner, res, parser, attrs, flags, outError);
                if (p == null) {
                    mParseError = PackageManager.INSTALL_PARSE_FAILED_MANIFEST_MALFORMED;
                    return false;
                }

                owner.providers.add(p);
            } else if (tagName.equals("activity-alias")) {
                Activity a = parseActivityAlias(owner, res, parser, attrs, flags, outError);
                if (a == null) {
                    mParseError = PackageManager.INSTALL_PARSE_FAILED_MANIFEST_MALFORMED;
                    return false;
                }
                owner.activities.add(a);

            } else if (parser.getName().equals("meta-data")) {
                // note: application meta-data is stored off to the side, so it can
                // remain null in the primary copy (we like to avoid extra copies because
                // it can be large)
                if ((owner.mAppMetaData = parseMetaData(res, parser, attrs, owner.mAppMetaData,
                        outError)) == null) {
                    mParseError = PackageManager.INSTALL_PARSE_FAILED_MANIFEST_MALFORMED;
                    return false;
                }
```

```
} else if (tagName.equals("library")) {
    sa = res.obtainAttributes(attrs,
            com.android.internal.R.styleable.AndroidManifestLibrary);

    // Note: don't allow this value to be a reference to a resource
    // that may change.
    String lname = sa.getNonResourceString(
            com.android.internal.R.styleable.AndroidManifestLibrary_name);

    sa.recycle();

    if (lname != null) {
        if (owner.libraryNames == null) {
            owner.libraryNames = new ArrayList<String>();
        }
        if (!owner.libraryNames.contains(lname)) {
            owner.libraryNames.add(lname.intern());
        }
    }

    XmlUtils.skipCurrentTag(parser);

} else if (tagName.equals("uses-library")) {
    sa = res.obtainAttributes(attrs,
            com.android.internal.R.styleable.AndroidManifestUsesLibrary);

    // Note: don't allow this value to be a reference to a resource
    // that may change.
    String lname = sa.getNonResourceString(
            com.android.internal.R.styleable.AndroidManifestUsesLibrary_name);
    boolean req = sa.getBoolean(
            com.android.internal.R.styleable.AndroidManifestUsesLibrary_required,
            true);

    sa.recycle();

    if (lname != null) {
        if (req) {
            if (owner.usesLibraries == null) {
                owner.usesLibraries = new ArrayList<String>();
            }
            if (!owner.usesLibraries.contains(lname)) {
                owner.usesLibraries.add(lname.intern());
            }
        } else {
            if (owner.usesOptionalLibraries == null) {
                owner.usesOptionalLibraries = new ArrayList<String>();
            }
            if (!owner.usesOptionalLibraries.contains(lname)) {
                owner.usesOptionalLibraries.add(lname.intern());
            }
        }
    }

    XmlUtils.skipCurrentTag(parser);

} else if (tagName.equals("uses-package")) {
    // Dependencies for app installers; we don't currently try to
    // enforce this.
    XmlUtils.skipCurrentTag(parser);

} else {
    if (!RIGID_PARSER) {
        Slog.w(TAG, "Unknown element under <application>: " + tagName
                + " at " + mArchiveSourcePath + " "
                + parser.getPositionDescription());
        XmlUtils.skipCurrentTag(parser);
        continue;
```

```
            } else {
                outError[0] = "Bad element under <application>: " + tagName;
                mParseError = PackageManager.INSTALL_PARSE_FAILED_MANIFEST_MALFORMED;
                return false;
            }
        }
    }

    return true;
}
```

通过上述函数的实现代码，对文件 AndroidManifest.xml 中的 application 标签进行了解析。常用到的 application 标签有 activity、service、receiver 和 provider 等，有关各个标签的具体含义信息，读者可以参考官方文档：http://developer.android.com/guide/topics/manifest/manifest-intro.html。

23.6 保存解析信息

当完成解析工作后，返回到前面的扫描函数 scanPackageLI 中，调用文件“frameworks/base/services/java/com/android/server/pm/PackageManagerService.java”中的函数 scanPackageLI 将解析后得到的应用程序信息保存下来。函数 PackageManagerService.scanPackageLI 的主要实现代码如下所示：

```
class PackageManagerService extends IPackageManager.Stub {
    ......

    // Keys are String (package name), values are Package.  This also serves
    // as the lock for the global state.  Methods that must be called with
    // this lock held have the prefix "LP".
    final HashMap<String, PackageParser.Package> mPackages =
        new HashMap<String, PackageParser.Package>();

    ......

    // All available activities, for your resolving pleasure.
    final ActivityIntentResolver mActivities =
    new ActivityIntentResolver();

    // All available receivers, for your resolving pleasure.
    final ActivityIntentResolver mReceivers =
        new ActivityIntentResolver();

    // All available services, for your resolving pleasure.
    final ServiceIntentResolver mServices = new ServiceIntentResolver();

    // Keys are String (provider class name), values are Provider.
    final HashMap<ComponentName, PackageParser.Provider> mProvidersByComponent =
        new HashMap<ComponentName, PackageParser.Provider>();

    ......

    private PackageParser.Package scanPackageLI(PackageParser.Package pkg,
            int parseFlags, int scanMode, long currentTime) {
        ......

        synchronized (mPackages) {
            ......

            // Add the new setting to mPackages
            mPackages.put(pkg.applicationInfo.packageName, pkg);

            ......

            int N = pkg.providers.size();
            int i;
```

```
        for (i=0; i<N; i++) {
            PackageParser.Provider p = pkg.providers.get(i);
            p.info.processName = fixProcessName(pkg.applicationInfo.processName,
                p.info.processName, pkg.applicationInfo.uid);
            mProvidersByComponent.put(new ComponentName(p.info.packageName,
                p.info.name), p);

            ......
        }

        N = pkg.services.size();
        for (i=0; i<N; i++) {
            PackageParser.Service s = pkg.services.get(i);
            s.info.processName = fixProcessName(pkg.applicationInfo.processName,
                s.info.processName, pkg.applicationInfo.uid);
            mServices.addService(s);

            ......
        }
        N = pkg.receivers.size();
        r = null;
        for (i=0; i<N; i++) {
            PackageParser.Activity a = pkg.receivers.get(i);
            a.info.processName = fixProcessName(pkg.applicationInfo.processName,
                a.info.processName, pkg.applicationInfo.uid);
            mReceivers.addActivity(a, "receiver");
            ......
        }

        N = pkg.activities.size();
        for (i=0; i<N; i++) {
            PackageParser.Activity a = pkg.activities.get(i);
            a.info.processName = fixProcessName(pkg.applicationInfo.processName,
                a.info.processName, pkg.applicationInfo.uid);
            mActivities.addActivity(a, "activity");

            ......
        }

        ......
    }

    ......
    return pkg;
  }

  ......
}
```

上述代码的功能是将前面解析应用程序得到的 package、provider、service、receiver 和 activity 等标签信息保存到 PackageManagerService 服务中。

到此为止，在 Android 系统启动时安装应用程序的过程全部讲解完毕。此时这些应用程序只是相当于在 PackageManagerService 服务中完成了注册，如果想要在 Android 桌面上看到这些应用程序，还需要通过 Home 应用程序从 PackageManagerService 服务中把这些安装好的应用程序取出来，并且以友好的方式在桌面上展现出来，例如通常以桌面快捷图标的方式进行展示。

第 24 章　ART 环境安装 APK 应用程序

在本书第 23 章的内容中，已经详细讲解了在 Dalvik VM 环境下安装 APK 应用程序的过程。Dalvik VM 环境和 ART 环境则不相同，因为在 ART 环境下运行的是本地机器码，所以必须在应用程序安装时，通过 dex2oat 工具将应用的 dex 字节码翻译成本地机器码。在本章的内容中，将首先详细分析在 ART 环境下安装 APK 应用程序的过程，为读者步入本书后面知识的学习打下基础。

24.1 Android 安装 APK 概述

在 Android 系统中，其实对于 APK 安装包来说，安装 APP 应用程序的主要方式如下所示。

- 系统启动时安装。
- adb 命令安装。
- Google Play 上下载安装。
- 通过 PackageInstaller 安装。

在上述安装方式中的最核心方法是 scanPackageLI()，上述安装方式最后都是调用这个函数完成主要的工作，不同安装方式的区别在于在此之前的处理过程不同。

在 Android 系统中，APK 安装包在装完后的内容会体现在如下所示的目录中。

- /data/app　// apk 包。
- /data/app-lib// native lib。
- /data/data //数据目录，其中的 lib 目录指向上面的/data/app-lib 目录。
- /data/dalvik-cache/data@app@<package-name>.apk@classes.dex //优化或编译后的 Java bytecode。

24.2 启动时安装

Android 启动时会把已有的 APP（应用程序）安装一遍，整个过程主要分如下所示的 3 部分。

- 读取安装信息。
- 扫描安装。
- 写回安装信息。

其中读取和写回主要是针对一些安装信息文件，这些信息保证了启动后 APP 与上一次的一致。关键步骤是扫描指定目录下的 APK 并安装。Android 中 APK 主要分布在以下几个目录，意味着启动时要扫描的主要是如下所示的目录。

- 系统核心应用目录：/system/priv-app。
- 系统 APP 目录：/system/app。
- 非系统 APP 目录：/data/app(安装于手机存储的一般 app)或/mnt/asec/<pkgname-number>/pkg.apk（sdcard 或 forward-locked）。

- ❑ 受 DRM 保护 APP 目录：/data/app-private。
- ❑ vendor-specific 的 APP：/vendor/app。
- ❑ 资源型 APP 目录：/system/framework。

整个启动时安装 APK 的流程如图 24-1 所示。

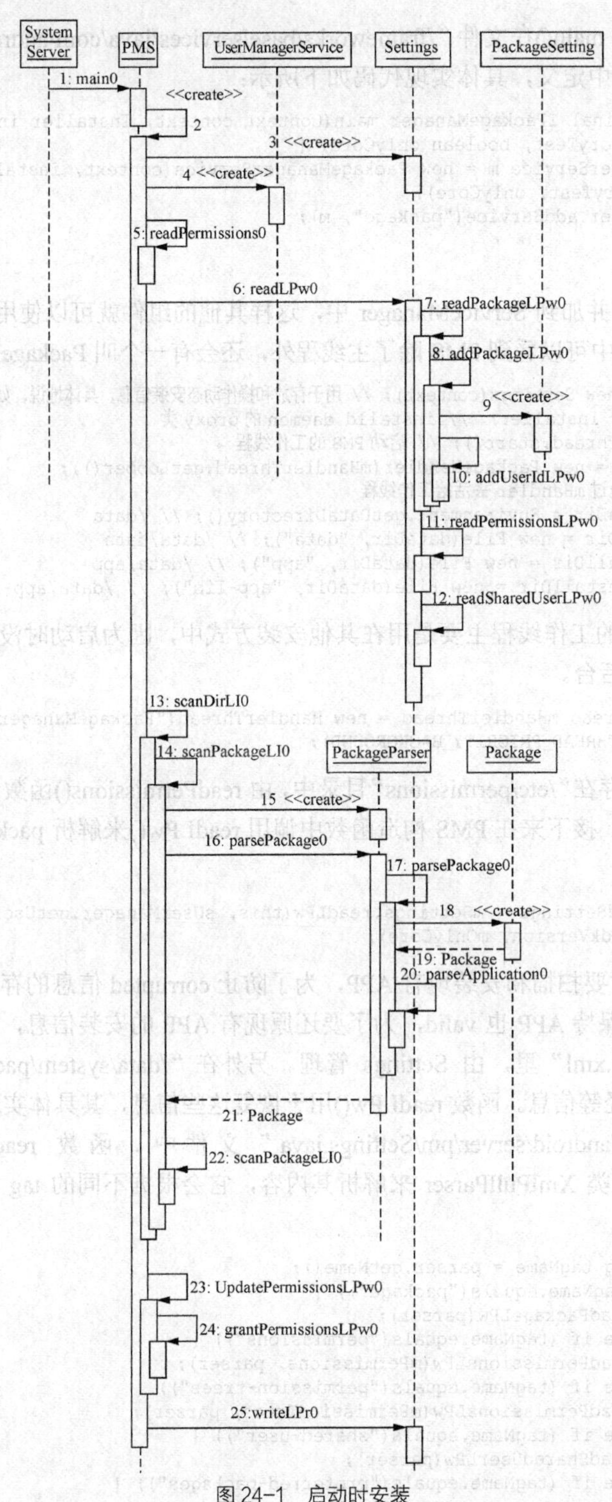

图 24-1 启动时安装

整个安装过程从 System Server 开始，在文件“/frameworks/base/services/java/com/android/server/SystemServer.java”中定义，具体实现代码如下所示：

```
pm = PackageManagerService.main(context, installer,
        factoryTest != SystemServer.FACTORY_TEST_OFF,
        onlyCore);
```

调用 PMS 的函数 main()在文件“/frameworks/base/services/java/com/android/server/pm/PackageManagerService.java”中定义，具体实现代码如下所示：

```
public static final IPackageManager main(Context context, Installer installer,
    boolean factoryTest, boolean onlyCore) {
    PackageManagerService m = new PackageManagerService(context, installer,
        factoryTest, onlyCore);
    ServiceManager.addService("package", m);
    return m;
}
```

在此构造了 PMS 并加到 ServiceManager 中，这样其他的组件就可以使用该服务了。在如下所示的 PMS 的构造函数中可以看到 PMS 除了主线程外，还会有一个叫 PackageManager 的工作线程：

```
mSettings = new Settings(context); // 用于存放和操作动态安装信息，具体地说，如uid, permission等。
mInstaller = installer;  // installd daemon 的proxy类
   mHandlerThread.start(); // 启动 PMS 的工作线程
   mHandler = new PackageHandler(mHandlerThread.getLooper());
  // PMS 会通过 mHandler 丢活给工作线程
   File dataDir = Environment.getDataDirectory(); // /data
   mAppDataDir = new File(dataDir, "data"); // /data/data
   mAppInstallDir = new File(dataDir, "app"); // /data/app
   mAppLibInstallDir = new File(dataDir, "app-lib"); // /data/app-lib
```

PackageManager 的工作线程主要是用在其他安装方式中，因为启动时没什么用户交互，基本上不需要把工作交给后台。

```
final HandlerThread mHandlerThread = new HandlerThread("PackageManager",
    Process.THREAD_PRIORITY_BACKGROUND);
```

具体权限信息保存在“/etc/permissions”目录中，由 readPermissions()函数读取，保存在 Settings 中的 mPermissions 中。接下来在 PMS 构造函数中调用 readLPw()来解析 packages.xml 文件得到上次的安装信息：

```
mRestoredSettings = mSettings.readLPw(this, sUserManager.getUsers(false),
        mSdkVersion, mOnlyCore);
```

Android 在启动时要扫描和安装现有 APP，为了防止 corrupted 信息的存在，为了能在升级系统或更新系统属性后保持 APP 也 valid，为了要还原现有 APP 的安装信息，这些信息被放在文件“/data/system/packages.xml”里，由 Settings 管理。另外在“/data/system/packages.list”中记录了 APP 的 uid 和数据路径等信息。函数 readLPw()用于恢复这些信息，其具体实现位于“/frameworks/base/services/java/com/android/server/pm/Settings.java”文件中。函数 readLPw()先打开文件 packages.xml，再通过类 XmlPullParser 来解析其内容，它会根据不同的 tag 调用相应的函数来读取信息：

```
String tagName = parser.getName();
if (tagName.equals("package")) {
   readPackageLPw(parser);
} else if (tagName.equals("permissions")) {
   readPermissionsLPw(mPermissions, parser);
} else if (tagName.equals("permission-trees")) {
   readPermissionsLPw(mPermissionTrees, parser);
} else if (tagName.equals("shared-user")) {
   readSharedUserLPw(parser);
} else if (tagName.equals("preferred-packages")) {
```

```
            // no longer used.
        } else if (tagName.equals("preferred-activities")) {
            // Upgrading from old single-user implementation;
            // these are the preferred activities for user 0.
            readPreferredActivitiesLPw(parser, 0);
        } else if (tagName.equals("updated-package")) {
            readDisabledSysPackageLPw(parser);
```

在文件 packages.xml 中，一个 APK 包对应的 package 结构大体如下：

```
<package name=""codePath="" nativeLibraryPath="" sharedUserId=""or userId="">
...
</package>
```

其中 readPackageLPw()调用 addPackageLPw()注册 APP 信息到变量 mPackages 中，uid 信息到 mUserIds/mOtherUserIds 中。也就是说，应该把这些信息从文件恢复回去。函数 addPackageLPw()会创建 PackageSetting 类，该类描述了该 APK 包的安装信息：

```
p = new PackageSetting(name, realName, codePath, resourcePath, nativeLibraryPathString,
        vc, pkgFlags);
p.appId = uid;
if (addUserIdLPw(uid, p, name)) {
    mPackages.put(name, p);
    return p;
}
```

在此需要注意，Android APP 中 uid 的范围区间是从 FIRST_APPLICATION_UID 到 LAST_APPLICATION_UID，即 10000 ～ 99999，FIRST_APPLICATION_UID 之下的给系统应用。

另外，有时会遇到多个 APP 为了共享权限而共享一个 uid 的情形。如果一个 APP 要用共享 uid，需要在 APK 的 AndroidManifest.xml 文件中申明 android:sharedUserId="..."，这时就要先为其创建 PendingPackage 并放到 mPendingPackages 等共享 uid 部分。在文件 packages.xml 中的共享用户被表示为下面的形式：

```
<shared-user name="" userId="">
...
</shared-user>
```

函数 readSharedUserLPw()用来处理 APP 安装信息中的共享用户部分，它会调用函数 addSharedUserLPw()来添加共享用户。函数 addSharedUserLPw()为共享用户创建 SharedUserSetting 类（SharedUserSetting 包含了一个 PackageSetting 的集合），再调用 addUserIdLPw()函数注册到 mUserIds 中（或 mOtherUserIds），如果成功会写入到 mSharedUsers 中：

```
s = new SharedUserSetting(name, pkgFlags);
s.userId = uid;
if (addUserIdLPw(uid, s, name)) {
    mSharedUsers.put(name, s);
    return s;
}
```

在文件 packages.xml 中保存了 APP 在之前安装时的配置信息。当一个 APP 卸载后会删除文件 packages.xml 中的该 APP 的信息。当卸载以后下一次安装同一个 APP 时会重新生成，uid 不会被保留。

回到 readLPw()函数，处理前面那些因为用了共享用户而待处理的 APP，也就是 mPendingPackages 中的内容。完成后再回到 PackageManagerService()。在 mSharedLibraries 中存放的是一些共享的 Java 库，在此会调用 dexopt()函数对它们进行优化：

```
if (dalvik.system.DexFile.isDexOptNeeded(lib)) {
    alreadyDexOpted.add(lib);
    mInstaller.dexopt(lib, Process.SYSTEM_UID, true);
    didDexOpt = true;
}
```

alreadyDexOpted 用于记录已经运行过 dexopt 的文件，像启动类和上面的共享库。下面的代码负责对 framework 中的包和类进行优化：

```
        for (int i=0; i<frameworkFiles.length; i++) {
...
            if (dalvik.system.DexFile.isDexOptNeeded(path)) {
            mInstaller.dexopt(path, Process.SYSTEM_UID, true);
```

接下来的代码监控“/system/framework”目录并扫描该目录：

```
    // Find base frameworks (resource packages without code).
    mFrameworkInstallObserver = new AppDirObserver(
        frameworkDir.getPath(), OBSERVER_EVENTS, true, false);
```

此处使用 Observer 模式来监视目录变动，这依赖于 Linux kernel 提供的 Inotify 机制。具体实现主要位于文件“/frameworks/base/core/java/android/os/FileObserver.java”和文件“/frameworks/base/core/jni/android_util_FileObserver.cpp”中。对于其继承类（如 AppDirObserver），只要实现 onEvent()函数来处理文件或目录的变动即可。例如函数 onEvent()在文件“/frameworks/base/services/java/com/android/server/pm/PackageManagerService.java”中实现，具体代码如下所示：

```
    public void onEvent(int event, String path) {
...
            p = scanPackageLI(fullPath, flags,
            SCAN_MONITOR | SCAN_NO_PATHS | SCAN_UPDATE_TIME,
            System.currentTimeMillis(), UserHandle.ALL);
...
            synchronized (mPackages) {
            updatePermissionsLPw(p.packageName, p,
            p.permissions.size() > 0 ? UPDATE_PERMISSIONS_ALL : 0);
            }
...
        synchronized (mPackages) {
        mSettings.writeLPr();
        }
```

扫描该目录的目的是要安装里边的 APP 包，主要实现函数是 scanDirLI()：

```
    scanDirLI(frameworkDir, PackageParser.PARSE_IS_SYSTEM
        | PackageParser.PARSE_IS_SYSTEM_DIR
        | PackageParser.PARSE_IS_PRIVILEGED,
        scanMode | SCAN_NO_DEX, 0);
```

对于其他几个目录（/system/priv-app，/system/app，/vendor/app，/data/app，/data/app-private）的处理过程也是一样的：

```
      mAppInstallObserver = new AppDirObserver(
          mAppInstallDir.getPath(), OBSERVER_EVENTS, false, false);
       mAppInstallObserver.startWatching();
       scanDirLI(mAppInstallDir, 0, scanMode, 0);
…
```

等全部安装完毕后，就可以更新权限信息并且写回安装信息：

```
    updatePermissionsLPw(null, null, UPDATE_PERMISSIONS_ALL
          | (regrantPermissions
                ? (UPDATE_PERMISSIONS_REPLACE_PKG|UPDATE_PERMISSIONS_REPLACE_ALL)
                : 0));
...
    // can downgrade to reader
    mSettings.writeLPr();
```

这样启动时安装 APP 主要工作就差不多完成了，其中目录的扫描和安装等核心工作是通过 scanDirLI()函数实现的：

```
scanDirLI()
    scanPackageLI(file, flags|PackageParser.PARSE_MUST_BE_APK, scanMode, currentTime, null);
```

```
    PackageParser pp = new PackageParser(scanPath);
    final PackageParser.Package pkg = pp.parsePackage(scanFile, scanPath, mMetrics, parseFlags);
        assmgr = new AssetManager();
        parser = assmgr.openXmlResourceParser(cookie, ANDROID_MANIFEST_FILENAME);
        pkg = parsePackage(res, parser, flags, errorText); // parse AndroidManifest.xml
    ...
    PackageParser.Package scannedPkg = scanPackageLI(pkg, parseFlags, scanMode | SCAN_
UPDATE_SIGNATURE, currentTime, user);
```

由此可以看到，scanPackageLI()和 parsePackage()皆有重载版本。内层的 parsePackage(res, ...)函数用于解析 AndroidManifest.xml 文件，此函数在文件“/frameworks/base/core/java/android/content/pm/PackageParser.java”中实现。而 AndroidManifest.xml 是 APK 中必不可少的配置文件，如果没有，APK 文件则无法在 Android 安装：

```
        String tagName = parser.getName();
        if (tagName.equals("application")) {
            if (foundApp) {
                if (RIGID_PARSER) {
                    outError[0] = "<manifest> has more than one <application>";
                    mParseError = PackageManager.INSTALL_PARSE_FAILED_MANIFEST_MALFORMED;
                    return null;
                } else {
                    Slog.w(TAG, "<manifest> has more than one <application>");
                    XmlUtils.skipCurrentTag(parser);
                    continue;
                }
            }

            foundApp = true;
            if (!parseApplication(pkg, res, parser, attrs, flags, outError)) {
                return null;
            }
        } else if (tagName.equals("keys")) {
            if (!parseKeys(pkg, res, parser, attrs, outError)) {
                return null;
            }
        } else if (tagName.equals("permission-group")) {
            if (parsePermissionGroup(pkg, flags, res, parser, attrs, outError) == null)
{
                return null;
            }
        } else if (tagName.equals("permission")) {
            if (parsePermission(pkg, res, parser, attrs, outError) == null) {
                return null;
            }
        } else if (tagName.equals("permission-tree")) {
            if (parsePermissionTree(pkg, res, parser, attrs, outError) == null) {
                return null;
            }
        } else if (tagName.equals("uses-permission")) {
            if (!parseUsesPermission(pkg, res, parser, attrs, outError)) {
                return null;
            }
```

函数 parseApplication()用于解析 application 标签里的东西，application 标签里包含了 Android 四大组件（Activity、Receiver、Service 和 Content Provider）的信息和库等信息。解析后返回结果 PackageParser.Package 对象 pkg，此类基本上就包含了 AndroidManifest.xml 文件中的信息。接下来 scanPackageLI(pkg, ...)被调用，前面返回的解析结果 pkg 被当作参数传入。函数 scanPackageLI (pkg, ...)用于处理共享用户和注册包信息，调用 NativeLibraryHelper 和 Installer 的相关函数进行安装等等。

如果该 APP 使用了共享用户，则调用 getSharedUserLPw()函数获取该共享 uid 的 SharedUserSetting。如果没有则新建，然后分配 uid。

```
SharedUserSetting getSharedUserLPw(String name,
        int pkgFlags, boolean create) {
    SharedUserSetting s = mSharedUsers.get(name);
    if (s == null)
    if (!create) {
        return null;
    }
        s = new SharedUserSetting(name, pkgFlags);
        s.userId = newUserIdLPw(s);
        Log.i(PackageManagerService.TAG, "New shared user " + name + ": id=" + s.userId);
        // < 0 means we couldn't assign a userid; fall out and return
        // s, which is currently null
        if (s.userId >= 0) {
            mSharedUsers.put(name, s);
        }
    }

    return s;
}
```

在函数 scanPackageLI(pkg, ...)中，调用函数 getPackageLPw()得到该 APK 包的 PackageSetting 对象：

```
        pkgSetting = mSettings.getPackageLPw(pkg, origPackage, realName, suid, destCodeFile,
                destResourceFile, pkg.applicationInfo.nativeLibraryDir,
                pkg.applicationInfo.flags, user, false);
```

getPackageLPw()实现在/frameworks/base/services/java/com/android/server/pm/Settings.java 中：

```
private PackageSetting getPackageLPw(String name, PackageSetting origPackage,
        String realName, SharedUserSetting sharedUser, File codePath, File resourcePath,
        String nativeLibraryPathString, int vc, int pkgFlags,
        UserHandle installUser, boolean add, boolean allowInstall) {
…
            p = new PackageSetting(name, realName, codePath, resourcePath,
                    nativeLibraryPathString, vc, pkgFlags);
...
                // Assign new user id
                p.appId = newUserIdLPw(p);
```

函数 newUserIdLPw()用于为 APP 分配 uid，这样该应用对应的 uid 就设置好了。

回到函数 scanPackageLI()开始设置进程名，进程名默认就是包名：

```
pkg.applicationInfo.processName = fixProcessName(
        pkg.applicationInfo.packageName,
        pkg.applicationInfo.processName,
        pkg.applicationInfo.uid);
```

对于大部分全新安装的一般应用而言，接下来需要为应用创建数据目录：

```
private int createDataDirsLI(String packageName, int uid, String seinfo) {
    int[] users = sUserManager.getUserIds();
    int res = mInstaller.install(packageName, uid, uid, seinfo);
...
    for (int user : users) {
        if (user != 0) {
            res = mInstaller.createUserData(packageName,
                    UserHandle.getUid(user, uid), user);
```

Installer 是一个代理类，它会和后台的 installd 通信并让 installd 完成具体工作。installd 的实现位于文件“/frameworks/native/cmds/installd/commands.c”中，函数 install()会创建 APP 目录(/data/data/<package-name>)，并做 lib 目录的软链接，如/data/data/xxx/lib -> /data/app-lib/xxx。

接下来设置的原生库目录为“/data/data/<package- name>/lib”，它指向“/data/app-lib”，然后再把原生库文件（如有）解压到该目录下：

```
if (pkgSetting.nativeLibraryPathString == null) {
    setInternalAppNativeLibraryPath(pkg, pkgSetting);
```

```
        } else {
            pkg.applicationInfo.nativeLibraryDir = pkgSetting.nativeLibraryPathString;
        }
    ...
    if (pkg.applicationInfo.nativeLibraryDir != null) {
        try {
            File nativeLibraryDir = new File(pkg.applicationInfo.nativeLibraryDir);
    ...
                try {
                    if  (copyNativeLibrariesForInternalApp(scanFile,  nativeLibraryDir)  !=
PackageManager.INSTALL_SUCCEEDED) {
                        Slog.e(TAG, "Unable to copy native libraries");
                        mLastScanError = PackageManager.INSTALL_FAILED_INTERNAL_ERROR;
                        return null;
```

上述代码调用了 copyNativeLibrariesForInternalApp()函数，此函数会调用 NativeLibraryHelper.copyNativeBinariesIfNeededLI()函数把 APK 里的原生库解压出来放到“/data/app-lib”的对应目录下。

接下来 PMS 调用 performDexOptLI()优化 Java 的 bytecode，即 dex 文件：

```
    if ((scanMode&SCAN_NO_DEX) == 0) {
        if (performDexOptLI(pkg, forceDex, (scanMode&SCAN_DEFER_DEX) != 0, false)
                == DEX_OPT_FAILED) {
            mLastScanError = PackageManager.INSTALL_FAILED_DEXOPT;
            return null;
        }
    }
```

真正的安装工作还是在后台 installd 进程中完成， installd 中的 dexopt()函数会根据当前是运行 dalvik 还是 art 虚拟机来选择调用 run_dexopt()或 run_dex2oat()：

```
    if (strncmp(persist_sys_dalvik_vm_lib, "libdvm", 6) == 0) {
        run_dexopt(zip_fd, out_fd, apk_path, out_path, dexopt_flags);
    } else if (strncmp(persist_sys_dalvik_vm_lib, "libart", 6) == 0) {
        run_dex2oat(zip_fd, out_fd, apk_path, out_path, dexopt_flags);
    } else {
        exit(69);   /* Unexpected persist.sys.dalvik.vm.lib value */
    }
```

前者适用于 Dalvik 环境，将 dex 优化成 odex 文件；后者适用于 ART 环境，将 dex 直接一步到位编译成 oat 文件（也就是可执行代码）了。由于下层是执行了“/system/bin/dexopt”或“/system/bin/dex2oat”文件，所以 dexopt()将之放到子进程去做，自己作为父进程等待它结束。在 ART 模式下，“/system/bin/dex2oat”被调用（实现位于“/art/dex2oat/dex2oat.cc”）。输出的 elf 文件放在/data/dalvik-cache/data@app@<package-name>@classes.dex。读者在此需要注意，对于 Dalvik 和 ART，这个文件名称相同但性质截然不同。在 Dalvik 环境下，该文件为优化后的 bytecode。而在 ART 环境下，这个就是可执行文件了。

如果更新已有的 APP，还要让 ActivityManager 调用 killApplication()函数把进程杀掉。当 APP 都更新后，老的还保留在内存中运行是不合适的：

```
        killApplication(pkg.applicationInfo.packageName,
                pkg.applicationInfo.uid, "update pkg");
```

然后将安装的 APP 注册到 PMS 的 mPackages 中。mPackges 是一个 Hash 表，保存了从包名到 PackageParser.Package 的映射。在 Settings 里也有一个 mPackages，在里面保存的是包名到 PackageSetting 的映射。前者主要是 APP 配置文件中的信息，而后者是安装过程中的信息。可以粗略理解为一个是静态信息，一个是动态信息：

```
    mSettings.insertPackageSettingLPw(pkgSetting, pkg);
     // PackageSetting <= PackageParser.Package
     addPackageSettingLPw(p, pkg.packageName, p.sharedUser)
         mPackages.put(name, p);
         // Add the new setting to mPackages
         mPackages.put(pkg.applicationInfo.packageName, pkg);
```

接下来把 APP 中的组件信息（content provider、service、receiver 和 activity）记录到系统中，另外根据前面 APP 配置文件中的权限信息进行初始化操作：

```
        int N = pkg.providers.size();
        StringBuilder r = null;
        int i;
        for (i=0; i<N; i++) {
           PackageParser.Provider p = pkg.providers.get(i);
           p.info.processName = fixProcessName(pkg.applicationInfo.processName,
                   p.info.processName, pkg.applicationInfo.uid);
           mProviders.addProvider(p);
…
        N = pkg.services.size();
…
        N = pkg.receivers.size();
…
        N = pkg.activities.size();
…
        N = pkg.permissionGroups.size();
...
        N = pkg.permissions.size();
```

返回到 PMS 构造函数中进行收尾工作，主要包括更新共享库信息、更新权限信息以及写回安装信息。当所有包都解析完后，意味着所有共享库信息都已解析，下面就可以调用 updateAllSharedLibrariesLPw()函数为那些使用动态库的 APP 绑定动态库信息了。下面的 updatePermissionsLPw()函数用于赋予 APP 相应的权限：

```
private void updatePermissionsLPw(String changingPkg,
        PackageParser.Package pkgInfo, int flags) {
...
    // Now update the permissions for all packages, in particular
    // replace the granted permissions of the system packages.
    if ((flags&UPDATE_PERMISSIONS_ALL) != 0) {
        for (PackageParser.Package pkg : mPackages.values()) {
            if (pkg != pkgInfo) {
               grantPermissionsLPw(pkg, (flags&UPDATE_PERMISSIONS_REPLACE_ALL) != 0);
            }
        }
    }
```

在文件 AndroidManifest.xml 中 APP 会申请一些权限，比如读取位置信息、读取联系人、操作摄像头等。文件 AndroidManifest.xml 中的格式如下所示：

```
<uses-permission android:name="permission_name">
```

这里的 permission_name 被放到 requestedPermission 中，表示该 APP 申请该权限。经过 grantPermissionsLPw()函数来判断能否给予相应权限，如果允许则授予权限（即把权限对应的 gid 加到 app 的 gid 列表中，因为权限在 Linux 中对应物就是 group，由 gid 表示），并把该权限加到 grantedPermissions 中，代表已授予该权限：

```
private void grantPermissionsLPw(PackageParser.Package pkg, boolean replace) {
...
    final int N = pkg.requestedPermissions.size();
    for (int i=0; i<N; i++) {
...
               if (!gp.grantedPermissions.contains(perm)) {
                   changedPermission = true;
                   gp.grantedPermissions.add(perm);
                   gp.gids = appendInts(gp.gids, bp.gids);
               } else if (!ps.haveGids) {
                   gp.gids = appendInts(gp.gids, bp.gids);
               }
```

最后，函数 writeLPr()将安装信息写回到 packages.xml 文件，这也是一开始 readLPw()函数读的那个文件，这样在下次启动时就可以按照这里边的信息重新安装了：

```
void writeLPr() {
...
        serializer.startTag(null, "permission-trees");
        for (BasePermission bp : mPermissionTrees.values()) {
            writePermissionLPr(serializer, bp);
        }
        serializer.endTag(null, "permission-trees");

        serializer.startTag(null, "permissions");
        for (BasePermission bp : mPermissions.values()) {
            writePermissionLPr(serializer, bp);
        }
        serializer.endTag(null, "permissions");

        for (final PackageSetting pkg : mPackages.values()) {
            writePackageLPr(serializer, pkg);
        }
```

到此为止，整个启动时安装 APP 的过程全部讲解完毕，整个过程主要用到了如下所示的类。

- PackageManagerService：APK 包安装服务。
- Settings：管理 APP 的安装信息。
- PackageSetting：APP 的动态安装信息。
- SharedUserSetting：共享 Linux 用户。
- PackageParser.Package：APP 的静态配置信息。
- Pm：pm 命令实现类。
- Installer：installd daemon 代理类。
- HandlerThread：PMS 工作线程。

24.3 ART 安装

在 Android 系统中，PackageManagerService 负责 APK 包的管理以及应用程序的安装、卸载，同时提供 APK 包的相关信息查询功能。DEX 到 OAT 的转换过程就是在 PackageManagerService 中完成的。PackageManagerService 由 SystemServer（源码路径为“/frameworks/base/services/java/com/android/server/SystemServer.java”）创建，相关代码如下所示：

```
1. public void initAndLoop(){
2. ......
3. Slog.i(TAG, "Waiting for installd to be ready.");
4. installer = new Installer();
5. installer.ping();
6. ......
7. pm = PackageManagerService.main(context, installer,
8. actoryTest != SystemServer.FACTORY_TEST_OFF, onlyCore);
9. ......
10. }
```

在上述代码中，第 4 行创建 Installer 实例，Installer 类主要负责和 installd 服务通过 socket 进行通信。第 7 行调用 PackageManagerService.main 函数，具体代码如下所示：

```
public static final IPackageManager main(Context context, Installer installer,
boolean factoryTest, boolean onlyCore) {
PackageManagerService m = new PackageManagerService(context, installer,
factoryTest, onlyCore);
ServiceManager.addService("package", m);
return m;
}
```

上述代码的功能是创建 PackageManagerService 的实例，并调用 ServiceManager.addService 函数将其添加到 ServiceManager 的相关数据结构中。PackageManagerService 的构造函数部分代码如下所示：

```
1. public PackageManagerService(Context context, Installer installer,
2. boolean factoryTest, boolean onlyCore) {
3. ......
4. mInstaller = installer;
5. ......
6. if (mSharedLibraries.size() > 0) {
7. Iterator<SharedLibraryEntry> libs = mSharedLibraries.values().iterator();
8. while (libs.hasNext()) {
9. String lib = libs.next().path;
10. ......
11. try {
12. if (dalvik.system.DexFile.isDexOptNeeded(lib)) {
13. alreadyDexOpted.add(lib);
14. mInstaller.dexopt(lib, Process.SYSTEM_UID, true);
15. didDexOpt = true;
16. }
18. }
19. }
20. }
21. ......
22. }
```

在此主要分析 DEX 到 OAT 格式的转化过程，对上述代码的具体说明如下所示。

第 4 行：将参数 installer 赋值给 mInstaller，installer 在 SystemServer 中完成初始化。

第 6 行：mSharedLibraries 是一个 HashMap 实例，保存了“/system/etc/permissions/platform.xml”文件中声明的系统库的信息，在 platform.xml 文件中包含了系统库信息，包括库名称以及库文件的具体路径。

第 9 行：获取库文件的路径。

第 12 行：调用 DexFile.isDexOptNeeded 函数判断 APK 或者 jar 文件是否需要进行优化处理，该函数代码如下，源文件路径为“/libcore/dalvik/src/main/java/dalvik/system/DexFile.java”：

```
native public static boolean isDexOptNeeded(String fileName)
throws FileNotFoundException, IOException;
```

由此可见，isDexOptNeeded 函数是一个 Native 函数，但是这个 Native 函数的定义则有两个，分别存在于文件“/dalvik/vm/native/dalvik_system_DexFile.cpp”和文件“/art/runtime/native/dalvik_system_DexFile.cc”中。从源码路径来看，前者是针对 Dalvik 运行环境的，而后者是 ART 运行环境的。在 ART 环境下，文件“/art/runtime/native/dalvik_system_DexFile.cc”中 Native 函数的注册过程可以参考 ART Runtime 中的 Native。

在文件 system_DexFile.cc 中对应 isDexOptNeeded 的 Native 函数是 DexFile_isDexOptNeeded，其部分代码如下所示，主要功能是根据一些规则来判断是否需要进行 DEX 文件的优化操作：

```
static jboolean DexFile_isDexOptNeeded(JNIEnv* env, jclass, jstring javaFilename)
{
......
ScopedUtfChars filename(env, javaFilename);
......
Runtime* runtime = Runtime::Current();
ClassLinker* class_linker = runtime->GetClassLinker();
const std::vector<const DexFile*>& boot_class_path = class_linker ->
GetBootClassPath();
for (size_t i = 0; i < boot_class_path.size(); i++) {

if (boot_class_path[i]->GetLocation() == filename.c_str()) {
return JNI_FALSE;
}
```

```
12. }
13. ......
14. std::string cache_location(GetDalvikCacheFilenameOrDie(filename.c_str()));
15. oat_file.reset(OatFile::Open(cache_location, filename.c_str(), NULL, false));
16. if (oat_file.get() == NULL) {
17. LOG(INFO) << "DexFile_isDexOptNeeded cache file " << cache_location
18. << " does not exist for " << filename.c_str();
19. return JNI_TRUE;
20. }
21. ......
22. }
```

- ❑ 第 5～11 行：判断文件是否存在于 bootclasspath 路径中，对于所有定义在 bootclasspath 中的文件默认都已经经过了优化处理，因此返回 false。
- ❑ 第 14～19 行：判断在 cache 目录下（/data/dalvik-cache/目录）是否存在已经优化过的 DEX 文件，若没有则打印 Log 信息并返回 true。

另外，函数 DexFile_isDexOptNeeded 还有一些其他的判断规则，例如检查是否存在对应的 odex 文件，classes.dex 文件的校验值等工作。

回到 PackageManagerService 的构造函数，在 12 行完成是否需要进行 DEX 优化的判断后，如果需要进行 DEX 优化，则首先将该文件路径加入到名为 alreadyDexOpted 的 HashSet 中，然后在第 14 行调用 Installer 类的 dexopt 函数，该函数的代码如下所示：

```
1. public int dexopt(String apkPath, int uid, boolean isPublic) {
2. StringBuilder builder = new StringBuilder("dexopt");
3. builder.append(' ');
4. builder.append(apkPath);
5. builder.append(' ');
6. builder.append(uid);
7. builder.append(isPublic ? " 1" : " 0");
8. return execute(builder.toString());
9. }
```

在上述代码中，第 2～7 行用于构造一段字符串，以系统库"/system/framework/javax.obex.jar"为例，则 apkPath 值为"/system/framework/javax.obex.jar"；uid 值为 Process.SYSTEM_UID，即 1000；isPublic 的值为 true。因此构造完的字符串为"dexopt /system/framework/javax.obex.jar 1000 1"。第 8 行调用 execute 执行传入的命令，函数 execute 的代码如下所示：

```
1. private int execute(String cmd) {
2. String res = transaction(cmd);
3. try {
4. return Integer.parseInt(res);
5. } catch (NumberFormatException ex) {
6. return -1;
7. }
8. }
```

在上述代码中，第 2 行调用了 transaction 函数，此函数的核心代码如下所示：

```
1. private synchronized String transaction(String cmd) {
2.if (!connect()) {
3. Slog.e(TAG, "connection failed");
4. return "-1";
5. }
6. if (!writeCommand(cmd)) {
7. Slog.e(TAG, "write command failed? reconnect!");
8. if (!connect() || !writeCommand(cmd)) {
9. return "-1";
10. }
11. }
12. ......
13. }
```

在上述代码中，因为类 Installer 主要负责和 installd 服务进行 socket 通信，所以第 2 行首先调用 connect 函数与 socket 建立连接，第 6 行调用 writeCommand 向 socket 发送数据。

接下来开始分析 installd 服务，其实现源代码位于文件“/frameworks/native/cmds/installd/installd.c”中，函数 main 的部分代码如下所示：

```
1. int main(const int argc, const char *argv[]) {
2. ......
3. for (;;) {
4. alen = sizeof(addr);
5. s = accept(lsocket, &addr, &alen);
6. ......
7. ALOGI("new connection\n");
8. for (;;) {
9. unsigned short count;
10. if (readx(s, &count, sizeof(count))) {
11. ALOGE("failed to read size\n");
12. break;
13. }
14. if ((count < 1) || (count >= BUFFER_MAX)) {
15. ALOGE("invalid size %d\n", count);
16. break;
17. }
18. if (readx(s, buf, count)) {
19. ALOGE("failed to read command\n");
20. break;
21. }
22. buf[count] = 0;
23. if (execute(s, buf)) break;
24.  }
25. ALOGI("closing connection\n");
26. close(s);
27. }
28.
29. return 0;
30. }
```

在上述代码中，当 installd 接收到一个 socket 请求后，首先会判断传入的消息字符串的长度（10～17 行）；接着将消息字符串读入到 buf 数组中（18～22 行）；最后调用函数 execute 执行命令。函数 execute 会解析命令字符串，并调用相应的函数执行，对于“dexopt /system/framework/javax.obex.jar1000 1”命令，最终会调用文件 installd.c 中的 do_dexopt 函数，而该函数只是去调用 dexopt 函数，具体代码如下所示：

```
static int do_dexopt(char **arg, char reply[REPLY_MAX])
{
return dexopt(arg[0], atoi(arg[1]), atoi(arg[2]));
}
```

在上述代码中，dexopt 传入的 3 个参数值分别为/system/framework/javax.obex.jar，1000 好人 1，该函数位于“/frameworks/native/cmds/installd/commands.c”源文件中，具体代码如下所示：

```
int dexopt(const char *apk_path, uid_t uid, int is_public)
{
struct utimbuf ut;
struct stat apk_stat, dex_stat;
char out_path[PKG_PATH_MAX];
char dexopt_flags[PROPERTY_VALUE_MAX];
char persist_sys_dalvik_vm_lib[PROPERTY_VALUE_MAX];
char *end;
int res, zip_fd=-1, out_fd=-1;
if (strlen(apk_path) >= (PKG_PATH_MAX - 8)) {
return -1;
}
property_get("persist.sys.dalvik.vm.lib", persist_sys_dalvik_vm_lib,
"libdvm.so");
sprintf(out_path, "%s%s", apk_path, ".odex");
if (stat(out_path, &dex_stat) == 0) {
return 0;
}
```

```
18. if (create_cache_path(out_path, apk_path)) {
19. return -1;
20. }
21. ......
22. pid_t pid;
23. pid = fork();
24. if (pid == 0) {
25. ......
26. if (strncmp(persist_sys_dalvik_vm_lib, "libdvm", 6) == 0) {
27. run_dexopt(zip_fd, out_fd, apk_path, out_path, dexopt_flags);
28. } else if (strncmp(persist_sys_dalvik_vm_lib, "libart", 6) == 0) {
29. run_dex2oat(zip_fd, out_fd, apk_path, out_path, dexopt_flags);
30. } else {
31. exit(69); /* Unexpected persist.sys.dalvik.vm.lib value */
32. }
33. exit(68); /* only get here on exec failure */
34. }
35. ......
36. }
```

对上述代码的具体说明如下所示。

- 10～12 行：对需要进行优化的文件路径长度进行判断。
- 13 行：获取名为 persist.sys.dalvik.vm.lib 的属性值，默认值为 libdvm.so，当然对于 ART 运行环境，该值为 libart.so。
- 14～17 行：判断是否已经存在对应的.odex 文件，如果有说明已经经过预处理，可以直接忽略，out_path 路径为"/system/framework/javax.obex.jar.odex"。
- 18 行：调用 create_cache_path 对 out_path 处理后其值最终为"/data/dalvik-cache/system@framerok@javax.obex.jar.odex"。
- 23 行：fork 一个子进程来完成 DEX 文件的优化。
- 26～29 行：根据运行环境调用不同的函数来行 DEX 文件的处理，若是 Dalvik，则调用 run_dexopt；若是 ART，则调用 run_dex2oat。

接着分析 run_dex2oat 函数，其具体代码如下所示：

```
1. static void run_dex2oat(int zip_fd, int oat_fd, const char* input_file_name,
2. const char* output_file_name, const char* dexopt_flags)
3. {
4. static const char* DEX2OAT_BIN = "/system/bin/dex2oat";
5. static const int MAX_INT_LEN = 12;
6. char zip_fd_arg[strlen("--zip-fd=") + MAX_INT_LEN];
7. char zip_location_arg[strlen("--zip-location=") + PKG_PATH_MAX];
8. char oat_fd_arg[strlen("--oat-fd=") + MAX_INT_LEN];
9. char oat_location_arg[strlen("--oat-name=") + PKG_PATH_MAX];
10.
11. sprintf(zip_fd_arg, "--zip-fd=%d", zip_fd);
12. sprintf(zip_location_arg, "--zip-location=%s", input_file_name);
13. sprintf(oat_fd_arg, "--oat-fd=%d", oat_fd);
14. sprintf(oat_location_arg, "--oat-location=%s", output_file_name);
15. execl(DEX2OAT_BIN, DEX2OAT_BIN,
16. zip_fd_arg, zip_location_arg,
17. oat_fd_arg, oat_location_arg,
18. (char*) NULL);
19. }
```

在上述代码中，DEX2OAT_BIN 指向 dex2oat 可执行文件的路径，即"/system/bin/dex2oat"，DEX 到 OAT 文件格式的转换最终需要通过该可执行文件完成；6～14 行完成参数的构造，"--zip-fd"表示待转换文件的文件描述符，"--zip-location"表示待转换文件的路径，值为"/system/framework/javax.obex.jar"，"--oat-fd"表示转换结果的文件描述符，"--oat-location"表示转换结果的输出文件路径，值为"/data/dalvik-cache/system@framerok@javax.obex.jar.odex"；第 15 行调用 execl 执行 dex2oat 命令。

24.4 实现 dex2oat 转换

/system/bin/dex2oat 对应的源码文件位于文件“/art/dex2oat/dex2oat.cc”，函数 main 的实现代码如下所示：

```
int main(int argc, char** argv) {
return art::dex2oat(argc, argv);
}
```

通过上述代码可知，直接调用 ART 命名空间下的 dex2oat 函数，虽然该函数有点长，但是有很大部分代码都是在解析参数。

24.4.1 参数解析

与参数解析相关的代码如下，大家可以参考源码，因为函数开头声明了很多变量。基本上后面遇到的所有变量都是在函数开头声明的，而且这些变量的值最终都是从传递过来的参数中获取的：

```
static int dex2oat(int argc, char** argv) {
......
#if defined(ART_USE_PORTABLE_COMPILER)
CompilerBackend compiler_backend = kPortable;
#else
CompilerBackend compiler_backend = kQuick;
#endif
#if defined(__arm__)
InstructionSet instruction_set = kThumb2;
#elif defined(__i386__)
InstructionSet instruction_set = kX86;
#elif defined(__mips__)
InstructionSet instruction_set = kMips;
#else
#error "Unsupported architecture"
#endif
......
for (int i = 0; i < argc; i++) {
const StringPiece option(argv[i]);
......
else if (option.starts_with("--zip-fd=")) {
const char* zip_fd_str = option.substr(strlen("--zip-fd=")).data();
if (!ParseInt(zip_fd_str, &zip_fd)) {
Usage("Failed to parse --zip-fd argument '%s' as an integer", zip_fd_str);
}
} else if (option.starts_with("--zip-location=")) {
zip_location = option.substr(strlen("--zip-location=")).data();
}
......
} else if (option.starts_with("--oat-fd=")) {
const char* oat_fd_str = option.substr(strlen("--oat-fd=")).data();
if (!ParseInt(oat_fd_str, &oat_fd)) {
Usage("Failed to parse --oat-fd argument '%s' as an integer", oat_fd_str);
}
}
......
} else if (option.starts_with("--oat-location=")) {
oat_location = option.substr(strlen("--oat-location=")).data();
}
......
}
if (oat_filename.empty() && oat_fd == -1) {
Usage("Output must be supplied with either --oat-file or --oat-fd");
}
if (!oat_filename.empty() && oat_fd != -1) {
Usage("--oat-file should not be used with --oat-fd");
}
......
```

```
49. if (oat_fd != -1 && !image_filename.empty()) {
50. Usage("--oat-fd should not be used with --image");
51. }
52. if (host_prefix.get() == NULL) {
53. const char* android_product_out = getenv("ANDROID_PRODUCT_OUT");
54. if (android_product_out != NULL) {
55. host_prefix.reset(new std::string(android_product_out));
56. }
57. }
58. if (android_root.empty()) {
59. const char* android_root_env_var = getenv("ANDROID_ROOT");
60. if (android_root_env_var == NULL) {
61. Usage("--android-root unspecified and ANDROID_ROOT not set");
62. }
63. android_root += android_root_env_var;
64. }
65. bool image = (!image_filename.empty());
66. if (!image && boot_image_filename.empty()) {
67. if (host_prefix.get() == NULL) {
68. boot_image_filename += GetAndroidRoot();
69. } else {
70. boot_image_filename += *host_prefix.get();
71. boot_image_filename += "/system";
72. }
73. boot_image_filename += "/framework/boot.art";
74. }
75. std::string boot_image_option;
76. if (!boot_image_filename.empty()) {
77. boot_image_option += "-Ximage:";
78. boot_image_option += boot_image_filename;
79. }
80. ......
81. if (dex_locations.empty()) {
82. for (size_t i = 0; i < dex_filenames.size(); i++) {
83. dex_locations.push_back(dex_filenames[i]);
84. }
85. } else if (dex_locations.size() != dex_filenames.size()) {
86. Usage("--dex-location arguments do not match --dex-file arguments");
87. }
88. ......
89. }
```

对上述代码的具体说明如下所示。

❑ 3～7 行：根据 ART_USE_PORTABLE_COMPILER 宏的值来确定 ART Compiler Driver 的工作方式。

❑ 8～16 行：根据设备的平台架构来确定指令集，包括 ARM、i386 以及 MIPS。

❑ 21～25 行：解析“--zip-fd”参数并转换为整型值赋值给“zip_fd”。

❑ 26～39 行：分别解析获得 zip_location、oat_fd 以及 oat_location。获得参数信息后，会对一些参数进行判断。

❑ 45～46 行：判断--oat-file 以及--oat-fd 参数是否同时被赋值，对于一些不正确的参数组合，会调用 Usage 函数打印 dex2oat 的使用方法并退出。

❑ 52 行：判断 host_prefix 是否为空，因为没有使用“--host-prefix”参数，所以进入 if 分支语句，获取 ANDROID_PRODUCT_OUT 环境变量并重新赋值给 host_prefix 变量。环境变量 ANDROID_PRODUCT_OUT 在 PC 上进行 Android 源码编译时设置的，一般为<ANDROID BASEDIR>/out/target/product/generic/。在 Android 手机设备上不存在此环境变量，host_prefix 值还是为 NULL。

❑ 58～64 行：获取 ANDROID_ROOT 环境变量，该值为/system，也就是 Android 系统下的 system 目录路径；

❑ 65 行：由于 image_filename 为空，因此布尔值 image 为 false，则会进入 66 行的 if 分支；

❑ 66 行：由于 host_prefix 为 NULL， boot_image_filename 值为“/system”（GetAndroidRoot

函数源码在“/art/runtime/utils.cc”文件中，仍然调用 getenv(“ANDROID_ROOT”)，不过会增加/system 目录是否存在的判断)，执行 73 行，最终 boot_iamge_filename 值为“/system/framework/boot.art”。

- ❑ 76 行：经过上述操作后，boot_image_filename 不为空，因此进入 if 分支语句，boot_image_option 最终值为“-Ximage:/system/framework/boot.art”；
- ❑ 81 行：dex_locations 为空，进入 if 分支，由于 dex_filenames 仍然为空，所以不会执行 for 循环。

24.4.2　创建 OAT 文件指针

当完成参数解析判断工作后会创建一个指向 oat_location 的文件指针，暂时没有真正的写入 OAT 格式的文件数据，具体代码如下所示：

```
1. UniquePtr<File> oat_file;
2. bool create_file = !oat_unstripped.empty();
3. if (create_file) {
4. oat_file.reset(OS::CreateEmptyFile(oat_unstripped.c_str()));
5. if (oat_location.empty()) {
6. oat_location = oat_filename;
7. }
8. } else {
9. oat_file.reset(new File(oat_fd, oat_location));
10. oat_file->DisableAutoClose();
11. }
12. if (oat_file.get() == NULL) {
13. PLOG(ERROR) << "Failed to create oat file: " << oat_location;
14. return EXIT_FAILURE;
15. }
16. if (create_file && fchmod(oat_file->Fd(), 0644) != 0) {
17. PLOG(ERROR) << "Failed to make oat file world readable: " << oat_location;
18. return EXIT_FAILURE;
19. }
```

在上述代码中，第 1 行声明了 File 指针变量 oat_file；第 2 行因为 oat_unstripped 为空，所以 create_file 布尔值为 flase；第 3 行判断 create_file 值，进入 else 分支；第 9 行根据 oat_fd 以及 oat_location 创建 File 实例并赋值给 oat_file；第 10 行调用 DisableAutoClose 函数禁止文件自动关闭；12～18 行会进行一些判断操作。上述代码只是创建了一个 File 实例，还没有真正写入任何数据。

24.4.3　dex2oat 准备工作

接下来开始完成 DEX 到 OAT 文件格式的转换工作，具体代码如下所示：

```
1. ......
2. Runtime::Options options;
3. options.push_back(std::make_pair("compiler", reinterpret_cast<void*>(NULL)));
4. std::vector<const DexFile*> boot_class_path
5. if (boot_image_option.empty()) {
6. ......
7. } else {
8. options.push_back(std::make_pair(boot_image_option.c_str(),
reinterpret_cast<void*>(NULL)));
9. }
10. ......
```

在上述代码中，第 2 行声明了 Runtime::Options 类型的变量 options，而 Runtime::Options 实际上是一个包含 pair（http://www.cplusplus.com/reference/utility/pair/）的 vector，在头文件“/art/runtime/runtime.h”中定义，具体代码如下所示：

```
class Runtime {
public:
typedef std::vector<std::pair<std::string, const void*> > Options;
......
}
```

在上述代码中，pair 中的两个数据一个是字符串，另一个为指针，类似于 HashMap 之类的结构。第 3 行在 options 变量中添加一个 pair 数据；第 4 行声明了包含 DexFile 的 vector 变量 boot_class_path，类 DexFile 在头文件“/art/runtime/dex_file.h”中定义，其实就是在 Dalvik 上关于对 dex 文件结构的定义；第 5 行由于 boot_image_option 不为空，所以执行 else 分支语句，跳到第 8 行，继续在 options 中增加一个 pair 数据，boot_image_option 值为-Ximage:/system/framework/boot.art。所以最终 options 中包含“compiler”和“-Ximage:/system/framework/boot.art”两项。

24.4.4 提取 classes.dex 文件

接下来始进入 OAT 文件的创建和转换工作，部分代码如下所示：

```
1. Dex2Oat* p_dex2oat;
2. if (!Dex2Oat::Create(&p_dex2oat, options, compiler_backend, instruction_set,
thread_count)) {
3. LOG(ERROR) << "Failed to create dex2oat";
4. return EXIT_FAILURE;
5. }
6. UniquePtr<Dex2Oat> dex2oat(p_dex2oat);
7. Thread* self = Thread::Current();
8. self->TransitionFromRunnableToSuspended(kNative);
9. WellKnownClasses::Init(self->GetJniEnv());
10. ......
11. std::vector<const DexFile*> dex_files;
12. if (boot_image_option.empty()) {
13. dex_files = Runtime::Current()->GetClassLinker()->GetBootClassPath();
14. } else {
15. if (dex_filenames.empty()) {
16. UniquePtr<ZipArchive> zip_archive(ZipArchive::OpenFromFd(zip_fd));
17. if (zip_archive.get() == NULL) {
18. LOG(ERROR) << "Failed to open zip from file descriptor for " << zip_location;
19. return EXIT_FAILURE;
20. }
21. const DexFile* dex_file = DexFile::Open(*zip_archive.get(), zip_location);
22. if (dex_file == NULL) {
23. ......
24. return EXIT_FAILURE;
25. }
26. dex_files.push_back(dex_file);
27. } else {
28. ......
29. }
30. for (const auto& dex_file : dex_files) {
31. if (!dex_file->EnableWrite()) {
32. ......
33. }
34. }
35. }
```

对上述代码的具体说明如下所示。

第 1 行：声明指向 Dex2Oat 的指针，类 Dex2Oat 定义在“/art/dex2oat/dex2oat.cc”文件中；

第 2 行：调用 Dex2Oat 的静态方法 Create，其中 options 参数已经在“dex2oat 准备工作”部分分析了，compiler_backend 在没有指定--compiler-backend 参数的情况下，会根据 ART_USE_PORTABLE_COMPILER 宏定义的情况取值，若该宏定义了则为 kPortable，否则为 kQuick；instruction_set 能够根据源码编译指定的目标平台使用相应的指令集。因为目前大部分 Android 设备使用的 ARM，所以 instruction_set 值为 kThumb2。thread_count 值在默认情况下通过 sysconf（_SC_NPROCESSORS_CONF）获取，也就是 CPU 的个数。函数 Dex2Oat::Create 的实现代码如下所示：

```
static bool Create(Dex2Oat** p_dex2oat,Runtime::Options& options,
CompilerBackend compiler_backend, InstructionSet instruction_set,
size_t thread_count)
SHARED_TRYLOCK_FUNCTION(true, Locks::mutator_lock_) {
if (!CreateRuntime(options, instruction_set)) {
```

```
6. *p_dex2oat = NULL;
7. return false;
8. }
9. *p_dex2oat = new Dex2Oat(Runtime::Current(), compiler_backend, instruction_set,
thread_count);
10. return true;
11. }
```

在上述代码中，第 5 行调用函数 CreateRuntime 获取 Runtime（/art/runtime/runtime.cc）类的实例，并进行相关的设置。因为 Runtime 使用单例模式，所以获取 Runtime 实例之前会先获取锁，如第 4 行所示，关于 Runtime 类实例的获取比较简单，是一个典型的单例设计模式；第 9 行创建 Dex2Oat 类的实例，Dex2Oat 的构造函数比较简单，只是一些类成员的赋值操作，具体代码如下所示，除了传入的一些参数外，还记录了实例化的时间保存在 start_ns_成员变量中：

```
explicit Dex2Oat(Runtime* runtime,
CompilerBackend compiler_backend,
InstructionSet instruction_set,
size_t thread_count)
: compiler_backend_(compiler_backend),
instruction_set_(instruction_set),
runtime_(runtime),
thread_count_(thread_count),
start_ns_(NanoTime()) {
}
```

继续回到 dex2oat 函数，第 6 行将 p_dex2oat 重新赋值给 dex2oat 变量；第 7 行调用 Thread::Current 得到当前线程，Thread 类定义在“/art/runtime/thread.h”头文件中；第 8 行将线程状态从 Runnable 切换到 Suspend，释放掉之前在调用 Dex2Oat::Create 中获取的锁，函数 TransitionFromRunnableToSuspend 在“/art/runtime/thread-inl.h”头文件中定义，是一个内联函数；第 9 行调用 WellKnownClasses::Init 函数完成一些 JNI 类、函数以及字段的初始化操作，主要是从 JNI 运行环境中查找一些类、函数等信息并返回，self->GetJniEnv()可以获得线程关联的 JNI 环境，Init 函数代码在文件“/art/runtime/well_known_classes.cc”中定义；11 行声明名为 dex_files 的 vector 变量；12 行由于 boot_image_option 非空，因此进入 else 分支；15 行 dex_filenames 为空，进入 if 分支；16 行调用 ZipArchive::OpenFromFd 函数创建 ZipArchive 类实例，主要完成 zip 文件到内存结构的映射。

在此展开分析类 ZipArchive（art/runtime/zip_archive.h 和/art/runtime/zip_archive.cc），函数 OpenFromFd 的具体实现代码如下所示：

```
1. ZipArchive* ZipArchive::OpenFromFd(int fd) {
2. ......
3. UniquePtr<ZipArchive> zip_archive(new ZipArchive(fd));
4. ......
5. if (!zip_archive->MapCentralDirectory()) {
6. zip_archive->Close();
7. return NULL;
8. }
9. if (!zip_archive->Parse()) {
10. zip_archive->Close();
11. return NULL;
12. }
13. return zip_archive.release();
14. }
```

在上述代码中，第 3 行初始化 ZipArchive 实例并赋值给 zip_archive 变量，类 ZipArchive 的构造函数比较简单，主要实现了成员变量的初始化工作，具体代码如下所示：

```
explicit ZipArchive(int fd) : fd_(fd), num_entries_(0), dir_offset_(0) {}
```

第 5 行调用 MapCentralDirectory 完成 Central Directory Header 的查找以及将所有 Central Directory Header 内容映射到内存，该函数的主要实现代码如下：

```
1. bool ZipArchive::MapCentralDirectory() {
2. off64_t file_length = lseek64(fd_, 0, SEEK_END);
3. ......
4. size_t read_amount = kMaxEOCDSearch;
5. if (file_length < off64_t(read_amount)) {
6. read_amount = file_length;
7. }
8. UniquePtr<uint8_t[]> scan_buf(new uint8_t[read_amount]);
9. ......
10. if (lseek64(fd_, 0, SEEK_SET) != 0) {
11. return false;
12. }
13. ssize_t actual = TEMP_FAILURE_RETRY(read(fd_, scan_buf.get(), sizeof(int32_t)));
14. ......
15. unsigned int header = Le32ToHost(scan_buf.get());
16. if (header != kLFHSignature) {
17. return false;
18. }
19. off64_t search_start = file_length - read_amount;
20. if (lseek64(fd_, search_start, SEEK_SET) != search_start) {
21. return false;
22. }
23. actual = TEMP_FAILURE_RETRY(read(fd_, scan_buf.get(), read_amount));
24. ......
25. int i;
26. for (i = read_amount - kEOCDLen; i >= 0; i--) {
27. if (scan_buf.get()[i] == 0x50 && Le32ToHost(&(scan_buf.get())[i]) ==
 kEOCDSignature) {
28. break;
29. }
30. }
31. ......
32. off64_t eocd_offset = search_start + i;
33. const byte* eocd_ptr = scan_buf.get() + i;
34. DCHECK(eocd_offset < file_length);
35. uint16_t disk_number = Le16ToHost(eocd_ptr + kEOCDDiskNumber);
36. uint16_t disk_with_central_dir = Le16ToHost(eocd_ptr + kEOCDDiskNumberForCD);
37. uint16_t num_entries = Le16ToHost(eocd_ptr + kEOCDNumEntries);
38. uint16_t total_num_entries = Le16ToHost(eocd_ptr + kEOCDTotalNumEntries);
39. uint32_t dir_size = Le32ToHost(eocd_ptr + kEOCDSize);
40. uint32_t dir_offset = Le32ToHost(eocd_ptr + kEOCDFileOffset);
41. uint16_t comment_size = Le16ToHost(eocd_ptr + kEOCDCommentSize);
42. ......
43. dir_map_.reset(MemMap::MapFile(dir_size, PROT_READ, MAP_SHARED, fd_,
dir_offset));
44. ......
45. num_entries_ = num_entries;
46. dir_offset_ = dir_offset;
47. return true;
48. }
```

对上述代码的具体说明如下所示。

❑ 第 1 行：通过 lseek64 函数将文件描述符定位到 ZIP 文件结尾处得到文件的长度。

❑ 第 3 行：将 kMaxEOCDSearch 赋值给 read_amount，kMaxEOCDSearch = (kMaxCommentLen + kEOCDLen) =65535 + 22，也就是 End of Central Directory 的最大长度，因为 Zip file comment 最大长度为 64 KB，所以 kMaxEOCDSearch 最大长度为 65535+22。

❑ 第 5 行：判断若整个 ZIP 文件的长度小于 read_amount，则将 read_amount 重新设置为 file_length 的值。

❑ 第 8 行：声明 scan_buf 指针用来存储读取的文件内容。

❑ 第 10 行：将文件描述符定位到 ZIP 文件的开头，因为在获取文件长度时将其定位到了文件结尾处，现在要重新进行定位。

❑ 第 13 行：从文件开始处连续读取 sizeof(int32_t)字节的内容到 scan_buf 中；15～18 行判断文件开头 4 个字节是否为 0x04034b50，也就是 Local File Header 的 Signature，若不是说明

该 ZIP 文件格式不正确。

- 第 19 行：计算 search_start 值，也就是后续 End of Central Directory 搜索的起始地址（相对于文件开头）。
- 第 20 行：将文件描述符定位到 search_start 处。
- 第 23 行：读取 search_start 之后的所有内容到 scan_buf。
- 第 26～30 行：开始搜索 End of Central Directory 的位置。
- 第 32 行：给 eocd_offset 赋值。
- 第 33 行：获得指向 EOCD 起始地址的指针。
- 第 35～41 行：获取 EOCD 中的字段值，其中 dir_size 表示所有 CentralDirectory Header 的大小，dir_offset 表示第一个 Central Directory Header 相对文件开始处的偏移地址；
- 第 43 行：将所有 Central Directory Header 的内容映射到内存；45～46 行给相应的成员变量赋值，保存解析到的 Central Directory Header 的个数及偏移量。

继续回到 OpenFromFd 函数，第 9 行调用函数 Parse 将 Central Directory Header 保存到名为 dir_entries_的类 SafeMap（/art/runtime/safe_map.h）中，其中 Key 为 StringPiece 实例（类 StringPiece 在头文件"/art/runtime /base/stringpiece.h"中定义），Value 为对应 Central Directory Header 映射到内存中的起始地址。

分析完 ZipArchive::OpenFromFd 后，继续回到 dex2oat 中，在 21 行调用 DexFile::Open 函数，该函数在文件"/art/runtime/dex_file.cc"中定义，注意有两个 DexFile::Open 函数，一个是 constDexFile* DexFile::Open(const std::string& filename, const std::string& location)，另一个是 constDexFile* DexFile::Open(const ZipArchive& zip_archive, const std::string& location)，这里显然调用的是后者，部分代码如下所示：

```
1. const DexFile* DexFile::Open(const ZipArchive& zip_archive, const std::string&
location) {
2. CHECK(!location.empty());
3. UniquePtr<ZipEntry> zip_entry(zip_archive.Find(kClassesDex));
4. ......
5. UniquePtr<MemMap> map(zip_entry->ExtractToMemMap(kClassesDex));
6. ......
7. UniquePtr<const DexFile> dex_file(OpenMemory(location, zip_entry->GetCrc32(),
map.release()));
8. ......
9. if (!DexFileVerifier::Verify(dex_file.get(), dex_file->Begin(),
dex_file->Size())) {
10. LOG(ERROR) << "Failed to verify dex file '" << location << "'";
11. return NULL;
12. }
13. if (!dex_file->DisableWrite()) {
14. LOG(ERROR)<< "Failed to make dex file read only '" << location << "'";
15. return NULL;
16. }
17. CHECK(dex_file->IsReadOnly()) << location;
18. return dex_file.release();
19. }
```

在上述代码中，第 3 行调用了类 ZipArchive 中的函数 Find，以文件名为 Key 找到对应 ZIP 文件中的文件并返回 ZipEntry 实例，该 ZipEntry 实例对应的内容为 Central Directory Header，其中 kClasssesDex 在文件 dex_file.cc 中的第 208 行定义，值为"classes.dex"，也就是从文件"/system/framework/javax.obex.jar"（JAR 文件实际上也是一个 ZIP 文件）中找到 classes.dex 的文件。第 5 行调用 ZipEntry 的 ExtractToMemMap 函数，该函数部分实现代码如下所示：

```
MemMap* ZipEntry::ExtractToMemMap(const char* entry_filename) {
std::string name(entry_filename);
name += " extracted in memory from ";
```

```
4. name += entry_filename;
5. UniquePtr<MemMap> map(MemMap::MapAnonymous(name.c_str(),NULL,
6. GetUncompressedLength(), PROT_READ | PROT_WRITE));
7. ......
8. bool success = ExtractToMemory(map->Begin(), map->Size());
9. ......
10. return map.release();
11. }
```

在上述代码中，第 2～4 行实现字符串拼接，最终字符串值为“classes.dex extracted in memory from classes.dex”；第 5 行调用类 MemMap 的 MapAnonymous(/art/runtime/mem_map.cc)创建 MemMap 的实例，其中函数 GetUncompressedLength 用于从 Central Directory Header 中获得 uncompressed size 的值，具体代码如下所示：

```
uint32_t ZipEntry::GetUncompressedLength() {
return Le32ToHost(ptr_ + ZipArchive::kCDEUncompLen);
}
```

返回到函数 ExtractToMemMap，第 8 行调用 ExtractToMemory，该函数的实现代码如下所示：

```
1. bool ZipEntry::ExtractToMemory(uint8_t* begin, size_t size) {
2. ......
3. off64_t data_offset = GetDataOffset();
4. ......
5. if (lseek64(zip_archive_->fd_, data_offset, SEEK_SET) != data_offset) {
6. PLOG(WARNING) << "Zip: lseek to data at " << data_offset << " failed";
7. return false;
8. }
9. switch (GetCompressionMethod()) {
10. case kCompressStored:
11. return CopyFdToMemory(begin, size, zip_archive_->fd_,
GetUncompressedLength());
12. case kCompressDeflated:
13. return InflateToMemory(begin, size, zip_archive_->fd_,
14. GetUncompressedLength(), GetCompressedLength());
15. default:
16. LOG(WARNING) << "Zip: unknown compression method " << std::hex <<
GetCompressionMethod();
17. return false;
18. }
19. }
```

在上述代码中，第 3 行调用 GetDataOffset 函数从 Central Directory Header 中找到 Local File Header 的偏移量，然后从 Local File Header 中定位到 File Data 的位置，该函数部分代码如下所示：

```
1. off64_t ZipEntry::GetDataOffset() {
2. ......
3. int64_t lfh_offset = Le32ToHost(ptr_ + ZipArchive::kCDELocalOffset);
4. ......
5. if (lseek64(zip_archive_->fd_, lfh_offset, SEEK_SET) != lfh_offset) {
6. PLOG(WARNING) << "Zip: failed seeking to LFH at offset " << lfh_offset;
7. return -1;
8. }
9. uint8_t lfh_buf[ZipArchive::kLFHLen];
10. ssize_t actual = TEMP_FAILURE_RETRY(read(zip_archive_->fd_, lfh_buf,
sizeof(lfh_buf)));
11. ......
12. off64_t data_offset = (lfh_offset + ZipArchive::kLFHLen
13. + Le16ToHost(lfh_buf + ZipArchive::kLFHNameLen)
14. + Le16ToHost(lfh_buf + ZipArchive::kLFHExtraLen));
15. ......
16. return data_offset;
17. }
```

在上述代码中，第 3 行获得该 Central Directory Header 对应的 File Local Header 的偏移量；第 5 行将 zip_archive_ 的 fd_ 定位到该偏移处；第 10 行从偏移处开始读取内容到 lfh_buf 中；第 12 行从 File Local Header 中获得 File Data 的偏移，具体过程可以参考 Local File Header 的结构来

看第 12 行代码；最后返回 data_offset 值。

返回到 ExtractToMemory，第 5 行根据 File Data 的偏移量将 ZIP 文件的文件描述符定位到 File Data 起始位置；第 9 行进入 switch 语句，GetCompressionMethod 从 Central Directory Header 中获得 compression method 字段值。因此 13 行的 InflateToMemory 函数会被调用，该函数的主要功能是完成 File Data 的解压工作。

返回到 DexFile::Open 函数中，至此已经完成了 classes.dex 文件的解压并映射到了内存中。第 7 行调用 OpenMemory 函数，该函数的主要实现代码如下所示：

```
1. const DexFile* DexFile::OpenMemory(const byte* base,size_t size,
2. const std::string& location,uint32_t location_checksum, MemMap* mem_map) {
3. CHECK_ALIGNED(base, 4);
4. UniquePtr<DexFile> dex_file(new DexFile(base, size, location, location_checksum,
mem_map));
5. if (!dex_file->Init()) {
6. return NULL;
7. } else {
8. return dex_file.release();
9. }
10. }
```

对上述代码的具体说明如下所示。

❑ 第 4 行：创建 DexFile 类的实例，完成部分成员变量的初始化，DexFile 的构造函数位于头文件“/art/runtime/dex_file.h”中。

❑ 第 5 行：调用 Init 函数完成初始化，如 DEX 文件的 StringId，TypeId 的初始化，以及 Dex 文件的一些合法性检查工作。

❑ 第 9 行：函数 DexFile::Open 调用函数 DexFileVerifier::Verify 对文件进行 DEX 合法性验证。

❑ 第 13 行：设置文件禁止写。

❑ 第 17 行：检查文件是否只读，也就是进一步判断设置禁止写是否成功。

返回到 dex2oat 中，在第 26 行调用 dex_files.push_back(dex_file)将创建的 DexFile 实例存入 vector 中；第 30～31 行将 DEX 文件设置为可写。

到此为止，主要完成了从 javax.obex.jar 文件中提取 classes.dex 并映射到内存的工作，并同时完成了 DEX 文件合法性的验证工作。

24.4.5　创建 OAT 文件

继续分析 dex2oat 的代码，接下来开始创建 OAT 文件，具体代码如下所示：

```
1. if (!image && (Runtime::Current()->GetCompilerFilter() !=
Runtime::kInterpretOnly)) {
2. size_t num_methods = 0;
3. for (size_t i = 0; i != dex_files.size(); ++i) {
4. const DexFile* dex_file = dex_files[i];
5. CHECK(dex_file != NULL);
6. num_methods += dex_file->NumMethodIds();
7. }
8. if (num_methods <= Runtime::Current()->GetNumDexMethodsThreshold()) {
9. Runtime::Current()->SetCompilerFilter(Runtime::kSpeed);
10. VLOG(compiler) << "Below method threshold, compiling anyways";
11. }
12. }
13. UniquePtr<const CompilerDriver>
compiler(dex2oat->CreateOatFile(boot_image_option,host_prefix.get(),android_roo
t,is_host, dex_files,oat_file.get(), bitcode_filename, image, image_classes,
dump_stats,timings));
```

在上述代码中，第 2～10 行主要从 DEX 文件中获得函数个数并判断是否达到阈值；第 13 行调用函数 CreateOatFile 创建 OAT 文件。其实 OAT 部分的内容是嵌在 ELF 格式的文件中的，Linux

下使用 file 命令来检验“/system/framework/javax.obex.jar” 转换后的文件“/data/dalvik-cache/system@framework@javax.objex.jar@classes.dex”。在文件“/art /compiler/elf_writer_quick.cc”中（当未定义 ART_USE_PORTABLE_COMPILER 宏时会进入 ElfWriterQuick::Create 函数）的 ElfWriteQuick::Write 函数中有一个文件格式的说明。

到此为止，完成了从 DEX 文件到 OAT 文件格式的转换工作，当然在 dex2oat 中还有一些其他的扫尾工作需要完成。

24.5 APK 文件的转换

经过本章前面内容的学习，已经了解了 dex2oat 针对系统库文件的转换操作过程。对于一般的 APK 文件而言，dex2oat 执行的流程可以从 PackageManagerService 中得到，下面以“/system/app”下的 APK 的转换为例，在文件 PackageManagerService.java 中的代码如下所示：

```
1. File systemAppDir = new File(Environment.getRootDirectory(), "app");
2. mSystemInstallObserver = new AppDirObserver(
3. systemAppDir.getPath(), OBSERVER_EVENTS, true, false);
4. mSystemInstallObserver.startWatching();
5. scanDirLI(systemAppDir, PackageParser.PARSE_IS_SYSTEM
6. | PackageParser.PARSE_IS_SYSTEM_DIR, scanMode, 0);
```

对上述代码的具体说明如下所示。

- 第 1 行：创建系统 APP 的目录路径，即“/system/app”。
- 第 2～4 行：实现监控文件夹目录。
- 第 5 行：调用 scanDirLI ()函数，此函数的功能是调用函数 scanPackageLI 对目录下的每个 APK 文件进行扫描。scanPackageLI 会调用 performDexOptLI 函数。在 performDexOptLI 中则调用 Installer.dexopt 函数，后续工作和前文分析“/system/framework/javax.obex.jar”的转换过程完全一样，在此将不再详细讲解。

这样当 system_server 启动后，会在 log 中显示“art”这个进程，通过 dex2oat 工具创建了一个巨大的“镜像”文件。在命令行参数映射中包含了如下所示的部分内容。

- 虚拟机的一些运行时参数。
- 将被编译成镜像的几个 dex 文件。
- 输出的镜像文件名。
- 输出的 oat 文件名。
- 一个包含应被编译成镜像的类的 jar 文件。
- 一个描述了哪些来自 jar 文件的类应该被使用的说明。

在创建镜像期间，所有已包含在内的 dex 文件会被编译。例如：

```
W/dex2oat ( 397): Verification of void org.ccil.cowan.tagsoup.HTMLSchema.<init>() took 187.968ms
```

由此可见，在 Dalvi 环境中的 Zygote 没有被取代，它依然存在于 ART 环境中，可以用与 Dalvik 一样的方式来处理它。在更早的启动过程中，包管理器在每一个已安装的应用中运行 dexopt。但是除此之外，dex2oat 编译器会把每一个已经产生的 dex 文件编译成 oat 文件。ART 使用了 Zygote，就像镜像一样，之后开始执行实际的应用程序。同时它也进行了大量的编译工作，甚至是对框架类的编译。这就意味着从字面上看，曾经熟知的 Android 整个系统都被改变了。ART 不仅仅是一个“更好的 Dalvik”，并且还意味着范例上的一个改变。

在 Android 系统中，ART 模式的核心是使用类 Instalerl 的成员函数 dexopt()对 APK 里面的 dex 字节码进行优化处理，所以拥有更高的处理效率和更好的用户体验。

读书笔记

读书笔记

读书笔记